# Lipids in Health and Disease

# SUBCELLULAR BIOCHEMISTRY



---

*Recent Volumes in this Series*

Peter J. Quinn · Xiaoyuan Wang
Editors

# Lipids in Health and Disease

*Editors*

Peter J. Quinn
King's College London
Biochemistry Department
150 Stamford Street
London SE1 9NH
United Kingdom
p.quinn@kcl.ac.uk

Xiaoyuan Wang
Jiangnan University
State Key Laboratory of Food Science and Technology
1800 Lihu Avenue
Wuxi 214122
China
xwang@jiangnan.edu.cn

ISBN: 978-1-4020-8830-8 e-ISBN: 978-1-4020-8831-5

Library of Congress Control Number: 2008931355



Printed on acid-free paper

9 8 7 6 5 4 3 2 1

springer.com

# Preface

Lipids are functionally versatile molecules. They have evolved from relatively simple hydrocarbons that serve as depot storages of metabolites and barriers to the permeation of solutes into complex compounds that perform a variety of signalling functions in higher organisms. This volume is devoted to the polar lipids and their constituents. We have omitted the neutral lipids like fats and oils because their function is generally to act as deposits of metabolizable substrates. The sterols are also outside the scope of the present volume and the reader is referred to volume 28 of this series which is the subject of cholesterol.

The polar lipids are comprised of fatty acids attached to either glycerol or sphingosine. The fatty acids themselves constitute an important reservoir of substrates for conversion into families of signalling and modulating molecules including the eicosanoids amongst which are the prostaglandins, thromboxanes and leucotrienes. The way fatty acid metabolism is regulated in the liver and how fatty acids are desaturated are subjects considered in the first part of this volume. This section also deals with the modulation of protein function and inflammation by unsaturated fatty acids and their derivatives. New insights into the role of fatty acid synthesis and eicosenoid function in tumour progression and metastasis are presented.

The phospholipids form the principal constituents of the lipid bilayer matrix of cell membranes. They constitute a range of lipid classes characterised by the substituents attached to the phosphate group. Each lipid class, in turn, consists of a range of molecular species characterised by the length, degree of unsaturation and position and type of attachment to the glycerol backbone. Cell membranes can be comprised of upwards of hundreds of individual molecular species of lipid. The proportion of each molecular species present in particular cell membranes of homeothermic organisms is preserved within relatively narrow limits by biochemical homeostatic mechanisms. The lipids found in the membranes of poikilothermic organisms are seen to change in response to environmental factors like temperature and salinity which infers that the biochemical changes in membrane lipids result in adaptive changes in the physical properties of the lipid matrix. One particular role of polar lipids in membranes is the integration and organization of intrinsic proteins into the matrix. This feature is examined by exploring how membrane lipids that form non-bilayer structures can influence the function of integral membrane proteins. Dysfunction in membranes has been associated with deficiencies of phospholipids that tend to form non-bilayer structures and the role of these lipids in protein folding and formation of oligomeric protein complexes is explored.

Disorders of lipid metabolism are believed to underlie a variety of organ pathologies and risk factors associated with circulatory diseases. Phospholipase hydrolysis of cardiac sarcolemmal membrane phospholipids generates products involved in the pathophysiology of heart disease. Signalling pathways augmented by these products are evaluated in the context of their potential for therapeutic intervention. One of the products of phospholipase $A_2$ hydrolysis, lysophospholipid, is known to be involved in activation of G-protein coupled receptors and one chapter is devoted to an overview of the role of lysophospholipids in normal and pathological conditions. A summary of recent findings in human and animal models is provided.

Oxidation of phospholipids may represent a general mechanism underlying a range of chronic inflammatory and autoimmune diseases. Oxidised phospholipids are also associated with many other conditions involving generation of reactive oxygen species. Deacylation and transacylation with lysophosphatide intermediates is an important process in membrane lipid homeostasis and the role of these lysolipids as cell modulators is now recognized. Free radical oxidation of membrane phospholipids results in release and activation of pro-apoptotic factors and the generation of "eat me" signals culminating in phagocytosis of the target cell.

Sphingolipids are also constituents of the lipid bilayer matrix. Some members form a group of complex glycosphingolipids many of which are surface antigens and are known to be involved in the social organization of cells in tissues. Sphingolipids and their metabolites act in a variety of regulatory roles including the metabolism of lipids in general and in the biosynthesis of sterols in particular. Although the precise mechanisms of how sphingolipids regulate lipid metabolism are not known, this relationship has important implications with regard to cellular lipid homeostasis, composition of lipoproteins and development of atherosclerosis. These functions have been examined together with how these lipids alter endothelial barrier functions and cellular immune responses. Sphingolipids are implicated in many disease states including metastatic conditions and apoptosis. The way sphingolipids act via membrane signalling platforms like rafts is discussed as well as how such actions may be targeted in the development of therapeutic strategies.

Finally, a complete understanding of the role of lipids in health and disease can only be achieved by detailed knowledge of the changes in molecular species of lipid in response to physiological or pathological states. This aspect of lipidology has received considerable impetus in the recent past by the combination of powerful separation and analytical techniques. An appreciation of the analytical power of current lipidomic techniques can be given by the fact that 100 nmoles of lipid, an amount that can be extracted from a conventional tissue culture flask containing one million cells, is sufficient for a complete lipidomic analysis. The application of such lipidomic analyses in discerning intracellular lipid traffic and monitoring disease is presented. It is anticipated that deployment of lipidomic methodology in wider fields will lead to a greater understanding of the role of lipids in health and disease. This volume is aimed to generate the necessary enthusiasm and curiosity to realize these ambitions.

London — Peter J. Quinn
Wuxi — Xiaoyuan Wang

# About the Editors

**Peter J Quinn** has been a Professor of Biochemistry at King's College London since 1989 and has held visiting Professorships at Pittsburgh, Nagoya and Tsinghua Universities. His primary research interest is biological membranes and their constituents. The approach in this research has been to apply a range of biophysical methods including real-time synchroron X-ray diffraction, neutron scattering, differential scanning calorimetry, freeze-fracture electron microscopy, nuclear magnetic resonance spectroscopy, laser flash photolysis and Fourier transform infrared spectroscopy to address questions concerned with relationships between biomembrane structure and function. Professor Quinn received his PhD from the University of Sydney and was awarded a DSc from the University of London in 1980.

**Xiaoyuan Wang** is the Cheung Kong Professor for Molecular Biology in the State Key Laboratory of Food Science and Technology at Jiangnan University. He received his PhD from University of London in 2000, and completed postdoctoral training in Dr. William Dowhan's laboratory at University of Texas Houston Medical School in 2002. Then he worked as a Research Associate in Dr. Christian Raetz's group at Duke University Medical School until 2007. During the years Dr. Wang has been using a combined molecular genetic and biochemical approach to study the structure, function, and assembly of phospholipids and lipid A in membranes of Gram-negative bacteria. In 2007 Dr. Wang joined Jiangnan University, and was appointed Cheung Kong Chair Professor by the Ministry of Education in the same year. His research interests now include food lipids, food toxins and food safety control.

# Contents

## Part II Phospholipids



## Part III Sphingolipids

# Contributors

Rao Muralikrishna Adibhatla
Department of Neurological Surgery, H4-330, Clinical Science Center, 600 Highland Avenue, University of Wisconsin School of Medicine and Public Health, Madison, WI 53792-3232, USA

Daniele Amadio
Department of Experimental Medicine & Biochemical Sciences, University of Rome "Tor Vergata", Rome, Italy

Claudia E. Banchio
Departamento de Ciencias Biológicas, Facultad de Ciencias Bioquímicas y Farmacéuticas, Universidad Nacional de Rosario, Suipacha 531, (S2002LRK) Rosario, Argentina

Thomas Baumruker
Novartis Institutes for BioMedical Research, Brunnerstrasse 59, A-1235 Vienna, Austria

Katrin Anne Becker
Dept. of Molecular Biology, University of Duisburg-Essen, Hufelandstrasse 55, 45122 Essen, Germany

Andreas Billich
Novartis Institutes for BioMedical Research, Brunnerstrasse 59, A-1235 Vienna, Austria

Mikhail Bogdanov
Department of Biochemistry and Molecular Biology, University of Texas-Houston, Medical School, 6431 Fannin St. Houston, TX 77030, USA

Stephanie D. Boomkamp
Glycobiology Institute, Department of Biochemistry, University of Oxford, Oxford, OX1 3QU, UK

Terry D. Butters
Glycobiology Institute, Department of Biochemistry, University of Oxford, Oxford, OX1 3QU, UK

Jerold Chun
Department of Molecular Biology, Helen L. Dorris Child and Adolescent Neuropsychiatric Disorder Institute, The Scripps Research Institute, 10550 North Torrey Pines Rd., ICND-118, La Jolla, CA 92037, USA

Sean S. Davies
Department of Pharmacology, Vanderbilt University, 506A RRB, 2222 Pierce Ave., Nashville, TN 37232, USA

Diego de Mendoza
Departamento de Microbiologia, Facultad de Ciencias Bioquímicas y Farmacéuticas, Universidad Nacional de Rosario, Suipacha 531, (S2002LRK) Rosario, Argentina

Chiara De Simone
Department of Experimental Medicine & Biochemical Sciences, University of Rome "Tor Vergata", Rome, Italy

William Dowhan
Department of Biochemistry and Molecular Biology, University of Texas-Houston, Medical School, 6431 Fannin St. Houston, TX 77030, USA

Filomena Fezza
Department of Experimental Medicine & Biochemical Sciences, University of Rome "Tor Vergata", Rome, Italy

Gilbert O. Fruhwirth
Institute of Biochemistry, Graz University of Technology, Petersgasse 12/2, A-8010 Graz, Austria

Beate Fuchs
University of Leipzig, Medical Department, Institute of Medical Physics and Biophysics, Härtelstr. 16/18, D-04107 Leipzig, Germany

Alexandra Gellhaus
Dept. of Molecular Biology, University of Duisburg-Essen, Hufelandstrasse 55, 45122 Essen, Germany

Hervé Guillou
Laboratoire de Pharmacologie et Toxicologie UR66, INRA, F-31000 Toulouse, France

Erich Gulbins
Dept. of Molecular Biology, University of Duisburg-Essen, Hufelandstrasse 55, 45122 Essen, Germany

J. F. Hatcher
Department of Neurological Surgery, H4-330, Clinical Science Center, 600 Highland Avenue, University of Wisconsin School of Medicine and Public Health, Madison, WI 53792-3232, USA

Albin Hermetter
The Richard Dimbleby Department of Cancer Research, King's College London, London SE1 1UL, United Kingdom

Kenneth V. Honn
Departments of Pathology, and Chemistry, Wayne State University School of Medicine, Detroit, MI 4822, USA

Jing X. Kang
Department of Medicine, Massachusetts General Hospital, Harvard Medical School, Boston, MA 02114, USA

Alexander Karakashian
University of Kentucky College of Medicine, Department of Physiology, Lexington, KY, USA

Steven J. Kridel
Department of Cancer Biology and Comprehensive Cancer Center, Wake Forest University School of Medicine, Medical Center Boulevard, Winston-Salem, NC 27157, USA

Sriram Krishnamoorthy
Departments of Pathology, and Chemistry, Wayne State University School of Medicine, Detroit, MI 4822, USA

Norbert Leitinger
Robert M. Berne Cardiovascular Research Center, 409 Lane Road, Charlottesville, VA 22908, USA

Joy L. Little
Department of Cancer Biology and Comprehensive Cancer Center, Wake Forest University School of Medicine, Medical Center Boulevard, Winston-Salem, NC 27157, USA

Natasha C. Lucki
School of Biology, Georgia Institute of Technology, 310 Ferst Drive, Atlanta, GA 30332-0230, USA

Mauro Maccarrone
Department of Biomedical Sciences, University of Teramo, Piazza A. Moro 45, 64100 Teramo, Italy

María Cecilia Mansilla
Departamento de Microbiologia, Facultad de Ciencias Bioquímicas y Farmacéuticas, Universidad Nacional de Rosario, Suipacha 531, (S2002LRK) Rosario, Argentina

Pascal G.P. Martin
Laboratoire de Pharmacologie et Toxicologie UR66, INRA, F-31000 Toulouse, France

Eugenia Mileykovskaya
Department of Biochemistry and Molecular Biology, University of Texas-Houston, Medical School, 6431 Fannin St. Houston, TX 77030, USA

Tetsuji Mutoh
Department of Molecular Biology, Helen L. Dorris Child and Adolescent Neuropsychiatric Disorder Institute, The Scripps Research Institute, 10550 North Torrey Pines Rd., ICND-118, La Jolla, CA 92037, USA

Mariana Nikolova-Karakashian
University of Kentucky College of Medicine, Department of Physiology, Lexington, KY, USA

Besim Ogretmen
Department of Biochemistry and Molecular Biology, Hollings Cancer Center, Medical University of South Carolina, Medical University of South Carolina, 173 Ashley Avenue, Charleston, SC 29425, USA

Thierry Pineau
Laboratoire de Pharmacologie et Toxicologie, Institut National de la Recherche Agronomique (INRA), 180 Chemin de Tournefeuille, BP 3, F 31931 Toulouse, Cedex 9, France

Peter J. Quinn
Department of Biochemistry, King's College London, 150 Stamford Street, London SE1 9NH, UK

Kristina Rutkute
University of Kentucky College of Medicine, Department of Physiology, Lexington, KY, USA

Sahar A. Saddoughi
Department of Biochemistry and Molecular Biology, Hollings Cancer Center, Medical University of South Carolina, Medical University of South Carolina, 173 Ashley Avenue, Charleston, SC 29425, USA

Jürgen Schiller
Universität Leipzig, Medizinische Fakultät, Institut für Medizinische Physik und Biophysik, Härtelstr. 16/18, D-04107 Leipzig, Germany

Marion B. Sewer
School of Biology, Georgia Institute of Technology, 310 Ferst Drive, Atlanta, GA 30332-0230, USA

Tushi Singal
Institute of Cardiovascular Sciences, St. Boniface General Hospital Research Centre & Department of Physiology, Faculty of Medicine, University of Manitoba, Winnipeg, Canada

Pengfei Song
Department of Biochemistry and Molecular Biology, Hollings Cancer Center, Medical University of South Carolina, Medical University of South Carolina, 173 Ashley Avenue, Charleston, SC 29425, USA

Paramjit S. Tappia
Institute of Cardiovascular Sciences, St. Boniface General Hospital Research Centre & Department of Human Anatomy & Cell Science, Faculty of Medicine, University of Manitoba, Winnipeg, Canada

Tilla S. Worgall
Dept. of Pathology, Columbia University, 168 W 168 St, BB 457, New York, N.Y. 10032, USA

Karsten H. Weylandt
Department of Gastroenterology, Rudolf Virchow Hospital, Charité University Medicine, Berlin, 13353, Germany

Elke Winterhager
Dept. of Molecular Biology, University of Duisburg-Essen, Hufelandstrasse 55, 45122 Essen, Germany

Claude Wolf
Mass Spectrometry Unit, INSERM U538, Faculte de Medecine P. et M. Curie, University Paris-6, 27 rue de Chaligny, Paris 75012, France

# Part I
# Fatty Acids

# Chapter 1
# Transcriptional Regulation of Hepatic Fatty Acid Metabolism

**Hervé Guillou, Pascal G.P. Martin and Thierry Pineau**

**Abstract** The liver is a major site of fatty acid synthesis and degradation. Transcriptional regulation is one of several mechanisms controlling hepatic metabolism of fatty acids. Two transcription factors, namely SREBP1-c and PPARα, appear to be the main players controlling synthesis and degradation of fatty acids respectively. This chapter briefly presents fatty acid metabolism. The first part focuses on SREBP1-c contribution to the control of gene expression relevant to fatty acid synthesis and the main mechanisms of activation for this transcriptional program. The second part reviews the evidence for the involvement of PPARα in the control of fatty acid degradation and the key features of this nuclear receptor. Finally, the third part aims at summarizing recent advances in our current understanding of how these two transcription factors fit in the regulatory networks that sense hormones or nutrients, including cellular fatty acids, and govern the transcription of genes implicated in hepatic fatty acid metabolism.

**Keywords** Lipogenesis · fatty acid · PPARalpha · SREBP-1c · LXR · PKB/AKT · liver · insulin · fasting · polyunsaturated fatty acids

**Abbreviations** ACC: Acetyl-CoA Carboxylase; ChREBP: Carbohydrate-responsive element-binding protein; FAS: Fatty acid synthase; FFA: Free fatty acid; FoxO: Forkhead box O; GSK-3: Glycogen synthase kinase 3; IR: Insulin Receptor; LXR: Liver X receptor; NEFA: Non-esterified fatty acid; PDK: Phosphoinositide dependent kinase; PGC: Peroxisome proliferator activated receptor general co-activator; PI3K: Phosphatidylinositol 3 kinase; PKA: Protein kinase A; PKB: Protein kinase B; PKC: Protein kinase C; PPAR: Peroxisome proliferator activated receptor; PUFA: Polyunsaturated fatty acid; RXR: Retinoid X receptors; SCD: Stearoyl-CoA Desaturase; SRE: Sterol regulatory element; SREBP: Sterol regulatory element-binding protein

T. Pineau
Laboratoire de Pharmacologie et Toxicologie, Institut National de la Recherche Agronomique (INRA), 180 Chemin de Tournefeuille, BP 3, F 31931, TOULOUSE, Cedex 9, France
e-mail: thierry.pineau@toulouse.inra.fr

P.J. Quinn, X. Wang (eds.), *Lipids in Health and Disease*,

## 1.1 Introduction

In mammals, the liver is a major site of fatty acid synthesis and degradation, triglyceride synthesis and energy homeostasis. Impaired balance between synthesis and degradation of fatty acids might result in increased triglyceride assembly and obesity. It can also contribute to the development of the metabolic syndrome as a consequence of ectopic triglycerides accumulation in tissues such as the liver, muscles and pancreatic β-cells. Thus, understanding the regulation of hepatic fatty acid metabolism is an important challenge in order to define nutritional or pharmacological approaches for better prevention and treatment of metabolic diseases.

In this chapter, an overview of pathways for fatty acid synthesis and degradation in the liver will be given in two independent parts. On the one hand, the sterol regulatory element-binding protein 1-c (SREBP1-c) controls the expression of hepatic genes involved in fatty acid and triglyceride synthesis (Liang et al., 2002; Shimomura et al., 1997). Therefore, main genes involved in fatty acid synthesis and their regulation by SREBP1-c are first presented. The different levels of activation for this transcription factor are then reviewed. On the other hand, there is clear evidence supporting a major contribution of the α isoform of the peroxisome proliferator activated receptors (PPARs) to the modulation of a transcriptional program determining the level of hepatic fatty acid degradation (Aoyama et al., 1998; Kersten et al., 1999; Kroetz et al., 1998). Hence, the second part of the chapter will attempt to concisely review our current knowledge of PPARα target genes involved in fatty acid degradation and what we know about PPARα itself.

Finally, examples of physiological and nutritional stimuli impacting on these transcriptional sensors will be provided. This will lead us to focus on SREBP1-c and PPARα within complex regulatory networks of dietary and hormonal regulations in the liver.

## 1.2 Fatty Acid Biosynthesis

Fatty acids are essential constituents of all biological membrane lipids and are important substrates for energy metabolism. They also contribute to the regulation of a wide variety of biological activities. In animal tissues, fatty acids are found as non-esterified (NEFA), as acyl-CoA and most predominantly as acyl chain in complex lipids through amide linkage (ceramide, sphingolipids) or esterified in cholesterol esters, mono- di- and triglycerides, and in phospholipids. Physical, chemical and biological properties of fatty acids largely depend on their length, the number and positions of double bonds.

The liver plays a central role in endogenously-synthesized fatty acid (de novo lipogenesis) and exogenously-derived (dietary) fatty acid metabolism. An overview of fatty acid biosynthesis is given in Fig. 1.1. As presented in this figure,

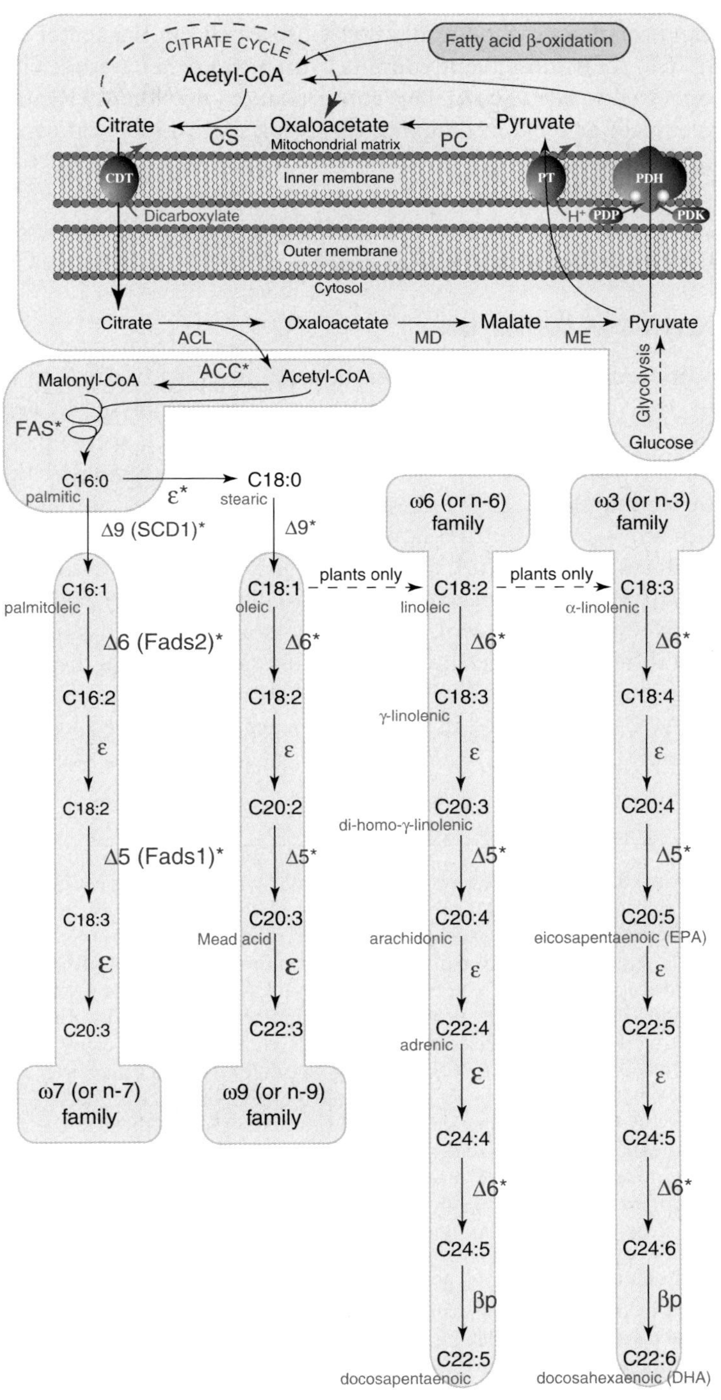

CITRATE CYCLE
Fatty acid β-oxidation
Acetyl-CoA
Citrate
CS
Oxaloacetate
PC
Pyruvate
Mitochondrial matrix
CDT
Inner membrane
PT
PDH
Dicarboxylate
H+
PDP
PDK
Outer membrane
Cytosol
Citrate
ACL
Oxaloacetate
MD
Malate
ME
Pyruvate
Glycolysis
Glucose
Malonyl-CoA
ACC*
Acetyl-CoA
FAS*
C16:0
palmitic
ε*
C18:0
stearic
Δ9 (SCD1)*
Δ9*
ω6 (or n-6) family
ω3 (or n-3) family
C16:1
palmitoleic
C18:1
oleic
plants only
C18:2
linoleic
plants only
C18:3
α-linolenic
Δ6 (Fads2)*
Δ6*
C16:2
C18:2
C18:3
γ-linolenic
C18:4
ε
C18:2
C20:2
C20:3
di-homo-γ-linolenic
C20:4
Δ5 (Fads1)*
Δ5*
C18:3
C20:3
Mead acid
C20:4
arachidonic
C20:5
eicosapentaenoic (EPA)
ε
C20:3
C22:3
C22:4
adrenic
C22:5
ω7 (or n-7) family
ω9 (or n-9) family
C24:4
C24:5
Δ6*
C24:5
C24:6
βp
C22:5
docosapentaenoic
C22:6
docosahexaenoic (DHA)

fatty acid synthesis is tightly linked to glucose metabolism. The full oxidation of glucose can provide mitochondrial acetyl-CoA which can then enter the citrate cycle (Fig. 1.1). This subsequently results in an increase in cytosolic citrate that can be converted to acetyl-coA. This conversion requires the ATP-citrate lyase (ACL), a critical enzyme in coupling of glycolysis to fatty acid synthesis by providing acetyl-CoA.

## 1.2.1 *De Novo Lipogenesis Enzymes*

### 1.2.1.1 Acetyl-CoA Carboxylase

De novo lipogenesis starts with the carboxylation of acetyl-CoA to malonyl-CoA (Fig. 1.1). This ATP-dependent carboxylation is catalysed by acetyl-coA carboxylase (ACC) (Tong and Harwood, 2006) which acts as the pace-setting enzyme of fatty acid synthesis. ACC activity is tightly regulated through a variety of stimuli (Kim, 1997). In particular, it is allosterically activated by citrate. Not only is ACC the rate-limiting enzyme in fatty acid biosynthesis, but it also produces malonyl-CoA an allosteric inhibitor of carnitine palmitoyl-transferase I (CPT-I) which is required for mitochondrial fatty acid uptake and oxidation (McGarry and Brown, 1997). ACC isoforms, ACC1 and ACC2, are encoded by different genes (Kim, 1997; Widmer et al., 1996), display distinct cellular distributions (Abu-Elheiga et al., 2000) and appear to have different functions. ACC1 is expressed in lipogenic tissues and localized in the cytosol. ACC2 is mostly expressed in oxidative tissues, such as heart and skeletal muscle,

**Fig. 1.1** (continued) Hepatic pathways of fatty acid biosynthesis. The upper panel illustrates the origin of cytosolic acetyl-CoA, the precursor for the de novo synthesis of palmitic acid which is illustrated in the middle panel. The lower part illustrates the synthesis of unsaturated FA. SREBP-1c target genes involved in fatty acid biosynthesis are marked with an asterisk (*). Mitochondrial acetyl-CoA, produced by mitochondrial FA oxidation and pyruvate oxidative decarboxylation (PDH: pyruvate dehydrogenase complex, regulated by phosphorylation through PDP: PDH phosphorylase and PDK: PDH kinase), is transported to the cytosol through its condensation as citrate (CS: citrate synthase) for which a transporter exist (CDT: citrate/dicarboxylate transporter). In the cytosol, the ATP-citrate lyase (ACL) reconverts citrate to acetyl-CoA which is then carboxylated to malonyl-CoA through the acetyl-CoA carboxylases (ACC). The fatty acid synthase (FAS) then synthesizes palmitic acid from malonyl-CoA through a series of reactions catalyzed by its different functional domains. The synthesis of unsaturated FAs from palmitic acid involves elongation (ε) and rate-limiting desaturation steps (additional requirements for βp: a peroxisomal β-oxidation round are indicated). FA desaturation is catalyzed by FA Δ9-, Δ6- and Δ5-desaturases (main hepatic isoforms: SCD1 for stearoyl-CoA desaturase 1, Fads2 for FA desaturase 2 and Fads1 respectively) which function on various substrates. Note that the synthesis of linoleic (C18:2ω6) and α-linolenic (C18:3ω3) is plant-specific. Thus, these essential precursors of the ω6 and ω3 families must be provided by dietary inputs. PT: pyruvate transporter; PC: pyruvate carboxylase; MD: malate dehydrogenase; ME: malic enzyme

and localizes at the mitochondrial membrane. Wakil and collaborators have investigated the respective roles of ACC isoforms in vivo. Transgenic mice lacking ACC1 are not viable (Abu-Elheiga et al., 2005). The liver-specific deletion of ACC1 results in a reduced triglyceride accumulation without affecting fatty acid oxidation (Mao et al., 2006). Mice lacking ACC2 show increased oxidation of fatty acids and decreased fat storage (Abu-Elheiga et al., 2001).

All together, these data suggest that the malonyl-CoA produced by ACC1 is primarily used for fatty acid synthesis while mitochondrial ACC2 produces a pool of malonyl-CoA that controls CPT-I activity and fatty acid β-oxidation.

#### 1.2.1.2 Fatty Acid Synthase

Malonyl-CoA is then used for fatty acid synthesis. This process requires Fatty Acid Synthase (FAS) (Chirala and Wakil, 2004) that uses acetyl-CoA as a primer, malonyl-CoA as a two-carbon donor, and NADPH as a reducing equivalent (Fig. 1.1). The predominant fatty acid produced by FAS is palmitic acid (C16:0). The structure of FAS has been extensively studied (Asturias et al., 2005; Maier et al., 2006). FAS is a multifunctional complex consisting of two identical monomers. However, only the dimeric form is active (Chirala et al., 2001). In animal FAS functions as a homodimeric complex of 250 kDa found in the cytoplasm. It harbours seven different enzymatic activities required for palmitic acid synthesis.

Transgenic mice lacking fatty acid synthase die in utero (Chirala et al., 2003) which suggest not only that this FAS activity is important during development but also that there is no alternative biosynthetic pathway. In mice with a liver-specific inactivation of FAS, lipid and glucose homeostasis is markedly impaired (Chakravarthy et al., 2005).

FAS represents a drug target not only for obesity-related diseases (Loftus et al., 2000) but also for treatment and prevention of human cancers (Kuhajda, 2006; Menendez and Lupu, 2007).

#### 1.2.1.3 Δ9-desaturase

Δ9-desaturase is generally considered as a lipogenic enzyme. It catalyses the introduction of a Δ9 double bound on the acyl-chain of saturated fatty acids such as palmitic acid, which can be produced by FAS. The hepatic Δ9-desaturase is well-known (Heinemann and Ozols, 2003) since it has been purified more than 30 years ago (Strittmatter et al., 1974). This Δ9-desaturase acts on both palmitic acid (C16:0) and stearic acid (C18:0) to produce pamitoleic acid (C16:1n-7) and oleic acid (C18:1n-9) respectively (Fig. 1.1). Under normal dietary conditions, C16:1 n-7 and C18:1n-9 are the main fatty acids of the n-7 and n-9 families in animals.

The protein lies in the endoplasmic reticulum where it co-localises with NADH cytochrome b5 reductase and cytochrome b5, both required for Δ9-desaturase activity (Jeffcoat et al., 1977; Strittmatter et al., 1974). Δ9-desaturase is also named stearoyl-CoA isomerase (SCD). At least four distinct SCD isoforms exist in mice (Miyazaki et al., 2003). These isoforms named SCD1, SCD2, SCD3 and SCD4 have distinct tissue distributions. In human only three isoforms have been identified. One human isoform, SCD1 (Zhang et al., 1999), is highly similar to mouse SCD1. SCD2 (Zhang et al., 2005) and another human isoform, named SCD5 (Wang et al., 2005), which is unique to primates, have been cloned.

SCD1 is, at least in rodents, the main hepatic isoform of SCD. Its physiological role has been extensively studied, in particular its role in the control of metabolism (Dobrzyn and Ntambi, 2005). SCD1 has been established as a critical player in the development of the metabolic syndrome. A very large body of evidence has been obtained in vivo showing that SCD1 deficiency protects against adiposity and several related pathologies. The different models used are the Asebia mouse, a natural mutant lacking SCD1 (Cohen et al., 2002; Zheng et al., 1999), a mouse with targeted disruption of SCD1 (Miyazaki et al., 2001; Ntambi et al., 2002) and more recently a mouse model with liver specific disruption of SCD1 (Miyazaki et al., 2007).

### *1.2.2 Elongases and PUFA-desaturases*

Elongases (marked ε on Fig. 1.1) and PUFA desaturases (marked Δ6 and Δ5 on Fig. 1.1) are not usually classified as "lipogenic" enzymes. However, they play a critical role in the synthesis of a variety of fatty acids either by acting on endogenously produced fatty acids (from lipogenesis) or on exogenous (dietary) fatty acids. Therefore, we chose to briefly introduce these enzymes.

#### 1.2.2.1 Elongases

Fatty acid elongation mainly occurs in the endoplasmic reticulum where elongases add two carbons to acyl-CoA substrates by using malonyl-CoA. Elongases are the rate-limiting protein acting in a multi proteic complex including a 3-keto acyl-CoA reductase and trans-2,3-enoylCoA reductase (Leonard et al., 2004; Moon et al., 2001; Prasad et al., 1986). The family of elongases consists of at least six members (Elovl-1, 2, 3, 4, 5 and 6) in mouse and human (Jakobsson et al., 2006) with tissue-specific, nutritional and developmental regulation (Wang et al., 2005). In the liver, the most abundant isoform is Elovl-5. Elovl-1, Elovl-2, and Elovl-6 are also expressed in the liver (Wang et al., 2006). Elovl-1 and Elovl-6 are saturated and monounsaturated fatty acid elongases while both Elovl-2 and Elovl-5 are likely to play a role in the PUFA biosynthetic pathway (Jakobsson et al., 2006). Elovl-6 particularly controls the balance between C18 and C16 fatty acids and has been recently shown to play a crucial role in obesity-induced insulin resistance (Matsuzaka et al., 2007).

#### 1.2.2.2 PUFA Desaturases

Two PUFA desaturases have been isolated in rodents and human. These are Δ6-desaturase (Cho et al., 1999b), encodef by the FADS2 gene, and Δ5-desaturase, encoded by the FADS1 gene (Cho et al., 1999a) which are highly expressed in the liver. In human, both genes belong to a gene cluster which includes FADS3, another putative desaturase which remains uncharacterized (Marquardt et al., 2000). Unlike Δ9-desaturases, both Δ6- and Δ5-desaturases contain a cytochrome b5 domain required for their activity (Guillou et al., 2004).

Under dietary conditions supplying both linoleic (C18:2n-6) and α-linolenic acid (C18:3n-3), Δ6-desaturase and Δ5-desaturase, together with elongases, synthesize very long chain PUFAs of the n-6 and n-3 series, such as arachidonic (C20:4n-6) and docosahexaenoic (C22:6n-3), which are involved in a number of pivotal biological functions. The clinical picture described in a human with a mutation in Δ6-desaturase promoter which results in a low expression of Δ6-desaturase emphasizes the importance of this pathway (Nwankwo et al., 2003). Δ6-desaturase has several substrates and acts twice in the pathways (Fig. 1.1), both on 18 carbon and 24-carbon PUFAs of n-6 and n-3 families (D'Andrea et al., 2002; Voss et al., 1991).

### *1.2.3 SREBP-1c a Key Regulator of Fatty Acid Synthesis*

Sterol regulatory element binding proteins (SREBPs) are a family of transcription factors originally discovered as involved in the regulation of genes controlling the cellular availability of cholesterol (Wang et al., 1994). SREBP-1c is the isoform which appears to be primarily involved in regulating the expression of lipogenic and fatty acid-metabolizing enzymes. SREBP-1c is also sometimes referred to as Adipocyte determination and differentiation-dependent factor 1 (ADD1) (Kim and Spiegelman, 1996; Tontonoz et al., 1993).

#### 1.2.3.1 The SREBP Family

The SREBP family of transcription factors control the expression of genes involved in cholesterol and fatty acid metabolism. They play critical roles during adipocyte differentiation and insulin-dependent gene expression (Bengoechea-Alonso and Ericsson, 2007; Eberle et al., 2004; Espenshade and Hughes, 2007; Ferre and Foufelle, 2007; Horton, 2002). The family consists of three different SREBP proteins, SREBP-1a, SREBP-1c and SREBP2. In human, SREBP-1a and SREBP-1c are produced from a single gene (Hua et al., 1995), SREBP-2 is produced from a separate gene (Miserez et al., 1997).

SREBPs are basic-helix-loop-helix-leucine zipper (bHLH-LZ) transcription factors synthesized as 1150 amino acids precursors bound to the endoplasmic reticulum (ER) membrane (Hua et al., 1996). All SREBPs precursors are organized into three distinct domains. The N-terminal domain contains the transactivation domain, a region rich in serine and proline and the bHLH-LZ region for DNA binding and dimerization. SREBPs contain two hydrophobic transmembrane domains and a short loop located in the ER lumen. The C-terminal end is a regulatory region. To be active, the N-terminal region of SREBPs must be released from the membrane and translocate to the nucleus. Unlike other bHLH-LZ transcription factors, instead of a well-conserved arginine residue in their basic domain, SREBPs contain a tyrosine residue. This feature allows SREBPs to bind not only to E-boxes (5′-CANNTG-3′, where N can be any base), like all bHLH proteins do, but also to sterol regulatory element (SRE) sequences (5′-TCACNCCCAC-3′) (Kim et al., 1995).

The relative abundance of SREBP isoforms differs in cultured cells and animal tissues. In cultured cells SREBP-1a is highly expressed (Shimano et al., 1997a). In adult animal tissues, SREBP-1c expression is approximately 10-times more abundant than SREBP-1a and twice as abundant as SREBP-1 (Shimomura et al., 1997).

### Regulation of Fatty Acid Synthesis by SREBP-1c

The relative function of SREBP isoforms has been investigated both in vitro (Pai et al., 1998) and in vivo (Horton et al., 1998b; Shimano et al., 1999). While SREBPs have overlapping function, studies performed with transgenic mice overexpressing nuclear forms of SREBPs (nSREBPs) in the liver have provided information on the specific target genes of individual isoforms.

SREBP-2 primarily regulates the expression of genes involved in cholesterol uptake and synthesis. The hepatic overexpression of nSREBP-2 in mice results in a marked increase in the expression of genes involved in cholesterol biosynthesis and a moderate induction of genes involved in fatty acid synthesis (Horton et al., 2003, 1998b). Overexpression of nSREBP-1a in mice liver strongly increases the expression of genes involved in cholesterol synthesis and fatty acid synthesis (Horton et al., 2003; Shimano et al., 1996). Interestingly, despite the overlap in target genes between SREBP-1a and SREBP-2, deletion of SREBP-2 in mice is lethal (Shimano et al., 1997b).

Finally, in mice overexpressing nSREBP-1c, the expression of lipogenic enzymes is elevated while no effect on cholesterol synthesis is reported (Shimano et al., 1997a). Moreover, the selective deletion of SREBP-1c isoform in mice leads to a reduction of mRNAs encoding enzymes involved in fatty acid synthesis (Liang et al., 2002). Altogether, these data showed that amongst SREBP isoforms, SREBP-1c is the isoform most specifically involved in the regulation of lipogenic genes such as ACC, FAS and SCD1. More recently Δ6-desaturase (Matsuzaka et al., 2002; Nara et al., 2002), Δ5-desaturase (Matsuzaka et al., 2002) and the Elovl6 elongase (Kumadaki et al., 2008; Moon et al., 2001) were reported to be SREBP-1c target genes.

## Mechanism of SREBP-1c Activation

### *Transcriptional Regulation of SREBP-1c*

Unlike other SREBP isoforms, SREBP-1c expression and nuclear abundance is not increased in case of low cholesterol availability (Sheng et al., 1995). In the liver, SREBP-1c expression is very sensitive to the nutritional status (Horton et al., 1998a). SREBP-1c mRNA level decreases during fasting and is increases markedly in animals refed a high carbohydrate diet. Experiments performed in hepatocytes (Foretz et al., 1999) showed that the transcription of SREBP1c is induced by insulin. In vivo, it has been shown that the 2.2 kb of the 5′-flanking sequence of SREBP-1c is sufficient for the fasting-refeeding regulation of its expression (Takeuchi et al., 2007). This induction of SREBP-1c transcription leads to a parallel increase in the level of the ER membrane-bound precursor and the nuclear form (Azzout-Marniche et al., 2000).

The signalling pathways involved in the control of SREBP-1c expression are not clearly understood. However, Protein Kinase B (PKB) (Fleischmann and Iynedjian, 2000; Porstmann et al., 2005) and Protein Kinase C λ (Matsumoto et al., 2003; Taniguchi et al., 2006) are likely to be involved (Fleischmann and Iynedjian, 2000). Moreover, the Liver X Receptor, a nuclear receptor highly expressed in the liver, also contributes to regulate SREBP-1c expression (Peet et al., 1998; Repa et al., 2000). In the last part of this chapter, we will see how PKB and LXR might be involved in the control of SREBP-1c expression in response to the dietary status.

### *Proteolytic Cleavage of SREBP-1c*

Several aspects of the mechanisms by which the cellular sterol concentration regulates SREBP-2 and SREBP-1a proteolytic cleavage have been partly characterized (Brown and Goldstein, 1997; Espenshade and Hughes, 2007). Under high cholesterol SREBP precursors are retained in the ER membranes through association with the SREBP cleavage activating protein (SCAP) and a protein of the insulin-induced gene (Insig) family (McPherson and Gauthier, 2004; Yang et al., 2002). Retention of the SREBP/SCAP complex in the ER is mediated by the sterol-mediated inhibition of the COPII proteins Sec 23/24 from binding to SCAP (Espenshade et al., 2002). Under low cholesterol, SCAP dissociates from Insig and escorts SREBP from the ER to the Golgi. This transport occurs through incorporation of SCAP/SREBP into COPII vesicles containing the subcomplex of COPII proteins consisting of the small GTPase Sar1 and the cargo recognition complex Sec 23/24 (Gurkan et al., 2006). Once in the Golgi, two distinct proteases, S1P and S2P, sequentially cleave the precursor form of SREBP releasing the mature form in the cytoplasm before it transfers to the nucleus (Duncan et al., 1998; Sakai et al., 1998). A representation of the sterol-sensitive pathway that regulates the proteolytic cleavage of SREBP-2 and SREBP-1a is given in Fig. 1.2. Studies performed in vivo have shown that sterol depletion does not influence the cleavage of SREBP-1c (Sheng et al., 1995).

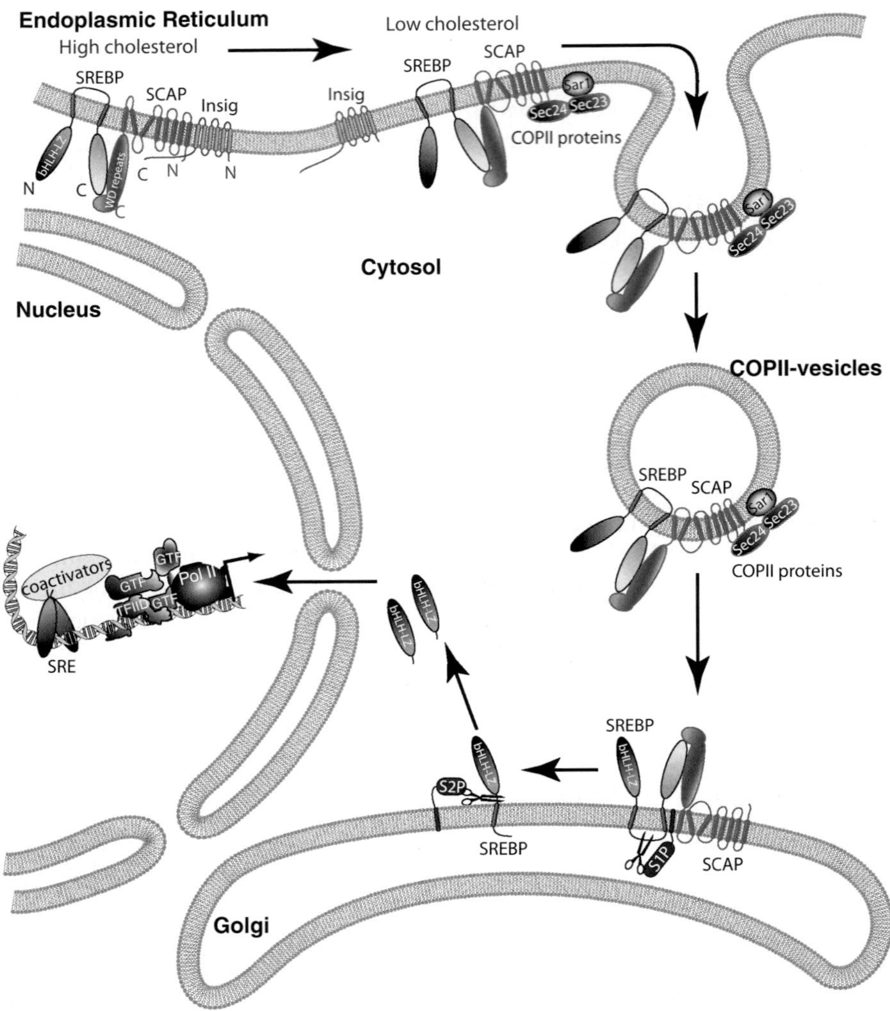


**Fig. 1.2** A schematic overview of the proteolytic processing of SREBPs. This pathway was described as a cholesterol-sensing process which applies to both SREBP-2 and SREBP-1a. While SCAP and Insig are involved in the proteolytic cleavage of SREBP-1c the detailed mechanisms by which insulin regulates SREBP-1c processing are not known

While it has been shown that SREBP-1c cleavage is under the control of insulin (Hegarty et al., 2005) the mechanism by which this cleavage occurs remains to be elucidated. However, both SCAP (Matsuda et al., 2001) and Insigs (Engelking et al., 2004; 2005, 2006) are involved in regulating SREBP-1c in vivo. One Insig isoform (Insig-2a) is highly expressed in the liver and it is selectively down-regulated by insulin (Yabe et al., 2003). Insig-2a might therefore be involved in a specific, insulin-sensitive and high cholesterol-insensitive, regulation of SREBP-1c.

*Nuclear Regulation of SREBP-1c Activity*

SREBPs interact with several co-activators and act in conjunction with several other transcription factors (Bengoechea-Alonso and Ericsson, 2007; Weber et al., 2004). This could lead to other signal-sensitive co-activation of SREBP activity. For instance, the peroxisome proliferator activated receptor general co-activator β (PGC-1β) interacts with and stimulates SREBP-1c activity in the liver of mice fed a diet rich in saturated fatty acids (Lin et al., 2005). Chromatin remodelling complexes have also recently been shown to interact with SREBP-1c and contribute to insulin sensitivity (Lee et al., 2007).

Nuclear SREBP activity is also controlled by several post-translational regulations including phosphorylation, acetylation, sumoylation and ubiquitination (Bengoechea-Alonso and Ericsson, 2007; Eberle et al., 2004). Several cross-talks with kinase-mediated signalling are likely to influence SREBP-1c in the liver in vivo. GSK3 is involved in the phosphorylation and subsequent ubiquitination of SREBP-1c (Punga et al., 2006; Sundqvist et al., 2005). PKA-dependent phosphorylation of SREBP-1c also reduces SREBP-1c activity (Lu and Shyy, 2006). Together with 26S proteasome, Erk-dependent pathways is involved in the reduction of nuclear SREBP-1 induced by docosahexaenoic acid (C22:6n-3) (Botolin et al., 2006).

SREBP-1c appears as a very important player in controlling the transcriptional program for lipogenesis and fatty acid synthesis. The synthesis of fatty acids in the liver is particularly intensive in response to feeding when insulin and glucose levels are high. Under those circumstances, particularly elevated insulin, the role of SREBP-1c is very clear. However, SREBP-1c deficiency does not fully abolish the expression of genes involved in lipogenesis (Liang et al., 2002) which implies that other transcription factors are involved. Amongst those, LXR and the carbohydrate-responsive element-binding protein (ChREBP) are likely to make a particularly significant contribution to the regulation of fatty acid synthesis in response to elevated glucose and insulin (Denechaud et al., 2008).

## 1.3 Fatty Acid Degradation

Fatty acid catabolism occurs in various organs throughout the body. Some organs such as the adult heart rely preferentially on fatty acid degradation as their energy source. The liver is also a major site of fatty acid catabolism and the pathways involved are strongly regulated in this organ. In the liver, fatty acid catabolism not only acts as an energy source but also as a source of substrates for the synthesis of ketone bodies (ketogenesis) which, under fasting conditions, can be used as fuels by extrahepatic tissues. In this chapter, we will first describe the different pathways involved in hepatic fatty acid catabolism and present the known hepatic regulations at the transcriptional level.

## 1.3.1 Metabolic Pathways in the Liver

Fatty acid breakdown is mainly accomplished *via* the mitochondrial and the peroxisomal β-oxidation pathways (Fig. 1.3). While short-chain and medium-chain fatty acids are mainly degraded in the mitochondria, very long-chain fatty acids (more than 20 carbons) are first shortened in the peroxisomes and then usually further oxidized in the mitochondria.

### 1.3.1.1 Import into Mitochondria

Before being oxidized in the mitochondrial matrix, long-chain fatty acids must be transported through the mitochondrial membrane *via* a tightly regulated transport system involving four different actors (Kerner and Hoppel, 2000). The first step is the activation of the free fatty acid (FFA) through its esterification with coenzyme A. This step is catalyzed by the long-chain acyl-CoA synthetase located at the mitochondrial outer membrane. Since the mitochondrial inner membrane is impermeable to long-chain acyl-CoAs, the mitochondrial carnitine-dependent transport system, involving three proteins, transports the acyl-CoA to the mitochondrial matrix. The carnitine palmitoyltransferase I (CPT-I), which is also located in the mitochondrial outer membrane, transesterifies the acyl-CoA to acylcarnitine. Then a carnitine/acylcarnitine translocase (CACT), an integral inner membrane protein, transports the acylcarnitine to the mitochondrial matrix together with carnitine, from the matrix to the intermembrane space. Finally, the carnitine palmitoyltransferase II (CPT-II), an enzyme associated with the inner leaflet of the mitochondrial inner membrane, catalyzes the reverse reaction as CPT-I and thus reconverts the acylcarnitine to its respective fatty acyl-CoA.

### 1.3.1.2 Mitochondrial β-Oxidation

Once into the mitochondrial matrix, fatty acyl-CoAs are β-oxidized through a pathway involving four sequential reactions. These are catalyzed by enzymes displaying chain length specificities. The first step involves acyl-CoA dehydrogenases (short, medium, long and very long-chain acyl-CoA dehydrogenases or SCAD, MCAD, LCAD and VLCAD) which produce enoyl-CoAs and use FAD as a cofactor. Enoyl-CoAs are then hydrated by at least two different enoyl-CoA hydratases (short chain enoyl-CoA hydratase -SCEH or Echs1- and long chain enoyl-CoA hydratase -LCEH-) which produce L-hydroxyacyl-CoAs. The third step is catalyzed by short chain 3-hydroxyacyl-CoA dehydrogenase (SCHAD or Hadh2, also known as 17β-hydroxysetroid dehydrogenase 10 or Hsd17b10) and by 3-hydroxyacyl-CoA dehydrogenase (HAD or Hadh) (Yang et al., 2005) which produce 3-ketoacyl-CoAs and use $NAD+$ as a cofactor. The final step is catalyzed by 3-ketoacyl-CoA thiolases amongst which acetyl-CoA acyltransferase 2 (Acaa2) and acetoacetyl-CoA thiolase (Acat1) have been best characterized. The latter also catalyzes the condensation of two acetyl-CoA which is the first reaction involved in ketogenesis. In

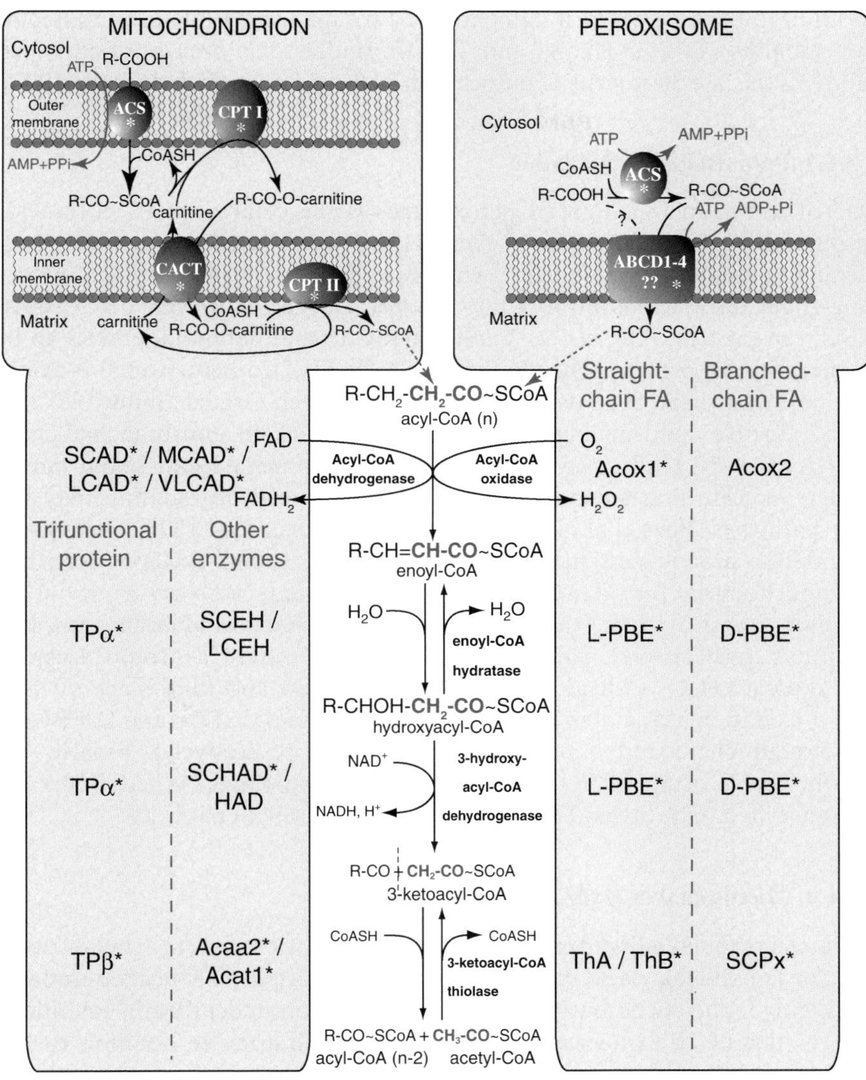


**Fig. 1.3** Hepatic pathways of mitochondrial and peroxisomal fatty acid catabolism. The upper part illustrates the systems of FA activation and entry into the mitochondrion and the peroxisome. The FA β-oxidation pathway is illustrated in the center and the specific cofactors and enzymes involved in the mitochondrion (*left*) and in the peroxisome (*right*) are presented on the sides. Genes for which a PPARα-dependent regulation has been reported in the literature are identified by an asterisk (*). For most of these genes, at least one functional PPRE has been characterized in their promoter. ACS: acyl-CoA synthesase; CPT: carnitine palmitoyl-transferase; CACT: carnitine:acylcarnitine translocase; ABCD1-4 : ATP-binding cassette transporter, sub-family D, members 1 to 4; S-, M-, L- and VLCAD: short, medium, long and very long chain acyl-CoA dehydrogenase; TPα (or β): trifunctional protein α (or β) subunit; SCEH and LCEH: short and long chain enoyl-CoA hydratase; (SC)HAD: (short chain) 3-hydroxyacyl-CoA dehydrogenase; Acaa2: acetyl-CoA acyltransferase 2; Acat1: acetyl-CoA acetyltransferase 1; Acox: acyl-CoA oxidase; L- and D-PBE : L- and D-peroxisomal bifunctional enzyme; ThA (or B): peroxisomal 3-ketoacyl-CoA thiolase A (or B); SCPx: sterol carrier protein X

addition to these enzymes, a mitochondrial trifunctional protein composed of two subunits (TPα or Hadha and TPβ or Hadhb) has been purified which catalyzes the last three steps of mitochondrial β-oxidation (Uchida et al., 1992).

#### 1.3.1.3 Peroxisomal β-Oxidation

One of the main functions of peroxisomes is the catabolism of fatty acids through β-oxidation and, for β-methyl fatty acids such as phytanic acid, through α-oxidation. Little is known about the mechanisms involved in very long-chain fatty acid transport into the peroxisome. However, peroxisomal ABC transporters (ABCD1, 2, 3 and 4 in mouse and human) are likely to be involved in this process (Theodoulou et al., 2006). Two peroxisomal β-oxidation pathways have been described which have been termed "inducible" and "non-inducible" and mainly function on straight-chain and branched-chain fatty acids respectively. Nevertheless, observations made in knockout mouse models indicate that some intermediates in fatty acid degradation may use both pathways (Baes et al., 2000; Jia et al., 2003; Qi et al., 1999). Peroxisomal β-oxidation proceeds essentially through pathways similar to those described for mitochondrial β-oxidaton except for the first step, catalyzed by acyl-CoA oxidases 1 and 2 (Acox1 and 2 for the straight-chain and branched-chain pathways, respectively) which use molecular oxygen as a proton acceptor and generate $H_2O_2$ instead of $FADH_2$. The second and third steps of fatty acid β-oxidation are catalyzed by bifunctional enzymes (L-PBE and D-PBE for the straight-chain and branched-chain pathways respectively). Finally, the fourth step is catalyzed by the 3-ketoacyl-CoA thiolases A and B for the straight-chain pathway and SCPx for the branched-chain pathway.

#### 1.3.1.4 Microsomal ω-Hydroxylation

While microsomal ω-hydroxylation is not believed to represent a major pathway for hepatic fatty acid catabolism under normal physiological conditions, this highly inducible pathway may become important under certain conditions such as fasting or exposure to peroxisome proliferators. In addition to FA catabolism, the cytochrome P450 4As (CYP4As) involved in this pathway catabolize leukotrienes and prostanoids and also generate bioactive molecules from arachidonic acid ω-hydroxylation.

#### 1.3.1.5 Ketogenesis

Although ketogenesis is not a pathway of fatty acid catabolism *per se*, it is activated in several situations of hepatic fatty acid breakdown such as fasting. Ketogenesis converts acetyl-CoA to ketone bodies (acetoacetate, acetone and β-hydroxybutyrate) which can be used as an energy source by certain tissues such as brain or muscles. The key enzyme of this pathway is the mitochondrial 3-hydroxy-3-methylglutaryl-CoA synthase or HMGCS2.

#### 1.3.1.6 Fatty Acid Catabolism Disorders and Knockout Models

In humans, several genetic disorders of fatty acid catabolism, such as the most common MCAD-deficiency, have been reported but their description falls beyond the scope of this review. Several recent reviews have described many of these diseases and their symptoms in more detail (Longo et al., 2006; Rinaldo et al., 2002; Wanders and Waterham, 2006; Yang et al., 2005). For mitochondrial fatty acid oxidation disorders the symptoms often develop at infancy during an episode of increased energy demand such as fasting, exercise or illness. Peroxisomal fatty acid oxidation enzyme deficiencies often involve neuropathy and retinopathy.

### *1.3.2 Transcriptional Regulation of Hepatic Fatty Acid Catabolism*

Various situations yield to induction of fatty acid catabolism in the liver such as birth which corresponds to a transition in energy source from a carbohydrate-rich to a lipid-rich regimen; fasting, which is associated with free fatty acid release in the blood due to lipolysis of the adipose tissue; as well as exposure to various xenobiotics. Several biochemical mechanisms exist to rapidly control the activity of the different enzymes involved in fatty acid oxidation. However, the transcriptional regulation of the expression of these enzymes is an additional level of control necessary for long-term maintenance of the organism energy balance.

#### 1.3.2.1 Identification of PPARα

One of the key transcriptional regulators of fatty acid catabolism which has been well characterized since its discovery in 1990 (Issemann and Green, 1990) is the nuclear receptor PPARα (Peroxisome Proliferator-Activated Receptor α). The nuclear receptor (NR) superfamily, which comprises 48 members in human and 49 in mouse, mainly includes transcription factors whose activity is primarily regulated by ligand binding. These receptors are organized in different functional domains. The poorly conserved amino-terminal domain (A/B domain) contains a ligand-independent transactivation function (AF-1) and may be involved in the interaction of the NR pathways with different signalling pathways though post-translational modifications of the NR. For PPARs, other important roles attributed to this domain include the interaction with coregulators (Tudor et al., 2007) and the control of the selectivity of target gene regulations among PPAR isoforms (Hummasti and Tontonoz, 2006). The central domain or DBD (DNA-binding domain) is highly conserved and interacts with the response elements located in the promoters of the target genes. Finally, the well-conserved carboxy-terminal ligand-binding domain (LBD) exhibits a ligand-dependent transactivation function which, upon ligand binding, interacts with different coregulators. The actions of these coregulators at the chromatin (remodelling and histone post-translational modifications) and at the basal transcriptional machinery (interactions with basal transcription factors) levels are eventually responsible for the regulation of the transcription of the target genes (Lonard and O'Malley B, 2007). X-ray crystal structures of the LBDs of many NRs have helped elucidating the main determinants of ligand binding, dimerization and cofactors interactions (Zoete et al., 2007).

#### 1.3.2.2 PPARα and its Target Genes

Three different PPARs have been identified: PPARα (NR1C1), PPARβ/δ (NR1C2) and PPARγ (NR1C3). The latter is almost exclusively expressed in adipose tissues where it is particularly involved in adipogenesis and mediates the effects of the thiazolidinedione drugs used as insulin sensitizers (Walczak and Tontonoz, 2002). PPARα is mainly expressed in tissues exhibiting high rates of fatty acid catabolism such as the liver, the heart, the skeletal muscle, the kidney or the brown adipose tissue (Desvergne and Wahli, 1999). Although in human, PPARα is expressed at much lower levels in the liver compared to rodent species (Palmer et al., 1998), it is generally considered the main mediator of the beneficial effects of the fibrate drugs, used for the treatment of dyslipidemias (Gervois et al., 2007; Zandbergen and Plutzky, 2007). Finally, PPARβ is more ubiquitously expressed. Although PPARβ has been less studied than the other two isoforms, it has recently emerged as another important regulator of fatty acid transport and catabolism, especially in muscle cells (Barish et al., 2006). All three PPARs function as heterodimers with the retinoid X receptors (RXRα, β or γ) and generally recognize response elements organized as direct repeats of the consensus motif AGGTCA separated by one nucleotide (DR1). Most direct PPARα target genes characterized to date (Mandard et al., 2004) are involved in fatty acid transport (lipoprotein metabolism, transport across plasma membrane), fatty acid intracellular trafficking (fatty acid or acyl-CoA binding proteins, mitochondrial and peroxisomal transport systems) and fatty acid oxidation (mitochondrial β-oxidation, peroxisomal α- and β-oxidation and microsomal ω-oxydation). In Fig. 1.3, several genes involved in fatty acid catabolism and known to be PPARα target genes have been marked.

The observations made in PPARα –/– mice (Lee et al., 1995) provide in vivo evidence for a major role of PPARα in regulating these enzymes. In PPARα –/– mice, Aoyama et al. (Aoyama et al., 1998) observed a marked decrease in the expression of genes involved in fatty acid-catabolizing enzymes. Consistent with an impaired ability to degrade fatty acid, PPARα–/– mice accumulate hepatic triacylglycerols (Costet et al., 1998).

#### 1.3.2.3 PPARα Ligands

The PPARα ligands, named peroxisome proliferators (PPs) include a wide range of structurally diverse molecules which induce, primarily in rodent species, hepatic peroxisome proliferation and the expression of PPARα target genes. Several early studies have identified fatty acids as ligands and activators of PPARα (Forman et al., 1997; Gottlicher et al., 1992; Kliewer et al., 1997; Krey et al., 1997; Murakami et al., 1999). Among the PPARs, PPARα appears to display the highest affinity for polyunsaturated fatty acids (PUFAs) and among the fatty acids, the saturated fatty acids appear as lower affinity ligands for the PPARs. Recent studies have further concluded that several fatty acids, especially PUFAs, display affinities in the nanomolar range for PPARα (Ellinghaus et al., 1999; Hostetler et al., 2005, 2006; Lin et al., 1999), which seems compatible with the circulating and more importantly with the estimated intracellular and intranuclear concentrations of these fatty acids. Additionally, it has been proposed that fatty acid-binding proteins (FABPs) may shuttle these ligands to the receptor located in the nucleus (Tan et al., 2002; Wolfrum et al., 2001). Some studies also suggest that fatty acyl-CoAs represent better ligands for PPARα than their corresponding free fatty acids (Forman et al., 1997; Hostetler et al., 2005, 2006). This hypothesis is further strengthened by the observations made by Reddy and collaborators on the Acox (Fan et al., 1998; Hashimoto et al., 1999, 2000) or bifunctional enzymes-deficient mice (Jia et al., 2003; Qi et al., 1999). In Acox1-deficient mice, the PPARα pathway is constitutively activated, leading to hepatic peroxisome proliferation, induction of PPARα target genes and eventually hepatocarcinogenesis (Fan et al., 1998; Hashimoto et al., 2000). In Acox1/PPARα-double knockout mice, spontaneous hepatic peroxisome proliferation and PPARα target genes induction are no longer observed (Hashimoto et al., 1999), suggesting that fatty acyl-CoA substrates of Acox1 may serve as endogenous PPARα ligands. In contrast, L-PBE-deficient mice do not display spontaneous peroxisome proliferation or the induction of PPARα target genes (Qi et al., 1999), which could further suggest that fatty acyl-CoA are the endogenous PPARα ligands. However, the young L-PBE-deficient mice do not display hepatic steatosis like the Acox1-deficient mice (Qi et al., 1999), suggesting that the substrates of L-PBE may also be metabolized by the D-PBE, which is consistent with the hepatic accumulation of long chain-fatty acids observed in D-PBE-deficient mice (Baes et al., 2000). This latter observation may explain the hepatic induction of the expression of PPARα target genes in D-PBE–/– mice, although it is less marked than in Acox1–/– mice (Baes et al., 2000; Jia et al., 2003). Interestingly, when treated with a pharmacological activator of PPARα, L-PBE-deficient mice do not display hepatic peroxisome proliferation despite an increased expression of PPARα target genes, suggesting a major role of L-PBE induction in the process of peroxisome proliferation (Qi et al., 1999). The L-PBE/D-PBE-double knockout mice show a spontaneous increased expression of PPARα target genes in the absence of peroxisome proliferation (Jia et al., 2003). Similarly, SCP2/SCPx-knockout mice also display an induction of PPARα-target genes but exhibit hepatic peroxisome proliferation (Seedorf et al., 1998). Altogether, these data suggest that in addition to fatty acyl-CoAs, other intermediates of the peroxisomal β-oxidation pathway may serve as endogenous PPARα ligands but further investigations are necessary to identify them. The data to date suggests that PPARα may act as a sensor of fatty acids and/or of intermediates of their oxidation. This hypothesis is further supported by results obtained from in vivo studies using PPARα-knockout mice challenged by fasting (Kersten et al., 1999; Kroetz et al., 1998), a high-fat diet (Patsouris et al., 2006), or low-fat diets with various dietary fatty acid profiles (Martin et al., 2007). Other specific fatty acids or derivatives such as conjugated linoleic acids (although many of their effects appear independent of PPARα activation, Clement et al., 2002; Moya-Camarena et al., 1999; Peters et al., 2001), oleylethanolamide (Fu et al., 2003) or palmitoylethanolamide (Lo Verme et al., 2005) also activate PPARα, but the physiological relevance of these putative ligands remains to be confirmed. Finally, some eicosanoids (20-carbon fatty acid derivatives such as prostaglandins, prostacyclins, thromboxanes, leukotrienes and lipoxins) may also bind to and activate PPARα as it has been shown for the leukotriene B4 (Devchand et al., 1996) or for the 8(S)hydroxyeicosatetraenoic acid or 8(S)HETE (Forman et al., 1997; Kliewer et al., 1997; Krey et al., 1997; Murakami et al., 1999). However, because these eicosanoids are usually short lived and thus act locally, it is likely that their relevance is limited to specific tissues or cell populations and cannot explain systemic effects in which the receptor is simultaneously activated in different tissues.

#### 1.3.2.4 Mechanism of Activation

Recently, some studies have started to unravel how the PPARα signalling pathway functions (Fig. 1.4). In the absence of exogenously added ligand, PPARα resides in the nucleus where it exhibits high mobility (Feige et al., 2005). In this situation, its mobility is mainly reduced by its association with RXR and with coregulators (Feige et al., 2005; Tudor et al., 2007). Since fatty acids, present in all cell types, are ligands of PPARα, it is difficult to define its "constitutive" activity. Even in the absence of exogenously added ligand, a transient transfection of PPARα generally exhibits a significant basal activity on a synthetic promoter (Hi et al., 1999). This was attributed, at least in part, to a strong AF-1 present in its amino-terminal domain (Hi et al., 1999; Juge-Aubry et al., 1999), a region involved in the association with coregulators (Tudor et al., 2007). In vivo, the invalidation of PPARα in mouse, reduces the constitutive hepatic expression of several genes involved in fatty acid catabolism (Aoyama et al., 1998). Thus, it is likely that in the absence of exogenous stimulation PPARα binds to at least some of its response elements and constitutively regulates (positively or negatively depending on its association with coactivators or corepressors, respectively) the expression of certain target genes. However, it cannot be ruled out that intracellular fatty acids indeed occupy the ligand binding pocket of PPARα in this situation. Although similar data is

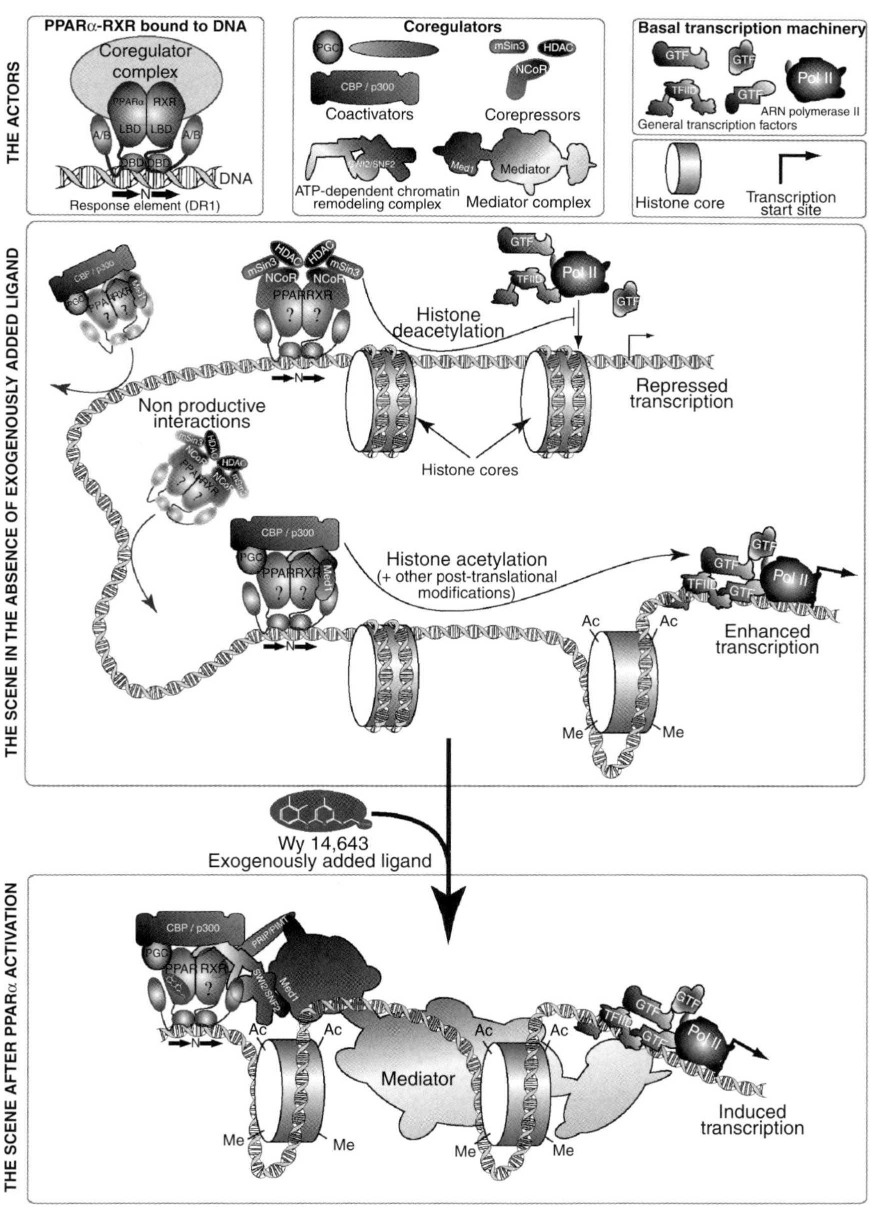


**Fig. 1.4** The PPARα pathway. The upper panels illustrate the different actors involved in this pathway. The large middle panel illustrates the potential actions of PPARα at the DNA level in the absence of exogenously added ligand. The lower panel illustrates the putative assembly of coactivators on the PPARα-RXR dimer bound to the promoter of a positive target gene following the exposure to an exogenously added activator ligand. The question marks (?) inside the NR LBDs illustrate the putative presence of endogenous ligands. In the absence of exogenously added ligand, PPARα, bound to RXR and to cofactors, is highly mobile in the nucleus and may transiently interact in a non-productive way with DNA. In this situation, a

lacking for the α isoform, it has recently been shown that PPARγ was able to "constitutively" inhibit the expression of some but not all of its target genes through DNA-binding on specific response elements and corepressors recruitment (Chui et al., 2005; Guan et al., 2005). Although significant, this basal activity of PPARα can undoubtedly be further increased through treatment with a peroxisome proliferator or through changes in the physiological state of the organism (fasting, high-fat diet, feeding with specific dietary fatty acids, streptozotocin-induced diabetes). Following the binding of an exogenously added activator to its LBD, the mobility of PPARα in the nucleus is reduced, mainly due to the recruitment of a larger complex of coactivators (Feige et al., 2005; Tudor et al., 2007). Such a complex has indeed been isolated from rat liver (Surapureddi et al., 2002) and several coactivators associated with PPARα-RXR have been identified such as CBP/p300 (Dowell et al., 1997; Tudor et al., 2007), PGC-1α (Vega et al., 2000), members of the SWI/SNF chromatin remodelling complexes (Surapureddi et al., 2002) and of Mediator complexes (Jia et al., 2004; Matsumoto et al., 2007). These coregulators mediate the effects of PPARα-RXR at the transcription level through their action on the chromatin and their interactions with the basal transcription machinery (Fig. 1.4). Although not developed in this review, it should finally be mentioned that a substantial part of the gene expression changes induced by PPARα activation, especially those negatively affecting inflammatory genes, are due to interactions of this NR with other transcription factor pathways (Gervois et al., 2007; Zandbergen and Plutzky, 2007).

Overall, PPARα emerges as a major regulator of hepatic fatty acid catabolism, whose activity is largely regulated by the binding of fatty acids and/or fatty acid derivatives to its LBD. However, an intricate network of factors, including the promoter context, PPARα expression level and post-translational modifications or the expression level, post-translational modifications, and cellular localization of its RXR partners, of coactivators and of interacting transcription factors are involved in fine-tuning the activity of this pathway and its specificity toward the regulation of its target genes.

◀

**Fig. 1.4** (continued) constitutive function of PPARα on certain promoter is likely but remains to be fully demonstrated. Such a constitutive function is illustrated in the middle panel through the association of PPARα-RXR (**a**) with corepressors acting through histone deacetylation (or other mechanisms not represented here) on the promoter of a negative PPARα target gene (*upper part*) and (**b**) with coactivators acting through histone acetylation (or other mechanisms not represented here) on the promoter of a positive PPARα target gene. Upon addition of an exogenous activator (here Wy14,643), a larger complex of coregulators is recruited to the PPARα-RXR heterodimer. It participates to the regulation of PPARα target genes through several mechanisms including histone modifications, chromatin remodeling and interactions with GTFs as illustrated in the lower panel for a positive PPARα target gene. The interactions of PPARα with other signalling pathways, leading for example to the down-regulation of several acute phase response genes in the liver, are not illustrated

## 1.4 SREBP-1c and PPARα in Nutritional Regulation of Hepatic Fatty Acid Metabolism

Here, a schematic overview of how SREBP-1c and PPARα respond to three challenges is described: feeding, fasting and dietary PUFAs exposure (Fig. 1.5). These different stimuli represent examples of regulation in which we aimed at integrating: (1) a coupling of fatty acid and glucose metabolism (2) the contribution of main signalling molecules and some other important transcription factors / co-regulators involved in these metabolic pathways.

### *1.4.1 SREBP-1c is a Central Integrator of the Hepatic Response to Feeding*

Under fed conditions, higher animals preferentially burn carbohydrates to generate ATP and surplus carbohydrate is converted to fatty acids, which are then stored as triacylglycerols in the adipose tissue.

In this paragraph, it is considered that "feeding" primarily implies elevation of both insulin and glucose concentration. We have attempted to review evidences linking high insulin and high glucose to changes in hepatic response, including recent advances in the regulation of both SREBP-1c and possibly PPARα downstream targets (Fig. 1.5a).

#### 1.4.1.1 PI3K/PKB Signalling and Insulin Effects

Insulin impacts on the expression of genes involved in the control of both glucose and fatty acid metabolism. It is generally considered that the hepatic response to insulin relies mainly, if not only, on activation of insulin receptor coupled to PI3 kinase (PI3K) signalling. In the liver, mechanisms linking PI3K signalling and changes in genes controlling metabolism involve the serine/threonine protein kinases PKB (also known as AKT) (Whiteman et al., 2002). The PKB family consists of three isoforms: PKBα (AKT1), PKBβ (AKT2) and PKBγ (AKT3). Isoform-specific knock-out mice have revealed distinct roles for two of the isoforms (PKBα and β). PKBβ deletion results in defects resembling diabetes mellitus, including elevated blood glucose (George et al., 2004).

PI3K activity leads to a rise in PtdIns(3,4,5)$P_3$ concentration at the plasma membrane which triggers membrane recruitment of signalling molecules containing PtdIns(3,4,5)$P_3$ specific PH domain (Hawkins et al., 2006). The phosphoinositide-dependent kinase (PDK) (Mora et al., 2004) and PKB both possess PH domains which bind PtdIns(3,4,5)$P_3$ (and one of its degradation product, PtdIns(3,4)$P_2$). After insulin-mediated increase in PI3K activity PDK and PKB translocate to the plasma membrane and this translocation leads to a

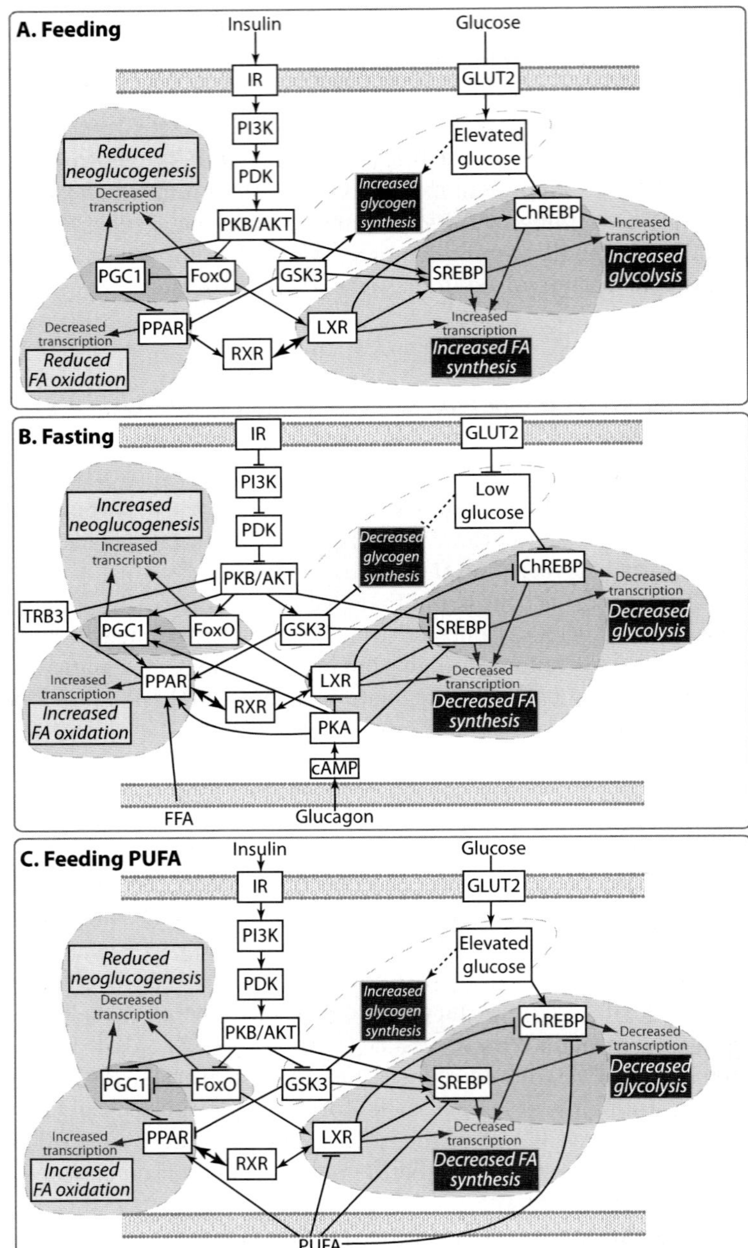


**Fig. 1.5** SREBP-1c and PPARα in nutritional regulation of fatty acid and glucose metabolism. (**A**) "Feeding". Insulin stimulates the PI3K/PKB signalling cascade and glucose enters the hepatocytes. (**B**) "Fasting". Insulin and glucose signals are switched off while glucagon stimulates the AMPc/PKA cascade and free fatty acids (FFA) enter the hepatocytes. (**C**) "Feeding PUFA". Dietary polyunsaturated fatty acids (PUFAs) activate PPARα while repressing SREBP-1c, ChREBP and possibly LXR

local increase in their relative concentration. Thus, PDK activates PKB through phosphorylation, initiating a large number of downstream cellular responses. Once activated, PKB dissociates from the plasma membrane and phosphorylates cytoplasmic and nuclear proteins such as those involved in regulating insulin-dependent processes. These phosphorylation events lead to activation or inactivation of a variety of molecules involved in the control of both glucose and fatty acid metabolism.

#### 1.4.1.2 PI3K/PKB Signalling and Insulin Effects on Glucose Homeostasis

Downstream of PI3K/PKB-dependent signalling cascade, The Forkhead box O (FoxO) transcription factor and the Glycogen Synthase Kinase (GSK-3) are two main effectors of glucose homeostasis. While these effectors are primarily involved in the direct control of glucose metabolism, they appear likely to also play a part in driving PPARα and SREBP-1c activity. Hence, we chose to briefly present these two signalling pathways.

##### FoxO and Glucose Homeostasis

The FoxO family of transcription factors (Barthel et al., 2005) contains three highly conserved PKB phosphorylation sites. Phosphorylation of these sites results in inactivation of FoxO transcriptional activity. Several mechanisms resulting in nuclear exclusion of the transcription factor are involved (Biggs et al., 1999; Brunet et al., 1999, 2001; Guo et al., 1999; Rena et al., 1999). Nuclear exclusion appears to involve the binding of protein 14-3-3 to the phosphorylation sites for PKB, thus masking the nuclear addressing signal (Cahill et al., 2001; Obsil et al., 2003; Obsilova et al., 2005).

FoxO is a regulator of the peroxisome proliferator activated receptor general co-activator α (PGC1-α) expression (Daitoku et al., 2003). FoxO also interacts with PGC1-α (Puigserver et al., 2003) and regulates the expression of target genes that include several key enzymes of neoglucogenesis (Barthel et al., 2005). Therefore, nuclear exclusion of FoxO in response to insulin contributes to reducing neoglucogenesis. On the contrary, hepatic expression of a constitutively active form of human FoxO1, in which PKB phosphorylation sites have been mutated, stimulates neoglucogenesis (Zhang et al., 2006).

##### Glycogen Synthase Kinase-3

In the liver, glycogen synthase kinase-3 (GSK-3) is another major target for PKB and a downstream effector of insulin action. GSK-3 is a serine/threonine kinase first identified as one of the primary regulator of glycogen synthase (GS) (Embi et al., 1980). It is now known that GSK-3 phosphorylates GS regulatory residues critical for GS inhibition and hence glycogen synthesis (Roach, 2002). GSK-3 exists as two homologs, GSK-3α and GSK-3β, both ubiquitously expressed (Woodgett, 1990). GSK-3 is constitutively active in resting cells but becomes inactivated by a variety of mechanisms upon stimulation. The most documented signalling pathway inhibiting GSK-3 is triggered by insulin, PKB activation, and subsequent phosphorylation of GSK-3 that leads to its inactivation (Frame et al., 2001), increase GS activity and glycogen synthesis. While insulin-mediated inhibition of GSK-3 supports glucose use for glycogen synthesis, GSK-3 is also likely to interfere with metabolic pathways through other substrates such as metabolic enzymes (ACC, ACL) and several transcription factors (Lee and Kim, 2007), including SREBP-1c (Kim et al., 2004) and possibly PPARα (Burns and Vanden Heuvel, 2007).

#### 1.4.1.3 Involvement of PI3K/PKB Signalling in Insulin Effects on Fatty Acid Metabolism

Recent studies strongly suggest a role of PKB in governing some aspects of hepatic fatty acid metabolism in response to insulin. The detailed mechanisms by which these regulations occur are not fully understood but increasing evidence suggests a cross-talk with both SREBP-1c and PPARα.

##### PI3K/PKB Signalling Controls SREBP1-c Activity

*Effect of Insulin on SREBP1-c Expression*

The primary regulation evidenced for SREBP-1c occurs at the transcriptional level. Insulin induces the transcription of the SREBP-1c gene, and this leads to a parallel increase in both the membrane-bound precursor and the mature nuclear form (Azzout-Marniche et al., 2000; Foretz et al., 1999; Shimomura et al., 1999). It appears that Liver X Receptor (LXR) is required for SREBP1-c expression (Repa et al., 2000) and the induction by insulin of SREBP1-c transcription (Chen et al., 2004). LXRα and LXRβ are nuclear receptors which, like PPARs, form active heterodimers with RXRs. The primary activators of LXRs are oxygenated intermediates in cholesterol synthesis (Janowski et al., 1996) and glucose (Mitro et al., 2007). It has been proposed that insulin may act through creating ligands for LXR itself or for its heterodimerization partner, RXR (Chen et al., 2004).

Wortmannin, an inhibitor of PI3K activity strongly impairs the insulin-mediated increase in SREBP-1c mRNA in human hepatocytes (Fleischmann and Iynedjian, 2000). This observation strongly suggests that PI3K signalling directly impacts on SREBP-1c mRNA level through a mechanism which remains to be established. One hypothesis has recently been proposed which is based on the observation that the hepatic overexpression of constitutively active version of FoxO1 in mice leads to decreased expression of SREBP-1c (Zhang et al., 2006). These authors have postulated that constitutively active FoxO1 might impair LXR activity. By contrast, under circumstances such as insulin stimulation, FoxO1 nuclear exclusion might be able to prevent a putative inhibitory effect on LXR-dependent transcription of SREBP-1c. This hypothesis is consistent with LXR being involved in mediating the effect of insulin on fatty acid biosynthesis (Tobin et al., 2002).

*Effect of Insulin on SREBP1-c Processing*

It has been shown in isolated hepatocytes and in vivo that insulin acutely controls the cleavage of the membrane-bound SREBP-1c precursor to its mature, nuclear form (Hegarty et al., 2005). This study also shows that PI3K inhibition by wortmannin, abolishes this insulin-stimulated increase in the abundance of the mature form. One possibility, which remains to be investigated, is that PI3K/PKB signalling governs key steps of SREBP-1c processing. Consistent with this proposal, it has been shown that in vitro the use of PI3K inhibitor and dominant negative version of PKB both disrupt the transport of SREBP escort protein, SCAP, from the endoplasmic reticulum to the Golgi (Du et al., 2006).

Other plausible mechanisms by which PI3K/PKB might influence SREBP-1c maturation involve control of the expression or activity of proteins required for the retention of SREBP-1c in the endoplasmic reticulum, its traffic to the Golgi, or its cleavage. A liver-specific mRNA for Insig-2 whose expression is down-regulated by insulin has been discovered (Yabe et al., 2003). To our knowledge, the hypothesis that this isoform specifically supports SREBP1-c processing in response to insulin has not yet been investigated.

*Effect of Insulin on SREBP1-c Degradation*

In the nucleus, SREBPs activity is directly and indirectly controlled by post-translational modification (Eberle et al., 2004). SREBPs are modified by the small ubiquitin-related modifier (SUMO)-1. Sumoylation of SREBPs represses the activity of the transcription factor without modifying its degradation (Hirano et al., 2003).

SREBPs are also target for ubiquitination and rapid degradation by the 26S proteasome (Hirano et al., 2001). It has been shown that GSK-3 negatively regulates the transcriptional activity of SREBP-1c by phosphorylation (Kim et al., 2004) and this repression process is likely to be relieved when GSK-3 is suppressed by the insulin signalling pathway. More recently, a phosphodegron for the SCF (Fbw7) ubiquitin ligase has been identified in SREBP1-a (Sundqvist et al., 2005). It is now demonstrated (Punga et al., 2006) that GSK-3-dependent phosphorylation of SREBP1 is enhanced in response to DNA binding. As a result for phosphorylation, the ubiquitin ligase Fbw7 is recruited to SREBPs associated with target promoters. Hence, it is proposed that inhibition of GSK-3 by insulin controls phosphorylation of SREBP-1c, its Fbw7-mediated ubiquitination after DNA binding, and the duration of SREBP-dependent transcription.

*Effect of Insulin on PPARα*

PPARs can be phosphorylated through various mechanisms (Burns and Vanden Heuvel, 2007). Insulin increases the phosphorylation of PPARα and this is associated with increased transcriptional activity (Shalev et al., 1996). However, the physiological relevance of this observation remains unclear. In vivo, the impact of insulin on PPARα target genes is not well known. The expression of CYP4A1 and AOX, two typical PPARα target genes, has been shown to be increased in mice made diabetic through streptozotocin injection which destroys pancreatic β–cells responsible for insulin production (Kroetz et al., 1998). By contrast, in the same work insulin injection had no effect on the relative expression of PPARα target genes. This suggests that PPARα phosphorylation in response to insulin might either not occur in vivo or not be associated with an increase in target gene expression.

In the liver, insulin stimulation impacts on co-factors important for PPARα activity. PGC1-α has been shown to be a downstream target for PI3K/PKB (Li et al., 2007). Insulin might repress PPARα activity by directly inhibiting PGC1-α.

#### 1.4.1.4 Effect of Glucose on Transcriptional Regulations

Glucose is both taken up and released in the bloodstream from the liver (Jurczak et al., 2007). In liver hepatocytes, glucose enters and exits the cell through a bidirectional transporter: GLUT2. GLUT2 is constitutively activated and localized in the plasma membrane in an insulin-independent manner. Therefore, a rise in blood glucose turns into elevation of intracellular glucose and facilitates its use (glycolysis) and storage (glycogen synthesis, lipogenesis). The hepatic response to elevated glucose is very difficult to dissociate from insulin signalling. However, a molecular pathway exists that allows increased glucose flux into hepatocytes to generate xylulose 5-phosphate. This signalling molecule triggers changes in glycolytic enzyme activities and nuclear import of the transcription factor ChREBP (Uyeda and Repa, 2006). ChREBP coordinates the regulation of enzymes channelling the end-products of glycolysis into lipogenesis.

The recent identification of LXR as a glucose sensor in vitro (Mitro et al., 2007) provided evidence for a possible alternative coupling between glycolysis and lipogenesis. Since LXR regulates the expression of both SREBP-1c (Repa et al., 2000) and ChREBP (Cha and Repa, 2007), LXR appeared likely to be a master switch for glycolysis, its coupling to lipogenesis, and lipogenesis itself, in response to elevated cellular glucose. A recent study (Denechaud et al., 2008) showed that in vivo the contribution of LXR to ChREBP function, and to glucose signalling is minor in the liver. However, in this work LXR appears essential for the expression of key lipogenic enzyme mRNAs such as SREBP-1c, FAS and SCD1, not ACC, in animals fed a high carbohydrate diet.

Interestingly, it has been reported in vitro that LXRs suppress lipid-degradation gene promoters through inhibition of PPAR signalling (Ide et al., 2003) suggesting a rate-limiting effect due to the availability of their common heterodimerization partner, RXRα. Whether such competition occurs in vivo, and better supports LXR/SREBP-mediated lipogenesis than PPARα-dependent pathway, remains to be established.

## *1.4.2 PPARα is a Central Integrator of the Hepatic Response to Fasting*

During fasting, low levels of both insulin and circulating glucose combine with elevation of glucagon and free fatty acids mobilized from the adipose stores (Fig. 1.5b). Under those circumstances the insulin signal is switched off, neoglucogenesis and glucose output are stimulated while free fatty acids are used to provide the cell with energy through oxidative pathways and ketogenesis. Under low insulin, the PI3K/PKB-mediated inhibition of FoxO and PGC1, we previously mentioned is abolished. This, in turn, results in increased expression of genes involved in neoglucogenesis. Similarly, low insulin also results in activation of other pathways that are not presented in Fig. 1.5b. For instance, low insulin activates the forkhead transcription factor Foxa2, which during fasting promotes fatty acid oxidation, ketogenesis and glycolysis (Wolfrum et al., 2004). Under high insulin, this pathway is repressed by PI3K/PKB (Wolfrum et al., 2003). Under low insulin, PGC1-β, a coactivator closely related to PGC1-α, is up-regulated (Lin et al., 2002) and coactivates Foxa2 (Wolfrum and Stoffel, 2006). Low insulin also ends the PI3K/PKB-mediated activation of SREBP-1c and contributes to reduce lipogenesis.

Other signalling pathways sense elevated glucagon and possibly free fatty acids. Evidence linking elevated glucagon and free fatty acids to both PPARα and SREBP-1c transcriptional response are presented next.

### 1.4.2.1 PKA, Central Effector of Glucagon Signalling

Glucagon is processed by the pancreatic islets in response to low blood glucose level. In the liver, glucagon binds to a membrane receptor coupled to GTP-bonding proteins, inducing an intracellular rise in cAMP, which subsequently activates the Protein kinase A (PKA) and downstream effectors.

PKA is a cAMP-dependent protein kinase that regulates numerous cellular functions by phosphorylating target proteins. It acts downstream of glucagon/adrenalin and counteracts some of the responses to insulin and glucose. It is considered to be one of the fasting signals that activates neoglucogenesis and β-oxidation (Desvergne et al., 2006). Its mechanism of action involves phosphorylation of several transcription factors, one of these transcription factors is the cAMP response relement binding protein (CREB), responsible for increased expression of gluconeogenic enzyme genes. CREB also regulates the expression of PGC1, a cofactor involved in several transcriptional regulation events, including neoglucogenesis via HNF4α and β-oxidation via PPARα.

Amongst the transcription factors targeted by PKA are PPARα, SREBP-1c and LXRα. PKA phosphorylates PPARα in vitro and activates its transcriptional activity (Lazennec et al., 2000). This implies that the PKA pathway might be an important modulator of fatty acid β-oxidation under conditions such as fasting. By contrast, it has also been recently reported that PKA phosphorylates SREBP-1c and suppresses its activity (Lu and Shyy, 2006). PKA activation also reduces SREBP-1c expression via phosphorylation of LXRα (Yamamoto et al., 2007). Therefore, PKA might be a downstream effector of glucagon and control the activation of β-oxidation and inhibition of lipogenesis through PPARα and SREBP-1c respectively.

#### 1.4.2.2 Effect of Adipose Triglyceride Mobilization

While SREBP-1c clearly plays a major part in the adaptation to feeding several starvation in vivo studies have provided strong evidence for the role of PPARα in the hepatic response to starvation (Kersten et al., 1999; Kroetz et al., 1998; Leone et al., 1999).

##### PPARα and FFA Influx

PPARα null mice are unable to sustain prolonged fasting and to enhance fatty acid oxidation. This results in severe hepatic steatosis in PPARα-null mouse (Hashimoto et al., 2000) and hypoketonemia (Kersten et al., 1999). This phenotype correlates with an impaired up-regulation of genes critical for fatty acid β-oxidation and ketogenesis observed in PPARα–/– mice. Moreover, during fasting, the expression of the mammalian tribbles homolog TRB-3 is increased (Koo et al., 2004). TRB-3 is a fasting-inducible inhibitor of PKB (Du et al., 2006) and a target for PPARα (Koo et al., 2004). Thus, during fasting, PPARα activation might subsequently result in further inhibition of the PI3K/PKB signalling pathway.

Food deprivation results in an increased expression of PPARα which, by itself, is likely to stimulate β-oxidation. The mechanism by which PPARα expression increases during fasting is unknown. However, as previously mentioned, it might involve PGC1 activation downstream of glucagon and glucocorticoïds (Louet et al., 2002; Yoon et al., 2001) together with low insulin.

Fasting also results in lipolysis of triglycerides from the adipose stores. Fatty acids activate PPARα (Jump et al., 2005; Sampath and Ntambi, 2005). The free fatty acids released in the blood flow during fasting enter the liver and may then act as ligands for PPARα. This hypothesis remains extremely challenging to investigate experimentally. Since PUFAs are better activators of PPARα, it can be speculated that the type of free fatty acids released during fasting could influence the extent to which PPARα target genes would be up-regulated. Kersten et al. (1999) have addressed this issue in vivo. They showed that the expression of L-FABP, a target gene for PPARα, is higher in fasted mice which have received a diet high in linoleic acid (C18:2n-6) as compared to a standard diet. However, this increase in L-FABP expression correlates to an increase in PPARα expression. These combined up-regulations of PPARα itself and its target gene makes the respective contribution of both mechanisms, i.e. PPARα expression and presence of its ligands, difficult to evaluate.

SREBP1-c and Free Fatty Acids Influx

During fasting, the hepatic level of SREBP-1c mRNA and the level of the nuclear protein markedly decrease in mice (Horton et al., 1998a). This is likely to be a direct consequence of the change in the hormonal status. Since both SREBP-1c and its regulator LXRα are fatty acid-sensitive transcription factors (Jump et al., 2005; Sampath and Ntambi, 2005), the effect of a change in the amount of cellular free fatty acids (FFA) is to be considered. However, while PUFAs may have an inhibitory effect on the expression of lipogenesis genes via LXRα-SREBP-1c (see below), saturated fatty acids have a pro-lipogenic effect on SREBP-1c (Lin et al., 2005). Hence, it is hardly conceivable that the inhibition of LXRα-SREBP-1c pathway occurs during fasting as a direct consequence of FFA influx.

Nevertheless, the increase in PPARα expression and activity increase during fasting might result in an indirect inhibition of LXRα activity and as a result lessen the expression of SREBP-1c and its target genes. Such PPARα-mediated suppression of SREBP-1c expression through inhibition of LXR signalling has been reported in vitro (Yoshikawa et al., 2003).

### *1.4.3 PPARα and SREBP-1c Transcriptional Sensors of Dietary PUFAs*

Dietary PUFAs (n-6 and n-3) have been shown to confer various health benefits. In particular, the effect of PUFAs on dyslipidemia and insulin signalling is well established. PUFAs might prevent the accumulation of lipotoxic triglycerides in insulin-sensitive tissues and improve insulin sensitivity.

PUFAs are generally considered to ameliorate obesity-related symptoms through both activation of PPARα and inhibition of SREBP-1c. All dietary PUFAs do not share the same "beneficial" impact on hepatic transcriptional regulations. It is very complex to establish the specific action of each type of fatty acids. Amongst the PUFAs influencing gene expression, long chain n-3 PUFAs are particularly prominent (Jump, 2002; Jump et al., 2005; Sampath and Ntambi, 2005). Fish oil consumption, providing DHA (C22:6n-3) and EPA (C20:5n-3), lead to several benefits in animal models such as increased fatty acid oxidation, decreased lipogenesis, decreased TG concentration, decreased VLDL secretion and prevention of insulin resistance. Presented next is our current understanding of how PUFAs act to influence the activity of PPARα and the abundance of SREBP-1c (Fig. 1.5c).

#### 1.4.3.1 PUFAs Stimulate PPARα-Responsive Fatty Acid Oxidation

PPARα was the first transcription factor identified as a fatty acid sensitive transcription factor (Gottlicher et al., 1992). Several studies using PPARα-deficient mice provided evidence that the receptor is required for the effect of dietary PUFAs on gene expression to occur. Some of these effects arise while animals have been fed with very moderate PUFAs intake (Martin et al., 2007). PPARα is required in vivo for the up-regulation of AOX (Ren et al., 1997) and other enzymes involved in fatty acid transport and catabolism in the liver of mice fed with PUFAs (Martin et al., 2007).

In vivo, PUFA intake up-regulates the expression of AOX without significantly influencing PPARα mRNA levels (Sekiya et al., 2003). Moreover, PPARα binds PUFAs (Xu et al., 1999a) and the mechanism by which n-3 PUFAs trigger fatty acid degradation in vivo is likely to be occurring through this binding and activation of PPARα. The reason for which some PUFAs and not other fatty acids activate PPARα might not only be a higher binding affinity. Certain types of PUFAs might also have a more specific metabolic fate and be predominant in the NEFA pool that can be targeted to the nucleus to exert their regulatory effect on nuclear receptors. This issue is extensively discussed in reviews by Jump and colleagues (Jump, 2002; Jump et al., 2005).

Amongst fatty acids, C20:5n-3 is considered as being a potent activator of PPARα in hepatocytes and in vivo, in the liver of animals fed n-3 PUFA diets. C20:5n-3 seem to have both the metabolic (Pawar and Jump, 2003) and structural (Jump, 2002) characteristics for being a PPARα activator. C20:5n-3 is a weak PPARα activator when compared to synthetic agonist such as fenofibrate. However, unlike fenofibrate, PUFAs also display other hypolipidemic effects on other transcription factors. For example, C20:5n-3 is known to suppress the activity of SREBP-1c (Sekiya et al., 2003).

#### 1.4.3.2 Role of SREBP-1c in the Repression of Lipogenesis by Dietary PUFAs

PUFA-containing diets have been shown to down-regulate the expression of several key lipogenic genes such as ACC, FAS, SCD1 and other desaturases. Inhibition of lipogenic enzymes through down-regulation of their mRNA level occurs through a sensitive mechanism that can take priority over the prolipogenic impact of feeding and insulin.

Most of the genes encoding enzymes involved in fatty acid synthesis are SREBP-1c targets. In vivo, the hepatic over-expression of a mature version of SREBP-1 impairs the PUFA-mediated inhibition of lipogenesis (Yahagi et al., 1999). This observation first suggested that the suppressive effect of PUFAs on lipogenic enzyme genes in the liver is caused by a decrease in the mature form of SREBP-1 protein. In contrast to PPARα, fatty acid regulation of SREBP-1c occurs through control of its nuclear abundance rather than through binding.

PUFAs suppress SREBP-1c abundance through several mechanisms. Firstly, PUFAs reduce SREBP-1c mRNA level. Both reduction in SREBP-1 mRNA stability (Xu et al., 2001) and decreased transcription of the SREBP-1 gene (Xu et al., 1999b) have been suggested. The mechanism by which PUFAs reduce SREBP-1 mRNA remains to be established. Several in vitro studies have shown that PUFAs can inhibit the lipogenic action of LXR, thereby inhibiting the expression of target genes, including SREBP-1c (Ntambi et al., 2002; Sampath and Ntambi, 2005). However, the putative role of LXR in mediating the effect of dietary PUFAs on SREBP-c expression and other genes involved in lipogenesis in the liver is very unlikely. Indeed, Pawar et al. (2003) showed no antagonism of C20:5n-3 on LXR activity or on the expression of LXR target genes, using doses that repressed SREBP-1c mRNA.

Secondly, PUFAs decrease the level of mature SREBP-1c by inhibition of the proteolytic processing (Yahagi et al., 1999) and stimulation of its degradation (Botolin et al., 2006). The mechanisms by which PUFAs exert a regulatory effect on SREBP-1c processing and degradation remain to be elucidated. It has been shown that 26S proteasome and ERK signalling are involved in controlling the nuclear abundance of SREBP-1c (Botolin et al., 2006).

PUFA-enriched diets, particularly those containing long chain n-3 fatty acids exhibit anti-obesity effects and also improve insulin sensitivity (Li et al., 2008). Long chain n-3 PUFA consumption is also shown to have a beneficial effect in preventing hepatic steatosis (Martin et al., 2007; Sekiya et al., 2003). Depending on the abundance of long chain PUFAs in the diet, the schematic overview of the effect of feeding we presented first should be revised. The two main revisions would be: a significant raise in the PPARα-responsive fatty acid oxidation and a moderation of the pro-lipogenic SREBP-1c dependent effect of insulin and glucose. Importantly, it has been shown that PUFAs also inhibit ChREBP nuclear translocation (Dentin et al., 2005). Together with inhibiting SREBP-1c, this contributes to reduce the expression of glycolytic and lipogenic genes.

## 1.5 Concluding Remarks

In this chapter we attempted to review the transcriptional regulation of fatty acid metabolism in the liver. Many transcription factors are involved in this process. We chose to focus on PPARα and SREBP-1c because of their established regulatory roles in the control of transcriptional programs that govern fatty acid degradation and synthesis, respectively. Moreover, we thought their distinct activation processes and sensitivity to various stimuli make them very complementary examples. Above all, their ability to sense molecules such as hormones or nutrients and, as a consequence, tune fatty acid homeostasis through very distinct mechanism is fascinating.

However, it must be said that most of the literature considered in this chapter refers to experimental work performed with cell lines or in vivo by means of rodent models. Both PPARα and SREBP-1c are very attractive drug targets. Therefore, it would be of crucial interest to document their respective gender and species-specificity. PPARα activators such as fibrates are currently widely prescribed drugs for human hyperlipidemia. To our knowledge, there is no current use of drugs targeting SREBP1c.

While being far from providing an exhaustive picture of very complex regulatory systems, we introduced evidence for cross-talks between PPARα or SREBP-1c and other signalling pathways involved in responses to dietary or hormonal stimuli. In this context, we have briefly introduced some molecular mechanisms that might associate PPARα and SREBP1-c to the insulin-sensitive signalosome.

Further characterization of the interactions with other signalling pathways and functional links to other hormonal, metabolic and xenobiotic sensors represents a very challenging issue. A refined picture of these networks would provide a better understanding of liver pathologies, including those related to the metabolic syndrome.

**Acknowledgments** We are grateful to Dr Suhasini Kulkarni for critical reading of the manuscript. This study was financially supported by ANR-PNRA PlastImpact.

## References

Abu-Elheiga, L., Brinkley, W. R., Zhong, L., Chirala, S. S., Woldegiorgis, G., and Wakil, S. J. The subcellular localization of acetyl-CoA carboxylase 2. Proc Natl Acad Sci U S A 97 (2000) 1444–1449.

Abu-Elheiga, L., Matzuk, M. M., Abo-Hashema, K. A., and Wakil, S. J. Continuous fatty acid oxidation and reduced fat storage in mice lacking acetyl-CoA carboxylase 2. Science 291 (2001) 2613–2616.

Abu-Elheiga, L., Matzuk, M. M., Kordari, P., Oh, W., Shaikenov, T., Gu, Z., and Wakil, S. J. Mutant mice lacking acetyl-CoA carboxylase 1 are embryonically lethal. Proc Natl Acad Sci U S A 102 (2005) 12011–12016.

Aoyama, T., Peters, J. M., Iritani, N., Nakajima, T., Furihata, K., Hashimoto, T., and Gonzalez, F. J. Altered constitutive expression of fatty acid-metabolizing enzymes in mice lacking the peroxisome proliferator-activated receptor alpha (PPARalpha). J Biol Chem 273 (1998) 5678–5684.

Asturias, F. J., Chadick, J. Z., Cheung, I. K., Stark, H., Witkowski, A., Joshi, A. K., and Smith, S. Structure and molecular organization of mammalian fatty acid synthase. Nat Struct Mol Biol 12 (2005) 225–232.

Azzout-Marniche, D., Becard, D., Guichard, C., Foretz, M., Ferre, P., and Foufelle, F. Insulin effects on sterol regulatory-element-binding protein-1c (SREBP-1c) transcriptional activity in rat hepatocytes. Biochem J 350 Pt 2 (2000) 389–393.

Baes, M., Huyghe, S., Carmeliet, P., Declercq, P. E., Collen, D., Mannaerts, G. P., and Van Veldhoven, P. P. Inactivation of the peroxisomal multifunctional protein-2 in mice impedes the degradation of not only 2-methyl-branched fatty acids and bile acid intermediates but also of very long chain fatty acids. J Biol Chem 275 (2000) 16329–16336.

Barish, G. D., Narkar, V. A., and Evans, R. M. PPAR delta: a dagger in the heart of the metabolic syndrome. J Clin Invest 116 (2006) 590–597.

Barthel, A., Schmoll, D., and Unterman, T. G. FoxO proteins in insulin action and metabolism. Trends Endocrinol Metab 16 (2005) 183–189.

Bengoechea-Alonso, M. T., and Ericsson, J. SREBP in signal transduction: cholesterol metabolism and beyond. Curr Opin Cell Biol 19 (2007) 215–222.

Biggs, W. H., 3rd, Meisenhelder, J., Hunter, T., Cavenee, W. K., and Arden, K. C. Protein kinase B/Akt-mediated phosphorylation promotes nuclear exclusion of the winged helix transcription factor FKHR1. Proc Natl Acad Sci U S A 96 (1999) 7421–7426.

Botolin, D., Wang, Y., Christian, B., and Jump, D. B. Docosahexaneoic acid (22:6,n-3) regulates rat hepatocyte SREBP-1 nuclear abundance by Erk- and 26S proteasome-dependent pathways. J Lipid Res 47 (2006) 181–192.

Brown, M. S., and Goldstein, J. L. The SREBP pathway: regulation of cholesterol metabolism by proteolysis of a membrane-bound transcription factor. Cell 89 (1997) 331–340.

Brunet, A., Bonni, A., Zigmond, M. J., Lin, M. Z., Juo, P., Hu, L. S., Anderson, M. J., Arden, K. C., Blenis, J., and Greenberg, M. E. Akt promotes cell survival by phosphorylating and inhibiting a Forkhead transcription factor. Cell 96 (1999) 857–868.

Brunet, A., Park, J., Tran, H., Hu, L. S., Hemmings, B. A., and Greenberg, M. E. Protein kinase SGK mediates survival signals by phosphorylating the forkhead transcription factor FKHRL1 (FOXO3a). Mol Cell Biol 21 (2001) 952–965.

Burns, K. A., and Vanden Heuvel, J. P. Modulation of PPAR activity via phosphorylation. Biochim Biophys Acta 1771 (2007) 952–960.

Cahill, C. M., Tzivion, G., Nasrin, N., Ogg, S., Dore, J., Ruvkun, G., and Alexander-Bridges, M. Phosphatidylinositol 3-kinase signaling inhibits DAF-16 DNA binding and function via 14-3-3-dependent and 14-3-3-independent pathways. J Biol Chem 276 (2001) 13402–13410.

Cha, J. Y., and Repa, J. J. The liver X receptor (LXR) and hepatic lipogenesis. The carbohydrate-response element-binding protein is a target gene of LXR. J Biol Chem 282 (2007) 743–751.

Chakravarthy, M. V., Pan, Z., Zhu, Y., Tordjman, K., Schneider, J. G., Coleman, T., Turk, J., and Semenkovich, C. F. "New" hepatic fat activates PPARalpha to maintain glucose, lipid, and cholesterol homeostasis. Cell Metab 1 (2005) 309–322.

Chen, G., Liang, G., Ou, J., Goldstein, J. L., and Brown, M. S. Central role for liver X receptor in insulin-mediated activation of Srebp-1c transcription and stimulation of fatty acid synthesis in liver. Proc Natl Acad Sci U S A 101 (2004) 11245–11250.

Chirala, S. S., Chang, H., Matzuk, M., Abu-Elheiga, L., Mao, J., Mahon, K., Finegold, M., and Wakil, S. J. Fatty acid synthesis is essential in embryonic development: fatty acid synthase null mutants and most of the heterozygotes die in utero. Proc Natl Acad Sci U S A 100 (2003) 6358–6363.

Chirala, S. S., Jayakumar, A., Gu, Z. W., and Wakil, S. J. Human fatty acid synthase: role of interdomain in the formation of catalytically active synthase dimer. Proc Natl Acad Sci U S A 98 (2001) 3104–3108.

Chirala, S. S., and Wakil, S. J. Structure and function of animal fatty acid synthase. Lipids 39 (2004) 1045–1053.

Cho, H. P., Nakamura, M., and Clarke, S. D. Cloning, expression, and fatty acid regulation of the human delta-5 desaturase. J Biol Chem 274 (1999a) 37335–37339.

Cho, H. P., Nakamura, M. T., and Clarke, S. D. Cloning, expression, and nutritional regulation of the mammalian Delta-6 desaturase. J Biol Chem 274 (1999b) 471–477.

Chui, P. C., Guan, H. P., Lehrke, M., and Lazar, M. A. PPARgamma regulates adipocyte cholesterol metabolism via oxidized LDL receptor 1. J Clin Invest 115 (2005) 2244–2256.

Clement, L., Poirier, H., Niot, I., Bocher, V., Guerre-Millo, M., Krief, S., Staels, B., and Besnard, P. Dietary trans-10,cis-12 conjugated linoleic acid induces hyperinsulinemia and fatty liver in the mouse. J Lipid Res 43 (2002) 1400–1409.

Cohen, P., Miyazaki, M., Socci, N. D., Hagge-Greenberg, A., Liedtke, W., Soukas, A. A., Sharma, R., Hudgins, L. C., Ntambi, J. M., and Friedman, J. M. Role for stearoyl-CoA desaturase-1 in leptin-mediated weight loss. Science 297 (2002) 240–243.

Costet, P., Legendre, C., More, J., Edgar, A., Galtier, P., and Pineau, T. Peroxisome proliferator-activated receptor alpha-isoform deficiency leads to progressive dyslipidemia with sexually dimorphic obesity and steatosis. J Biol Chem 273 (1998) 29577–29585.

D'Andrea, S., Guillou, H., Jan, S., Catheline, D., Thibault, J. N., Bouriel, M., Rioux, V., and Legrand, P. The same rat Delta6-desaturase not only acts on 18- but also on 24-carbon fatty acids in very-long-chain polyunsaturated fatty acid biosynthesis. Biochem J 364 (2002) 49–55.

Daitoku, H., Yamagata, K., Matsuzaki, H., Hatta, M., and Fukamizu, A. Regulation of PGC-1 promoter activity by protein kinase B and the forkhead transcription factor FKHR. Diabetes 52 (2003) 642–649.

Denechaud, P. D., Bossard, P., Lobaccaro, J. M., Millatt, L., Staels, B., Girard, J., and Postic, C. ChREBP, but not LXRs, is required for the induction of glucose-regulated genes in mouse liver. J Clin Invest 118 (2008) 956–964.

Dentin, R., Benhamed, F., Pegorier, J. P., Foufelle, F., Viollet, B., Vaulont, S., Girard, J., and Postic, C. Polyunsaturated fatty acids suppress glycolytic and lipogenic genes through the inhibition of ChREBP nuclear protein translocation. J Clin Invest 115 (2005) 2843–2854.

Desvergne, B., Michalik, L., and Wahli, W. Transcriptional regulation of metabolism. Physiol Rev 86 (2006) 465–514.

Desvergne, B., and Wahli, W. Peroxisome proliferator-activated receptors: nuclear control of metabolism. Endocr Rev 20 (1999) 649–688.

Devchand, P. R., Keller, H., Peters, J. M., Vazquez, M., Gonzalez, F. J., and Wahli, W. The PPARalpha-leukotriene B4 pathway to inflammation control. Nature 384 (1996) 39–43.

Dobrzyn, A., and Ntambi, J. M. The role of stearoyl-CoA desaturase in the control of metabolism. Prostaglandins Leukot Essent Fatty Acids 73 (2005) 35–41.

Dowell, P., Ishmael, J. E., Avram, D., Peterson, V. J., Nevrivy, D. J., and Leid, M. p300 functions as a coactivator for the peroxisome proliferator-activated receptor alpha. J Biol Chem 272 (1997) 33435–33443.

Du, X., Kristiana, I., Wong, J., and Brown, A. J. Involvement of Akt in ER-to-Golgi transport of SCAP/SREBP: a link between a key cell proliferative pathway and membrane synthesis. Mol Biol Cell 17 (2006) 2735–2745.

Duncan, E. A., Dave, U. P., Sakai, J., Goldstein, J. L., and Brown, M. S. Second-site cleavage in sterol regulatory element-binding protein occurs at transmembrane junction as determined by cysteine panning. J Biol Chem 273 (1998) 17801–17809.

Eberle, D., Hegarty, B., Bossard, P., Ferre, P., and Foufelle, F. SREBP transcription factors: master regulators of lipid homeostasis. Biochimie 86 (2004) 839–848.

Ellinghaus, P., Wolfrum, C., Assmann, G., Spener, F., and Seedorf, U. Phytanic acid activates the peroxisome proliferator-activated receptor alpha (PPARalpha) in sterol carrier protein 2-/ sterol carrier protein x-deficient mice. J Biol Chem 274 (1999) 2766–2772.

Embi, N., Rylatt, D. B., and Cohen, P. Glycogen synthase kinase-3 from rabbit skeletal muscle. Separation from cyclic-AMP-dependent protein kinase and phosphorylase kinase. Eur J Biochem 107 (1980) 519–527.

Engelking, L. J., Evers, B. M., Richardson, J. A., Goldstein, J. L., Brown, M. S., and Liang, G. Severe facial clefting in Insig-deficient mouse embryos caused by sterol accumulation and reversed by lovastatin. J Clin Invest 116 (2006) 2356–2365.

Engelking, L. J., Kuriyama, H., Hammer, R. E., Horton, J. D., Brown, M. S., Goldstein, J. L., and Liang, G. Overexpression of Insig-1 in the livers of transgenic mice inhibits SREBP processing and reduces insulin-stimulated lipogenesis. J Clin Invest 113 (2004) 1168–1175.

Engelking, L. J., Liang, G., Hammer, R. E., Takaishi, K., Kuriyama, H., Evers, B. M., Li, W. P., Horton, J. D., Goldstein, J. L., and Brown, M. S. Schoenheimer effect explained–feedback regulation of cholesterol synthesis in mice mediated by Insig proteins. J Clin Invest 115 (2005) 2489–2498.

Espenshade, P. J., and Hughes, A. L. Regulation of sterol synthesis in eukaryotes. Annu Rev Genet 41 (2007) 401–427.

Espenshade, P. J., Li, W. P., and Yabe, D. Sterols block binding of COPII proteins to SCAP, thereby controlling SCAP sorting in ER. Proc Natl Acad Sci U S A 99 (2002) 11694–11699.

Fan, C. Y., Pan, J., Usuda, N., Yeldandi, A. V., Rao, M. S., and Reddy, J. K. Steatohepatitis, spontaneous peroxisome proliferation and liver tumors in mice lacking peroxisomal fatty acyl-CoA oxidase. Implications for peroxisome proliferator-activated receptor alpha natural ligand metabolism. J Biol Chem 273 (1998) 15639–15645.

Feige, J. N., Gelman, L., Tudor, C., Engelborghs, Y., Wahli, W., and Desvergne, B. Fluorescence imaging reveals the nuclear behavior of peroxisome proliferator-activated receptor/retinoid X receptor heterodimers in the absence and presence of ligand. J Biol Chem 280 (2005) 17880–17890.

Ferre, P., and Foufelle, F. SREBP-1c transcription factor and lipid homeostasis: clinical perspective. Horm Res 68 (2007) 72–82.

Fleischmann, M., and Iynedjian, P. B. Regulation of sterol regulatory-element binding protein 1 gene expression in liver: role of insulin and protein kinase B/cAkt. Biochem J 349 (2000) 13–17.

Foretz, M., Guichard, C., Ferre, P., and Foufelle, F. Sterol regulatory element binding protein-1c is a major mediator of insulin action on the hepatic expression of glucokinase and lipogenesis-related genes. Proc Natl Acad Sci U S A 96 (1999) 12737–12742.

Forman, B. M., Chen, J., and Evans, R. M. Hypolipidemic drugs, polyunsaturated fatty acids, and eicosanoids are ligands for peroxisome proliferator-activated receptors alpha and delta. Proc Natl Acad Sci U S A 94 (1997) 4312–4317.

Frame, S., Cohen, P., and Biondi, R. M. A common phosphate binding site explains the unique substrate specificity of GSK3 and its inactivation by phosphorylation. Mol Cell 7 (2001) 1321–1327.

Fu, J., Gaetani, S., Oveisi, F., Lo Verme, J., Serrano, A., Rodriguez De Fonseca, F., Rosengarth, A., Luecke, H., Di Giacomo, B., Tarzia, G., and Piomelli, D. Oleylethanolamide regulates feeding and body weight through activation of the nuclear receptor PPAR-alpha. Nature 425 (2003) 90–93.

George, S., Rochford, J. J., Wolfrum, C., Gray, S. L., Schinner, S., Wilson, J. C., Soos, M. A., Murgatroyd, P. R., Williams, R. M., Acerini, C. L., et al. A family with severe insulin resistance and diabetes due to a mutation in AKT2. Science 304 (2004) 1325–1328.

Gervois, P., Fruchart, J. C., and Staels, B. Drug Insight: mechanisms of action and therapeutic applications for agonists of peroxisome proliferator-activated receptors. Nat Clin Pract Endocrinol Metab 3 (2007) 145–156.

Gottlicher, M., Widmark, E., Li, Q., and Gustafsson, J. A. Fatty acids activate a chimera of the clofibric acid-activated receptor and the glucocorticoid receptor. Proc Natl Acad Sci U S A 89 (1992) 4653–4657.

Guan, H. P., Ishizuka, T., Chui, P. C., Lehrke, M., and Lazar, M. A. Corepressors selectively control the transcriptional activity of PPARgamma in adipocytes. Genes Dev 19 (2005) 453–461.

Guillou, H., D'Andrea, S., Rioux, V., Barnouin, R., Dalaine, S., Pedrono, F., Jan, S., and Legrand, P. Distinct roles of endoplasmic reticulum cytochrome b5 and fused cytochrome b5-like domain for rat Delta6-desaturase activity. J Lipid Res 45 (2004) 32–40.

Guo, S., Rena, G., Cichy, S., He, X., Cohen, P., and Unterman, T. Phosphorylation of serine 256 by protein kinase B disrupts transactivation by FKHR and mediates effects of insulin on insulin-like growth factor-binding protein-1 promoter activity through a conserved insulin response sequence. J Biol Chem 274 (1999) 17184–17192.

Gurkan, C., Stagg, S. M., Lapointe, P., and Balch, W. E. The COPII cage: unifying principles of vesicle coat assembly. Nat Rev Mol Cell Biol 7 (2006) 727–738.

Hashimoto, T., Cook, W. S., Qi, C., Yeldandi, A. V., Reddy, J. K., and Rao, M. S. Defect in peroxisome proliferator-activated receptor alpha-inducible fatty acid oxidation determines the severity of hepatic steatosis in response to fasting. J Biol Chem 275 (2000) 28918–28928.

Hashimoto, T., Fujita, T., Usuda, N., Cook, W., Qi, C., Peters, J. M., Gonzalez, F. J., Yeldandi, A. V., Rao, M. S., and Reddy, J. K. Peroxisomal and mitochondrial fatty acid beta-oxidation in mice nullizygous for both peroxisome proliferator-activated receptor alpha and peroxisomal fatty acyl-CoA oxidase. Genotype correlation with fatty liver phenotype. J Biol Chem 274 (1999) 19228–19236.

Hawkins, P. T., Anderson, K. E., Davidson, K., and Stephens, L. R. Signalling through Class I PI3Ks in mammalian cells. Biochem Soc Trans 34 (2006) 647–662.

Hegarty, B. D., Bobard, A., Hainault, I., Ferre, P., Bossard, P., and Foufelle, F. Distinct roles of insulin and liver X receptor in the induction and cleavage of sterol regulatory element-binding protein-1c. Proc Natl Acad Sci U S A 102 (2005) 791–796.

Heinemann, F. S., and Ozols, J. Stearoyl-CoA desaturase, a short-lived protein of endoplasmic reticulum with multiple control mechanisms. Prostaglandins Leukot Essent Fatty Acids 68 (2003) 123–133.

Hi, R., Osada, S., Yumoto, N., and Osumi, T. Characterization of the amino-terminal activation domain of peroxisome proliferator-activated receptor alpha. Importance of alpha-helical structure in the transactivating function. J Biol Chem 274 (1999) 35152–35158.

Hirano, Y., Murata, S., Tanaka, K., Shimizu, M., and Sato, R. Sterol regulatory element-binding proteins are negatively regulated through SUMO-1 modification independent of the ubiquitin/26 S proteasome pathway. J Biol Chem 278 (2003) 16809–16819.

Hirano, Y., Yoshida, M., Shimizu, M., and Sato, R. Direct demonstration of rapid degradation of nuclear sterol regulatory element-binding proteins by the ubiquitin-proteasome pathway. J Biol Chem 276 (2001) 36431–36437.

Horton, J. D. Sterol regulatory element-binding proteins: transcriptional activators of lipid synthesis. Biochem Soc Trans 30 (2002) 1091–1095.

Horton, J. D., Bashmakov, Y., Shimomura, I., and Shimano, H. Regulation of sterol regulatory element binding proteins in livers of fasted and refed mice. Proc Natl Acad Sci U S A 95 (1998a) 5987–5992.

Horton, J. D., Shah, N. A., Warrington, J. A., Anderson, N. N., Park, S. W., Brown, M. S., and Goldstein, J. L. Combined analysis of oligonucleotide microarray data from transgenic and knockout mice identifies direct SREBP target genes. Proc Natl Acad Sci U S A 100 (2003) 12027–12032.

Horton, J. D., Shimomura, I., Brown, M. S., Hammer, R. E., Goldstein, J. L., and Shimano, H. Activation of cholesterol synthesis in preference to fatty acid synthesis in liver and adipose tissue of transgenic mice overproducing sterol regulatory element-binding protein-2. J Clin Invest 101 (1998b) 2331–2339.

Hostetler, H. A., Kier, A. B., and Schroeder, F. Very-long-chain and branched-chain fatty acyl-CoAs are high affinity ligands for the peroxisome proliferator-activated receptor alpha (PPARalpha). Biochemistry 45 (2006) 7669–7681.

Hostetler, H. A., Petrescu, A. D., Kier, A. B., and Schroeder, F. Peroxisome proliferator-activated receptor alpha interacts with high affinity and is conformationally responsive to endogenous ligands. J Biol Chem 280 (2005) 18667–18682.

Hua, X., Sakai, J., Brown, M. S., and Goldstein, J. L. Regulated cleavage of sterol regulatory element binding proteins requires sequences on both sides of the endoplasmic reticulum membrane. J Biol Chem 271 (1996) 10379–10384.

Hua, X., Wu, J., Goldstein, J. L., Brown, M. S., and Hobbs, H. H. Structure of the human gene encoding sterol regulatory element binding protein-1 (SREBF1) and localization of SREBF1 and SREBF2 to chromosomes 17p11.2 and 22q13. Genomics 25 (1995) 667–673.

Hummasti, S., and Tontonoz, P. The peroxisome proliferator-activated receptor N-terminal domain controls isotype-selective gene expression and adipogenesis. Mol Endocrinol 20 (2006) 1261–1275.

Ide, T., Shimano, H., Yoshikawa, T., Yahagi, N., Amemiya-Kudo, M., Matsuzaka, T., Nakakuki, M., Yatoh, S., Iizuka, Y., Tomita, S., et al. Cross-talk between peroxisome proliferator-activated receptor (PPAR) alpha and liver X receptor (LXR) in nutritional regulation of fatty acid metabolism. II. LXRs suppress lipid degradation gene promoters through inhibition of PPAR signaling. Mol Endocrinol 17 (2003) 1255–1267.

Issemann, I., and Green, S. Activation of a member of the steroid hormone receptor superfamily by peroxisome proliferators. Nature 347 (1990) 645–650.

Jakobsson, A., Westerberg, R., and Jacobsson, A. Fatty acid elongases in mammals: their regulation and roles in metabolism. Prog Lipid Res 45 (2006) 237–249.

Janowski, B. A., Willy, P. J., Devi, T. R., Falck, J. R., and Mangelsdorf, D. J. An oxysterol signalling pathway mediated by the nuclear receptor LXR alpha. Nature 383 (1996) 728–731.

Jeffcoat, R., Brawn, P. R., Safford, R., and James, A. T. Properties of rat liver microsomal stearoyl-coenzyme A desaturase. Biochem J 161 (1977) 431–437.

Jia, Y., Qi, C., Kashireddi, P., Surapureddi, S., Zhu, Y. J., Rao, M. S., Le Roith, D., Chambon, P., Gonzalez, F. J., and Reddy, J. K. Transcription coactivator PBP, the peroxisome proliferator-activated receptor (PPAR)-binding protein, is required for PPARalpha-regulated gene expression in liver. J Biol Chem 279 (2004) 24427–24434.

Jia, Y., Qi, C., Zhang, Z., Hashimoto, T., Rao, M. S., Huyghe, S., Suzuki, Y., Van Veldhoven, P. P., Baes, M., and Reddy, J. K. Overexpression of peroxisome proliferator-activated receptor-alpha (PPARalpha)-regulated genes in liver in the absence of peroxisome proliferation in mice deficient in both L- and D-forms of enoyl-CoA hydratase/dehydrogenase enzymes of peroxisomal beta-oxidation system. J Biol Chem 278 (2003) 47232–47239.

Juge-Aubry, C. E., Hammar, E., Siegrist-Kaiser, C., Pernin, A., Takeshita, A., Chin, W. W., Burger, A. G., and Meier, C. A. Regulation of the transcriptional activity of the peroxisome proliferator-activated receptor alpha by phosphorylation of a ligand-independent trans-activating domain. J Biol Chem 274 (1999) 10505–10510.

Jump, D. B. The biochemistry of n-3 polyunsaturated fatty acids. J Biol Chem 277 (2002) 8755–8758.

Jump, D. B., Botolin, D., Wang, Y., Xu, J., Christian, B., and Demeure, O. Fatty acid regulation of hepatic gene transcription. J Nutr 135 (2005) 2503–2506.

Jurczak, M. J., Danos, A. M., Rehrmann, V. R., and Brady, M. J. The role of protein translocation in the regulation of glycogen metabolism. J Cell Biochem (2007) *In press.*

Kerner, J., and Hoppel, C. Fatty acid import into mitochondria. Biochim Biophys Acta 1486 (2000) 1–17.

Kersten, S., Seydoux, J., Peters, J. M., Gonzalez, F. J., Desvergne, B., and Wahli, W. Peroxisome proliferator-activated receptor alpha mediates the adaptive response to fasting. J Clin Invest 103 (1999) 1489–1498.

Kim, J. B., and Spiegelman, B. M. ADD1/SREBP1 promotes adipocyte differentiation and gene expression linked to fatty acid metabolism. Genes Dev 10 (1996) 1096–1107.

Kim, J. B., Spotts, G. D., Halvorsen, Y. D., Shih, H. M., Ellenberger, T., Towle, H. C., and Spiegelman, B. M. Dual DNA binding specificity of ADD1/SREBP1 controlled by a single amino acid in the basic helix-loop-helix domain. Mol Cell Biol 15 (1995) 2582–2588.

Kim, K. H. Regulation of mammalian acetyl-coenzyme A carboxylase. Annu Rev Nutr 17 (1997) 77–99.

Kim, K. H., Song, M. J., Yoo, E. J., Choe, S. S., Park, S. D., and Kim, J. B. Regulatory role of glycogen synthase kinase 3 for transcriptional activity of ADD1/SREBP1c. J Biol Chem 279 (2004) 51999–52006.

Kliewer, S. A., Sundseth, S. S., Jones, S. A., Brown, P. J., Wisely, G. B., Koble, C. S., Devchand, P., Wahli, W., Willson, T. M., Lenhard, J. M., and Lehmann, J. M. Fatty acids and eicosanoids regulate gene expression through direct interactions with peroxisome proliferator-activated receptors alpha and gamma. Proc Natl Acad Sci U S A 94 (1997) 4318–4323.

Koo, S. H., Satoh, H., Herzig, S., Lee, C. H., Hedrick, S., Kulkarni, R., Evans, R. M., Olefsky, J., and Montminy, M. PGC-1 promotes insulin resistance in liver through PPAR-alpha-dependent induction of TRB-3. Nat Med 10 (2004) 530–534.

Krey, G., Braissant, O., L'Horset, F., Kalkhoven, E., Perroud, M., Parker, M. G., and Wahli, W. Fatty acids, eicosanoids, and hypolipidemic agents identified as ligands of peroxisome proliferator-activated receptors by coactivator-dependent receptor ligand assay. Mol Endocrinol 11 (1997) 779–791.

Kroetz, D. L., Yook, P., Costet, P., Bianchi, P., and Pineau, T. Peroxisome proliferator-activated receptor alpha controls the hepatic CYP4A induction adaptive response to starvation and diabetes. J Biol Chem 273 (1998) 31581–31589.

Kuhajda, F. P. Fatty acid synthase and cancer: new application of an old pathway. Cancer Res 66 (2006) 5977–5980.

Kumadaki, S., Matsuzaka, T., Kato, T., Yahagi, N., Yamamoto, T., Okada, S., Kobayashi, K., Takahashi, A., Yatoh, S., Suzuki, H., et al. Mouse Elovl-6 promoter is an SREBP target. Biochem Biophys Res Commun 368 (2008) 261–266.

Lazennec, G., Canaple, L., Saugy, D., and Wahli, W. Activation of peroxisome proliferator-activated receptors (PPARs) by their ligands and protein kinase A activators. Mol Endocrinol 14 (2000) 1962–1975.

Lee, J., and Kim, M. S. The role of GSK3 in glucose homeostasis and the development of insulin resistance. Diabetes Res Clin Pract 77 Suppl 1 (2007) S49–S57.

Lee, S. S., Pineau, T., Drago, J., Lee, E. J., Owens, J. W., Kroetz, D. L., Fernandez-Salguero, P. M., Westphal, H., and Gonzalez, F. J. Targeted disruption of the alpha isoform of the peroxisome proliferator-activated receptor gene in mice results in abolishment of the pleiotropic effects of peroxisome proliferators. Mol Cell Biol 15 (1995) 3012–3022.

Lee, Y. S., Sohn, D. H., Han, D., Lee, H. W., Seong, R. H., and Kim, J. B. Chromatin remodeling complex interacts with ADD1/SREBP1c to mediate insulin-dependent regulation of gene expression. Mol Cell Biol 27 (2007) 438–452.

Leonard, A. E., Pereira, S. L., Sprecher, H., and Huang, Y. S. Elongation of long-chain fatty acids. Prog Lipid Res 43 (2004) 36–54.

Leone, T. C., Weinheimer, C. J., and Kelly, D. P. A critical role for the peroxisome proliferator-activated receptor alpha (PPARalpha) in the cellular fasting response: the PPARalpha-null mouse as a model of fatty acid oxidation disorders. Proc Natl Acad Sci U S A 96 (1999) 7473–7478.

Li, J. J., Huang, C. J., and Xie, D. Anti-obesity effects of conjugated linoleic acid, docosahexaenoic acid, and eicosapentaenoic acid. Mol Nutr Food Res 52 (2008) *In press*.

Li, X., Monks, B., Ge, Q., and Birnbaum, M. J. Akt/PKB regulates hepatic metabolism by directly inhibiting PGC-1alpha transcription coactivator. Nature 447 (2007) 1012–1016.

Liang, G., Yang, J., Horton, J. D., Hammer, R. E., Goldstein, J. L., and Brown, M. S. Diminished hepatic response to fasting/refeeding and liver X receptor agonists in mice with selective deficiency of sterol regulatory element-binding protein-1c. J Biol Chem 277 (2002) 9520–9528.

Lin, J., Puigserver, P., Donovan, J., Tarr, P., and Spiegelman, B. M. Peroxisome proliferator-activated receptor gamma coactivator 1beta (PGC-1beta ), a novel PGC-1-related transcription coactivator associated with host cell factor. J Biol Chem 277 (2002) 1645–1648.

Lin, J., Yang, R., Tarr, P. T., Wu, P. H., Handschin, C., Li, S., Yang, W., Pei, L., Uldry, M., Tontonoz, P., et al. Hyperlipidemic effects of dietary saturated fats mediated through PGC-1beta coactivation of SREBP. Cell 120 (2005) 261–273.

Lin, Q., Ruuska, S. E., Shaw, N. S., Dong, D., and Noy, N. Ligand selectivity of the peroxisome proliferator-activated receptor alpha. Biochemistry 38 (1999) 185–190.

Lo Verme, J., Fu, J., Astarita, G., La Rana, G., Russo, R., Calignano, A., and Piomelli, D. The nuclear receptor peroxisome proliferator-activated receptor-alpha mediates the anti-inflammatory actions of palmitoylethanolamide. Mol Pharmacol 67 (2005) 15–19.

Loftus, T. M., Jaworsky, D. E., Frehywot, G. L., Townsend, C. A., Ronnett, G. V., Lane, M. D., and Kuhajda, F. P. Reduced food intake and body weight in mice treated with fatty acid synthase inhibitors. Science 288 (2000) 2379–2381.

Lonard, D. M., and O'Malley B, W. Nuclear receptor coregulators: judges, juries, and executioners of cellular regulation. Mol Cell 27 (2007) 691–700.

Longo, N., Amat di San Filippo, C., and Pasquali, M. Disorders of carnitine transport and the carnitine cycle. Am J Med Genet C Semin Med Genet 142 (2006) 77–85.

Louet, J. F., Hayhurst, G., Gonzalez, F. J., Girard, J., and Decaux, J. F. The coactivator PGC-1 is involved in the regulation of the liver carnitine palmitoyltransferase I gene expression by cAMP in combination with HNF4 alpha and cAMP-response element-binding protein (CREB). J Biol Chem 277 (2002) 37991–38000.

Lu, M., and Shyy, J. Y. Sterol regulatory element-binding protein 1 is negatively modulated by PKA phosphorylation. Am J Physiol Cell Physiol 290 (2006) C1477–1486.

Maier, T., Jenni, S., and Ban, N. Architecture of mammalian fatty acid synthase at 4.5 A resolution. Science 311 (2006) 1258–1262.

Mandard, S., Muller, M., and Kersten, S. Peroxisome proliferator-activated receptor alpha target genes. Cell Mol Life Sci 61 (2004) 393–416.

Mao, J., DeMayo, F. J., Li, H., Abu-Elheiga, L., Gu, Z., Shaikenov, T. E., Kordari, P., Chirala, S. S., Heird, W. C., and Wakil, S. J. Liver-specific deletion of acetyl-CoA carboxylase 1 reduces hepatic triglyceride accumulation without affecting glucose homeostasis. Proc Natl Acad Sci U S A 103 (2006) 8552–8557.

Marquardt, A., Stohr, H., White, K., and Weber, B. H. cDNA cloning, genomic structure, and chromosomal localization of three members of the human fatty acid desaturase family. Genomics 66 (2000) 175–183.

Martin, P. G., Guillou, H., Lasserre, F., Dejean, S., Lan, A., Pascussi, J. M., Sancristobal, M., Legrand, P., Besse, P., and Pineau, T. Novel aspects of PPARalpha-mediated regulation of lipid and xenobiotic metabolism revealed through a nutrigenomic study. Hepatology 45 (2007) 767–777.

Matsuda, M., Korn, B. S., Hammer, R. E., Moon, Y. A., Komuro, R., Horton, J. D., Goldstein, J. L., Brown, M. S., and Shimomura, I. SREBP cleavage-activating protein (SCAP) is required for increased lipid synthesis in liver induced by cholesterol deprivation and insulin elevation. Genes Dev 15 (2001) 1206–1216.

Matsumoto, K., Yu, S., Jia, Y., Ahmed, M. R., Viswakarma, N., Sarkar, J., Kashireddy, P. V., Rao, M. S., Karpus, W., Gonzalez, F. J., and Reddy, J. K. Critical role for transcription coactivator peroxisome proliferator-activated receptor (PPAR)-binding protein/TRAP220 in liver regeneration and PPARalpha ligand-induced liver tumor development. J Biol Chem 282 (2007) 17053–17060.

Matsumoto, M., Ogawa, W., Akimoto, K., Inoue, H., Miyake, K., Furukawa, K., Hayashi, Y., Iguchi, H., Matsuki, Y., Hiramatsu, R., et al. PKClambda in liver mediates insulin-induced

SREBP-1c expression and determines both hepatic lipid content and overall insulin sensitivity. J Clin Invest 112 (2003) 935–944.

Matsuzaka, T., Shimano, H., Yahagi, N., Amemiya-Kudo, M., Yoshikawa, T., Hasty, A. H., Tamura, Y., Osuga, J., Okazaki, H., Iizuka, Y., et al. Dual regulation of mouse Delta(5)- and Delta(6)-desaturase gene expression by SREBP-1 and PPARalpha. J Lipid Res 43 (2002) 107–114.

Matsuzaka, T., Shimano, H., Yahagi, N., Kato, T., Atsumi, A., Yamamoto, T., Inoue, N., Ishikawa, M., Okada, S., Ishigaki, N., et al. Crucial role of a long-chain fatty acid elongase, Elovl6, in obesity-induced insulin resistance. Nat Med 13 (2007) 1193–1202.

McGarry, J. D., and Brown, N. F. The mitochondrial carnitine palmitoyltransferase system. From concept to molecular analysis. Eur J Biochem 244 (1997) 1–14.

McPherson, R., and Gauthier, A. Molecular regulation of SREBP function: the Insig-SCAP connection and isoform-specific modulation of lipid synthesis. Biochem Cell Biol 82 (2004) 201–211.

Menendez, J. A., and Lupu, R. Fatty acid synthase and the lipogenic phenotype in cancer pathogenesis. Nat Rev Cancer 7 (2007) 763–777.

Miserez, A. R., Cao, G., Probst, L. C., and Hobbs, H. H. Structure of the human gene encoding sterol regulatory element binding protein 2 (SREBF2). Genomics 40 (1997) 31–40.

Mitro, N., Mak, P. A., Vargas, L., Godio, C., Hampton, E., Molteni, V., Kreusch, A., and Saez, E. The nuclear receptor LXR is a glucose sensor. Nature 445 (2007) 219–223.

Miyazaki, M., Flowers, M. T., Sampath, H., Chu, K., Otzelberger, C., Liu, X., and Ntambi, J. M. Hepatic stearoyl-CoA desaturase-1 deficiency protects mice from carbohydrate-induced adiposity and hepatic steatosis. Cell Metab 6 (2007) 484–496.

Miyazaki, M., Jacobson, M. J., Man, W. C., Cohen, P., Asilmaz, E., Friedman, J. M., and Ntambi, J. M. Identification and characterization of murine SCD4, a novel heart-specific stearoyl-CoA desaturase isoform regulated by leptin and dietary factors. J Biol Chem 278 (2003) 33904–33911.

Miyazaki, M., Man, W. C., and Ntambi, J. M. Targeted disruption of stearoyl-CoA desaturase1 gene in mice causes atrophy of sebaceous and meibomian glands and depletion of wax esters in the eyelid. J Nutr 131 (2001) 2260–2268.

Moon, Y. A., Shah, N. A., Mohapatra, S., Warrington, J. A., and Horton, J. D. Identification of a mammalian long chain fatty acyl elongase regulated by sterol regulatory element-binding proteins. J Biol Chem 276 (2001) 45358–45366.

Mora, A., Komander, D., van Aalten, D. M., and Alessi, D. R. PDK1, the master regulator of AGC kinase signal transduction. Semin Cell Dev Biol 15 (2004) 161–170.

Moya-Camarena, S. Y., Vanden Heuvel, J. P., Blanchard, S. G., Leesnitzer, L. A., and Belury, M. A. Conjugated linoleic acid is a potent naturally occurring ligand and activator of PPARalpha. J Lipid Res 40 (1999) 1426–1433.

Murakami, K., Ide, T., Suzuki, M., Mochizuki, T., and Kadowaki, T. Evidence for direct binding of fatty acids and eicosanoids to human peroxisome proliferators-activated receptor alpha. Biochem Biophys Res Commun 260 (1999) 609–613.

Nara, T. Y., He, W. S., Tang, C., Clarke, S. D., and Nakamura, M. T. The E-box like sterol regulatory element mediates the suppression of human Delta-6 desaturase gene by highly unsaturated fatty acids. Biochem Biophys Res Commun 296 (2002) 111–117.

Ntambi, J. M., Miyazaki, M., Stoehr, J. P., Lan, H., Kendziorski, C. M., Yandell, B. S., Song, Y., Cohen, P., Friedman, J. M., and Attie, A. D. Loss of stearoyl-CoA desaturase-1 function protects mice against adiposity. Proc Natl Acad Sci U S A 99 (2002) 11482–11486.

Nwankwo, J. O., Spector, A. A., and Domann, F. E. A nucleotide insertion in the transcriptional regulatory region of FADS2 gives rise to human fatty acid delta-6-desaturase deficiency. J Lipid Res 44 (2003) 2311–2319.

Obsil, T., Ghirlando, R., Anderson, D. E., Hickman, A. B., and Dyda, F. Two 14-3-3 binding motifs are required for stable association of Forkhead transcription factor FOXO4 with 14-3-3 proteins and inhibition of DNA binding. Biochemistry 42 (2003) 15264–15272.

Obsilova, V., Vecer, J., Herman, P., Pabianova, A., Sulc, M., Teisinger, J., Boura, E., and Obsil, T. 14-3-3 Protein interacts with nuclear localization sequence of forkhead transcription factor FoxO4. Biochemistry 44 (2005) 11608–11617.

Pai, J. T., Guryev, O., Brown, M. S., and Goldstein, J. L. Differential stimulation of cholesterol and unsaturated fatty acid biosynthesis in cells expressing individual nuclear sterol regulatory element-binding proteins. J Biol Chem 273 (1998) 26138–26148.

Palmer, C. N., Hsu, M. H., Griffin, K. J., Raucy, J. L., and Johnson, E. F. Peroxisome proliferator activated receptor-alpha expression in human liver. Mol Pharmacol 53 (1998) 14–22.

Patsouris, D., Reddy, J. K., Muller, M., and Kersten, S. Peroxisome proliferator-activated receptor alpha mediates the effects of high-fat diet on hepatic gene expression. Endocrinology 147 (2006) 1508–1516.

Pawar, A., Botolin, D., Mangelsdorf, D. J., and Jump, D. B. The role of liver X receptor-alpha in the fatty acid regulation of hepatic gene expression. J Biol Chem 278 (2003) 40736–40743.

Pawar, A., and Jump, D. B. Unsaturated fatty acid regulation of peroxisome proliferator-activated receptor alpha activity in rat primary hepatocytes. J Biol Chem 278 (2003) 35931–35939.

Peet, D. J., Janowski, B. A., and Mangelsdorf, D. J. The LXRs: a new class of oxysterol receptors. Curr Opin Genet Dev 8 (1998) 571–575.

Peters, J. M., Park, Y., Gonzalez, F. J., and Pariza, M. W. Influence of conjugated linoleic acid on body composition and target gene expression in peroxisome proliferator-activated receptor alpha-null mice. Biochim Biophys Acta 1533 (2001) 233–242.

Porstmann, T., Griffiths, B., Chung, Y. L., Delpuech, O., Griffiths, J. R., Downward, J., and Schulze, A. PKB/Akt induces transcription of enzymes involved in cholesterol and fatty acid biosynthesis via activation of SREBP. Oncogene 24 (2005) 6465–6481.

Prasad, M. R., Nagi, M. N., Ghesquier, D., Cook, L., and Cinti, D. L. Evidence for multiple condensing enzymes in rat hepatic microsomes catalyzing the condensation of saturated, monounsaturated, and polyunsaturated acyl coenzyme A. J Biol Chem 261 (1986) 8213–8217.

Puigserver, P., Rhee, J., Donovan, J., Walkey, C. J., Yoon, J. C., Oriente, F., Kitamura, Y., Altomonte, J., Dong, H., Accili, D., and Spiegelman, B. M. Insulin-regulated hepatic gluconeogenesis through FOXO1-PGC-1alpha interaction. Nature 423 (2003) 550–555.

Punga, T., Bengoechea-Alonso, M. T., and Ericsson, J. Phosphorylation and ubiquitination of the transcription factor sterol regulatory element-binding protein-1 in response to DNA binding. J Biol Chem 281 (2006) 25278–25286.

Qi, C., Zhu, Y., Pan, J., Usuda, N., Maeda, N., Yeldandi, A. V., Rao, M. S., Hashimoto, T., and Reddy, J. K. Absence of spontaneous peroxisome proliferation in enoyl-CoA Hydratase/L-3-hydroxyacyl-CoA dehydrogenase-deficient mouse liver. Further support for the role of fatty acyl CoA oxidase in PPARalpha ligand metabolism. J Biol Chem 274 (1999) 15775–15780.

Ren, B., Thelen, A. P., Peters, J. M., Gonzalez, F. J., and Jump, D. B. Polyunsaturated fatty acid suppression of hepatic fatty acid synthase and S14 gene expression does not require peroxisome proliferator-activated receptor alpha. J Biol Chem 272 (1997) 26827–26832.

Rena, G., Guo, S., Cichy, S. C., Unterman, T. G., and Cohen, P. Phosphorylation of the transcription factor forkhead family member FKHR by protein kinase B. J Biol Chem 274 (1999) 17179–17183.

Repa, J. J., Liang, G., Ou, J., Bashmakov, Y., Lobaccaro, J. M., Shimomura, I., Shan, B., Brown, M. S., Goldstein, J. L., and Mangelsdorf, D. J. Regulation of mouse sterol regulatory element-binding protein-1c gene (SREBP-1c) by oxysterol receptors, LXRalpha and LXRbeta. Genes Dev 14 (2000) 2819–2830.

Rinaldo, P., Matern, D., and Bennett, M. J. Fatty acid oxidation disorders. Annu Rev Physiol 64 (2002) 477–502.

Roach, P. J. Glycogen and its metabolism. Curr Mol Med 2 (2002) 101–120.

Sakai, J., Nohturfft, A., Goldstein, J. L., and Brown, M. S. Cleavage of sterol regulatory element-binding proteins (SREBPs) at site-1 requires interaction with SREBP cleavage-activating protein. Evidence from in vivo competition studies. J Biol Chem 273 (1998) 5785–5793.

Sampath, H., and Ntambi, J. M. Polyunsaturated fatty acid regulation of genes of lipid metabolism. Annu Rev Nutr 25 (2005) 317–340.

Seedorf, U., Raabe, M., Ellinghaus, P., Kannenberg, F., Fobker, M., Engel, T., Denis, S., Wouters, F., Wirtz, K. W., Wanders, R. J., et al. Defective peroxisomal catabolism of branched fatty acyl coenzyme A in mice lacking the sterol carrier protein-2/sterol carrier protein-x gene function. Genes Dev 12 (1998) 1189–1201.

Sekiya, M., Yahagi, N., Matsuzaka, T., Najima, Y., Nakakuki, M., Nagai, R., Ishibashi, S., Osuga, J., Yamada, N., and Shimano, H. Polyunsaturated fatty acids ameliorate hepatic steatosis in obese mice by SREBP-1 suppression. Hepatology 38 (2003) 1529–1539.

Shalev, A., Siegrist-Kaiser, C. A., Yen, P. M., Wahli, W., Burger, A. G., Chin, W. W., and Meier, C. A. The peroxisome proliferator-activated receptor alpha is a phosphoprotein: regulation by insulin. Endocrinology 137 (1996) 4499–4502.

Sheng, Z., Otani, H., Brown, M. S., and Goldstein, J. L. Independent regulation of sterol regulatory element-binding proteins 1 and 2 in hamster liver. Proc Natl Acad Sci U S A 92 (1995) 935–938.

Shimano, H., Horton, J. D., Hammer, R. E., Shimomura, I., Brown, M. S., and Goldstein, J. L. Overproduction of cholesterol and fatty acids causes massive liver enlargement in transgenic mice expressing truncated SREBP-1a. J Clin Invest 98 (1996) 1575–1584.

Shimano, H., Horton, J. D., Shimomura, I., Hammer, R. E., Brown, M. S., and Goldstein, J. L. Isoform 1c of sterol regulatory element binding protein is less active than isoform 1a in livers of transgenic mice and in cultured cells. J Clin Invest 99 (1997a) 846–854.

Shimano, H., Shimomura, I., Hammer, R. E., Herz, J., Goldstein, J. L., Brown, M. S., and Horton, J. D. Elevated levels of SREBP-2 and cholesterol synthesis in livers of mice homozygous for a targeted disruption of the SREBP-1 gene. J Clin Invest 100 (1997b) 2115–2124.

Shimano, H., Yahagi, N., Amemiya-Kudo, M., Hasty, A. H., Osuga, J., Tamura, Y., Shionoiri, F., Iizuka, Y., Ohashi, K., Harada, K., et al. Sterol regulatory element-binding protein-1 as a key transcription factor for nutritional induction of lipogenic enzyme genes. J Biol Chem 274 (1999) 35832–35839.

Shimomura, I., Bashmakov, Y., Ikemoto, S., Horton, J. D., Brown, M. S., and Goldstein, J. L. Insulin selectively increases SREBP-1c mRNA in the livers of rats with streptozotocin-induced diabetes. Proc Natl Acad Sci U S A 96 (1999) 13656–13661.

Shimomura, I., Shimano, H., Horton, J. D., Goldstein, J. L., and Brown, M. S. Differential expression of exons 1a and 1c in mRNAs for sterol regulatory element binding protein-1 in human and mouse organs and cultured cells. J Clin Invest 99 (1997) 838–845.

Strittmatter, P., Spatz, L., Corcoran, D., Rogers, M. J., Setlow, B., and Redline, R. Purification and properties of rat liver microsomal stearyl coenzyme A desaturase. Proc Natl Acad Sci U S A 71 (1974) 4565–4569.

Sundqvist, A., Bengoechea-Alonso, M. T., Ye, X., Lukiyanchuk, V., Jin, J., Harper, J. W., and Ericsson, J. Control of lipid metabolism by phosphorylation-dependent degradation of the SREBP family of transcription factors by SCF(Fbw7). Cell Metab 1 (2005) 379–391.

Surapureddi, S., Yu, S., Bu, H., Hashimoto, T., Yeldandi, A. V., Kashireddy, P., Cherkaoui-Malki, M., Qi, C., Zhu, Y. J., Rao, M. S., and Reddy, J. K. Identification of a transcriptionally active peroxisome proliferator-activated receptor alpha -interacting cofactor complex in rat liver and characterization of PRIC285 as a coactivator. Proc Natl Acad Sci U S A 99 (2002) 11836–11841.

Takeuchi, Y., Yahagi, N., Nakagawa, Y., Matsuzaka, T., Shimizu, R., Sekiya, M., Iizuka, Y., Ohashi, K., Gotoda, T., Yamamoto, M., et al. In vivo promoter analysis on refeeding response of hepatic sterol regulatory element-binding protein-1c expression. Biochem Biophys Res Commun 363 (2007) 329–335.

Tan, N. S., Shaw, N. S., Vinckenbosch, N., Liu, P., Yasmin, R., Desvergne, B., Wahli, W., and Noy, N. Selective cooperation between fatty acid binding proteins and peroxisome proliferator-activated receptors in regulating transcription. Mol Cell Biol 22 (2002) 5114–5127.

Taniguchi, C. M., Kondo, T., Sajan, M., Luo, J., Bronson, R., Asano, T., Farese, R., Cantley, L. C., and Kahn, C. R. Divergent regulation of hepatic glucose and lipid metabolism by phosphoinositide 3-kinase via Akt and PKClambda/zeta. Cell Metab 3 (2006) 343–353.

Theodoulou, F. L., Holdsworth, M., and Baker, A. Peroxisomal ABC transporters. FEBS Lett 580 (2006) 1139–1155.

Tobin, K. A., Ulven, S. M., Schuster, G. U., Steineger, H. H., Andresen, S. M., Gustafsson, J. A., and Nebb, H. I. Liver X receptors as insulin-mediating factors in fatty acid and cholesterol biosynthesis. J Biol Chem 277 (2002) 10691–10697.

Tong, L., and Harwood, H. J., Jr. Acetyl-coenzyme A carboxylases: versatile targets for drug discovery. J Cell Biochem 99 (2006) 1476–1488.

Tontonoz, P., Kim, J. B., Graves, R. A., and Spiegelman, B. M. ADD1: a novel helix-loop-helix transcription factor associated with adipocyte determination and differentiation. Mol Cell Biol 13 (1993) 4753–4759.

Tudor, C., Feige, J. N., Pingali, H., Lohray, V. B., Wahli, W., Desvergne, B., Engelborghs, Y., and Gelman, L. Association with coregulators is the major determinant governing peroxisome proliferator-activated receptor mobility in living cells. J Biol Chem 282 (2007) 4417–4426.

Uchida, Y., Izai, K., Orii, T., and Hashimoto, T. Novel fatty acid beta-oxidation enzymes in rat liver mitochondria. II. Purification and properties of enoyl-coenzyme A (CoA) hydratase/3-hydroxyacyl-CoA dehydrogenase/3-ketoacyl-CoA thiolase trifunctional protein. J Biol Chem 267 (1992) 1034–1041.

Uyeda, K., and Repa, J. J. Carbohydrate response element binding protein, ChREBP, a transcription factor coupling hepatic glucose utilization and lipid synthesis. Cell Metab 4 (2006) 107–110.

Vega, R. B., Huss, J. M., and Kelly, D. P. The coactivator PGC-1 cooperates with peroxisome proliferator-activated receptor alpha in transcriptional control of nuclear genes encoding mitochondrial fatty acid oxidation enzymes. Mol Cell Biol 20 (2000) 1868–1876.

Voss, A., Reinhart, M., Sankarappa, S., and Sprecher, H. The metabolism of 7,10,13,16,19-docosapentaenoic acid to 4,7,10,13,16,19-docosahexaenoic acid in rat liver is independent of a 4-desaturase. J Biol Chem 266 (1991) 19995–20000.

Walczak, R., and Tontonoz, P. PPARadigms and PPARadoxes: expanding roles for PPAR-gamma in the control of lipid metabolism. J Lipid Res 43 (2002) 177–186.

Wanders, R. J., and Waterham, H. R. Peroxisomal disorders: the single peroxisomal enzyme deficiencies. Biochim Biophys Acta 1763 (2006) 1707–1720.

Wang, J., Yu, L., Schmidt, R. E., Su, C., Huang, X., Gould, K., and Cao, G. Characterization of HSCD5, a novel human stearoyl-CoA desaturase unique to primates. Biochem Biophys Res Commun 332 (2005) 735–742.

Wang, X., Sato, R., Brown, M. S., Hua, X., and Goldstein, J. L. SREBP-1, a membrane-bound transcription factor released by sterol-regulated proteolysis. Cell 77 (1994) 53–62.

Wang, Y., Botolin, D., Xu, J., Christian, B., Mitchell, E., Jayaprakasam, B., Nair, M. G., Peters, J. M., Busik, J. V., Olson, L. K., and Jump, D. B. Regulation of hepatic fatty acid elongase and desaturase expression in diabetes and obesity. J Lipid Res 47 (2006) 2028–2041.

Weber, L. W., Boll, M., and Stampfl, A. Maintaining cholesterol homeostasis: sterol regulatory element-binding proteins. World J Gastroenterol 10 (2004) 3081–3087.

Whiteman, E. L., Cho, H., and Birnbaum, M. J. Role of Akt/protein kinase B in metabolism. Trends Endocrinol Metab 13 (2002) 444–451.

Widmer, J., Fassihi, K. S., Schlichter, S. C., Wheeler, K. S., Crute, B. E., King, N., Nutile-McMenemy, N., Noll, W. W., Daniel, S., Ha, J., et al. Identification of a second human acetyl-CoA carboxylase gene. Biochem J 316 (Pt 3) (1996) 915–922.

Wolfrum, C., Asilmaz, E., Luca, E., Friedman, J. M., and Stoffel, M. Foxa2 regulates lipid metabolism and ketogenesis in the liver during fasting and in diabetes. Nature 432 (2004) 1027–1032.

Wolfrum, C., Besser, D., Luca, E., and Stoffel, M. Insulin regulates the activity of forkhead transcription factor Hnf-3beta/Foxa-2 by Akt-mediated phosphorylation and nuclear/cytosolic localization. Proc Natl Acad Sci U S A 100 (2003) 11624–11629.

Wolfrum, C., Borrmann, C. M., Borchers, T., and Spener, F. Fatty acids and hypolipidemic drugs regulate peroxisome proliferator-activated receptors alpha - and gamma-mediated gene expression via liver fatty acid binding protein: a signaling path to the nucleus. Proc Natl Acad Sci U S A 98 (2001) 2323–2328.

Wolfrum, C., and Stoffel, M. Coactivation of Foxa2 through Pgc-1beta promotes liver fatty acid oxidation and triglyceride/VLDL secretion. Cell Metab 3 (2006) 99–110.

Woodgett, J. R. Molecular cloning and expression of glycogen synthase kinase-3/factor A. Embo J 9 (1990) 2431–2438.

Xu, H. E., Lambert, M. H., Montana, V. G., Parks, D. J., Blanchard, S. G., Brown, P. J., Sternbach, D. D., Lehmann, J. M., Wisely, G. B., Willson, T. M., et al. Molecular recognition of fatty acids by peroxisome proliferator-activated receptors. Mol Cell 3 (1999a) 397–403.

Xu, J., Nakamura, M. T., Cho, H. P., and Clarke, S. D. Sterol regulatory element binding protein-1 expression is suppressed by dietary polyunsaturated fatty acids. A mechanism for the coordinate suppression of lipogenic genes by polyunsaturated fats. J Biol Chem 274 (1999b) 23577–23583.

Xu, J., Teran-Garcia, M., Park, J. H., Nakamura, M. T., and Clarke, S. D. Polyunsaturated fatty acids suppress hepatic sterol regulatory element-binding protein-1 expression by accelerating transcript decay. J Biol Chem 276 (2001) 9800–9807.

Yabe, D., Komuro, R., Liang, G., Goldstein, J. L., and Brown, M. S. Liver-specific mRNA for Insig-2 down-regulated by insulin: implications for fatty acid synthesis. Proc Natl Acad Sci U S A 100 (2003) 3155–3160.

Yahagi, N., Shimano, H., Hasty, A. H., Amemiya-Kudo, M., Okazaki, H., Tamura, Y., Iizuka, Y., Shionoiri, F., Ohashi, K., Osuga, J., et al. A crucial role of sterol regulatory element-binding protein-1 in the regulation of lipogenic gene expression by polyunsaturated fatty acids. J Biol Chem 274 (1999) 35840–35844.

Yamamoto, T., Shimano, H., Inoue, N., Nakagawa, Y., Matsuzaka, T., Takahashi, A., Yahagi, N., Sone, H., Suzuki, H., Toyoshima, H., and Yamada, N. Protein kinase A suppresses sterol regulatory element-binding protein-1C expression via phosphorylation of liver X receptor in the liver. J Biol Chem 282 (2007) 11687–11695.

Yang, S. Y., He, X. Y., and Schulz, H. 3-Hydroxyacyl-CoA dehydrogenase and short chain 3-hydroxyacyl-CoA dehydrogenase in human health and disease. Febs J 272 (2005) 4874–4883.

Yang, T., Espenshade, P. J., Wright, M. E., Yabe, D., Gong, Y., Aebersold, R., Goldstein, J. L., and Brown, M. S. Crucial step in cholesterol homeostasis: sterols promote binding of SCAP to INSIG-1, a membrane protein that facilitates retention of SREBPs in ER. Cell 110 (2002) 489–500.

Yoon, J. C., Puigserver, P., Chen, G., Donovan, J., Wu, Z., Rhee, J., Adelmant, G., Stafford, J., Kahn, C. R., Granner, D. K., et al. Control of hepatic gluconeogenesis through the transcriptional coactivator PGC-1. Nature 413 (2001) 131–138.

Yoshikawa, T., Ide, T., Shimano, H., Yahagi, N., Amemiya-Kudo, M., Matsuzaka, T., Yatoh, S., Kitamine, T., Okazaki, H., Tamura, Y., et al. Cross-talk between peroxisome proliferator-activated receptor (PPAR) alpha and liver X receptor (LXR) in nutritional regulation of fatty acid metabolism. I. PPARs suppress sterol regulatory element binding protein-1c promoter through inhibition of LXR signaling. Mol Endocrinol 17 (2003) 1240–1254.

Zandbergen, F., and Plutzky, J. PPARalpha in atherosclerosis and inflammation. Biochim Biophys Acta 1771 (2007) 972–982.

Zhang, L., Ge, L., Parimoo, S., Stenn, K., and Prouty, S. M. Human stearoyl-CoA desaturase: alternative transcripts generated from a single gene by usage of tandem polyadenylation sites. Biochem J 340 (Pt 1) (1999) 255–264.

Zhang, S., Yang, Y., and Shi, Y. Characterization of human SCD2, an oligomeric desaturase with improved stability and enzyme activity by cross-linking in intact cells. Biochem J 388 (2005) 135–142.

Zhang, W., Patil, S., Chauhan, B., Guo, S., Powell, D. R., Le, J., Klotsas, A., Matika, R., Xiao, X., Franks, R., et al. FoxO1 regulates multiple metabolic pathways in the liver: effects on gluconeogenic, glycolytic, and lipogenic gene expression. J Biol Chem 281 (2006) 10105–10117.

Zheng, Y., Eilertsen, K. J., Ge, L., Zhang, L., Sundberg, J. P., Prouty, S. M., Stenn, K. S., and Parimoo, S. Scd1 is expressed in sebaceous glands and is disrupted in the asebia mouse. Nat Genet 23 (1999) 268–270.

Zoete, V., Grosdidier, A., and Michielin, O. Peroxisome proliferator-activated receptor structures: ligand specificity, molecular switch and interactions with regulators. Biochim Biophys Acta 1771 (2007) 915–925.

# Chapter 2
# Modulation of Protein Function by Isoketals and Levuglandins

**Sean S. Davies**

**Abstract** Oxidative stress, defined as an increase in reactive oxygen species, leads to peroxidation of polyunsaturated fatty acids and generates a vast number of biologically active molecules, many of which might contribute in some way to health and disease. This chapter will focus on one specific class of peroxidation products, the levuglandins and isoketals (also called isolevuglandins). These γ-ketoaldehydes are some of the most reactive products derived from the peroxidation of lipids and exert their biological effects by rapidly adducting to primary amines such as the lysyl residues of proteins. The mechanism of their formation and remarkable reactivity will be described, along with evidence for their increased formation in disease conditions linked with oxidative stress and inflammation. Finally, the currently known effects of these γ-ketoaldehydes on cellular function will then be discussed and when appropriate compared to the effects of α,β-unsaturated fatty aldehydes, in order to illustrate the significant differences between these two classes of peroxidation products that modify proteins.

**Keywords** Aldehydes · isoketals · levuglandins · lipid peroxidation · protein modification

## 2.1 Introduction

### *2.1.1 Mechanisms of Isoketal and Levuglandin Formation*

Conversion of arachidonic acid to an eicosanoid γ-ketoaldehyde can proceed by two separate pathways, one driven enzymatically by cyclooxygenases and the other driven non-enzymatically by free radicals. The two pathways differ only in the mechanism used to generate the key intermediate, a bicyclic endoperoxide with two aliphatic side chains, which can undergo non-enzymatic rearrangement to

S.S. Davies
Department of Pharmacology, Vanderbilt University, 506A RRB, 2222 Pierce Ave, Nashville, TN, 37232, USA

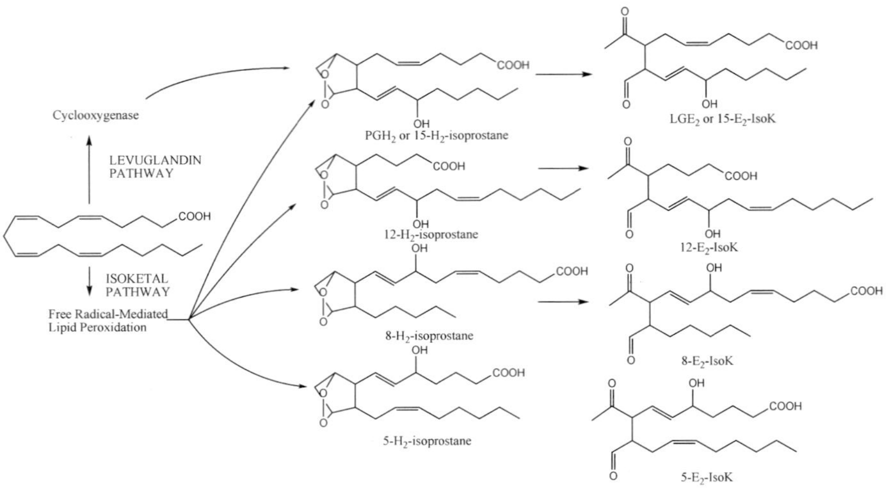

**Fig. 2.1** Formation of levuglandins (LG) and isoketals (IsoK) from arachidonic acid. Arachidonic acid can be converted to a γ-ketoaldehyde by two distinct pathways. In the levuglandin pathway, cyclooxygenase enzymes convert arachidonic acid to $PGH_2$, which then rearranges non-enzymatically to form LGs. In the isoketal pathway, free radicals mediated lipid peroxidation forms four different regioisomers of $H_2$-isoprostanes, that also rearrange non-enzymatically to form IsoKs. The different regioisomers are identified by the position of their hydroxyl group. 15-$H_2$-isoprostane is the same regioisomer as $PGH_2$, so that one of the eight stereoisomers of 15-$E_2$-IsoK is identical to $LGE_2$

form the γ-ketoaldehyde. In the enzymatic pathway, cyclooxgenases convert non-esterified arachidonic acid to the bicyclic endoperoxide, prostaglandin $H_2$ ($PGH_2$) (Fig. 2.1). In the non-enzymatic pathway, free radical mediated peroxidation of esterified and non-esterified arachidonic acid generates four $PGH_2$-like regioisomers ($H_2$-isoprostanes) (Morrow et al., 1990). The specific regioisomer formed depends on which of the bis-allylic hydrogen the free radical initially attacks and the different $H_2$-isoprostane regioisomers are denoted by the carbon number of their hydroxyl group. These bicyclic endoperoxides have 4 stereocenters, so that each radical generated regioisomer includes 16 stereoisomers for a total of 64 $H_2$-isoprostane isomers compared to the single stereoisomer, $PGH_2$, generated by the cyclooxygenases.

After formation of the bicyclic endoperoxides, formation of the γ-ketoaldehydes proceeds by an identical non-enzymatic concerted scission rearrangement for both pathways (Fig. 2.2) (Salomon et al., 1984). Because base-catalyzed rearrangement can be initiated on either side of the nearly symmetrical bridgehead, both $E_2$- or $D_2$- isomers of γ-ketoaldehyde form. $E_2$-isomers have the ketone group adjacent to the carboxylate side chain and the aldehyde adjacent to the second side chain, while $D_2$-isomers have the aldehyde group adjacent to the carboxylate side chain and the ketone group adjacent to the second side chain. In the case of $PGH_2$ rearrangement, the two isomers

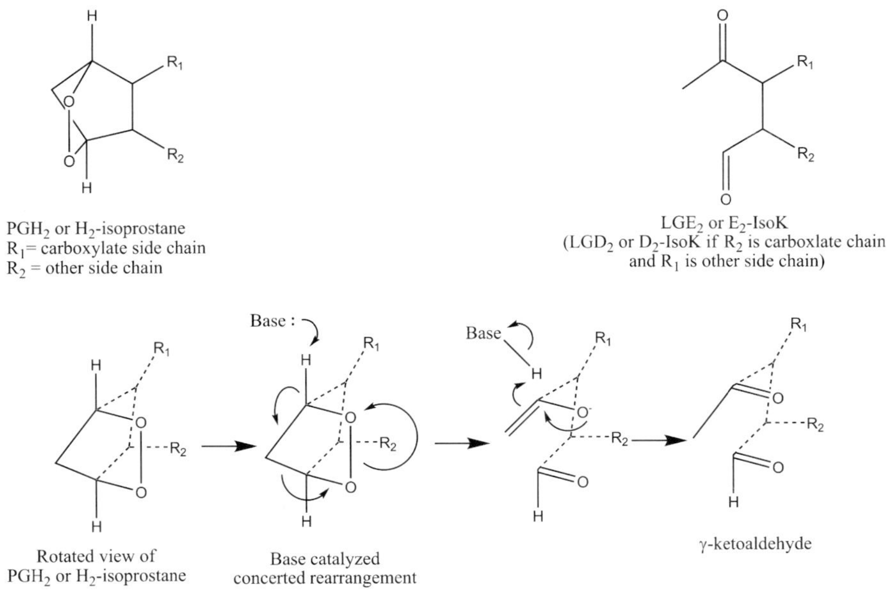

$PGH_2$ or $H_2$-isoprostane
$R_1$= carboxylate side chain
$R_2$ = other side chain

$LGE_2$ or $E_2$-IsoK
($LGD_2$ or $D_2$-IsoK if $R_2$ is carboxlate chain and $R_1$ is other side chain)

Rotated view of $PGH_2$ or $H_2$-isoprostane

Base catalyzed concerted rearrangement

γ-ketoaldehyde

**Fig. 2.2** Rearrangement of bicyclic endoperoxides to form γ-ketoaldehyde. Proposed mechanism for formation of γ-ketoaldehydes from bicyclic endoperoxide such as $PGH_2$ and $H_2$-isoprostanes by base-catalyzed concerted rearrangement (Salomon et al., 1984). Attack of base is shown at the top of the bridgehead, which results in formation of $E_2$-isomers. Attack at the symmetrical position below the bridgehead forms the $D_2$-isomers

are called levuglandin (LG) $E_2$ and $D_2$. With $H_2$-isoprostane rearrangement, the resulting collective set of 64 regio- and stereo-isomers are given the trivial name $E_2$- and $D_2$-isoketals (IsoK) or alternatively isolevuglandins (Brame et al., 1999). As with $H_2$-isoprostanes, $D_2$- and $E_2$-IsoK regioisomers are denoted by the position of the hydroxyl group, so that 15-$E_2$-IsoK represents the same regioisomer as $LGE_2$, and in fact, one of the eight stereoisomers of 15-$E_2$-IsoK is structurally identical to $LGE_2$. Finally, it should be noted that peroxidation of docosahexanoic acid (DHA) generates similar γ-ketoaldehydes designated as neuroketals (Bernoud-Hubac et al., 2001).

Incubation of $PGH_2$ in phosphate buffer yields 22% LG (Salomon et al., 1984). Incubation of $PGH_2$ in DMSO, which may better model the hydrophobic environment near cyclooxygenases, yields 70% LG. The non-enzymatic rearrangement of $PGH_2$ to LG occurs even in cells with active prostaglandin synthases. For instance, stimulated platelet produce prodigious amounts of thromboxane through the cyclooxgenase/thromboxane synthase pathway, yet platelet stimulation also produces significant amounts of LG (Boutaud et al., 2003). LG formation can be enhanced in this system by inhibition of thromboxane synthase. Because prostaglandin synthases do not act on $H_2$-isoprostanes, the relative yield of IsoK derived from $H_2$-isoprostane formed in cells is expected to be similar to that found in vitro.

In summary, γ-ketoaldehydes can be formed by two pathways whose products can only be distinguished by the number of isomers formed and by the potential presence of esterified forms for IsoKs. So far, no experimental data has shown that the biological effects of the γ-ketoaldehydes are specific only to a particular regio- or stereo-isomers of the γ-ketoaldehydes, so that the blanket term IsoK/LG will be used in this review even when a specific effect was demonstrated using a particular LG or IsoK species.

### *2.1.2 Cardinal Features of IsoK/LG Reaction with Proteins*

Interest in the biological activity of IsoK/LGs stems from the cardinal features of IsoK/LG biochemistry: (1) their extremely rapid adduction to proteins, (2) their proclivity to crosslink proteins, and (3) their propensity to disrupt protein function.

#### 2.1.2.1 IsoK/LG Rapidly Adduct to Proteins

IsoK/LGs react nearly instantaneously with primary amines such as the lysyl residues of proteins. Although 4-hydroxynonenal (HNE), an α,β-unsaturated aldehyde that also forms by lipid peroxidation, is often considered to be highly reactive, it reacts at a very pedestrian rate when compared to IsoK/LG. For instance, when IsoK/LG or HNE is added to human serum albumin, the half-life of unadducted IsoK/LG is less than 20 s, while the half-life of HNE is about 60 min (Brame et al., 1999). In practical terms, the difference in reactivity between IsoK/LG and HNE means that only adducted IsoK/LG can be found in vivo, while unadducted HNE can be readily measured in tissues and plasma. Additionally, the reactivity of IsoK/LG precludes significant diffusion, so that only the proteins nearest the sites of IsoK/LG formation, such as membrane-associated proteins, are likely to be adducted.

The rapid reaction rate of IsoK/LG is driven by the stability of the pyrrole adducts formed (Fig. 2.3). In common with all aldehydes including HNE, an initial nucleophilic attack by the lysyl nitrogen forms a hemiaminal adduct which dehydrates to an imine (Schiff base) adduct. This highly reversible reaction product can usually only be measured after conversion to a reduced Schiff base by a strong reducing agent such as sodium borohydride. What makes IsoK/LGs so much more reactive than ordinary aldehydes? With γ-ketoaldehydes, formation of the initial hemiaminal adduct positions the second carbonyl group to also undergo nucleophilic attack. This pyrrolidine adduct then quickly undergoes dehydration to form a pyrrole, making the reaction essentially irreversible. For other aldehydes such as HNE, no secondary nucleophilic attack is possible, so the unstable hemiaminal adduct

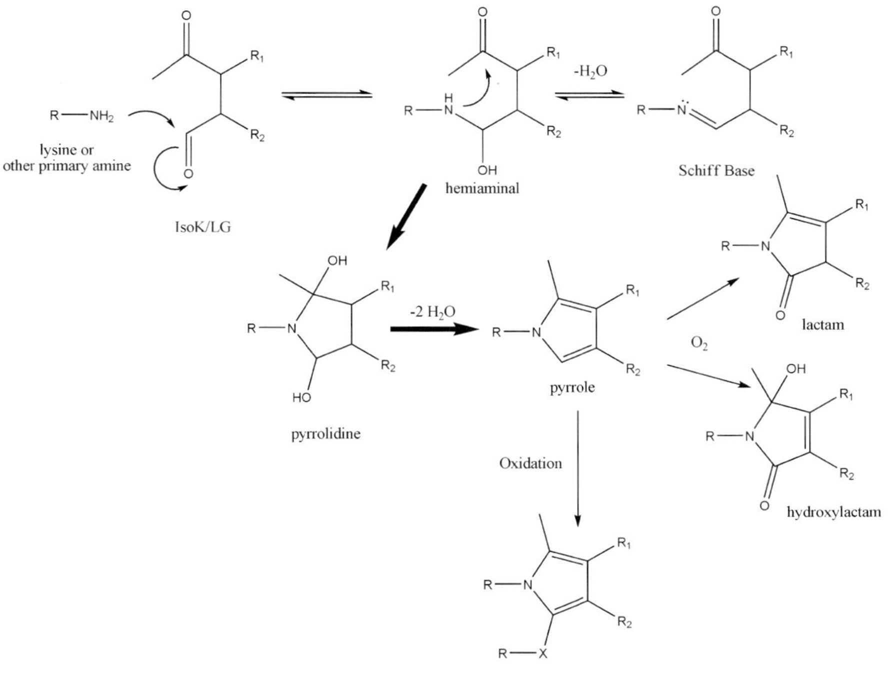

**Fig. 2.3** Reaction of IsoK/LG with primary amines to form stable adducts. Primary amines including lysine react with IsoK/LGs to form a hemiaminal adduct. Unlike most aldehydes which can only form the highly reversible Schiff base adduct, the hemiaminal adduct of γ-ketoaldehydes can undergo a second nucleophilic attack to form a pyrrolidine adduct which dehydrates to form an irreversible pyrrole adduct. In the presence of oxygen, the pyrrole is converted to lactam and hydroxylactam adducts. Oxidation of the pyrrole leads to formation of stable crosslinked species

simply reverses back to unadducted HNE and lysine. Stable adduction of protein by HNE and related α,β-unsaturated aldehydes generally occurs through a Michael addition reaction, for which thiols are more reactive nucleophiles than lysines.

The rate of pyrrole adduct formation can be determined by addition of the Ehrlich reagent that reacts with pyrroles to form a visibly purple product (DiFranco et al., 1995). In the presence of oxygen, the ability of adduct to react with the Ehrlich reagent diminishes over time. Analysis by mass spectrometry found that in the presence of oxygen, the pyrrole adduct goes on to form highly stable lactam and hydroxylactam adducts (Fig. 2.3) (Brame et al., 1999). Therefore, quantification of lactam adducts, rather than pyrrole adducts, is probably most useful except in artificial conditions when oxygen can be completely excluded.

#### 2.1.2.2 IsoK/LG Crosslink Proteins

Another cardinal feature of IsoK/LG biochemistry is the proclivity to form crosslinked protein aggregrates. This feature can be readily appreciated by dose curves from treatment of a model protein such as chicken egg ovalbumin with increasing molar equivalents of IsoK/LG (Davies et al., 2002; Iyer et al., 1989). At low molar equivalents of IsoK/LG, there is a small proportion of protein that migrates on SDS-PAGE with the apparent mass of dimers and trimers that indicate intermolecular crosslinking. At ten or greater molar equivalents, most of the protein runs as highly oligimerized forms of the protein. Because each molecule of ovalbumin is estimated to have twenty surface lysines, saturation of available surface lysines is not required to generate extensive crosslinking.

The mechanism that underlie this proclivity to crosslink is believed to be oxidation of the pyrrole adduct to form electrophiles that can readily react with nucleophiles including thiols, amines, or unoxidized pyrroles (Fig. 2.3) (Amarnath et al., 1994). Therefore, IsoK/LG can readily crosslink proteins not only to adjacent proteins but to DNA or polyamines as well (Boutaud et al., 2001; Murthi et al., 1993). Detection of IsoK/LG intermolecular crosslinks currently relies on visualization of co-migrating species on SDS-PAGE. Unfortunately, the exact molecular species of the crosslinks has eluded characterization by mass spectrometry or NMR. One reason for this failure is the difficulty of working with highly crosslinked material. The purple-to-brownish crosslinked material typically forms insoluble aggregates that do not pass through solid phase extraction cartridges and HPLC columns and that poorly ionize in mass spectrometers.

The conditions that facilitate the formation of lactams versus crosslinks are poorly characterized but are likely to be mediated by the proximity of lysine and other nucleophiles. Increasing the molar equivalents of IsoK/LG versus lysine residues significantly increases the extent of intermolecular crosslinking visualized by SDS-PAGE, but also linearly increases the amount of lactam adduct measured (Davies et al., 2007). Thus measurement of IsoK-lysyl-lactam appears to be a reasonable surrogate marker for all other IsoK/LG adducts formed, and increases in IsoK-lysyl-lactam in tissues or cells most likely also indicates increases in IsoK/LG crosslinked proteins as well.

#### 2.1.2.3 Disruption of Protein Function

A final cardinal feature of IsoK/LG biochemistry is their propensity to disrupt protein function. Modification of proteins by IsoK/LG could theoretically alter protein function by several mechanisms. Adduction of IsoK/LG to lysine converts a short, positively charged group to a bulky, hydrophobic, negatively charged group. Modifications of lysyl residues in the active site of enzymes will therefore eliminate catalytic activity. Catalytic activity may be lost even if the modified residue is simply adjacent to the active site, as the bulky IsoK/LG adduct may sterically hinder substrate binding or product release. Protein modification by IsoK/LG may also disrupt protein function by altering protein conformation. Crosslinking of nearby cysteine and lysyl residues could significantly deform the conformation and lock the protein in an active or inactive conformation. Intermolecular crosslinking that initiates protein aggregates may not only alter conformation, but potentially initiate cellular stress responses.

In addition to directly altering catalytic activity, IsoK/LG modification may alter protein function by changing interactions with other regulatory proteins. For example, modification of lysines required for protein-protein interactions will disrupt this interaction. Depending on the specific protein, such disruption could be inhibiting or activating. Similarly, addition of a bulky hydrophobic group could also cause the adducted protein to more strongly partition to membranes, thus altering interaction with normal binding partners and creating novel partners. Finally, adduction of IsoK/LG to lysyl groups may alter the degradation rate or pathways of adducted proteins. In doing so, it may significantly prolong or shorten the half-life of the adducted protein and thereby lead to dysregulation.

### 2.1.3 IsoK/LG Protein Adducts form in Various Disease Conditions

Because of the myriad ways that IsoK/LG adduction might disrupt normal physiological proteins, a rational exploration of how IsoK/LGs contribute in disease processes first requires defining conditions where IsoK/LG adducts are increased. Currently, there are two complimentary methods for quantifying IsoK/LG adducts in vivo. The first method utilizes the sensitivity and specificity of electrospray ionization tandem mass spectrometry. Quantitative measurement of IsoK-lysyl-lactam adducts can be made by complete enzymatic proteolysis of tissue followed by partial purification on solid phase extraction cartridges and HPLC and then analysis by mass spectrometry (Davies et al., 2007). A heavy isotope labeled internal standard is added to the sample for quantification. The second method utilizes antibody based approaches such as ELISA, Western blotting, or immunohistochemistry. Anti-IsoK/LG antibodies are made by immunizing animals with IsoK/LG adducted proteins or by screening single-chain antibody libraries with IsoK/LG adducted peptides.

Both quantitative approaches have drawbacks. For instance, the current mass spectrometric method does not distinguish between the various isomers of IsoK/LG adducts and thus provides no information on the pathway leading to adducts. Because complete proteolyis is required, it is also not possible to specifically determine the identity of the adducted protein. The method is also very time-consuming and limited to labs with electrospray ionization tandem mass spectrometers and associated expertise. In contrast to mass spectrometry methods, antibody based methods are relatively easy and inexpensive; however, in comparison to anti-protein antibodies, anti-lipid antibodies tend to suffer from relatively poor affinity and greater non-specificity. The poor performance of anti-lipid antibodies is likely inherent to the chemical properties of the antigen, as lipid antigens lack the high number of rigid structures that are typically required for high affinity, high specificity antigen binding. The only rigid structure in the IsoK/LG adducts are the pyrrole or lactam ring, structures shared with other biologically relevant compounds such as porphyrins. Nevertheless, the presence of aliphatic side chains in IsoK/LG adducts appear to confer reasonable specificity as antibodies with selectivity for separate regioisomers of IsoK/LG adducts have been reported (DiFranco et al., 1995; Poliakov et al., 2004; Salomon et al., 1997b; Salomon et al., 1999; Salomon et al., 2000). Ideally, both mass spectrometric and antibody methods would be used to confirm increases in specific disease conditions.

IsoK/LG adducts increase in a number of conditions related to oxidative stress and inflammation. The first published report of IsoK/LG adducted proteins in vivo was in very small groups of patients with documented atherosclerosis or end-stage renal disease using ELISA measurements (Salomon et al., 1997b). $LGE_2$ adducted keyhole limpet hemocyanin was the immunizing antigen for the antibody utilized in this study, therefore both the IsoK and LG pathways might have contributed to detected immunoreactivity. A follow-up study also used a second antibody that recognized an IsoK regioisomer adduct that could only be derived from the IsoK pathway. In this study, both of the antibodies measured an approximately two-fold change in plasma adduct levels in the atherosclerotic and renal disease patients compared to controls (Salomon et al., 2000). This finding implicates the IsoK pathway, rather than the LG pathway, as the major source of adducts in these diseases.

The first study demonstrating IsoK/LG adducts in vivo using mass spectrometric methods used carbon chloride treated rats (Brame et al., 2004). Carbon tetrachloride ($CCl_4$) is converted to trichloromethyl free radical by cyctochrome P450s in the liver, leading to a massive increase in lipid peroxidation. After four hours of treatment with either $CCl_4$ or vehicle, the rats were sacrificed and a portion of the recovered livers reacted with sodium borohydride to allow measurement of reduced Schiff Base adducts. Additionally, a portion of liver protein extracts were base hydrolyzed in order to measure total (both esterified and non-esterified) IsoK adducts. In livers from untreated animals, no reduced Schiff base adducts was detected, but lactam adduct was measured to be 3.5±0.2 ng/g tissue. In $CCl_4$ treated animals, the total reduced Schiff base adduct was 21±4 ng/g tissue, with only 0.5±0.1 ng/g tissue being non-esterified. Non-esterified lactam adduct was 6.4±0.3 ng/g and similar levels of total lactam adduct were found. This result is consistent with the notion that immediately following $CCl_4$ treatment, IsoKs initially form in situ on phospholipids and are still esterified when they react with proteins to form Schiff base adducts. The presence of lactam adduct even in untreated animals suggests that the lactam adduct is stable and can thus accumulate during basal rates of lipid peroxidation in vivo. That only non-esterified lactam adduct is found even after $CCl_4$ treatment suggests the presence of a phospholipase that can act on either the pyrrole or lactam adduct to hydrolyze the modified phospholipid. While no further studies have been carried out to confirm the existence of this phospholipase, an attractive candidate would be platelet-activating factor acetylhydrolase, which hydrolyzes other phospholipid-esterified isoprostane products.

Increased tissue levels of IsoK/LG protein adducts have been found in a number of other conditions associated with oxidative stress and inflammation including ischemic heart (Fukuda et al., 2005), hyperoxic lung (Davies et al., 2004), Alzheimer's Disease brain (Zagol-Ikapitte et al., 2005), experimental sepsis plasma (Poliakov et al., 2003), allergic inflammation lung (Talati et al., 2006), and glaucomatous trabecular meshwork (Govindarajan et al., 2008). From this limited sampling, it seems reasonable to expect increases in other conditions where oxidative stress or inflammation occur. While the demonstration of increased IsoK/LG is consistent with their contribution to the disease process, these results by themselves do not provide conclusive evidence as to whether IsoK/LG adducts have any real pathophysiological significance or are simply tombstones of more important events. This important question must be addressed in two ways: first, by demonstrating in cellular assays plausible mechanisms for IsoK/LG adduction to contribute to pathophysiology and second, by demonstrating that reducing IsoK/LG adducts, preferably without altering other oxidative events, significantly ameliorates pathophysiology.

## *2.1.4 Other Primary Amines Compete with Lysines for Adduction to IsoK/LGs*

### 2.1.4.1 Other Endogenous Amines React with IsoK/LGs

Although the lysyl residues of proteins are probably the most important target of IsoK/LG in the cell, IsoK/LG reacts with a wide range of primary amines. Another abundant primary amine in cells is the ethanolamine head group of phosphatidylethanolamine (PE). Because IsoKs form on membrane phospholipids, they would be well-positioned to react with PE. Incubation of one molar equivalent of IsoK/LG and 1-palmitoyl, 2-linoleolyl-PE for two hours produced a stable IsoK/LG-PE pyrrole adduct (Bernoud-Hubac et al., 2004). No evidence of a lactam or hydroxylactam adduct was found under these conditions, likely because the reaction conditions were designed to exclude oxygen. The formation of IsoK/LG-PE is of interest because oxidatively modified phospholipids have been implicated in certain autoimmune diseases such as lupus. Whether this reaction actually occurs in vivo has yet to be determined.

IsoK/LG can also react with arginine to form a bis-IsoK/LG-arginine adduct which then undergoes decomposition to a bis-IsoK/LG-urea (BLU)

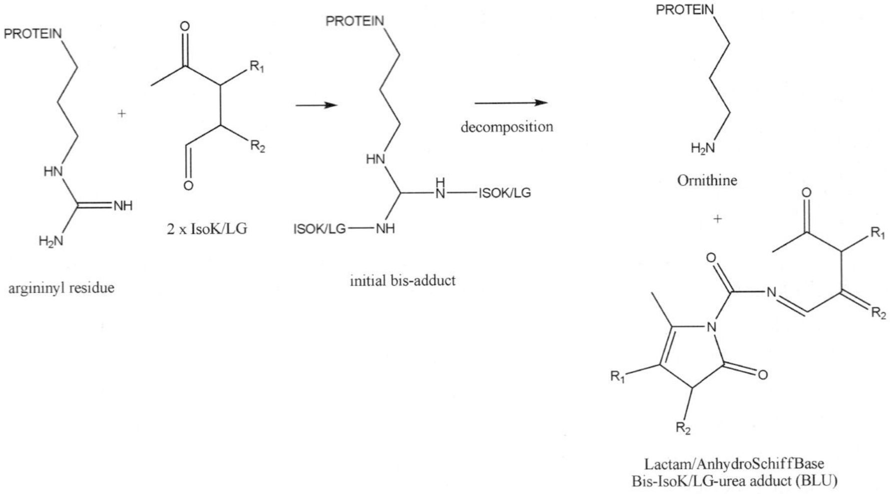

Fig. 2.4 Reaction of IsoK/LG with arginine. According to proposed mechanism (Zagol-Ikapitte et al., 2004) two IsoK/LGs react with the two amine groups of arginine to form bis-IsoK/LG adducts that then undergoes decomposition to a bis-IsoK/LG urea (BLU) adduct and ornithine

adduct and ornithine (Zagol-Ikapitte et al., 2004) (Fig. 2.4). To test the relative reactivity of IsoK/LG for lysine versus arginine, a polyglycine peptide also featuring one lysine and one arginine residue was reacted with 1 or 3 molar equivalents of IsoK/LG. When one molar equivalent of IsoK/LG was added, only the lysyl residue of the peptide was adducted. When three molar equivalents of IsoK/LG was added to the peptide, BLU adduct could be readily detected as well as the ornithine variant of the IsoK/LG lysyl adducted peptide. This result is consistent with the argininyl residue being substantially less reactive than the lysyl residue. The relevance of the reaction with arginine in vivo, especially its role in the formation of orthinine is currently under investigation.

Besides PE and arginine, other abundant endogenous primary amines include the polyamines. Polyamines such as spermidine and spermine have been postulated to protect DNA from reactive carbonyls and their high concentration in cytoplasm makes them a plausible agent for preventing the adduction of cytoplasmic proteins. Although polyamines have been demonstrated to be crosslinked to proteins in the presence of IsoK/LG (Boutaud et al., 2001), there have been no published characterizations of IsoK/LG-spermine or IsoK/LG-spermidine only adducts. If formed, these adducts might be useful surrogate markers for the extent of IsoK/LG formation under various oxidative conditions, as these adducts would not require proteolysis to be quantified by mass spectrometry making their measurement more straightforward than protein adducts.

#### 2.1.4.2 Pyridoxamine Analogs are IsoK/LG Scavengers

That different primary amines reacted with IsoK/LG at different rates suggested the possibility that novel amines with even greater reactivity than lysyl residues could be identified. These novel amines could then be used as IsoK/LG scavengers to protect proteins from inactivation and to determine the impact on pathophysiology of specifically reducing the levels of IsoK/LG protein adduct. Screening a series of primary amines identified that pyridoxamine, a vitamin $B_6$ vitamer, potently competed with lysine for reaction to IsoK/LGs (Amarnath et al., 2004). To determine the structure-activity relationship for γ-ketoaldehyde scavenging, a series of pyridoxamine analogs were reacted with 4-oxopentanal (OPA), a model γ-ketoaldehyde, and the rate of pyrrole formation determined using Ehrlich reagent (Fig. 2.5). The second order reaction rate of pyridoxamine was found to be 2,309-fold faster than that of lysine. The key structural requirements for scavenging by pyridoxamine appear to be an aminomethyl group with an adjacent hydroxyl on an aromatic ring. Reaction of γ-ketoaldehyde with the aminomethyl group of pyridoxamine forms a hemiaminal adduct whose ketone group hydrogens bonds with the phenolic hydroxyl group of pyridoxamine. This bonding facilitates the nucleophilic attack on the ketone group required for ring closure and pyrrole formation. Methylation of the hydroxyl group prevents this hydrogen bonding and accounts for the highly diminished reactivity of 2-methoxybenzylamine compared to salicylamine.

The identification of the key features for IsoK/LG scavenging led to the synthesis of other related phenolic amines that also potently scavenge IsoK/LG in vitro such as pentyl-pyridoxamine (Davies et al., 2006). Lipophilic scavengers such as salicylamine and pentyl-pyridoxamine localize to the membranes where

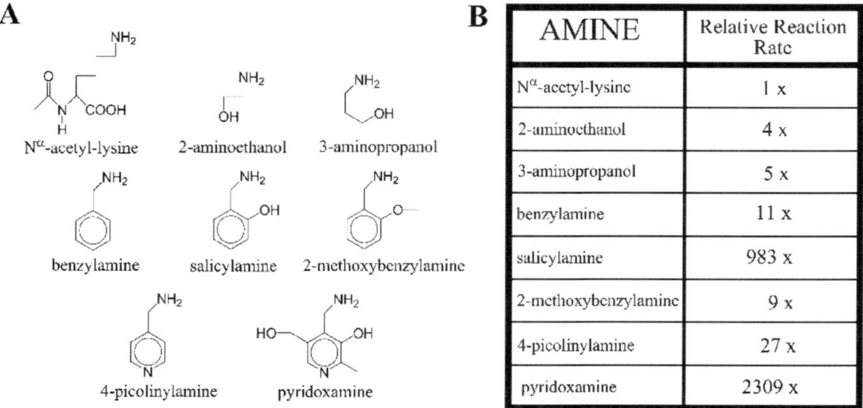


| AMINE | Relative Reaction Rate |
|---|---|
| $N^{\alpha}$-acetyl-lysine | 1 x |
| 2-aminoethanol | 4 x |
| 3-aminopropanol | 5 x |
| benzylamine | 11 x |
| salicylamine | 983 x |
| 2-methoxybenzylamine | 9 x |
| 4-picolinylamine | 27 x |
| pyridoxamine | 2309 x |

**Fig. 2.5** Structural requirements for IsoK/LG reactivity. Second order reaction rate of various pyridoxamine analogs for the formation of pyrrole adduct were determined relative to $N^{\alpha}$-acetyl-lysine (Amarnath et al., 2004). The reaction rate of pyridoxamine was 2,309x faster than $N^{\alpha}$-acetyl-lysine

IsoK/LG form and should therefore be more effective in vivo than hydrophilic compounds such as pyridoxamine. The greater efficacy of salicylamine and pentyl-pyridoxamine compared to pyridoxamine was borne out in studies where platelets were treated with individual scavengers and then activated with arachidonic acid to make LGs. While all three scavengers significantly inhibited the formation of LG protein adducts, salicylamine was the most potent followed by pentyl-pyridoxamine and then pyridoxamine (Davies et al., 2006).

How selective are pyridoxamine and lipophilic pyridoxamine analogs for scavenging γ-ketoaldehydes compared to other reactive lipid peroxidation products? The reactivity of pyridoxamine with α,β-unsaturated aldehydes is completely trivial, and pyridoxamine does not protect proteins from HNE adduction (Amarnath et al., 2004; Davies et al., 2006). Pyridoxamine does react with α-ketoaldehydes such as methylglyoxal that form from the oxidative decomposition of carbohydrates and lipids (Voziyan et al., 2002), and these adducts can be detected in the urine of rodents fed pyridoxamine in their drinking water (Metz et al., 2003). However, the reaction rate of pyridoxamine with γ-ketoaldehyde is 187 times greater than with methylglyoxal (Amarnath et al., 2004). To address the specificity of phenolic amine scavenging during lipid peroxidation, aliquots of lysine and iron-oxidized arachidonic acid were incubated with either pyridoxamine, pentyl-pyridoxamine, salicylamine, or vehicle and the resulting α- and γ-ketoaldehyde adducts measured by mass spectrometry (Davies et al., 2006). All three phenolic amines significantly reduced the levels of IsoK-lysyl-lactam adduct compared to vehicle. Levels of IsoK-phenolic amine adduct were nearly proportional to the decrease in lysyl-lactam adducts as would be expected for scavenging of IsoK. Importantly, IsoK-phenolic amine adduct were formed in significantly greater abundance than those derived from α-ketoaldehydes. Pentyl-pyridoxamine appeared to be most selective, with a nearly 50-fold greater yield of IsoK-pentyl-pyridoxamine adduct than the most abundant α-ketoaldehyde adduct. Salicylamine was the least selective phenolic amine with about a 7-fold greater abundance of IsoK-salicylamine adduct than the most abundant α-ketoaldehyde adduct. The discovery of these efficient and relatively selective IsoK/LG scavengers makes possible their future use in cells and in vivo to examine the contribution of IsoK/LG to cellular and organ dysfunction in the complex environment of oxidative stress.

## 2.2 Effects of IsoK/LGs on Cellular Function

### *2.2.1 IsoK/LG are Some of the Most Higly Cytotoxic Products of Lipid Peroxidation*

Perhaps the most striking biological effect of IsoK/LG is their potent cytoxicity. The first demonstration of this effect came from direct injection of 100 nmol of

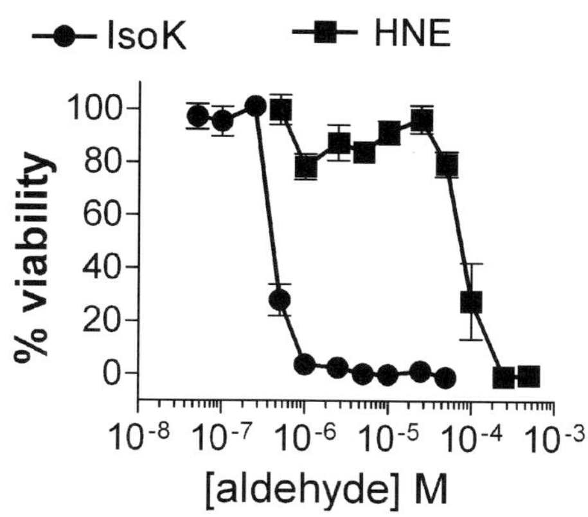


**Fig. 2.6** IsoK is considerably more cytotoxic than HNE. P19 neuroglial cells were treated with various concentration of 15-$E_2$-IsoK or HNE for 24 h and the viability of cells determined by the MTT conversion assay. Percent viability was calculated relative to untreated cells. (Mean ± SEM $n=8$)

IsoK/LG into the substance of the cerebral hemisphere of rats (Schmidley et al., 1992). The injected area showed pallor and loss of cellular constituents typical of necrotic cells and the marginal area became hypercellular because of the infiltration by macrophages. Evans Blue extravasation measurements also demonstrated a dose-dependent loss of blood brain barrier integrity. Adding nanomolar concentrations of IsoK/LG to cultured neuroglial cells directly induces cytotoxicity (Davies et al., 2002). IsoK/LG is several orders of magnitude more cytotoxic than HNE (Fig. 2.6), making it one of the most potent cytotoxic lipid peroxidation species known. Nevertheless, because toxicity in situ is a function of both the concentration of the product generated by peroxidation and of its potency, these experiments alone do not define to what extent each of the various peroxidation products contribute to the cytotoxicity induced by oxidative stress.

Because the IsoK/LG scavengers do not scavenge HNE and related electrophiles, these scavengers can be used to compare the contribution of IsoK/LG and HNE to the cytotoxicity of cells treated with hydrogen peroxide to induce oxidative stress. Pretreatment of HepG2 cells with IsoK/LG scavengers significantly inhibits the cytotoxicity induced by hydrogen peroxide (Fig. 2.7) (Davies et al., 2006). Only the lipophilic IsoK/LG scavengers, salicylamine and pentylpyridoxamine, provided protection, while the hydrophilic IsoK/LG scavenger, pyridoxamine, failed to provide significant protection. This result suggests that formation of IsoK adduct are a critical component of cell death induced by oxidative stress and that the critical targets of adduction are located near membranes, although lipophilic scavengers may also penetrate into the cells better than hydrophilic scavengers.

The exact mechanisms whereby IsoK/LG potently induces cell death are currently under investigation. IsoK/LG does not need to adduct to intracellular proteins to be toxic. Incubation of IsoK/LG with amyloid beta peptide causes oligimerization of amyloid beta (Davies et al., 2002) and IsoK/LG crosslinked amyloid $\beta_{1-42}$ peptide is highly neurotoxic (Boutaud et al., 2006). IsoK/LG or

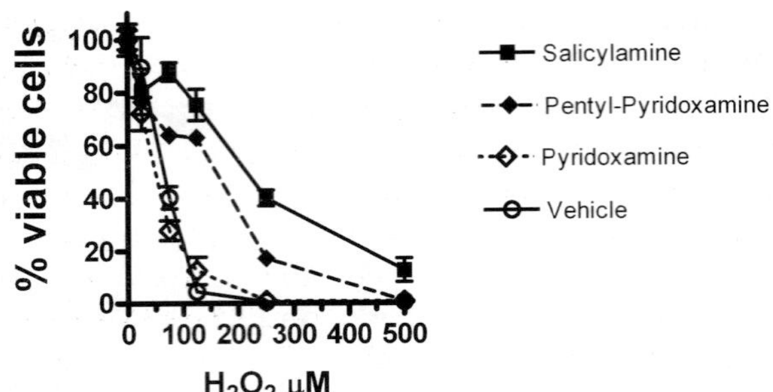


**Fig. 2.7** Lipophilic IsoK/LG scavengers protect against cytotoxicity induced by oxidative stress. HepG2 cells were incubated with vehicle, pyridoxamine, pentyl-pyridoxamine, or salicylamine for 30 min prior to treatment with various concentrations of hydrogen peroxide (Davies et al., 2006). Viability was determined by detection of ATP using ATPlite luminescence assay and percent viability calculated relative to untreated cells (Mean ± SEM $n = 8$)

amyloid $\beta_{1-42}$ alone did not produce similar neurotoxicity, so that adduction and oligimerization is required for this potent toxicity. Amyloid-derived diffusible ligands are thought to be critical neurotoxic products in Alzheimer's disease and antibodies to amyloid-derived diffusible ligands recognize IsoK/LG adducted amyloid $\beta_{1-42}$. Whether IsoK/LG contributes to the formation of amyloid-derived diffusible ligands in vivo remains to be determined. Another important question is whether the neurotoxic effect of IsoK/LG induced oligimerization can be generalized to other amyloid forming peptides.

### *2.2.2 Proteasome Inhibition may be an Important Mechanism of Cytotoxicity*

The presence of oligimerized amyloid peptides does not, of course, account for the toxicity of IsoK/LG in culture cells that lack amyloid-like peptides. Nevertheless, studies with IsoK/LG adducted amyloid β may provide some insight into the generalized mechanisms that underlie IsoK/LG toxicity. For instance, IsoK/LG adducted amyloid β is a potent competitive inhibitor of the proteasome (Davies et al., 2002). The proteasome rids the cell of unneeded, incorrectly folded, or oxidatively damaged proteins. Pharmacological inhibition of the proteasome results in cell death, probably because of the accumulation of undesirable proteins. Proteolysis by the proteasome requires passing an unfolded loop of the protein down the barrel of the proteasome into the trypsin-, chymotrypsin-, and post-glutamyl peptidase-like catalytic sites buried deep within the multiple subunit complex. Incubating a model protein, ovalbumin, with equimolar IsoK/LG to form adducted protein decreased proteasomal degradation of the ovalbumin by about 50% (Davies et al., 2002). Incubating ovalbumin with ten molar equivalents of IsoK/LG completely prevented proteasomal degradation. It should be noted that in these studies, the 20S proteasome was used rather than the 26S proteasome, so that blocking ubiquination by adducting the relevant lysines does not account for the decreased proteolysis. Instead, it is easy to imagine that adduction of lysine blocks the ability of the trypsin-like activity to hydrolyze the protein, so that the processing of IsoK/LG adducted proteins stalls inside the proteasome's barrel. These undigestable proteins would then inhibit access by other proteins targeted for degradation. Adduction of peptides with high affinity for the proteasome, such as amyloid β, would be particularly likely to inhibit proteasome activity. This mechanism is supported by a recent report that ubiquinated, IsoK/LG modified calpain-1 protein inhibited 26S proteasome activity while unmodified calpain-1 had no effect (Govindarajan et al., 2008). IsoK/LG can also directly act on the proteasome to inhibit its activity, although the exact mechanism for inhibition is unknown (Davies et al., 2002). Similarly, IsoK/LG can act on calpain-1, another significant pathway of protein degradation, to inhibit its activity (Govindarajan et al., 2008). Thus, formation of even a small amount of IsoK/LG or IsoK/LG adducted protein may lead to significant accumulation of undesired proteins within the cells and thus to cell death.

Accumulation of aggregated proteins is a hallmark of several neurodegenerative diseases, including Alzheimer's, Huntington's, and Parkinson's Disease. Proteasome activity, but not protein mass, is reduced in Alzheimer's disease. A recent report suggested that there is increased modification of proteasome by neuroketals in Alzheimer's disease brain (Cecarini et al., 2007). Although the specific epitope recognized by this antibody was poorly defined, if substantiated this finding would suggest proteasomal inhibition by IsoK/LG and neuroketal adducted proteins may very well play a central role in the neurodegeneration associated with these diseases.

### 2.2.3 IsoK/LG in Cardiovascular Disease

Another area of active interest into the effects of IsoK/LG is the role they may play in cardiovascular disease. While the increased levels of IsoK/LG adducts found in the circulation of atherosclerotic subjects suggests that formation of IsoK/LG may be important, this finding alone provides little insight into either the mechanisms of their formation or their role in the disease. Several plausible mechanisms may account for increased formation of IsoK/LG in atherosclerosis. Myeloperoxidase and cyclooxygenase levels increase in the vascular wall during atherosclerosis, and both enzymes generate prodigious amounts of IsoK/LGs, at least in vitro (Boutaud et al., 2001; Poliakov et al., 2003). Additionally, ischemia is well known to generate reactive oxygen species and ligation of the coronary descending artery in dogs for five days induced about a three-fold increase in the levels of IsoK/LG adducts, measured by mass spectrometry, in the infarct border zone (Fukuda et al., 2005). Using immunohistochemistry, IsoK/LG adducts were found to localize to the epicardium and the myocardial core of the border zone after infarction.

#### 2.2.3.1 IsoK/LG Adduction to Sodium Channel is Proarrhythmic

The localization of adducts to the epicardial border zone suggested the possibility that IsoK/LG adducts contribute to cardiac arrhythmias. Ventricular tachycardia/fibrillation following myocardial infarction is a major cause of sudden cardiac death. Arrhythmias in ischemic myocardium arise from sodium channel blockade. Sodium channels are hypothesized to cycle between three conformational states: a deactivated closed state, an activated open state, and an inactivated closed state. Upon depolarization, the deactivated state converts to the activated state and sodium current flows for a brief time before the channel enters the inactive state. The channel only converts from the inactive state to the deactivated state when the membrane repolarizes during the falling phase of the action potential. Changes in the ability to convert from the inactive to the deactivated state are critical to the initiation and perpetuation of arrhythmias.

Treatment of HEK-293 cells expressing the human cardiac sodium channel (Nav1.5-HEK) with an oxidant, tert-butyl-hydroperoxide (tBHP), results in a negative shift in the voltage dependence of inactivation (Fukuda et al., 2005). If translated in vivo, this negative shift would be proarrhythmic. In contrast to its effects on inactivation, tBHP treatment did not alter voltage dependent activation significantly. Thus oxidative stress does not simply destroy sodium channel function, but rather alters its susceptibility to inactivation. A similar effect of tBHP treatment occurs in HL-1 mouse atrial cell, which endogenously express both the cardiac sodium channel and its accessory beta subunits. Adding thiols such as DTT or glutathione, in order to scavenge α,β-unsaturated carbonyls such as HNE, did not protect sodium channel function during tBHP treatment. Additionally, treating Nav1.5-HEK cells with HNE did not induce voltage-dependent inactivation. Therefore, HNE and similar α,β-unsaturated carbonyls do not appear to contribute to sodium channel inactivation. In contrast to HNE, addition of 10 μM IsoK/LG to Nav1.5-HEK caused voltage-dependent inactivation of the sodium channel in a similar manner as tBHP. Addition of 15-$F_{2t}$-isoprostane, which is structurally related to IsoK/LG, but cannot adduct to proteins, had no effect on sodium channel function. IsoK/LG also caused sodium channel inactivation in HL-1 cells. Interestingly, treatment with IsoK/LG appeared to have overlapping as well as synergistic effects as flecainide, a Nav1.5 sodium channel blocker. An outstanding question is whether IsoK/LG scavengers blocked the effect of tBHP on the sodium channel. A preliminary conference report indicated that 10 μM salicylamine completely mitigated the effect of tBHP in Nav1.5-HEK cells and 100 μM salicylamine blocked the effect in HL-1 cells. If this report is substantiated, these results would strongly implicate IsoK/LG as important effectors of sodium channel dysfunction after oxidative stress and would suggest the possibility that IsoK/LG scavengers might be effective antiarrhythmic agents. Studies are currently underway to test the efficacy of IsoK/LG scavengers in preventing arrhythmias after myocardial infarction in animal models. These studies should provide a clearer picture not only of the contribution of IsoK/LG to arrhythmias, but also to the efficacy of the scavengers in vivo.

#### 2.2.3.2 Effect of IsoK/LG on Other Ion Channels

Besides the sodium channel, IsoK/LG has effects on other ion channels as well. For instance, addition of synthetic IsoK/LG to an atrial tumor myocyte cell line, AT-1, resulted in a pronounced dose-dependent inhibition of the inward rectifying potassium current induced by a –40 mV voltage step (Brame et al., 2004). Both activating and deactivating currents were suppressed, so the effect on potassium channels differs in this regard from sodium channels and suggests a more wholesale destruction of channel function. The $IC_{50}$ for inhibition of $I_{Kr}$ was 2.2 μM and full inhibition was only achieved after incubating the cells for 60 min. Washing the cells after this period of time did not revert $I_{Kr}$ current to normal, consistent with covalent modification by IsoK/LG inducing the inhibition. The delayed time course of inactivation is consistent with the need to crosslink the channel in order to inactivate it, but no further attempt to characterize the mechanism of inhibition or the sites of adduction were reported.

Investigations into the effect of IsoK/LG modification on calcium channels or associated receptors that activate calcium release have only recently begun. Provocative preliminary results suggest that IsoK/LG modification may lead to activation of calcium currents, but additional experiments are needed to confirm this result and identify the mechanisms responsible.

#### 2.2.3.3 Oxidized Lipoproteins

Although modifications of ion channels could contribute to the late stages of cardiovascular disease, formation of IsoK/LG adducts may also play a role early in atherosclerosis. Atherosclerosis is initiated when macrophages take up oxidized low density lipoprotein (oxLDL) via scavenger receptors such as SR-A and CD36 to form foam cells. In vitro oxidation of LDL results in the formation of IsoK adducts on the particle (Brame et al., 1999; Salomon et al., 1997a; 1999), suggesting that IsoK modification could potentially mediate this process. Addition of increasing concentration of IsoK/LG to native LDL results in a dose-dependent shift in the mobility of the LDL during electrophoresis, reminiscent of what is observed with oxLDL (Hoppe et al., 1997). Importantly, IsoK/LG modification of native LDL also dose-dependently increased the binding and uptake of the LDL by cultured mouse peritoneal macrophages to a similar extent as found with oxLDL. Acetylated LDL, a substrate for SR-A, did not compete with IsoK/LG-modified LDL for uptake by these macrophages, but oxLDL completely competed off the binding and uptake of IsoK/LG-modified LDL. Therefore, IsoK/LG-modified LDL must be taken up by scavenger receptors other than SR-A. While these results suggest a role for IsoK/LG-modified LDL in atherosclerosis, they leave unanswered a number of important questions. For instance, do the receptors responsible for IsoK/LG-modified LDL uptake also facilitate the uptake of other IsoK/LG-modified proteins? Can IsoK/LG-modified LDL stimulate monocyte/macrophage chemotaxis and cytokine secretion in a similar manner as oxLDL? Hopefully, these and other questions will be addressed by future studies.

### *2.2.4 Effect of IsoK/LG on Other Generalized Cellular Functions*

#### 2.2.4.1 Tubulin/Microtubules

Microtubules play an integral part in a large number of cellular processes including mitosis. Polymerization and depolymization of tubulin, the major component of microtubules and one of the most abundant proteins in the cell, is critical to these functions and its polymerization is often regulated by membrane associated proteins. To test the effect of IsoK/LG adduction on microtubule function, IsoK/LG was added to sea urchin eggs. IsoK/LG dose-dependently ($IC_{50}$ 15 μM) inhibited cell division of fertilized eggs, with lower doses ($<$ 10 μM) inducing abnormal cleavage (Murthi et al., 1990). Whether similar inhibition of microtubule function occurs in mammalian cells is unclear as unpublished studies in cultured neuronal cells found that exogenous addition of IsoK/LG did not have significant effects either on neurite outgrowth or microtubule organization. Perhaps this is because insufficient IsoK/LG is able to penetrate through the membrane of neuronal cells, because incubation of purified GTP-depleted microtubule protein preparations from bovine brains with IsoK/LG dose-dependently inhibited microtubule assembly stimulated by addition of GTP (Murthi et al., 1990). Complete inhibition of tubulin polymerization required two molecules of IsoK/LG for each molecule of tubulin in the microtubule preparation. Interestingly, adduction of already assembled spindles by IsoK/LG did not lead to disassembly of the spindles. Similar concentrations of related lipid molecules, such as arachidonic acid or $PGE_2$ had no effect on microtubule assembly.

#### 2.2.4.2 Histones and DNA

Addition of IsoK/LG to V79 Chinese hamster lung fibroblasts or nuclei caused formation of DNA-protein crosslinks (Murthi et al., 1993). These crosslinks were demonstrated indirectly by the amount of DNA retained on a nitrocelluse filter after passing lysates from treated cells or nuclei through the filter. As the nitrocellulose filter normally traps only protein, but not DNA, the presence of DNA was indicative of a covalent bound with proteins. Treatment of the lysate with proteinase K prior to addition to the filter prevented DNA binding, confirming the requirement for covalent binding to protein. Formation of DNA-protein crosslinks was time-dependent, with 60 min or more required for maximal crosslinking, which is in keeping with the time required for formation of protein-protein crosslinks. The identity of the proteins involved in the DNA-protein crosslink was not determined. However, the lysine-rich histones readily undergo adduction in the presence of IsoK/LG (Boutaud et al., 2001) and DNA tightly coils around histones in cells, making these proteins the most likely protein candidate for DNA-IsoK/LG-protein crosslinks. What, if any, impact the formation of DNA-IsoK/LG-protein crosslinks has on the cell, or even if they form under biologically relevant conditions, has not been determined. However, it is not difficult to imagine that DNA-histone crosslinking might significantly disruption normal transcription and transcriptional regulation.

## 2.3 Concluding Remarks

Much of the early work relating to IsoK/LG has necessarily focused on understanding the basic chemistry of their reaction with proteins and other biological amines, developing methods to quantify IsoK/LG adducts in vivo, and developing selective inhibitors of their formation. From these initial studies comes tantalizing evidence for an important contribution of IsoK/LG adduction to pathophysiology. Not only are IsoK/LG adducts increased in disease conditions associated with oxidative stress and inflammation, but their application to cultured cells induce effects highly relevant to these conditions. The use of IsoK/LG scavengers improves cellular viability and function under oxidative stress. While these early results are encouraging, a large number of outstanding questions remain. For instance, are the levels of IsoK/LG modification measured in vivo sufficient to induce pathophysiology? Does treatment with IsoK/LG scavengers significantly lower the levels of IsoK/LG adducts in vivo and do these scavengers protect against disease? What are the mechanisms underlying IsoK/LG induced toxicity and ion channel dysfunction? Which specific proteins (or other biologically important amines) are adducted during oxidative stress and which ones are protected during treatment with IsoK/LG scavengers? Answering these questions may not only provide insight into the pathogenesis of diseases related to oxidative stress and inflammation, but also into the feasibility of using IsoK/LG scavengers as novel therapeutic agents for these conditions. Thus, the study of IsoK/LG may be fruitful field of endeavor for many years to come.

## References

Amarnath, V., Valentine, W.M., Amarnath, K., Eng, M.A., and Graham, D.G. The mechanism of nucleophilic substitution of alkylpyrroles in the presence of oxygen. Chem Res Toxicol 7 (1994) 56–61.

Amarnath, V., Amarnath, K., Amarnath, K., Davies, S., and Roberts, L.J., Pyridoxamine: an extremely potent scavenger of 1,4-dicarbonyls. Chem Res Toxicol 17 (2004) 410–415.

Bernoud-Hubac, N., Davies, S.S., Boutaud, O., Montine, T.J., and Roberts, L.J., Formation of highly reactive gamma-ketoaldehydes (neuroketals) as products of the neuroprostane pathway. J Biol Chem 276 (2001) 30964–30970.

Bernoud-Hubac, N., Fay, L.B., Armarnath, V., Guichardant, M., Bacot, S., Davies, S.S., Roberts, L.J., 2nd, and Lagarde, M. Covalent binding of isoketals to ethanolamine phospholipids. Free Radic Biol Med 37 (2004) 1604–1611.

Boutaud, O., Montine, T.J., Chang, L., Klein, W.L., and Oates, J.A. PGH2-derived levuglandin adducts increase the neurotoxicity of amyloid beta1-42. J Neurochem 96 (2006) 917–923.

Boutaud, O., Li, J., Zagol, I., Shipp, E.A., Davies, S.S., Roberts, L.J., 2nd, and Oates, J.A. Levuglandinyl adducts of proteins are formed via a prostaglandin H2 synthase-dependent pathway after platelet activation. J Biol Chem 278 (2003) 16926–16928.

Boutaud, O., Brame, C.J., Chaurand, P., Li, J., Rowlinson, S.W., Crews, B.C., Ji, C., Marnett, L.J., Caprioli, R.M., Roberts, L.J., 2nd, and Oates, J.A. Characterization of the lysyl adducts of prostaglandin H-synthases that are derived from oxygenation of arachidonic acid. Biochemistry 40 (2001) 6948–6955.

Brame, C.J., Salomon, R.G., Morrow, J.D., and Roberts, L.J. Identification of extremely reactive gamma-ketoaldehydes (isolevuglandins) as products of the isoprostane pathway and characterization of their lysyl protein adducts. J Biol Chem 274 (1999) 13139–13146.

Brame, C.J., Boutaud, O., Davies, S.S., Yang, T., Oates, J.A., Roden, D., and Roberts, L.J. Modification of proteins by isoketal-containing oxidized phospholipids. J Biol Chem 279 (2004) 13447–13451.

Cecarini, V., Ding, Q., and Keller, J.N. Oxidative inactivation of the proteasome in Alzheimer's disease. Free Radic Res 41 (2007) 673–680.

Davies, S.S., Amarnath, V., Brame, C.J., Boutaud, O., and Roberts, L.J. Measurement of chronic oxidative and inflammatory stress by quantification of isoketal/levuglandin gamma-ketoaldehyde protein adducts using liquid chromatography tandem mass spectrometry. Nat Protoc 2 (2007) 2079–2091.

Davies, S.S., Amarnath, V., Montine, K.S., Bernoud-Hubac, N., Boutaud, O., Montine, T.J., and Roberts, L.J., Effects of reactive gamma-ketoaldehydes formed by the isoprostane pathway (isoketals) and cyclooxygenase pathway (levuglandins) on proteasome function. Faseb J 16 (2002) 715–717.

Davies, S.S., Talati, M., Wang, X., Mernaugh, R.L., Amarnath, V., Fessel, J., Meyrick, B.O., Sheller, J., and Roberts, L.J., Localization of isoketal adducts in vivo using a single-chain antibody. Free Radic Biol Med 36 (2004) 1163–1174.

Davies, S.S., Brantley, E.J., Voziyan, P.A., Amarnath, V., Zagol-Ikapitte, I., Boutaud, O., Hudson, B.G., Oates, J.A., and Ii, L.J. Pyridoxamine Analogues Scavenge Lipid-Derived gamma-Ketoaldehydes and Protect against H(2)O(2)-Mediated Cytotoxicity. Biochemistry 45 (2006) 15756–15767.

DiFranco, E., Subbanagounder, G., Kim, S., Murthi, K., Taneda, S., Monnier, V.M., and Salomon, R.G. Formation and stability of pyrrole adducts in the reaction of levuglandin E2 with proteins. Chem Res Toxicol 8 (1995) 61–67.

Fukuda, K., Davies, S.S., Nakajima, T., Ong, B.H., Kupershmidt, S., Fessel, J., Amarnath, V., Anderson, M.E., Boyden, P.A., Viswanathan, P.C., Roberts, L.J., 2nd, and Balser, J.R. Oxidative mediated lipid peroxidation recapitulates proarrhythmic effects on cardiac sodium channels. Circ Res 97 (2005) 1262–1269.

Govindarajan, B., Laird, J., Salomon, R.G., and Bhattacharya, S.K. 2008. Isolevuglandin-Modified Proteins, Including Elevated Levels of Inactive Calpain-1, Accumulate in Glaucomatous Trabecular Meshwork. Biochemistry in press.

Hoppe, G., Subbanagounder, G., O'Neil, J., Salomon, R.G., and Hoff, H.F. Macrophage recognition of LDL modified by levuglandin E2, an oxidation product of arachidonic acid. Biochim Biophys Acta 1344 (1997) 1–5.

Iyer, R.S., Ghosh, S., and Salomon, R.G. 1989. Levuglandin E2 crosslinks proteins. Prostaglandins 37 (1989) 471–480.

Metz, T.O., Alderson, N.L., Chachich, M.E., Thorpe, S.R., and Baynes, J.W. Pyridoxamine traps intermediates in lipid peroxidation reactions in vivo: evidence on the role of lipids in chemical modification of protein and development of diabetic complications. J Biol Chem 278 (2003) 42012–42019.

Morrow, J.D., Harris, T.M., and Roberts, L.J., Noncyclooxygenase oxidative formation of a series of novel prostaglandins: analytical ramifications for measurement of eicosanoids. Anal Biochem 184 (1990) 1–10.

Murthi, K.K., Salomon, R.G., and Sternlicht, H. Levuglandin E2 inhibits mitosis and microtubule assembly. Prostaglandins 39 (1990) 611–622.

Murthi, K.K., Friedman, L.R., Oleinick, N.L., and Salomon, R.G. Formation of DNA-protein cross-links in mammalian cells by levuglandin E2. Biochemistry 32 (1993) 4090–4097.

Poliakov, E., Meer, S.G., Roy, S.C., Mesaros, C., and Salomon, R.G. Iso[7]LGD2-protein adducts are abundant in vivo and free radical-induced oxidation of an arachidonyl phospholipid generates this D series isolevuglandin in vitro. Chem Res Toxicol 17 (2004) 613–622.

Poliakov, E., Brennan, M.L., Macpherson, J., Zhang, R., Sha, W., Narine, L., Salomon, R.G., and Hazen, S.L. Isolevuglandins, a novel class of isoprostenoid derivatives, function as integrated sensors of oxidant stress and are generated by myeloperoxidase in vivo. Faseb J 17 (2003) 2209–2220.

Salomon, R.G., Miller, D.B., Zagorski, M.G., and Coughlin, D.J. Solvent Induced Fragmentation of Prostaglandin Endoperoxides. New Aldehyde Products from PGH2 and Novel Intramolecular 1,2-Hydride Shift During Endoperoxide Fragmentation in Aqueous Solution. J Am Chem Soc 106 (1984) 6049–6060.

Salomon, R.G., Subbanagounder, G., Singh, U., O'Neil, J., and Hoff, H.F. Oxidation of low-density lipoproteins produces levuglandin-protein adducts. Chem Res Toxicol 10 (1997a) 750–759.

Salomon, R.G., Subbanagounder, G., O'Neil, J., Kaur, K., Smith, M.A., Hoff, H.F., Perry, G., and Monnier, V.M. Levuglandin E2-protein adducts in human plasma and vasculature. Chem Res Toxicol 10 (1997b) 536–545.

Salomon, R.G., Sha, W., Brame, C., Kaur, K., Subbanagounder, G., O'Neil, J., Hoff, H.F., and Roberts, L.J., Protein adducts of iso[4]levuglandin E2, a product of the isoprostane pathway, in oxidized low density lipoprotein. J Biol Chem 274 (1999) 20271–20280.

Salomon, R.G., Batyreva, E., Kaur, K., Sprecher, D.L., Schreiber, M.J., Crabb, J.W., Penn, M.S., DiCorletoe, A.M., Hazen, S.L., and Podrez, E.A. Isolevuglandin-protein adducts in humans: products of free radical-induced lipid oxidation through the isoprostane pathway. Biochim Biophys Acta 1485 (2000) 225–235.

Schmidley, J.W., Dadson, J., Iyer, R.S., and Salomon, R.G. Brain tissue injury and blood-brain barrier opening induced by injection of LGE2 or PGE2. Prostaglandins Leukot Essent Fatty Acids 47 (1992) 105–110.

Talati, M., Meyrick, B., Peebles, R.S., Jr., Davies, S.S., Dworski, R., Mernaugh, R., Mitchell, D., Boothby, M., Roberts, L.J., and Sheller, J.R. Oxidant stress modulates murine allergic airway responses. Free Radic Biol Med 40 (2006) 1210–1219.

Voziyan, P.A., Metz, T.O., Baynes, J.W., and Hudson, B.G. A post-Amadori inhibitor pyridoxamine also inhibits chemical modification of proteins by scavenging carbonyl intermediates of carbohydrate and lipid degradation. J Biol Chem 277 (2002) 3397–3403.

Zagol-Ikapitte, I., Bernoud-Hubac, N., Amarnath, V., Roberts, L.J., 2nd, Boutaud, O., and Oates, J.A. Characterization of bis(levuglandinyl) urea derivatives as products of the reaction between prostaglandin H2 and arginine. Biochemistry 43 (2004) 5503–5510.

Zagol-Ikapitte, I., Masterson, T.S., Amarnath, V., Montine, T.J., Andreasson, K.I., Boutaud, O., and Oates, J.A. Prostaglandin H(2)-derived adducts of proteins correlate with Alzheimer's disease severity. J Neurochem 94 (2005) 4 :1140–1145.

# Chapter 3
# Signalling Pathways Controlling Fatty Acid Desaturation

**María Cecilia Mansilla, Claudia Banchio and Diego de Mendoza**

**Abstract** Microorganisms, plants and animals regulate the synthesis of unsaturated fatty acids (UFAs) during changing environmental conditions as well as in response to nutrients. Unsaturation of fatty acid chains has important structural roles in cell membranes: a proper ratio of saturated to UFAs contributes to membrane fluidity. Alterations in this ratio have been implicated in various disease states including cardiovascular diseases, immune disorders, cancer and obesity. They are also the major components of triglycerides and intermediates in the synthesis of biologically active molecules such as eicosanoids, which mediates fever, inflammation and neurotransmission. UFAs homeostasis in many organisms is achieved by feedback regulation of fatty acid desaturases gene transcription. Here, we review recently discovered components and mechanisms of the regulatory machinery governing the transcription of fatty acid desaturases in bacteria, yeast and animals.

**Keywords** Desaturase · gene regulation · membrane fluidity · signal transduction · unsaturated fatty acids

**Abbreviations** a-BFAs: anteiso-branched fatty acids; ACP: acyl carrier protein; BHLH: active soluble domain of SREBP; CNS: central nervous system; D5D: delta 5 desaturase; D6D: delta 6 desaturase; ER: endoplasmic reticulum; Hik: histidine kinase; HUFA: highly unsaturated fatty acid; LXR: liver X receptor; MUFA: monounsaturated fatty acid; NF-Y: nuclear factor Y; PLs: phospholipids; PP: peroxisome proliferator; PPAR: PP-activated receptor; PPRE: PP response element; PUFA: polyunsaturated fatty acid; PUFA-BP: PUFA binding protein; PUFA-RE: PUFA response element; RXR: retinoid X receptor; SCAP: SREBP cleavage-activating protein; SCD: steraroyl CoA desaturase; SFA: saturated fatty acid; SRE: sterol response

D. de Mendoza
Instituto de Biología Molecular y Celular de Rosario (IBR), Consejo Nacional de Investigaciones Científicas y Técnicas and Departamento de Microbiología, Suipacha 531, (S2002LRK) Rosario, Argentina

P.J. Quinn, X. Wang (eds.), *Lipids in Health and Disease*,

element; SREBP: sterol-regulatory element binding protein; SSD: sterol sensing domain; TGs: triacylglycerols; TMS: transmembrane segment; UFA: unsaturated fatty acid

## 3.1 Introduction

### *3.1.1 Role of Unsaturated Fatty Acids*

*cis*-Unsaturated fatty acids (UFAs) have crucial roles in membrane biology and signalling processes in organisms ranging from bacteria to man. UFAs have a much lower transition temperature than saturated fatty acids (SFAs) because the steric hindrance imparted by the rigid kink of the *cis*-double bond results in much poorer packing of the acyl chains. Thus, UFAs are key molecules in the regulation of cellular membrane fluidity (Cronan and Gelmann, 1973) and major determinants of the melting temperature of triglycerides (TGs) in animals. In addition to their structural role, UFAs have recently been recognized as signalling molecules involved in several essential cellular processes, such as cell differentiation and DNA replication (for recent reviews see Heird and Lapillonne, 2005; Mansilla and de Mendoza, 2005). Furthermore, alterations in UFA biosynthesis have been implicated in various disease states, including cardiovascular disease, obesity, non-insulin dependent diabetes mellitus, hypertension, neurological diseases and cancer (Nakamura and Nara, 2004; Sampath and Ntambi, 2005; Thijssen and Mensink, 2005).

### *3.1.2 General Features of Desaturases*

The desaturases encompass a superfamily of iron-dependent enzymes that have the function of introducing double bounds into fatty acyl chains. All these enzymes utilize molecular oxygen and reducing equivalents obtained from an electron transport chain and are categorized according to their substrate specificity and regioselectivity, which designates the preferred position for substrate modification. They are present in all groups of organisms, i.e., bacteria, fungi, plants and animals, and play a key role in the maintenance of the proper structure and functioning of biological membranes (Pereira et al., 2003; Sperling et al., 2003; Tocher et al., 1998).

According to its solubility these enzymes can be classified into two non evolutively related groups: the soluble acyl carrier protein (ACP) desaturases and the membrane-bound desaturases, which includes the acyl-lipid desaturases and the acyl-CoA desaturases. The soluble ACP desaturases introduce double bonds into fatty acids esterified to ACP, and are found in the stroma of plant plastids (Shanklin and Cahoon, 1998) and some bacteria, as *Mycobacterium* and *Streptomyces* (Phetsuksiri et al., 2003). The acyl-lipid desaturases, that introduce double bonds in fatty acids esterified in glycerolipids, are membrane bound-enzymes associated with the endoplasmic reticulum (Tocher et al., 1998), the plant chloroplast membrane (Ohlrogge and Browse, 1995), the cytoplasmic membrane of some bacteria (Aguilar et al., 1998), or the plasmatic and thylakoid membranes of cyanobacteria (Mustardy et al., 1996). The acyl-CoA desaturases introduce double bonds into fatty acids esterified to CoA (Enoch et al., 1976), which are associated to the endoplasmic reticulum membrane of animals and fungi (Tocher et al., 1998).

The amino acid residues involved in binding of the di-iron complex in the enzymes belonging to the soluble class form two characteristic D/EXXH motifs, as revealed by the X-ray structure of the castor $\Delta^9$stearoyl-acyl carrier protein desaturase (Lindqvist et al., 1996). The electrons required for acyl group desaturation are delivered from ferredoxin. A different consensus motif for the putative active site is present in the membrane-bound desaturases. It is composed of three histidine-rich regions ("HXXXH", "HXXHH", and "HXXHH"), which are presumably involved in iron binding, predicted to be exposed on the cytoplasmic face of the membrane (Shanklin et al., 1994). Since all the substrates for these enzymes are highly hydrophobic, they will likely partition into the lipid bilayer. In contrast, the electron donors for these enzymes are either soluble proteins or peripheral membrane proteins as ferredoxin (cyanobacterial and plastidial enzymes) or cytochrome b5 (either in free or fused form). Unfortunately, the structural information on membrane desaturases is scarce, due to the technical limitations in obtaining large quantities of purified protein and the intrinsic difficulties in obtaining crystals from membrane proteins. Two topological models have been proposed for the membrane-bound desaturases. A theorycal one was initially proposed by Stukey et al., based on the sequences of rat and yeast $\Delta^9$ stearoyl-CoA desaturases (Stukey et al., 1990). In this model the desaturases were anchored to the membrane through two long hydrophobic regions of 50 residues, predicted to form two transmembrane segments (TMSs). Most of the protein, including the three clusters of histidine residues, would be exposed on the cytosolic surface of the endoplasmic reticulum membrane. The topological features of the rat $\Delta^9$desaturase have been recently confirmed experimentally, using functional recombinant constructs that contain internal epitope tags (Man et al., 2006). The second model is based in the experimental analysis of the topology of the *Bacillus subtilis* acyl-lipid desaturase, a polytopic membrane protein with six TMSs and one membrane-associated domain, which might be involved in keeping the catalytic site in close proximity to the membrane (Diaz et al., 2002). Moreover, the presently available hydrophobicity profile of many acyl-lipid desaturases suggests that this group of enzymes contains a new TM domain that might play a critical role in the desaturation of fatty acids esterified in glycerolipids.

Each desaturase introduces an unsaturated bond at a specific position in a fatty acyl chain. There appear to be three modes of regioselectivity. The $\Delta^x$ desaturases introduce a double bond between Cx and C(x + 1) in the fatty acid moiety of the substrate. The $\omega^x$ desaturases catalyze a reaction that introduces a double bond between the x and (x + 1) positions from the methyl end of fatty acids. There is a third mode of regioselectivity called + x in which the double bond is introduced relative to an existing double bond ("front end" desaturases).

Some acyl-lipid desaturases recognize specific polar head groups, as well as the *sn*-position of the glycerol backbone to which the fatty acid is esterifed (Murata and Wada, 1995). However, most acyl-lipid desaturases are insensitive to head groups and to *sn*-positions.

## 3.2 Unsaturated Fatty Acids Synthesis in Bacteria

There are two major mechanisms by which bacteria synthesize UFAs: mostly of them, including *Escherichia coli*, synthesize UFAs anaerobically (Mansilla et al., 2004) whereas some prokaryotes such as cyanobacteria, bacilli, mycobacteria and pseudomonads use an oxygen-dependent fatty acid desaturation pathway (Mansilla and de Mendoza, 2005; Phetsuksiri et al., 2003; Zhu et al., 2006).

In *E. coli*, UFAs are generated through the activity of FabA, which anaerobically introduces the double bond into a 10-carbon intermediate formed in the fatty acid biosynthetic pathway (Bloch, 1963, for a recent review see Mansilla et al., 2004). However, other bacteria lacking *fabA* synthesize UFAs under anoxic conditions. For example, *Streptococcus pneumoniae* compensates FabA absence with an enzyme called FabM, that is capable of isomerising the *trans* unsaturated bond of the key 10-carbon intermediate to its *cis*-isomer (Marrakchi et al., 2002). Nevertheless, FabM seems to be specific for streptococci indicating that there are new anaerobic pathways of UFAs synthesis to be discovered.

The other known pathway of UFAs synthesis, by action of fatty acid desaturases, functions only in aerobic organisms, where desaturation of full-length fatty acids to unsaturated derivatives can occur (Tocher et al., 1998).

### *3.2.1 Bacterial Desaturases*

ORFs with significant similarity to fatty acid desaturase genes have been found in the genomes of many bacteria (Sperling et al., 2003). Although their function in most cases has not yet experimentally demonstrated, fatty acid desaturases may be more widespread than originally thought. In this section we will focus on bacterial desaturases whose mechanisms of regulation are being deciphered.

*Bacillus* cells respond to a decrease in the ambient temperature by increasing the synthesis of UFAs (Fulco, 1969). *B. subtilis* contains a sole desaturase, encoded by the *des* gene (Aguilar et al., 1998), that catalyses the introduction of a *cis*-double bond at the Δ5 position of existing SFAs attached to membrane phospholipids (PLs). Thus, this protein is an acyl-lipid desaturase and was named Δ5-Des (Altabe et al., 2003). Its topology model consisting in six TMSs and one membrane-associated domain constitutes the first experimental evidence for the topology of a plasma membrane desaturase (Diaz et al., 2002).

*Pseudomonas aeruginosa* is also capable of regulating its UFAs biosynthesis in response to changes in the growth conditions. The predominant UFA synthetic pathway in *P. aeruginosa* is the anaerobic one, through the FabA dehydratase/isomerase of the bacterial II fatty acid synthase. However this bacterium also possesses two oxygen-dependent desaturases, DesA and DesB (Zhu et al., 2006). DesB is an inducible $\Delta^9$desaturase that introduces double bonds into acyl-CoAs that are produced from exogenous fatty acids. The DesA acyl-lipid desaturase introduces double bonds at the Δ9 position of fatty acyl chains attached to the 2-position of existing PLs, produced from both the *de novo* fatty acid synthesis and exogenous SFAs, thus, cells can quickly modify their membrane biophysical properties using this mechanism to adapt to changes in growth conditions.

The effects of the unsaturation of fatty acids on membrane fluidity has also been extensively studied in two strains of cyanobacteria, *Synechocystis* sp. PCC 6803 and *Synechococcus* sp. PCC 7942 (Los and Murata, 2004). The genes for four specific acyl-lipid desaturases, designated *desA*, *desB*, *desC* and *desD*, have been cloned from *Synechocystis*. These desaturases introduce double bonds at the Δ12 (Sakamoto and Bryant, 1997), ω3 (Sakamoto et al., 1994a), Δ9 (Sakamoto and Bryant, 1997; Sakamoto et al., 1994b), and Δ6 (Reddy et al., 1993) positions of fatty acids, respectively. The order in which desaturases operates is very strictly determined: the first double bond is introduced by the $\Delta^9$desaturase into stearic acid; and the $\Delta^{12}$ and $\Delta^6$ desaturases introduce a double bond into fatty acids that have a double bond at the Δ9 position (Higashi and Murata, 1993). The $\omega^3$ desaturase introduces a double bond into fatty acids that have a double bond at the Δ12 position (Higashi and Murata, 1993). Unlike *Synechocystis*, *Synechococcus* has only a single $\Delta^9$desaturase and it synthesizes mono-unsaturated fatty acids (MUFAs) exclusively.

### *3.2.2 Modulation of Membrane Fluidity*

When poikilothermic organisms such as bacteria, plants and fish are exposed to temperatures below those of their normal conditions, the lipids of their membrane become rigidified leading to a suboptimal functioning of cellular activities (Phadtare, 2004; Mansilla and de Mendoza, 2005; Al-Fageeh and Smales, 2006). These organisms can acclimate to such new conditions by remodelling its membrane lipid composition. This can be achieved through an increase in the desaturation of the acyl chains of membrane PLs, because PLs containing UFAs have a much lower transition temperature than those lipids made of

SFAs (Cronan and Gelmann, 1973). As a result, the fluidity of membrane lipids returns to their original state, or close to it, with restoration of normal cell activity at the lower temperature.

#### 3.2.2.1 The Des Pathway of *Bacillus subtilis*

In *B. subtilis* the transcription of the *des* gene increases in response to a decrease in temperature (Aguilar et al., 1999). The induction of UFAs synthesis in this bacterium depends on the extent of the shift in temperature and not on absolute temperature (Grau and de Mendoza, 1993). A canonical two-component regulatory system comprising the histidine kinase DesK and the response regulator DesR regulates *des* expression (Aguilar et al., 2001). The *B. subtilis* DesK protein features five TMSs that define the sensor domain and a long cytoplasmic C-terminal tail, which harbours the kinase domain, DesKC. DesKC undergoes autophosphorylation in the presence of ATP in the conserved His 188, which is the target residue of its autokinase activity (Albanesi et al., 2004). Autophosphorylated DesKC transfers the phosphoryl group to the effector protein DesR that becomes phosphorylated in the predicted Asp 54 residue (DesR-P) (Albanesi et al., 2004). Phosphorylation of the regulatory domain of dimeric DesR promotes, in a cooperative fashion, the hierarchical occupation of two adjacent, non identical, DesR-P DNA binding sites, so that there is a shift in the equilibrium toward the tetrameric active form of the response regulator (Cybulski et al., 2004). This results in the recruitment of RNA polymerase to the *des* promoter and activation of *des* transcription, as demonstrated by *in vitro* transcription experiments (Cybulski et al., 2004). Genetic and biochemical experiments demonstrated that the level of phosphorylation of DesR is determined by the balance of the two activities possessed by DesK, a phosphate donor for DesR and a phosphatase of DesR-P (Aguilar et al., 2001; Albanesi et al., 2004). These activities would be reciprocally regulated by changes in growth temperature that, in turn, adjust the fluidity of membrane PLs.

Evidence that membrane fluidity, rather than growth temperature, controls transcription of the *des* gene was obtained by experiments in which the proportion of anteiso-branched chain fatty acids (a-BFAs) of *B. subtilis* membranes was varied controlling the provision of exogenous fatty acid precursors (Cybulski et al., 2002). The a-BFAs, which are synthesized using ketoacids derived from isoleucine as primers (Kaneda, 1991), are essential to decrease the transition temperature of *B. subtilis* membrane PLs to maintain the appropriate fluidity. Limiting the supply of isoleucine dramatically reduces the amount of a-BFAs of plasma membrane lipids (Klein et al., 1999), resulting in ordered membrane lipids. Growth of cells in the absence of isoleucine results in activation of *des* transcription under isothermal conditions, using a DesK/DesR-dependent mechanism (Cybulski et al., 2002). Thus, a decrease in the content of membrane isoleucine-derived fatty acids at constant temperature mimics a drop in growth temperature, and both stimuli can be sensed by DesK as an increase in the order of membrane lipids, leading to induction of UFAs synthesis.

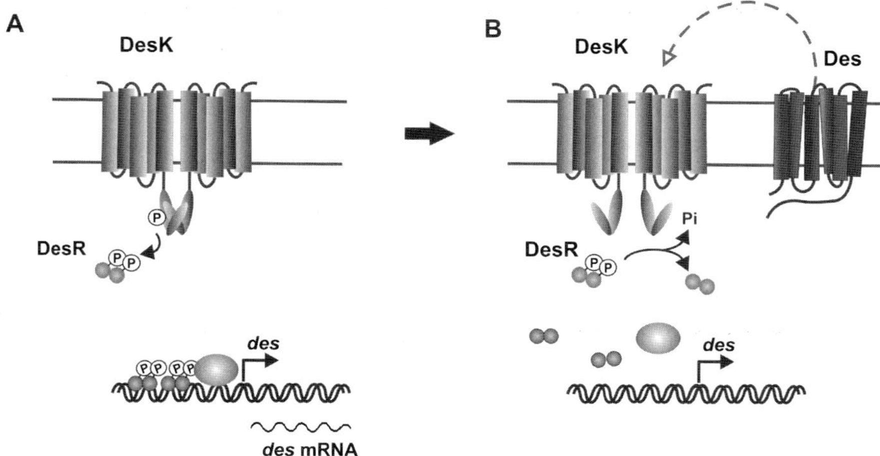


**Fig. 3.1** Model for the signal transduction pathway leading to membrane fluidity optimization in *B. subtilis*. (**A**) A kinase dominant state of DesK predominates upon an increase in the proportion of ordered membrane lipids. The phosphate group is transferred to DesR. Two DesR-P dimers interact with the *des* promoter and RNA polymerase, resulting in transcriptional activation of *des*. (**B**) Δ5-Des is synthesized and desaturates the acyl chains of membrane phospholipids. The decrease in membrane lipids order favors the phosphatase-dominant state of DesK. DesR is dephosphorylated, resulting in decreased transcription of the *des* gene

How does DesK recognize changes in membrane fluidity? One or more of the five TMSs in DesK could undergo a conformational change induced by a change in the physical state of the membrane lipid bilayer, and transmit this information to the cytoplasmic domain of DesK, thereby altering its activity (Aguilar et al., 2001). Direct support for the functional importance of the TM helices comes from studies of *B. subtilis* strains producing the soluble DesKC domain. In these strains, the *des* gene is constitutively expressed and its transcription is affected neither by growth temperature nor by the presence of UFAs (Albanesi et al., 2004). Moreover, when DesKC is anchored to the membrane by only one TMS, the expression of *des* is not induced by cold-shock (Cybulski et al, unpublished results). Therefore, the TM segments of DesK directly participate in signal transduction recognition and play an essential role in the end-product feedback regulation of *des* transcription.

A model that accounts for our present knowledge on the DesK/DesR-mediated regulatory pathway, termed the Des pathway, is presented in Fig. 3.1.

#### 3.2.2.2 Two Aerobics Pathways of Desaturation in *Pseudomona aeruginosa*

The anaerobic UFA biosynthetic pathway in *P. aeruginosa* is supplemented by two inducible aerobic desaturases, DesA and DesB. The expression of *desA* is induced under anaerobic conditions but is not regulated significantly following an abrupt temperature shift (Zhu et al., 2006). Rather than being regulated by a thermosensor like in *B. subtilis, P. aeruginosa desA* expression is most highly regulated by changes in oxygen availability, which is reminiscent of the induction of *OLE1* expression by Mga2p under hypoxic conditions (See section 3.3; Jiang et al., 2001; Vasconcelles et al., 2001). However, the mechanism of regulation of *desA* by oxygen is not known. Therefore the position of the double bond and the lack of temperature regulation distinguish the acyl-lipid desaturase of *P. aeruginosa* from that of the Δ5Des in *B. subtilis* (Aguilar et al., 1998).

The second aerobic pathway of UFA synthesis in *P. aeruginosa* requires DesB which can desaturate exogenous stearate and palmitate, and the principal products are 18:1Δ9 and 16:1Δ9, respectively (Zhu et al., 2006). The *desB* gene is in an operon with *desC*, a gene encoding an oxidoreductase that is predicted to function in the electron transport coupled with the DesB desaturation reaction. The expression of *desCB* operon is controlled by the DesT repressor, which binds to a palindrome in the *desT–desCB* intergenic region (Zhang et al., 2007). DesT is able to respond to the changes in the composition of cellular acyl-CoA pool modifying its binding to DNA. In this way the expression of the acyl-CoA $\Delta^9$desaturase is modulated to adjust fatty acid desaturation activity accordingly (Zhang et al., 2007). Briefly, under normal growth conditions, DesT exists in equilibrium between the free and DNA-bound forms, and a basal level of *desCB* transcription is maintained (Fig. 3.2A). The presence of SFAs in the growth environment leads to an increase of the intracellular saturated acyl-CoA levels. The binding of saturated long-chain acyl-CoAs, such as palmitoyl-CoA, to DesT releases the protein from the *desCB* promoter and derepresses *desCB* transcription (Fig. 3.2A). As a result, the protein levels of DesB and DesC increase, saturated acyl-CoAs are converted to Δ9-unsaturated acyl-CoAs and incorporated into membrane PLs. As unsaturated acyl- CoAs accumulate in the cell, from either the action of the DesB enzyme or uptake and activation of exogenous UFAs by acyl-CoA synthetase (FadD), these acyl-CoAs also bind to DesT. However, the complex of DesT with UFA-CoA has increased affinity for the *desCB* promoter and represses *desCB* transcription more efficiently than DesT alone (Fig. 3.2B). Thus, the presence of unsaturated acyl-CoA represses *desCB* transcription to levels below basal. The presence of another palindrome downstream of the *desT* transcription initiation, suggests the existence of another yet unknown regulator that controls *desT* expression in response to different regulatory signals other than acyl-CoAs (Zhang et al., 2007).

*P. aeruginosa* grows on fatty acids or alkanes with different chain length as the sole carbon and energy source (van Beilen et al., 1994). The regulation of *desB* expression by DesT is important for *Pseudomonas* growth under conditions where fatty acids are an environmental component. In cystic fibrosis lungs, pulmonary surfactant PLs are degraded by lipases and phospholipases secreted by the colonizing *P. aeruginosa*, releasing free fatty acids (Beatty et al., 2005). Also, the DesT–DesCB system allows the bacterium to dispense with the

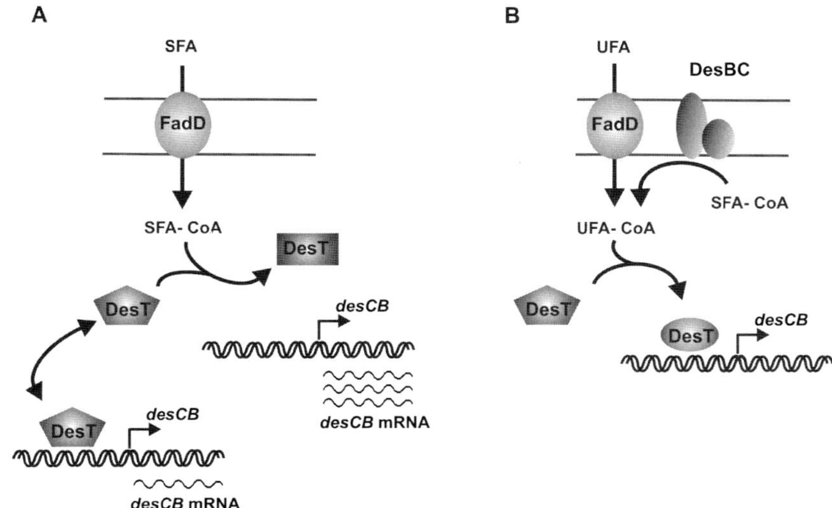


**Fig. 3.2** Model for regulation of *desCB* expression in *P. aeruginosa*. (**A**) Under normal growth conditions, DesT exists in equilibrium between the free and DNA-bound forms, and a basal level of *desCB* transcription is maintained. When SFAs are present in the growth environment, they are taken up by the cell and activated by acyl-CoA synthetase (FadD). This increased level of saturated acyl-CoA in the cell allows the formation of a complex DesT-SFA-CoA that prevents DesT binding to the *desCB* promoter. As a result, the repression of *desCB* transcription is released, the protein levels of DesB and DesC increase, and SFA-CoAs are desaturated. (**B**) As cellular UFA-CoAs accumulate, either from the action of DesB enzyme or from activation of exogenous UFAs, the DesT·UFA-CoA complex is formed, which binds tightly to the *desCB* promoter, leading to repression of *desCB* transcription

energy-intensive fatty acid synthesis and provides appropriately modified acyl-CoAs to the phospholipid biosynthetic pathway.

#### 3.2.2.3 A Multi-Stress Sensor in Cyanobacteria

Three of the four genes for desaturases in *Synechocystis* (*desA*, *desB*, and *desD*) are cold-inducible (Los et al., 1993; Los et al., 1997). The enhanced synthesis *de novo* of these three fatty-acid desaturases under cold stress and the subsequent introduction of additional double bonds into the fatty-acyl chains of membrane lipids are involved in the maintenance of membrane fluidity in the liquid-crystalline phase and prevent the membranes from undergoing phase transition to the lethal gel phase (Hazel, 1995). From these results it might be expected that the desaturation of fatty acids could also compensate for the rigidification of membrane lipids in cells exposed to hyperosmotic stress. However, genomewide analysis of transcription in *Synechocystis*, using DNA microarrays, indicated that hyperosmotic stress does not activate the transcription of genes for desaturases (Kanesaki et al., 2002). Moreover, in *B. subtilis* hyperosmotic stress decreases the fluidity of cell membranes and subsequently increases the levels of

UFAs in membrane lipids (Lopez et al., 2000), but this phenomena is not due to enhanced expression of the gene for the desaturase (Mansilla, unpublished results).

The histidine kinase 33 (Hik33), is involved in the cold-inducible expression of *desB* and *desD* genes, which encode the $\omega^3$ and the $\Delta^6$ desaturases of *Synechocystis*, but is not involved in the expression of the *desA* gene (Suzuki et al., 2000) (Fig. 3.3). By screening of a library of *Synechocystis* cells with random mutations that affected regulation of the transcription of the *desB* gene, the response regulator Rer1was identified (Suzuki et al., 2000). However, this protein is not involved in the expression of *desD*, and their regulator is still unknown. A second histidine kinase, Hik19, would serve as an intermediate between the Hik33 and its cognate response regulators (Suzuki et al., 2000). Although the phosphorylation relay between Hik33 and Rer1 and the subsequent regulation of transcription of the *desB* gene by Rer1 remain to be characterized, the cold-regulated induction of gene expression in *Synechocystis* seems to be similar to the pathway controlled by DesK/DesR in *B. subtilis*. As DesK, it is very likely that Hik33 recognizes changes in membrane fluidity as the primary signal of cold stress, therefore, the mechanism of regulation of UFAs synthesis mediated by a histidine kinase and its cognate response regulator appears to be found in several prokaryotes using the desaturation of fatty acids for adaptation to low temperatures. In contrast, while the *B. subtilis*

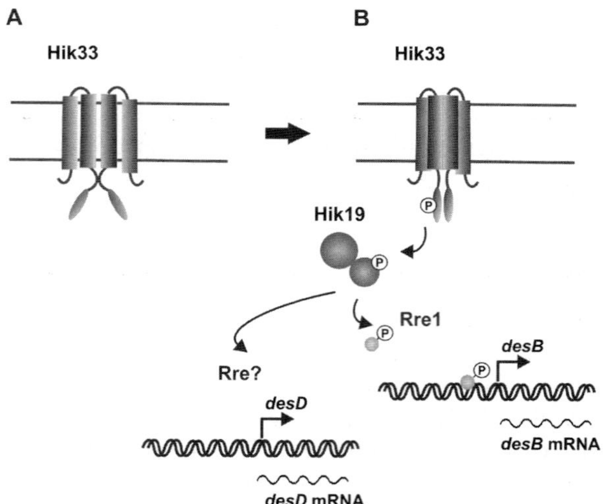


**Fig. 3.3** Hypothetical scheme for the pathway for low-temperature signal transduction in *Synechocystis*. An inactive form of Hik33 is present when membrane lipids are disordered (**A**). A decrease in temperature rigidifies the membrane that would force the transmembrane segments to move close together, leading to changes in the linker conformation and autophosphorylation of the histidine kinase domains (**B**). A phosphate group would be transferred to Hik19, and finally to Rer1, to activate *desB* transcription. The response regulator involved in *desD* transcription has not yet been identified

system seems to regulate the expression of the *des* gene only (Beckering et al., 2002), Hik33 is a global cold sensor involved in the expression of most of the genes that are induced at low temperature (Murata and Suzuki, 2006). In addition, genome-wide analysis of transcription in *Synechocystis* with DNA microarrays indicated that Hik33 is also involved in the sensing of hyperosmotic stress (Mikami et al., 2002), salt stress (Shoumskaya et al., 2005) and is the main sensor of peroxydative stress in this bacterium (Kanesaki et al., 2007). It is remarkable that Hik33 and its cognate regulator/s regulate the expression of distinct respective sets of genes under different kinds of stresses, and this finding cannot be explained by the currently accepted model of two-component systems. It seems reasonable to postulate the presence of some unknown factor(s) that provides each two-component system with strict specificity that is related to the specific nature of the stress. Future work will need biochemical, structural and biophysical approaches to gain insight into the molecular mechanism of signal perception of sensor kinases involved in the expression of desaturase genes.

## 3.3 Regulation of Long Chain Unsaturated Fatty Acids Biosynthesis in Yeast

A large number of lower eukaryotes such as fungi, algae and protozoa are known to produce large amounts of polyunsaturated fatty acids (PUFAs) with chain lengths of C20 or greater (Singh and Ward, 1997); thus, these organisms contain the complete array of biosynthetic enzymes including desaturases and elongases (for a recent review see Uttaro, 2006). In addition, some lower eukaryotes display an alternative pathway for C20 PUFA production involving the $\Delta^4$and $\Delta^8$ desaturases (Wallis and Browse, 1999).

*Sacharomyces cerevisiae* and *Schizosaccharomyces pombe* are unusual fungi, however, that form only MUFAs (for review see Martin et al., 2007). *S. cerevisiae* possesses a single fatty acid desaturase named Ole1p, which introduces a double bound in the D9 position of fatty acids esterified to CoA (Stukey et al., 1990). Ole1p localizes into the endoplasmic reticulum (ER), where most of the lipid biosynthetic machinery resides. In this organelle, saturated C-16:0 (palmitic acid) and C-18:0 (stearic acid) acyl-CoA precursors are desaturated, yielding C-16:1 (palmitoleic acid) and C-18:1 (oleic acid) respectively, which are then distributed throughout the membranes of the cell systems and comprise more than 70% of the total fatty acids. Although *Sacharomyces* form only MUFAs via a single enzyme system, it is a highly regulated function. The Ole1p desaturase activity must respond to a number of environmental and nutritional signals in order to provide essential precursors for membrane assembly as cells grow and adapt to different physiological and metabolic conditions. Like the *B. subtilis des* gene, *OLE1* transcription is transiently activated at low temperature (Nakagawa et al., 2002). It is also induced under hypoxic conditions (Kwast et al., 1998). As the desaturation reaction utilizes oxygen as an electron acceptor *OLE1* induction under hypoxia might be a response to UFA depletion under such limiting substrate conditions (Nakagawa et al., 2001). Two related homologous genes of the mammalian transcription factor NF-kB, *SPT23* and *MGA2*, are required for *OLE1* transcription. Disruption of either *SPT23* or *MGA2* has little effect on growth or UFA synthesis, whereas the simultaneous gene disruption results in synthetic auxotrophy for UFAs due to loss of *OLE1* mRNA (Zhang et al., 1999b). Spt23p and Mga2p are initially synthesized as inactive precursors that are anchored to the ER membrane via their single C-terminal TM spans (Hoppe et al., 2000). An ubiquitin/proteasome-dependent process cleaves both precursors within their TM regions and the soluble N-terminal domains are transported into the nucleus to promote *OLE1* transcription (Hitchcock et al., 2001) (Fig. 3.4A). Under optimal growth conditions, both *spt23*D and *mga2*D single mutants activate *OLE1* transcription to similar extents, indicating that each transcription factor by itself is sufficient for *OLE1* expression (Zhang et al., 1999b; Jiang et al., 2001). However, *MGA2* is essential, whereas *SPT23* is dispensable for *OLE1* transcriptional induction under both hypoxic and cold-shock conditions (Jiang et al., 2001; Nakagawa et al., 2002). In agreement with the preferential role of *MGA2* on *OLE1* transcriptional induction, Mga2p processing is induced during $O_2$ depletion (Jiang et al., 2001). Taken together, these results indicate that environmental signals that perturb membrane fluidity induce *OLE1* expression using a mechanism that involves proteolytic processing of TM domains of ER-anchored transcription factors.

Early work showed that fatty acid desaturation in yeasts is strongly inhibited by addition of UFAs to the growth media (Bossie and Martin, 1989). Furthermore, it was demonstrated that *OLE1* transcription is repressed by a variety of UFAs, the extent of inhibition increasing as the melting point of the added UFA declines (McDonough et al., 1992; Fujiwara et al., 1998). UFAs also increase *OLE1* mRNA destabilization by a mechanism that is independent of the nonsense-decay pathway, which requires the *OLE1* mRNA 5′ untranslated region and seems to be mediated by exosome mRNA degradation (Gonzalez and Martin, 1996; Vemula et al., 2003; Kandasamy et al., 2004). Remarkably, Spt23p and Mga2p cleavage, and therefore generation of the competent transcription factors, are also affected by UFAs. Under normal growth conditions, UFA addition almost completely blocks Spt23p cleavage, whereas Mga2p processing seems to be mildly affected (Hoppe et al., 2000; Jiang et al., 2002). However, Mga2p cleavage, which is strongly induced by hypoxia, is counteracted by exposure of cells to UFAs (Jiang et al., 2002). The synthesis of a soluble N-terminal fragment of Mga2p in a *spt23*D *mga2*D double mutant strain is sufficient to promote *OLE1* transcription, but expression of *OLE1* remains sensitive to inhibition by UFAs. This finding indicates that UFA-mediated repression of *OLE1* can also act downstream of Mga2p proteolytic cleavage (Chellappa et al., 2001). On the other hand, *OLE1* was also expressed but UFA-mediated repression was not observed when a soluble version of Spt23p was

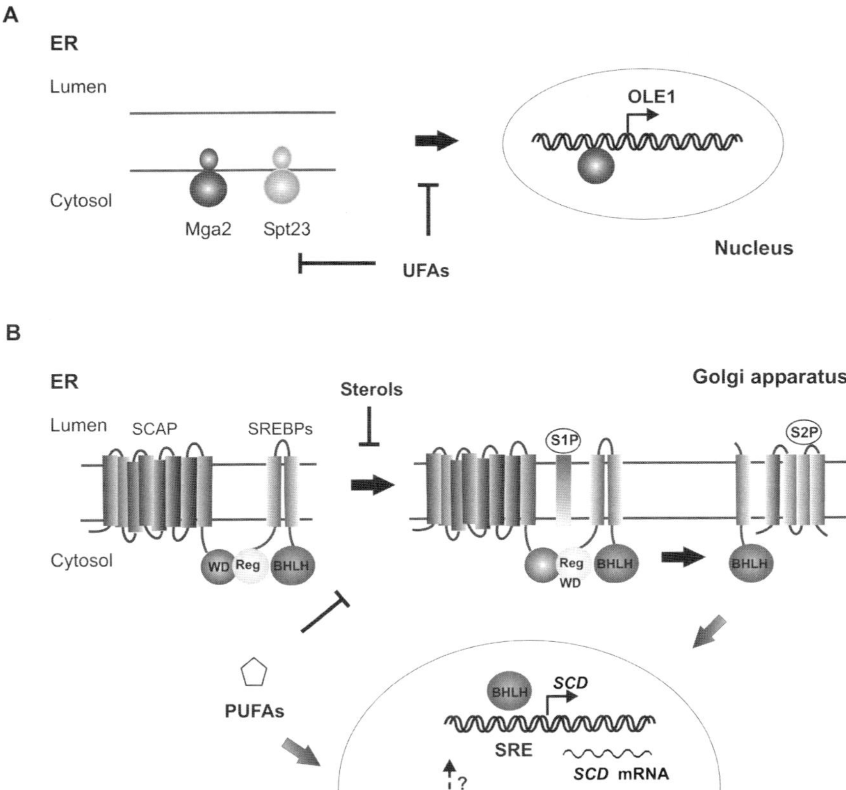


**Fig. 3.4** Model for signal transduction pathway controlling fatty acid desaturation in different eukaryotes. **A**. Budding yeast desaturase gene expression requires proteolytic processing in either ER transcription factors Mga2p or Spt23p. UFAs negatively regulates OLE1 expression at multiple levels, including repression of intramembrane proteolytic processing. **B**. Model for the sterol-mediated proteolytic release of SREBPs from membranes. SREBP cleavage-activating protein (SCAP) is a sensor of sterols and an escort of SREBPs. When cells are depleted of sterols, SCAP transports SREBPs from the ER to the Golgi apparatus, where two proteases, Site-1 (S1P) and Site-2 (S2P), act sequentially to release the NH2-terminal BHLH domain from the membrane. The BHLH domain enters the nucleus and bind to a sterol response element (SRE) in the enhancer/promoter region of target genes, activating their transcription. Cholesterol represses the maturation of SREBP via the SCAP while PUFAs inhibits this process by an unknown mechanism (see text). PUFAs either from the diet or membrane phospholipids independently bind the putative binding protein (PUFA-BP) and the PUFA and PUFA-BP complex repress transcription of the SCD gene by binding to the PUFA response elements (PUFA-RE). Crosstalk between the two pathways is designed by the dashed arrow

produced in the *spt23*D *mga2*D mutant strain, indicating that processing is the key step in Spt23p regulation by UFAs (Chellappa et al., 2001). Regulation of proteolytic processing of Spt23p and Mga2p by UFAs and different environmental stresses that affect the physical state of cellular membrane lipids resembles the regulated intramembrane proteolysis (RIP) pathway. The differences between Mga2p and Spt23p suggest that both proteins have evolved complementary, overlapping roles in the regulation of *OLE1*, and perhaps on other genes that remain to be identified.

## 3.4 Polyunsaturated Fatty Acids Synthesis in Animals

### *3.4.1 Animal Desaturases*

Three desaturases are known in animals. Stearoyl CoA desaturases (SCDs, also called $\Delta^9$desaturases) catalyze the synthesis of oleic acid (18:1), which is mostly esterified into TGs as an energy reserve. $\Delta^6$ Desaturase (D6D) and $\Delta^5$ desaturase (D5D) are the key enzymes for the synthesis of highly unsaturated fatty acids (HUFAs) such as arachidonic acid (20:4 ω6) and docosahexanoic acid (22:6 ω3), that are incorporated into PLs and perform essential physiological functions including eicosanoid signaling (Funk, 2001), pinocytosis, ion channel modulation (Kang and Leaf, 1996), and regulation of gene expression (Clarke and Jump, 1994).

The stearoyl CoA ($\Delta^9$) desaturase is one of the best-studied desaturases to date. This enzyme catalyzes the synthesis of MUFAs by introducing a double bond at the 9,10 position from the carboxyl end of the fatty acids. This enzyme catalyzes the D9 desaturation of fatty acyl-CoAs with 12 to 19 carbon chains (Ntambi, 1999). It is a microsomal-membrane-bound protein and functions in conjunction with cytochrome *b5* and NADH-dependent cytochrome *b5* reductase. The genes for SCD have been cloned from different species including rat (Mihara, 1990), mice (Kaestner et al., 1989; Ntambi et al., 1988) and human (Zhang et al., 1999a; Zhang et al., 2001). Four isoforms have been identified in mice (SCD-1,-2,-3,-4), whereas only two in human (SCD-1 and SCD-2) (Miyazaki et al., 2000; Ntambi and Miyazaki, 2003). The main product, oleic acid (18:1 Δ9), is ubiquitously present in all tissues. In adipose tissue, TGs mainly consist of long chain fatty acids with 16 and 18 carbons but certain amount of UFAs (18:1 and 18:2) are required to maintain physical properties at the body temperature of mammals. Thus, 18:1 Δ9 is the major species in adipose TGs (Phinney, 1990). Therefore, SCD is an essential component for the *de novo* lipogenesis to store excess energy as TG. Indeed, disruption of the SCD-1 gene in mice affects energy metabolism and makes animals resistant to developing obesity (Cohen et al., 2002; Ntambi and Miyazaki, 2003; Ntambi et al., 2002b). Otherwise, in various tissues and serum, the significant amount of 18:1 Δ9 present in PLs contributes to maintenance of biological membrane fluidity. SCD is also required for cholesterol ester synthesis in liver (Napier et al., 1998), and is induced by dietary cholesterol. Another function of SCD was discovered in SCD-1 null mice, which showed defects in lipids synthesis and secretion from skin and eyelid (Miyazaki et al., 2001). The involvement of SCD in these multiple metabolic pathways requires complex regulation of the gene encoding for this desaturase by various nutrients, as will be discussed in the next section.

D6Ds are membrane-bound acyl-CoA desaturases found in the endoplasmic reticulum of animals. They catalyze the rate-limiting step for the synthesis of PUFAs. D6D is classified as a front-end desaturase because it introduces a double bond between the pre-existing double bond and the carboxyl end of the fatty acid. These enzymes differ from the other desaturases in that they contain a fused cytochrome *b5* domain at the N-terminus, which play a role as an electron donor during desaturation. The first gene encoding a D6D was cloned from *Synechocystis* using a gain-of-function cloning approach (Reddy et al., 1993). Based in this sequence, later, other D6Ds were cloned from humans (Cho et al., 1999a), mice (Cho et al., 1999b), and rats (Aki et al., 1999). However, up to date, not much progress has been made in terms of biochemical characterization of these enzymes.

D5Ds, another front-end desaturases present in animals, catalyzes HUFA synthesis. After desaturation and elongation by D6D and elongase, respectively, D5Ds introduce another double bond at the $\Delta5$ position of 20-carbon fatty acids 20:3 ω6 and 20:4 ω3. D5D genes have been cloned from several organisms including human (Cho et al., 1999a) and rat (Zolfaghari et al., 2001).

Most vertebrates, unlike plants, do not have $\omega^3$desaturases, and are incapable of desaturating C18 acyl chains at the D12 and D15 positions and consequently must obtain ω3 PUFAs from their diet. The nutritional importance of specific fatty acid was first revealed through the pioneering work of Burr and Burr in 1929 (Burr, 1981). They fed rats with a fat-free diet and observed retarded growth, scaly skin, tail necrosis and eventual death which were reversed by feeding specific fat. Linoleic acid (18:2 ω6) and linolenic (18:3 ω3) were recognized as active agent and the term "essential fatty acid" was coined. The ω3 and ω6 PUFAs have been known to confer various health benefits, including increase insuline signalling, enhanced immune response (Ntambi et al., 2002a) and decrease incidence of lung and coronary diseases (Schwartz, 2000; Skerrett and Hennekens, 2003).

## *3.4.2 Regulation of Desaturation Through Sensors and Receptors*

### 3.4.2.1 The Sterol Regulatory Element Binding Proteins (SREBPs)

In mammals, the liver is the organ most active in lipids synthesis, and in this organelle, the transcription factors SREBPs (sterol-regulatory element binding protein) play crucial roles in lipids homeostasis. In addition to cholesterol, the liver synthesizes large amounts of fatty acids, not only for its own membranes, but also for export to other tissues in lipoproteins. SREBPs control both processes (Horton et al., 2002). The two prominent isoforms of SREBP in liver, SREBP-2 and SREBP-1c, have divergent but partially overlapping functions. SREBP-2 is primarily involved in stimulating cholesterol synthesis, whereas SREBP-1c primarily stimulates fatty acid synthesis. SREBP-regulated cleavage is best understood for the control of cholesterol homeostasis. In the presence of sterols, SREBP and SREBP cleavage-activating protein (SCAP) form a complex that remains in the ER. Retention of the SREBP-SCAP complex in the endoplasmic reticulum requires interaction with INSIG-1 or -2. In the absence of sterols, SCAP undergoes a conformational change, through its sterol-sensing domain (SSD), which detaches the SCAP-SREBP complex from INSIG-1/2. The SCAP-SREBP complex is then transported to the Golgi, where SREBP is proteolitically processed, releasing an active soluble domain (BHLH). Processed SREBP enter to the nucleus and binds to the SRE (sterol regulatory elements) located in the 5̀ flanking regions of more than 20 genes involved in lipids synthesis and uptake (Goldstein et al., 2006), including the SCD (Brown and Goldstein, 1999; Ericsson et al., 1999; Magana et al., 1997) (see Fig. 3.4 B). In transgenic mice, truncated forms of SREBP give a 3- to 4- fold increase in SCD gene expression, and overexpressed SREBP-1 increases the levels of the D5D and D6D mRNAs (Matsuzaka et al., 2002). The proteolytic processing of SREBPs is under feedback control by cholesterol. Thus, when sterols accumulate in cells, the SCAP-SREBP complex fails to move to the Golgi, and SREBPs are not processed (Nohturfft et al., 2000). The nuclear SREBPs are rapidly degraded by proteosomal process, and the synthesis of sterols and fatty acids, primarily 18:1 unsaturates, declines.

## Role of SREBP-1c in Desaturases Regulation by PUFAs

Several studies have shown that the expression of desaturases is highly regulated by dietary factors. A high carbohydrate fat-free diet induces SCD1 mRNA in mouse liver, whereas the supplementation of PUFAs suppress the expression of genes involved in fatty acid synthesis (Clarke and Jump, 1994) and also the D6D, D5D and SCD1 genes (Holloway and Holloway, 1975; Ntambi et al., 1996). The repression of the SCD1 gene by PUFAs has been shown to be dependent on SREBPs. In fact, several studies indicate that UFAs inhibit the cleavage of the precursor form of SREBP, resulting in decreased levels of nuclear SREBP and concomitant reduction of SREBP target gene transcription (Sampath and Ntambi, 2005). Other studies also indicate that, in addition to decrease the level of the processed form of the transcription factor, UFAs markedly reduces the mRNA levels of SREBP (Hannah et al., 2001).

Several lines of evidence indicate that a *cis*-acting PUFA binding protein element (PUFA-RE) exists in the promoter of the SCD genes and hypothesized that a putative transcription factor would bind and block the transcription of these genes (Tabor et al., 1999; Waters et al., 1997). This *cis*- element was mapped in the SCD promoter by deletion analysis, and by EMSA assays it was demonstrated the binding of nuclear proteins to the PUFA-RE. Two of these proteins were identified as SREBPs and the CCAAT-binding factor/nuclear factor Y or NF-Y. Since SREBPs are transcriptional activators of SCD, the repression of SCD through the PUFA-RE element could be attributed to an indirect effect exerted by SREBPs involving transcriptional complexes. Taken together, PUFAs appear to bind to a not yet identified PUFA-binding protein (PUFA-BP) forming part of the complex that repress SCD expression. Therefore, several transcription factors seem to be crucial to input from pathways involved in the expression of SCD (see Fig. 3.4B).

The SREPB also mediates the suppression of D6D promoter activity by HUFAs (Nara et al., 2002). HUFAs suppress the D6D-target gene transcription by reducing the active form of SREBP-1c by several mechanisms: (i) by reducing the nuclear form of SREBP-1 (Xu et al., 1999); (ii) decreasing the stability of SREBP-1 mRNA (Xu et al., 2001) and (iii) suppressing the SREBP-1c transcription in cell lines by a mechanism dependent of the liver X receptor (LXR) (Kim et al., 2002), although dietary PUFAs did not suppressed D6D gene transcription in rats (Yoshikawa et al., 2002). The identification of SREBP-1c as a key regulator of D6D poses an interesting question: what is the physiological role that SREBP-1c plays in liver? The primarily role of *de novo* fatty acid synthesis in liver is considered to be the production and secretion of TGs (Towle et al., 1997). If this were the case, the major role of SREBP-1c would be the conversion of excess glucose into TGs for storage because SREBP-1c activates genes for fatty acid synthesis in liver (Horton et al., 2002). However, as we describe above, SREBP-1c plays a key regulatory role in HUFA synthesis by inducing D5D and D6D expression. Because HUFAs are mainly incorporated into PLs, and are poor substrate for TG synthesis, the primary role of SREBP in liver may be the regulation of fatty acid synthesis for PLs rather than for TGs. Interestingly, it was demonstrated in Drosophila cells that phosphatidylethanolamine controls maturation of SREBP, and proposed the hypothesis that the primary role of SREBP in animals is to monitor and maintain cell membrane composition (Dobrosotskaya et al., 2002). Furthermore, transgenic mice overexpressing the mature form of SREBP-1 in liver developed steatosis but did not show an elevation of HUFA incorporation into PLs (Shimomura et al., 1997) which suggest the presence of another mechanism that may limit the HUFAs incorporation into PLs in addition to the regulation of HUFAs synthesis by SREBP-1.

### Regulation of Stearoyl CoA Desaturase Expression by Cholesterol

Regulation of SCD gene expression by cholesterol is opposite depending if the studies are performed *in vivo* or using cell culture. It has been shown that in Chinese hamster ovary cells cholesterol represses the expression of SCD genes and enzyme activities by a similar mechanism to that mediated by PUFAs. Conversely, the *in vivo* studies showed that high levels of cholesterol can induce SCD gene expression in liver, increasing the oleic acid synthesis and enhancing its incorporation into cholesteryl esters. This last mechanism of regulation is not well understood but appears to involve the LXRα receptor. LXRs are transcription factors of the nuclear receptor family that form heterodimer with RXR (retinoid X receptor) and bind the repeat 4 elements present in target gene promoters such us SREBP1-c. Cholesterol metabolites and oxysterols are its natural ligands and activate LXRs and gene expression. Indeed, mice lacking LXRα receptor expressed low levels of SCD mRNA in the presence or absence of cholesterol feeding, while the wild type animals fed with a diet supplemented with 2% cholesterol had their SCD mRNA levels up-regulated four-fold (Chin and Chang, 1982). These *in vivo* studies would be consistent with the possibility that the liver, when challenged with excess of cholesterol, increases SCD activity to provide oleyl-CoA as a substrate for cholesterol esterification. The cholesteryl esters then will be secreted and transported to other tissues. It is not clear however, whether cholesterol or its metabolites induce SCD transcription through the LXRα or indirectly through the regulation of SREBPs.

Role of SREBP-1c in Induction of Desaturases by Insulin

Accumulating evidence indicates that SREBP-1c mediates the effect of insulin on transcriptional activation of genes involved in fatty acid synthesis (Horton et al., 2002; Shimomura et al., 1999; Foretz et al., 1999). When mice are fed with a fat free high-carbohydrate diet, the liver SCD mRNA is induced about 50 fold (Ntambi, 1995; Sessler and Ntambi, 1998), however, the induction is blunted when SREBP1-c is disrupted. Because SDC and D6D mRNA expression were diminished in diabetic rats, and restored by insulin administration (Waters and Ntambi, 1994), the insulin effect on the desaturase genes is likely to be mediated by SREBP-1c. Studies with primary culture rat hepatocytes showed that the expression of a dominant negative SREBP1-c blocked the effect of insulin on the transcriptional activation of SCD, whereas expression of a dominant positive SREBP1-c mimicked the insulin effect (Foretz et al., 1999). Further studies are necessary to clarify the mechanism by which insulin induces SREBP expression.

#### 3.4.2.2 Peroxisomal Proliferators-Activated Receptor (PPARs)

Another common feature among desaturases is an induction of these enzymes by peroxisome proliferators (PPs) (Kawashima et al., 1983; Kawashima et al., 1990). Fatty acids and their oxygenated derivates are examples of PPs known to be physiological activators of members of the PPAR family nuclear receptors (Willson et al., 2001). PP-activated receptor α (PPARα) is a transcription factor primarily expressed in the liver in which it has been shown to promote β-oxidation of fatty acids (Schoonjans et al., 1996). However, PPARγ is a critical modulator of fat cell differentiation and function, thus providing a direct link between the intracellular levels of fatty acid and the regulation of gene transcription in adipocytes. PPs bind to the nuclear receptor PPARα, which them activates transcription of target genes by binding as a heterodimer to a cis-element, PP response element (PPRE) (Forman et al., 1997). The mouse SCD-1 (Miller and Ntambi, 1996) and D6D promoters have a PPRE site, although the mechanism of activation is unknown. Both, SCD1 and D6D mRNA are induced after PP administration by a mechanism dependent of protein synthesis (Kawashima et al., 1990). However, the intracellular levels of HUFAs (the product of D5D, D6D pathway) remain constant rather than be increased (Miller and Ntambi, 1996). This suggests that the mechanism of desaturase activation by PP is not direct. Instead, the induction of desaturase expression could be a compensatory response to an accelerated degradation of UFAs due to the induction of β-oxidation and an increased demand of HUFAs for PLs biosynthesis, required for peroxisome proliferation caused by PP administration.

#### 3.4.2.3 Cross-Talk Between PPARs and SREBPs

In conclusion, mammalian fatty acyl desaturases share common regulatory mechanism. Most notably, SREBP-1c activates the transcription of SCD-1, D6D and probably D5D genes, and its activation is suppressed by HUFAs. PPs induce both SCD1 and D6D probably due to the increase in HUFAs requirement as a consequence of PPs induction of peroxisome proliferation and fatty acid oxidation. Fatty acid desaturases are the only known genes regulated by both PPARα and SREPBs (Nakamura and Nara, 2002) which regulate oppositely fatty acid metabolic pathways. PPRAα is a general inducer of genes involved in fatty acid β-oxidation, while SREBP-1 induces genes of fatty acid synthesis. The cross-talk between PPARs and SREBPs could play a crucial role in the regulation of fatty acid metabolism but the interaction between these nuclear factors and their regulatory action on desaturase expression is complex and possess many paradoxes that need to be resolved (Yoshikawa et al., 2003).

### *3.4.3 Age-Related, Dietary and Hormonal Regulation of Δ9 Desaturase*

#### 3.4.3.1 Dietary Regulation of Desaturases by Dietary PUFAs in Liver

Extreme response to dietary alterations is a remarkable feature of $\Delta^9$ desaturase (SCD-1). Liver SCD-1 mRNA is dramatically altered by dietary changes, when rats are starved for 12–72 h, the $\Delta^9$ desaturase activity declines to less than 5% of control values. After refeeding, $\Delta^9$ desaturase activity increases dramatically to more than 2-fold above the normal levels. This phenomenon has been termed "super-induction" as levels of enzyme activity can be more than 100-fold above the fasted state. For example, this "super-induction" occurs in rat pups nursed by mothers on an essential fatty acid deficient diet; and a 45-fold increase occurs in liver upon re-feeding fasted mice with a fat-free, high carbohydrate diet (Jeffcoat and James, 1978; Ntambi, 1992). Responses of liver enzyme to dietary intake probably explain the "circadian change" in $\Delta^9$ desaturase, where liver activities can fluctuate 4-fold over a 24 h period; highest activity (around midnight) correspond to maximal food intake in nocturnal rats.

On the other hand, brain $\Delta^9$ desaturase (SCD-2) is almost no altered by dietary restrictions ensuring continuing activity during crucial stages of brain development. In contrast to the extreme changes in liver SCD-1, brain SCD-2 mRNA increases only 2-fold in neonates that are suckling mothers on an essential fatty acid deficient diet. Brain $\Delta^9$ desaturase activity is greatest during the perinatal and suckling periods in rats and is generally higher than in liver. However, after weaned, brain $\Delta^9$ desaturase activity slowly declines (DeWillie and Farmer, 1992).

#### 3.4.3.2 Hormonal Regulation of Desaturases

Regulation of Stearoyl CoA Desaturases by Insulin

Hormonal regulation of $\Delta^9$ desaturase is complex and not fully understood. Rats with genetic diabetes, or made diabetic by destruction of pancreatic β-cells, have depressed $\Delta^9$ desaturase activity in liver, mammary gland and adipose tissue which is restored *in vivo* by insulin. Insulin appears essential for basal transcription of the SCD1 gene and, as described before, SREPB appears to be the key regulator (Rimoldi et al., 2001; Waters and Ntambi, 1994).

Regulation of Stearoyl CoA Desaturases by Leptin

SCD activity has been shown to be elevated in the adipose tissue of various animal models of obesity (Cohen et al., 2002) and a positive correlation between SCD activity in skeletal muscle and the percentage of body weight has been reported in humans (Jones et al., 1996). Leptin deficient mice (*ob*/*ob*) are characterized by a five-fold higher deposition of body fat than their counterparts, and the consistent change in the fatty acid composition in these mice is an increase in MUFAs as a result of increased SCD activity (Cohen et al., 2002). SCD1 mRNA levels, enzymatic activity, and the levels of MUFAs are markedly increased in *ob*/*ob* mice and are all specifically reduced by leptin administration. These data suggested the possibility that leptin specific downregulation of

SCD1 mRNA and enzyme activity might mediate some of its metabolic effect. Supporting this hypothesis, double mutants *ab*$^J$/*ab*$^J$,*ob*/*ob* mice, where the SCD1 is deficient, showed a dramatic reduction of body weight and level of TGs. These mice consumed more food than *ob*/*ob* littermates, suggesting that SDC1 deficiency may modulate central nervous system (CNS) pathways that regulate food intake, but have a complete correction of their hypometabolic phenotype, with energy expenditure equivalent to, or even greater than, wild type. These data show that SDC1 is required for the development of obese phenotype of *ob*/*ob* mice and further suggest that a significant proportion of leptins metabolic effects may result from inhibition of this enzyme. The metabolic effect of leptin on SCD1 in liver, however, are likely to be the results of central action, as mice lacking leptin receptor in brain enlarged fatty livers, while livers from mice with liver specific knockout of the leptin receptor appear normal. Leptin also reduced hepatic SCD1 activity when administrated intracerebroventriculary. However, the nature of the CNS signals that modulate liver metabolism in response to leptin is unknown. Thus, in addition to play a role in lipids metabolism, the SCD1 is a promising therapeutic target in the treatment of the metabolic syndrome.

## 3.5 Conclusions and Perspectives

The data reviewed here clearly show that desaturase expression is tightly regulated by complex signaling pathways in organisms ranging from bacteria to humans. An interesting unresolved problem is how the sensors of membrane fluidity, found in bacteria, operate at the molecular level. These sensors can adopt alternative signaling states that are regulated by the biophysical properties of the membrane and are coupled to transcription factors that control membrane lipid homeostasis. A second unanswered question relates to the mechanism of signal recognition that allows activation of yeast desaturase expression by the ubiquitin-mediated proteolysis of the ER-associated transcription factors, Spt23p and Mga2p. This represents an interesting and novel regulation system responding to a wide range of physiological nutritional and metabolic conditions, which appears to have arisen early in the evolution of eukaryotes.

A final major unanswered question relates to the way in which the sophisticated set of controls, involving multiple sensors and signaling pathways regulate the expression of mammalian desaturases. This field is not without discrepancies, and it is probable that novel transcription factors and signaling mechanisms will be identified as our understanding of the topics grows. Gaining knowledge in this complex regulatory network will facilitate the future biomedical investigation of obesity, and lipids-related disorders such as diabetes and cardiovascular disease.

**Acknowledgments** This work was supported by a grant from Agencia Nacional de Promoción Científica y Tecnológica (FONCYT). C. E. Banchio, M. C. Mansilla and D. de Mendoza are Career Investigators from Consejo Nacional de Investigaciones Científicas y Técnicas (CONICET). D. de Mendoza is an International Research Scholar from Howard Hughes Medical Institute.

## References

Aguilar, P.S., Cronan, J.E., Jr. and de Mendoza, D. A *Bacillus subtilis* gene induced by cold shock encodes a membrane phospholipid desaturase. *J Bacteriol*, 180 (1998) 2194–2200.

Aguilar, P.S., Lopez, P. and de Mendoza, D. Transcriptional control of the low-temperature-inducible des gene, encoding the delta5 desaturase of *Bacillus subtilis*. *J Bacteriol*, 181 (1999) 7028–7033.

Aguilar, P.S., Hernandez-Arriaga, A.M., Cybulski, L.E., Erazo, A.C. and de Mendoza, D. Molecular basis of thermosensing: a two-component signal transduction thermometer in *Bacillus subtilis*. *Embo J*, 20 (2001)1681–1691.

Aki, T., Shimada, Y., Inagaki, K., Higashimoto, H., Kawamoto, S., Shigeta, S., Ono, K. and Suzuki, O. Molecular cloning and functional characterization of rat delta-6 fatty acid desaturase. *Biochem Biophys Res Commun*, 255 (1999) 575–579.

Al-Fageeh, M.B. and Smales, C.M. Control and regulation of the cellular responses to cold shock: the responses in yeast and mammalian systems. *Biochem J*, 397(2006) 247–259.

Albanesi, D., Mansilla, M.C. and de Mendoza, D. The membrane fluidity sensor DesK of *Bacillus subtilis* controls the signal decay of its cognate response regulator. *J Bacteriol*, 186 (2004) 2655–2663.

Altabe, S.G., Aguilar, P., Caballero, G.M. and de Mendoza, D. The *Bacillus subtilis* acyl lipid desaturase is a delta5 desaturase. *J Bacteriol*, 185 (2003) 3228–3231.

Beatty, A.L., Malloy, J.L. and Wright, J.R. *Pseudomonas aeruginosa* degrades pulmonary surfactant and increases conversion in vitro. *Am J Respir Cell Mol Biol*, 32 (2005)128–134.

Beckering, C.L., Steil, L., Weber, M.H., Volker, U. and Marahiel, M.A. Genomewide transcriptional analysis of the cold shock response in *Bacillus subtilis*. *J Bacteriol*, 184 (2002) 6395–6402.

Bloch, K. The biological synthesis of unsaturated fatty acids. *Biochem Soc Symp*, 24 (1963) 1–16.

Bossie, M.A. and Martin, C.E. Nutritional regulation of yeast delta-9 fatty acid desaturase activity. *J Bacteriol*, 171 (1989) 6409–6413.

Brown, M.S. and Goldstein, J.L. A proteolytic pathway that controls the cholesterol content of membranes, cells, and blood. *Proc Natl Acad Sci U S A*, 96 (1999) 11041–11048.

Burr, G.O. The essential fatty acids fifty years ago. *Prog Lipid Res*, 20 (1981) xxvii–xxix.

Clarke, S.D. and Jump, D.B. Dietary polyunsaturated fatty acid regulation of gene transcription. *Annu Rev Nutr*, 14 (1994) 83–98.

Cohen, P., Miyazaki, M., Socci, N.D., Hagge-Greenberg, A., Liedtke, W., Soukas, A.A., Sharma, R., Hudgins, L.C., Ntambi, J.M. and Friedman, J.M. Role for stearoyl-CoA desaturase-1 in leptin-mediated weight loss. *Science*, 297 (2002) 240–243.

Cronan, J.E., Jr. and Gelmann, E.P. An estimate of the minimum amount of unsaturated fatty acid required for growth of *Escherichia coli*. *J Biol Chem*, 248 (1973) 1188–1195.

Cybulski, L.E., Albanesi, D., Mansilla, M.C., Altabe, S., Aguilar, P.S. and de Mendoza, D. Mechanism of membrane fluidity optimization: isothermal control of the *Bacillus subtilis* acyl-lipid desaturase. *Mol Microbiol*, 45 (2002) 1379–1388.

Cybulski, L.E., del Solar, G., Craig, P.O., Espinosa, M. and de Mendoza, D. *Bacillus subtilis* DesR functions as a phosphorylation-activated switch to control membrane lipid fluidity. *J Biol Chem*, 279 (2004) 39340–39347.

Chellappa, R., Kandasamy, P., Oh, C.S., Jiang, Y., Vemula, M. and Martin, C.E. The membrane proteins, Spt23p and Mga2p, play distinct roles in the activation of *Saccharomyces cerevisiae* OLE1 gene expression. Fatty acid-mediated regulation of Mga2p activity is independent of its proteolytic processing into a soluble transcription activator. *J Biol Chem*, 276 (2001) 43548–43556.

Chin, J. and Chang, T.Y. Further characterization of a Chinese hamster ovary cell mutant requiring cholesterol and unsaturated fatty acid for growth. *Biochemistry*, 21 (1982) 3196–3202.

Cho, H.P., Nakamura, M. and Clarke, S.D. Cloning, expression, and fatty acid regulation of the human delta-5 desaturase. *J Biol Chem*, 274 (1999a) 37335–37339.

Cho, H.P., Nakamura, M.T. and Clarke, S.D. Cloning, expression, and nutritional regulation of the mammalian Delta-6 desaturase. *J Biol Chem* (1999b) 274, 471–477.

DeWillie, J.W. and S.J. Farmer. Postnatal dietary fat influences mRNAs involved in mylenation. *Dev Neurosci.*14 (1992) 61–68.

Diaz, A.R., Mansilla, M.C., Vila, A.J. and de Mendoza, D. Membrane topology of the acyl-lipid desaturase from *Bacillus subtilis*. *J Biol Chem*, 277 (2002) 48099–48106.

Dobrosotskaya, I.Y., Seegmiller, A.C., Brown, M.S., Goldstein, J.L. and Rawson, R.B. Regulation of SREBP processing and membrane lipid production by phospholipids in Drosophila. *Science*, 296 (2002) 879–883.

Enoch, H.G., Catala, A. and Strittmatter, P. Mechanism of rat liver microsomal stearyl-CoA desaturase. Studies of the substrate specificity, enzyme-substrate interactions, and the function of lipid. *J Biol Chem*, 251 (1976) 5095–5103.

Ericsson, J., Usheva, A. and Edwards, P.A. YY1 is a negative regulator of transcription of three sterol regulatory element-binding protein-responsive genes. *J Biol Chem*, 274 (1999) 14508–14513.

Foretz, M., Pacot, C., Dugail, I., Lemarchand, P., Guichard, C., Le Liepvre, X., Berthelier-Lubrano, C., Spiegelman, B., Kim, J.B., Ferre, P. and Foufelle, F. ADD1/SREBP-1c is required in the activation of hepatic lipogenic gene expression by glucose. *Mol Cell Biol*, 19 (1999) 3760–3768.

Forman, B.M., Chen, J. and Evans, R.M. Hypolipidemic drugs, polyunsaturated fatty acids, and eicosanoids are ligands for peroxisome proliferator-activated receptors alpha and delta. *Proc Natl Acad Sci U S A*, 94 (1997) 4312–4317.

Fujiwara, D., Yoshimoto, H., Sone, H., Harashima, S. and Tamai, Y. Transcriptional co-regulation of *Saccharomyces cerevisiae* alcohol acetyltransferase gene, ATF1 and delta-9 fatty acid desaturase gene, OLE1 by unsaturated fatty acids. *Yeast* 14 (1998), 711–721.

Fulco, A.J. The biosynthesis of unsaturated fatty acids by bacilli. I. Temperature induction of the desaturation reaction. *J Biol Chem*, 244 (1969) 889–895.

Funk, C.D. Prostaglandins and leukotrienes: advances in eicosanoid biology. *Science*, 294 (2001) 1871–1875.

Goldstein, J.L., DeBose-Boyd, R.A. and Brown, M.S. Protein sensors for membrane sterols. *Cell*, 124 (2006) 35–46.

Gonzalez, C.I. and Martin, C.E. Fatty acid-responsive control of mRNA stability. Unsaturated fatty acid-induced degradation of the *Saccharomyces* OLE1 transcript. *J Biol Chem*, 271 (1996) 25801–25809.

Grau, R. and de Mendoza, D. Regulation of the synthesis of unsaturated fatty acids by growth temperature in *Bacillus subtilis*. *Mol Microbiol*, 8 (1993) 535–542.

Hannah, V.C., Ou, J., Luong, A., Goldstein, J.L. and Brown, M.S. Unsaturated fatty acids down-regulate srebp isoforms 1a and 1c by two mechanisms in HEK-293 cells. *J Biol Chem*, 276 (2001) 4365–4372.

Hazel, J.R. Thermal adaptation in biological membranes: is homeoviscous adaptation the explanation? *Annu Rev Physiol*, 57 (1995) 19–42.

Heird, W.C. and Lapillonne, A. The role of essential fatty acids in development. *Annu Rev Nutr*, 25 (2005) 549–571.

Higashi, S. and Murata, N. An in vivo study of substrate specificities of acyl-lipid desaturases and acyltransferases in lipid synthesis in *Synechocystis* PCC6803. *Plant Physiol*, 102 (1993) 1275–1278.

Hitchcock, A.L., Krebber, H., Frietze, S., Lin, A., Latterich, M. and Silver, P.A. The conserved npl4 protein complex mediates proteasome-dependent membrane-bound transcription factor activation. *Mol Biol Cell*, 12 (2001) 3226–3241.

Holloway, C.T. and Holloway, P.W. Stearyl coenzyme a desaturase activity in mouse liver microsomes of varying lipid composition. *Arch Biochem Biophys*, 167 (1975) 496–504.

Hoppe, T., Matuschewski, K., Rape, M., Schlenker, S., Ulrich, H.D. and Jentsch, S. Activation of a membrane-bound transcription factor by regulated ubiquitin/proteasome-dependent processing. *Cell*, 102 (2000) 577–586.

Horton, J.D., Goldstein, J.L. and Brown, M.S. SREBPs: activators of the complete program of cholesterol and fatty acid synthesis in the liver. *J Clin Invest*, 109 (2002) 1125–1131.

Jeffcoat, R. and James, A.T. The control of stearoyl-CoA desaturase by dietary linoleic acid. *FEBS Lett*, 85 (1978) 114–118.

Jiang, Y., Vasconcelles, M.J., Wretzel, S., Light, A., Gilooly, L., McDaid, K., Oh, C.S., Martin, C.E. and Goldberg, M.A. Mga2p processing by hypoxia and unsaturated fatty acids in *Saccharomyces cerevisiae*: impact on LORE-dependent gene expression. *Eukaryot Cell*, 1 (2002) 481–490.

Jiang, Y., Vasconcelles, M.J., Wretzel, S., Light, A., Martin, C.E. and Goldberg, M.A. MGA2 is involved in the low-oxygen response element-dependent hypoxic induction of genes in *Saccharomyces cerevisiae*. *Mol Cell Biol*, 21 (2001) 6161–6169.

Jones, B.H., Maher, M.A., Banz, W.J., Zemel, M.B., Whelan, J., Smith, P.J. and Moustaid, N. Adipose tissue stearoyl-CoA desaturase mRNA is increased by obesity and decreased by polyunsaturated fatty acids. *Am J Physiol*, 271 (1996) E44–49.

Kaestner, K.H., Ntambi, J.M., Kelly, T.J., Jr. and Lane, M.D. Differentiation-induced gene expression in 3T3-L1 preadipocytes. A second differentially expressed gene encoding stearoyl-CoA desaturase. *J Biol Chem*, 264 (1989) 14755–14761.

Kandasamy, P., Vemula, M., Oh, C.S., Chellappa, R. and Martin, C.E. Regulation of unsaturated fatty acid biosynthesis in *Saccharomyces*: the endoplasmic reticulum membrane protein, Mga2p, a transcription activator of the OLE1 gene, regulates the stability of the OLE1 mRNA through exosome-mediated mechanisms. *J Biol Chem*, 279 (2004) 36586–36592.

Kaneda, T. Iso- and anteiso-fatty acids in bacteria: biosynthesis, function, and taxonomic significance. *Microbiol Rev*, 55 (1991) 288–302.

Kanesaki, Y., Suzuki, I., Allakhverdiev, S.I., Mikami, K. and Murata, N. Salt stress and hyperosmotic stress regulate the expression of different sets of genes in *Synechocystis* sp. PCC 6803. *Biochem Biophys Res Commun*, 290 (2002) 339–348.

Kanesaki, Y., Yamamoto, H., Paithoonrangsarid, K., Shoumskaya, M., Suzuki, I., Hayashi, H. and Murata, N. Histidine kinases play important roles in the perception and signal transduction of hydrogen peroxide in the cyanobacterium *Synechocystis* sp. PCC 6803. *Plant J*, 49 (2007) 313–324.

Kang, J.X. and Leaf, A. Antiarrhythmic effects of polyunsaturated fatty acids. Recent studies. *Circulation*, 94 (1996)1774–1780.

Kawashima, Y., Hanioka, N., Matsumura, M. and Kozuka, H. Induction of microsomal stearoyl-CoA desaturation by the administration of various peroxisome proliferators. *Biochim Biophys Acta*, 752 (1983) 259–264.

Kawashima, Y., Musoh, K. and Kozuka, H. Peroxisome proliferators enhance linoleic acid metabolism in rat liver. Increased biosynthesis of omega 6 polyunsaturated fatty acids. *J Biol Chem*, 265 (1990) 9170–9175.

Kim, H.I., Cha, J.Y., Kim, S.Y., Kim, J.W., Roh, K.J., Seong, J.K., Lee, N.T., Choi, K.Y., Kim, K.S. and Ahn, Y.H. Peroxisomal proliferator-activated receptor-gamma upregulates glucokinase gene expression in beta-cells. *Diabetes*, 51 (2002) 676–685.

Klein, W., Weber, M.H. and Marahiel, M.A. Cold shock response of *Bacillus subtilis*: isoleucine-dependent switch in the fatty acid branching pattern for membrane adaptation to low temperatures. *J Bacteriol*, 181 (1999) 5341–5349.

Kwast, K.E., Burke, P.V. and Poyton, R.O. Oxygen sensing and the transcriptional regulation of oxygen-responsive genes in yeast. *J Exp Biol*, 201 (1998) 1177–1195.

Lindqvist, Y., Huang, W., Schneider, G. and Shanklin, J. Crystal structure of delta9 stearoyl-acyl carrier protein desaturase from castor seed and its relationship to other di-iron proteins. *Embo J*, 15 (1996) 4081–4092.

Lopez, C.S., Heras, H., Garda, H., Ruzal, S., Sanchez-Rivas, C. and Rivas, E. Biochemical and biophysical studies of *Bacillus subtilis* envelopes under hyperosmotic stress. *Int J Food Microbiol*, 55 (2000) 137–142.

Los, D.A. and Murata, N. Membrane fluidity and its roles in the perception of environmental signals. *Biochim Biophys Acta*, 1666 (2004) 142–157.

Los, D., Horvath, I., Vigh, L. and Murata, N. The temperature-dependent expression of the desaturase gene desA in *Synechocystis* PCC6803. *FEBS Lett*, 318 (1993) 57–60.

Los, D.A., Ray, M.K. and Murata, N. Differences in the control of the temperature-dependent expression of four genes for desaturases in *Synechocystis* sp. PCC 6803. *Mol Microbiol*, 25 (1997) 1167–1175.

Magana, M.M., Lin, S.S., Dooley, K.A. and Osborne, T.F. Sterol regulation of acetyl coenzyme A carboxylase promoter requires two interdependent binding sites for sterol regulatory element binding proteins. *J Lipid Res*, 38 (1997) 1630–1638.

Man, W.C., Miyazaki, M., Chu, K. and Ntambi, J.M. Membrane topology of mouse stearoyl-CoA desaturase 1. *J Biol Chem*, 281 (2006) 1251–1260.

Mansilla, M.C. and de Mendoza, D. The *Bacillus subtilis* desaturase: a model to understand phospholipid modification and temperature sensing. *Arch Microbiol*, 183 (2005) 229–235.

Mansilla, M.C., Cybulski, L.E., Albanesi, D. and de Mendoza, D. Control of membrane lipid fluidity by molecular thermosensors. *J Bacteriol*, 186 (2004) 6681–6688.

Marrakchi, H., Choi, K.H. and Rock, C.O. A new mechanism for anaerobic unsaturated fatty acid formation in *Streptococcus pneumoniae*. *J Biol Chem*, 277 (2002) 44809–44816.

Martin, C.E., Oh, C.S. and Jiang, Y. Regulation of long chain unsaturated fatty acid synthesis in yeast. *Biochim Biophys Acta*, 1771 (2007) 271–285.

Matsuzaka, T., Shimano, H., Yahagi, N., Amemiya-Kudo, M., Yoshikawa, T., Hasty, A.H., Tamura, Y., Osuga, J., Okazaki, H., Iizuka, Y., Takahashi, A., Sone, H., Gotoda, T., Ishibashi, S. and Yamada, N. Dual regulation of mouse Delta(5)- and Delta(6)-desaturase gene expression by SREBP-1 and PPARalpha. *J Lipid Res*, 43 (2002) 107–114.

McDonough, V.M., Stukey, J.E. and Martin, C.E. Specificity of unsaturated fatty acid-regulated expression of the *Saccharomyces cerevisiae* OLE1 gene. *J Biol Chem*, 267 (1992) 5931–5936.

Mihara, K. Structure and regulation of rat liver microsomal stearoyl-CoA desaturase gene. *J Biochem*, 108 (1990) 1022–1029.

Mikami, K., Kanesaki, Y., Suzuki, I. and Murata, N. The histidine kinase Hik33 perceives osmotic stress and cold stress in *Synechocystis* sp PCC 6803. *Mol Microbiol*, 46 (2002) 905–915.

Miller, C.W. and Ntambi, J.M. Peroxisome proliferators induce mouse liver stearoyl-CoA desaturase 1 gene expression. *Proc Natl Acad Sci U S A*, 93 (1996) 9443–9448.

Miyazaki, M., Kim, Y.C., Gray-Keller, M.P., Attie, A.D. and Ntambi, J.M. The biosynthesis of hepatic cholesterol esters and triglycerides is impaired in mice with a disruption of the gene for stearoyl-CoA desaturase 1. *J Biol Chem*, 275 (2000) 30132–30138.

Miyazaki, M., Man, W.C. and Ntambi, J.M. Targeted disruption of stearoyl-CoA desaturase1 gene in mice causes atrophy of sebaceous and meibomian glands and depletion of wax esters in the eyelid. *J Nutr*, 131 (2001) 2260–2268.

Murata, N. and Wada, H. Acyl-lipid desaturases and their importance in the tolerance and acclimatization to cold of cyanobacteria. *Biochem J*, 308 ( Pt 1) (1995) 1–8.

Murata, N. and Suzuki, I. Exploitation of genomic sequences in a systematic analysis to access how cyanobacteria sense environmental stress. *J Exp Bot*, 57, (2006) 235–247.

Mustardy, L., Los, D.A., Gombos, Z. and Murata, N. Immunocytochemical localization of acyl-lipid desaturases in cyanobacterial cells: evidence that both thylakoid membranes and cytoplasmic membranes are sites of lipid desaturation. *Proc Natl Acad Sci U S A*, 93 (1996) 10524–10527.

Nakagawa, Y., Sugioka, S., Kaneko, Y. and Harashima, S. O2R, a novel regulatory element mediating Rox1p-independent O(2) and unsaturated fatty acid repression of OLE1 in *Saccharomyces cerevisiae*. *J Bacteriol*, 183 (2001) 745–751.

Nakagawa, Y., Sakumoto, N., Kaneko, Y. and Harashima, S. Mga2p is a putative sensor for low temperature and oxygen to induce OLE1 transcription in *Saccharomyces cerevisiae*. *Biochem Biophys Res Commun*, 291 (2002) 707–713.

Nakamura, M.T. and Nara, T.Y. Gene regulation of mammalian desaturases. *Biochem Soc Trans*, 30 (2002) 1076–1079.

Nakamura, M.T. and Nara, T.Y. Structure, function, and dietary regulation of delta6, delta5, and delta9 desaturases. *Annu Rev Nutr*, 24 (2004) 345–376.

Napier, J.A., Hey, S.J., Lacey, D.J. and Shewry, P.R. Identification of a *Caenorhabditis elegans* delta6-fatty-acid-desaturase by heterologous expression in *Saccharomyces cerevisiae*. *Biochem J*, 330 (Pt 2) (1998) 611–614.

Nara, T.Y., He, W.S., Tang, C., Clarke, S.D. and Nakamura, M.T. The E-box like sterol regulatory element mediates the suppression of human Delta-6 desaturase gene by highly unsaturated fatty acids. *Biochem Biophys Res Commun*, 296 (2002) 111–117.

Nohturfft, A., Yabe, D., Goldstein, J.L., Brown, M.S. and Espenshade, P.J. Regulated step in cholesterol feedback localized to budding of SCAP from ER membranes. *Cell*, 102 (2000) 315–323.

Ntambi, J.M. Dietary regulation of stearoyl-CoA desaturase 1 gene expression in mouse liver. *J Biol Chem*, 267 (1992) 10925–10930.

Ntambi, J.M. The regulation of stearoyl-CoA desaturase (SCD). *Prog Lipid Res*, 34 (1995) 139–150.

Ntambi, J.M. Regulation of stearoyl-CoA desaturase by polyunsaturated fatty acids and cholesterol. *J Lipid Res*, 40 (1999) 1549–1558.

Ntambi, J.M. and Miyazaki, M. Recent insights into stearoyl-CoA desaturase-1. *Curr Opin Lipidol*, 14 (2003) 255–261.

Ntambi, J.M., Sessler, A.M. and Takova, T. A model cell line to study regulation of stearoyl-CoA desaturase gene 1 expression by insulin and polyunsaturated fatty acids. *Biochem Biophys Res Commun*, 220 (1996) 990–995.

Ntambi, J.M., Buhrow, S.A., Kaestner, K.H., Christy, R.J., Sibley, E., Kelly, T.J., Jr. and Lane, M.D. Differentiation-induced gene expression in 3T3-L1 preadipocytes. Characterization of a differentially expressed gene encoding stearoyl-CoA desaturase. *J Biol Chem*, 263 (1988) 17291–17300.

Ntambi, J.M., Choi, Y., Park, Y., Peters, J.M. and Pariza, M.W. Effects of conjugated linoleic acid (CLA) on immune responses, body composition and stearoyl-CoA desaturase. *Can J Appl Physiol*, 27 (2002a) 617–628.

Ntambi, J.M., Miyazaki, M., Stoehr, J.P., Lan, H., Kendziorski, C.M., Yandell, B.S., Song, Y., Cohen, P., Friedman, J.M. and Attie, A.D. Loss of stearoyl-CoA desaturase-1 function protects mice against adiposity. *Proc Natl Acad Sci U S A*, 99 (2002b) 11482–11486.

Ohlrogge, J. and Browse, J. Lipid biosynthesis. *Plant Cell*, 7 (1995) 957–970.

Pereira, S.L., Leonard, A.E. and Mukerji, P. Recent advances in the study of fatty acid desaturases from animals and lower eukaryotes. *Prostaglandins Leukot Essent Fatty Acids*, 68 (2003) 97–106.

Phadtare, S. Recent developments in bacterial cold-shock response. *Curr Issues Mol Biol*, 6 (2004) 125–136.

Phetsuksiri, B., Jackson, M., Scherman, H., McNeil, M., Besra, G.S., Baulard, A.R., Slayden, R.A., DeBarber, A.E., Barry, C.E., 3rd, Baird, M.S., Crick, D.C. and Brennan, P.J. Unique mechanism of action of the thiourea drug isoxyl on *Mycobacterium tuberculosis*. *J Biol Chem*, 278 (2003) 53123–53130.

Phinney, S.D. British radiation study. *Science*, 248 (1990) 1595.

Reddy, A.S., Nuccio, M.L., Gross, L.M. and Thomas, T.L. Isolation of a delta 6-desaturase gene from the cyanobacterium *Synechocystis* sp. strain PCC 6803 by gain-of-function expression in Anabaena sp. strain PCC 7120. *Plant Mol Biol*, 22 (1993) 293–300.

Rimoldi, O.J., Finarelli, G.S. and Brenner, R.R. Effects of diabetes and insulin on hepatic delta6 desaturase gene expression. *Biochem Biophys Res Commun*, 283 (2001) 323–326.

Sakamoto, T. and Bryant, D.A. Temperature-regulated mRNA accumulation and stabilization for fatty acid desaturase genes in the cyanobacterium *Synechococcus* sp. strain PCC 7002. *Mol Microbiol*, 23 (1997)1281–1292.

Sakamoto, T., Los, D.A., Higashi, S., Wada, H., Nishida, I., Ohmori, M. and Murata, N. Cloning of omega 3 desaturase from cyanobacteria and its use in altering the degree of membrane-lipid unsaturation. *Plant Mol Biol*, 26 (1994a) 249–263.

Sakamoto, T., Wada, H., Nishida, I., Ohmori, M. and Murata, N. delta 9 Acyl-lipid desaturases of cyanobacteria. Molecular cloning and substrate specificities in terms of fatty acids, sn-positions, and polar head groups. *J Biol Chem*, 269 (1994b) 25576–25580.

Sampath, H. and Ntambi, J.M. Polyunsaturated fatty acid regulation of genes of lipid metabolism. *Annu Rev Nutr*, 25 (2005) 317–340.

Schoonjans, K., Staels, B. and Auwerx, J. Role of the peroxisome proliferator-activated receptor (PPAR) in mediating the effects of fibrates and fatty acids on gene expression. *J Lipid Res*, 37 (1996) 907–925.

Schwartz, J. Role of polyunsaturated fatty acids in lung disease. *Am J Clin Nutr*, 71 (2000) 393S–396S.

Sessler, A.M. and Ntambi, J.M. Polyunsaturated fatty acid regulation of gene expression. *J Nutr*, 128 (1998) 923–926.

Shanklin, J. and Cahoon, E.B. Desaturation and related modifications of fatty acids1. *Annu Rev Plant Physiol Plant Mol Biol*, 49 (1998) 611–641.

Shanklin, J., Whittle, E. and Fox, B.G. Eight histidine residues are catalytically essential in a membrane-associated iron enzyme, stearoyl-CoA desaturase, and are conserved in alkane hydroxylase and xylene monooxygenase. *Biochemistry*, 33 (1994) 12787–12794.

Shimomura, I., Bashmakov, Y., Shimano, H., Horton, J.D., Goldstein, J.L. and Brown, M.S. Cholesterol feeding reduces nuclear forms of sterol regulatory element binding proteins in hamster liver. *Proc Natl Acad Sci U S A*, 94 (1997) 12354–12359.

Shimomura, I., Bashmakov, Y. and Horton, J.D. Increased levels of nuclear SREBP-1c associated with fatty livers in two mouse models of diabetes mellitus. *J Biol Chem*, 274 (1999) 30028–30032.

Shoumskaya, M.A., Paithoonrangsarid, K., Kanesaki, Y., Los, D.A., Zinchenko, V.V., Tanticharoen, M., Suzuki, I. and Murata, N. Identical Hik-Rre systems are involved in perception and transduction of salt signals and hyperosmotic signals but regulate the expression of individual genes to different extents in synechocystis. *J Biol Chem*, 280 (2005) 21531–21538.

Singh A., and Ward OP. Microbial production of docosahexaenoic acid (DHA, C22:6). *Adv Appl Microbiol*, 45 (1997) 271–312.

Skerrett, P.J. and Hennekens, C.H. Consumption of fish and fish oils and decreased risk of stroke. *Prev Cardiol*, 6 (2003) 38–41.

Sperling, P., Ternes, P., Zank, T.K. and Heinz, E. The evolution of desaturases. *Prostaglandins Leukot Essent Fatty Acids*, 68 (2003) 73–95.

Stukey, J.E., McDonough, V.M. and Martin, C.E. The OLE1 gene of *Saccharomyces cerevisiae* encodes the delta 9 fatty acid desaturase and can be functionally replaced by the rat stearoyl-CoA desaturase gene. *J Biol Chem*, 265 (1990)20144–20149.

Suzuki, I., Los, D.A., Kanesaki, Y., Mikami, K. and Murata, N. The pathway for perception and transduction of low-temperature signals in *Synechocystis*. *Embo J*, 19 (2000) 1327–1334.

Tabor, D.E., Kim, J.B., Spiegelman, B.M. and Edwards, P.A. Identification of conserved cis-elements and transcription factors required for sterol-regulated transcription of stearoyl-CoA desaturase 1 and 2. *J Biol Chem*, 274 (1999) 20603–20610.

Thijssen, M.A. and Mensink, R.P. Fatty acids and atherosclerotic risk. *Handb Exp Pharmacol*. (2005)165–194.

Tocher, D.R., Leaver, M.J. and Hodgson, P.A. Recent advances in the biochemistry and molecular biology of fatty acyl desaturases. *Prog Lipid Res*, 37 (1998) 73–117.

Towle, H.C., Kaytor, E.N. and Shih, H.M. Regulation of the expression of lipogenic enzyme genes by carbohydrate. *Annu Rev Nutr*, 17 (1997) 405–433.

Uttaro, A.D. Biosynthesis of polyunsaturated fatty acids in lower eukaryotes. *IUBMB Life*, 58 (2006) 563–571.

van Beilen, J.B., Wubbolts, M.G. and Witholt, B. Genetics of alkane oxidation by *Pseudomonas oleovorans*. *Biodegradation*, 5 (1994) 161–174.

Vasconcelles, M.J., Jiang, Y., McDaid, K., Gilooly, L., Wretzel, S., Porter, D.L., Martin, C.E. and Goldberg, M.A. Identification and characterization of a low oxygen response element involved in the hypoxic induction of a family of *Saccharomyces cerevisiae* genes. Implications for the conservation of oxygen sensing in eukaryotes. *J Biol Chem*, 276 (2001) 14374–14384.

Vemula, M., Kandasamy, P., Oh, C.S., Chellappa, R., Gonzalez, C.I. and Martin, C.E. Maintenance and regulation of mRNA stability of the *Saccharomyces cerevisiae* OLE1 gene requires multiple elements within the transcript that act through translation-independent mechanisms. *J Biol Chem*, 278 (2003) 45269–45279.

Wallis, J.G. and Browse, J. The Delta8-desaturase of *Euglena gracilis*: an alternate pathway for synthesis of 20-carbon polyunsaturated fatty acids. *Arch Biochem Biophys*, 365 (1999) 307–316.

Waters, K.M. and Ntambi, J.M. Insulin and dietary fructose induce stearoyl-CoA desaturase 1 gene expression of diabetic mice. *J Biol Chem*, 269 (1994) 27773–27777.

Waters, K.M., Miller, C.W. and Ntambi, J.M. Localization of a polyunsaturated fatty acid response region in stearoyl-CoA desaturase gene 1. *Biochim Biophys Acta*, 1349 (1997) 33–42.

Willson, T.M., Lambert, M.H. and Kliewer, S.A. Peroxisome proliferator-activated receptor gamma and metabolic disease. *Annu Rev Biochem*, 70 (2001) 341–367.

Xu, J., Nakamura, M.T., Cho, H.P. and Clarke, S.D. Sterol regulatory element binding protein-1 expression is suppressed by dietary polyunsaturated fatty acids. A mechanism for the coordinate suppression of lipogenic genes by polyunsaturated fats. *J Biol Chem*, 274, (1999) 23577–23583.

Xu, J., Teran-Garcia, M., Park, J.H., Nakamura, M.T. and Clarke, S.D. Polyunsaturated fatty acids suppress hepatic sterol regulatory element-binding protein-1 expression by accelerating transcript decay. *J Biol Chem*, 276 (2001) 9800–9807.

Yoshikawa, T., Shimano, H., Yahagi, N., Ide, T., Amemiya-Kudo, M., Matsuzaka, T., Nakakuki, M., Tomita, S., Okazaki, H., Tamura, Y., Iizuka, Y., Ohashi, K., Takahashi, A., Sone, H., Osuga Ji, J., Gotoda, T., Ishibashi, S. and Yamada, N. Polyunsaturated fatty acids suppress sterol regulatory element-binding protein 1c promoter activity by inhibition of liver X receptor (LXR) binding to LXR response elements. *J Biol Chem*, 277 (2002) 1705–1711.

Yoshikawa, T., Ide, T., Shimano, H., Yahagi, N., Amemiya-Kudo, M., Matsuzaka, T., Yatoh, S., Kitamine, T., Okazaki, H., Tamura, Y., Sekiya, M., Takahashi, A., Hasty, A.H., Sato, R., Sone, H., Osuga, J., Ishibashi, S. and Yamada, N. Cross-talk between peroxisome proliferator-activated receptor (PPAR) alpha and liver X receptor (LXR) in nutritional regulation of fatty acid metabolism. I. PPARs suppress sterol regulatory element binding protein-1c promoter through inhibition of LXR signaling. *Mol Endocrinol*, 17 (2003) 1240–1254.

Zhang, L., Ge, L., Parimoo, S., Stenn, K. and Prouty, S.M. Human stearoyl-CoA desaturase: alternative transcripts generated from a single gene by usage of tandem polyadenylation sites. *Biochem J*, 340 (Pt 1) (1999a) 255–264.

Zhang, L., Ge, L., Tran, T., Stenn, K. and Prouty, S.M. Isolation and characterization of the human stearoyl-CoA desaturase gene promoter: requirement of a conserved CCAAT cis-element. *Biochem J*, 357 (2001) 183–193.

Zhang, S., Skalsky, Y. and Garfinkel, D.J. MGA2 or SPT23 is required for transcription of the delta9 fatty acid desaturase gene, OLE1, and nuclear membrane integrity in *Saccharomyces cerevisiae*. *Genetics*, 151 (1999b) 473–483.

Zhang, Y.M., Zhu, K., Frank, M.W. and Rock, C.O. A *Pseudomonas aeruginosa* transcription factor that senses fatty acid structure. *Mol Microbiol*, 66 (2007) 622–632.

Zhu, K., Choi, K.H., Schweizer, H.P., Rock, C.O. and Zhang, Y.M. Two aerobic pathways for the formation of unsaturated fatty acids in *Pseudomonas aeruginosa*. *Mol Microbiol*, 60 (2006) 260–273.

Zolfaghari, R., Cifelli, C.J., Banta, M.D. and Ross, A.C. Fatty acid delta(5)-desaturase mRNA is regulated by dietary vitamin A and exogenous retinoic acid in liver of adult rats. *Arch Biochem Biophys*, 391 (2001) 8–15.

# Chapter 4
# Fatty Acid Amide Hydrolase: A Gate-Keeper of the Endocannabinoid System

**Filomena Fezza, Chiara De Simone, Daniele Amadio and Mauro Maccarrone**

**Abstract** The family of endocannabinoids contains several polyunsaturated fatty acid amides such as anandamide (AEA), but also esters such as 2-arachidonoylglycerol (2-AG). These compounds are the main endogenous agonists of cannabinoid receptors, able to mimic several pharmacological effects of $\Delta^9$-tetrahydrocannabinol ($\Delta^9$-THC), the active principle of *Cannabis sativa* preparations like hashish and marijuana. The activity of AEA at its receptors is limited by cellular uptake, through a putative membrane transporter, followed by intracellular degradation by fatty acid amide hydrolase (FAAH). Growing evidence demonstrates that FAAH is the critical regulator of the endogenous levels of AEA, suggesting that it may serve as an attractive therapeutic target for the treatment of human disorders. In particular, FAAH inhibitors may be next generation therapeutics of potential value for the treatment of pathologies of the central nervous system, and of peripheral tissues. Investigations into the structure and function of FAAH, its biological and therapeutic implications, as well as a description of different families of FAAH inhibitors, are the topic of this chapter.

**Keywords** Cannabinoids · endocannabinoid system · FAAH · gene expression · inhibitor

**Abbreviations** AEA: anandamide; 2-AG: 2-arachidonoylglycerol; AMT: anandamide membrane transporter; AS: amidase signature; AD: Alzheimer's disease; CB1R: type 1 cannabinoid receptor; CB2R: type 2 cannabinoid receptor; DAG: 1-acyl-2-arachidonoylglycerol; DAGLs: diacylglycerol lipases; EAE: autoimmune encephalomyelitis; ECS: endocannabinoid system; ETTH: episodic tension-type headache; FAAH: fatty acid amide hydrolase; FAAs: fatty acid amides; HAEAs: hydroxyanandamides; HD: Huntington's

M. Maccarrone
Department of Biomedical Sciences, University of Teramo, Teramo, Italy and European Center for Brain Research (CERC)/IRCCS S. Lucia Foundation, Rome, Italy

disease; MAE2: malonamidase; MAFP: methoxy arachidonoyl fluorophosphonate; MAGL: monoacylglycerol lipase; MoA: migraine without aura; MS: multiple sclerosis; NAGly: *N*-arachidonoyl-glycine; NAPE-PLD: *N*-acylphosphatidylethanolamide-phospholipase D; NArPE: *N*-arachidonoylphosphatidylethanolamine; NAT: *N*-acyltransferase; PAM: C-terminal peptide amidase; PMSF: phenylmethylsulfonyl fluoride; SNPs: single nucleotide polymorphisms; $\Delta^9$-THC: $\Delta^9$-tetrahydrocannabinol; TM: N-terminal transmembrane; TRPV1, transient receptor potential vanilloid 1

## 4.1 Introduction

### *4.1.1 (Endo)cannabinoids*

The recreational value of *Cannabis sativa* preparations is known to most people, largely as a result of the explosion in its use in the late 1960s; indeed, marijuana is still one of the most widespread illicit drugs of abuse in the world (Adams and Martin, 1996). The plant contains about 60 cannobinoid compounds (Ross and Elsohly, 1996), among which $\Delta^9$-tetrahydrocannabinol ($\Delta^9$-THC) (Fig. 4.1) is the primary psychoactive component and is thought to mediate most of the physiological effects associated with marijuana smoking (Dewey, 1986). $\Delta^9$-THC was used in folklore medicine long before the discovery of its mechanism of action.

The stringent structural characteristics that cannabinoid compounds must possess in order to exert their psychotropic effects, and the key observation that cannabinoids inhibit adenylate cyclase, supported the presence of a specific, high-affinity binding site for these lipidic substances (Howlett and Fleming, 1984). Shortly afterwards, the first membrane receptor for $\Delta^9$-THC was identified in rat brain (Devane et al., 1988). Its distribution was consistent with the pharmacological properties of psycotropic cannabinoids, and therefore it was

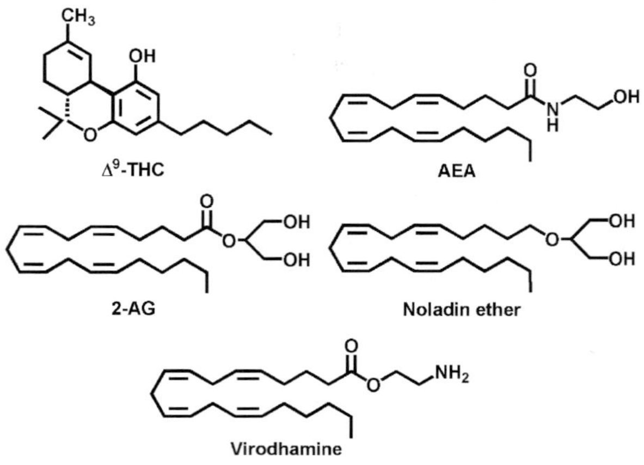

**Fig. 4.1** Chemical structures of various (endo)cannabinoids

designated type 1 cannabinoid receptor (CB1R) (Devane et al., 1988). A peripheral cannabinoid-binding receptor was identified a few years later in spleen and immune cells, and was called type 2 cannabinoid receptor (CB2R) (Munro et al., 1993). Since then, a number of endogenous agonists of CB receptors were characterized, i.e. amides, esters and ethers of long chain polyunsaturated fatty acids collectively termed 'endocannabinoids'. Remarkably, these compounds are structurally different from $\Delta^9$-THC or other plant cannabinoids (Mechoulam et al., 2002; Piomelli, 2003; De Petrocellis et al., 2004). In fact two arachidonate derivatives, the amide *N*-arachidonoylethanolamine (anandamide, AEA) (Devane et al., 1992) and the ester 2-arachidonoylglycerol (2-AG) (Mechoulam et al., 1995; Sugiura et al., 1995), both shown in Fig. 4.1, are the most biologically active endocannabinoids described to date (Piomelli, 2003; De Petrocellis et al., 2004). Also the ether 2-arachidonoylglyceryl-ether (noladin ether) (Fig. 4.1) has been shown to act as an endocannabinoid (Hanus et al., 2001), but its actual physiological relevance remains a matter of debate (Oka et al., 2003). Furthermore an 'inverted anandamide', *O*-arachidonoylethanolamine (virodhamine) (Fig. 4.1), has been shown to behave as a partial agonist or as a full agonist at CB1 or CB2 receptors, respectively (Porter et al., 2002).

Instead, the amides *N*-oleoylethanolamine (OEA), *N*-palmitoylethanolamine (PEA) and oleamide are better considered 'endocannabinoid-like' compounds, because they do not activate directly CB receptors, but rather prolong the activity of true endocannabinoids within the cell by an 'entourage effect' (De Petrocellis et al., 2004). AEA and 2-AG are present in the central nervous system (CNS) and also in peripheral tissues (Sugiura et al., 2002), but exhibit important differences in their quantitative distribution; 2-AG is more abundant than AEA in the brain and behaves as a full agonist for CB1R and CB2R, while AEA acts as partial agonist for CB1R and as a weak partial agonist for CB2R (Sugiura et al., 2000a). AEA levels may vary by 4–6-fold in different regions of the rat brain, with the highest levels in the striatum and brainstem and the lowest levels in the cerebellum and cortex (Bisogno et al., 1999a; Yang et al., 1999). AEA was found in regions of both rat and human brains that contain high densities of CB1R (e.g., hippocampus, cerebellum, and striatum) and also in a region that is sparse in CB1R like the thalamus (Felder et al., 1996). It is clear from these data that for AEA the relative regional abundance in the brain does not correlate with the distribution of CB1R. AEA levels in the brain are equivalent to those of other neurotransmitters such as dopamine and serotonin, but at least 10-fold lower than the levels reported for GABA and glutamate. AEA has also been found in peripheral tissues such as human and rat spleen, which expresses high levels of CB2R. Small amounts of AEA were also detected in human serum, plasma, and cerebrospinal fluid (Felder et al., 1996).

The concentration of 2-AG can be up to ~200-fold higher than that of AEA in the brain (Bisogno et al., 1999a). Yet, there are reports showing much lower 2-AG:AEA ratios in rat striatum (~10), substantia nigra (~3), and globus pallidus (~4) (Di Marzo et al., 2000; Gubellini et al., 2002). These differences may arise from different methodologies, for instance killing the animals by decapitation without immediate freezing instead of soaking in liquid nitrogen can increase 2-AG levels by ~15-fold (Sugiura et al., 2002). Discrepancies may also be a consequence of the high sensitivity of endocannabinoids to environmental factors like animal diets, caging and bedding systems, viral load, water quality, and pathogen infections. A recent example of the dramatic effect of these factors on endocannabinoid levels has been recently reported (Guo et al., 2005). In this context, it seems noteworthy that a recent study has shown that the extracellular concentrations of AEA and 2-AG are both in the nanomolar range (Caillè et al., 2007), suggesting that these two compounds have a similar availability for their CBR-mediated biological actions. On the other hand, the spatial distribution of the two endocannabinoids is similar in different regions of the brain. In fact, the highest concentrations of 2-AG were found in the brainstem, medulla, limbic forebrain, striatum, and hippocampus, and the lowest in the cortex, diencephalons, mesencephalon, hypothalamus, and cerebellum (Sugiura et al., 2002). Therefore, much alike AEA, no correlation was found between 2-AG concentrations and CB1R distribution. 2-AG was also detected in the peripheral nervous system, i.e. in the sciatic nerve, lumbar spinal cord, and lumbar dorsal root ganglion cells (Sugiura et al., 2002).

In just one decade, endocannabinoids have been shown to play manifold roles, both in the CNS and in the periphery. In particular, it is now widely accepted that the biological activity of AEA and 2-AG is largely dependent on a 'metabolic control', that modulates the effects of these substances by modulating their in vivo concentration (or endogenous tone) (Cravatt and Lichtman, 2002).

### *4.1.2 Overview of the Endocannabinoid System*

Investigations of the pathways involved in the metabolism of endocannabinoids have grown exponentially in recent years following the discovery of cannabinoid receptors. As other lipid mediators, AEA and 2-AG are released from cells 'on demand' by stimulus-dependent cleavage of membrane phospolipid precursors (Di Marzo et al., 1994).

AEA biosynthesis has been shown to occur through several pathways mediated by *N*-acylphosphatidylethanolamide-phospholipase D (NAPE-PLD), a secretory $PLA_2$ and PLC. 2-AG is generated through the action of selective enzymes such as phosphatidic acid phsophohydrolase, diacylglycerol lipase (DAGL), phosphoinositide-specific PLC (PI-PLC) and lyso-PLC. A putative membrane transporter, catalyzing a facilitated diffusion process, is involved in the cellular uptake or release of endocannabinoids. AEA is metabolized by fatty acid amidohydrolase (FAAH) and 2-AG is metabolized by monoacylglycerol lipase (MAGL), and to a lesser extent by FAAH.

Taken together AEA and 2-AG, their congeners and metabolic enzymes, their purported transporters and molecular targets form the 'endocannabinoid system (ECS)'.

#### 4.1.2.1 Molecular Targets

Endocannabinoids act primarily at cannabinoid receptors. These are seven trans-membrane spanning receptors that include type-1 cannabinoid receptors (CB1R), which are present mainly in the CNS but are also expressed in peripheral tissues and cells like lymphocytes (Börner et al., 2007), and type-2 cannabinoid receptors (CB2R), expressed predominantly by astrocytes, spleen and immune cells (Lunn et al., 2006), but also present in the brainstem (Van Sickle et al., 2005; Aguado et al., 2007 ). CB1R and CB2R belong to the rhodopsin family of G protein-coupled receptors (GPCRs), particularly those of the Gi/o family (Howlett et al., 2002). The binding of endocannabinoids to these receptors induces several biological actions, such as the inhibition of adenylate cyclase (AC), the regulation of ionic currents (inhibition of voltage-gated L, N and P/Q-type $Ca^{2+}$ channels, activation of $K^{+}$ channels), the activation of focal adhesion kinase, of mitogen-activated protein kinase (MAPK), and of cytosolic phospholipase $A_2$, and the activation (CB1R) or the inhibition (CB2R) of nitric oxide synthetase (NOS). Additionally, a recent report has shown an unprecedented coupling of CB1R to Gq/11 proteins, suggesting further diversity of CB1R signaling pathways (Lauckner et al., 2005). Furthermore, there is some evidence that endocannabinoids induce a biological activity via other CB receptors, like a purported CB3 (GPR55) receptor (Sawzdargo et al., 1999; Baker et al., 2006; McPartland et al., 2006; Ryberg et al., 2007), via non-CB1/non-CB2 receptors, or via non-cannabinoid receptors. In the latter group, transient receptor potential vanilloid 1 (TRPV1) has emerged as an important target of AEA, but remarkably not of 2-AG. TRPV1 is a six transmembrane spanning protein with intracellular N- and C-terminals, and a pore-loop between the fifth and sixth transmembrane helices (Jung et al., 1999). This ligand-gated and non-selective cationic channel is activated by molecules derived from plants, such as the pungent component of 'hot' red peppers capsaicin, by noxious stimuli like heat and protons (Jordt and Julius, 2002), and by peptides contained in spider toxins (Siemens et al., 2006). Also AEA is considered a true 'endovanilloid' (van der Stelt et al., 2004; Starowicz et al., 2007), that behaves as an authentic (though weak) ligand of TRPV1.

#### 4.1.2.2 Biosynthesis of Endocannabinoids

The main route for AEA biosynthesis occurs by two enzymatic steps involving the sequential action of a calcium dependent *N*-acyltransferase (NAT) and of a NAPE-specific phospholipase D (NAPE-PLD) (Okamoto et al., 2004). In the first step, NAT catalyzes direct transfer of arachidonic acid from the *sn*-1 position of phosphatidylcholine (PC), generating

*N*-arachidonoylphosphatidylethanolamine (NArPE), the AEA precursor (Fig. 4.2). This biosynthetic pathway is in agreement with the observation that AEA levels are generally lower than those of the other NAEs in most of the tissue analyzed so far, because the arachidonic acid levels in position 1 of phospholipids are very low.

In the last step, NArPE is hydrolyzed by NAPE-PLD which releases AEA and phosphatidic acid (PA). This enzyme has been cloned and purified from rat heart and classified as a member of the zinc metallo-hydrolase family of the β-lactamase fold (Okamoto et al., 2004). NAPE-PLD does not recognize phosphatidylcholine and phoshatidylethanolamine as substrates, and it is widely distributed in mouse organs, with highest concentrations in brain, kidney and testis (Okamoto et al., 2004). The same group who characterized NAPE-PLD also suggested that several $PLA_1/A_2$ isozymes can generate *N*-arachidonoyl-lysoPE (NAr-lysoPE) from NArPE, and that a lysoPLD may release AEA from NAr-lysoPE. Therefore, the sequential action of $PLA_1/A_2$ and lysoPLD may represent an alternative biosynthetic pathway for NAEs, including AEA (Sun et al., 2004) (Fig. 4.2).

Recently, it has been shown that in RAW264-7 macrophages, the lipopolysaccharide-induced anadamide production appears to depend mainly on a pathway whereby NAPE is hydrolysed to yield a phosphor-AEA, which is

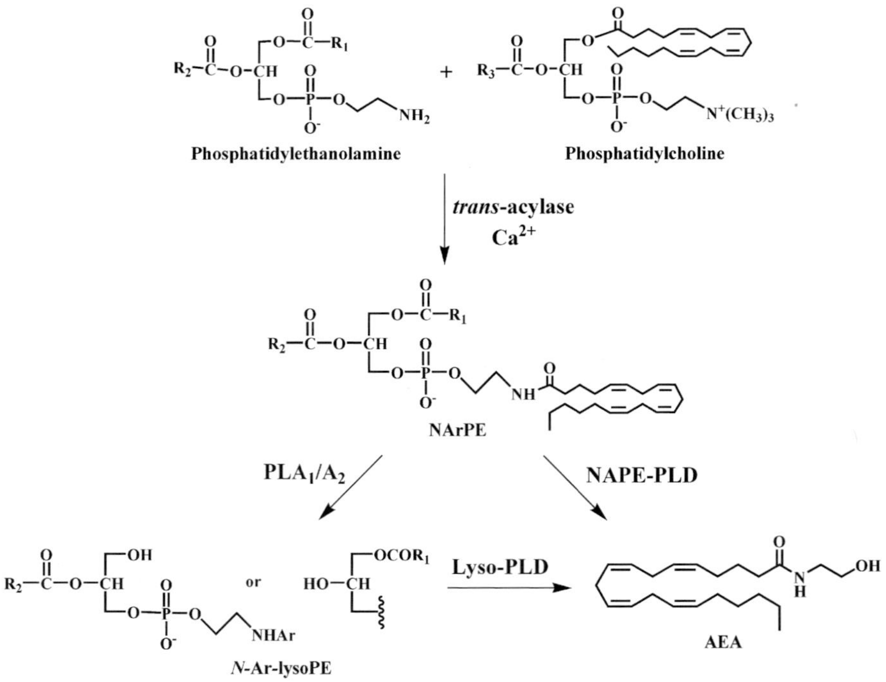

**Fig. 4.2** Major biosynthetic pathways of AEA. See text for details

then dephosphorylated (Liu et al., 2006). Furthermore, a non exclusive role of NAPE–PLD in the conversion of NAPE to AEA is clearly indicated by the unchanged brain levels of AEA in NAPE–PLD knockout mice (Leung et al., 2006). In fact, an independent pathway may occur through a double-deacylation of NAPE to generate *lyso*-NAPE and then glycerophospho-NAE, that is rapidly cleaved to release the corresponding NAE. This novel route is driven by the sequential action of a fluorophosphonate-sensitive serine hydrolase and a metal-dependent phosphodiesterase (Simon and Cravatt, 2006).

The levels of 2-AG in tissues and cells are usually much higher than those of AEA, and in principle they are sufficient to activate both cannabinoid receptor subtypes (Sugiura et al., 1995). At any rate, the 2-AG found in cells and tissues is probably not uniquely used to stimulate cannabinoid receptors, as 2-AG is at the crossroads of several metabolic pathway. It is likely that a particular pool of 2-AG is produced via a special pathway only for the purpose of functioning as endocannabinoid. In line with this hypothesis, the extracellular levels of 2-AG are close to those of AEA, and are in the nanomolar range (Caillè et al., 2007).

A biosynthetic pathway for 2-AG provides for quick hydrolysis of inositol phospholipids by a specific PLC, generating 1-acyl-2-arachidonoylglycerol (DAG) (Di Marzo et al., 1999; Piomelli et al., 1998). DAG is then converted to 2-AG by a *sn*-1-DAG lipase (Stella et al., 1997; Bisogno et al., 2003). Another pathway for 2-AG formation involves the hydrolysis of phosphatidylinositol (PI) by $PLA_1$ into lysoPI, followed by hydrolysis by phospholipase C (PLC) to produce 2-AG (Sugiura and Waku, 2000b). Furthermore, 2-AG has been shown to be produced also by PLC-independent pathways (Bisogno et al., 1999b). Very recently, two *sn*-1-specific DAG lipases ($\alpha$ and $\beta$) responsible for the synthesis of 2-AG have been cloned by comparing human genome with Penicillium DAGL sequence. Both DAGL $\alpha$ and $\beta$ are associated with the cell membrane and are stimulated by high concentrations of $Ca^{2+}$ and, remarkably, by physiological concentrations of glutathione (Bisogno et al., 2003).

#### 4.1.2.3 Degradation of Endocannabinoids

The endocannabinoid actions are relatively short-lasteing, due to the presence of effective mechanisms for their cellular removal and subsequent degradation. Because they are lipophilic compounds, endocannabinoids can diffuse through the cell membrane. However, in order to be rapid, selective and subject to regulation, the diffusion process needs to be facilitated by a carrier, or to be driven by a mechanism capable of rapidly reducing the intracellular concentration of endocannabinoids, or both.

Indeed, AEA appears to be taken up by several cells via a facilitated transport mechanism, possibly mediated by a purported anandamide membrane transporter (AMT) (Fig. 4.3). In fact, cellular uptake of AEA is saturable, temperature-dependent and sensitive to synthetic inhibitors, as expected for a protein-mediated process (Maccarrone et al., 1998, Bisogno et al., 2001a). However, some authors have reported evidence against the existence of AMT,

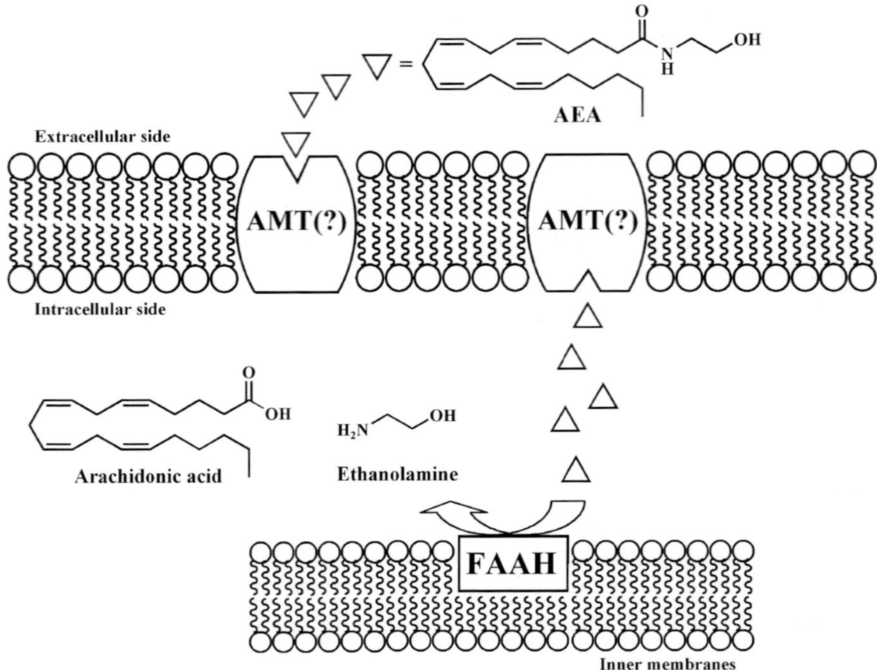


**Fig. 4.3** Pathways of AEA inactivation. Once taken up by a purported transporter on the plasma membrane (AMT), AEA is rapidly cleaved by endomembrane-bound FAAH, realesing arachidonic acid (AA) and ethanolamine (EA)

suggesting that the enzyme mostly responsible for AEA hydrolysis, fatty acid amide hydrolase (FAAH) (Fig. 4.3), may be the sole responsible of AEA cellular uptake, by reducing its intracellular concentration (Bracey et al., 2002; Glaser et al., 2003).

On the other hand, several data are in agreement with a facilitated transport of AEA independent of FAAH. In fact, different cells that do not express FAAH are still able to rapidly take up AEA (Day et al., 2001); compounds that inhibit AEA cellular uptake without affecting FAAH activity have been synthesized (López-Rodríguez et al., 2001; Ortar et al., 2003); saturable AEA accumulation can be still observed in synaptosomes and cells prepared from FAAH-null mice (Ligresti et al., 2004; Fegley et al., 2004). Overall, from the available data it is possible to conclude that FAAH activity can contribute to facilitated AEA transport, yet it is not necessary; other mechanisms different from intracellular hydrolysis may also enhance the rate of endocannabinoid uptake. In line with this, a new model for AEA transport has been proposed, that might engage a caveolae/lipid rafts-related endocytic process (McFarland et al., 2004; Bari et al., 2005). On the other hand, it has been suggested that the 2-AG membrane transporter may be the same used by anandamide, i.e. AMT (Beltramo and Piomelli, 2000; Bisogno et al., 2001b).

Once inside the cell, endocannabinoids are degraded through mechanisms depending on their chemical nature. FAAH has been identified as the main responsible for AEA hydrolysis to arachidonic acid and ethanolamine (Cravatt et al., 1996; Cravatt and Lichtman, 2002) (Fig. 4.3). Although FAAH can catalyze also the hydrolysis of 2-AG (Di Marzo and Deutsch, 1998), the levels of the latter substance, unlike those of AEA, are not increased in FAAH 'knockout' mice (Lichtman et al., 2002). This observation is in agreement with the existence of other enzymes catalyzing 2-AG degradation (Di Marzo et al., 1999; Goparaju et al., 1999a). In fact, monoacylglycerol lipase (MAGL) is a cytosolic enzyme that cleaves efficiently 2-AG (Ben-Shabat et al., 1998; Di Marzo and Deutsch, 1998). In rat brain, MAGL is more abundant in regions where also CB1 receptors are highly expressed (hippocampus, cortex, anterior thalamus and cerebellum). Furthermore, immunohistochemical studies in the hippocampus suggested a presynaptic localization of MAGL, supporting its role in the degradation of 2-AG as retrograde messenger. Interestingly, recent studies have confirmed a sort of 'complementary localization' of MAGL and FAAH in the brain, pre-synaptic and post-synaptic respectively, suggesting different roles for AEA and 2-AG in endocannabinoid signaling within the CNS (Gulyas et al., 2004). Incidentally, the data on MAGL localization supplement previous observations showing that the diacylglycerol lipases (DAGLs) responsible for 2-AG production are instead post-synaptic in the adult brain (Dinh et al., 2002; Bisogno et al., 2003).

## 4.2 Properties of Fatty Acid Amide Hydrolase

The actual enzymes involved in fatty acid amides (FAAs) metabolism remained unknown until the late 1990s, when a rat liver oleamide hydrolase activity was affinity-purified and its cDNA was cloned (Cravatt et al., 1996). Oleamide amidase was connected to AEA hydrolysis, because AEA and oleamide were catalyzed by the same enzyme, called fatty acid amide hydrolase (FAAH; *N*-arachidonoylethanolamine amidohydrolase, EC 3.5.1.4) (Maurelli et al., 1995).

Later on, FAAH was cloned from human and mouse liver (Giang and Cravatt, 1997), and from porcine brain (Goparaju et al., 1999b). All these enzymes are composed of 579 amino acids, and their molecular weights are ~63 kDa. The porcine enzyme shows 80, 81, and 85% identity of the deduced amino acid sequence to mouse, rat and human FAAH respectively (Goparaju et al., 1999b).

In rats, FAAH is mainly distributed in liver, small intestine, brain, testis, uterus, kidney, ocular tissues and spleen, but not in skeletal muscle or heart (Deutsch and Chin, 1993; Desarnaud et al., 1995; Ueda et al., 1995). In humans, the distribution is different: FAAH is mainly detected in pancreas, brain, kidney, skeletal muscle (Giang and Cravatt, 1997), placenta (Park et al., 2003) and is less abundant in liver (Giang and Cravatt, 1997). FAAH activity can also be detected in mouse uterus (Paria et al., 1996), and its expression is regulated during pregnancy (Paria et al., 1999; Maccarrone et al., 2000a).

FAAH is a membrane-bound serine hydrolase, that shows its maximal activity at pH 9 (Cravatt et al., 1996). This enzyme belongs to a protein family called 'amidase signature (AS)' (Chebrou et al., 1996), whose members share a common, conserved amino acid sequence comprising ~130 residues, the so-called 'amidase-signature sequence'. The AS family of enzymes is mainly represented among bacteria and fungi, and FAAH was, until recently, the only known representative of this class of proteins in mammals.

Further studies were conducted with the aim of identifying the primary-sequence of FAAH, to unravel the properties of the region that allows anchoring to the membrane. Although the amino acids in positions 9–29 were predicted with the aid of a sequence-analysis software to constitute the FAAH transmembrane domain, deletion of this segment did not release FAAH from lipid bilayers (Arreaza and Deutsch, 1999; Patricelli et al., 1998). Noteworthy, this so-called transmembrane domain, while not necessary for hydrolase activity, seems to be involved in the self-association of FAAH, because a mutant lacking the first 30 amino acids showed a reduced tendency to form oligomers (Arreaza and Deutsch, 1999).

## 4.2.1 Structural Features

FAAH has been crystallized in complex with an irreversible active site-directed inhibitor, the methoxy arachidonyl fluoro-phosphonate (MAFP), and its three-dimensional structure has been analyzed at a 2.8 Å resolution (Bracey et al., 2002). To obtain the crystalline structure, a catalytically active mutant was generated (ΔTM-FAAH), where the first 29 amino acids were deleted (Patricelli et al., 1998). ΔTM-FAAH is soluble and homogeneous in detergent-containing buffers, opening the avenue to the in vitro mechanistic and structural studies, and is still able to bind membranes (Fig. 4.3). The X-ray structure of this mutant confirmed that FAAH is an integral membrane enzyme with a globular shape: the enzyme crystallized as a homodimer, indicating that it is at least a dimer in solution (McKinney and Cravatt, 2005).

More than 100 members of the AS family of enzymes have been reported in the literature, but only for malonamidase (MAE2) (Shin et al., 2002) and C-terminal peptide amidase (PAM) (Labahn et al., 2002), two soluble bacterial enzymes, structural data are available. All three resolved structures of AS enzymes (FAAH, MAE2, and PAM) revealed a common core, consisting of a twisted β-sheet of 11 mixed strands, surrounded by a large number of α-helices (those of FAAH are shown in Fig. 4.3). Compared to other AS enzymes, which are mostly soluble proteins, FAAH displays two distinguished features: i) integration into membranes, and ii) strong preference for hydrophobic substrates.

Furthermore, three well-defined domains have been identified in FAAH: i) a transmembrane domain at the N-terminus which directs protein oligomerization, ii) a serine- and glycine-rich domain, and iii) a proline-rich domain.

FAAH has several elements of secondary structure: the twisted β-sheet consisting of 11 mixed strands (accounting for ~17% of the whole protein structure) is surrounded by 28 α-helices of various lengths (accounting for ~53% of the whole protein structure). Recently, the stability of ΔTM-FAAH has been studied as a function of chemical (guanidinium hydrochloride) or physical (high hydrostatic pressure) denaturation (Mei et al., 2007). The unfolding transition of the enzyme was observed to be complex and required a fitting procedure based on a three-state process with a monomeric intermediate. The first transition was characterized by dimer dissociation, with a free energy change of ~11 kcal/mol that accounted for ~80% of the total stabilization energy. This process was also paralleled by a large change in the solvent-accessible surface area, because of the hydration occurring both at the dimeric interface and within the monomers. As a consequence, the isolated subunits were found to be much less stable (ΔG ~3 kcal/mol). The addition of MAFP enhanced the stability of the dimer by ~2 kcal/mol, toward denaturant- and pressure-induced unfolding. FAAH inhibition by MAFP also reduced the ability of the protein to bind to the membranes. Taken together, these findings suggest that local conformational changes at the level of the active site might induce a tighter interaction between the subunits of FAAH, thus affecting the enzymatic activity and the interaction with membranes (Mei et al., 2007).

ΔTM-FAAH appears to bind membrane lipids via helices α-18 and α-19 (amino acid 410–438), which form a helix-turn-helix motif. This motif interrupts the AS fold and is comprised mainly of hydrophobic residues (with few basic amino acid) that are likely to constitute a membrane binding surface of FAAH. In addition, a predicted N-terminal transmembrane (TM) domain (amino acids 9–29) forms a membrane binding helix that strengthens the interactions of the α-18 and α-19 helices with membranes (McKinney and Cravatt, 2005). Remarkably, sequence comparisons revealed that this domain is not present in other AS enzymes (Cravatt et al., 1996).

The two monomers of FAAH have a parallel alignment, that allows both subunits to function concomitantly by recruiting substrates from the same membrane. The parallel orientation is required to have the α-18 and α-19 membrane cap on the same face of the dimer, thus enhancing membrane binding (McKinney and Cravatt, 2005). Noteworthy, the intimate relationship between the membrane binding surface and the active site of FAAH resembles the membrane-binding domains of two other integral membrane enzymes, like squalene cyclase (Wendt et al., 1997) and prostaglandin $H_2$ synthase (Picot et al., 1994). Also these enzymes act on lipid-soluble substrates and have hydrophobic caps surrounding the entrance of the corresponding active sites. These three enzymes share no sequence or fold homology, indicating that they have evolved independently similar strategies for membrane integration (Bracey et al., 2004). However, all three enzymes are dimeric proteins, with the active site capped by a hydrophobic domain, that is surrounded by basic amino acids in order to interact with negatively charged phospholipids (McKinney and Cravatt, 2005).

It has been suggested that FAAH may have different structural alterations, allowing direct access from the cytosolic and the membrane side to its active site. In fact, X-ray analysis revealed several unusual features of the enzyme: the resolved crystal structure confirms that FAAH has different key regions, including a remarkable collection of channels that form a 'cytosolic port' and a 'membrane port' to facilitate substrate recognition, binding, hydrolysis and product release (thus improving the catalytic turnover). These ports might grant the simultaneous access to both membrane and cytosolic compartments of the cell, useful for substrate entry and/or product exit during the catalytic reaction (Cravatt and Lichtman, 2003).

A potential substrate entryway (which presents anphipathic residues possibly to accommodate polar substrate head groups towards the FAAH active site) has been identified next to α-18 and α-19 helices, and it may indicate direct connection between the FAAH active site and the hydrophobic membrane bilayer. The mode for membrane binding of FAAH may facilitate movement of the FAA substrates directly from the bilayer to the active site, with no need for transport of these lipids through the aqueous cytosol. In this model, the substrate would first enter via the membrane to the active site; following hydrolysis, the released fatty acid (hydrophobic) and amine (hydrophilic) products would then exit through the membrane-access and cytosolic-access channels, respectively. Moreover, the cytoplasmic port may serve the additional function of providing a way for a water molecule required for deacylation of the FAA-FAAH acyl-enzyme intermediate, which has been already characterized by LC-MS (Patricelli and Cravatt., 1999).

### 4.2.2 Catalytic Mechanism

FAAH presents unique biochemical properties due to an unusual serine-serine-lysine (Ser241-Ser217-Lys142) catalytic triad. In fact, differently from the substrate selectivity displayed by most serine hydrolases, which react with esters at rates several orders of magnitude faster than amides, FAAH reacts with esters and amides at equivalent rates. It has been demonstrated that this unusual property depends on a single lysine residue (Lys142), since its mutation to alanine greatly reduces the amidase activity of FAAH, without affecting the esterase activity (Patricelli and Cravatt., 1999). Many investigations have been focused to clarify the catalytic mechanism of FAAH. A number of mutagenesis, kinetic and chemical labeling studies have revealed that the FAAH nucleophile is Ser241 (Patricelli et al., 1999). Mutagenesis studies also invoked the participation of additional residues in the catalytic mechanism of FAAH: in particular, a serine residue (Ser217) mutated to alanine produced a mutant FAAH with a significant reduction of hydrolytic activity (~2000 fold), reduction that was much less severe than that observed with mutants lacking either the serine nucleophile (Ser241) or the lysine base/acid (Lys142). Remarkably, the unusual catalytic core of FAAH is highly conserved among the AS family members. Lys142 appears to play a critical role as both base and acid in the hydrolytic cycle. In fact, several lines of experiments show that Lys142 plays a role as a base that activates the Ser241 nucleophile in FAAH, whereas other kinetic data seem to support a role for Lys142 as an acid that participates in the protonation of the substrate leaving group (Patricelli and Cravatt., 1999). The relative importance of acid-catalyzed leaving group protonation for amide hydrolysis compared to ester hydrolysis has been emphasized previously in semi-empirical studies (Fersht, 1971): consistent with these predictions, a tight coupling of base-catalyzed nucleophile activation and acid-catalyzed leaving group protonation might explain the ability of FAAH to normalize the acylation/hydrolysis rates of an amide or an ester substrate (McKinney and Cravatt, 2005). This hypothetical mechanism assumes that Lys142 would be deprotonated in the absence of bound substrate, leading to a constitutively activated nucleophile (Ser241). The structural arrangement analysis of catalytic residues indicates that in FAAH the impact of Lys142 on Ser241 nucleophile strands and the leaving group protonation likely occurs indirectly, via the bridging Ser217 of the triad; the latter may act as a 'proton shuttle'. In this model, FAAH would force protonation of the substrate leaving group early in the transition state of acylation, concomitantly with the nucleophile attack on the substrate carbonyl group (McKinney and Cravatt, 2005). The overall catalytic cycle of FAAH is shown in Fig. 4.4.

It should be pointed out that the comparable hydrolysis rate for amide and ester bonds has a biological meaning: FAAH must bind and hydrolyze its FAA substrates against a background of a large excess of structurally related esters such as monoacylglycerols (Mechoulam et al., 1995). To reach this goal, the

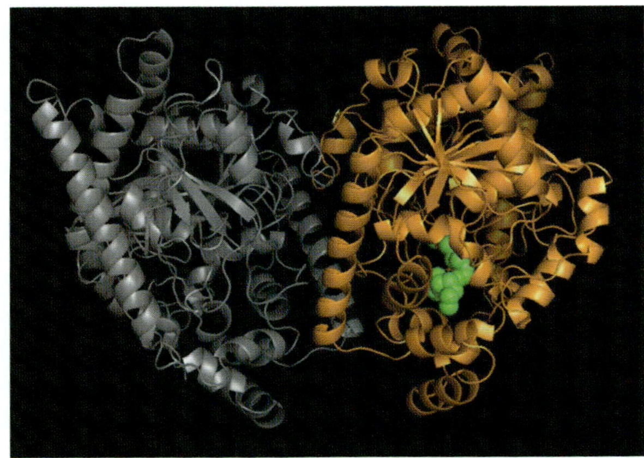

**Fig. 4.4** Schematic representation of rat ΔTM-FAAH, complexed with the irreversible inhibitor MAFP (*in green*) (Protein Data Bank file 1MT5)

active site of FAAH has specifically evolved and adapted to hydrolysis of FAA substrates in a cellular environment with high concentration of fatty acid esters. Therefore, the unique biochemical proprieties of FAAH permit to this enzyme to act as a lipid amidase in vivo.

### *4.2.3 Synthetic and Natural Inhibitors*

Growing evidence demonstrates that FAAH is the critical regulator of the endogenous levels of AEA, suggesting that it may serve as an attractive therapeutic target for the treatment of human disorders (Maccarrone, 2006). Unfortunately, AEA is rapidly inactivated by FAAH, which prevents its therapeutic exploitation. Yet, inhibitors of FAAH that block degradation of AEA and related endocannabinoids might be useful to tackle pathologies in which endocannabinoid levels are reduced.

The first non-specific inhibitor reported for FAAH was the serine protease inhibitor phenylmethylsulfonyl fluoride (PMSF) (Table 4.1) (Deutsch and Chin, 1993). Since this compound was not selective for FAAH, there was a growing interest to design more potent and selective inhibitors. New compounds were obtained from the derivatization of various fatty acids with functional groups, previously reported to react and form covalent adducts with catalitycally active serine and cysteine residues. This method allowed the discovery of novel FAAH inhibitors, like diazomethylarachidonoyl ketone (Edgemond et al., 1998), stearylsulfonyl fluoride (Deutsch et al., 1997a), methyldodecyl fluorophosphonate (Martin et al., 2000), arachidonylsulfonyl fluoride (Segall et al., 2003), and the most potent methoxy arachidonoyl fluorophosphonate (MAFP) (Table 4.1) (Deutsch et al.,1997b; De Petrocellis et al., 1997). All these compounds are potent irreversible inhibitors of FAAH, however, they also have remarkable affinity for the CB1 receptor.

More recently, a series of irreversible aryl-carbamates inhibitors were described (Mor et al., 2004). URB597 (cyclohexyl carbamic acid 3'-carbamoylbiphenyl-3-yl ester) (Table 4.1), the most potent member of this family, inhibited FAAH activity with an $IC_{50}$ value of 4.6 nM in rat brain extracts (Mor et al., 2004), and of 0.5 nM in intact neurons (Piomelli et al., 2006), without affecting other serine hydrolases. In addition, introduction of small polar groups in meta-position of the distal phenyl ring, and in para-position of the proximal phenyl ring, were found to improve inhibition (Mor et al., 2004; Tarzia et al., 2006). These carbamates inhibit FAAH activity through irreversible interaction based on nucleophilic attack of Ser241 in the active site. Biochemical evidence (Alexander and Cravatt, 2005) showed that these inhibitors covalently modify the active site by adopting an orientation opposite of that originally predicted from modeling (Mor et al., 2004). Indeed, the *O*-biaryl substituents would reside in the cytoplasmic-access channel (rather than in the acyl-chain-binding channel), where they would be susceptible to enzyme-catalyzed protonation to

**Table 4.1** Chemical structures of relevant-FAAH inhibitors

| | |
|---|---|
| 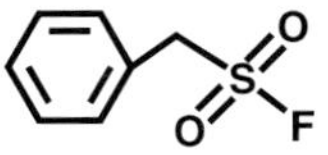 **PMSF** | 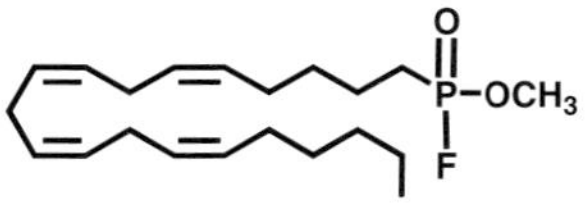 **MAFP** |
| 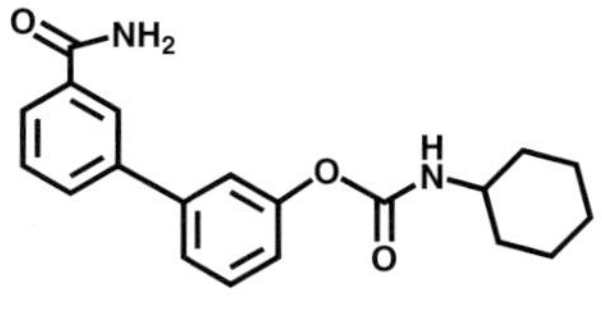 **URB597** | 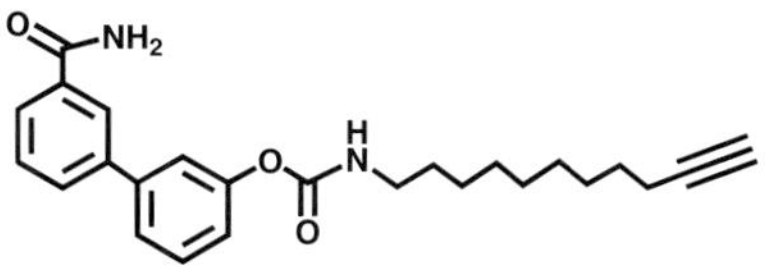 **JP-104** |
| 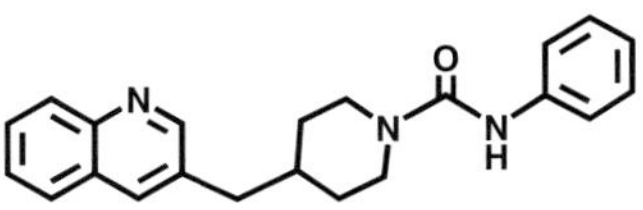 **PF-750** | 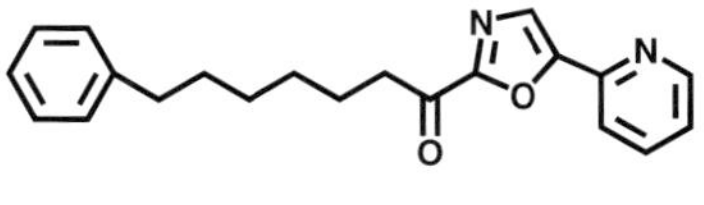 **OL-135** |
| 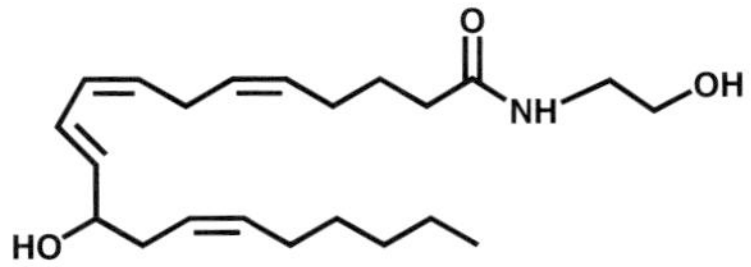 **12-OH-AEA** | 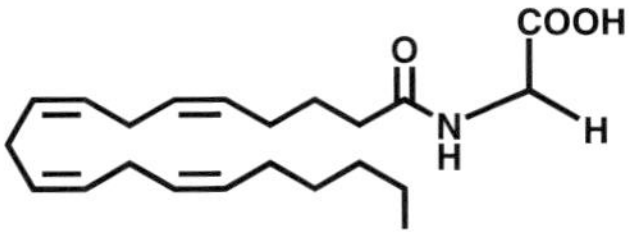 **NAGLy** |

enhance their function as leaving groups. Based on these results, a series of carbamates were designed, in which the *N*-cyclohexyl unit was replaced with various *N*-alkyl groups mimicking the acyl chains of anandamide. These compounds, of which JP-104 (Undec-10-ynyl-carbamic acid 3'-carbamoyl-biphenyl-3-yl ester) is a prototype member (Table 4.1), generally exhibited enhanced potency (Alexander and Cravatt, 2005). More recently, Ahn and coworkers have described a new series of 'PF' urea-based inhibitors with piperidine/piperazine groups [see PF-750 (N-phenyl-4-(quinolin-3-ylmethyl)piperidine-1-carboxamide) in Table 4.1]. These compounds have been shown to covalently inactivate FAAH via carbamylation of the serine nucleophile in the active site, and did not show any detectable activity against other serine hydrolases in mammalian proteomes (Ahn et al., 2007).

In general, an irreversible mechanism of inhibition might reduce the versatility of a drug for in vivo applications. Thus, a major challenge for the ongoing pharmaceutical research is the development of potent and selective, but reversible, inhibitors of FAAH. Based on α-ketoheterocycle protease inhibitors (Edwards et al., 1995), potent reversible competitive inhibitors were developed, combining an unsaturated acyl chain and an α-keto-*N*4-oxazolopyridine, with incorporation of a second weakly basic N-atom. This class of compounds showed potency in the subnanomolar range, with Ki values falling below 200 pM (Boger et al., 2000). The inhibition potency was strongly dependent on the hydrophobicity of the flexible acyl chain, and on the degree of α-substitution (Boger et al., 2001). For example, the compound OL-135 (1-oxo-1[5-(2-pyridyl)-2-yl]-7-phenylheptane) (Table 4.1) displayed an exceptional combination of high potency (Ki = 4.7 nM towards rat-recombinant FAAH) and high selectivity in vivo (Lichtman et al., 2004; Boger et al., 2005).

Several inhibitors have been tested in vivo and their ability to inactivate FAAH was shown to elicit pain and anxiety (Kathuria et al., 2003), without the side effects (hypomotility, hypothermia and catalepsy) that usually accompany activation of CB1 receptors by exogenous cannabinoids like $\Delta^9$-THC (Piomelli et al., 2000; Fowler, 2003; Cravatt and Lichtman, 2003). To maintain this lack of 'cannabinoid side effects', FAAH inhibitors must be devoid of affinity for the cannabinoid receptors. The most studied of the FAAH inhibitors, URB597, does not exhibit affinity for cannabinoid receptors. Thus, at doses that inhibit FAAH and substantially raise brain levels of AEA, but not of 2-AG, this compound did not induce common side effects of typical CB1 agonists, suggesting that they it might be exploited as an innovative anti-anxiety therapeutic (Gaetani et al., 2003).

Also OL-135 (see above) allowed a profound increase of anandamide levels in the brain and spinal cord, and displayed CB1-dependent antinociceptive effects in the hot-plate, tail-immersion, and formalin tests (Lichtman et al., 2004).

Besides the design of synthetic molecules, several papers reported the presence of specific enzymatic reactions able to produce also in the cells compounds able to act as reversible FAAH inhibitors (Maccarrone and Finazzi-Agrò, 2004a). In particular, it has been found that oxidative metabolites of AEA generated by various lipoxygenases, i.e. the hydroxyanandamides (HAEAs; see 12-OH-AEA in Table 4.1), are powerful inhibitors of FAAH (van der Stelt et al., 2002). Instead, derivatives of AEA generated by cyclooxygenase-2, and termed prostamides, have been recently shown to be ineffective on FAAH activity (Matias et al., 2004).

Of interest is the fact that all HAEAs are reversible competitive inhibitors of FAAH (van der Stelt et al., 2002). In addition, the fact that various lipoxygenases (i.e., 5-, 12-, and 15-LOXs) generate different HAEAs with different inhibition profiles towards the proteins of the ECS, suggests that cells with different LOXs might contribute different selectivity to networks regulating endocannabinoids' actions. These compounds may be the 'physiological' inhibitors of FAAH, of potential utility in the control of emotional states and of those disorders whose onset or symptoms are associated with defective production or excessive degradation of AEA and congeners. More in general, it should be pointed out that HAEAs represent one of the newest paradigms of the ability of cells to make their own tools to regulate key targets like FAAH, and they seem to do it more simply than is done in the laboratory.

Moreover, the natural occurrence in mammalian tissues of several arachidonoylated amino acids prompted different research groups to evaluate the effects of these compounds as FAAH inhibitors. It has been shown that these derivatives may act as additional regulatory factors for FAAH. A typical example of these mediators is the *N*-arachidonoyl-glycine (NAGly) shown in Table 4.1 (Huang et al., 2001). NAGly has been shown to exert both analgesic and anti-inflammatory effects, despite its lack of activity at both CB1 and CB2 receptors, through inhibition of FAAH (Huang et al., 2001; Burstein et al., 2002).

## 4.2.4 Subcellular Localization

FAAH has been found mainly in microsomal and mitochondrial fractions of rat brain and liver (Deutsch and Chin, 1993; Desarnaud et al., 1995), and of porcine brain (Ueda et al., 1995). Recent studies performed with confocal microscopy, showed that FAAH is localized intracellularly as a vesicular-like staining, that has no association with the plasma membranes and is partially co-localized with the endoplasmic reticulum (Fig. 4.5). These morphological data were corroborated by biochemical assays of FAAH activity in subcellular fractions, showing that AEA hydrolysis was primarily confined to the endomembrane compartment (Oddi et al., 2005). Moreover, by means of reconstituted vesicles derived from purified membrane fractions, it was demonstrated that transport activity is retained by plasma membrane vesicles devoid of FAAH, thereby indicating that AEA hydrolase activity is not necessary for AEA membrane transport. Overall, by means of confocal microscopy, subcellular fractionation, and

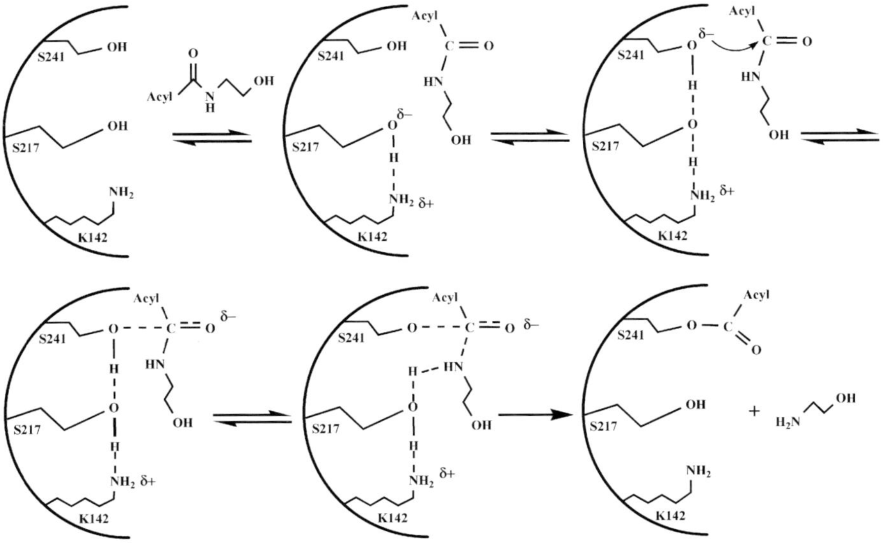

**Fig. 4.5** Proposed mechanism for amide hydrolysis by FAAH. See text for details

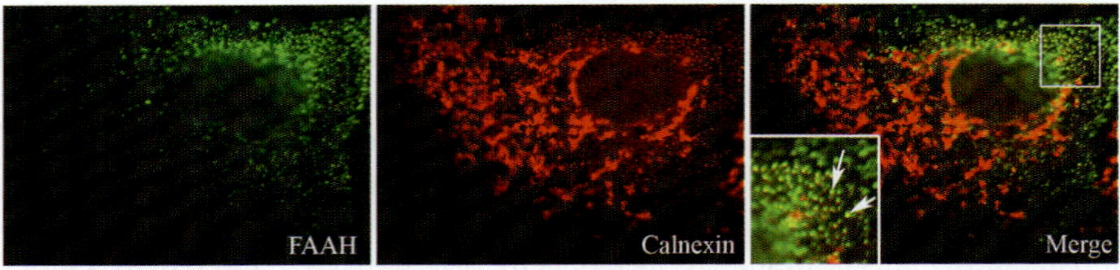


**Fig. 4.6** Cellular localization of FAAH in human keratinocytes. Co-localization of FAAH with calnexin (marker to endoplasmic reticulum). Human keratinocytes (HaCaT cells) were co-stained with anti-FAAH (*in green*) and anti-calnexin (*in red*) antibodies. Superimposition of the two stainings (*merge*) revealed a vesicular region of the endoplasmic reticulum where FAAH and calnexin largely overlapped (*yellow*). Dot structures, where FAAH and calnexin co-localized, are indicated by the white arrows in the inset at the bottom of the merge panel. The remaining part of the reticulum, with lamellar appearence, did not display any co-localization of the two proteins. Courtesy of Dr. Sergio Oddi (University of Teramo, Italy)

biochemical analysis it can be demonstrated, at least in some cell types, that transport and hydrolysis of AEA are uncoupled also in cells with a normal genetic background for FAAH. Therefore, it can be concluded that the transport and the hydrolysis steps are two spatially and functionally independent events of the AEA inactivation pathway.

### *4.2.5 Regulation of Gene Expression*

The genomic organization of mouse and human *FAAH* genes was reported in 1998 (Wan et al., 1998). In humans, the *FAAH* gene is localized to chromosome 1p, while in mice it is on chromosome 4 (Wan et al., 1998). The genomic configuration of the human and mouse *FAAH* exons is highly conserved, with 15 exons ranging in size from 40–207 bp. Each splice donor and acceptor sites are conserved and, with the exception of two introns (2 and 7), even the intron size of human and mouse *FAAH* gene is conserved (Wan et al., 1998). With the mouse genomic organization completed, another DNA region amenable to study was the *FAAH* promoter, and in fact a number of investigators turned their attention to understanding how *FAAH* gene expression is regulated.

A number of studies showed that in the periphery, estrogen and progesterone would in part regulate *FAAH* gene expression (Paria et al., 1996; Maccarrone et al., 2000a; Maccarrone et al., 2001). In 2001, a mouse *FAAH* promoter analysis using neuronal and muscle cell lines suggested that a tissue-specific expression of *FAAH* is accomplished via elements within 700 bp of the *FAAH* initiation codon ATG (Puffenbarger et al., 2001). In 2002, further mouse *FAAH* promoter analysis was published and while these first two studies did not mark identical sites for the start site of mouse *FAAH* mRNA (+1 of transcription), neither group identified a TATA box which might explain the variation of transcription initiation sites in brain and liver (Puffenbarger et al., 2001; Waleh et al., 2002). Human *FAAH* promoter studies have been performed in human T lymphocytes, where leptin and progesterone activation of *FAAH* transcription was shown to occur via STAT3 (signal transduction and activator of transcription-3) and Ikaros transcription factors, respectively (Maccarrone et al., 2003a; Maccarrone et al., 2003b). Neither the human nor mouse *FAAH* promoters seem to have an active TATA box, while both contain several SP1 binding sites (Maccarrone et al., 2003b). The essential elements of human *versus* mouse *FAAH* promoter are schematically represented in Fig. 4.7. Interestingly, later on the human *FAAH* promoter was examined in lymphoma U937 *versus* neuroblastoma CHP100 cells (Maccarrone et al., 2004b). It was found that, while leptin and progesterone strongly enhanced *FAAH* promoter activity in lymphoma cells, neither leptin nor progesterone (alone or in combination) significantly changed *FAAH* expression in neuroblastoma cells, suggesting significant differences in the response of *FAAH* promoter sequences along the neuroimmune axis (Maccarrone et al., 2004b). Further work will be needed to understand the tissue-specific regulation of *FAAH* gene, and to determine which transcription factors drive FAAH expression within the CNS. Nowadays, with *FAAH* promoter sequences outlined and its genomic organization completed, another avenue of research could be to 'knockout' or silence *FAAH* gene expression, by using homologous recombination to target the *FAAH* gene. It seems noteworthy that tissue extracts from *FAAH*(−/−) mice displayed 50–100-fold lower hydrolysis rates towards AEA and other FAAs, indicating that FAAH is indeed the primary enzyme responsible for the hydrolytic degradation of these lipids in vivo. Consistent with this premise, the pharmacological administration of AEA produced greatly exaggerated behavioral effects in *FAAH*(−/−) mice, compared to wild-type littermates, including hypomotility, analgesia, hypothermia, and catalepsy. All of the effects of AEA in *FAAH* (−/−) mice were blocked by a CB1R antagonist, indicating that this substance acts as a selective CB1R ligand in these animal models (Lichtman et al., 2002).

### 4.2.6 *FAAH-1* versus *FAAH-2*

Recent proteomic data suggest the existence of a second mammalian AS enzyme with FAAH activity, called FAAH-2 (Wei et al., 2006). The FAAH-2

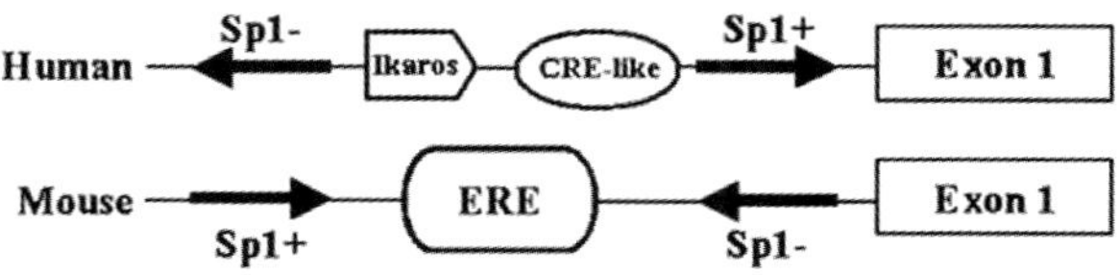


**Fig. 4.7** The *FAAH* promoter. Comparison of the upstream transcription factor binding sites of human and mouse FAAH promoters (not drawn to scale). Arrows indicate SP-1 binding sites on the − (←) or + (→) strand. An Ikaros (Ik) binding site, a cyclicAMP responsive-like element (CRE), and an estrogen-responsive element (ERE) are schematically represented in boxes

gene was found in primates and in distantly related vertebrates but not in rodents like mice and rats. This enzyme exhibits an overlapping but distinct tissue distribution, substrate selectivity, and inhibitor sensitivity compared to the original FAAH enzyme (FAAH-1, discussed above). Both FAAH-1 and FAAH-2 share 20% amino acid sequence identity. Similarly to FAAH-1, FAAH-2 possesses an N-terminal transmembrane domain and an AS sequence containing the serine-serine-lysine catalytic triad, along with other amino acid residues required for enzyme activity (Wei et al., 2006). Interestingly, the C-terminal catalytic domains of FAAH-1 and FAAH-2 would be located in the cytoplasmic and luminal compartments of the cell, respectively, suggesting that the opposite relative orientation of these enzymes within the membrane could influence their respective access to specific FAA substrates in the cell, especially if these lipids show preferential localization to the inner or outer leaflet of the membrane bilayer (Wei et al., 2006).

Comparison of the enzymatic properties of FAAH-1 and FAAH-2 revealed that FAAH-1 has much higher hydrolytic activity than FAAH-2, with AEA (C20:4) as substrate. This differential activity contrasted with the similar rates of hydrolysis displayed by the two enzymes with oleamide (C18:10) and linoleamide (C18:1) FAAs. FAAH-2 thus appears to prefer monounsaturated over polyunsaturated acyl chains, while FAAH-1 exhibits the opposite selectivity.

These observations indicate that FAAH-2 may be important for the regulation of monounsaturated lipid amides in the CNS and peripheral tissues; however, further investigation is needed (Wei et al., 2006). To date, in addition to FAAH-1 and FAAH-2, other enzymes have been shown to be involved in the termination of endocannabinoid signaling, like the above-discussed MAGL and the *N*-acylethanolamine acid amidase (NAAA) (Tsuboi et al., 2005). The involvement of these additional hydrolases in regulating endocannabinoid tone needs to be clarified, and raises a question about the biological meaning of such a diversity of metabolic pathways (Vandevoorde et al., 2005; Dinh et al., 2004; Muccioli et al., 2007). Indeed, these proteins are able to hydrolyze, at least in vitro, a wide and overlapping panel of endocannabinoids, but the precise role played by each enzyme in vivo still remains unclear. Unfortunately, the lack of selective inhibitors to be used as pharmacological tools, as well as of knockout mice models, impairs a thorough characterization of the patho-physiological roles of these enzymes.

## 4.3 Involvement of Faah in Health and Disease

Experimental observations suggest that altered levels of endocannabinoids are associated with several physiopathological conditions, therefore the role of the endocannabinoid system is currently the subject of intense investigation. In particular several studies have provided strong evidence that FAAH, due to its broad distribution, represents an attractive therapeutic target for the treatment of many diseases in both the CNS and the periphery (Maccarrone, 2006).

In the periphery, endocannabinoid signaling was soon identified as crucial for uterine receptivity for embryo implantation in mouse, with high AEA levels associated to an impairment of the implantation process. This effect was inhibited by the antagonist SR141716, indicating that anandamide is acting through CB1 receptors (Paria et al., 1996). The relevance of the effects of AEA on early pregnancy and on neuroendocrine function underlined the possible leading role played by FAAH in reproduction. In humans, decreased FAAH expression and activity, as well as elevated AEA concentrations in peripheral lymphocytes, are correlated with spontaneous abortion (Maccarrone et al., 2000b). Accordingly, the recent article on *FAAH(−/−)* mice outlined the capital importance of FAAH in the (pre-)implantation process and fertility (Wang et al., 2006). Indeed, an increase in AEA levels in knockout mice resulted in altered oviductal embryo transport and expression of genes required for differentiation and blastocyst implantation, ultimately leading to impairment of fertility. Based on the available data, it can be proposed that drugs that are able to enhance FAAH activity (e.g., by mimicking Ikaros or STAT3) might become useful therapeutic tools to correct defects in human fertility (Maccarrone and Finazzi-Agrò, 2004a).

Besides lymphocytes, other blood cells express an ECS that plays an important role in human pathologies. For instance, in platelets isolated from humans with headache, Cupini et al. (2006) found an increase in the activity of FAAH in two groups of headache subjects: migraine without aura (MoA) or episodic tension-type headache (ETTH) patients. In particular, this FAAH dysfunction was found in female but not male migraineurs. Furthermore, in a recent paper it has been reported a decrease of AEA and 2-AG levels in other two types of headache disorders (Rossi et al., 2007).

The alteration of the endocannabinoid system is implicated also in a number of human behaviors that seems to be, at least in part, determined by genetics. In fact, several articles have shown an association of genetic polymorphisms of FAAH with disease conditions (Norrod and Puffenbarger, 2007). Genetic polymorphisms are variations in DNA sequences from person to person. Polymorphic sequences may be single nucleotide polymorphisms (SNPs) or larger changes, including differences in the number of trinucleotide repeats.

In particular, the first mutations in the human FAAH sequence were discovered in 2002 (Sipe et al., 2002). In this study, it was described a natural SNP in the human gene that encodes for FAAH, that in homozygous form is strongly associated with risk factors for problem alcohol and drug use. This single nucleotide polymorphism results in a missense mutation (385C→385A), that converts a conserved proline residue to threonine (Pro129→Thr), producing a FAAH variant that displays normal catalytic properties but an enhanced sensitivity to proteolytic degradation. In the same line, studies on human T lymphocytes from genotyped blood donors, have revealed that 385A/385A donors who would have only Pro129→Thr type FAAH had less than half the

FAAH activity found in 385C/385C wild-type donors. Very interestingly, in a study of Japanese methamphetamine users, no significant linkage was found between the 385C→A and methamphetamine use (Chiang et al., 2004). Thus, it remains to be seen how the FAAH 385C→A polymorphism might affect the risk of drug and alcohol abuse in other ethnic groups.

The FAAH polymorphism has been associated also with eating disorders in subjects of multiple ethnic backgrounds. In fact, the homozygous FAAH 385A/A genotype was significantly associated with overweight and obesity in white subjects and in black subjects, but not in a small group of Asians. The median Body Mass Index (BMI) for all subjects was significantly greater in the FAAH 385A/A genotype group compared to heterozygote and wild-type groups. In white subjects, there was an increasing frequency of the FAAH 385A/A genotype with increasing BMI categories of overweight and obese, with the same trend in black subjects (Sipe et al., 2005).

In another study a group of 451 obese and dyslipidaemic participants were genotyped, and their biometric and metabolic outcome after a 6 week low fat diet was observed. Carriers of the P129T mutation in FAAH had a significantly great decrease in triglycerides and total cholesterol, as compared to wild-type. These findings remain to be elucidated, however, a hepatic down-regulation of endocannabinoid tone may contribute to the observed outcome in studied subjects (Aberle et al., 2007).

Several lines of evidence suggest that the endocannabinoid system is profoundly involved in neurodegenerative diseases, like Alzheimer's disease (AD), Huntington's disease (HD) and multiple sclerosis (MS). In fact, Benito and co-workers reported that FAAH expression and activity, and CB2 receptor expression are selectively enhanced in glial cells that are linked to the inflammatory process that accompanies AD. This is the first observation in a human tissue that suggests a role for the ECS in the progression of this neurodegenerative disease. FAAH exhibits upregulation in glial cells associated with senile plaques and its expression appears to be restricted to reactive astrocytes. While CB2 receptors are expressed only in activated microglial cells (Benito et al., 2003).

Alterations in FAAH activity seem to be implicated also in HD. Peripheral lymphocytes of HD patients are known to express mutant huntingtin (htt) (Ide et al., 1995), and to replicate some of the transcriptional abnormalities found in HD brain (Borovecki et al., 2005). According to that in a recent paper lymphocytes were used and it has been observed that FAAH activity (but not its expression) was dramatically decreased in HD compared to healthy subjects. In addition, FAAH activity was also decreased to approximately 50% of controls in specimens from HD brains, arguably suggesting that diminished FAAH activity in lymphocytes and brains are correlated (Battista et al., 2007).

A recent study has demonstrated that AEA, but not 2-AG, is increased in the cerebrospinal fluid of MS patients (Centonze et al., 2007a). Remarkably, increased synthesis, reduced degradation and increased levels of AEA were also observed in lymphocytes of MS patients, indicating that an important source of AEA in the CNS of these subjects may be represented by activated inflammatory cells invading the brain from the periphery. Furthermore, by means of neurophysiological recordings from single neurons, it was confirmed that excitatory transmission is inhibited by CB1 receptor activation in mice with experimental autoimmune encephalomyelitis (EAE), a preclinical model of MS, and that GABA transmission becomes conversely insensitive to CB1 receptor stimulation (Centonze et al., 2007a). Overall, together with previous experimental findings, these results suggest that during immuno-mediated attack of the CNS, the activation of ECS represents a protective mechanism aimed at reducing both neurodegenerative and inflammatory damage through various and partially converging mechanisms that involve neuronal and immune cells (Centonze et al., 2007b).

## 4.4 Conclusions and Future Perspectives

In this chapter we have outlined the structural properties of FAAH, its catalytic mechanism, its gene expression and its pathophysiological roles, against the wider background of the endocannabinoid system to which FAAH belongs. Since there is a general consensus that endocannabinoid tone and activity are under a 'metabolic control', and that FAAH is the key-player in this process (McKinney and Cravatt, 2005; Maccarrone, 2006), it can be proposed that this enzyme could be viewed as a possible important target for the development of new drugs.

It seems that modulating endocannabinoid metabolism, rather than agonizing or antagonizing cannabinoid and noncannabinoid receptors, might be the way to better understand the pathophysiological implications of these bioactive lipids, and to exploit them for therapeutic purposes. In this context, it can be suggested that not only inhibitors of FAAH, but also drugs able to enhance its activity might become useful therapeutic tools for the treatment of human diseases. If not as therapeutic agents *per se*, FAAH inhibitors or activators could be used together with AEA analogues to lower the doses or to shorten the treatment necessary in vivo to observe an effect, and hence to minimize the possible psychotropic side effects of endocannabinoids when they are used as pharmaceuticals.

On a final note, it seems necessary to remind that: i) several endogenous endocannabinoid(-like) compounds, whose functions are not yet understood, are present in our body, and their biological activity might be affected in unexpected ways by drugs that modulate FAAH or other known proteins of the endocannabinoid system; ii) new metabolic enzymes have been recently identified, that catalyze the hydrolysis and synthesis of AEA or 2-AG, and it remains to be elucidated how these multiple pathways may contribute to the overall tone and biological activity of endocannabinoids.

## References

Adams, I. B. and Martin, B. R., Cannabis: pharmacology and toxicology in animals and humans, Addiction, 91 (1996) 1585–1614.

Aberle, J., Fedderwitz, I., Klages, N., George, E. and Beil, F. U., Genetic variation in two proteins of the endocannabinoid system and their influence on body mass index and metabolism under low fat diet, Horm Metab Res, 39 (2007) 395–397.

Aguado, T., Carracedo, A., Julien, B., Velasco, G., Milman, G., Mechoulam, R., Alvarez, L., Guzmán, M. and Galve-Roperh, I., Cannabinoids induce glioma stem-like cell differentiation and inhibit gliomagenesis, J Biol Chem, 282 (2007) 6854–6862.

Ahn, K., Johnson, D. S., Fitzgerald, L. R., Liimatta, M., Arendse, A., Stevenson, T., Lund, E. T., Nugent, R. A., Nomanbhoy, T. K., Alexander, J. P. and Cravatt, B. F., Novel mechanistic class of fatty acid amide hydrolase inhibitors with remarkable selectivity, Biochemistry, 46 (2007) 13019–13030.

Alexander, J. P. and Cravatt, B. F., Mechanism of carbamate inactivation of FAAH: implications for the design of covalent inhibitors and in vivo functional probes for enzymes, Chem Biol, 12 (2005) 1179–1187.

Arreaza, G. and Deutsch, D. G., Deletion of a proline-rich region and a transmembrane domain in fatty acid amide hydrolase, FEBS Lett, 454 (1999) 57–60.

Baker, D., Pryce, G., Davies, W. L. and Hiley, C. R., In silico patent searching reveals a new cannabinoid receptor, Trends Pharmacol Sci, 1 (2006) 1–4.

Bari, M., Battista, N., Fezza, F., Finazzi Agrò, A. and Maccarrone, M., Lipid rafts control signaling of type-1 cannabinoid receptors in neuronal cells. Implications for anandamide-induced apoptosis, J Biol Chem, 280 (2005) 12212–12220.

Battista, N., Bari, M., Tarditi, A., Mariotti, C., Bachoud-Lévi, A. C., Zuccato, C., Finazzi-Agrò, A., Genitrini, S., Peschanski, M., Di Donato, S., Cattaneo, E. and Maccarrone, M., Severe deficiency of the fatty acid amide hydrolase (FAAH) activity segregates with the Huntington's disease mutation in peripheral lymphocytes, Neurobiol Dis, 27 (2007) 108–116.

Beltramo, M. and Piomelli, D., Carrier-mediated transport and enzymatic hydrolysis of the endogenous cannabinoid 2-arachidonylglycerol, Neuroreport, 11 (2000) 1231–1235.

Benito, C., Núñez, E., Tolón, R. M., Carrier, E. J., Rábano, A., Hillard, C. J. and Romero, J., Cannabinoid CB2 receptors and fatty acid amide hydrolase are selectively overexpressed in neuritic plaque-associated glia in Alzheimer's disease brains, J Neurosci, 23 (2003) 11136–11141.

Ben-Shabat, S., Fride, E., Sheskin, T., Tamiri, T., Rhee, M. H., Vogel, Z., Bisogno, T., De Petrocellis, L., Di Marzo, V. and Mechoulam, R., An entourage effect: inactive endogenous fatty acid glycerol esters enhance 2-arachidonoyl-glycerol cannabinoid activity, Eur J Pharmacol, 353 (1998) 23–31.

Bisogno, T., Berrendero, F., Ambrosino, G., Cebeira, M., Ramos, J. A., Fernandez-Ruiz, J. J. and Di Marzo, V., Brain regional distribution of endocannabinoids: implications for their biosynthesis and biological function, Biochem Biophys Res Commun, 256 (1999a) 377–380.

Bisogno, T., Melck, D., De Petrocellis, L. and Di Marzo, V., Phosphatidic acid as the biosynthetic precursor of the endocannabinoid 2-arachidonoylglycerol in intact mouse neuroblastoma cells stimulated with ionomycin, J Neurochem, 72 (1999b) 2113–2119.

Bisogno, T., Hanus, L., De Petrocellis, L., Tchilibon, S., Ponde, D. E., Brandi, I., Moriello, A. S., Davis, J. B., Mechoulam, R. and Di Marzo, V., Molecular targets for cannabidiol and its synthetic analogues: effect on vanilloid VR1 receptors and on the cellular uptake and enzymatic hydrolysis of anandamide, Br J Pharmacol, 134 (2001a) 845–852.

Bisogno, T., Maccarrone, M., De Petrocellis, L., Jarrahian, A., Finazzi-Agro, A., Hillard, C. and Di Marzo, V., The uptake by cells of 2-arachidonoylglycerol, an endogenous agonist of cannabinoid receptors, Eur J Biochem, 268 (2001b) 1982–1989.

Bisogno, T., Howell, F., Williams, G., Minassi, A., Cascio, M. G., Ligresti, A., Matias, I., Schiano-Moriello, A., Paul, P., Williams, E. J., Gangadharan, U., Hobbs, C., Di Marzo, V. and Doherty, P., Cloning of the first sn1-DAG lipases points to the spatial and temporal regulation of endocannabinoid signaling in the brain, J Cell Biol, 163 (2003) 463–468.

Boger, D. L., Sato, H., Lerner, A. E., Hedrick, M. P., Fecik, R. A., Miyauchi, H., Wilkie, G. D., Austin, B. J., Patricelli, M. P. and Cravatt, B. F., Exceptionally potent inhibitors of fatty acid amide hydrolase: the enzyme responsible for degradation of endogenous oleamide and anandamide, Proc Natl Acad Sci U S A, 97 (2000) 5044–5049.

Boger, D. L., Miyauchi, H. and Hedrick, M. P., Alpha-Keto heterocycle inhibitors of fatty acid amide hydrolase: carbonyl group modification and alpha-substitution, Bioorg Med Chem Lett, 11 (2001) 1517–1520.

Boger, D.L., Miyauchi, H., Du, W., Hardouin, C., Fecik, R. A., Cheng, H., Hwang, I., Hedrick, M. P., Leung, D., Acevedo, O., Guimarães, C. R., Jorgensen, W. L. and Cravatt, B. F., Discovery of a potent, selective, and efficacious class of reversible alpha-ketoheterocycle inhibitors of fatty acid amide hydrolase effective as analgesics, J Med Chem, 48 (2005) 1849–1856.

Börner, C., Höllt, V., Sebald, W. and Kraus, J., Transcriptional regulation of the cannabinoid receptor type 1 gene in T cells by cannabinoids, J Leukoc Biol, 81 (2007) 336–343.

Borovecki, F., Lovrecic, L., Zhou, J., Jeong, H., Then, F., Rosas, H. D., Hersch, S. M., Hogarth, P., Bouzanou, B., Jensen, R. V., Krainc, D., Genome-wide expression profiling of human blood reveals biomarkers for Huntington's disease, Proc Natl Acad Sci U S A, 102 (2005) 11023–11028.

Bracey, M. H., Hanson, M. A., Masuda, K. R., Stevens, R. C. and Cravatt, B. F., Structural adaptations in a membrane enzyme that terminates endocannabinoid signaling, Science, 298 (2002) 1793–1796.

Bracey, M. H., Cravatt, B. F. and Stevens, R. C., Structural commonalities among integral membrane enzymes, FEBS Lett, 567 (2004) 159–165.

Burstein, S. H., Huang, S. M., Petros, T. J., Rossetti, R. G., Walker, J. M. and Zurier, R. B., Regulation of anandamide tissue levels by N-arachidonylglycine, Biochem Pharmacol, 64 (2002) 1147–1150.

Caillé, S., Alvarez-Jaimes, L., Polis, I., Stouffer, D. G. and Parsons, L. H., Specific alterations of extracellular endocannabinoid levels in the nucleus accumbens by ethanol, heroin, and cocaine self-administration, J Neurosci, 27 (2007) 3695–3702.

Centonze, D., Bari, M., Rossi, S., Prosperetti, C., Furlan, R., Fezza, F., De Chiara, V., Battistini, L., Bernardi, G., Bernardini, S., Martino, G. and Maccarrone, M., The endocannabinoid system is dysregulated in multiple sclerosis and in experimental autoimmune encephalomyelitis, Brain, 130 (2007a) 2543–2553.

Centonze, D., Finazzi-Agrò, A., Bernardi, G. and Maccarrone, M., The endocannabinoid system in targeting inflammatory neurodegenerative diseases, Trends Pharmacol Sci, 28 (2007b) 180–187.

Chebrou, H., Bigey, F., Arnaud, A. and Galzy, P., Study of the amidase signature group. Biochim Biophys Acta, 1298 (1996) 285–293.

Chiang, K. P., Gerber, A. L., Sipe, J. C. and Cravatt, B. F., Reduced cellular expression and activity of the P129T mutant of human fatty acid amide hydrolase: evidence for a link between defects in the endocannabinoid system and problem drug use, Hum Mol Genet, 13 (2004) 2113–2119.

Cravatt, B. F., Giang, D. K., Mayfield, S. P., Boger, D. L., Lerner, R. A. and Gilula, N. B., Molecular characterization of an enzyme that degrades neuromodulatory fatty-acid amides, Nature, 384 (1996) 83–87.

Cravatt, B. F. and Lichtman, A. H., The enzymatic inactivation of the fatty acid amide class of signaling lipids, Chem Phys Lipids, 121 (2002) 135–148.

Cravatt, B. F. and Lichtman, A. H., Fatty acid amide hydrolase: an emerging therapeutic target in the endocannabinoid system, Curr Opin Chem Biol, 7 (2003) 469–475.

Cupini, L. M., Bari, M., Battista, N., Argirò, G., Finazzi-Agrò, A., Calabresi, P. and Maccarrone, M., tBiochemical changes in endocannabinoid system are expressed in platelets of female but not male migraineurs, Cephalalgia, 26 (2006) 277–281.

Day, T. A., Rakhshan, F., Deutsch, D. G. and Barker, E. L., Role of fatty acid amide hydrolase in the transport of the endogenous cannabinoid anandamide, Mol Pharmacol, 59 (2001) 1369–1375.

De Petrocellis, L., Melck, D., Ueda, N., Maurelli, S., Kurahashi, Y., Yamamoto, S., Marino, G. and Di Marzo, V., Novel inhibitors of brain, neuronal and basophilic anandamide amidohydrolase, Biochem Biophys Res Commun, 231 (1997) 82–88.

De Petrocellis, L., Cascio, M. G. and Di Marzo, V., The endocannabinoid system: a general view and latest additions, Br J Pharmacol, 141 (2004) 765–774.

Desarnaud, F., Cadas, H. and Piomelli, D., Anandamide amidohydrolase activity in rat brain microsomes. Identification and partial characterization, J Biol Chem, 270 (1995) 6030–6035.

Deutsch, D. G. and Chin, S. A., Enzymatic synthesis and degradation of anandamide, a cannabinoid receptor agonist, Biochem, 46 (1993) 791–796.

Deutsch, D. G., Lin, S., Hill, W. A., Morse, K. L., Salehani, D., Arreaza, G., Omeir, R. L. and Makriyannis, A., Fatty acid sulfonyl fluorides inhibit anandamide metabolism and bind to the cannabinoid receptor, Biochem Biophys Res Commun, 231 (1997a) 217–221.

Deutsch, D. G., Omeir, R., Arreaza, G., Salehani, D., Prestwich, G. D., Huang, Z. and Howlett, A., Methyl arachidonyl fluorophosphonate: a potent irreversible inhibitor of anandamide amidase, Biochem Pharmacol, 53 (1997b) 255–260.

Devane, W. A., Dysarz, F. A., Johnson, M. R., Melvin, L. S. and Howlett, A. C., Determination and characterisation of a cannabinoid receptor in rat brain, Mol Pharmacol, 34 (1988) 605–613.

Devane, W. A., Hannus, L., Breuer, A., Pertwee, R. G., Stevenson, L. A., Griffin, G., Gibson, D., Mandelbaum, A., Etinger, A. and Mechoulam, R., Isolation and structure of a brain constituent that binds to the cannabinoid receptor, Science, 258 (1992) 1946–1949.

Dewey, W. L., Cannabinoid pharmacology, Pharmacol Rev, 38 (1986) 151–178.

Di Marzo, V., Fontana, A., Cadas, H., Schinelli, S., Cimino, G., Schwartz, J. C. and Pomelli, D., Formation and inactivation of endogenous cannabinoid anandamide in central neurons, Nature, 372 (1994) 686–691.

Di Marzo, V. and Deutsch, D. G., Biochemistry of the endogenous ligands of cannabinoid receptors, Neurobiol Dis, 5 (1998) 386–404.

Di Marzo, V., Bisogno, T., De Petrocellis, L., Melck, D., Orlando, P. and Wagner, J. A., Biosynthesis and inactivation of the endocannabinoid 2-arachidonoylglycerol in circulating and tumoral macrophages, Eur J Biochem, 264 (1999) 258–267.

Di Marzo, V., Hill, M. P., Bisogno, T., Crossman, A. R. and Brotchie, J. M., Enhanced levels of endogenous cannabinoids in the globus pallidus are associated with a reduction in movement in an animal model of Parkinson's disease, FASEB J, 14 (2000) 1432–1438.

Dinh, T. P., Carpenter, D., Leslie, F. M., Freund, T. F., Katona, I., Sensi, S. L., Kathuria, S. and Piomelli, D., Brain monoglyceride lipase participating in endocannabinoid inactivation, Proc Natl Acad Sci USA, 99 (2002) 10819–10824.

Dinh, T. P., Kathuria, S. and Piomelli, D., RNA interference suggests a primary role for monoacylglycerol lipase in the degradation of the endocannabinoid 2-arachidonoylglycerol, Mol Pharmacol, 66 (2004) 1260–1264.

Edgemond, W. S., Greenberg, M. J., McGinley, P. J., Muthian, S., Campbell, W. B. and Hillard, C. J., Synthesis and characterization of diazomethylarachidonyl ketone: an irreversible inhibitor of N-arachidonylethanolamine amidohydrolase, J Pharmacol Exp Ther, 286 (1998) 184–190.

Edwards, P. D., Zottola, M. A., Davis, M., Williams, J. and Tuthill, P. A., Peptidyl alpha-ketoheterocyclic inhibitors of human neutrophil elastase. 3. In vitro and in vivo potency of a series of peptidyl alpha-ketobenzoxazoles, J Med Chem, 38 (1995) 3972–3982.

Fegley, D., Kathuria, S., Mercier, R., Li, C., Goutopoulos, A., Makriyannis, A. and Piomelli, D., Anandamide transport is independent of fatty-acid amide hydrolase activity and is blocked by the hydrolysis-resistant inhibitor AM1172, Proc Natl Acad Sci U S A, 101 (2004) 8756–8761.

Felder, C. C., Nielsen, A., Briley, E. M., Palkovits, M., Priller, J., Axelrod, J., Nguyen, D. N., Richardson, J. M., Riggin, R. M., Koppel, G. A., Paul, S. M. and Becker, G. W., Isolation and measurement of the endogenous cannabinoid receptor agonist, anandamide, in brain and peripheral tissues of human and rat, FEBS Lett, 393 (1996) 231–235.

Fersht, A. R., Acyl-transfer reactions of amides and esters with alcohols and thiols. A reference system for the serine and cysteine proteinases, Concerning the N protonation of amides and amide-imidate equilibria, J Am Chem Soc, 93 (1971) 3504–3515.

Fowler, C. J., Plant-derived, synthetic and endogenous cannabinoidsas neuroprotective agents. Non-psychoactive cannabinoids,'entourage' compounds and inhibitors of N-acyl ethanolamine breakdown as therapeutic strategies to avoid psychotropic effects, Brain Res Rev, 41 (2003) 26–43.

Gaetani, S., Cuomo, V. and Piomelli, D., Anandamide hydrolysis: a new target for anti-anxiety drugs?, Trends Mol Med 9 (2003) 474–478.

Giang, D. K. and Cravatt, B. F., Molecular characterization of human and mouse fatty acid amide hydrolases, Proc Natl Acad Sci U S A, 94 (1997) 2238–2242.

Glaser, S. T., Abumrad, N. A., Fatade, F., Kaczocha, M., Studholme, K. M. and Deutsch, D. G., Evidence against the presence of an anandamide transporter, Proc Natl Acad Sci U S A, 100 (2003) 4269–4274.

Goparaju, S. K., Ueda, N., Taniguchi, K. and Yamamoto, S., Enzymes of porcine brain hydrolyzing 2-arachidonoylglycerol, an endogenous ligand of cannabinoid receptors, Biochem Pharmacol, 57 (1999a) 417–423.

Goparaju, S. K., Kurahashi, Y., Suzuki, H., Ueda, N. and Yamamoto, S., Anandamide amidohydrolase of porcine brain:cDNA cloning, functional expression and site-directed mutagenesis, Biochim Biophys Acta, 1441 (1999b) 77–84.

Gubellini, P., Picconi, B., Bari, M., Battista, N., Calabresi, P., Centonze, D., Bernardi, G., Finazzi-Agrò, A. and Maccarrone, M., Experimental parkinsonism alters endocannabinoid degradation: implications for striatal glutamatergic transmission, J Neurosci, 22 (2002) 6900–6907.

Gulyas, A. I., Cravatt, B. F., Bracey, M. H., Dinh, T. P., Piomelli, D., Boscia, F. and Freund, T. F., Segregation of two endocannabinoid-hydrolyzing enzymes into preand postsynaptic compartments in the rat hippocampus, cerebellum and amygdale, Eur J Neurosci, 20 (2004) 441–458.

Guo, Y., Wang, H., Okamoto, Y., Ueda, N., Kingsley, P. J., Marnett, L. J., Schmid, H. H., Das, S. K. and Dey, S. K., N-acylphosphatidylethanolamine-hydrolyzing phospholipase D is an important determinant of uterine anandamide levels during implantation, J Biol Chem, 280 (2005) 23429–23432.

Hanus, L., Abu-Lafi, S., Fride, E., Breuer, A., Vogel, Z., Shalev, D. E., Kustanovich, I. and Mechoulam, R. 2-Arachidonyl glyceryl ether, an endogenous agonist of the cannabinoid $CB_1$ receptor, Proc Natl Acad Sci U S A, 98 (2001) 3662–3665.

Howlett, A. C. and Fleming, R. M., Cannabinoid inhibition of adenylate cyclase. Pharmacology of the response in neuroblastoma cell membranes, Mol Pharmacol, 26 (1984) 532–528.

Howlett, A. C., Barth, F., Bonner, T. I., Cabral, G., Casellas, P., Devane, W. A., Felder, C. C., Herkenham, M., Mackie, K., Martin, B. R., Mechoulam, R. and Pertwee, R. G., International Union of Pharmacology. XXVII. Classification of cannabinoid receptors, Pharmacol Rev, 54 (2002) 161–202.

Huang, S. M., Bisogno, T., Petros, T. J., Chang, S. Y., Zavitsanos, P. A., Zipkin, R. E., Sivakumar, R., Coop, A., Maeda, D. Y., De Petrocellis, L., Burstein, S., Di Marzo, V. and Walker, J. M., Identification of a new class of molecules, the arachidonyl amino acids, and characterization of one member that inhibits pain, J Biol Chem, 276 (2001) 42639–42644.

Ide, K., Nukina, N., Masuda, N., Goto, J., Kanazawa, I., Abnormal gene product identified in Huntington's disease lymphocytes and brain, Biochem Biophys Res Commun, 209 (1995) 1119–1125.

Jordt, S. E. and Julius, D., Molecular basis for species-specific sensitivity to "hot" chili peppers, Cell, 108 (2002) 421–430.

Jung, J., Hwang, S. W., Kwak, J., Lee, S. Y., Kang, C. J., Kim, W. B., Kim, D. and Oh, U., Capsaicin binds to the intracellular domain of the capsaicin-activated ion channel, J Neurosci, 19 (1999) 529–538.

Kathuria, S., Gaetani, S., Fegley, D., Valiño, F., Duranti, A., Tontini, A., Mor, M., Tarzia, G., La Rana, G., Calignano, A., Giustino, A., Tattoli, M., Palmery, M., Cuomo, V. and Piomelli, D., Modulation of anxiety through blockade of anandamide hydrolysis, Nat Med, 9 (2003) 76–81.

Labahn, J., Neumann, S., Büldt, G., Kula, M. R. and Granzin, J., An alternative mechanism for amidase signature enzymes, J Mol Biol, 322 (2002) 1053–1064.

Lauckner, J. E., Hille, B. and Mackie, K., The cannabinoid agonist WIN55, 212-2 increases intracellular calcium via CB1 receptor coupling to Gq/11 G proteins, Proc Natl Acad Sci U S A, 102 (2005) 19144–19149.

Leung, D., Saghatelian, A., Simon, G. M. and Cravatt, B. F., Inactivation of N-acyl phosphatidylethanolamine phospholipase D reveals multiple mechanisms for the biosynthesis of endocannabinoids, Biochemistry, 45 (2006) 4720–4726.

Lichtman, A. H., Hawkins, E. G., Griffin, G. and Cravatt, B. F., Pharmacological activity of fatty acid amides is regulated, but not mediated, by fatty acid amide hydrolase in vivo, J Pharmacol Exp Ther, 302 (2002) 73–79.

Lichtman, A. H., Leung, D., Shelton, C. C., Saghatelian, A., Hardouin, C., Boger, D. L. and Cravatt, B. F., Reversible inhibitors of fatty acid amide hydrolase that promote analgesia: evidence for an unprecedented combination of potency and selectivity, J Pharmacol Exp Ther, 311 (2004) 441–448.

Ligresti, A., Morera, E., Van Der Stelt, M., Monory, K., Lutz, B., Ortar, G. and Di Marzo, V., Further evidence for the existence of a specific process for the membrane transport of anandamide, Biochem J, 380 (2004) 265–272.

Liu, J., Wang, L., Harvey-White, J., Osei-Hyiaman, D., Razdan, R., Gong, Q., Chan, A. C., Zhou, Z., Huang, B. X., Kim, H. Y. and Kunos, G., A biosynthetic pathway for anandamide, Proc Natl Acad Sci U S A, 103 (2006) 13345–13350.

López-Rodríguez, M. L., Viso, A., Ortega-Gutiérrez, S., Lastres-Becker, I., González, S., Fernández-Ruiz, J. and Ramos, J. A., Design, synthesis and biological evaluation of novel arachidonic acid derivatives as highly potent and selective endocannabinoid transporter inhibitors, J Med Chem, 44 (2001) 4505–4508.

Lunn, C. A., Reich, E. P. and Bober, L., Targeting the CB2 receptor for immune modulation, Expert Opin Ther Targets, 5 (2006) 653–663.

Maccarrone, M., Fatty acid amide hydrolase: a potential target for next generation therapeutics, Curr Pharm Des, 12 (2006) 759–772.

Maccarrone, M. and Finazzi-Agrò, A., Anandamide hydrolase: a guardian angel of human reproduction?, Trends Pharmacol Sci, 25 (2004a) 353–357.

Maccarrone, M., van der Stelt, M., Rossi, A., Veldink, G. A., Vliegenthart, J. F. and Finazzi-Agrò A., Anandamide hydrolysis by human cells in culture and brain, J Biol Chem, 273 (1998) 32332–32339.

Maccarrone, M., De Felici, M., Bari, M., Klinger, F., Siracusa, G. and Finazzi-Agro, A., Down-regulation of anandamide hydrolase in mouse uterus by sex hormones, Eur J Biochem, 267 (2000a) 2991–2997.

Maccarrone, M., Valensise, H., Bari, M., Lazzarin, N., Romanini, C. and Finazzi-Agrò, A., Relation between decreased anandamide hydrolase concentrations in human lymphocytes and miscarriage, Lancet, 355 (2000b) 1326–1329.

Maccarrone, M., Valensise, H., Bari, M., Lazzarin, N., Romanini, C. and Finazzi-Agrò, A., Progesterone up-regulates anandamide hydrolase in human lymphocytes: role of cytokines and implications for fertility, J Immunol, 166 (2001) 7183–7189.

Maccarrone, M., Di Rienzo, M., Finazzi-Agrò, A. and Rossi, A., Leptin activates the anandamide hydrolase promoter in human T lymphocytes through STAT3, J Biol Chem, 278 (2003a) 13318–13324.

Maccarrone, M., Bari, M., Di Rienzo, M., Finazzi-Agro, A. and Rossi, A., Progesterone activates fatty acid amide hydrolase (FAAH) promoter in human T lymphocytes through the transcription factor Ikaros, Evidence for a synergistic effect of leptin, J Biol Chem, 278 (2003b) 32726–32732.

Maccarrone, M., Gasperi, V., Fezza, F., Finazzi-Agro, A. and Rossi, A., Differential regulation of fatty acid amide hydrolase promoter in human immune cells and neuronal cells by leptin and progesterone, Eur J Biochem, 271 (2004b) 4666–4676.

Martin, B. R., Beletskaya, I., Patrick, G., Jefferson, R., Winckler, R., Deutsch, D. G., Di Marzo, V., Dasse, O., Mahadevan, A. and Razdan, R. K., Cannabinoid properties of methylfluorophosphonate analogs, J Pharmacol Exp Ther, 294 (2000) 1209–1218.

Matias, I., Chen, J., De Petrocellis, L., Bisogno, T., Ligresti, A., Fezza, F., Krauss, A. H., Shi, L., Protzman, C. E., Li, C., Liang, Y., Nieves, A. L., Kedzie, K. M., Burk, R. M., Di Marzo, V. and Woodward, D. F., Prostaglandin ethanolamides (prostamides): in vitro pharmacology and metabolism, J Pharmacol Exp Ther, 309 (2004) 745–757.

Maurelli, S., Bisogno, T., De Petrocellis, L., Di Luccia, A., Marino, G. and Di Marzo, V., Two novel classes of neuroactive fatty acid amides are substrates for mouse neuroblastoma 'anandamide amidohydrolase', FEBS Lett, 377 (1995) 82–86.

McFarland, M. J., Porter, A. C., Rakhshan, F. R., Rawat, D. S., Gibbs, R. A. and Barker, E. L., A role for caveolae/lipid rafts in the uptake and recycling of the endogenous cannabinoid anandamide, J Biol Chem, 279 (2004) 41991–41997.

McKinney, M. K. and Cravatt, B. F., Structure and function of fatty acid amide hydrolase, Annu Rev Biochem, 74 (2005) 411–432.

McPartland, J. M., Matias, I., Di Marzo, V. and Glass, M., Evolutionary origins of the endocannabinoid system, Gene, 370 (2006) 64–74.

Mechoulam, R., Ben-Shabat, S., Hanus, L., Ligumsky, M., Kaminski, N. E., Schatz, A. R., Gopher, A., Almog, S., Martin, B. R., Compton, D. R., Pertwee, R. G., Griffine, G., Bayewitchf, M., Bargf, J. and Vogelf, Z., Identification of an endogenous 2-monoglyceride, present in canine gut, that binds to cannabinoid receptors, Biochem Pharmacol, 50 (1995) 83–90.

Mechoulam, R., Panikashvili, D. and Shohami, E., Cannabinoids and brain injury: therapeutic implications, Trends Mol Med, 8 (2002) 58–61.

Mei, G., Di Venere, A., Gasperi, V., Nicolai, E., Masuda, K. R., Finazzi-Agrò, A., Cravatt, B. F. and Maccarrone, M., Closing the gate to the active site: effect of the inhibitor methoxyarachidonyl fluorophosphonate on the conformation and membrane binding of fatty acid amide hydrolase, J Biol Chem, 282 (2007) 3829–3836.

Mor, M., Rivara, S., Lodola, A., Plazzi, P. V., Tarzia, G., Duranti, A., Tontini, A., Piersanti, G., Kathuria, S. and Piomelli, D., Cyclohexylcarbamic acid 3'- or 4'-substituted biphenyl-3-yl esters as fatty acid amide hydrolase inhibitors: Synthesis, quantitative structure-activity relationships, and molecular modeling studies, J Med Chem, 47 (2004) 4998–5008.

Muccioli, G. G., Xu, C., Odah, E., Cudaback, E., Cisneros, J. A., Lambert, D. M., López Rodríguez, M. L., Bajjalieh, S. and Stella, N., Identification of a novel endocannabinoid-hydrolyzing enzyme expressed by microglial cells, J Neurosci, 27 (2007) 2883–2889.

Munro, S., Thomas, K. L. and Abu-Shaar, M., Molecular characterization of a peripheral receptor for cannabinoids, Nature, 365 (1993) 61–65.

Norrod, A. G. and Puffenbarger, R. A., Genetic polymorphisms of the endocannabinoid system, Chem Biodivers, 4 (2007) 1926–1932.

Oddi, S., Bari, M., Battista, N., Barsacchi, D., Cozzani, I. and Maccarrone, M., Confocal microscopy and biochemical analysis reveal spatial and functional separation between anandamide uptake and hydrolysis in human keratinocytes, Cell Mol Life Sci, 62 (2005) 386–395.

Oka, S., Tsuchie, A., Tokumura, A., Muramatsu, M., Suhara, Y., Takayama, H., Waku, K., and Sugiura, T., Ether-linked analogue of 2-arachidonoylglycerol (noladin ether) was not detected in the brains of various mammalian species, J Neurochem, 85 (2003) 1374–1381.

Okamoto, Y., Morishita, J., Tsuboi, K., Tonai, T. and Ueda, N., Molecular characterization of a phospholipase D generating anandamide and its congeners, J Biol Chem, 279 (2004), 5298–5305.

Ortar, G., Ligresti, A., De Petrocellis, L., Morera, E. and Di Marzo, V., Novel selective and metabolically stable inhibitors of anandamide cellular uptake, Biochem Pharmacol, 65 (2003) 1473–1481.

Paria, B. C., Deutsch, D. G. and Dey, S. K., The uterus is a potential site for anandamide synthesis and hydrolysis: differential profiles of anandamide synthase and hydrolase activities in the mouse uterus during the preiimplantation period, Mol Reprod Dev, 45 (1996) 183–192.

Paria, B. C., Zhao, X., Wang, J., Das, S. K. and Dey, S. K., Fatty-acid amide hydrolase is expressed in the mouse uterus and embryo during the preiimplantation period, Biol Repro, 60 (1999) 1151–1157.

Park, B., Gibbons, H. M., Mitchell, M. D. and Glass, M., Identification of the CB1 cannabinoid receptor and fatty acid amide hydrolase (FAAH) in the human placenta, Placenta, 24 (2003) 990–995.

Patricelli, M. P. and Cravatt, B. F., Fatty acid amide hydrolase competitively degrades bioactive amides and esters through a non conventional catalytic mechanism, Biochemistry, 38 (1999) 14125–14130.

Patricelli, M. P., Lashuel, H. A., Giang, D. K., Kelly, J. W. and Cravatt, B.F., Comparative characterization of a wild type and transmembrane domain-deleted fatty acid amide hydrolase: identification of the transmembrane domain as a site for oligomerization, Biochemistry, 37(43): (1998) 15177–15187.

Patricelli, M. P., Lovato, M. A. and Cravatt, B. F., Chemical and mutagenic investigations of fatty acid amide hydrolase: evidence for a family of serine hydrolases with distinct catalytic properties, Biochemistry, 38 (1999) 9804–9812.

Picot, D., Loll, P. J. and Garavito, R. M., The X-ray crystal structure of the membrane protein prostaglandin H2 synthase-1, Nature, 367 (1994) 243–249.

Piomelli, D., The molecular logic of endocannabinoid signalling, Nature Rev Neurosci, 4 (2003) 873–884.

Piomelli, D., Beltramo, M., Giuffrida, A. and Stella, N., Endogenous cannabinoid signaling, Neurobiol Dis, 6 (1998) 462–473.

Piomelli, D., Giuffrida, A., Calignano, A. and de Fonseca, F. R., The endocannabinoid system as a target for therapeutic drugs, Trends Pharmacol Sci, 21 (2000) 218–224.

Piomelli, D., Tarzia, G., Duranti, A., Tontini, A., Mor, M., Compton, T. R., Dasse, O., Monaghan, E. P., Parrott, J. A. and Putman, D., Pharmacological profile of the selective FAAH inhibitor KDS-4103 (URB597), CNS Drug Rev, 12 (2006) 21–38.

Porter, A. C., Sauer, J. M., Knierman, M. D., Becker, G. W., Berna, M. J., Bao, J., Nomikos, G. G., Carter, P., Bymaster, F. P., Leese, A. B. and Felder, C. C., Characterization of a novel endocannabinoid, virodhamine, with antagonist activity at the CB1 receptor, J Pharmacol Exp Ther, 301 (2002) 1020–1024.

Puffenbarger, R. A., Kapulina, O., Howell, J. M. and Deutsch, D. G., Characterization of the 5'-sequence of the mouse fatty acid amide hydrolase, Neurosci Lett, 314 (2001) 21–24.

Ross, S. A. and Elsohly, M. A., The volatile oil composition of fresh and air-dried buds of Cannabis sativa, J Nat Prod, 59 (1996) 49–51.

Rossi, C., Pini, L. A., Cupini, M. L., Calabresi, P. and Sarchielli, P., Endocannabinoids in platelets of chronic migraine patients and medication-overuse headache patients: relation with serotonin levels, Eur J Clin Pharmacol, 64 (2008) 1–8.

Ryberg, E., Larsson, N., Sjögren, S., Hjorth, S., Hermansson, N. O., Leonova, J., Elebring, T., Nilsson, K., Drmota, T. and Greasley, P. J., The orphan receptor GPR55 is a novel cannabinoid receptor, Br J Pharmacol, 152 (2007) 1092–1101.

Sawzdargo, M., Nguyen, T., Lee, D. K., Lynch, K. R., Cheng, R., Heng, H. H., George, S. R. and O'Dowd, B. F., Identification and cloning of three novel human G protein-coupled receptor genes GPR52, PsiGPR53 and GPR55: GPR55 is extensively expressed in human brain, Brain Res Mol Brain Res, 64 (1999) 193–198.

Segall, Y., Quistad, G. B., Nomura, D. K. and Casida, J. E., Arachidonylsulfonyl derivatives as cannabinoid CB1 receptor and fatty acid amide hydrolase inhibitors, Bioorg Med Chem Lett, 13 (2003) 3301–3303.

Shin, S., Lee, T. H., Ha, N. C., Koo, H. M., Kim, S. Y., Lee, H. S., Kim, Y. S. and Oh, B. H., Structure of malonamidase E2 reveals a novel Ser-Ser-Lys catalytic triad in a new serine hydrolase fold that is prevalent in nature, EMBO J, 21 (2002) 2509–2516.

Siemens, J., Zhou, S., Piskorowski, R., Nikai, T., Lumpkin, E. A., Basbaum, A. I., King, D. and Julius, D., Spider toxins activate the capsaicin receptor to produce inflammatory pain, Nature, 444 (2006) 208–212.

Simon, G. M. and Cravatt, B. F., Endocannabinoid biosynthesis proceeding through glycerophospho-*N*-acyl ethanolamine and a role for alpha/beta-hydrolase 4 in this pathway, J Biol Chem, 281 (2006) 26465–26472.

Sipe, J. C., Chiang, K., Gerber, A. L., Beutler, E. and Cravatt, B. F., A missense mutation in human fatty acid amide hydrolase associated with problem drug use, Proc Natl Acad Sci U S A, 99 (2002) 8394–8399.

Sipe, J. C., Waalen, J., Gerber, A. and Beutler, E., Overweight and obesity associated with a missense polymorphism in fatty acid amide hydrolase (FAAH), Int J Obesity, 29 (2005) 755–799.

Starowicz, K., Nigam, S. and Di Marzo, V., Biochemistry and pharmacology of endovanilloids, Pharmacol Ther, 114 (2007) 13–33.

Stella, N., Schweitzer, P. and Piomelli, D., A second endogenous cannabinoid that modulates long-term potentiation, Nature, 388 (1997) 773–778.

Sugiura, T. and Waku, K., 2-Arachidonoylglycerol and the cannabinoid receptors, Chem Phys Lipids, 108 (2000b) 89–106.

Sugiura, T., Kondo, S., Sukagawa, A., Nakane, S., Shinoda, A., Itoh, K., Yamashita, A. and Waku, K., Arachidonoylglycerol: a possible endogenous cannabinoid receptor ligand in brain, Biochem Biophys Res Commun, 215 (1995) 89–95.

Sugiura, T., Kondo, S., Kishimoto, S., Miyashita, T., Natane, S., Kodaka, T., Suhara, Y., Takayama, H. and Waku, K., Evidence that 2-arachidonoylglycerol but not N-palmitoylethanolamine or anandamide is the physiological ligand for the cannabinoid CB2 receptor. Comparison of the agonistic activities of various cannabinoid receptor ligands in HL-60 cells, J Biol Chem, 275 (2000a) 605–612.

Sugiura, T., Kobayashi, Y., Oka, S. and Waku, K., Biosynthesis and degradation of anandamide and 2-arachidonoylglycerol and their possible physiological significance, Prostaglandins Leukot Essent Fatty Acids, 66 (2002) 173–192.

Sun, Y. X., Tsuboi, K., Okamoto, Y., Tonai, T., Murakami, M., Kudo, I. and Ueda, N., Biosynthesis of anandamide and N-palmitoylethanolamine by sequential actions of phospholipase A2 and lysophospholipase D, Biochem J, 380 (2004) 749–756.

Tarzia, G., Duranti, A., Gatti, G., Piersanti, G., Tontini, A., Rivara, S., Lodola, A., Plazzi, P. V., Mor, M., Kathuria, S. and Piomelli, D., Synthesis and structure-activity relationships of FAAH inhibitors: cyclohexylcarbamic acid biphenyl esters with chemical modulation at the proximal phenyl ring, Chem Med Chem, 1 (2006) 130–139.

Tsuboi, K., Sun, Y. X., Okamoto, Y., Araki, N., Tonai, T. and Ueda, N., Molecular characterization of N-acylethanolamine-hydrolyzing acid amidase, a novel member of the choloylglycine hydrolase family with structural and functional similarity to acid ceramidase, J Biol Chem, 280 (2005) 11082–11092.

Ueda, N., Kurahashi, Y., Yamamoto, S. and Tokunaga, T., Partial purification and characterization of the porcine brain enzymehydrolyzing and synthesizing anandamide, J Biol Chem, 270 (1995) 23823–23827.

van der Stelt, M. and Di Marzo, V., Endovanilloids. Putative endogenous ligands of transient receptor potential vanilloid 1 channels, Eur J Biochem, 271 (2004) 1827–1834.

van der Stelt, M., van Kuik, J. A., Bari, M., van Zadelhoff, G., Leeflang, B. R., Veldink, G. A., Finazzi-Agrò, A., Vliegenthart, J. F. and Maccarrone, M., Oxygenated metabolites of anandamide and 2-arachidonoyl-glycerol: conformational analysis and interaction with cannabinoid receptors, membrane transporter and fatty acid amide hydrolase, J Med Chem, 45 (2002) 3709–3720.

Van Sickle, M. D., Duncan, M., Kingsley, P. J., Mouihate, A., Urbani, P., Mackie, K., Stella, N., Makriyannis, A., Piomelli, D., Davison, J. S., Marnett, L. J., Di Marzo, V., Pittman, Q. J., Patel, K. D. and Sharkey, K. A., Identification and functional characterization of brainstem cannabinoid CB2 receptors, Science, 310 (2005) 329–332.

Vandevoorde, S., Saha, B., Mahadevan, A., Razdan, R. K., Pertwee, R. G., Martin, B. R. and Fowler, C. J., Influence of the degree of unsaturation of the acyl side chain upon the interaction of analogues of 1-arachidonoylglycerol with monoacylglycerol lipase and fatty acid amide hydrolase, Biochem Biophys Res Commun, 337 (2005) 104–109.

Waleh, N. S., Cravatt, B. F., Apte-Deshpande, A., Terao, A. and Kilduff, T. S., Transcriptional regulation of the mouse fatty acid amide hydrolase gene, Gene, 291 (2002) 203–210.

Wan, M., Cravatt, B. F., Ring, H. Z., Zhang, X. and Francke, U., Conserved chromosomal location and genomic structure of human and mouse fatty-acid amide hydrolase genes and evaluation of clasper as a candidate neurological mutation, Genomics, 54 (1998) 408–414.

Wang, H., Xie, H., Guo, Y., Zhang, H., Takahashi, T., Kingsley, P. J., Marnett, L. J., Das, S. K., Cravatt, B. F. and Dey, S. K., Fatty acid amide hydrolase deficiency limits early pregnancy events, J Clin Invest, 116 (2006) 2122–2131.

Wei, B. Q., Mikkelsen, T. S., McKinney, M. K., Lander, E. S. and Cravatt, B. F., A second fatty acid amide hydrolase with variable distribution among placental mammals, J Biol Chem, 281 (2006) 36569–36578.

Wendt, K. U., Poralla, K. and Schulz, G.E., Structure and function of a squalene cyclase, Science, 277 (1997) 1811–1815.

Yang, H. Y., Karoum, F., Felder, C., Badger, H., Wang, T. C. and Markey, S. P., GC/MS analysis of anandamide and quantification of N-arachidonoylphosphatidylethanolamides in various brain regions, spinal cord, testis, and spleen of the rat, J Neurochem, 72 (1999) 1959–1968.

# Chapter 5
# Modulation of Inflammatory Cytokines by Omega-3 Fatty Acids

**Jing X. Kang and Karsten H. Weylandt**

**Abstract** Many human diseases have been linked to inflammation, which is mediated by a number of chemical molecules including lipid mediators and cytokines. Polyunsaturated fatty acids (omega-6 and omega-3 fatty acids) are the precursors of the lipid mediators and play an important role in regulation of inflammation. Generally, omega-6 fatty acids (e.g. arachidonic acid) promote inflammation whereas omega-3 fatty acids (e.g. eicosapentaenoic acid and docosahexaenoic acid) have anti-inflammatory properties. Omega-3 fatty acids dampen inflammation through multiple pathways. On the one hand, omega-3 fatty acids inhibit the formation of omega-6 fatty acids-derived pro-inflammatory eicosanoids (e.g. $PGE_2$ and $LTB_4$), and on the other hand these fatty acids can form several potent anti-inflammatory lipid mediators (e.g. resolvins and protectins). These together directly or indirectly suppress the activity of nuclear transcription factors, such as NFκB, and reduce the production of pro-inflammatory enzymes and cytokines, including COX-2, tumor necrosis factor (TNF)-α, and interleukin (IL)-1β. This chapter focuses on the evidence from recent studies using new experimental models.

**Keywords** Omega-3 fatty acids · omega-6 fatty acids · lipid mediators · cytokines · inflammation

**Abbreviations** AA: arachidonic acid; ALA: alpha-linolenic acid; AP-1: activator protein 1; COX: cyclooxygenase; DHA: docosahexaenoic acid; EPA: eicosapentaenoic acid; IL: interleukin; LA: linoleic acid; LOX: lipoxygenase; LPS: lipopolysaccaride; LT: leukotriene; NF-κB: nuclear factor-kappa B, PG: prostaglandin; PPARs: peroxisome proliferator-activated receptors; PUFA: polyunsaturated fatty acids, RvE1: resolvin E1, TNFα: tumor necrosis factor alpha, TX: thromboxane

---

J.X. Kang
Department of Medicine, Massachusetts General Hospital, Harvard Medical School, Boston, MA, 02114, USA

P.J. Quinn, X. Wang (eds.), *Lipids in Health and Disease*,

# 5.1 Introduction

## *5.1.1 Inflammation and Chronic Diseases*

Inflammation is the activation of the immune system in response to infection, irritation, or injury, characterized by an influx of white blood cells, redness, heat, swelling, pain, and dysfunction of the organs. It has important functions in both defense and pathophysiological events maintaining the dynamic homeostasis of a host organism including its tissues, organs and individual cells. However, when inflammation persists, known as chronic inflammation, it can lead to chronic diseases. In fact, abnormalities associated with inflammation comprise a large, unrelated group of disorders which underly a variety of human diseases, including cardiovascular disease, cancer, diabetes and neurodegenerative disease. Thus, information is recently considered as a common mechanism of disease (Libby, 2007).

Inflammation involves various immune-system cells and numerous mediators. Recruitment of blood leukocytes characterizes the initiation of inflammatory response. The migrated or activated immune cells generate and release a variety of mediators that control the progression and resolution of information. Among the numerous inflammatory mediators are cytokines and lipid mediators.

## *5.1.2 Cytokines and Inflammation*

Cytokines are small proteins ranging in molecular weight from 8 to 30 kDa. They are important mediators regulating the development of acute or chronic inflammation. Different cytokines are produced by various cells (particularly activated tissue macrophages) and have a wide range of different biological activities. The key cytokines IL-1, TNF-$\alpha$ and IL-6 exhibit redundant and pleiotropic effects that together contribute to the inflammatory response. Some of the effects mediated by these cytokines include increased vascular permeability, increased adhesion molecules on vascular endothelium, chemokine induction, T-cell and B-cell activation, chemoattraction of leukocytes and induction of cell death (Dinarello, 2000). Nuclear factor-kappa B (NF-$\kappa$B), a nuclear transcription factor, is involved in regulating expression of these cytokines (Hayden and Ghosh, 2008).

## *5.1.3 Polyunsaturated Fatty Acids, Lipid Mediators and Inflammation*

Some lipid metabolites, derived from polyunsaturated fatty acids (PUFA), act as inflammatory mediators. There are two kinds of PUFA: omega-6 and

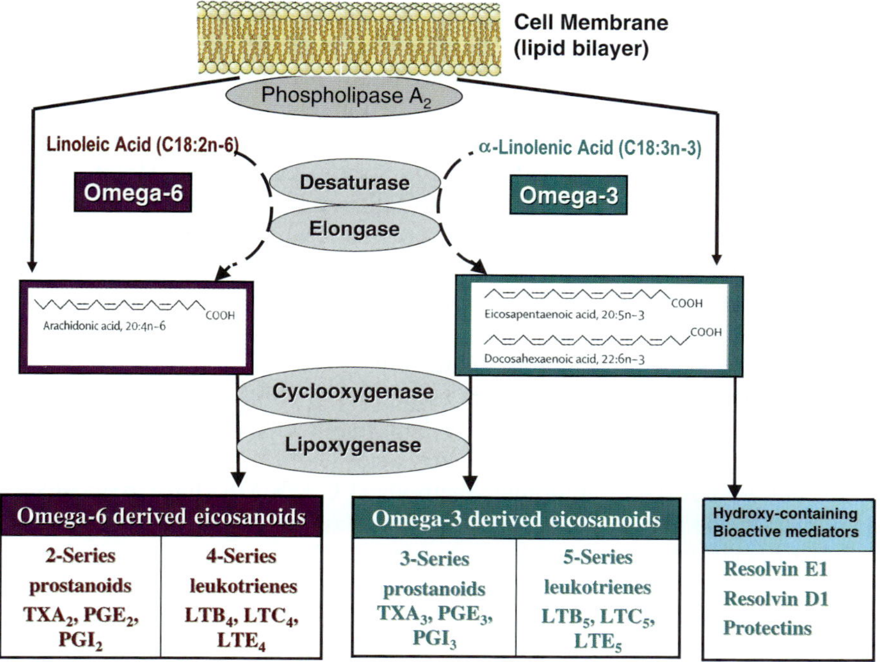


**Fig. 5.1** Metabolic pathways for the production of lipid mediators from omega-6 and omega-3 fatty acids

omega-3 (relating to the position of the first double bond in their hydrocarbon chains) (Fig. 5.1). The omega-6 PUFA that can be metabolized to form inflammatory mediators is arachidonic acid (AA). The omega-3 PUFA that can be converted to lipid mediators are primarily eicosapentaenoic acid (EPA) and docosahexaenoic acid (DHA), which mainly found in fish and fish oil.

Following cell activation by inflammatory stimuli, PUFA in membrane phospholipids of various cell types are released by phospholipase A2 and converted to eicosanoids by cyclooxygenases (COX) and the lipoxygenases (LOX) (Fig. 5.1). Metabolism of the omega-6 arachidonic acid by the COX pathway produces prostaglandins (PGs) and thromboxanes (TX). The common PGs generated by immune cells include $PGE_2$, $PGF_2$ and $PGD_2$. These prostaglandins have diverse physiological effects, including increased vascular permeability, increased vascular dilation, induction of neutrophil chemotaxis, and stimulation of smooth muscle cell migration and proliferation (Richard et al., 2000). The thromboxanes (e.g. $TXA_2$) cause platelet aggregation and constriction of blood vessels (Sellers and Stallone, 2008). Arachidonic acid is also metabolized by the LOX pathway to yield the four leukotrienes (LT): $LTC_4$, $LTB_4$, $LTD_4$, and $LTE_4$. Three of these ($LTC_4$, $LTD_4$, and $LTE_4$) together make up what was formerly called slow-reacting substance of anaphylaxis

(SRS-A); these mediators induce smooth-muscle contraction. $LTB_4$ is a potent chemoattractant of neutrophils (Peters-Golden et al. 2005). Thus, many eicosanoids derived from the omega-6 AA are highly pro-inflammatory.

The omega-3 fatty acid EPA is metabolised in mammalian cells through the same COX and LOX pathways to form 3-series prostaglandins and thromboxanes ($PGD_3$, $PGF_3$, $PGE_3$ and $TXA_3$) and 5-series leukotrienes ($LTB_5$, $LTC_5$, $LTD_5$ and $LTE_5$), which are much less pro-inflammatory or even have opposing effects as compared to their counterparts derived from AA (Simopoulos, 2002; Calder, 2006). For example, $PGE_3$ and $LTB_5$ (unlike $PGE_2$ and $LTB_4$ which can induce violent contracture of heart cells) have little effect on myocyte contraction (Li et al., 1997), and they both are potent in inhibiting mitogen-induced lymphocyte proliferation (Shapiro et al., 1993). Thus, inhibition of the formation of AA-derived pro-inflammatory mediators by competing with AA for the COX and LOX enzymes is thought to be a major mechanism underlying the anti-inflammatory effect of omega-3 fatty acids (James et al., 2000).

Recent studies using lipidomics methods (LC-MS/MS) have been able to demonstrate the generation of potent anti-inflammatory mediators from the n-3 fatty acids EPA and DHA (Serhan et al., 2008). The omega-3 derived mediators have been implicated in the resolution of inflammation and therefore termed resolvins and protectins (Schwab et al., 2007). The resolvin formed from EPA is resolvin E1 (RvE1). Those derived from DHA metabolites are named resolvin D (RvD1–RvD6) and the protectins (protectin D1, PD1). RvE1 is the first omega-3 fatty acid derived lipid mediator, for which a distinct receptor, namely ChemR23, has been identified to mediate its antiinflammatory action (Arita et al., 2005a). The protective effect of the DHA-derived lipid mediators has been examined in the model of ischemic brain injury, showing a significant neuroprotective effect by inhibition of NFkB activity, leukocyte migration and COX-2 induction (Hong et al., 2003). Animal study using colitis as an inflammatory disease model showed that pretreatment with resolvin E1 protected mice from TNBS-induced colitis, a hapten based colitis model (Arita et al., 2005b). Thus, these studies demonstrate that derivates (resolvins and protectins) of the n-3 fatty acids EPA and DHA are bioactive mediators with potent anti-inflammatory properties.

The relative abundance or ratio between the long-chain omega-6 (AA) and omega-3 (EPA and DHA) fatty acids in cell membranes depends on their dietary supplies and the conversion (elongation and desaturation) of their precursors linoleic acid (LA) and alpha-linolenic acid (ALA), respectively (Fig. 5.1). In mammals, the conversion of LA to AA is much more efficient than the conversion of ALA to EPA/DHA. LA and ALA are "essential" because the body cannot make them and we must take in via the food we eat. In addition to the competition for the COX and LOX enzymes, omega-6 and omega-3 fatty acids also compete for the enzymes of elongation and desaturation as well as incorporation into phospholipids (Fig. 5.1). When one type of fatty acid predominates in cell membrane phospholipids, it takes up most of these enzymes, leaving behind little for the other. For example, flush of the omega-3 ALA or EPA into cells would dramatically reduce the omega-6 LA and AA contents in cellular phospholipids. Thus, dietary supplies of these fatty acids can influence their composition in cell membrane phospholipids and modulate the production of pro- or anti-inflammatory mediators and thereby the inflammatory status (Calder, 2006). For this reason, maintaining a balance of omega-6 and omega-3 PUFA is important for optimal biochemical balance in the body.

## 5.2 Effect of Omega-3 Fatty Acids on NF-$\kappa$B Activation and Cytokine Production

The anti-inflammatory effects of the long-chain omega-3 polyunsaturated fatty acids EPA and DHA were amongst their earliest identified properties. During the last decades, many studies in both animals and humans have been done to evaluate the beneficial effects of these fatty acids on a variety of inflammatory diseases (Calder, 2006, 2007; Sijben and Calder, 2007; Simopoulos, 2002). The outcomes of theses studies have been reviewed frequently and, therefore, are not the focus of this chapter. Noticeably, the mechanism underlying the anti-inflammatory effects of omega-3 fatty acids remains to be fully elucidated.

The original view that omega-3 fatty acids exert anti-inflammatory effects only by blocking arachidonic acid metabolism (production of pro-inflammatory eicosanoids) was too simplistic. It appears that they also alter expression of inflammatory genes, particularly those encoding cytokines. Although this possible mechanism has been investigated for many years, the study results of the effect on cytokine production were inconsistent for different species and for different clinical conditions (Blok et al., 1996). These differences could be due to the occurrence of confounding factors of diet in the studies using dietary supplementation or a poorly controlled experimental system. In this context, results from well-controlled studies, free of dietary confounding factors, are critical for addressing the effect of omega-3 fatty acids on cytokine production. Here, we review the results on cytokine production from the recent studies using a novel mouse model.

### *5.2.1 The Fat-1 Transgenic Mouse Model*

A well-controlled experimental model that can eliminate or minimize the confounding factors of diet is critical for addressing nutrient-gene interaction. The newly generated fat-1 transgenic mouse was genetically engineered to carry a gene, namely *fat-1,* from the round worm *C. elegans* and is capable of converting n-6 to n-3 fatty acids (which is naturally impossible in mammals), leading to an increase in n-3 fatty acid content with a balanced n-6/n-3 fatty acid ratio in all tissues, without the need of dietary n-3 supply (Kang et al., 2004). Feeding an identical diet (high in n-6) to the transgenic and wild type littermates can produce different fatty acid profiles in these animals. Thus, this model allows well-controlled studies to be performed, without the interference of the potential confounding factors of diet, ideal for studying the benefits of n-3 fatty acids and the molecular mechanisms of their action (Kang, 2007).

## 5.2.2 *NFκB*

Nuclear factor-kappa B (NF-κB) is a widely expressed inducible transcription factor and is involved in the induction of several pro-inflammatory cytokines and enzymes that are critically involved in the pathogenesis of chronic inflammatory diseases. NF-κB is composed of homodimers and heterodimers, the most abundant and best-studied form in mammalian cells consisting of the p65 and p50 subunits. Activation of NF-κB typically involves the phosphorylation of cytoplasmic IκB by the IκB kinase (IKK) complex, resulting in IκB degradation via the proteosomal system. The degradation of IκB releases the NF-κB heterodimers to translocate to the nucleus where they bind to κB motifs in the promoters of pro-inflammatory genes such as TNF-α, IL-1β, IL-6 and COX-2 leading to their induction (Hayden et al., 2006).

Using the fat-1 mouse model, Bhattacharya et al. have shown a significant reduction of NF-κB (p65/p50) activity in LPS-stimulated splenocytes from fat-1 transgenic mice rich in omega-3 fatty acids (EPA and DHA) either on a normal or a calorie restricted diet, accompanied by a lower IL-6 and TNFα secretion from the LPS-treated splenocytes when compared to those in the wild type mice high in omega-6 and low in omega-3 fatty acids (Bhattacharya et al., 2006). These findings are consistent with the observation in a murine macrophage cell culture model showing that omega-3 PUFA could decrease NFκB activity by inhibition of IkB phosphorylation (Novak et al., 2003). Along the line, we also observed a significant decrease in NFκB activity (as measured by a p65 activation assay) in the colons of the omega-3 enriched fat-1 mice with chemically induced colitis (Hudert et al., 2006). In that study the inflammation of colon, in terms of both clinical manifestation and pathology, was significantly less severe in fat-1 transgenic mice than that in wild type littermates. The ratio of long-chain omeg-6 fatty acid to long-chain omega-3 fatty acids was 1.7 in *fat-1* transgenics and 30.1 in wild type mice, accompanied with increased formation of the derivates of n-3 fatty acids (RvE1 and RvD3, and PD1) (Hudert et al., 2006). A similar inhibitory effect on NFκB activity was also found in the fat-1 transgenic mice with colitis-associated colon tumors (Nowak et al., 2007).

As to the potential link of omega-3 PUFA-derived mediators to NFκB activity, Arita et al. showed, by using a NFκB luciferase activation assay, that the EPA-derived resolvin E1 (RvE1) was able to inhibit TNF-α induced NFκB activation (Arita et al., 2005a). This suggests that omega-3 PUFA metabolites may be able to directly regulate the expression of NFκB associated genes.

### 5.2.3 TNFα

Tumor necrosis factor alpha (TNFα) plays a major pathogenic role in many inflammatory diseases. The expression of TNFα can be induced by the activation of NFκB (Collart et al., 1990). Of interest, TNFα itself is a potent inducer of NFκB activity (Hayden et al., 2006). Increased production of TNFα has been shown to contribute to the development of inflammatory bowel disease (Papadakis and Targan, 2000), acute liver inflammation (Sass et al., 2002; Tilg et al., 2003), retinopathy (Connor et al., 2007) and many other inflammatory conditions. Many previous studies have shown that n-3 PUFA can decrease TNFα production in vitro and in vivo (Babcock et al., 2002; Endres et al., 1989; Novak et al., 2003). Consistent with these observations, our recent studies using the fat-1 mouse model showed that the production of TNFα was significant lower in splenocytes following LPS treatment (Bhattacharya et al., 2006), colons with DSS-induced inflammation (Hudert et al., 2006), livers with D-GalN/LPS-induced hepatitis (Schmocker et al., 2007) and retinas after hypoxia-induced injury (Connor et al., 2007), accompanied with a reduced severity of inflammation in the fat-1 mice rich in omega-3 PUFA when compared to that in wild type mice high in omega-6 PUFA.

### 5.2.4 IL-1β

Interleukin 1β (IL-1β) (a 17 kD protein) has many functions on many different cells and is secreted by a number of cells including macrophages, monocytes and dendritic cells. IL-1 induces fever and pain and activates various immune cells. It can also induce expression of many inflammatory genes including $PLA_2$ and $COX_2$, leading to increased production of pro-inflammatory eicosanoids (White et al., 2008). IL-1β acts via binds to two trans-membrane IL-1 receptors (IL-1R type 1 and 2). Activation of the IL-1R type leads to NFκB activation (Martin and Wesche, 2002). Similar to TNFα , it can be induced by NFκB (Hiscott et al., 1993). Together with TNFα, IL-1β plays an important role in the promotion of inflammatory response. In the fat-1 transgenic mice, we found that increased tissue content of omega-3 PUFA suppressed IL-1β expression in the target tissues of several inflammatory conditions, including DSS-induced colitis (Hudert et al., 2006), D-GalN/LPS-induced hepatitis (Schmocker et al., 2007) and cerulein-induced pancreatitis (unpublished results).

### 5.2.5 IL-6

Interleukin-6 (IL-6) is another important mediator of fever and inflammation and has been shown to be a prognostic indicator in human pancreatitis (Mayer et al., 2000; Stimac et al., 2006) as well as an indicator of disease severity in animal models of hepatitis (Sass et al., 2002). Our recent studies with fat-1 transgenic mice showed a significantly decreased expression of IL-6 in *fat-1* mice with hepatitis (Schmocker et al., 2007), lower IL-6 serum levels in the fat-1 mice with pancreatitis (unpublished results) and a reduced secretion of IL-6 from splenocytes from the fat-1 mice following LPS treatment (Bhattacharya et al., 2006).

These studies demonstrate a role played by an increased tissue status of n-3 fatty acids and decreased n-6/n-3 ratio in modulation of cytokine production (mainly NFκB, TNFα, IL-1β and IL-6).

## 5.3 Discussion and Conclusions

It is evident that omega-3 polyunsaturated fatty acids can modulate both the synthesis of lipid mediators and the production of cytokines. Their effects on lipid mediator formation include a reduction of pro-inflammatory eicosanoids (derived from the omega-6 AA) and an increase in anti-inflammatory mediators (derived from the omega-3 EPA and DHA themselves). The inhibitory effect on cytokine production might be subsequent to the changes in the types and/or amount of lipid mediators formed, or due to a direct effect of the fatty acids themselves on gene expression. Thus, it seems that omega-3 PUFA exert anti-inflammatory effects through a wider variety of metabolic pathways than thought previously. However, the relative contribution of each of them to the anti-inflammatory effect and the relationships among them (action sequence/pathway) remain to be further elucidated. Noticeably, there are looping feed-backs or interplays among cytokines, transcriptional factors and lipid mediators. For example, TNFα activates NFκB, which in turn induces TNFα and COX-2, leading to increased production of $PGE_2$, and consequently further activation of NFκB by $PGE_2$, and so on. Since NF-κB seems to be the core factor mediating all the effects/interplays, inhibition of NF-κB may be the key target for the omega-3 PUFA's effects on cytokine production and inflammation. Of interest, omega-3 PUFA appear to act also on some other targets in addition to NFκB. These multiple effects of omega-3 PUFA render them an effective natural anti-inflammatory agent. The possible interactions and working pathways are schemed in Fig. 5.2. However, further investigations are warranted to validate these potential effects and pathways by using qualified experimental models together with state-of-the-art technologies (e.g. lipidomics, genomics and proteomics). The understanding of the anti-inflammatory molecules derived from omega-3 PUFA and their regulatory signaling pathways provides new insights into the molecular pathophysiology of chronic diseases and opportunities for the design of therapeutic strategies.

In summary, omega-3 fatty acids not only modulate the production of lipid mediators but also alter expression of inflammatory genes. This broad spectrum of anti-inflammatory effects explains why omega-3 fatty acid can help in the

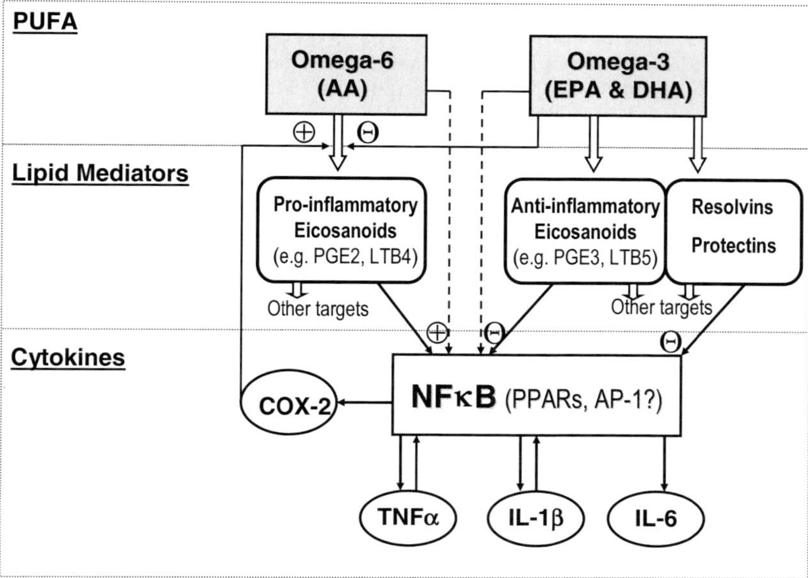


**Fig. 5.2** Possible mechanisms by which omega-3 fatty acids modulate the cytokine production and potential interplays of PUFA, lipid mediators and cytokines. ⊕, activation or up-regulation;Θ, inhibition or down-regulation

prevention, and possibly the amelioration of diseases with an inflammatory component, including diabetes, Alzheimer's, cancer, and cardiovascular disease (Calder, 2006; Simopoulos, 2002). Given the safety and multiple health benefits of omega-3 fatty acids, increased intake of the nutrients may be a promising approach to disease prevention, or an alternative therapy for inflammatory diseases.

**Acknowledgments** This work was supported by grants from the NIH (R01CA-113605) (to J.X.K.) and from an MSD scholarship award (to K.H.W.).

## References

Arita, M., Bianchini, F., Aliberti, J., Sher, A., Chiang, N., Hong, S., Yang, R., Petasis, N. A. and Serhan, C. N. Stereochemical assignment, antiinflammatory properties, and receptor for the omega-3 lipid mediator resolvin E1. J Exp Med **201** (2005a) 713–722.

Arita, M., Yoshida, M., Hong, S., Tjonahen, E., Glickman, J. N., Petasis, N. A., Blumberg, R. S. and Serhan, C. N. Resolvin E1, an endogenous lipid mediator derived from omega-3 eicosapentaenoic acid, protects against 2,4,6-trinitrobenzene sulfonic acid-induced colitis. Proc Natl Acad Sci USA **102** (2005b) 7671–7676.

Babcock, T. A., Helton, W. S., Hong, D. and Espat, N. J. Omega-3 fatty acid lipid emulsion reduces LPS-stimulated macrophage TNF-alpha production. Surg Infect (Larchmt) **3** (2002): 145–149.

Bhattacharya, A., Chandrasekar, B., Rahman, M. M., Banu, J., Kang, J. X. and Fernandes, G. Inhibition of inflammatory response in transgenic fat-1 mice on a calorie-restricted diet. Biochem Biophys Res Commun **349** (2006) 925–930.

Blok, W.L., Katan, M.B., and van der Meer, J.W.M. Modulation of inflammation and cytokine production by dietary n-3 fatty acids. J Nutr **126** (1996) 1515–1533.

Calder P.C. n–3 Polyunsaturated fatty acids, inflammation, and inflammatory disease. Am J Clin Nutr. **83** (2006)1505S–1519S.

Calder, P.C. Immunomodulation by omega-3 fatty acids. Prostaglandins Leukot Essent Fatty Acids **77** (2007) 327–335.

Collart, M.A., Baeuerle, P. and Vassalli, P. Regulation of tumor necrosis factor alpha transcription in macrophages: involvement of four kappa B-like motifs and of constitutive and inducible forms of NF-kappa B. Mol Cell Biol **10** (1990) 1498–1506.

Connor, K.M., SanGiovanni, J.P., Lofqvist, C., Aderman, C.M., Chen, J., Higuchi, A., Hong, S., Pravda, E.A., Majchrzak, S., Carper, D., Hellstrom, A., Kang, J.X., Chew, E.Y., Salem, N., Jr., Serhan, C.N. and Smith, L.E. Increased dietary intake of omega-3-polyunsaturated fatty acids reduces pathological retinal angiogenesis. Nat Med **13** (2007) 868–873.

Dinarello, C. A. Proinflammatory cytokines. Chest **118** (2000) 503–508.

Endres, S., Ghorbani, R., Kelley, V. E., Georgilis, K., Lonnemann, G., van der Meer, J.W., Cannon, J.G., Rogers, T. S., Klempner, M.S., Weber, P.C. and et al. The effect of dietary supplementation with n-3 polyunsaturated fatty acids on the synthesis of interleukin-1 and tumor necrosis factor by mononuclear cells. N Engl J Med **320** (1989) 265–271.

Hayden, M. S. and Ghosh, S. Shared principles in NF-kappaB signaling. Cell **132** (2008) 344–362.

Hayden, M. S., West, A. P. and Ghosh, S. NF-kappaB and the immune response. Oncogene **25** (2006) 6758–6780.

Hiscott, J., Marois, J., Garoufalis, J., D'Addario, M., Roulston, A., Kwan, I., Pepin, N., Lacoste, J., Nguyen, H., Bensi, G. and et al. Characterization of a functional NF-kappa B site in the human interleukin 1 beta promoter: evidence for a positive autoregulatory loop. Mol Cell Biol **13** (1993) 6231–6240.

Hong, S., Gronert, K., Devchand, P. R., Moussignac, R. L. and Serhan, C. N. Novel docosatrienes and 17S-resolvins generated from docosahexaenoic acid in murine brain, human blood, and glial cells. Autacoids in anti-inflammation. J Biol Chem **278** (2003) 14677–14687.

Hudert, C. A., Weylandt, K. H., Lu, Y., Wang, J., Hong, S., Dignass, A., Serhan, C. N. and Kang, J. X. Transgenic mice rich in endogenous omega-3 fatty acids are protected from colitis. Proc Natl Acad Sci USA **103** (2006) 11276–11281.

James, M. J., Gibson, R. A. and Cleland, L. G. Dietary polyunsaturated fatty acids and inflammatory mediator production. Am J Clin Nutr **71** (2000) 343S–348S.

Kang, J.X. Fat-1 transgenic mice: a new model for omega-3 research. Prostaglandins Leukot Essent Fatty Acids **77** (2007) 263–267.

Kang J.X., Wang, J., Wu, L., Kang, Z.B. *Fat-1* transgenic mice convert n-6 to n-3 fatty acids. Nature **427** (2004) 504.

Li, Y., Kang, J.K., and Leaf, A. Differential effects of various eicosanoids on the production or prevention of arrhythmias in cultured neonatal rat cardiac myocytes. Prostaglandins **54** (1997) 511–530.

Libby, P. Inflammatory mechanisms: the molecular basis of inflammation and disease. Nutr Rev **65** (2007) S140–S146.

Martin, M. U. and Wesche, H. Summary and comparison of the signaling mechanisms of the Toll/interleukin-1 receptor family. Biochim Biophys Acta **1592** (2002) 265–280.

Mayer, J., Rau, B., Gansauge, F. and Beger, H. G. Inflammatory mediators in human acute pancreatitis: clinical and pathophysiological implications. Gut **47** (2000) 546–552.

Novak, T. E., Babcock, T. A., Jho, D. H., Helton, W. S. and Espat, N. J. NF-kappa B inhibition by omega-3 fatty acids modulates LPS-stimulated macrophage TNF-alpha transcription. Am J Physiol Lung Cell Mol Physiol **284** (2003) L84–L89.

Nowak, J., Weylandt, K. H., Habbel, P., Wang, J., Dignass, A., Glickman, J. N. and Kang, J. X. Colitis-associated colon tumorigenesis is suppressed in transgenic mice rich in endogenous n-3 fatty acids. Carcinogenesis **28** (2007) 1991–1995.

Papadakis, K. A. and Targan, S. R. Role of cytokines in the pathogenesis of inflammatory bowel disease. Annu Rev Med **51** (2000) 289–298.

Peters-Golden, M., Canetti, C., Mancuso, P. and Coffey, M. J. Leukotrienes: underappreciated mediators of innate immune responses. J Immunol **174** (2005) 589–594.

Richard, A.G., Thomas, J.K., Barbara, A.O. and Janis, K. (Editors) *Immunology* (5th Edition, W. H. Freeman and Company, New York) 2000; pp 338–360.

Sass, G., Heinlein, S., Agli, A., Bang, R., Schumann, J. and Tiegs, G. Cytokine expression in three mouse models of experimental hepatitis. Cytokine **19** (2002) 115–120.

Schmocker, C., Weylandt, K. H., Kahlke, L., Wang, J., Lobeck, H., Tiegs, G., Berg, T. and Kang, J. X. Omega-3 fatty acids alleviate chemically induced acute hepatitis by suppression of cytokines. Hepatology **45** (2007) 864–869.

Schwab, J. M., Chiang, N., Arita, M. and Serhan, C. N. Resolvin E1 and protectin D1 activate inflammation-resolution programmes. Nature **447** (2007) 869–74.

Sellers, M. M. and Stallone, J. N. Sympathy for the Devil: The Role of Thromboxane in the Regulation of Vascular Tone and Blood Pressure. Am J Physiol Heart Circ Physiol. (2008) (Epub ahead of print– PMID: 18310512).

Serhan, C. N. Yacoubian, S., and Yang, Y. Anti-Inflammatory and Proresolving Lipid Mediators. Annu Rev Pathol. **28** (2008) 279–312.

Shapiro, A.C., Wu, D., and Meydani, S.N. Eicosanoids derived from arachidonic and eicosapentaenoic acids inhibit T cell proliferative response. Prostaglandins **45** (1993) 229–240.

Sijben, J. W. and Calder, P. C. Differential immunomodulation with long-chain n-3 PUFA in health and chronic disease. Proc Nutr Soc **66** (2007) 237–259.

Simopoulos, A. P. Omega-3 fatty acids in inflammation and autoimmune diseases. J Am Coll Nutr **21** (2002) 495–505.

Stimac, D., Fisic, E., Milic, S., Bilic-Zulle, L. and Peric, R. Prognostic values of IL-6, IL-8, and IL-10 in acute pancreatitis. J Clin Gastroenterol **40** (2006) 209–212.

Tilg, H., Jalan, R., Kaser, A., Davies, N. A., Offner, F. A., Hodges, S. J., Ludwiczek, O., Shawcross, D., Zoller, H., Alisa, A., Mookerjee, R. P., Graziadei, I., Datz, C., Trauner, M., Schuppan, D., Obrist, P., Vogel, W. and Williams, R. Anti-tumor necrosis factor-alpha monoclonal antibody therapy in severe alcoholic hepatitis. J Hepatol **38** (2003) 419–425.

White, K. E., Ding, Q., Moore, B. B., Peters-Golden, M., Ware, L. B., Matthay, M. A. and Olman, M. A. Prostaglandin E2 mediates IL-1beta-related fibroblast mitogenic effects in acute lung injury through differential utilization of prostanoid receptors. J Immunol **180** (2008) 637–646.

# Chapter 6
# Eicosanoids in Tumor Progression and Metastasis

**Sriram Krishnamoorthy and Kenneth V. Honn**

**Abstract** Eicosanoids and the enzymes responsible for their generation in living systems are involved in the mediation of multiple physiological and pathophysiological responses. These bioactive metabolites are part of complex cascades that initiate and perpetuate several disease processes such as atherosclerosis, arthritis, neurodegenerative conditions, and cancer. The intricate role played by each of these metabolites in the initiation, progression, and metastasis of solid tumors has been a subject of intense research in the scientific community. This review summarizes some of the key aspects of eicosanoids and the associated enzymes, and the pathways they mediate in promoting tumor progression and metastasis.

**Keywords** Eicosanoids · cyclooxygenase · lipoxygenase · platelets · metastasis

## 6.1 Introduction

Eicosanoids are 20-carbon lipid molecules derived from the enzymatic breakdown of membrane lipid precursors, chiefly arachidonic acid. Mediators in this family are generally, local and short acting metabolites, that have potent and stereospecific action in a multitude of host physiological and pathological processes like inflammation, asthma, and cancer (Funk, 2001). The three major enzyme families that are actively involved in catalyzing the conversion of arachidonic acid into the bioactive eicosanoids are the cyclooxygenases (COX), the lipoxygenases (LOX), and the epoxygenases (cytochrome p450 enzyme family). Lipid mediators generated by these enzymes are involved in a wide variety of cellular and molecular pathways, including but not limited to apoptosis, cell survival, proliferation, chemotaxis, senescence etc. This review paper will chiefly focus on the molecular mechanisms involved in the regulation of tumor progression and metastasis by these mediators importantly the COX and LOX generated metabolites.

---

S. Krishnamoorthy
Departments of Pathology, and Chemistry, Wayne State University School of Medicine, Detroit, MI 48202

P.J. Quinn, X. Wang (eds.), *Lipids in Health and Disease*,

## 6.2 COX and LOX Pathways of Arachidonic Acid Metabolism

Prostaglandins, bioactive metabolites generated from arachidonic acid, were first isolated from seminal fluid in the year 1935 by a Swedish physiologist, Ulf von Euler. This discovery came decades after the analytical synthesis of aspirin, a pain killer, derived from acetyl salicylic acid, the active principle in willow bark. Aspirin and other NSAIDs commonly used in the clinic, are potent inhibitors of the COX enzyme. This enzyme is involved in the conversion of arachidonic acid into prostaglandins (Fig. 6.1A). Cellular phospholipases ($PLA_2$) cleave membrane phospholipids at the sn-2 position to liberate arachidonic acid which is acted upon by either COX or LOX enzymes. In the COX pathway, the arachidonic acid thus released is presented to the enzyme prostaglandin H synthase (PGHS or COX), which converts it to $PGH_2$. $PGH_2$ undergoes further metabolism to downstream products which are formed in a cell type specific fashion, depending on

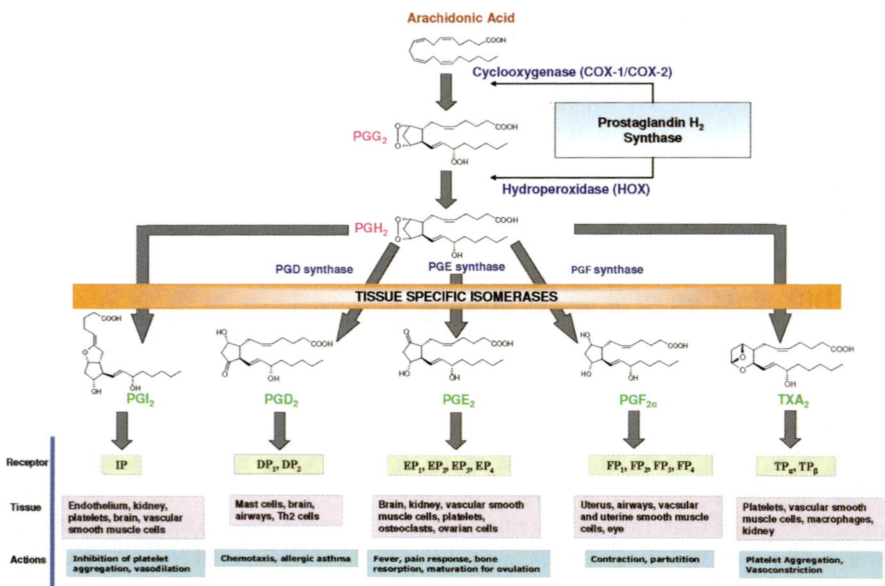


**Fig. 6.1** (**A**) Cyclooxygenase pathway of arachidonic acid metabolism. Arachidonic acid generated from cell membrane phospholipids, is acted upon by COX-1 or COX-2 to generate $PGG_2$ and subsequently $PGH_2$. Tissue specific isomerases take up $PGH_2$, and convert it into respective lipid mediators, for example $PGE_2$ synthase converts $PGH_2$ to $PGE_2$. Each of these lipid mediators have specific cell surface receptor(s), which relay signals to the interior of the cell resulting in physiological or pathophysiological consequences. (**B**) Lipoxygenase pathway of arachidonic acid metabolism. In this pathway, arachidonic acid is converted chiefly to the HETEs or the 5-LOX products, leukotrienes, in a cell or tissue specific manner. For example, platelets are abundant in 12-LOX, and produce 12-HETE. Resulting lipid mediators, have several cell type specific actions in both physiological and pathophysiological contexts

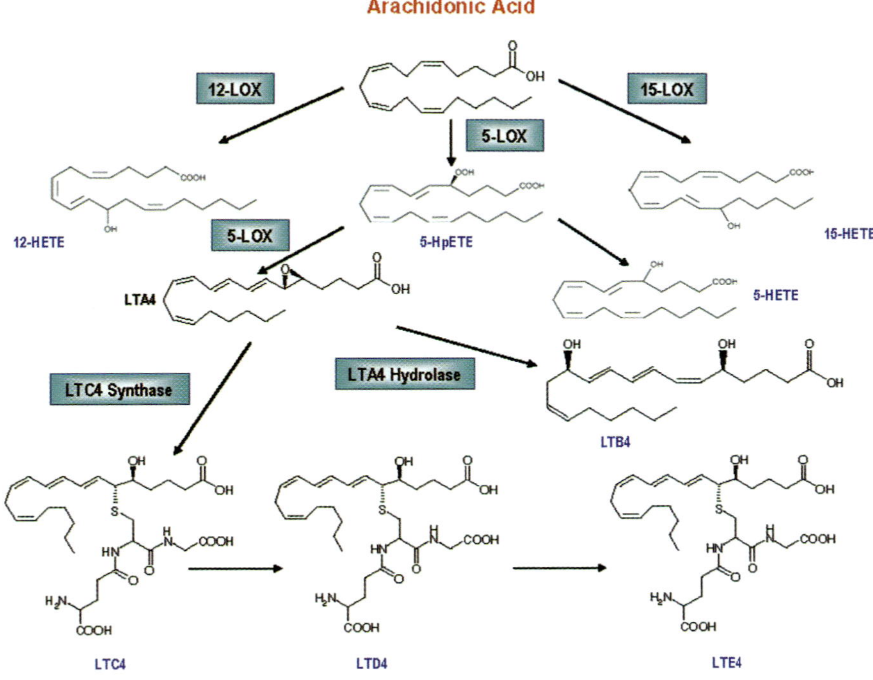


**Fig. 6.1** (Continued)

the presence of the enzyme in these cell types. For example, mast cells and brain convert $PGH_2$ to $PGD_2$ using the PGD synthase; platelets harbor the enzyme thromboxane synthase and produce thromboxanes from $PGH_2$; endothelial cells utilize arachidonic acid to form prostacyclin, catalyzed by prostacyclin synthase; uterine cells form $PGF_2\alpha$. Microsomal $PGE_2$ synthase enzyme is responsible for the formation of $PGE_2$ from $PGH_2$.

Conversely, arachidonic acid is also metabolized by various enzymes of the LOX pathway (Fig. 6.1B), to produce bioactive lipid mediators. The chief enzymes of the LOX pathway are the 5-LOX, 12-LOX, and the 15-LOX enzymes, whose names are derived from the position in which molecular oxygen is inserted by the enzyme species into the arachidonic acid backbone. The major bioactive products of the 5-LOX pathway are the leukotrienes (LT). Leukotrienes are formed from arachidonic acid presented to the 5-LOX enzyme by the 5-lipoxygenase-activating protein (FLAP). 5-LOX converts arachidonic acid to LTA4, which can be hyrdrolyzed by LTA4 hydrolase to LTB4. LTA4 can also give rise to glutathione conjugates by the action of LTC4 synthase, giving rise to the cysteinyl leukotrienes, namely LTC4, LTD4, and LTE4. Other enzymes, 12-LOX and 15-LOX give rise to 12-HETE and 15-HETE respectively, upon their action on arachidonic acid.

## 6.3 Cyclooxygenases, Metabolites and Tumor Progression

The COX enzymes, chiefly COX-2, have been linked to development and progression of multiple neoplasias. In the early 1980s, it was clinically observed that NSAIDs may inhibit tumor progression in patients with familial adenomatous polyposis (Waddell & Loughry, 1983). This was followed by the finding that administration of NSAIDs regressed intestinal polyps in an autochthonous rat model of intestinal cancer (Pollard et al., 1983). Epidemiological observations revealed that frequent intake of NSAIDs prevented cancers in humans. In a large prospective study, it was reported that people with aspirin intake had a reduced risk of fatal colon cancer (Thun et al., 1991). A few years after these findings, the expression of COX-2 at the message and protein levels was demonstrated to be upregulated in human colorectal adenomas and adenocarcinomas (Eberhart et al., 1994; Kargman et al., 1995) . Subsequent studies elaborated that, chemical inhibition or gene knockout of COX-2 in $Apc^{\Delta 716}$ knockout mice, a model of human familial adenomatous polyposis, led to a dramatic reduction in the number and size of intestinal polyps, which are precursor lesions for colon cancer (Oshima et al., 1996). Among the COX metabolites, prostaglandin $E_2$ ($PGE_2$) has been widely implicated in various steps of cancer development and progression such as angiogenesis, cell survival, proliferation, and chronic inflammation (Marnett & DuBois, 2002) (Fig. 6.2). In colorectal carcinomas, $PGE_2$ levels are maintained in a steady state by the actions of two pathways, i.e., the biosynthetic pathway catalyzed by the PG synthases and the enzymatic breakdown into the 15-keto $PGE_2$ metabolite by the enzyme 15-hydroxy prostaglandin dehydrogenase (15-PGDH). Loss of 15PGDH expression has been demonstrated in colorectal carcinomas, colorectal cancer cell lines, breast cancer, and lung cancer (Backlund et al., 2005). COX-2 overexpression also has been detected in prostate adenocarcinoma (Gupta et al., 2000; Yoshimura et al., 2000). The expression of COX-2 in tumors, was found to be upregulated by various oncogenes such as Her-2 or *ras* and downregulated by tumor suppressor genes like p53 (Subbaramaiah et al., 1999; Subbaramaiah et al., 2002). Several COX-2 inhibitors such as celecoxib and NS-398, have been demonstrated to induce apoptosis in prostate cancer cell lines (Hsu et al., 2000; Liu et al., 1998).

### 6.3.1 *Prostaglandin $E_2$*

Prostaglandin $E_2$ ($PGE_2$) is formed by the action of specific prostaglandin E synthases (PGES) which convert $PGH_2$ to $PGE_2$. Three isoforms of PGES exist, ie mPGES-1, mPGES-2, and cPGES. COX-2 and/or mPGES-1 have been found to be upregulated in several epithelial cancers (Muller-Decker & Furstenberger, 2007). The expression of mPGES-1 was found to be upregulated in 80% of non-small cell lung cancers and was localized to neoplastic epithelial cells The same study was able to identify that TNF-$\alpha$, a pro-inflammatory cytokine, was able

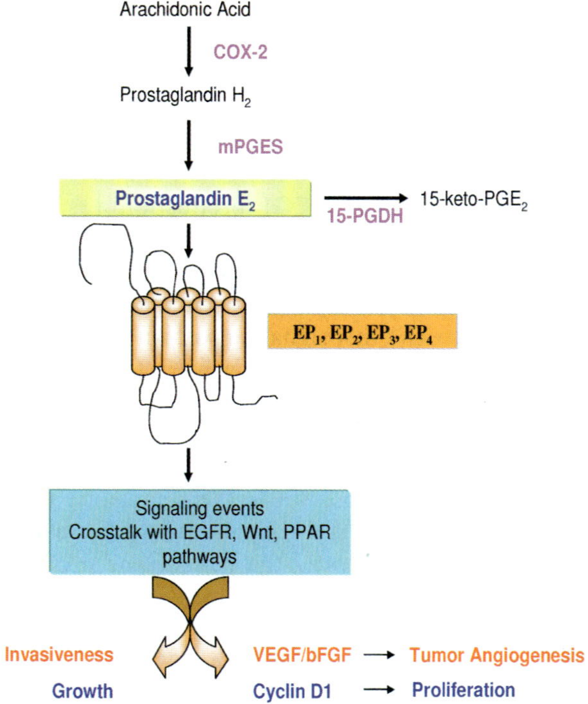


**Fig. 6.2** Prostaglandin $E_2$ – Formation, Degradation, and Role in Cancer progression. $PGE_2$ is formed by the conversion of arachidonic acid to $PGH_2$ by COX, which is later converted to the active $PGE_2$ by mPGES. $PGE_2$ is metabolised to the inactive keto form by PGDH. $PGE_2$ mediates its cellular actions by binding to one or more of the specific EP receptors on the cell surface of target cells resulting in the activation of one or more signaling pathways. Studies have suggested that $PGE_2$ crosstalks with other cell signaling pathways (see text for details), leading to upregulation of target genes of these pathways, which are vital for tumor progression and metastasis. Overexpression of COX-2 and / or mPGES have been reported in various cancers and cancer cell lines, leading to increased production of $PGE_2$, while downregulation or loss of PGDH in cancers results in poor metabolic inactivation of $PGE_2$. These events are critical in regulating the availability and actions of $PGE_2$ in the tumor microenvironmental milieu

to upregulate mPGES-1 in NSCLC cell lines, but failed to do so in a non-tumorigenic bronchial epithelial cell line (Yoshimatsu et al., 2001). Similarly overexpression of mPGES-1 was found in 70% of cases of invasive breast cancer by immunohistochemistry, but was undetectable in normal breast epithelial cells (Mehrotra et al., 2006).

$PGE_2$ exerts its molecular action by binding to unique cell surface G-protein coupled receptors namely EP1, EP2, EP3, and EP4. EP2 and EP4 signal via intracellular activation of cAMP through the $G\alpha s$ protein; EP1 signals through Gi raising intracellular calcium concentration; EP3 decreases cAMP formation via Gi. All four receptors have been demonstrated to be associated with tumor formation and progression, by independent studies (Cha & DuBois, 2007). For example, *Apc*$^{\Delta 716}$ mice with knockdown of the $EP_2$ receptor exhibited a decrease in the size and number of polyps. These polyps also showed a decrease in the secretion of the vascular endothelial growth factor (VEGF) linking $PGE_2$ signaling and angiogenesis in this model (Sonoshita et al., 2001). Growth of xenografts derived from sarcoma 180 and Lewis Lung carcinomas was significantly hampered in EP3 null mice. VEGF production was reduced in tumors lacking the EP3 receptor in this model, thus verifying the findings from the previous study (Amano et al., 2003). EP receptor signaling activated by $PGE_2$ is an intricate mechanism, in that the pathways emanating from the cell surface get involved in cross-talk with other signaling pathways which are important in tumor progression, especially in colorectal cancer. It has been demonstrated that $PGE_2$ treatment results in the activation of the Wnt signaling pathway elements, resulting in the translocation of β-catenin to the nucleus, and upregulation of Wnt target genes such as VEGF and cyclin D (Castellone et al., 2005; Shao et al., 2005). These observations lend credence to earlier findings emphasizing the significance of NSAIDs in inhibiting colon cancer progression. $PGE_2$ induced signaling also activates the peroxisome proliferator activated receptor δ (PPARδ) in colorectal cancer, which signals to activate the Wnt cascade, thus reinforcing the already existing crosstalk via the Wnt pathway. This phenomenon was demonstrated by Wang et al, in *Apc*$^{\Delta 716}$ mice, where $PGE_2$ activates PPARδ in a PI3K-Akt dependent fashion. The effect of $PGE_2$ in this system was abolished when PPARδ was knocked out in these mice (Wang et al., 2004). In addition to these pathways, $PGE_2$ stimulated signaling via the EP4 receptor, has also been shown to crosstalk with EGFR activated tyrosine kinase pathways through βarrestin 1/c-src signaling complex, resulting in downstream pathophysiological actions such as cancer cell migration and invasion, which is thought to play a crucial role in the metastatic spread of colorectal cancer to liver (Buchanan et al., 2006).

$PGE_2$ also plays a critical role in the modulation of the tumor angiogenic response. Several studies have clearly demonstrated that $PGE_2$ upregulates the production of pro-angiogenic factors VEGF and bFGF. In co-culture systems, it was found that COX-2 overexpressing colon cancer cells activated endothelial cell migration and tube formation, by virtue of the secretion of pro-angiogenic factors. These responses were inhibited by treating the cancer cells with aspirin, whereas, treatment of COX-2 negative cells did not affect the endothelial cell responses (Tsujii et al., 1998). It was also demonstrated that $PGE_2$ upregulates the chemokine receptor CXCR4, via VEGF and bFGF, on microvascular endothelial cells (Salcedo et al., 2003). The molecular mechanism by which $PGE_2$ upregulates VEGF in human colon carcinoma cells was found to be dependent on the induction of HIF-1α. The expression of VEGF mRNA in these cells, when exposed to $PGE_2$, was mediated by the transcriptional activator HIF-1α. $PGE_2$ mediated activation of HIF-1α was found to be signaled via activation of ERK phosphorylation and c-Src kinase activity, which demonstrates the requirement of multiple signaling pathway in the regulation of angiogenesis by this eicosanoid (Fukuda et al., 2003).

### *6.3.2 Thromboxane $A_2$*

Thromboxanes are biologically active lipid mediators, first identified in the year 1975, from washed human platelets incubated with arachidonic acid or $PGG_2$ for 30 s, which led to formation of a highly unstable factor, that induced irreversible platelet aggregation (Hamberg et al., 1975). Thromboxane $A_2$ ($TXA_2$), a short-lived lipid mediator, is formed by the action of thromboxane synthase on $PGH_2$. $TXA_2$ is extremely labile and breaks down to $TXB_2$, which is the biologically inactive form. Physiological functions for $TXA_2$ include platelet activation, aggregation, and thrombosis (Needleman et al., 1976). Human prostate cancer PC-3 cells express functionally active TX synthase enzyme and are able to biosynthesize $TXA_2$. TX synthase mRNA expression was found to be higher in prostate, renal and breast carcinomas compared to matched normal tissues. Immunohistochemical analysis demonstrated that normal prostate tissues showed weak expression of the enzyme compared to prostate carcinoma. The degree of expression of TX synthase was found to correlate with the severity of prostate carcinoma lesions, with advanced stages and poorly differentiated forms having the highest expression levels. Within the cancer tissue, expression of TX synthase was localized to areas of perineural invasion. The enzyme was found to be involved in motility, but not proliferation or survival, of prostate cancer cells (Nie et al., 2004). Overexpression of thromboxane synthase and/or increased $TXB_2$ levels were found also in lung cancer tissues, benign and malignant papillary thyroid carcinoma, renal carcinoma, and larynx squamous cell carcinoma (Bryant, 1994; Casey et al., 2004; Chen et al., 2006; Kajita et al., 2005; Pinto et al., 1993). Thromboxane synthase was also found to be over-expressed in several forms of bladder cancers such as transitional cell carcinoma, squamous cell carcinoma, and adenocarcinoma compared to non-tumor tissue. Patients with more than 4-fold increase in TXS expression were found to have a poor prognosis. Treatment of bladder cancer cell lines with TXS inhibitors and $TXA_2$ receptor antagonists resulted in a decrease in cell survival, migration, and invasion, and the opposite effects were seen with receptor agonists (Moussa et al., 2005). $TXA_2$ was also shown to mediate endothelial functions such as migration and angiogenesis. Pro-angiogenic factors bFGF and VEGF was demonstrated to enhance $TXA_2$ synthesis in endothelial cells by 3-5 fold. Inhibition of TXS activity resulted in a decrease in the endothelial cell migration response to VEGF and bFGF. Similar changes were seen with $TXA_2$ receptor antagonists (Nie et al., 2000a).

$TXA_2$ mediates its cellular functions by binding to its receptors -TP$\alpha$ and TP$\beta$. These are GPCRs which couple to Gq, G11, and G12/13 and elevate intracellular calcium levels, by signaling via PLC dependent inositol phosphate generation, leading to vasoconstriction and platelet aggregation (Breyer et al., 2001). TPα and TPβ are alternatively spliced variants that differ in amino acid sequences distal to Arg-328 at the C-terminal end of the receptor. mRNAs of both the receptors are expressed widely in organs such as lungs, kidneys, liver, uterus, heart, etc. (Miggin & Kinsella, 1998). Overexpression of the TP receptor has been identified in human breast cancer specimens compared to normal breast tissues. TP overexpression was seen in aggressive tumors and linked with poor prognosis. The same study also found that TXAS expression was significantly low in high grade tumors and in patients with a poor prognosis (Watkins et al., 2005). Recent studies from our group have clearly demonstrated that prostate carcinoma express functionally active TP receptors. Signaling via these receptors were found to regulate prostate cancer cell motility and migration in a Rho dependent fashion and subsequent reorganization of the cytoskeleton (Nie et al., 2008).

### *6.3.3 Prostacyclin*

Prostacyclin was discovered by Sir John Vane's group in the year 1976 as a substance generated by arterial vessel walls mediating relaxation of mesenteric and celiac arteries and inhibiting platelet aggregation (Bunting et al., 1976). The structure of prostacyclin was deduced in the same year (Whittaker et al., 1976). Prostacyclin or $PGI_2$ is an eicosanoid lipid mediator synthesized chiefly in endothelial cells, by the action of prostacyclin synthase on $PGH_2$. The major physiological action of $PGI_2$ is to oppose the functions of thromboxanes, inhibit platelet aggregation, and promote vasodilation. Together, prostacyclins and thromboxanes play a pivotal role in maintaining cardiovascular homeostasis (Wu & Liou, 2005). Prostacyclin mediates its cellular actions by binding to the cell surface $PGI_2$ receptor, named IP, a seven transmembrane GPCR (Narumiya et al., 1999). Knockout models of the IP receptor in mice have demonstrated the occurrence of thrombosis, reperfusion injury, intimal hyperplasia, and restenosis (Cheng et al., 2002; Murata et al., 1997; Xiao et al., 2001). Prostacyclin, in concert with thromboxanes play a major role in regulating metastasis of cancers *(see below in section on role of eicosanoids in platelet-tumor interactions)*.

## 6.4 Lipoxygenases and Tumor Progression

Lipoxygenases (LOX) constitute a family of lipid peroxidizing enzymes, which are distributed widely in the plant and animal kingdoms. These enzymes preferentially metabolize substrates that are polyunsaturated fatty acids containing a series of *cis* double bonds, which are the essential fatty acids for human beings (Kuhn & Thiele, 1999). Lipoxygenases are dioxygenases in nature and catalyze the stereospecific insertion of molecular oxygen into polyunsaturated fatty acids (PUFA). The primary products of the lipoxygenase reaction are the hydroperoxy fatty acids. Based on the currently used nomenclature, lipoxygenases are classified with respect to their positional specificity of arachidonic acid oxygenation. In mammalian cells, there are three major types of lipoxygenases – 5-LOX, 12-LOX, and 15-LOX. The major end product of LOX mediated enzymatic breakdown of arachidonic acid is hydroxyeicosa 5, 8, 10, 14-tetraenoic acid (HETE). 12-LOX introduces molecular oxygen at carbon 12 and forms 12-HpETE which is converted to 12-HETE nonenzymatically. Similarly 5-LOX and 15-LOX catalyze the conversion of arachidonic acid into 5-HETE and 15-HETE respectively (Marks & Furstenberger, 1999). In case of the 5-LOX enzyme, 5-HETE is further metabolized in the presence of the 5-Lipoxygenase Activating Protein (FLAP), to the epoxide intermediate, leukotriene A4 (LTA4). LTA4 could be subsequently converted enzymatically by LTA4 hydrolase to leukotriene B4 (LTB4). LTA4 can also undergo enzymatic conjugation to glutathione, generating the cysteinyl leukotrienes LTC4, LTD4, and LTE4. Each LOX is predominantly expressed in a specific tissue. For example, 5-LOX is abundantly expressed in polymorphonuclear leukocytes (Funk & FitzGerald, 1991; Funk et al., 1989). Similarly, 15-LOX represents one of the major proteins besides hemoglobin in reticulocytes during anemia (Fleming et al., 1989; Rapoport et al., 1979; Turk et al., 1982) whereas platelets constitutively express 12-LOX (Chen & Funk, 1993).

### *6.4.1 5-LOX and metabolites*

5-LOX enzymatic activity was first described in 1976 by Borgeat et al in rabbit polymorphonuclear leukocytes (Borgeat et al., 1976). 5-LOX catalyzes the conversion of arachidonic acid to 5-HETE or LTA4, which generates LTB4 and the cysteinyl leukotrienes. Majority of the 5-LOX metabolites such as LTB4 and the cysteinyl leukotrienes play crucial roles in the inflammatory processes of the host. LTB4 is a potent neutrophil chemotactic factor and also stimulates transendothelial migration of these cells. Cys-LTs are chiefly involved in mounting the allergic inflammatory response. Specific receptors for LTB4 and Cys-LTs have been identified. Two receptors for LTB4 have been cloned – BLT1 and BLT2. The Cys-LTs mediate their physiological actions by binding to the Cys-LT1 and Cys-LT2 receptors (Funk, 2001).

Several reports have suggested an association between 5-LOX, 5-(S)HETE formation and carcinogenesis. The expression of 5-LOX has been documented in several cancers including prostate, colon, lung, breast, pancreas, bone, brain, and mesothelium (Romano & Claria, 2003). Inhibition of 5-LOX enzymatic activity by specific chemical inhibitors blocked the stimulatory effect of arachidonic acid on the growth of prostate cancer cells. Conversely, addition of 5-HETE but not leukotrienes promoted the growth of prostate cancer cells, suggesting the involvement of this metabolite in modulating growth stimulatory effects in prostate cancer (Ghosh & Myers, 1997). Similarly, 5-LOX and 5-HETE were found to mediate growth of lung cancer cells, and inhibition of this pathway led to growth arrest and apoptosis of these cells (Avis et al., 1996). In the case of pancreatic cancer and mesothelioma, low or undetectable levels of 5-LOX were observed in normal cells and tissues, compared to the tumor cells, which had high expression and activity of this enzyme (Hennig et al., 2002; Romano et al., 2001). Specific inhibition of 5-LOX led to apoptosis in prostate cancer cells and blocked proliferation in human leukemia cell lines (Ghosh & Myers, 1998; Tsukada et al., 1986). Overexpression of LTB4 receptor was also identified in human pancreatic cancer tissues by immunohistochemistry (Hennig et al., 2002). Upregulation of LTD4 receptor, Cys-LT1 has been detected in colorectal adenocarcinoma (Ohd et al., 2003). In a human malignant mesothelioma model, it was reported that 5-LOX and LTA4 but not LTB4, upregulated the expression and secretion of the proangiogenic factor VEGF. Selective inhibition of VEGF, a prosurvival factor for mesothelioma cells, brought about by 5-LOX antisense or inhibitors resulted in apoptotic cell death, suggesting the involvement of 5-LOX and its metabolites in promoting tumor cell survival (Romano et al., 2001).

### *6.4.2 12-LOX and 12(S)-HETE*

The conversion of arachidonic acid to 12S-hydroxy-5,8,10,14 eicosatetraenoic acid 12(S)-HETE was first demonstrated in human and bovine platelets (Yoshimoto & Takahashi, 2002). The three predominant forms of 12-LOX are platelet-type 12-LOX, epidermis-type 12-LOX, and leukocyte-type 12-LOX (Limor et al., 2001; Siebert et al., 2001; Yamamoto et al., 1997). These are distinct enzymes by sequence, catalytic properties, and function. 12-LOX and its product 12-HETE have been shown to be involved in a variety of cancers. The mRNA of 12-LOX has been detected in erythroleukemia, colon carcinoma, epidermoid carcinoma A431 cells, human glioma, prostate, and breast cancer cells. Additionally, this enzyme also has been detected in smooth muscle cells (Kim et al., 1995), keratinocytes (Krieg et al., 1995), and endothelial cells (Funk et al., 1992). The product of 12-LOX in amelanotic melanoma cells has been found to be the S enantiomer (12(S)-HETE) by GC-MS spectral analysis (Liu et al., 1994). Production of endogenous 12S-HETE has been documented in human colon carcinoma, rat Walker carcinosarcoma, mouse melanoma and lung carcinoma (Chen et al., 1994). In human prostate carcinoma, the level of platelet-type 12-LOX expression was correlated with the tumor stage and grade (Gao et al., 1995; Timar et al., 2000). Platelet-type 12-LOX mRNA was shown to be increased in prostate cancer tissues, and the expression correlated with clinical stage of the disease (Timar et al., 2000). Studies using inhibitors of platelet-type 12-LOX have shown that the blocking 12-LOX activity results in the arrest of cell cycle progression and induction of apoptosis in prostate cancer cell lines (Pidgeon et al., 2002). These studies have demonstrated that the 12-LOX pathway plays an important role in regulating prostate cancer progression and apoptosis. Urinary levels of 12(S)-HETE, the metabolite produced by 12-LOX, in prostate cancer patients are significantly elevated when compared to normal individuals and removal of the prostate gland results in a significant decrease in the urinary concentration of this eicosanoid (Nithipatikom et al., 2006) In contrast, other HETEs (i.e. 5-and 15-HETE), although detected remain unchanged following radical prostatectomy (Nithipatikom et al., 2006). Approximately 38% of the 138 prostate cancer patients studied exhibited an elevated expression of 12-LOX at the mRNA level in prostate tumor tissues compared to matched normal tissues. This elevated 12-LOX mRNA expression was found to have a positive correlation with advanced stage and poor differentiation of prostate cancer (Gao et al., 1995). 12-LOX and 12-HETE have been shown to be important determinants of tumor cell survival and apoptosis (Honn et al., 1996).

12-LOX has been found to be an important marker for cancer progression within the melanoma system, and therefore could be a useful biomarker and therapeutic target for melanoma chemoprevention (Winer et al., 2002). It also has been found that 12(S)-HETE is involved in the proliferation of pancreatic carcinoma cells (Ding et al., 2001) and inhibition of the 12-LOX has been shown to cause apoptosis of these cells (Tong et al., 2002). 12-LOX and 12(S)-HETE were also found to enhance proliferation and survival of gastric cancer cell lines (Wong et al., 2001).

Overexpression of the platelet-type 12-LOX in human prostate cancer PC-3 cells stimulated growth by enhanced tumor angiogenesis (Nie et al., 1998) and 12(S)-HETE, the sole and stable end product of arachidonic acid metabolism by the platelet-type 12-LOX, has been shown to protect tumor cells from apoptosis and induce invasion, motility, and angiogenesis (Gao & Honn, 1995; Honn et al., 1994a; Nie et al., 2000c) as well as surface expression of $\alpha v\beta 3$ integrin (Tang et al., 1993b). Promotion of such divergent biological functions by 12(S)-HETE is indicative of a complex signaling mechanism leading to metastasis and survival of tumor cells. Attempts at understanding the 12(S)-HETE signaling mechanisms revealed several interesting features involving G-proteins. While specific 12(S)-HETE receptor(s) is yet to be identified, both high and low affinity binding sites have been identified on B16a murine melanoma cells (Liu et al., 1995), keratinocytes (Arenberger et al., 1993), and A431 cells (Szekeres et al., 2000a). We have recently elucidated some of the signaling events down stream of the putative 12(S)-HETE receptor (Szekeres et al., 2000a; Szekeres et al., 2000b). 12(S)-HETE stimulates phosphorylation of PLCγ1, which in turn is responsible for the activation of PKCα. PKCα plays a significant, but not exclusive role in ERK1/2 activation. Further, we have demonstrated the 12(S)-HETE induced activation of Src family kinases, and the subsequent phosphorylation of adapter proteins Shc and Grb2, which lead to activation of ERK1/2 *via* Ras. Our work also suggested that protein tyrosine phosphatases are involved in 12(S)-HETE signaling in A431 cells. Additionally, 12(S)-HETE also was shown to activate PI3 kinase which is rate limiting for ERK1/2 activation (Szekeres et al., 2000b). PI3 kinase mediated activation of ERK by 12(S)-HETE involves PKCζ. Thus, 12(S)-HETE activates (through its putative receptor) PKCα *via* PLCγ1 and stimulates PKCζ *via* inositide kinase. Both PKC isoforms contribute to phosphorylation of the Raf/MEK/ERK cascade. The natural convergence point for this vast array of mitogenic signaling mechanisms is the extremely variable transcriptional machinery.

One possible integrator of these 12(S)-HETE signaling mechanisms in tumor cells is the pleiotropic transcription factor NF-κB, which plays an important role in the control of cell proliferation and apoptosis. Using PC-3 prostate cancer cells, we have shown that either overexpression of the platelet-type 12-LOX or exogenously added 12(S)-HETE activates NF-κB (Kandouz et al., 2003). NFκB is normally sequestered in the cytosol where it is bound to IκBα proteins (Ghosh et al., 1998). 12(S)-HETE induced the degradation of IκBα, which resulted in the nuclear translocation of NF-κB and enhanced transcriptional activity. Among the activators and regulators of NF-κB complex are several members of the MEK kinase family (MEKK1, 2, and 3 and NIK), TAK1, PKCζ, and S6 kinase (Baumann et al., 2000; Lee et al., 1998; Nakano et al., 1998; Nemoto et al., 1998; Pearson et al., 2001). Given the role of 12(S)-HETE in the activation of the MAP kinase signaling pathway to induce Raf, PLC, PKCζ, as well as PI3 kinase (Szekeres et al., 2000a; Szekeres et al., 2000b) and the role of MAP kinase cascade in the activation of NF-κB, it is highly likely that the 12(S)-HETE activation of NF-κB proceeds *via* the MAP kinase cascade. A recent report on the activation of NF-κB by MEKK1 in Raf mediated cell transformation suggests the potential for a 12(S)-HETE induced Raf participation in the signaling pathway (Baumann et al., 2000).

The effect of 12-LOX on tumor growth in vivo, has been positively correlated with its ability to increase tumor angiogenesis (Nie & Honn, 2002). Studies have revealed that prostate cancer cells expressing high levels of platelet-type 12-LOX are more angiogenic than those expressing none or low levels of 12-LOX. The same study also revealed that the increased angiogenicity of the 12-LOX overexpressing cells is directly related to the ability of the metabolite 12(S)-HETE to stimulate endothelial cell migration (Nie et al., 1998). It was identified by two separate groups that 12-LOX induced stimulation of angiogenesis involved the upregulation of the pro-angiogenic factor VEGF (McCabe et al., 2006; Nie et al., 2006). The 12-LOX inhibitor BMD 122 (Nbenzyl-N-hydroxy-5-phenyl-pentanamide) was found to reduce endothelial cell proliferation stimulated by basic fibroblast growth factor (bFGF) or by vascular endothelial growth factor (VEGF). This inhibition could be partly restored by the addition of 12(S)-HETE. The same inhibitor also blocked *in vitro* blood vessel formation by rat vascular endothelial cells. These findings taken together have proved that arachidonic acid metabolism in endothelial cells through the 12-LOX pathway plays a critical role in angiogenesis (Nie et al., 2000b).

Overexpression of 12-LOX in breast cancer cells has resulted in enhanced tumor angiogenesis and growth in a fat pad animal model (Connolly & Rose, 1998).

Many studies have demonstrated that 12(S)-HETE is a proangiogenic agent. 12(S)HETE has a variety of effects on endothelial cells. This metabolite has been shown to upregulate the surface expression of $\alpha v\beta 3$ integrin in rat aorta endothelial cells and murine pulmonary vascular endothelial cells (Tang et al., 1993a, b, 1994, 1995a). 12(S)HETE has also been found to act as a mitogen for microvascular endothelial cells (Tang et al., 1995b). Taken together, the above findings clearly illustrate a definitive role for 12-LOX in tumor angiogenesis.

### 6.4.3 15-Lipoxygenase and Metabolites

Two isoforms of 15-LOX exist – 15-LOX-1 and 15-LOX-2. 15-LOX-1 was originally described in rabbit reticulocytes in the year 1975 as an enzyme that oxidizes phospholipids in intact mitochondria and cell membrane (Schewe et al., 1975). In 1988, the human ortholog of 15-LOX-1 was purified from eosinophilic leukocytes (Sigal et al., 1988). In 1997, a second isoform of 15-LOX, namely 15-LOX-2 was identified from human hair roots. This enzyme has only a low degree of sequence similiarity with 15-LOX-1, and was found to be expressed in prostate, skin, lung, and cornea (Brash et al., 1997). The 15-LOX-1 isoform is capable of utilizing both arachidonic acid and linoleic acid as substrates. It inserts molecular oxygen at C-15 of arachidonic acid to generate 15-HPETE, and at the C-13 of linoleic acid to produce 13-HODE (Kuhn et al., 2002). Conversely, 15-LOX-2 does not act efficiently on linoleic acid, and converts arachidonic acid to 15(S)HETE. This enzyme shares the highest homology with the mouse 8-LOX enzyme (Tang et al., 2007).

The role of 15-LOX-1 in carcinogenesis is controversial, and both pro and anti-tumor properties have been assigned to this enzyme by various researchers. Ikawa et al, identified stronger immunostaining for 15-LOX-1 in tumors compared to normal tissue, in 18 of 21 matched pairs of colorectal tumor samples and adjacent normal tissues (Ikawa et al., 1999). In contrast, a study by Shureiqi et al., reported weak immunohistochemical scores for 15-LOX-1 and lower levels of 13HODE in colorectal cancer when compared with normal tissue. It was concluded that 15-LOX-1 expression and 13-HODE generation may have tumor suppressor properties in colorectal cancer, by suppressing cell proliferation and promoting apoptosis (Shureiqi et al., 1999). Later, it was identified that COX inhibitors upregulated 15-LOX-1 expression in colon cancer cells, and this upregulation may be related to triggering apoptotic pathways in these cells (Shureiqi et al., 2000). Similarly, 15-LOX-1 expression and 13-HODE production was found to be downregulated in esophageal cancer cells, and inhibition of 15-LOX-1 activity was found to dampen the cytotoxic effects of NSAIDs in these cells, which was rescued by exogenous addition of 13(S)-HODE, suggesting the tumor suppressor roles for these molecules (Shureiqi et al., 2001). Spindle et al, have identified that the levels of 13(S)-HODE, measured by enzyme immunoassay, is elevated in human prostate carcinoma. In another study, 15LOX-1 was found to be coexpressed with a mutant p53 isoform, in 48 prostatectomy specimens with a high degree of statistical significance ($p < 0.001$), and enzyme expression correlated with Gleason staging of prostate cancer (Kelavkar et al., 2000). It was later identified that 15-LOX-1 is overexpressed in PC-3 prostate cancer cell lines and prostate adenocarcinoma. Lower levels of 15LOX-1 were identified in normal prostate tissues compared to 15-LOX-2, whereas 15-LOX-1 overexpression and concomitant 15-LOX-2 downregulation was observed in cancerous prostate tissues suggesting the involvement of 15-LOX-1 in prostate cancer (Kelavkar et al., 2006). On the contrary, 15-LOX-1 expression and activity was found to exert anti-tumor properties in human pancreatic cancer. Immunohistochemical analysis demonstrated that pancreatic cancer cell lines show a weak expression of this enzyme, and poor staining was seen in tumor samples compared to normal. Restitution of 15-LOX-1 activity in these cancer cells resulted in decreased cell growth suggesting a tumor suppressor role for this enzyme in pancreatic cancer (Hennig et al., 2007).

The second isoform of 15-LOX, namely 15-LOX-2 is considered to be a tumor suppressor, based on studies conducted in various cancers. The expression of 15-LOX-2 is restricted primarily to the prostate and also to the skin, lung, and cornea (Brash et al., 1997). The expression and activity of this enzyme is highly diminished in high grade prostatic intraepithelial neoplasia (HGPIN) and prostate cancer (Shappell et al., 1999; 2001b). Downregulation of 15LOX-2 was also observed in benign and neoplastic sebaceous glands, esophageal cancer and lung cancer (Gonzalez et al., 2004; Shappell et al., 2001a; Xu et al., 2003). The exact molecular mechanisms leading to the loss of 15-LOX-2 expression in prostate cancer cells is still under investigation and many hypotheses have been tested. Studies conducted in normal human prostate epithelial cells (NHP) have revealed that 15-LOX-2 has multiple alternatively spliced isoforms. In these cells 15-LOX-2 functions as a negative regulator of cell cycle progression. The expression of 15-LOX-2 in the NHP cells is regulated by Sp1 and is repressed by Sp3 (Tang et al., 2007).

## 6.5 Role of Eicosanoids in Platelet-Tumor Interactions

The involvement of platelets in assisting hematogenous spread of metastatic tumor cells and the interactions between platelets, cancer cells, and the blood vessel wall were proposed decades ago. This was confirmed in experimental model systems of thrombocytopenia which showed inhibition of metastasis (Gasic et al., 1973; Kimoto et al., 1993). Honn et al, proposed the first hypothesis on the involvement of bioactive lipid mediators, specifically $TXA_2$

and $PGI_2$, produced by the blood vessel wall, platelets, and tumor cells to play a complex and intricate role in the metastatic spread of cancer cells via the blood stream (Fig. 6.3). It was hypothesized that an intricate balance exists between $TXA_2$ and $PGI_2$ in mediating cancer cell metastasis in patients, and any pathophysiological disruption in this equilibrium favoring dominance of $TXA_2$ over $PGI_2$, would promote metastasis of tumor cells from the primary site (Honn & Meyer, 1981). In agreement with this hypothesis, it was demonstrated that $PGI_2$ is a potent antimetastatic agent. In this study, experimental metastasis of B16 melanoma cells to the lungs was dramatically inhibited by $PGI_2$, a response, which was not reproduced by other lipid mediators like $PGE_2$, which is a vasodilator but not a platelet anti-aggregating factor. $PGD_2$ which also functions similar to $PGI_2$ in preventing platelet aggregation, was able to mount this antimetastatic response, but was less potent compared to the latter. Intracellular actions of $PGI_2$ are mediated by the signaling molecule cAMP. Thus, combination of $PGI_2$ treatment with a phosphodiesterase inhibitor, which prolongs the half-life of intracellular cAMP, potentiated the actions of $PGI_2$ in inhibiting platelet aggregation and reducing tumor metastasis (Honn et al., 1981). In a subsequent study, it was revealed that $PGI_2$ blocks platelet tumor cell aggregation promoted by cathepsin B or calpains, whose actions are similar to cathepsin B (Honn et al., 1982). Yet another study demonstrated that $PGI_2$ is effective in inhibiting both tumor cell

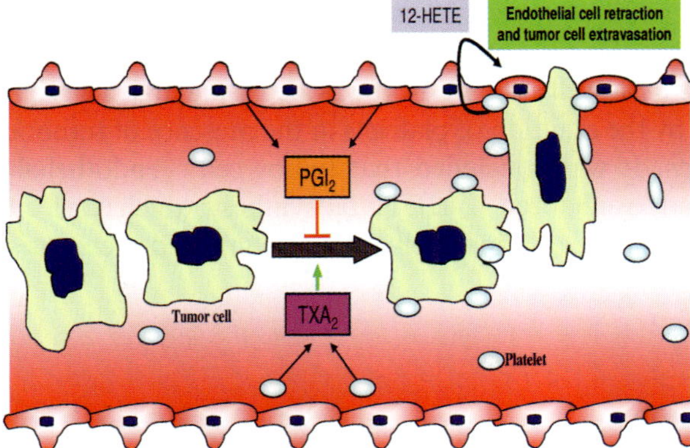


**Fig. 6.3** Eicosanoids and Tumor-platelet interactions in metastasis and the role of 12-HETE in tumor cell extravasation. Studies have clearly shown that hematogenous route of metastasis spread of cancer cells, involves interactions with platelets. Tumor-platelet interactions and subsequent aggregation is critically controlled by a delicate balance between the level of endothelium derived $PGI_2$ and platelet or tumor derived $TXA_2$. Elevated $TXA_2$ levels in the circulation can tip the balance towards platelet aggregation and tumor metastasis to distant organs, whereas increases in $PGI_2$ levels can block this interaction preventing spread of cancer cells. Shown in this illustration is a schematic of a blood vessel, with metastatic tumor cells interacting with platelets. Interactions of tumor cells with platelets and endothelial cells have been demonstrated to induce 12(S)-HETE production, which leads to retraction of endothelial cell layers enabling metastatic tumor cells to extravasate and set up secondary colonies of metastasis

induced platelet aggregation (TCIPA) and platelet facilitated tumor cell adhesion in vitro. In this study, $PGI_2$ was 100-fold more potent that $PGE_1$ or $PGD_2$, and 1000-fold more potent than its non-enzymatic metabolite 6-keto $PGF_2\alpha$. Interestingly, $PGE_2$ did not inhibit TCIPA but blocked TCIPA induced by $PGI_2$. It was also demonstrated that platelets promoted the adhesion of W256 cells to the culture dish, and $PGI_2$ was able to block this platelet induced tumor cell adhesion, suggesting that $PGI_2$ has antimetastatic actions in vivo (Menter et al., 1984). In agreement with these findings, it was demonstrated that patients with bone tumors had extremely low plasma levels of 6-keto-$PGF_{1\alpha}$, the stable hydrolysis product of $PGI_2$, whereas plasma concentrations of $TXA_2$ were in the normal range. Patients with malignant bone tumors were also found to have deficiency of plasma factors repsonsible for stabilization of $PGI_2$. Arterial tissues from patients with malignant disease were found to generate very low concentrations of $PGI_2$ compared to normal individuals (reviewed by Mehta, P.) (Mehta, 1984). Inhibition of TX synthase activity and treatment with a $TXA_2$ receptor antagonist were found to inhibit metastasis of B16 melanoma cells injected into the tail vein of mice (Honn, 1983). Later, it was reported that W256 cells can produce $TXA_2$ and 12-HETE concomitant to inducing platelet aggregation. The role of these eicosanoids in promoting TCIPA was verified by using COX and LOX inhibitors, combination treatment of which led to ablation of platelet aggregation, in repsonse to tumor cells (Honn et al., 1987). Subsequent studies revealed that 12-HETE treatment led to retraction of CD3 endothelial cell monolayers, as assessed by quantitative binding assays and by phase contrast microscopy. Platelet enhanced endothelial cell retraction was blocked by treating either tumor cells or platelets with lipoxygenase inhibitors as well as by $PGI_2$. It was also found that 12-HETE biosynthesis was enhanced by the addition of platelets in the presence of tumor cell endothelial cell interactions. It was concluded that tumor cell – endothelial cell – platelet interactions lead to enhanced 12-HETE biosynthesis which can lead to endothelial cell retraction facilitating tumor cell extravasation and metastasis (Honn et al., 1994b).

## 6.6 Conclusion

Eicosanoids are bioactive lipid mediators which have clearly defined roles in mediating tumor progression and metastasis. Both arms of the arachidonic acid metabolism pathway ie the COX and the LOX pathways, have been demonstrated to be pivotal in promoting the malignant phenotype. Some of these metabolites have also been demonstrated to have an anti-tumor effect. Exploiting these pathways will help in better understanding the intricacies of the metastatic cascade and to develop newer therapeutic agents aimed at blocking the actions of these metabolites. Inclusion of specific and safe inhibitors of these pathways along with the usual regimen of anti-tumor drugs, can be beneficial in better control of tumor progression and metastasis.

**Acknowledgments** Work described in this review contributed by the authors is supported by NIH grant R01CA114051 (K.V. Honn) and National Institutes of Health Grant CA-29997 (K.V. Honn).

## References

Amano H, Hayashi I, Endo H, Kitasato H, Yamashina S, Maruyama T, Kobayashi M, Satoh K, Narita M, Sugimoto Y, Murata T, Yoshimura H, Narumiya S, Majima M. Host prostaglandin E(2)-EP3 signaling regulates tumor-associated angiogenesis and tumor growth. J Exp Med. 197 (2003) 221–232.

Arenberger P, Kemeny L, Ruzicka T. Characterization of high-affinity 12(S)-hydroxyeicosatetraenoic acid (12(S)-HETE) binding sites on normal human keratinocytes. Epithelial Cell Biol. 2 (1993) 1–6.

Avis IM, Jett M, Boyle T, Vos MD, Moody T, Treston AM, Martinez A, Mulshine JL. Growth control of lung cancer by interruption of 5-lipoxygenase-mediated growth factor signaling. J Clin Invest. 97 (1996) 806–813.

Backlund MG, Mann JR, Holla VR, Buchanan FG, Tai HH, Musiek ES, Milne GL, Katkuri S, DuBois RN. 15-Hydroxyprostaglandin dehydrogenase is down-regulated in colorectal cancer. J Biol Chem. 280 (2005) 3217–3223.

Baumann B, Weber CK, Troppmair J, Whiteside S, Israel A, Rapp UR, Wirth T. Raf induces NF-kB by membrane shuttle kinase MEKK1, a signaling pathway critical for transformation. Proc.Natl.Acad.Sci.U.S.A. 97 (2000) 4615–4620.

Borgeat P, Hamberg M, Samuelsson B. Transformation of arachidonic acid and homo-gammalinolenic acid by rabbit polymorphonuclear leukocytes. Monohydroxy acids from novel lipoxygenases. J Biol Chem. 251 (1976) 7816–7820.

Brash AR, Boeglin WE, Chang MS. Discovery of a second 15S-lipoxygenase in humans. Proc Natl Acad Sci U S A. 94 (1997) 6148–6152.

Breyer RM, Bagdassarian CK, Myers SA, Breyer MD. Prostanoid receptors: subtypes and signaling. Annu Rev Pharmacol Toxicol. 41 (2001) 661–690.

Bryant J. Correlation between metastatic patterns in renal cell carcinoma and tissue distribution of thromboxane synthetase. Acta Oncol. 33 (1994) 708–709.

Buchanan FG, Gorden DL, Matta P, Shi Q, Matrisian LM, DuBois RN. Role of beta-arrestin 1 in the metastatic progression of colorectal cancer. Proc Natl Acad Sci U S A. 103 (2006) 1492–1497.

Bunting S, Gryglewski R, Moncada S, Vane JR. Arterial walls generate from prostaglandin endoperoxides a substance (prostaglandin X) which relaxes strips of mesenteric and coeliac ateries and inhibits platelet aggregation. Prostaglandins. 12 (1976) 897–913.

Casey MB, Zhang S, Jin L, Kajita S, Lloyd RV. Expression of cyclooxygenase-2 and thromboxane synthase in non-neoplastic and neoplastic thyroid lesions. Endocr Pathol. 15 (2004) 107–116.

Castellone MD, Teramoto H, Williams BO, Druey KM, Gutkind JS. Prostaglandin E2 promotes colon cancer cell growth through a Gs-axin-beta-catenin signaling axis. Science. 310 (2005) 1504–1510.

Cha YI, DuBois RN. NSAIDs and cancer prevention: targets downstream of COX-2. Annu Rev Med. 58 (2007) 239–252.

Chen GG, Lee TW, Yip JH, Xu H, Lee IK, Mok TS, Warner TD, Yim AP. Increased thromboxane B(2) levels are associated with lipid peroxidation and Bcl-2 expression in human lung carcinoma. Cancer Lett. 234 (2006) 193–198.

Chen X-S, Funk CD. Structure-function properties of human platelet 12-lipoxygenase: Chimeric enzyme and in vitro mutagenesis studies. FASEB J. 7 (1993) 694–701.

Chen YQ, Duniec ZM, Liu B, Hagmann W, Gao X, Shimoji K, Marnett LJ, Johnson CR, Honn KV. Endogenous 12(S)-HETE production by tumor cells and its role in metastasis. Cancer Res. 54 (1994) 1574–1579.

Cheng Y, Austin SC, Rocca B, Koller BH, Coffman TM, Grosser T, Lawson JA, FitzGerald GA. Role of prostacyclin in the cardiovascular response to thromboxane A2. Science. 296 (2002) 539–541.

Connolly JM, Rose DP. Enhanced angiogenesis and growth of 12-lipoxygenase gene-transfected MCF-7 human breast cancer cells in athymic nude mice. Cancer Lett. 132 (1998) 107–112.

Ding XZ, Tong WG, Adrian TE. 12-lipoxygenase metabolite 12(S)-HETE stimulates human pancreatic cancer cell proliferation via protein tyrosine phosphorylation and ERK activation. Int J Cancer. 94 (2001) 630–636.

Eberhart CE, Coffey RJ, Radhika A, Giardiello FM, Ferrenbach S, DuBois RN. Up-regulation of cyclooxygenase 2 gene expression in human colorectal adenomas and adenocarcinomas. Gastroenterology. 107 (1994) 1183–1188.

Fleming I, Thiel BA, Chester J, O'Prey J, Janetzki S, Aitken A, Anton IA, Rapoport SM, Harrison PR. The complete sequence of the rabbit erythroid cell-specific 15-lipoxygenase mRNA: Comparison of the predicted amino acid sequence of the erythrocyte lipoxygenase with other lipoxygenases. Gene. 79 (1989) 181–188.

Fukuda R, Kelly B, Semenza GL. Vascular endothelial growth factor gene expression in colon cancer cells exposed to prostaglandin E2 is mediated by hypoxia-inducible factor 1. Cancer Res. 63 (2003) 2330–2334.

Funk CD. Prostaglandins and leukotrienes: advances in eicosanoid biology. Science. 294 (2001) 1871–1875.

Funk CD, FitzGerald GA. Eicosanoid forming enzyme mRNA in human tissues. J.Biol. Chem. 266 (1991) 12508–12513.

Funk CD, Funk LB, FitzGerald GA, Samuelsson B. Characterization of human 12-lipoxygenase genes. Proc Natl Acad Sci U S A. 89 (1992) 3962–3966.

Funk CD, Hoshiko S, Matsumoto T, Rådmark O, Samuelsson B. Characterization of the human 5lipoxygenase gene. Proc.Natl.Acad.Sci.USA. 86 (1989) 2587–2591.

Gao X, Grignon DJ, Chbihi T, Zacharek A, Chen YQ, Sakr W, Porter AT, Crissman JD, Pontes JE, Powell IJ, Honn KV. Elevated 12-lipoxygenase mRNA expression correlated with advanced stage and poor differentiation of human prostate cancer. Urology. 46 (1995) 227–237.

Gao X, Honn KV. Biological properties of 12(S)-HETE in cancer metastasis. Adv.Prostaglandin Thromboxane Leukot.Res. 23 (1995) 439–444.

Gasic GJ, Gasic TB, Galanti N, Johnson T, Murphy S. Platelet-tumor-cell interactions in mice. The role of platelets in the spread of malignant disease. Int J Cancer. 11 (1973) 704–718.

Ghosh J, Myers CE. Arachidonic acid stimulates prostate cancer cell growth: critical role of 5lipoxygenase. Biochem Biophys Res Commun. 235 (1997) 418–423.

Ghosh J, Myers CE. Inhibition of arachidonate 5-lipoxygenase triggers massive apoptosis in human prostate cancer cells. Proc Natl Acad Sci U S A. 95 (1998) 13182–13187.

Ghosh S, May MJ, Kopp EB. NF-κB and Rel proteins: evolutionarily conserved mediators of immune responses. Annu.Rev.Immunol. 16 (1998) 225–260.

Gonzalez AL, Roberts RL, Massion PP, Olson SJ, Shyr Y, Shappell SB. 15-Lipoxygenase-2 expression in benign and neoplastic lung: an immunohistochemical study and correlation with tumor grade and proliferation. Hum Pathol. 35 (2004) 840–849.

Gupta S, Srivastava M, Ahmad N, Bostwick DG, Mukhtar H. Over-expression of cyclooxygenase-2 in human prostate adenocarcinoma. Prostate. 42 (2000) 73–78.

Hamberg M, Svensson J, Samuelsson B. Thromboxanes: a new group of biologically active compounds derived from prostaglandin endoperoxides. Proc Natl Acad Sci U S A. 72 (1975) 2994–2998.

Hennig R, Ding XZ, Tong WG, Schneider MB, Standop J, Friess H, Buchler MW, Pour PM, Adrian TE. 5-Lipoxygenase and leukotriene B(4) receptor are expressed in human pancreatic cancers but not in pancreatic ducts in normal tissue. Am J Pathol. 161 (2002) 421–428.

Hennig R, Kehl T, Noor S, Ding XZ, Rao SM, Bergmann F, Furstenberger G, Buchler MW, Friess H, Krieg P, Adrian TE. 15-lipoxygenase-1 production is lost in pancreatic cancer and overexpression of the gene inhibits tumor cell growth. Neoplasia. 9 (2007) 917–926.

Honn KV. Inhibition of tumor cell metastasis by modulation of the vascular prostacyclin/thromboxane A2 system. Clin Exp Metastasis. 1 (1983) 103–114.

Honn KV, Aref A, Chen YQ, Cher ML, Crissman JD, Forman JD, Gao X, Grignon D, Hussain M, Porter AT, Pontes EJ, Powell II, Redman B, Sakr W, Severson R, Tang DG, Wood DP, Jr. Prostate Cancer -Old Problems and New Approaches. (Part II. Diagnostic and Prognostic Markers, Pathology and Biological Aspects). Pathol Oncol Res. 2 (1996) 191–211.

Honn KV, Cavanaugh P, Evens C, Taylor JD, Sloane BF. Tumor cell-platelet aggregation: induced by cathepsin B-like proteinase and inhibited by prostacyclin. Science. 217 (1982) 540–542.

Honn KV, Cicone B, Skoff A. Prostacyclin: a potent antimetastatic agent. Science. 212 (1981) 1270–1272.

Honn KV, Meyer J. Thromboxanes and prostacyclin: positive and negative modulators of tumor growth. Biochem Biophys Res Commun. 102 (1981) 1122–1129.

Honn KV, Steinert BW, Moin K, Onoda JM, Taylor JD, Sloane BF. The role of platelet cyclooxygenase and lipoxygenase pathways in tumor cell induced platelet aggregation. Biochem Biophys Res Commun. 145 (1987) 384–389.

Honn KV, Tang DG, Gao X, Butovich IA, Liu B, Timar J, Hagmann W. 12-lipoxygenases and 12(S)HETE: role in cancer metastasis. Cancer Metastasis Rev. 13 (1994a) 365–396.

Honn KV, Tang DG, Grossi IM, Renaud C, Duniec ZM, Johnson CR, Diglio CA. Enhanced endothelial cell retraction mediated by 12(S)-HETE: a proposed mechanism for the role of platelets in tumor cell metastasis. Exp Cell Res. 210 (1994b) 1–9.

Hsu AL, Ching TT, Wang DS, Song X, Rangnekar VM, Chen CS. The cyclooxygenase-2 inhibitor celecoxib induces apoptosis by blocking Akt activation in human prostate cancer cells independently of Bcl-2. J Biol Chem. 275 (2000) 11397–11403.

Ikawa H, Kamitani H, Calvo BF, Foley JF, Eling TE. Expression of 15-lipoxygenase-1 in human colorectal cancer. Cancer Res. 59 (1999) 360–366.

Kajita S, Ruebel KH, Casey MB, Nakamura N, Lloyd RV. Role of COX-2, thromboxane A2 synthase, and prostaglandin I2 synthase in papillary thyroid carcinoma growth. Mod Pathol. 18 (2005) 221–227.

Kandouz M, Nie D, Pidgeon GP, Krishnamoorthy S, Maddipati KR, Honn KV. Platelet-type 12lipoxygenase activates NF-kappaB in prostate cancer cells. Prostaglandins Other Lipid Mediat. 71 (2003) 189–204.

Kargman SL, O'Neill GP, Vickers PJ, Evans JF, Mancini JA, Jothy S. Expression of prostaglandin G/H synthase-1 and -2 protein in human colon cancer. Cancer Res. 55 (1995) 2556–2559.

Kelavkar U, Lin Y, Landsittel D, Chandran U, Dhir R. The yin and yang of 15-lipoxygenase-1 and delta-desaturases: dietary omega-6 linoleic acid metabolic pathway in prostate. J Carcinog. 5 (2006) 9–13.

Kelavkar UP, Cohen C, Kamitani H, Eling TE, Badr KF. Concordant induction of 15-lipoxygenase-1 and mutant p53 expression in human prostate adenocarcinoma: correlation with Gleason staging. Carcinogenesis. 21 (2000) 1777–1787.

Kim JA, Gu JL, Natarajan R, Berliner JA, Nadler JL. A leukocyte type of 12-lipoxygenase is expressed in human vascular and mononuclear cells. Evidence for upregulation by angiotensin II. Arterioscler Thromb Vasc Biol. 15 (1995) 942–948.

Kimoto M, Ando K, Koike S, Matsumoto T, Jibu T, Moriya H, Kanegasaki S. Significance of platelets in an antimetastatic activity of bacterial lipopolysaccharide. Clin Exp Metastasis. 11 (1993) 285–292.

Krieg P, Kinzig A, Ress-Loschke M, Vogel S, Vanlandingham B, Stephan M, Lehmann WD, Marks F, Furstenberger G. 12-Lipoxygenase isoenzymes in mouse skin tumor development. Mol Carcinog. 14 (1995) 118–129.

Kuhn H, Thiele BJ. The diversity of the lipoxygenase family. Many sequence data but little information on biological significance. FEBS Lett. 449 (1999) 7–11.

Kuhn H, Walther M, Kuban RJ. Mammalian arachidonate 15-lipoxygenases structure, function, and biological implications. Prostaglandins Other Lipid Mediat. 68–69 (2002) 263–290.

Lee FS, Peters RT, Dang LC, Maniatis T. MEKK1 activates both IkB kinase a and IkB kinase beta. Proc.Natl.Acad.Sci.U.S.A. 95 (1998) 9319–9324.

Limor R, Weisinger G, Gilad S, Knoll E, Sharon O, Jaffe A, Kohen F, Berger E, Lifschizt-Mercer B, Stern N. A novel form of platelet-type 12-lipoxygenase mRNA in human vascular smooth muscle cells. Hypertension. 38 (2001) 864–871.

Liu B, Khan WA, Hannun YA, Timar J, Taylor JD, Lundy S, Butovich I, Honn KV. 12(S)hydroxyeicosatetraenoic acid and 13(S)-hydroxyoctadecadienoic acid regulation of protein kinase C-a in melanoma cells: role of receptor-mediated hydrolysis of inositol phospholipids. Proc.Natl.Acad.Sci.U.S.A. 92 (1995) 9323–9327.

Liu B, Marnett LJ, Chaudhary A, Ji C, Blair IA, Johnson CR, Diglio CA, Honn KV. Biosynthesis of 12(S)-hydroxyeicosatetraenoic acid by B16 amelanotic melanoma cells is a determinant of their metastatic potential. Lab Invest. 70 (1994) 314–323.

Liu XH, Yao S, Kirschenbaum A, Levine AC. NS398, a selective cyclooxygenase-2 inhibitor, induces apoptosis and down-regulates bcl-2 expression in LNCaP cells. Cancer Res. 58 (1998) 4245–4249.

Marks F, Furstenberger G. *Prostaglandins, Leukotrienes and Other Eicosanoids*. Weinheim: Wiley-VCH, 1999, 388p.

Marnett LJ, DuBois RN. COX-2: a target for colon cancer prevention. Annu Rev Pharmacol Toxicol. 42 (2002) 55–80.

McCabe NP, Selman SH, Jankun J. Vascular endothelial growth factor production in human prostate cancer cells is stimulated by overexpression of platelet 12-lipoxygenase. Prostate. 66 (2006) 779–787.

Mehrotra S, Morimiya A, Agarwal B, Konger R, Badve S. Microsomal prostaglandin $E_2$ synthase-1 in breast cancer: a potential target for therapy. J Pathol. 208 (2006) 356–363.

Mehta P. Potential role of platelets in the pathogenesis of tumor metastasis. Blood. 63 (1984) 55–63.

Menter DG, Onoda JM, Taylor JD, Honn KV. Effects of prostacyclin on tumor cell-induced platelet aggregation. Cancer Res. 44 (1984) 450–456.

Miggin SM, Kinsella BT. Expression and tissue distribution of the mRNAs encoding the human thromboxane A2 receptor (TP) alpha and beta isoforms. Biochim Biophys Acta. 1425 (1998) 543–559.

Moussa O, Yordy JS, Abol-Enein H, Sinha D, Bissada NK, Halushka PV, Ghoneim MA, Watson DK. Prognostic and functional significance of thromboxane synthase gene overexpression in invasive bladder cancer. Cancer Res. 65 (2005) 11581–11587.

Muller-Decker K, Furstenberger G. The cyclooxygenase-2-mediated prostaglandin signaling is causally related to epithelial carcinogenesis. Mol Carcinog. 46 (2007) 705–710.

Murata T, Ushikubi F, Matsuoka T, Hirata M, Yamasaki A, Sugimoto Y, Ichikawa A, Aze Y, Tanaka T, Yoshida N, Ueno A, Oh-ishi S, Narumiya S. Altered pain perception and inflammatory response in mice lacking prostacyclin receptor. Nature. 388 (1997) 678–682.

Nakano H, Shindo M, Sakon S, Nishinaka S, Mihara M, Yagita H, Okumura K. Differential regulation of IkB kinase a and beta by two upstream kinases, NF-kB-inducing kinase and mitogen-activated protein kinase/ERK kinase kinase-1. Proc.Natl.Acad.Sci.U.S.A. 95 (1998) 3537–3542.

Narumiya S, Sugimoto Y, Ushikubi F. Prostanoid receptors: structures, properties, and functions. Physiol Rev. 79 (1999) 1193–1226.

Needleman P, Minkes M, Raz A. Thromboxanes: selective biosynthesis and distinct biological properties. Science. 193 (1976) 163–165.

Nemoto S, DiDonato JA, Lin A. Coordinate regulation of IkB kinases by mitogen-activated protein kinase kinase kinase 1 and NF-kB-inducing kinase. Mol.Cell.Biol. 18 (1998) 7336–7343.

Nie D, Che M, Zacharek A, Qiao Y, Li L, Li X, Lamberti M, Tang K, Cai Y, Guo Y, Grignon D, Honn KV. Differential expression of thromboxane synthase in prostate carcinoma: role in tumor cell motility. Am J Pathol. 164 (2004) 429–439.

Nie D, Guo Y, Yang D, Tang Y, Chen Y, Wang MT, Zacharek A, Qiao Y, Che M, Honn KV. Thromboxane A2 receptors in prostate carcinoma: expression and its role in regulating cell motility via small GTPase Rho. Cancer Res. 68 (2008) 115–121.

Nie D, Hillman GG, Geddes T, Tang K, Pierson C, Grignon DJ, Honn KV. Platelet-type 12lipoxygenase in a human prostate carcinoma stimulates angiogenesis and tumor growth. Cancer Res. 58 (1998) 4047–4051.

Nie D, Honn KV. Cyclooxygenase, lipoxygenase and tumor angiogenesis. Cell Mol Life Sci. 59 (2002) 799–807.

Nie D, Krishnamoorthy S, Jin R, Tang K, Chen Y, Qiao Y, Zacharek A, Guo Y, Milanini J, Pages G, Honn KV. Mechanisms regulating tumor angiogenesis by 12-lipoxygenase in prostate cancer cells. J Biol Chem. 281 (2006) 18601–18609.

Nie D, Lamberti M, Zacharek A, Li L, Szekeres K, Tang K, Chen Y, Honn KV. Thromboxane A(2) regulation of endothelial cell migration, angiogenesis, and tumor metastasis. Biochem Biophys Res Commun. 267 (2000a) 245–251.

Nie D, Tang K, Diglio C, Honn KV. Eicosanoid regulation of angiogenesis: role of endothelial arachidonate 12-lipoxygenase. Blood. 95 (2000b) 2304–2311.

Nie D, Tang K, Szekeres K, Li L, Honn KV. Eicosanoid regulation of angiogenesis in human prostate carcinoma and its therapeutic implications. Ann.N.Y.Acad.Sci. 905 (2000c) 165–176.

Nithipatikom K, Isbell MA, See WA, Campbell WB. Elevated 12-and 20-hydroxyeicosatetraenoic acid in urine of patients with prostatic diseases. Cancer Lett. 233 (2006) 219–225.

Ohd JF, Nielsen CK, Campbell J, Landberg G, Lofberg H, Sjolander A. Expression of the leukotriene D4 receptor CysLT1, COX-2, and other cell survival factors in colorectal adenocarcinomas. Gastroenterology. 124 (2003) 57–70.

Oshima M, Dinchuk JE, Kargman SL, Oshima H, Hancock B, Kwong E, Trzaskos JM, Evans JF, Taketo MM. Suppression of intestinal polyposis in Apc delta716 knockout mice by inhibition of cyclooxygenase 2 (COX-2). Cell. 87 (1996) 803–809.

Pearson G, English JM, White MA, Cobb MH. ERK5 and ERK2 cooperate to regulate NF-kB and cell transformation. J.Biol.Chem. 276 (2001) 7927–7931.

Pidgeon GP, Kandouz M, Meram A, Honn KV. Mechanisms controlling cell cycle arrest and induction of apoptosis after 12-lipoxygenase inhibition in prostate cancer cells. Cancer Res. 62 (2002) 2721–2727.

Pinto S, Gori L, Gallo O, Boccuzzi S, Paniccia R, Abbate R. Increased thromboxane A2 production at primary tumor site in metastasizing squamous cell carcinoma of the larynx. Prostaglandins Leukot Essent Fatty Acids. 49 (1993) 527–530.

Pollard M, Luckert PH, Schmidt MA. The suppressive effect of piroxicam on autochthonous intestinal tumors in the rat. Cancer Lett. 21 (1983) 57–61.

Rapoport SM, Schewe T, Wiesner R, Halangk W, Ludwig P, Janicke-Hohne M, Tannert C, Hiebsch C, Klatt D. The lipoxygenase of reticulocytes. Purification, characterization and biological dynamics of the lipoxygenase; its identity with the respiratory inhibitors of the reticulocyte. Eur J Biochem. 96 (1979) 545–561.

Romano M, Catalano A, Nutini M, D'Urbano E, Crescenzi C, Claria J, Libner R, Davi G, Procopio A. 5-lipoxygenase regulates malignant mesothelial cell survival: involvement of vascular endothelial growth factor. FASEB J. 15 (2001) 2326–2336.

Romano M, Claria J. Cyclooxygenase-2 and 5-lipoxygenase converging functions on cell proliferation and tumor angiogenesis: implications for cancer therapy. FASEB J. 17 (2003) 1986–1995.

Salcedo R, Zhang X, Young HA, Michael N, Wasserman K, Ma WH, Martins-Green M, Murphy WJ, Oppenheim JJ. Angiogenic effects of prostaglandin E2 are mediated by up-regulation of CXCR4 on human microvascular endothelial cells. Blood. 102 (2003) 1966–1977.

Schewe T, Halangk W, Hiebsch C, Rapoport SM. A lipoxygenase in rabbit reticulocytes which attacks phospholipids and intact mitochondria. FEBS Lett. 60 (1975) 149–152.

Shao J, Jung C, Liu C, Sheng H. Prostaglandin E2 Stimulates the beta-catenin/T cell factor-dependent transcription in colon cancer. J Biol Chem. 280 (2005) 26565–26572.

Shappell SB, Boeglin WE, Olson SJ, Kasper S, Brash AR. 15-lipoxygenase-2 (15-LOX-2) is expressed in benign prostatic epithelium and reduced in prostate adenocarcinoma. Am J Pathol. 155 (1999) 235–245.

Shappell SB, Keeney DS, Zhang J, Page R, Olson SJ, Brash AR. 15-Lipoxygenase-2 expression in benign and neoplastic sebaceous glands and other cutaneous adnexa. J Invest Dermatol. 117 (2001a) 36–43.

Shappell SB, Manning S, Boeglin WE, Guan YF, Roberts RL, Davis L, Olson SJ, Jack GS, Coffey CS, Wheeler TM, Breyer MD, Brash AR. Alterations in lipoxygenase and cyclooxygenase-2 catalytic activity and mRNA expression in prostate carcinoma. Neoplasia. 3 (2001b) 287–303.

Shureiqi I, Chen D, Lotan R, Yang P, Newman RA, Fischer SM, Lippman SM. 15-Lipoxygenase-1 mediates nonsteroidal anti-inflammatory drug-induced apoptosis independently of cyclooxygenase-2 in colon cancer cells. Cancer Res. 60 (2000) 6846–6850.

Shureiqi I, Wojno KJ, Poore JA, Reddy RG, Moussalli MJ, Spindler SA, Greenson JK, Normolle D, Hasan AA, Lawrence TS, Brenner DE. Decreased 13-S-hydroxyoctadecadienoic acid levels and 15lipoxygenase-1 expression in human colon cancers. Carcinogenesis. 20 (1999) 1985–1995.

Shureiqi I, Xu X, Chen D, Lotan R, Morris JS, Fischer SM, Lippman SM. Nonsteroidal antiinflammatory drugs induce apoptosis in esophageal cancer cells by restoring 15-lipoxygenase-1 expression. Cancer Res. 61 (2001) 4879–4884.

Siebert M, Krieg P, Lehmann WD, Marks F, Furstenberger G. Enzymic characterization of epidermis-derived 12-lipoxygenase isoenzymes. Biochem.J. 355 (2001) 97–104.

Sigal E, Grunberger D, Cashman JR, Craik CS, Caughey GH, Nadel JA. Arachidonate 15lipoxygenase from human eosinophil-enriched leukocytes: partial purification and properties. Biochem Biophys Res Commun. 150 (1988) 376–383.

Sonoshita M, Takaku K, Sasaki N, Sugimoto Y, Ushikubi F, Narumiya S, Oshima M, Taketo MM. Acceleration of intestinal polyposis through prostaglandin receptor EP2 in Apc(-Delta 716) knockout mice. Nat Med. 7 (2001) 1048–1051.

Subbaramaiah K, Altorki N, Chung WJ, Mestre JR, Sampat A, Dannenberg AJ. Inhibition of cyclooxygenase-2 gene expression by p53. J Biol Chem. 274 (1999) 10911–10915.

Subbaramaiah K, Norton L, Gerald W, Dannenberg AJ. Cyclooxygenase-2 is overexpressed in HER-2/neu-positive breast cancer: evidence for involvement of AP-1 and PEA3. J Biol Chem. 277 (2002) 18649–18657.

Szekeres CK, Tang K, Trikha M, Honn KV. Eicosanoid activation of extracellular signal-regulated kinase1/2 in human epidermoid carcinoma cells. J.Biol.Chem. 275 (2000a) 38831–38841.

Szekeres CK, Trikha M, Nie D, Honn KV. Eicosanoid 12(S)-HETE activates phosphatidylinositol 3kinase. Biochem.Biophys.Res.Commun. 275 (2000b) 690–695.

Tang DG, Bhatia B, Tang S, Schneider-Broussard R. 15-lipoxygenase 2 (15-LOX2) is a functional tumor suppressor that regulates human prostate epithelial cell differentiation, senescence, and growth (size). Prostaglandins Other Lipid Mediat. 82 (2007) 135–146.

Tang DG, Chen YQ, Diglio CA, Honn KV. Protein kinase C-dependent effects of 12(S)-HETE on endothelial cell vitronectin receptor and fibronectin receptor. J Cell Biol. 121 (1993a) 689–704.

Tang DG, Diglio CA, Bazaz R, Honn KV. Transcriptional activation of endothelial cell integrin alpha v by protein kinase C activator 12(S)-HETE. J Cell Sci. 108 ( Pt 7) (1995a) 2629–2644.

Tang DG, Diglio CA, Honn KV. Activation of microvascular endothelium by eicosanoid 12(S)hydroxyeicosatetraenoic acid leads to enhanced tumor cell adhesion via up-regulation of surface expression of alpha v beta 3 integrin: a posttranscriptional, protein kinase C-and cytoskeleton-dependent process. Cancer Res. 54 (1994) 1119–1129.

Tang DG, Grossi IM, Chen YQ, Diglio CA, Honn KV. 12(S)-HETE promotes tumor-cell adhesion by increasing surface expression of a V beta 3 integrins on endothelial cells. Int J Cancer. 54 (1993b) 102–111.

Tang DG, Renaud C, Stojakovic S, Diglio CA, Porter A, Honn KV. 12(S)-HETE is a mitogenic factor for microvascular endothelial cells: its potential role in angiogenesis. Biochem Biophys Res Commun. 211 (1995b) 462–468.

Thun MJ, Namboodiri MM, Heath CW, Jr. Aspirin use and reduced risk of fatal colon cancer. N Engl J Med. 325 (1991) 1593–1596.

Timar J, Raso E, Dome B, Li L, Grignon D, Nie D, Honn KV, Hagmann W. Expression, subcellular localization and putative function of platelet-type 12-lipoxygenase in human prostate cancer cell lines of different metastatic potential. Int J Cancer. 87 (2000) 37–43.

Tong WG, Ding XZ, Witt RC, Adrian TE. Lipoxygenase inhibitors attenuate growth of human pancreatic cancer xenografts and induce apoptosis through the mitochondrial pathway. Mol Cancer Ther. 1 (2002) 929–935.

Tsujii M, Kawano S, Tsuji S, Sawaoka H, Hori M, DuBois RN. Cyclooxygenase regulates angiogenesis induced by colon cancer cells. Cell. 93 (1998) 705–716.

Tsukada T, Nakashima K, Shirakawa S. Arachidonate 5-lipoxygenase inhibitors show potent antiproliferative effects on human leukemia cell lines. Biochem Biophys Res Commun. 140 (1986) 832–836.

Turk J, Maas RL, Brash AR, Roberts LJ, II, Oates JA. Arachidonic acid 15-lipoxygenase products from human eosinophils. J.Biol.Chem. 257 (1982) 7068–7076.

Waddell WR, Loughry RW. Sulindac for polyposis of the colon. J Surg Oncol. 24 (1983) 83–87.

Wang D, Wang H, Shi Q, Katkuri S, Walhi W, Desvergne B, Das SK, Dey SK, DuBois RN. Prostaglandin E(2) promotes colorectal adenoma growth via transactivation of the nuclear peroxisome proliferator-activated receptor delta. Cancer Cell. 6 (2004) 285–295.

Watkins G, Douglas-Jones A, Mansel RE, Jiang WG. Expression of thromboxane synthase, TBXAS1 and the thromboxane A2 receptor, TBXA2R, in human breast cancer. Int Semin Surg Oncol. 2 (2005) 23–30.

Whittaker N, Bunting S, Salmon J, Moncada S, Vane JR, Johnson RA, Morton DR, Kinner JH, Gorman RR, McGuire JC, Sun FF. The chemical structure of prostaglandin X (prostacyclin). Prostaglandins. 12 (1976) 915–928.

Winer I, Normolle DP, Shureiqi I, Sondak VK, Johnson T, Su L, Brenner DE. Expression of 12lipoxygenase as a biomarker for melanoma carcinogenesis. Melanoma Res. 12 (2002) 429–434.

Wong BC, Wang WP, Cho CH, Fan XM, Lin MC, Kung HF, Lam SK. 12-Lipoxygenase inhibition induced apoptosis in human gastric cancer cells. Carcinogenesis. 22 (2001) 1349–1354.

Wu KK, Liou JY. Cellular and molecular biology of prostacyclin synthase. Biochem Biophys Res Commun. 338 (2005) 45–52.

Xiao CY, Hara A, Yuhki K, Fujino T, Ma H, Okada Y, Takahata O, Yamada T, Murata T, Narumiya S, Ushikubi F. Roles of prostaglandin I(2) and thromboxane A(2) in cardiac ischemia-reperfusion injury: a study using mice lacking their respective receptors. Circulation. 104 (2001) 2210–2215.

Xu XC, Shappell SB, Liang Z, Song S, Menter D, Subbarayan V, Iyengar S, Tang DG, Lippman SM. Reduced 15S-lipoxygenase-2 expression in esophageal cancer specimens and cells and upregulation in vitro by the cyclooxygenase-2 inhibitor, NS398. Neoplasia. 5 (2003) 121–127.

Yamamoto S, Suzuki H, Ueda N. Arachidonate 12-lipoxygenases. Prog.Lipid Res. 36 (1997) 23–41.

Yoshimatsu K, Altorki NK, Golijanin D, Zhang F, Jakobsson PJ, Dannenberg AJ, Subbaramaiah K. Inducible prostaglandin E synthase is overexpressed in non-small cell lung cancer. Clin Cancer Res. 7 (2001) 2669–2674.

Yoshimoto T, Takahashi Y. Arachidonate 12-lipoxygenases. Prostaglandins Other Lipid Mediat. 68–69 (2002) 245–262.

Yoshimura R, Sano H, Masuda C, Kawamura M, Tsubouchi Y, Chargui J, Yoshimura N, Hla T, Wada S. Expression of cyclooxygenase-2 in prostate carcinoma. Cancer. 89 (2000) 589–596.

# Chapter 7
# Fatty Acid Synthase Activity in Tumor Cells

**Joy L. Little and Steven J. Kridel**

**Abstract** While normal tissues are tightly regulated by nutrition and a carefully balanced system of glycolysis and fatty acid synthesis, tumor cells are under significant evolutionary pressure to bypass many of the checks and balances afforded normally. Cancer cells have high energy expenditure from heightened proliferation and metabolism and often show increased lipogenesis. Fatty acid synthase (FASN), the enzyme responsible for catalyzing the ultimate steps of fatty acid synthesis in cells, is expressed at high levels in tumor cells and is mostly absent in corresponding normal cells. Because of the unique expression profile of FASN, there is considerable interest not only in understanding its contribution to tumor cell growth and proliferation, but also in developing inhibitors that target FASN specifically as an anti-tumor modality. Pharmacological blockade of FASN activity has identified a pleiotropic role for FASN in mediating aspects of proliferation, growth and survival. As a result, a clearer understanding of the role of FASN in tumor cells has been developed.

**Keywords** Cancer · fatty acid synthase · lipogenesis

**Abbreviations** FASN, fatty acid synthase ACC, acetyl-CoA-carboxylase ACL, ATP-citrate lyase NADPH, nicotinamide adenine dinucleotide phosphate MAT, malonyl acetyl transferases KS, ketoacyl synthase KR, β-ketoacyl reductase DH, β-hydroxyacyl dehydratase ER, enoyl reductase TE, thioesterase ACP, acyl carrier protein VLCFA, very long chain fatty acids ELOVL, elongation of very long chain fatty acids SCD1, stearoyl-CoA desaturase-1 AMPK, AMP-activated kinase ME, malic enzyme FASKOL, liver-specific deletion of FAS PPARα, Peroxisome Proliferator-Activating Receptor alpha HMG-CoA, 3-hydroxy-3-methyl-glutaryl-CoA SREBP, sterol response element binding protein S1P, site-one protease S2P, site-two

S.J. Kridel
Department of Cancer Biology and Comprehensive Cancer Center, Wake Forest University School of Medicine, Medical Center Boulevard, Winston-Salem, NC 27157, USA

P.J. Quinn, X. Wang (eds.), *Lipids in Health and Disease*, 

protease RIPCre, Cre-recombinase under the control of rat insulin 2 promoter CPT1, carnitine palmityl transferase 1 MCD, malonyl-CoA desaturase SCAP, SREBP cleavage activating protein NF-Y, nuclear factor Y SP1, stimulatory protein 1 RNAi, RNA interference PI3K, phosphatidylinositol-3 kinase KGF, keratinocyte growth factor EGF, epidermal growth factor JNK, cJun N-terminal kinase RTK, receptor tyrosine kinase AR, androgen receptor PR, progesterone receptor USP2a, ubiquitin-specific protease 2a EGCG, epigallocatechin-3-gallate TOFA, 5-(tetradecyloxy)-2-furoic acid FDA, food and drug administration.

## 7.1 Fatty Acid Synthesis

### *7.1.1 The FASN Enzyme*

One of the metabolic hallmarks of a tumor cell is increased lipogenesis (Kuhajda, 2006; Swinnen et al., 2006). In fact, in many instances the vast majority of fatty acids in tumors are synthesized *de novo* (Ookhtens et al., 1984). In mammalian cells, fatty acid synthase (FASN) is the central enzyme of long chain fatty acid synthesis. FASN is a multifunctional polypeptide that is comprised of seven separate functional domains (Fig. 7.1A). The individual domains of FASN work in concert to catalyze thirty-two different reactions to synthesize the sixteen carbon fatty acid palmitate, using acetyl-CoA and malonyl-CoA as substrates and nicotinamide adenine dinucleotide phosphate (NADPH) as an electron donor. The fatty acid synthesis reaction mechanism can be separated into three functional groupings: (1) to bind and condense the substrates, (2) to reduce the intermediates and (3) to release the final saturated long chain fatty acid palmitate (Fig. 7.1B). The malonyl acetyl transferase (MAT) domain binds malonyl-CoA and acetyl-CoA, while the ketoacyl synthase (KS) domain acts to condense the acyl chain (Fig. 7.1B). This β-ketoacyl moiety is then reduced in steps by the β-ketoacyl reductase (KR), β-hydroxyacyl dehydratase (DH), and enoyl reductase (ER) domains to a saturated acyl intermediate. This derivative can then be elongated by repeating the reactions catalyzed by the five previous enzyme activities for seven cycles until the thioesterase (TE) domain cleaves the final product, the sixteen carbon fatty acid palmitate. Throughout the entire synthesis of palmitate, the acyl carrier protein (ACP) acts as a coenzyme to bind intermediates by a 4'-phosphopantetheine group (Fig. 7.1B). In total, approximately 30 intermediates are involved in the process, but it is the high specificity of the TE domain for a 16 carbon fatty acid, as well as the MAT specificity for malonyl-CoA, that are responsible for preventing leakage of intermediates (Wakil, 1989). The overall FASN reaction is as follows:

$$\text{Acetyl-CoA} + 7\text{Malonyl-CoA} + 14\text{NADPH} + 14\text{H}^+ \rightarrow$$
$$\text{Palmitic-acid} + 7\text{CO}_2 + 8\text{CoA} + 14\text{NADP}^+ + 6\text{H}_2\text{O}$$

A.

B.

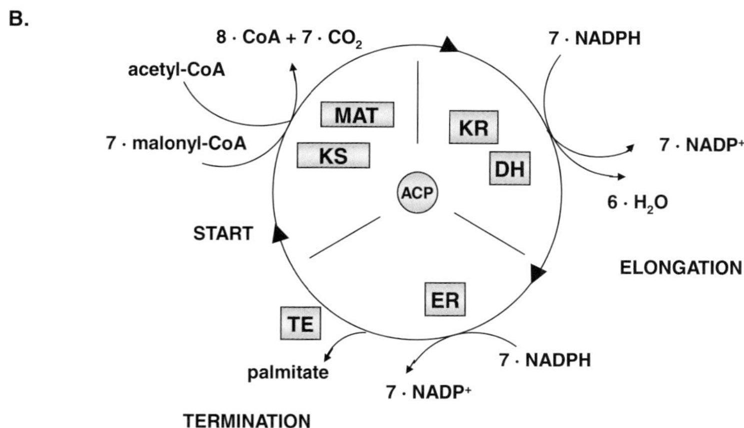


**Fig. 7.1** The FASN Enzyme. **A**. The FASN polypeptide comprises seven functional domains: the ketoacyl synthase (KS), malonyl acetyl transferase (MAT), β-hydroxyacyl dehydratase (DH), enoyl reductase (ER), β-ketoacyl reductase (KR), the acyl carrier protein (ACP), and thioesterase (TE) domains. **B**. The FASN reaction mechanism. The MAT domain of the enzyme binds malonyl-CoA and acetyl-CoA, while the KS domain acts to condense the growing acyl chain. The resulting β-ketoacyl moiety is then reduced in steps by the KR, DH, and ER to a saturated acyl intermediate. This process is repeated in seven cycles, after which, the TE domain releases the sixteen carbon fatty acid palmitate

The structure of FASN has yet to be definitively characterized, as there are two distinct models (Smith, 2006). Early complementation studies suggest that FASN functions as a homodimer in head-to-tail conformation with two simultaneous reactions beginning in one subunit and finishing in the other (Wakil, 1989; Smith et al., 2003; Rangan et al., 1998; Rangan et al., 2001). However a more recent crystal structure analysis of porcine FASN challenges this historical model. The 4.5 Å structure reveals FASN as an intertwined dimer in a conformation resembling an “X” with one central core region with two arms and two legs (Maier et al., 2006). However, at this lower resolution, the definitive placement of the flexible TE domain and ACP is not possible. It is also unclear whether the body of the FASN complex can be identified as two distinct monomers. In this model, the KS domain is near the bottom of the central core of the complex and two MAT domains are in the “legs” of the X shape. The DH domains are located in the top half of the central region just under the ER domains. Adjacent to the ER domains are the KR domains that comprise the “arms” of this X complex. The study equates the reaction pockets of this structure as having "double hot dog" folds but observes asymmetry of the two sides of the reaction chambers that may reveal hinge regions that allow different conformations of the FASN complex (Maier, et al., 2006; Smith, 2006).

### 7.1.2 Other Players in the Fatty Acid Synthesis Pathway

While FASN is the central enzyme of fatty acid synthesis, other enzymes and pathways upstream of FASN are required to generate and supply substrates. Glucose enters the cell and is converted through glycolysis to pyruvate which is then shuttled into the mitochondria to enter the citric acid cycle. Citrate is shuttled out of the mitochondria, where ATP-citrate lyase (ACL) catalyzes the conversion of citrate to oxaloacetate and acetyl-CoA. Acetyl-CoA Carboxylase (ACC) catalyzes the conversion of acetyl-CoA to malonyl-CoA in the rate limiting and first committed step of lipogenesis. Unlike FASN, which is primarily regulated transcriptionally, ACC is negatively regulated by post-translational phosphorylation at serine 79 by AMP-activated kinase (AMPK). Energy deficiency stimulates AMPK to regulate energy consumption of cells, specifically by regulating ACC among other enzymes. Fatty acid synthesis requires NADPH, which is provided through the hexose monophosphate shunt and malic enzyme (ME) to donate electrons (Wakil et al., 1983). Recent findings also suggest that glutamine metabolism can generate sufficient NADPH in glycolytic tumor cells, as well as act as a carbon source for fatty acid synthesis (Deberardinis et al., 2007).

After fatty acid synthesis, downstream enzymes can further modify palmitate for various cellular functions. In the endoplasmic reticulum, the 16 carbon fatty acid can be modified to fatty acids with eighteen or more carbons known as very long chain fatty acids (VLCFA), such as stearate (18:0) by a family of elongase enzymes called elongation of very long chain fatty acids (ELOVL1-6) (Jakobsson et al., 2006). Palmitate and stearate can also be desaturated by stearoyl-CoA desaturase-1 (SCD1) at the cis-9 carbon to palmitoleate (16:1) and oleate (18:1), respectively (Sampath and Ntambi, 2005).

## 7.2 FASN Expression

### 7.2.1 FASN Expression in Normal Cells

In normal tissue, FASN is expressed and active in cells that have a high lipid metabolism, such as liver and adipose tissues, to generate triglycerides in response to excess caloric intake (Jayakumar et al., 1995; Volpe and Marasa,

1975; Wakil et al., 1983). FASN is also expressed in a niche-specific manner in specialized tissues such as lactating mammary glands (Kusakabe et al., 2000; Thompson and Smith, 1985) cycling endometrium (Pizer et al., 1997; Kusakabe et al., 2000), and various other cell types including type II alveolar cells to produce lung surfactant (Buechler and Rhoades, 1980; Kusakabe et al., 2000), brain cells (Kusakabe et al., 2000; Jayakumar et al., 1995), and seminal vesicles to produce seminal fluid (Kusakabe et al., 2000). FASN is only weakly detectable, if at all, in other rapidly dividing normal tissues such as the intestinal epithelium, stomach epithelium, and hematopoietic cells in adults and is not detectable in most other adult tissues (Kusakabe et al., 2000).

Despite the low expression profile in most adult tissues, FASN is critical for developing embryos and is highly expressed in proliferative fetal cells (Kusakabe et al., 2000). The importance of FASN in development is underscored by the fact that mice with homozygous deletions of the *FASN* gene display an embryonic lethal phenotype (Chirala et al., 2003). *FASN* $^{-/-}$ mice die before implantation around embryonic day 3.5, most likely because developing embryos are unable to acquire enough fatty acids from the mother for adequate membrane biogenesis. The importance of FASN during development is further highlighted by the fact that the majority of heterozygotes are also resorbed after implantation. Those that survive do not live long beyond birth, indicating that one *FASN* allele is usually insufficient for embryogenesis, implantation, and developing tissues (Chirala et al., 2003). The importance of the fatty acid synthesis pathway in development is further supported by the demonstration that deletion of *ACC1* in mice also results in an embryonic lethal phenotype (Abu-Elheiga et al., 2005).

Mice harboring tissue-specific deletions of *FASN* have been generated to facilitate understanding of the role of FASN in normal tissue. To date *FASN* has been deleted in liver, β-cells, and hypothalamus (Chakravarthy et al., 2005, 2007). To knock out *FASN* in the liver, mice with a "floxed" *FASN* allele were crossed with mice harboring an allele of Cre driven by a rat albumin promoter. Although this liver-specific deletion of *FASN* (FASKOL) leaves animals viable without severe physiological effects, it is not without consequence. When FASKOL mice are fed a diet containing zero fat or are fasted for prolonged periods, they develop symptoms similar to those seen in mice engineered to lack Peroxisome Proliferator-Activating Receptor alpha (PPARα) (Kersten et al., 1999). Both *PPARα* knockout and FASKOL mice become hypoglycemic, develop steatosis (fatty liver) that correlates with reduced serum and liver cholesterol, reduced expression of 3-hydroxy-3-methyl-glutaryl-CoA (HMG-CoA) reductase, decreased cholesterol biosynthesis activity, and elevated sterol response element binding protein 2 (SREBP-2) expression. While the hypoglycemia and fatty liver may be reversed with dietary fat, all effects including cholesterol biosynthesis, HMG-CoA reductase and SREBP2 levels, as well as cholesterol levels in the serum and liver are rescued by administration of a PPARα agonist. This reveals distinct levels of metabolic regulation between *de novo* and dietary fat and indicates that products downstream of FASN activity regulate cholesterol, glucose, and fatty-acid homeostasis in the liver through activation of PPARα (Chakravarthy et al., 2005). Interestingly, mice with a liver-specific knockout of *ACC1* are still able to undergo fatty acid synthesis, but this discrepancy can be attributed to compensatory production of malonyl-CoA by the ACC2 isoform (Harada et al., 2007).

To determine whether FASN plays a role in pancreatic β-cell function, a knockout of *FASN* was generated. Crossing floxed *FASN* mice with mice harboring Cre under the control of rat insulin 2 promoter (RIPCre) causes specific deletion of *FASN* in pancreatic β-islet cells, as well as the hypothalamus, a region of the brain known for controlling motivational states, such as feeding. The resulting *FASN* knockout (FASKO) mice exhibit reduced feeding behavior and are highly active, even while maintained on a high fat diet (Chakravarthy et al., 2007). This correlates with studies showing the small molecule FASN inhibitor C75 acts in the hypothalamus to stimulate fatty acid oxidation via carnitine palmityl transferase 1 (CPT1) and induces a reversible anorexic phenotype (see Section 7.4.2). Interestingly, the β-cells lacking FASN are unaffected as loss of FASN does not alter insulin or glucose levels during glucose tolerance testing or stimulation either *in vivo* or *in vitro*. Therefore, the fasting phenotype of FASKO mice appears to be solely attributable to the effects on the hypothalamus. As a matter of fact, this observation is in agreement with a recent study showing FASN is not required for normal insulin secretion of β-cells *in vitro* (Joseph et al., 2007). Intracerebroventricular injection of FASKO mice with a small molecule drug Wy14,643 to activate PPARα restores feeding and weight gain, indicating that FASN controls PPARα activation in the hypothalamus. Pharmacological activation of PPARα in these mice also restores expression of CPT-1 and malonyl-CoA desaturase (MCD) that control cellular levels of malonyl-CoA by controlling the rate of transfer of fatty acids into the mitochondria for β-oxidation and malonyl-CoA stability, respectively (Chakravarthy et al., 2007). These studies elucidate the importance of FASN in energy homeostasis and provide a mechanism through which FASN can regulate its effects.

### *7.2.2 FASN Expression in Tumor Cells*

As discussed above, FASN has historically been studied in relation to normal physiology and as a central mediator of energy balance. In the last few decades, however, it has become clear that FASN is associated with tumor development. Accordingly, high FASN expression has been identified in many tumor types (Kuhajda, 2000, 2006). Haptoglobin-related protein (Hpr) was demonstrated to correlate with breast cancer stage, prognosis, as well as recurrence and patient survival (Kuhajda, et al., 1989a,b). Shortly after this observation, Hpr, or oncogenic antigen (OA-519) protein was identified as FASN (Kuhajda et al.,

1994). Since these discoveries, FASN upregulation has been demonstrated in every type of solid tumor. An initial retrospective study showed FASN expression correlated with staining of the proliferation marker MIB-1 to predict survival of breast cancer patients (Jensen et al., 1995). Subsequent studies confirmed the association of FASN with breast cancer recurrence, as well as shorter overall and disease-free survival in early breast cancer patients (Alo et al., 1996, 1999b). Breast cancer is not the only tumor type with elevated FASN levels. FASN expression is associated with prostate cancer prognosis, progression, and stage (Shurbaji, et al., 1992, 1996; Epstein et al., 1995). As a matter of fact, FASN is upregulated in androgen-independent prostate tumors and expression correlates with disease stage, as the highest levels of FASN expression are in androgen independent metastases (Pizer et al., 2001; Rossi et al., 2003). FASN expression correlates with poor prognosis, advanced progression, and/or decreased survival in a number of other cancers of different origins including: ovarian (Gansler et al., 1997; Alo et al., 2000), melanoma (Innocenzi et al., 2003; Kapur et al., 2005), nephroblastoma (Wilms tumor) (Camassei et al., 2003b), retinoblastoma (Camassei et al., 2003a), bladder (Visca et al., 2003), pancreas (Alo et al., 2007), soft tissue sarcoma (Takahiro et al., 2003), non-small cell lung cancer (Visca et al., 2004), endometrium (Sebastiani et al., 2004), and Paget's disease of the vulva (Alo et al., 2005). While FASN expression correlates with decreased survival and/or poor prognosis in a large number of tumor types, there are tumor types that show elevated FASN expression but no correlation with patient survival or disease stage (Rashid et al., 1997; Nemoto et al., 2001; Silva et al., 2008). In addition, there are several tumor types that show increased FASN expression, but correlation with disease progression or patient survival has not been investigated or published at this time. These tumors include hyperplastic parathyroid (Alo et al., 1999a), stomach carcinoma (Kusakabe et al., 2002), mesothelioma (Gabrielson et al., 2001), glioma (Zhao et al., 2006), and hepatocellular carcinoma (Yahagi et al., 2005).

Increased FASN expression in tumors is an early, common event (Swinnen et al., 2002; Myers et al., 2001) and its correlation with reduced survival and increased recurrence rationalizes the potential for anti-FASN tumor therapeutics (Kuhajda, 2000, 2006; Kridel et al., 2007). As evidence that lipogenesis as a whole is important in cancer, many of the enzymes upstream of FASN show altered expression patterns in human tumor cells, as well. For instance, ACL is overexpressed in cancer cells of breast and bladder (Szutowicz et al., 1979; Turyn et al., 2003). ACC is overexpressed in breast and prostate cancer cells (Milgraum et al., 1997; Swinnen et al., 2000b, 2006; Heemers et al., 2003). Interestingly, the tumor suppressor breast cancer susceptibility gene 1 (BRCA1) can bind the phosphorylated inactive ACC to prevent re-activation (Moreau et al., 2006). In addition, squamous cell carcinomas of the lung show lower immunohistochemical staining of phosphorylated inactive ACC than adenocarcinoma with poor prognosis (Conde et al., 2007). The strong functional correlation between upstream mediators of fatty acid synthesis and cancer underscores the importance of this pathway in tumor biology.

## 7.3 FASN Regulation

### 7.3.1 FASN Regulation in Normal Cells

In nonmalignant tissues, FASN expression is primarily regulated at the transcriptional level (Fig. 7.2A) (Hillgartner et al., 1995). There is a single *FASN* gene and the signals in normal cells that stimulate *FASN* transcription are numerous but strictly defined (Amy et al., 1990). Transcription of *FASN* is stimulated by dietary carbohydrate, glucose, insulin, amino acids, sterols and cyclic-AMP through specific response elements (Paulauskis and Sul, 1988; Rufo et al., 2001; Foufelle et al., 1992; Moustaid et al., 1994; Wang and Sul, 1998; Wakil et al., 1983; Rangan et al., 1996; Wakil, 1989). Hormones such as the thyroid hormone triiodothyronine (T3) (Moustaid and Sul, 1991), progesterone (Lacasa et al., 2001), androgen (Heemers et al., 2003) and adrenal glucocorticoids (Volpe and Marasa, 1975) can also upregulate FASN in liver and adipose tissues. *FASN* transcription is mediated by multiple transcription factors.

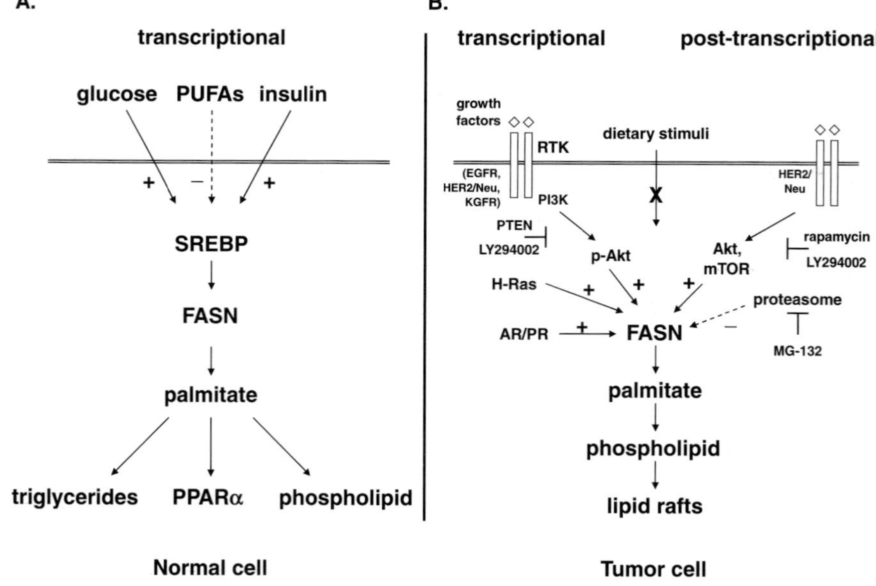


**Fig. 7.2** Regulation of FASN Expression in Normal and Tumor Cells. **A.** In normal cells (hepatocytes and adipocytes) FASN expression is primarily regulated through transcriptional mechanisms. **B.** In tumor cells, FASN expression is regulated by transcriptional and non-transcriptional mechanisms via multiple pathways

Upstream stimulatory factors (USFs) are required for insulin mediation of FASN expression, but other factors such as nuclear factor Y (NF-Y) and stimulatory protein 1 (SP1) can also play a role in *FASN* transcription (Teran-Garcia et al., 2007; Bennett et al., 1995). However, the vast majority of FASN-regulatory signals act through a family of transcription factors known as sterol response element binding proteins (SREBPs) that control lipid homeostasis and bind to various elements in the *FASN* promoter. There are three SREBP family members: SREBP-1a, SREBP-1c, and SREBP-2. SREBP-1a and SREBP-1c have been most widely linked to regulation of lipogenic gene transcription, while SREBP-2 is most linked to cholesterol metabolism. The SREBPs exist as endoplasmic reticulum membrane bound precursors that are activated after proteolytic processing by site-one and site-two proteases (S1P, S2P). When sterol levels are low, S1P cleaves the SREBP molecule to release the N terminal portion from the endoplasmic reticulum (Sakai et al., 1998). SREBP then binds to the SREBP cleavage activating protein (SCAP) and is translocated to the Golgi where S2P further processes the molecule so that the transcription factor is activated. The processed SREBP then translocates to the nucleus to bind specific E box motifs and sterol response elements (SREs) (Magana and Osborne, 1996). There is evidence that dietary factors stimulate the expression of FASN in a manner mediated through signaling pathways such as the phosphoinositide-3 kinase (PI3K) pathway. For instance, nonmalignant 3T3-L1 adipocytes regulate insulin-mediated FASN expression through Akt in a manner independent of both mitogen activated protein kinase (MAPK) and P70 S6 kinase, but dependent on SREBPs (Wang and Sul, 1998; Porstmann et al., 2005).

Expression of FASN is tightly controlled so that transcription does not continue unabated under typical circumstances. Polyunsaturated fatty acids (PUFAs) (Xu et al., 1999; Moon et al., 2002; Jump et al., 1994), sterols (Adams et al., 2004; Bennett et al., 1995), and leptin (Fukuda et al., 1999) all act to repress FASN transcription and do so by specifically down-regulating SREBP-1 in hepatocytes (Worgall et al., 1998; Teran-Garcia et al., 2007). This highly complex organization of checks and balances for FASN expression is necessary to supply the cell with essential *de novo* fatty acids for cellular function and growth (Fig. 7.2A). Just as importantly, controls keep the cell from continuing unnecessary lipogenesis.

### *7.3.2 FASN Regulation in Tumor Cells*

While FASN expression is tightly controlled through dietary and hormonal stimuli in nonmalignant cells, tumor cells ignore these restrictions and increase FASN beyond typical levels (Fig. 7.2B). In fact, an early study of orthotopic hepatomas revealed that while low-fat, high-fat, and high-cholesterol diets all affected rates of fatty acid synthesis in the normal liver, the rates of hepatoma fatty acid synthesis were unchanged (Sabine et al., 1967). It has since been discovered that deregulation of upstream signals drive FASN expression in a manner that is largely transcriptional in tumors (Fig. 7.2B) (Swinnen et al., 2006).

Overexpression of FASN in tumor cells is induced at the transcriptional level by receptor tyrosine kinase (RTK)-stimulation of Ras and Akt (Fig. 7.2B). Keratinocyte growth factor (KGF) can induce the Akt- and cJun N-terminal kinase (JNK)-dependent expression of FASN in pulmonary cancer cells (Chang et al., 2005). Epidermal growth factor (EGF) has also been shown to increase FASN in prostate cancer cells (Swinnen et al., 2000a).

In addition to growth factor signaling, activation of the RTK HER2/Neu is linked with *FASN* expression in tumor cells. HER2/Neu upregulates PI3K-dependent *FASN* transcription in breast cancer cells (Kumar-Sinha et al., 2003; Yoon et al., 2007). Interestingly, blocking HER2/Neu with Herceptin decreases FASN expression (Kumar-Sinha et al., 2003). In fact, there appears to be a crosstalk between these pathways, as inhibition of FASN activity leads to the downregulation of HER2/NEU (Menendez et al., 2004). While HER2/Neu is primarily associated with breast cancer progression, HER2/Neu and FASN expression correlate in squamous cell carcinomas of the tongue, as well (Silva et al., 2008). Surprisingly, HER2/Neu can also regulate FASN expression in prostate cancer cells (Yeh et al., 1999). These data suggest there is a coordinate regulation of activated HER2/Neu and FASN upregulation in tumor cells.

Downstream of RTK signaling, the PI3K/Akt pathway has been shown to upregulate FASN. Loss of PTEN is a frequent transformation event in cancer, that leads to a gain of function in Akt signaling (Mulholland et al., 2006; Blanco-Aparicio et al., 2007). In prostate cancer cells, this signaling cascade drives androgen receptor (AR)-mediated oncogenic transcription and progression to metastatic disease (Wang et al., 2003; Mulholland et al., 2006). The PTEN-null LNCaP tumor cell line has high levels of FASN. Reintroducing PTEN or using the PI3K inhibitor LY294002 can decrease FASN expression, whereas introducing constitutively active Akt can restore FASN expression (Van de Sande et al., 2002). The connection between FASN expression and PI3K activity is further observed in prostate carcinoma samples with high Gleason scores, where high FASN expression correlates with phosphorylated Akt that is localized to the nucleus (Van de Sande et al., 2005). Moreover, a crosstalk between these pathways has been identified. In ovarian cancer cell lines, phosphorylated Akt correlates with and drives FASN expression. Conversely, inhibiting FASN results in decreased Akt phosphorylation (Wang et al., 2005). These data suggest that PI3K signaling through Akt is an important mediator of *FASN* transcription in tumor cells.

In addition to RTK-driven stimulation of Akt, there is evidence that the small GTP-ase protein Ras can influence FASN expression in tumors. Constitutively active H-ras induces increased PI3K and MAPK-dependent FASN expression in MCF-10A cells (Yang et al., 2002). Consistent with this notion, the expression of activated K-ras correlates with FASN expression in human colorectal cancer samples (Ogino, et al., 2006, 2007). Altogether, these data suggest that RTK signaling, Ras, and PI3K-Akt pathways can drive transcriptional up-regulation of FASN expression in tumor cells (Fig. 7.2B).

Not surprisingly, hormones are another common factor driving FAS expression in tumor cells (Fig. 7.2B). Progestins stimulate FASN expression in breast cancer cells (Chalbos et al., 1987; Lacasa et al., 2001; Menendez et al., 2005a). Consistent with this finding, increased FASN expression in endometrial carcinoma correlates with expression of both estrogen and progesterone receptors (PR) (Pizer et al., 1998b). In prostate cancer, FASN expression can be regulated by androgens in prostate cancer through upregulation of transcription factors such as S14 and SREBPs (Swinnen et al., 1997a,b; Heemers et al., 2000, 2001). In addition, HER2/Neu can drive activation of AR in prostate cancer cells to increase MAPK-dependent induction of FASN in the absence of androgen (Yeh et al., 1999).

While the main mechanism of FASN overexpression in tumors is through transcriptional upregulation, there is also evidence that FASN is regulated by post-transcriptional mechanisms (Fig. 7.2B). For instance, HER2/Neu driven expression of both FASN and ACC can be regulated at the translational level through Akt, PI3K, and mTOR-dependent mechanisms (Yoon et al., 2007). FASN stabilization is tightly linked with the de-ubiquitinating enzyme ubiquitin-specific protease 2a (USP2A) in prostate cancer cells. USP2A is androgen regulated and is not only upregulated similarly to FASN, but actually interacts with FASN to enhance FASN stability (Graner et al., 2004). Treating prostate tumor cells with the proteasome inhibitor MG-132, also increases FASN expression, further supporting evidence that FASN is regulated by the proteasome (Graner et al., 2004). Interestingly, yeast studies provided early evidence of FASN regulation by proteasomal degradation (Egner et al., 1993). It is also worth mentioning that FASN can also be upregulated in cancer cells by *FASN* gene amplification (Shah et al., 2006). The fact that numerous mechanisms act to increase FASN expression in tumor cells highlights the importance of FASN in tumor progression.

### *7.3.3 Palmitate Utilization in Normal and Tumor Cells*

Upregulation of FASN activity causes the increased production of fatty acids, particularly palmitate. While the mechanisms that drive FASN expression are different in tumors as compared to normal cells, the utilization of its products differs, as well. Fatty acids are used for a variety of cellular functions. In nonmalignant adipose and hepatic tissue, palmitate is incorporated into triglycerides for secretion and storage to be ultimately used as an energy source through β-oxidation (Thupari et al., 2002). Fatty acids such as palmitate can also comprise a regulatory pool that activates energy mediators such as PPARα in the liver and hypothalamus (Chakravarthy, et al., 2005, 2007). In addition, key signaling molecules, such as Ras and Hedgehog, can be palmitoylated to target these proteins to cellular membranes (Resh, 2006). So far, a link between protein palmitoylation and FASN activity has not been established though. In development, fatty acids can segregate into phospholipids to create cellular membranes (Chirala et al., 2003). Similarly, tumor FASN-derived palmitate segregates into phospholipid microdomains known as lipid rafts (Fig. 7.2B) (Swinnen et al., 2003). Lipid rafts are involved in a number of key biological functions including signal transduction, polarization, trafficking, and migration (Freeman et al., 2005, 2007). Considering that palmitate can ultimately be used for a number of cellular processes, including being elongated and desaturated for subsequent events, it is apparent that FASN occupies an important niche in tumor cells.

## 7.4 Inhibiting FASN Activity

### *7.4.1 Small Molecule Inhibitors of FASN*

Because of the unique expression of FASN in tumors, much emphasis has been put toward the development of pharmacological agents that inhibit FASN activity and, therefore, inhibit tumor growth and progression. Historically, a *Cephalosporium caerulens* mycotoxin metabolite known as cerulenin [(2S, 3R)-2,3-epoxy-4-oxo-7,10-dodecadienoylamide] has been the primary FASN inhibitor used in biological studies. Cerulenin covalently binds the β-ketoacyl synthase domain in FASN that is responsible for binding and condensing the substrates (Funabashi et al., 1989). More recently, C75 was formulated as a synthetic analog of cerulenin due to instability and poor systemic availability of cerulenin (Kuhajda et al., 2000). C93 is the newest generation of C75 analogues (Zhou et al., 2007). Both C75 and C93 target the β-ketoacyl synthase activity of FASN (Kuhajda et al., 2000; Zhou et al., 2007). Recently, orlistat (Xenical©), a FDA-approved drug for obesity that targets gastrointestinal lipases, was described as a novel inhibitor of FASN thioesterase activity (Kridel et al., 2004). There also exists a growing body of literature showing that various natural products such as the green tea polyphenolic component epigallocatechin-3-gallate (EGCG) can inhibit FASN activity (Tian, 2006).

### *7.4.2 Effects* In Vivo

To date, all small molecule inhibitors of FASN have demonstrated ability to block tumor growth *in vivo*. Cerulenin greatly increases survival and delays progression of ovarian cancer xenografts without significantly affecting fatty acid synthesis in the liver (Pizer et al., 1996b). C75 reduces growth of several tumor xenograft models, including prostate, breast, ovarian and mesothelioma

(Pizer et al., 2000, 2001; Wang et al., 2005; Gabrielson et al., 2001). C93 and C75 both reduce ovarian and lung cancer xenograft growth (Zhou et al., 2007; Orita et al., 2007). The novel FASN inhibitor orlistat has also been shown to inhibit prostate tumor xenograft growth (Kridel et al., 2004). FASN inhibitors also work in genetic models of tumorigenesis, including the *Neu-N* murine mammary transgenic model (Hennigar et al., 1998; Pflug et al., 2003; Alli et al., 2005). While FASN inhibitors are not typically given orally due to poor bioavailability, recent work shows that C93 can work *in vivo* after oral administration (Orita et al., 2007). Surprisingly, cerulenin, C75 and related compounds induce a reversible anorexic phenotype that is associated with β-oxidation in the hypothalamus. This phenotype is mimicked in mice with *FASN* deleted in the hypothalamus (see Section 7.2.1) (Loftus et al., 2000; Thupari et al., 2004; Tu et al., 2005; Orita et al., 2007; Chakravarthy et al., 2007). Interestingly, the anorexic effect of FASN inhibitors has been overcome with newer generation drugs like C93 that can reduce tumor growth with no anorexic effect (Orita et al., 2007). The discrepancy between the knockout studies and pharmacological findings has yet to be explained.

### *7.4.3 Cell Cycle Effects* In Vitro

To determine the cellular consequences of FASN inhibition, numerous studies have focused on the *in vitro* anti-tumor effects of these inhibitors. Many studies have linked FASN inhibitors with cell cycle and growth arrest (Fig. 7.3). Cerulenin acts *in vitro* to inhibit fatty acid synthesis-mediated growth of breast carcinoma cells that can be rescued with palmitate (Kuhajda et al., 1994). Cerulenin induces a block at the G2/M cell cycle checkpoint in an androgen-independent

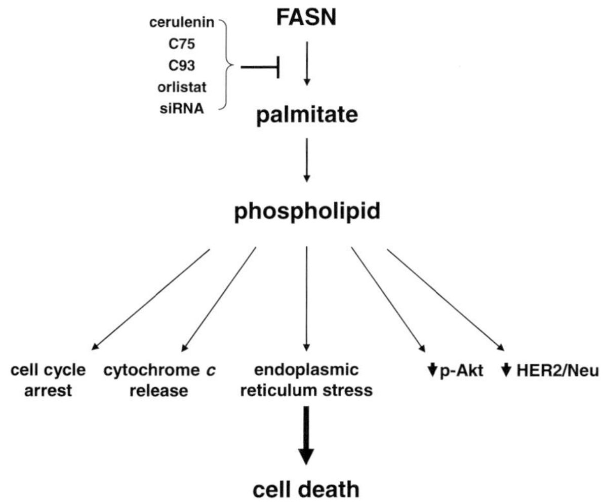


**Fig. 7.3** Inhibiting FASN in Tumor Cells. Several small molecule drugs can inhibit FASN activity. Blockade of FASN activity leads to a reduction in lipogenesis and phospholipid content in tumor cells. Inhibiting FASN also induces cycle arrest, cytochrome *c* release, and endoplasmic reticulum stress. In addition, FASN inhibitors can reduce the activation and expression of Akt and HER2/Neu

prostate cancer cell line that correlates with an induction of cyclin-dependent kinase inhibitors p21 and p27 (Furuya et al., 1997). However, glioma cells accumulate in S phase after cerulenin treatment (Zhao et al., 2006). Different hepatocellular carcinoma cell lines treated with C75 undergo either G1 or G2 cell cycle arrest independent of p53 status (Gao et al., 2006). In melanoma A-375 cells, cerulenin induces accumulation of cells in S phase, while C75 induces accumulation of G2/M phase cells (Ho et al., 2007). RKO colorectal cancer cells treated with either cerulenin or C75 show a transient accumulation of cells in S and G2/M phases, but accumulation in G1 and G2/M phases later (Li et al., 2001). Both cerulenin and C75 induce S phase arrest and inhibit DNA replication in breast, colorectal, and promyelocytic leukemia cancer cells (Pizer et al., 1998a). Orlistat induces cell cycle arrest by downregulating Skp2, a deubiquinating enzyme, leading to decreased turnover of p27/kip1, therefore blocking prostate tumor cells from entering S phase (Knowles et al., 2004). Orlistat has also been shown to induce an accumulation of breast cancer cells in S phase (Menendez et al., 2005b). Use of RNA interference (RNAi) to mediate knockdown of both the FASN and ACC$\alpha$ genes induces a decrease in S phase cells, further supporting the role of fatty acid synthesis in progression to or in S phase (Brusselmans et al., 2005). The data show there is little consensus on the phase that tumor cells arrest growth after inhibition of FASN in various tumor cells, which may be attributed to different tumor cell types. It is likely that a lack of *de novo* fatty acid synthesis in tumor cells impacts phospholipid synthesis required for proper DNA synthesis and cell cycle progression (Jackowski, 1994).

### 7.4.4 Cell Signaling Effects

The effects of FASN inhibitors are also mediated through key tumor signaling pathways. For example, it has been demonstrated that pharmacological inhibition of FASN activity results in reduced Akt activation in multiple tumor cell lines (Fig. 7.3) (Wang et al., 2005; Liu et al., 2006). As mentioned previously, it has been demonstrated that PI3K and Akt can drive FASN expression in tumor cells (Fig. 7.2B) (Van de Sande et al., 2002; Wang et al., 2005). The demonstration that reduced FASN activity negatively affects Akt activation identifies a feedback between the two pathways. Not surprisingly, inhibiting the PI3K pathway synergizes with cell death induced by genetic and pharmacological inhibition of *FASN* (Bandyopadhyay et al., 2005; Wang et al., 2005; Liu et al., 2006).

In addition to the PI3K pathway, HER2/Neu has also been linked with FASN expression in breast and prostate cancer cells (Kumar-Sinha et al., 2003; Yoon et al., 2007; Yeh et al., 1999). Inhibiting FASN with cerulenin and C75 reduces expression of Her2/neu expression in breast cancer cell lines (Fig. 7.3) (Menendez et al., 2004; Kumar-Sinha et al., 2003). Additionally, inhibiting Her2/Neu with Herceptin synergizes with FASN inhibitors to induce cell death (Menendez et al., 2004). Altogether, these data indicate that the very pathways that drive FASN expression in malignant cells are also affected when FASN activity is blocked. Moreover, tumor cell killing can be potentiated when FASN inhibitors are combined with inhibitors of these signaling pathways. The reason for this crosstalk has not been clearly defined, but it is tempting to speculate that inhibition of FASN activity directly impacts on lipid raft function, which results in reduced kinase signaling.

### *7.4.5* **In Vitro** ***Tumor Cell Death***

In addition to cell cycle arrest, all FASN inhibitors induce cell death in tumor cells (Pizer et al., 1996a, 1998a; Kridel et al., 2004; Zhou et al., 2007). Cerulenin induces breast and prostate cancer cell death that correlates with DNA fragmentation and morphology characteristic of apoptosis (Pizer, et al., 1996a, 2000; Furuya, et al., 1997). The mitochondria have also been linked to facilitation of cell death induced by cerulenin. For instance, the pro-apoptotic mitochondrial factor Bax is induced in cells treated with cerulenin. (Heiligtag et al., 2002). This correlation between cerulenin and the mitochondrial pathway of apoptosis is further supported by the induction of cytochrome *c* release (Fig. 7.3) (Heiligtag et al., 2002). FASN inhibition has been linked to p53 status of tumor cells, but whether p53 plays any role in FASN-expressing cells is unclear, as FASN is expressed in tumors independent of p53 status. FASN is strongly and significantly associated with p53 expression in hyperplastic parathyroids (Alo et al., 1999a). In various cancer cells, blocking p 53 activity with a dominant negative construct potentiates FASN inhibitor-induced cell death (Li et al., 2001). Conversely, others have reported that FASN inhibitors work equally well in tumors independent of p53 status (Heiligtag et al., 2002).

Cell death induced by FASN inhibitors could be a result of the cell lacking fatty acid for membrane biogenesis. Inhibiting FASN and ACC reduces incorporation of fatty acid into membrane phospholipids, which occurs in the endoplasmic reticulum (Zhou et al., 2003). Inhibiting FASN incorporation into phospholipids corresponds to a decrease in cell volume and other morphological changes ultimately leading to apoptosis (De Schrijver et al., 2003). Inhibiting FASN with small molecules (cerulenin, C75, orlistat), or with siRNAs induces endoplasmic reticulum stress and activation of the unfolded protein response (UPR) (Little et al., 2007). The UPR is able to induce cell death if homeostasis is not restored and, therefore, FASN inhibitors may be inducing cell death that is mediated by the UPR (Fig. 7.3) (Little et al., 2007).

When FASN is inhibited malonyl-CoA accumulates (Pizer et al., 2000). One hypothesis for the mechanism of FASN inhibitor-induced cell death is attributed to this accumulation of malonyl-CoA and, potentially, its interaction with CPT-1, the enzyme responsible for transferring fatty acids into the mitochondria for oxidation. Malonyl-CoA acts as a natural inhibitor of CPT-1 activity to prevent fatty acids being simultaneously synthesized and then oxidized (McGarry et al., 1983). Driving this hypothesis is a study showing that co-treating breast or ovarian cancer cells with the ACC inhibitor 5-(tetradecyloxy)-2-furoic acid (TOFA) partially rescues cell death induced by FASN inhibitors C75 and cerulenin (Pizer et al., 2000; Zhou et al., 2003). However, C75 alone can increase CPT-1 activity and directly compete with malonyl-CoA (Thupari et al., 2002; Yang et al., 2005). Therefore, it is important to note that MCF-7 cells co-treated with C75 and the CPT-1 inhibitor etomoxir show no effect on C75-induced cell death (Zhou et al., 2003). Hence, malonyl-CoA accumulation, not CPT-1 activation, is mediating death induced by FASN inhibitors (Fig. 7.3). In addition, siRNA-mediated knockdown of FASN induces accumulation of ceramide and malonyl-CoA that leads to inhibition of CPT-1 and induction of apoptotic genes *BNIP3*, *TRAIL*, and *DAPK2* (Bandyopadhyay et al., 2006).

Upstream lipogenesis mediators ACL and ACC are also important in maintaining tumor cell survival. RNAi-mediated knockdown or chemical inhibition of ACL in human tumor cells decreases proliferation and induces cell death *in vitro* and limits tumor growth by stimulating differentiation of tumor cells *in vivo* (Hatzivassiliou et al., 2005). ACL inhibition can also can impair Akt-mediated tumorigenesis and induce tumor cell death (Bauer et al., 2005). In addition, silencing ACC using RNAi induces apoptosis in breast and prostate cancer cells (Brusselmans et al., 2005; Chajes et al., 2006). Chemical inhibition of ACC can also induce tumor cell death (Beckers et al., 2007). While the effects of FASN inhibitors on tumor cells are clearly pleiotropic, and in some cases maybe even specific to the tumor type, it is evident that many of the effects can ultimately be tied to decreases in de novo synthesized fatty acids which can be extended to phospholipid synthesis. Whatever the mechanisms may be, the data clearly suggest that FASN occupies an important regulatory position in tumor cells to facilitate the processes that lead to tumor cell proliferation and survival.

## 7.5 Concluding Remarks

In summary, FASN is upregulated in multiple tumor types and correlates with poor patient prognosis and reduced survival. Correspondingly, a body of literature has demonstrated a requirement of FASN activity for tumor cell viability. Phospholipids synthesized from FASN-derived palmitate are important for cell cycle progression, lipid raft signaling, and endoplasmic reticulum homeostasis, all of which contribute to tumor cell survival, thereby, underscoring the importance of FASN. These findings signify a central role for fatty acid synthesis in critical cellular processes. In addition, tumor cells have developed feedback mechanisms to mediate crosstalk between FASN and signaling pathways like PI3-kinase and Her2/Neu. The discovery and development of pharmacological agents that block FASN activity suggest that FASN can be targeted for anti-tumor therapy. So far, anti-FASN drugs have successfully inhibited tumor growth in several tumor models with minimal side effects. Therefore, FASN represents a highly tractable anti-tumor target with significant clinical potential.

**Acknowledgments** The authors would like to apologize to those whose work we were unable to cite. We would also like to thank Isabelle Berquin and members of the Kridel laboratory for critical reading of the manuscript. Work in the Kridel laboratory is supported by the NIH/NCI (CA114104) and Department of Defense Prostate Cancer Research Program (W81XWH-07-1-0024).

## References

Abu-Elheiga, L., Matzuk, M. M., Kordari, P., Oh, W., Shaikenov, T., Gu, Z. and Wakil, S. J., Mutant mice lacking acetyl-CoA carboxylase 1 are embryonically lethal, *Proc Natl Acad Sci USA* 102 (2005) 12011–12016.

Adams, C. M., Reitz, J., De Brabander, J. K., Feramisco, J. D., Li, L., Brown, M. S. and Goldstein, J. L., Cholesterol and 25-Hydroxycholesterol Inhibit Activation of SREBPs by Different Mechanisms, Both Involving SCAP and Insigs, *J Biol Chem* 279 (2004) 52772–52780.

Alli, P. M., Pinn, M. L., Jaffee, E. M., McFadden, J. M. and Kuhajda, F. P., Fatty acid synthase inhibitors are chemopreventive for mammary cancer in neu-N transgenic mice, *Oncogene* 24 (2005) 39–46.

Alo, P. L., Amini, M., Piro, F., Pizzuti, L., Sebastiani, V., Botti, C., Murari, R., Zotti, G. and Di Tondo, U., Immunohistochemical expression and prognostic significance of fatty acid synthase in pancreatic carcinoma, *Anticancer Res* 27 (2007) 2523–2527.

Alo, P. L., Galati, G. M., Sebastiani, V., Ricci, F., Visca, P., Mariani, L., Romagnoli, F., Lombardi, G. and Tondo, U. D., Fatty acid synthase expression in Paget's disease of the vulva, *Int J Gynecol Pathol* 24 (2005) 404–408.

Alo, P. L., Visca, P., Framarino, M. L., Botti, C., Monaco, S., Sebastiani, V., Serpieri, D. E. and Di Tondo, U., Immunohistochemical study of fatty acid synthase in ovarian neoplasms, *Oncol Rep* 7 (2000) 1383–1388.

Alo, P. L., Visca, P., Marci, A., Mangoni, A., Botti, C. and Di Tondo, U., Expression of fatty acid synthase (FAS) as a predictor of recurrence in stage I breast carcinoma patients, *Cancer* 77 (1996) 474–482.

Alo, P. L., Visca, P., Mazzaferro, S., Serpieri, D. E., Mangoni, A., Botti, C., Monaco, S., Carboni, M., Zaraca, F., Trombetta, G. and Di Tondo, U., Immunohistochemical study of fatty acid synthase, Ki67, proliferating cell nuclear antigen, and p53 expression in hyperplastic parathyroids, *Ann Diagn Pathol* 3 (1999a) 287–293.

Alo, P. L., Visca, P., Trombetta, G., Mangoni, A., Lenti, L., Monaco, S., Botti, C., Serpieri, D. E. and Di Tondo, U., Fatty acid synthase (FAS) predictive strength in poorly differentiated early breast carcinomas, *Tumori* 85 (1999b) 35–40.

Amy, C. M., Williams-Ahlf, B., Naggert, J. and Smith, S., Molecular cloning of the mammalian fatty acid synthase gene and identification of the promoter region, *Biochem J* 271 (1990) 675–679.

Bandyopadhyay, S., Pai, S. K., Watabe, M., Gross, S. C., Hirota, S., Hosobe, S., Tsukada, T., Miura, K., Saito, K., Markwell, S. J., Wang, Y., Huggenvik, J., Pauza, M. E., Iiizumi, M. and Watabe, K., FAS expression inversely correlates with PTEN level in prostate cancer and a PI 3-kinase inhibitor synergizes with FAS siRNA to induce apoptosis, *Oncogene* 24 (2005) 5389–5395.

Bandyopadhyay, S., Zhan, R., Wang, Y., Pai, S. K., Hirota, S., Hosobe, S., Takano, Y., Saito, K., Furuta, E., Iiizumi, M., Mohinta, S., Watabe, M., Chalfant, C. and Watabe, K., Mechanism of Apoptosis Induced by the Inhibition of Fatty Acid Synthase in Breast Cancer Cells, *Cancer Res* 66 (2006) 5934–5940.

Bauer, D. E., Hatzivassiliou, G., Zhao, F., Andreadis, C. and Thompson, C. B., ATP citrate lyase is an important component of cell growth and transformation, *Oncogene* 24 (2005) 6314–6322.

Beckers, A., Organe, S., Timmermans, L., Scheys, K., Peeters, A., Brusselmans, K., Verhoeven, G. and Swinnen, J. V., Chemical inhibition of acetyl-CoA carboxylase induces growth arrest and cytotoxicity selectively in cancer cells, *Cancer Res* 67 (2007) 8180–8187.

Bennett, M. K., Lopez, J. M., Sanchez, H. B. and Osborne, T. F., Sterol regulation of fatty acid synthase promoter. Coordinate feedback regulation of two major lipid pathways, *J Biol Chem* 270 (1995) 25578–25583.

Blanco-Aparicio, C., Renner, O., Leal, J. F. M. and Carnero, A., PTEN, more than the AKT pathway, *Carcinogenesis* 28 (2007) 1379–1386.

Brusselmans, K., De Schrijver, E., Verhoeven, G. and Swinnen, J. V., RNA Interference-Mediated Silencing of the Acetyl-CoA-Carboxylase-{alpha} Gene Induces Growth Inhibition and Apoptosis of Prostate Cancer Cells, *Cancer Res* 65 (2005) 6719–6725.

Buechler, K. F. and Rhoades, R. A., Fatty acid synthesis in the perfused rat lung, *Biochim Biophys Acta* 619 (1980) 186–195.

Camassei, F. D., Cozza, R., Acquaviva, A., Jenkner, A., Rava, L., Gareri, R., Donfrancesco, A., Bosman, C., Vadala, P., Hadjistilianou, T. and Boldrini, R., Expression of the lipogenic enzyme fatty acid synthase (FAS) in retinoblastoma and its correlation with tumor aggressiveness, *Invest Ophthalmol Vis Sci* 44 (2003a) 2399–2403.

Camassei, F. D., Jenkner, A., Rava, L., Bosman, C., Francalanci, P., Donfrancesco, A., Alo, P. L. and Boldrini, R., Expression of the lipogenic enzyme fatty acid synthase (FAS) as a predictor of poor outcome in nephroblastoma: an interinstitutional study, *Med Pediatr Oncol* 40 (2003b) 302–308.

Chajes, V., Cambot, M., Moreau, K., Lenoir, G. M. and Joulin, V., Acetyl-CoA Carboxylase {alpha} Is Essential to Breast Cancer Cell Survival, *Cancer Res* 66 (2006) 5287–5294.

Chakravarthy, M. V., Pan, Z., Zhu, Y., Tordjman, K., Schneider, J. G., Coleman, T., Turk, J. and Semenkovich, C. F., "New" hepatic fat activates PPARalpha to maintain glucose, lipid, and cholesterol homeostasis, *Cell Metab* 1 (2005) 309–322.

Chakravarthy, M. V., Zhu, Y., Lopez, M., Yin, L., Wozniak, D. F., Coleman, T., Hu, Z., Wolfgang, M., Vidal-Puig, A., Lane, M. D. and Semenkovich, C. F., Brain fatty acid synthase activates PPARalpha to maintain energy homeostasis, *J Clin Invest* 117 (2007) 2539–2552.

Chalbos, D., Chambon, M., Ailhaud, G. and Rochefort, H., Fatty acid synthetase and its mRNA are induced by progestins in breast cancer cells, *J Biol Chem* 262 (1987) 9923–9926.

Chang, Y., Wang, J., Lu, X., Thewke, D. P. and Mason, R. J., KGF induces lipogenic genes through a PI3K and JNK/SREBP-1 pathway in H292 cells, *J Lipid Res* 46 (2005) 2624–2635.

Chirala, S. S., Chang, H., Matzuk, M., Abu-Elheiga, L., Mao, J., Mahon, K., Finegold, M. and Wakil, S. J., Fatty acid synthesis is essential in embryonic development: fatty acid synthase null mutants and most of the heterozygotes die in utero, *Proc Natl Acad Sci USA* 100 (2003) 6358–6363.

Conde, E., Suarez-Gauthier, A., Garcia-Garcia, E., Lopez-Rios, F., Lopez-Encuentra, A., Garcia-Lujan, R., Morente, M., Sanchez-Verde, L. and Sanchez-Cespedes, M., Specific pattern of LKB1 and phospho-acetyl-CoA carboxylase protein immunostaining in human normal tissues and lung carcinomas, *Hum Pathol* 38 (2007) 1351–1360.

De Schrijver, E., Brusselmans, K., Heyns, W., Verhoeven, G. and Swinnen, J. V., RNA interference-mediated silencing of the fatty acid synthase gene attenuates growth and induces morphological changes and apoptosis of LNCaP prostate cancer cells, *Cancer Res* 63 (2003) 3799–3804.

Deberardinis, R. J., Mancuso, A., Daikhin, E., Nissim, I., Yudkoff, M., Wehrli, S. and Thompson, C. B., Beyond aerobic glycolysis: Transformed cells can engage in glutamine metabolism that exceeds the requirement for protein and nucleotide synthesis, *Proc Natl Acad Sci USA* 104 (2007) 19345–19350.

Egner, R., Thumm, M., Straub, M., Simeon, A., Schuller, H. J. and Wolf, D. H., Tracing intracellular proteolytic pathways. Proteolysis of fatty acid synthase and other cytoplasmic proteins in the yeast Saccharomyces cerevisiae, *J Biol Chem* 268 (1993) 27269–27276.

Epstein, J. I., Carmichael, M. and Partin, A. W., OA-519 (fatty acid synthase) as an independent predictor of pathologic state in adenocarcinoma of the prostate, *Urology* 45 (1995) 81–86.

Foufelle, F., Gouhot, B., Pegorier, J. P., Perdereau, D., Girard, J. and Ferre, P., Glucose stimulation of lipogenic enzyme gene expression in cultured white adipose tissue. A role for glucose 6-phosphate, *J Biol Chem* 267 (1992) 20543–20546.

Freeman, M. R., Cinar, B., Kim, J., Mukhopadhyay, N. K., Di Vizio, D., Adam, R. M. and Solomon, K. R., Transit of hormonal and EGF receptor-dependent signals through cholesterol-rich membranes, *Steroids* 72 (2007) 210–217.

Freeman, M. R., Cinar, B. and Lu, M. L., Membrane rafts as potential sites of nongenomic hormonal signaling in prostate cancer, *Trends Endocrinol Metabol* 16 (2005) 273–279.

Fukuda, H., Iritani, N., Sugimoto, T. and Ikeda, H., Transcriptional regulation of fatty acid synthase gene by insulin/glucose, polyunsaturated fatty acid and leptin in hepatocytes and adipocytes in normal and genetically obese rats, *Eur J Biochem* 260 (1999) 505–511.

Funabashi, H., Kawaguchi, A., Tomoda, H., Omura, S., Okuda, S. and Iwasaki, S., Binding site of cerulenin in fatty acid synthetase, *J Biochem (Tokyo)* 105 (1989) 751–755.

Furuya, Y., Akimoto, S., Yasuda, K. and Ito, H., Apoptosis of androgen-independent prostate cell line induced by inhibition of fatty acid synthesis, *Anticancer Res* 17 (1997) 4589–4593.

Gabrielson, E. W., Pinn, M. L., Testa, J. R. and Kuhajda, F. P., Increased fatty acid synthase is a therapeutic target in mesothelioma, *Clin Cancer Res* 7 (2001) 153–157.

Gansler, T. S., Hardman, W., 3rd, Hunt, D. A., Schaffel, S. and Hennigar, R. A., Increased expression of fatty acid synthase (OA-519) in ovarian neoplasms predicts shorter survival, *Hum Pathol* 28 (1997) 686–692.

Gao, Y., Lin, L. P., Zhu, C. H., Chen, Y., Hou, Y. T. and Ding, J., Growth arrest induced by C75, A fatty acid synthase inhibitor, was partially modulated by p38 MAPK but not by p53 in human hepatocellular carcinoma, *Cancer Biol Ther* 5 (2006) 978–985.

Graner, E., Tang, D., Rossi, S., Baron, A., Migita, T., Weinstein, L. J., Lechpammer, M., Huesken, D., Zimmermann, J., Signoretti, S. and Loda, M., The isopeptidase USP2a regulates the stability of fatty acid synthase in prostate cancer, *Cancer Cell* 5 (2004) 253–261.

Harada, N., Oda, Z., Hara, Y., Fujinami, K., Okawa, M., Ohbuchi, K., Yonemoto, M., Ikeda, Y., Ohwaki, K., Aragane, K., Tamai, Y. and Kusunoki, J., Hepatic de novo lipogenesis is present in liver-specific ACC1-deficient mice, *Mol Cell Biol* 27 (2007) 1881–1888.

Hatzivassiliou, G., Zhao, F., Bauer, D. E., Andreadis, C., Shaw, A. N., Dhanak, D., Hingorani, S. R., Tuveson, D. A. and Thompson, C. B., ATP citrate lyase inhibition can suppress tumor cell growth, *Cancer Cell* 8 (2005) 311–321.

Heemers, H., Maes, B., Foufelle, F., Heyns, W., Verhoeven, G. and Swinnen, J. V., Androgens stimulate lipogenic gene expression in prostate cancer cells by activation of the sterol regulatory element-binding protein cleavage activating protein/sterol regulatory element-binding protein pathway, *Mol Endocrinol* 15 (2001) 1817–1828.

Heemers, H., Vanderhoydonc, F., Heyns, W., Verhoeven, G. and Swinnen, J. V., Progestins and androgens increase expression of Spot 14 in T47-D breast tumor cells, *Biochem Biophys Res Commun* 269 (2000) 209–212.

Heemers, H., Vanderhoydonc, F., Roskams, T., Shechter, I., Heyns, W., Verhoeven, G. and Swinnen, J. V., Androgens stimulate coordinated lipogenic gene expression in normal target tissues in vivo, *Mol Cell Endocrinol* 205 (2003) 21–31.

Heiligtag, S. J., Bredehorst, R. and David, K. A., Key role of mitochondria in cerulenin-mediated apoptosis, *Cell Death Differ* 9 (2002) 1017–1025.

Hennigar, R. A., Pochet, M., Hunt, D. A., Lukacher, A. E., Venema, V. J., Seal, E. and Marrero, M. B., Characterization of fatty acid synthase in cell lines derived from experimental mammary tumors, *Biochim Biophys Acta* 1392 (1998) 85–100.

Hillgartner, F. B., Salati, L. M. and Goodridge, A. G., Physiological and molecular mechanisms involved in nutritional regulation of fatty acid synthesis, *Physiol Rev* 75 (1995) 47–76.

Ho, T. S., Ho, Y. P., Wong, W. Y., Chi-Ming Chiu, L., Wong, Y. S. and Eng-Choon Ooi, V., Fatty acid synthase inhibitors cerulenin and C75 retard growth and induce caspase-dependent apoptosis in human melanoma A-375 cells, *Biomed Pharmacother* 61 (2007) 578–587.

Innocenzi, D., Alo, P. L., Balzani, A., Sebastiani, V., Silipo, V., La Torre, G., Ricciardi, G., Bosman, C. and Calvieri, S., Fatty acid synthase expression in melanoma, *J Cutan Pathol* 30 (2003) 23–28.

Jackowski, S., Coordination of membrane phospholipid synthesis with the cell cycle, *J Biol Chem* 269 (1994) 3858–3867.

Jakobsson, A., Westerberg, R. and Jacobsson, A., Fatty acid elongases in mammals: their regulation and roles in metabolism, *Prog Lipid Res* 45 (2006) 237–249.

Jayakumar, A., Tai, M. H., Huang, W. Y., al-Feel, W., Hsu, M., Abu-Elheiga, L., Chirala, S. S. and Wakil, S. J., Human fatty acid synthase: properties and molecular cloning, *Proc Natl Acad Sci USA* 92 (1995) 8695–8699.

Jensen, V., Ladekarl, M., Holm-Nielsen, P., Melsen, F. and Soerensen, F. B., The prognostic value of oncogenic antigen 519 (OA-519) expression and proliferative activity detected by antibody MIB-1 in node-negative breast cancer, *J Pathol* 176 (1995) 343–352.

Joseph, J. W., Odegaard, M. L., Ronnebaum, S. M., Burgess, S. C., Muehlbauer, J., Sherry, A. D. and Newgard, C. B., Normal flux through ATP-citrate lyase or fatty acid synthase is not required for glucose-stimulated insulin secretion, *J Biol Chem* 282 (2007) 31592–31600.

Jump, D. B., Clarke, S. D., Thelen, A. and Liimatta, M., Coordinate regulation of glycolytic and lipogenic gene expression by polyunsaturated fatty acids, *J Lipid Res* 35 (1994) 1076–1084.

Kapur, P., Rakheja, D., Roy, L. C. and Hoang, M. P., Fatty acid synthase expression in cutaneous melanocytic neoplasms, *Mod Pathol* 18 (2005) 1107–1112.

Kersten, S., Seydoux, J., Peters, J. M., Gonzalez, F. J., Desvergne, B. and Wahli, W., Peroxisome proliferator-activated receptor alpha mediates the adaptive response to fasting, *J Clin Invest* 103 (1999) 1489–1498.

Knowles, L. M., Axelrod, F., Browne, C. D. and Smith, J. W., A fatty acid synthase blockade induces tumor cell-cycle arrest by down-regulating Skp2, *J Biol Chem* 279 (2004) 30540–30545.

Kridel, S. J., Axelrod, F., Rozenkrantz, N. and Smith, J. W., Orlistat is a novel inhibitor of fatty acid synthase with antitumor activity, *Cancer Res* 64 (2004) 2070–2075.

Kridel, S. J., Lowther, W. T. and Pemble, C. W., Fatty acid synthase inhibitors: new directions for oncology, *Expert Opin Investig Drugs* 16 (2007) 1817–1829.

Kuhajda, F. P., Fatty-acid synthase and human cancer: new perspectives on its role in tumor biology, *Nutrition* 16 (2000) 202–208.

Kuhajda, F. P., Fatty acid synthase and cancer: new application of an old pathway, *Cancer Res* 66 (2006) 5977–5980.

Kuhajda, F. P., Jenner, K., Wood, F. D., Hennigar, R. A., Jacobs, L. B., Dick, J. D. and Pasternack, G. R., Fatty acid synthesis: a potential selective target for antineoplastic therapy, *Proc Natl Acad Sci USA* 91 (1994) 6379–6383.

Kuhajda, F. P., Katumuluwa, A. I. and Pasternack, G. R., Expression of haptoglobin-related protein and its potential role as a tumor antigen, *Proc Natl Acad Sci USA* 86 (1989a) 1188–1192.

Kuhajda, F. P., Piantadosi, S. and Pasternack, G. R., Haptoglobin-related protein (Hpr) epitopes in breast cancer as a predictor of recurrence of the disease, *N Engl J Med* 321 (1989b) 636–641.

Kuhajda, F. P., Pizer, E. S., Li, J. N., Mani, N. S., Frehywot, G. L. and Townsend, C. A., Synthesis and antitumor activity of an inhibitor of fatty acid synthase, *Proc Natl Acad Sci USA* 97 (2000) 3450–3454.

Kumar-Sinha, C., Ignatoski, K. W., Lippman, M. E., Ethier, S. P. and Chinnaiyan, A. M., Transcriptome analysis of HER2 reveals a molecular connection to fatty acid synthesis, *Cancer Res* 63 (2003) 132–139.

Kusakabe, T., Maeda, M., Hoshi, N., Sugino, T., Watanabe, K., Fukuda, T. and Suzuki, T., Fatty acid synthase is expressed mainly in adult hormone-sensitive cells or cells with high lipid metabolism and in proliferating fetal cells, *J Histochem Cytochem* 48 (2000) 613–622.

Kusakabe, T., Nashimoto, A., Honma, K. and Suzuki, T., Fatty acid synthase is highly expressed in carcinoma, adenoma and in regenerative epithelium and intestinal metaplasia of the stomach, *Histopathology* 40 (2002) 71–79.

Lacasa, D., Le Liepvre, X., Ferre, P. and Dugail, I., Progesterone stimulates adipocyte determination and differentiation 1/sterol regulatory element-binding protein 1c gene expression. potential mechanism for the lipogenic effect of progesterone in adipose tissue, *J Biol Chem* 276 (2001) 11512–11516.

Li, J. N., Gorospe, M., Chrest, F. J., Kumaravel, T. S., Evans, M. K., Han, W. F. and Pizer, E. S., Pharmacological inhibition of fatty acid synthase activity produces both cytostatic and cytotoxic effects modulated by p53, *Cancer Res* 61 (2001) 1493–1499.

Little, J. L., Wheeler, F. B., Fels, D. R., Koumenis, C. and Kridel, S. J., Inhibition of fatty acid synthase induces endoplasmic reticulum stress in tumor cells, *Cancer Res* 67 (2007) 1262–1269.

Liu, X., Shi, Y., Giranda, V. L. and Luo, Y., Inhibition of the phosphatidylinositol 3-kinase/Akt pathway sensitizes MDA-MB468 human breast cancer cells to cerulenin-induced apoptosis, *Mol Cancer Ther* 5 (2006) 494–501.

Loftus, T. M., Jaworsky, D. E., Frehywot, G. L., Townsend, C. A., Ronnett, G. V., Lane, M. D. and Kuhajda, F. P., Reduced Food Intake and Body Weight in Mice Treated with Fatty Acid Synthase Inhibitors, *Science* 288 (2000) 2379–2381.

Magana, M. M. and Osborne, T. F., Two tandem binding sites for sterol regulatory element binding proteins are required for sterol regulation of fatty-acid synthase promoter, *J Biol Chem* 271 (1996) 32689–32694.

Maier, T., Jenni, S. and Ban, N., Architecture of mammalian fatty acid synthase at 4.5 A resolution, *Science* 311 (2006) 1258–1262.

McGarry, J. D., Mills, S. E., Long, C. S. and Foster, D. W., Observations on the affinity for carnitine, and malonyl-CoA sensitivity, of carnitine palmitoyltransferase I in animal and human tissues. Demonstration of the presence of malonyl-CoA in non-hepatic tissues of the rat, *Biochem J* 214 (1983) 21–28.

Menendez, J. A., Oza, B. P., Colomer, R. and Lupu, R., The estrogenic activity of synthetic progestins used in oral contraceptives enhances fatty acid synthase-dependent breast cancer cell proliferation and survival, *Int J Oncol* 26 (2005a) 1507–1515.

Menendez, J. A., Vellon, L. and Lupu, R., Antitumoral actions of the anti-obesity drug orlistat (XenicalTM) in breast cancer cells: blockade of cell cycle progression, promotion of apoptotic cell death and PEA3-mediated transcriptional repression of Her2/neu (erbB-2) oncogene, *Ann Oncol* 16 (2005b) 1253–1267.

Menendez, J. A., Vellon, L., Mehmi, I., Oza, B. P., Ropero, S., Colomer, R. and Lupu, R., Inhibition of fatty acid synthase (FAS) suppresses HER2/neu (erbB-2) oncogene overexpression in cancer cells, *Proc Natl Acad Sci USA* 101 (2004) 10715–10720.

Milgraum, L. Z., Witters, L. A., Pasternack, G. R. and Kuhajda, F. P., Enzymes of the fatty acid synthesis pathway are highly expressed in in situ breast carcinoma, *Clin Cancer Res* 3 (1997) 2115–2120.

Moon, Y. S., Latasa, M. J., Griffin, M. J. and Sul, H. S., Suppression of fatty acid synthase promoter by polyunsaturated fatty acids, *J Lipid Res* 43 (2002) 691–698.

Moreau, K., Dizin, E., Ray, H., Luquain, C., Lefai, E., Foufelle, F., Billaud, M., Lenoir, G. M. and Venezia, N. D., BRCA1 Affects Lipid Synthesis through Its Interaction with Acetyl-CoA Carboxylase, *J Biol Chem* 281 (2006) 3172–3181.

Moustaid, N., Beyer, R. S. and Sul, H. S., Identification of an insulin response element in the fatty acid synthase promoter, *J Biol Chem* 269 (1994) 5629–5634.

Moustaid, N. and Sul, H. S., Regulation of expression of the fatty acid synthase gene in 3T3-L1 cells by differentiation and triiodothyronine, *J Biol Chem* 266 (1991) 18550–18554.

Mulholland, D. J., Dedhar, S., Wu, H. and Nelson, C. C., PTEN and GSK3beta: key regulators of progression to androgen-independent prostate cancer, *Oncogene* 25 (2006) 329–337.

Myers, R. B., Oelschlager, D. K., Weiss, H. L., Frost, A. R. and Grizzle, W. E., Fatty acid synthase: an early molecular marker of progression of prostatic adenocarcinoma to androgen independence, *J Urol* 165 (2001) 1027–1032.

Nemoto, T., Terashima, S., Kogure, M., Hoshino, Y., Kusakabe, T., Suzuki, T. and Gotoh, M., Overexpression of fatty acid synthase in oesophageal squamous cell dysplasia and carcinoma, *Pathobiology* 69 (2001) 297–303.

Ogino, S., Brahmandam, M., Cantor, M., Namgyal, C., Kawasaki, T., Kirkner, G., Meyerhardt, J. A., Loda, M. and Fuchs, C. S., Distinct molecular features of colorectal carcinoma with signet ring cell component and colorectal carcinoma with mucinous component, *Mod Pathol* 19 (2006) 59–68.

Ogino, S., Kawasaki, T., Ogawa, A., Kirkner, G. J., Loda, M. and Fuchs, C. S., Fatty acid synthase overexpression in colorectal cancer is associated with microsatellite instability, independent of CpG island methylator phenotype, *Hum Pathol* 38 (2007) 842–849.

Ookhtens, M., Kannan, R., Lyon, I. and Baker, N., Liver and adipose tissue contributions to newly formed fatty acids in an ascites tumor, *Am J Physiol Regul Integr Comp Physiol* 247 (1984) R146–R153.

Orita, H., Coulter, J., Lemmon, C., Tully, E., Vadlamudi, A., Medghalchi, S. M., Kuhajda, F. P. and Gabrielson, E., Selective inhibition of fatty acid synthase for lung cancer treatment, *Clin Cancer Res* 13 (2007) 7139–7145.

Paulauskis, J. D. and Sul, H. S., Cloning and expression of mouse fatty acid synthase and other specific mRNAs. Developmental and hormonal regulation in 3T3-L1 cells, *J Biol Chem* 263 (1988) 7049–7054.

Pflug, B. R., Pecher, S. M., Brink, A. W., Nelson, J. B. and Foster, B. A., Increased fatty acid synthase expression and activity during progression of prostate cancer in the TRAMP model, *Prostate* 57 (2003) 245–254.

Pizer, E. S., Chrest, F. J., DiGiuseppe, J. A. and Han, W. F., Pharmacological inhibitors of mammalian fatty acid synthase suppress DNA replication and induce apoptosis in tumor cell lines, *Cancer Res* 58 (1998a) 4611–4615.

Pizer, E. S., Jackisch, C., Wood, F. D., Pasternack, G. R., Davidson, N. E. and Kuhajda, F. P., Inhibition of fatty acid synthesis induces programmed cell death in human breast cancer cells, *Cancer Res* 56 (1996a) 2745–2747.

Pizer, E. S., Kurman, R. J., Pasternack, G. R. and Kuhajda, F. P., Expression of fatty acid synthase is closely linked to proliferation and stromal decidualization in cycling endometrium, *Int J Gynecol Pathol* 16 (1997) 45–51.

Pizer, E. S., Lax, S. F., Kuhajda, F. P., Pasternack, G. R. and Kurman, R. J., Fatty acid synthase expression in endometrial carcinoma: correlation with cell proliferation and hormone receptors, *Cancer* 83 (1998b) 528–537.

Pizer, E. S., Pflug, B. R., Bova, G. S., Han, W. F., Udan, M. S. and Nelson, J. B., Increased fatty acid synthase as a therapeutic target in androgen-independent prostate cancer progression, *Prostate* 47 (2001) 102–110.

Pizer, E. S., Thupari, J., Han, W. F., Pinn, M. L., Chrest, F. J., Frehywot, G. L., Townsend, C. A. and Kuhajda, F. P., Malonyl-coenzyme-A is a potential mediator of cytotoxicity induced by fatty-acid synthase inhibition in human breast cancer cells and xenografts, *Cancer Res* 60 (2000) 213–218.

Pizer, E. S., Wood, F. D., Heine, H. S., Romantsev, F. E., Pasternack, G. R. and Kuhajda, F. P., Inhibition of fatty acid synthesis delays disease progression in a xenograft model of ovarian cancer, *Cancer Res* 56 (1996b) 1189–1193.

Porstmann, T., Griffiths, B., Chung, Y. L., Delpuech, O., Griffiths, J. R., Downward, J. and Schulze, A., PKB/Akt induces transcription of enzymes involved in cholesterol and fatty acid biosynthesis via activation of SREBP, *Oncogene* 24 (2005) 6465–6481.

Rangan, V. S., Joshi, A. K. and Smith, S., Fatty acid synthase dimers containing catalytically active beta-ketoacyl synthase or malonyl/acetyltransferase domains in only one subunit can support fatty acid synthesis at the acyl carrier protein domains of both subunits, *J Biol Chem* 273 (1998) 34949–34953.

Rangan, V. S., Joshi, A. K. and Smith, S., Mapping the functional topology of the animal fatty acid synthase by mutant complementation in vitro, *Biochemistry* 40 (2001) 10792–10799.

Rangan, V. S., Oskouian, B. and Smith, S., Identification of an inverted CCAAT box motif in the fatty-acid synthase gene as an essential element for modification of transcriptional regulation by cAMP, *J Biol Chem* 271 (1996) 2307–2312.

Rashid, A., Pizer, E. S., Moga, M., Milgraum, L. Z., Zahurak, M., Pasternack, G. R., Kuhajda, F. P. and Hamilton, S. R., Elevated expression of fatty acid synthase and fatty acid synthetic activity in colorectal neoplasia, *Am J Pathol* 150 (1997) 201–208.

Resh, M. D., Trafficking and signaling by fatty-acylated and prenylated proteins, *Nat Chem Biol* 2 (2006) 584–590.

Rossi, S., Graner, E., Febbo, P., Weinstein, L., Bhattacharya, N., Onody, T., Bubley, G., Balk, S. and Loda, M., Fatty acid synthase expression defines distinct molecular signatures in prostate cancer, *Mol Cancer Res* 1 (2003) 707–715.

Rufo, C., Teran-Garcia, M., Nakamura, M. T., Koo, S. H., Towle, H. C. and Clarke, S. D., Involvement of a unique carbohydrate-responsive factor in the glucose regulation of rat liver fatty-acid synthase gene transcription, *J Biol Chem* 276 (2001) 21969–21975.

Sabine, J. R., Abraham, S. and Chaikoff, I. L., Control of lipid metabolism in hepatomas: insensitivity of rate of fatty acid and cholesterol synthesis by mouse hepatoma BW7756 to fasting and to feedback control, *Cancer Res* 27 (1967) 793–799.

Sakai, J., Nohturfft, A., Goldstein, J. L. and Brown, M. S., Cleavage of Sterol Regulatory Element-binding Proteins (SREBPs) at Site-1 Requires Interaction with SREBP Cleavage-activating Protein. Evidence from in vivo competition studies, *J Biol Chem* 273 (1998) 5785–5793.

Sampath, H. and Ntambi, J. M., The fate and intermediary metabolism of stearic acid, *Lipids* 40 (2005) 1187–1191.

Sebastiani, V., Visca, P., Botti, C., Santeusanio, G., Galati, G. M., Piccini, V., Capezzone de Joannon, B., Di Tondo, U. and Alo, P. L., Fatty acid synthase is a marker of increased risk of recurrence in endometrial carcinoma, *Gynecol Oncol* 92 (2004) 101–105.

Shah, U. S., Dhir, R., Gollin, S. M., Chandran, U. R., Lewis, D., Acquafondata, M. and Pflug, B. R., Fatty acid synthase gene overexpression and copy number gain in prostate adenocarcinoma, *Hum Pathol* 37 (2006) 401–409.

Shurbaji, M. S., Kalbfleisch, J. H. and Thurmond, T. S., Immunohistochemical detection of a fatty acid synthase (OA-519) as a predictor of progression of prostate cancer, *Hum Pathol* 27 (1996) 917–921.

Shurbaji, M. S., Kuhajda, F. P., Pasternack, G. R. and Thurmond, T. S., Expression of oncogenic antigen 519 (OA-519) in prostate cancer is a potential prognostic indicator, *Am J Clin Pathol* 97 (1992) 686–691.

Silva, S. D., Perez, D. E., Alves, F. A., Nishimoto, I. N., Pinto, C. A. L., Kowalski, L. P. and Graner, E., ErbB2 and fatty acid synthase (FAS) expression in 102 squamous cell carcinomas of the tongue: Correlation with clinical outcomes, *Oral Oncol* 44(5) (2008) 484–490.

Smith, S., Architectural options for a fatty acid synthase, *Science* 311 (2006) 1251–1252.

Smith, S., Witkowski, A. and Joshi, A. K., Structural and functional organization of the animal fatty acid synthase, *Prog Lipid Res* 42 (2003) 289–317.

Swinnen, J. V., Brusselmans, K. and Verhoeven, G., Increased lipogenesis in cancer cells: new players, novel targets, *Curr Opin Clin Nutr Metab Care* 9 (2006) 358–365.

Swinnen, J. V., Esquenet, M., Goossens, K., Heyns, W. and Verhoeven, G., Androgens stimulate fatty acid synthase in the human prostate cancer cell line LNCaP, *Cancer Res* 57 (1997a) 1086–1090.

Swinnen, J. V., Heemers, H., Deboel, L., Foufelle, F., Heyns, W. and Verhoeven, G., Stimulation of tumor-associated fatty acid synthase expression by growth factor activation of the sterol regulatory element-binding protein pathway, *Oncogene* 19 (2000a) 5173–5181.

Swinnen, J. V., Roskams, T., Joniau, S., Van Poppel, H., Oyen, R., Baert, L., Heyns, W. and Verhoeven, G., Overexpression of fatty acid synthase is an early and common event in the development of prostate cancer, *Int J Cancer* 98 (2002) 19–22.

Swinnen, J. V., Ulrix, W., Heyns, W. and Verhoeven, G., Coordinate regulation of lipogenic gene expression by androgens: evidence for a cascade mechanism involving sterol regulatory element binding proteins, *Proc Natl Acad Sci USA* 94 (1997b) 12975–12980.

Swinnen, J. V., Van Veldhoven, P. P., Timmermans, L., De Schrijver, E., Brusselmans, K., Vanderhoydonc, F., Van de Sande, T., Heemers, H., Heyns, W. and Verhoeven, G., Fatty acid synthase drives the synthesis of phospholipids partitioning into detergent-resistant membrane microdomains, *Biochem Biophys Res Commun* 302 (2003) 898–903.

Swinnen, J. V., Vanderhoydonc, F., Elgamal, A. A., Eelen, M., Vercaeren, I., Joniau, S., Van Poppel, H., Baert, L., Goossens, K., Heyns, W. and Verhoeven, G., Selective activation of the fatty acid synthesis pathway in human prostate cancer, *Int J Cancer* 88 (2000b) 176–179.

Szutowicz, A., Kwiatkowski, J. and Angielski, S., Lipogenetic and glycolytic enzyme activities in carcinoma and nonmalignant diseases of the human breast, *Br J Cancer* 39 (1979) 681–687.

Takahiro, T., Shinichi, K. and Toshimitsu, S., Expression of fatty acid synthase as a prognostic indicator in soft tissue sarcomas, *Clin Cancer Res* 9 (2003) 2204–2212.

Teran-Garcia, M., Adamson, A. W., Yu, G., Rufo, C., Suchankova, G., Dreesen, T. D., Tekle, M., Clarke, S. D. and Gettys, T. W., Polyunsaturated fatty acid suppression of fatty acid synthase (FASN): evidence for dietary modulation of NF-Y binding to the Fasn promoter by SREBP-1c, *Biochem J* 402 (2007) 591–600.

Thompson, B. J. and Smith, S., Biosynthesis of fatty acids by lactating human breast epithelial cells: an evaluation of the contribution to the overall composition of human milk fat, *Pediatr Res* 19 (1985) 139–143.

Thupari, J. N., Kim, E. K., Moran, T. H., Ronnett, G. V. and Kuhajda, F. P., Chronic C75 treatment of diet-induced obese mice increases fat oxidation and reduces food intake to reduce adipose mass, *Am J Physiol Endocrinol Metab* 287 (2004) E97–E104.

Thupari, J. N., Landree, L. E., Ronnett, G. V. and Kuhajda, F. P., C75 increases peripheral energy utilization and fatty acid oxidation in diet-induced obesity, *Proc Natl Acad Sci USA* 99 (2002) 9498–9502.

Tian, W. X., Inhibition of fatty acid synthase by polyphenols, *Curr Med Chem* 13 (2006) 967–977.

Tu, Y., Thupari, J. N., Kim, E. K., Pinn, M. L., Moran, T. H., Ronnett, G. V. and Kuhajda, F. P., C75 alters central and peripheral gene expression to reduce food intake and increase energy expenditure, *Endocrinology* 146 (2005) 486–493.

Turyn, J., Schlichtholz, B., Dettlaff-Pokora, A., Presler, M., Goyke, E., Matuszewski, M., Kmiec, Z., Krajka, K. and Swierczynski, J., Increased activity of glycerol 3-phosphate dehydrogenase and other lipogenic enzymes in human bladder cancer, *Horm Metab Res* 35 (2003) 565–569.

Van de Sande, T., De Schrijver, E., Heyns, W., Verhoeven, G. and Swinnen, J. V., Role of the phosphatidylinositol 3′-kinase/PTEN/Akt kinase pathway in the overexpression of fatty acid synthase in LNCaP prostate cancer cells, *Cancer Res* 62 (2002) 642–646.

Van de Sande, T., Roskams, T., Lerut, E., Joniau, S., Van Poppel, H., Verhoeven, G. and Swinnen, J. V., High-level expression of fatty acid synthase in human prostate cancer tissues is linked to activation and nuclear localization of Akt/PKB, *J Pathol* 206 (2005) 214–219.

Visca, P., Sebastiani, V., Botti, C., Diodoro, M. G., Lasagni, R. P., Romagnoli, F., Brenna, A., De Joannon, B. C., Donnorso, R. P., Lombardi, G. and Alo, P. L., Fatty acid synthase (FAS) is a marker of increased risk of recurrence in lung carcinoma, *Anticancer Res* 24 (2004) 4169–4173.

Visca, P., Sebastiani, V., Pizer, E. S., Botti, C., De Carli, P., Filippi, S., Monaco, S. and Alo, P. L., Immunohistochemical expression and prognostic significance of FAS and GLUT1 in bladder carcinoma, *Anticancer Res* 23 (2003) 335–339.

Volpe, J. J. and Marasa, J. C., Hormonal regulation of fatty acid synthetase, acetyl-CoA carboxylase and fatty acid synthesis in mammalian adipose tissue and liver, *Biochim Biophys Acta* 380 (1975) 454–472.

Wakil, S. J., Fatty acid synthase, a proficient multifunctional enzyme, *Biochemistry* 28 (1989) 4523–4530.

Wakil, S. J., Stoops, J. K. and Joshi, V. C., Fatty acid synthesis and its regulation, *Annu Rev Biochem* 52 (1983) 537–579.

Wang, D. and Sul, H. S., Insulin stimulation of the fatty acid synthase promoter is mediated by the phosphatidylinositol 3-kinase pathway. Involvement of protein kinase B/Akt, *J Biol Chem* 273 (1998) 25420–25426.

Wang, H. Q., Altomare, D. A., Skele, K. L., Poulikakos, P. I., Kuhajda, F. P., Di Cristofano, A. and Testa, J. R., Positive feedback regulation between AKT activation and fatty acid synthase expression in ovarian carcinoma cells, *Oncogene* 24 (2005) 3574–3582.

Wang, S., Gao, J., Lei, Q., Rozengurt, N., Pritchard, C., Jiao, J., Thomas, G. V., Li, G., Roy-Burman, P., Nelson, P. S., Liu, X. and Wu, H., Prostate-specific deletion of the murine Pten tumor suppressor gene leads to metastatic prostate cancer, *Cancer Cell* 4 (2003) 209–221.

Worgall, T. S., Sturley, S. L., Seo, T., Osborne, T. F. and Deckelbaum, R. J., Polyunsaturated Fatty Acids Decrease Expression of Promoters with Sterol Regulatory Elements by Decreasing Levels of Mature Sterol Regulatory Element-binding Protein, *J Biol Chem* 273 (1998) 25537–25540.

Xu, J., Nakamura, M. T., Cho, H. P. and Clarke, S. D., Sterol regulatory element binding protein-1 expression is suppressed by dietary polyunsaturated fatty acids. A mechanism for the coordinate suppression of lipogenic genes by polyunsaturated fats, *J Biol Chem* 274 (1999) 23577–23583.

Yahagi, N., Shimano, H., Hasegawa, K., Ohashi, K., Matsuzaka, T., Najima, Y., Sekiya, M., Tomita, S., Okazaki, H., Tamura, Y., Iizuka, Y., Ohashi, K., Nagai, R., Ishibashi, S., Kadowaki, T., Makuuchi, M., Ohnishi, S., Osuga, J.-i. and Yamada, N., Co-ordinate activation of lipogenic enzymes in hepatocellular carcinoma, *Eur J Cancer* 41 (2005) 1316–1322.

Yang, N., Kays, J. S., Skillman, T. R., Burris, L., Seng, T. W. and Hammond, C., C75 [4-Methylene-2-octyl-5-oxo-tetrahydro-furan-3-carboxylic Acid] Activates Carnitine Palmitoyltransferase-1 in Isolated Mitochondria and Intact Cells without Displacement of Bound Malonyl CoA, *J Pharmacol Exp Ther* 312 (2005) 127–133.

Yang, Y. A., Han, W. F., Morin, P. J., Chrest, F. J. and Pizer, E. S., Activation of fatty acid synthesis during neoplastic transformation: role of mitogen-activated protein kinase and phosphatidylinositol 3-kinase, *Exp Cell Res* 279 (2002) 80–90.

Yeh, S., Lin, H.-K., Kang, H.-Y., Thin, T. H., Lin, M.-F. and Chang, C., From HER2/Neu signal cascade to androgen receptor and its coactivators: A novel pathway by induction of androgen target genes through MAP kinase in prostate cancer cells, *Proc Natl Acad Sci USA* 96 (1999) 5458–5463.

Yoon, S., Lee, M. Y., Park, S. W., Moon, J. S., Koh, Y. K., Ahn, Y. H., Park, B. W. and Kim, K. S., Up-regulation of acetyl-CoA carboxylase alpha and fatty acid synthase by human epidermal growth factor receptor 2 at the translational level in breast cancer cells, *J Biol Chem* 282 (2007) 26122–26131.

Zhao, W., Kridel, S., Thorburn, A., Kooshki, M., Little, J., Hebbar, S. and Robbins, M., Fatty acid synthase: a novel target for antiglioma therapy, *Br J Cancer* 95 (2006) 869–878.

Zhou, W., Han, W. F., Landree, L. E., Thupari, J. N., Pinn, M. L., Bililign, T., Kim, E. K., Vadlamudi, A., Medghalchi, S. M., El Meskini, R., Ronnett, G. V., Townsend, C. A. and Kuhajda, F. P., Fatty acid synthase inhibition activates AMP-activated protein kinase in SKOV3 human ovarian cancer cells, *Cancer Res* 67 (2007) 2964–2971.

Zhou, W., Simpson, P. J., McFadden, J. M., Townsend, C. A., Medghalchi, S. M., Vadlamudi, A., Pinn, M. L., Ronnett, G. V. and Kuhajda, F. P., Fatty Acid Synthase Inhibition Triggers Apoptosis during S Phase in Human Cancer Cells, *Cancer Res* 63 (2003) 7330–7337.

# Part II
# Phospholipids

# Chapter 8
# Lipids in the Assembly of Membrane Proteins and Organization of Protein Supercomplexes: Implications for Lipid-Linked Disorders

**Mikhail Bogdanov, Eugenia Mileykovskaya and William Dowhan**

**Abstract** Lipids play important roles in cellular dysfunction leading to disease. Although a major role for phospholipids is in defining the membrane permeability barrier, phospholipids play a central role in a diverse range of cellular processes and therefore are important factors in cellular dysfunction and disease. This review is focused on the role of phospholipids in normal assembly and organization of the membrane proteins, multimeric protein complexes, and higher order supercomplexes. Since lipids have no catalytic activity, it is difficult to determine their function at the molecular level. Lipid function has generally been defined by affects on protein function or cellular processes. Molecular details derived from genetic, biochemical, and structural approaches are presented for involvement of phosphatidylethanolamine and cardiolipin in protein organization. Experimental evidence is presented that changes in phosphatidylethanolamine levels results in misfolding and topological misorientation of membrane proteins leading to dysfunctional proteins. Examples are presented for diseases in which proper protein folding or topological organization is not attained due to either demonstrated or proposed involvement of a lipid. Similar changes in cardiolipin levels affects the structure and function of individual components of the mitochondrial electron transport chain and their organization into supercomplexes resulting in reduced mitochondrial oxidative phosphorylation efficiency and apoptosis. Diseases in which mitochondrial dysfunction has been linked to reduced cardiolipin levels are described. Therefore, understanding the principles governing lipid-dependent assembly and organization of membrane proteins and protein complexes will be useful in developing novel therapeutic approaches for disorders in which lipids play an important role.

**Keywords** Membrane protein folding · phosphatidylethanolamine · cardiolipin · mitochondria · oxidative phosphorylation

---

W. Dowhan
Department of Biochemistry and Molecular Biology, University of Texas-Houston, Medical School, 6431 Fannin St., Houston, TX 77030, USA

**Abbreviations** CL, cardiolipin; CFTR, cystic fibrosis transmembrane conductance regulator; Complex I, NADH:ubiquinone oxidoreductase; Complex II, sucinate:ubiquinone oxidoreductase; Complex III, cytochrome $bc_1$ complex (ubiquinone:cytochrome *c* oxidoreductase); Complex IV, cytochrome *c* oxidase; Complex V, $F_1F_0$-ATP synthase; DGlcDG, diglucosyldiacylglycerol; ER, endoplasmic reticulum; GPI, glycosaminylated phosphatidylinositol; LacY, lactose permease; NKT, natural killer T; NBD-1, first nucleotide binding domain; MGlcDG, monoglucosyldiacylglycerol; PC, phosphatidylcholine; PE, phosphatidylethanolamine; PG, phosphatidylglycerol; PI, phosphatidylinositol; PS, phosphatidylserine TM, transmembrane domain.

## 8.1 Introduction

Deciphering the role of lipids in normal and dysfunctional cell function is a more complex and challenging task than defining similar roles for specific proteins. Roles for individual lipids and lipid classes are diverse and widespread throughout a cell where individual proteins have narrow and well-defined functions that usually are localized to discrete cellular locations. Lipids define the essential membrane permeability barrier of cells and internal organelles. The membrane lipid bilayer is a dynamic non-covalent supermolecular organization of individual lipid molecules whose combined physical and chemical properties define the matrix within which membrane proteins are organized. Lipids govern the folding, organization, and final structure of all membrane proteins. Lipids directly influence and modulate the function of membrane proteins and a large number of amphitropic proteins that reversibly interact with the membrane surface. They act as metabolic signaling molecules and are the substrates for posttranslation modification of proteins.

Lipids exert their influence on cellular processes through their diverse chemical spectrum and the cooperative physical properties of lipid mixtures of variable composition (Dowhan, 1997; Dowhan et al., 2008). The complexity of the lipidome is often not fully appreciated. Eukaryotic cells contain phospholipids, sphingolipids, and glycolipids with a wide variation in the hydrophilic headgroups as well as diverse fatty acid compositions. Add to this complexity other lipids such as steroids and fatty acid derived cellular products, and the complexity of the lipidome equals or exceeds that of the proteome. While the basic characteristics and structure of the proteome among all organisms are largely conserved, lipid complexity increases significantly when all forms of life are considered. Through the efforts of LIPID MAPS in the United States (Schmelzer et al., 2007) and the European Lipodomics Initiative (van Meer et al., 2007), the composition of the lipidome is rapidly approaching the level of detail available for the proteome. This type of structural information will be essential for complete understanding of the molecular basis for normal lipid function and lipid-related cellular dysfunction in disease.

Lipids are not covalently bound in membranes but rather interact dynamically to form transient arrangements with asymmetry both perpendicular and parallel to the plane of the lipid bilayer. The fluidity, supermolecular-phase propensity, lateral pressure and surface charge of the bilayer matrix is largely determined by the collective properties of the complex mixture of individual lipid species, some of which are shown in Fig. 8.1. Lipids also interact with and bind to proteins in stiochiometric amounts affecting protein structure and function. The broad range of lipid properties coupled with the dynamic organization of lipids in membranes multiplies their functional diversity in modulating the environment and therefore the function of membrane proteins.

The pleiotropic function of even a single lipid and its widespread distribution within a given cell poses difficult challenges to defining roles for a lipid in cellular processes. Unlike proteins lipids display no inherent catalytic activity and are not encoded by genes so that many of the initial clues to lipid function

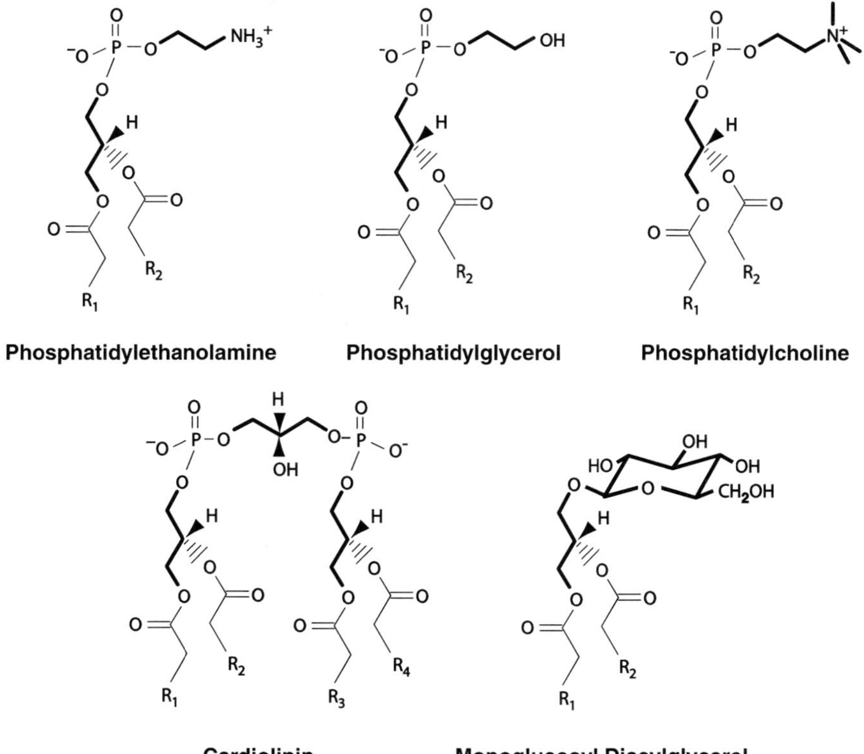


**Fig. 8.1** Structure of glycerol-based lipids. Stick drawing represents carbon backbone of headgroup (*bold*) and fatty acid chains. "R" represents hydrocarbon chains ranging from 10 to 22 in length with and without double bonds. Headgroups are attached to the *sn*-3 position of glycerol and the nature chirality is noted for the *sn*-2 position

outside of forming membrane bilayers have been derived from *in vitro* studies. Lipid function is many times defined by an effect on a biological process or enzyme reaction elicited by addition of a lipid. Often not considered in such experiments is the effect of the physical properties of a lipid since within any give set of lipids with the same hydrophilic domains exists a large array of possible hydrophobic domains (such as the diverse fatty acid composition of phospholipids). Although there is a large body of information describing the physical properties (phase properties, supermolecular organization, fluidity, etc.) of individual lipids and lipid mixtures, it is still not clear how to translate this information into a biological function or mechanism. Is the affect of a lipid on a process studied *in vitro* due to its physical properties, its chemical properties or an artifact due to adding a lipid with the wrong properties? Therefore, studying lipid-protein interaction only *in vitro* has serious limitations. However, studies *in vivo* pose additional obstacles. Genetic approaches are indirect since mutations must be made in genes that encode for biosynthetic enzymes along a pathway leading to a final lipid product. Such mutations result in changes in multiple lipid intermediates many of which may have important functions. In many cases such mutations are lethal causing loss of membrane integrity before more specific functions are affected and identified. The cumulative effect of loss of diverse functions often results in cell death and in complex phenotypes, especially in eukaryotic cells containing multiple organelles. However, in many cases viable lipid mutants in culture and even in whole animals have been generated. Defining lipid function requires biochemical dissection of biological processes as influenced by lipids and must be based on full knowledge of the physical and chemical properties of lipids. The *in vitro* properties must be verified by and be consistent with the properties of cells whose lipid metabolism and composition have been subjected to molecular genetic manipulation. Molecular genetic manipulation of cellular lipid metabolism has been most widely applied to bacteria and *S. cerevisiae* (Dowhan et al., 2004), but such approaches are increasingly successful in mammalian cell culture and whole animals.

Membrane proteins represent at least 30% of the all currently sequenced genomes and represent 60 percent of drug targets. In addition at least an equal number of proteins transiently interact with the membrane surface. More than half the drug targets pursued by pharmaceutical companies are related to membranes or membrane bound proteins (Drews, 2006). Effective drug design is dependent on understanding membrane protein structure and the rules that govern the folding and assembly of native and mutant membrane proteins as well as the principles governing interaction of "soluble" proteins with the membrane. In the past decade major advances have been made towards understanding the mechanisms by which polytopic membrane proteins fold and assemble in cellular membranes. However, the role that lipids play in the folding and assembly of membrane proteins, in the higher order organization of molecular machines, and in stabilizing final functional organization of proteins has only recently received attention. Through genetic manipulation of cellular lipid composition, it is now clear that membrane lipid composition is a determinant in the folding and topological organization of membrane proteins. Recent advances in detailed structural analysis of membrane proteins coupled with genetic manipulation of lipid composition has demonstrated that lipids play a specific role as integral components of multisubunit membrane protein complexes and higher order organization of complexes into molecular machines. Therefore, how lipids influence folding, assembly and function of proteins will be useful in developing novel therapeutic approaches for disorders involving proteins that associate with the membrane.

This review will focus on a combination of molecular genetic and biochemical studies on the role of primarily phosphatidylethanolamine (PE) and cardiolipin (CL) in the folding and organization of individual membrane proteins and multicomponent supercomplexes. The results of such studies will be related to the known and possible involvement of lipids in diseases resulting from lack of proper organization of membrane proteins. Rather than being an inclusive review of protein-lipid interactions, the aim is to select specific well-documented examples of lipid-protein interactions to illustrate the broader role of lipids in determining cellular function.

## 8.2 Lipid-Assisted Protein Folding

### *8.2.1 Experimental Evidence for Lipid-Assisted Folding of Proteins*

Molecular chaperones, traditionally proteins, facilitate the folding of proteins by interacting non-covalently with non-native folding intermediates and not with either the native or totally unfolded protein. When folding is complete, molecular chaperones are not required to maintain proper conformation. However, molecular chaperone function is not restricted to proteins (Bogdanov et al., 1996; Ellis, 1997). Specific lipids are able to interact with partially folded proteins in a transient manner in either *de novo* protein folding or protein renaturation *in vitro* similar to that of protein molecular chaperones.

The most compelling evidence for a specific role of phospholipids in membrane protein folding is the requirement for PE in the folding of the integral membrane protein lactose permease (LacY) of *Escherichia coli* (Bogdanov and Dowhan, 1999). LacY is organized within the inner cytoplasmic membrane as twelve transmembrane domains (TMs) connected by alternating solvent exposed cytoplasmic and periplasmic domains with both the N- and C-terminus oriented inward (Fig. 8.2). Normal assembly of LacY occurs into *E. coli* membranes containing an abundance of PE (70% with the remainder being about 20% phosphatidylglycerol (PG) and 5% CL) so that a separation between phospholipid-assisted and unassisted folding pathways cannot be distinguished *in vivo*. Studies of the lipid-assisted folding of periplasmic domain P7 connecting segments TMVII and TMVIII of LacY was made possible by a conformation specific monoclonal antibody directed to domain P7, viable

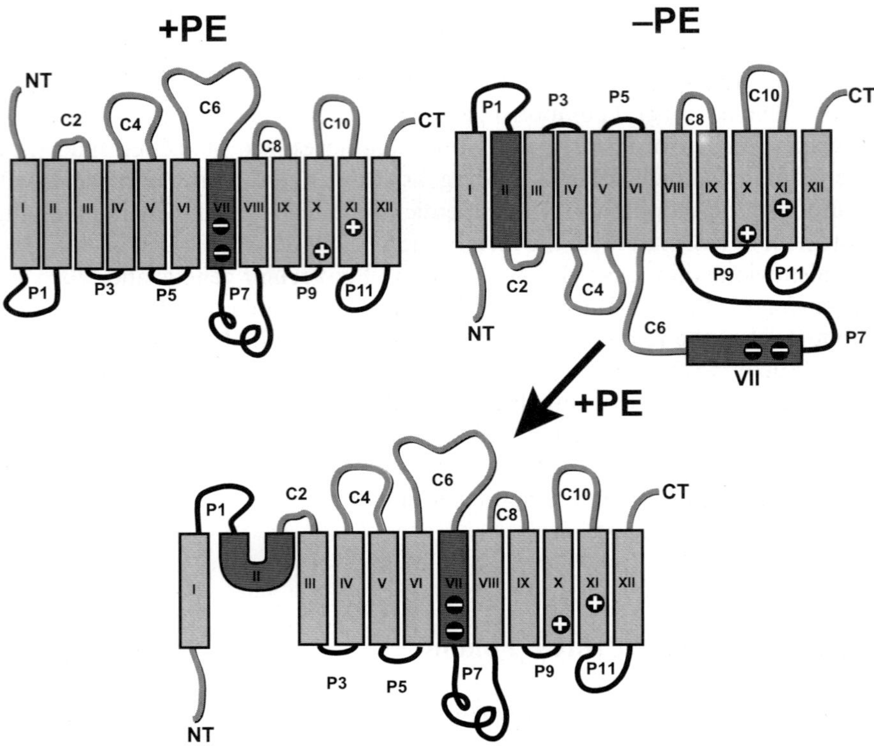


**Fig. 8.2** Topological models for LacY as a function of lipid composition. Topology of LacY in PE-containing (+PE) and PE-lacking (–PE) *E. coli* cells is illustrated in the upper left and right diagrams, respectively. The topology of LacY after initial assembly in cells lacking PE followed by post-assembly synthesis of PE is shown in the bottom diagram. The sequences within the rectangular shaded areas (TMs) define the amino acids that lie within the 27 Å hydrocarbon core of the bilayer excluding the lipid head groups. The darkly shaded TMs are those that undergo rearrangements as a function of membrane lipid composition (see text). TMs (Roman numerals), extramembrane domains (P in *black* for periplasmic and C in *gray* for cytoplasmic domains as in PE-containing cells), N-terminus (NT) and C-terminus (CT) are shown. The minus signs in TMVII represent the negative amino acids that salt bridge in PE-containing cells with positive residues in TMX and TMXI

mutants of *E. coli* lacking PE, and the development of a modification of the Western blotting procedure (Eastern-Western blot) that allowed renaturation of proteins in the presence of lipids on a solid support (Bogdanov et al., 1996, 1999). The loss of uphill energy-dependent transport function of LacY in mutant cells lacking PE was initially correlated with misfolding of domain P7 that is crucial for uphill energy dependent transport of substrate (Sun et al., 1996). Sodium dodecyl sulfate polyacrylamide gel electrophoresis (which only partially denatures LacY) followed by Western blotting analysis (using the conformation specific monoclonal antibody) of LacY from PE-containing cells demonstrated that the protein could be separated from PE (as determined by absence of radiolabeled phospholipid) and still retain its native structure with respect to domain P7. However, LacY from cells lacking PE was not detected by the conformation specific monoclonal antibody but was detected by a polyclonal antibody. Thus LacY is "denatured" *in vivo* with respect to the P7 domain by assembly in a non-native lipid environment. Therefore, once information was imparted during folding of LacY *in vivo* in the presence of PE, this lipid was no longer required for maintenance of proper conformation of the protein.

The misfolding of LacY due to assembly in PE-lacking cells was corrected by employing the Eastern-Western blotting technique in which LacY was exposed to hydrated phospholipids during renaturation from sodium dodecyl sulfate on a solid support followed by probing with a conformation specific monoclonal antibody (Bogdanov and Dowhan, 1999; Bogdanov et al., 1996, 1999). LacY regained the native conformation of domain P7, which was absent during *in vivo* assembly, after renaturation in the presence of specifically PE. Anionic lipids PG and CL and "foreign" lipids such as phosphatidylcholine (PC) did not support proper refolding. Proper refolding did not occur with protein extensively denatured by sodium dodecyl sulfate-urea treatment. In addition monoclonal antibody recognition of LacY assembled either *in vitro* (Bogdanov and Dowhan, 1998) or *in vivo* (Bogdanov et al., 2002) in the absence of PE could be restored after its insertion and assembly by the initiation of PE synthesis in the absence of new synthesis of LacY.

The interaction between lipids and partially folded LacY during refolding was found to be structurally specific. Both the chemical properties of the individual lipid molecules and the collective properties of phospholipid mixtures were determinants supporting proper protein folding (Bogdanov et al., 1999). Minimal requirements for refolding were lipid mixtures in the bilayer state containing diacyl phospholipids with an ionizable primary amine (i.e., PE, phosphatidylserine (PS), and mono- and dimethyl-PE, but not PC). Non-bilayer prone forms of PE were only effective if mixed with an excess of bilayer forming lipids. Stereoisomers of amino-containing lipids with either unnatural backbone or head group configuration did not support proper refolding. Therefore, there is a specific requirement for an ionizable amine-containing phospholipid of natural chirality and preference for bilayer organization to facilitate proper folding of LacY into its native conformation. Thus PE appears to facilitate *in vitro* refolding and *in vivo* folding into a fully native conformation by interacting with LacY via a transient non-covalent interaction with a folding intermediate and fulfills the minimum requirements of a molecular chaperone.

Is lipid-assisted folding a widespread phenomenon and possibly applicable to soluble proteins? The erythrocyte membrane contains about 20-mole % of PE that is almost exclusively localized in the inner leaflet and is in contact with highly concentrated heme-containing proteins. The refolding of the denatured soluble and heme-containing enzyme horseradish peroxidase (HRP) was followed in the presence and absence of liposomes made up of different phospholipids (Debnath et al., 2003). Remarkably, dimyristoyl-PE (a bilayer-forming

PE) was able to drastically increase the yield of renatured enzyme relative to refolding in the absence of liposomes. However, dioleoyl-PE, which does not favor bilayer organization, did not support proper refolding. PCs containing a wide range of fatty acids were either non-supportive of refolding or inhibited refolding relative to folding in the absence of liposomes. Moreover Trp117 quenching through energy transfer with the heme moiety indicated that the denatured protein after dimyristoyl PE-assisted folding assumed an overall conformation similar to that of the native protein with the heme moiety in a native-like conformation. Therefore, LacY and the peroxidase share common requirements for proper refolding dependent on lipids.

Based on all these results lipids can function as non-protein molecular chaperones or lipochaperones that specifically mediate the folding of proteins thereby extending the definition of chaperones to other biomolecules in addition to proteins (Bogdanov and Dowhan, 1999).

## *8.2.2 Lipochaperones and Protein Folding Disorders*

### 8.2.2.1 Alzheimer's and Scapies Diseases

Even under healthy conditions the energetic balance between folding and misfolding pathways of membrane proteins is very delicate and fragile (Sanders and Myers, 2004). Therefore, membrane protein folding can be re-routed in the pathophysiological direction by both mutations and changes in protein environment including the membrane lipid composition. Errors in insertion (Milenkovic et al., 2007), folding (Lin and Liu, 2006), localization and intracellular trafficking (Aridor and Hannan, 2000), processing (Fadiel et al., 2007) or degradation (Gelman and Kopito, 2003), and turnover (Sambamurti et al., 2006) of integral transmembrane proteins are responsible for numerous diseases including neurodegenerative cerebral amyloidoses, cystic fibrosis and others (Harrison et al., 2007). Misfolded proteins fail to be degraded and become prone to formation of toxic aggregates. Diverse disorders such as Alzheimer's disease and prion/scrapie disease (Monaco et al., 2006), cystic fibrosis (Mendoza and Thomas, 2007) and even cataract formation (Crabbe, 1998) and type 2 diabetes (Hayden et al., 2005) arise from protein misfolding and are grouped together under the category of conformational diseases.

Prion and scrapie diseases are linked with the conformational transition of normally monomeric α-helical cellular prion protein, $PrP^{c}$, to a ß-sheet-rich pathogenic form, $PrP^{SC}$, which is prone to aggregation. A similar conformational transition of the normal cellular form of α–helical amyloid peptide (αAP (1-40)) into the disease-specific largely ß-sheet form of amyloid peptide (ßAP (1-40)) occurs in Alzheimer's disease, which results in amyloid deposits (Monaco et al., 2006). So far more than 19 different mutations in the human PrP gene have been linked with inherited prion diseases (Monaco et al., 2006). However, the molecular event triggering the spontaneous conversion of wild-type and mutant $PrP^C$ forms to the infectious $PrP^{SC}$ isoform is still unknown. Since αAP is normally produced as a soluble peptide, the question arises as to what conditions induce conversion to the ßAP form and aggregation of the peptide? Thus the search for molecules interacting with cellular forms of PrP or Alzheimer amyloid peptide is a major effort in the study of transmissible amyloidoses.

An abundance of evidence exists suggesting that specific membrane lipids might serve as templates or nucleation sites, which play an important role in these two diseases by promoting the pathological folding of these proteins *in vivo*. Syrian hamster prion protein PrP has a high affinity for negatively charged phospholipid membranes, but it does not bind to membranes made of only zwitterionic PC (Sanghera and Pinheiro, 2002). The association of human (Morillas et al., 1999) or Syrian hamster PrP (Sanghera and Pinheiro, 2002) to negatively charged membranes is accompanied by an increase in ß-sheet structure, which results mostly from electrostatic lipid-protein interactions while binding of PrP to zwitterionic membranes composed of saturated PC mixed with cholesterol and sphingomyelin in a raft-mimicking ratio leads to a stabilization of α-helical structure through predominant hydrophobic lipid-protein interactions. Therefore electrostatic lipid-PrP interactions appear to promote ß-sheet formation, while hydrophobic lipid-protein interactions seem to preserve α-helix formation (Sanghera and Pinheiro, 2002).

A growing number of observations indicate that at least some pathological effects of the ßAP (1-40), the major component of Alzheimer plaques, could be mediated by peptide-lipid interactions. This peptide binds (Yanagisawa et al., 1995) to ganglioside GM1 containing membranes and upon binding undergoes a rapid conformational transition from random coil to an ordered conformation rich in ß-sheet structure which was not detected upon binding to ganglioside-free liposomes composed of zwitterionic phospholipids (PC), acidic phospholipids (PG and PS ) or the isolated oligosaccharide moiety of the ganglioside (Choo-Smith and Surewicz, 1997). Furthermore, GM1 tends to accumulate in certain regions of neurons forming ganglioside-rich domains in neural membranes. Since GM1-bound ßAP (1–40) is associated with early diffuse plaques, it was suggested that GM1-bound peptide may act as a template that facilitates self-aggregation of ßAP peptides as a precursor to formation of a mature amyloid plaques (Yanagisawa, 2005). Therefore, ganglioside appears to act as an anti-chaperone by inducing a misfolding event. Taken together, these data suggest that structural propensities of AP and PrP peptides are determined by the nuances of specific lipid environments and major conformational changes in prion and Alzheimer proteins may involve lipids as auxiliary molecules in the pathogenesis of these diseases.

#### 8.2.2.2 Lipid Basis for Cystic Fibrosis

Lack of Phe508 in the first nucleotide binding domain (NBD-1) of the cystic fibrosis transmembrane conductance regulator (CFTR) is the molecular basis for the most common form of cystic fibrosis (Mendoza and Thomas, 2007).

Isolated NBD-1 lacking Phe508 displays a kinetic defect in folding as evidenced by a dramatic reduction in the refolding efficiency starting from an unfolded state (Qu et al., 1997). This mutant form of CFTR fails to reach the plasma membrane and is retained in the endoplasmic reticulum (ER) membrane due to either illicit interactions between the mutated domain and molecular chaperones or other cellular factors in the ER (Cheng et al., 1990). The ability of acrylamide to quench the fluorescence of Trp496 located within the wild-type NBD-1 was drastically suppressed in the presence of PS while PC failed to protect this residue from acrylamide quenching (Eidelman et al., 2002). However, the changes in secondary structure due to a Phe508 mutation resulted in protection of Trp496 in the presence of both PC and PS. Thus the wild-type NBD-1 interacts selectively with PS while NBD-1 carrying the Phe508 mutation loses the ability to discriminate between these two phospholipids (Eidelman et al., 2002). This observation demonstrates that specific phospholipid-protein interactions are critical in maintaining the defined structure of the NBD-1 domain and that mutations that change the dynamics of this interaction may be the molecular basis for misfolding of the protein. The authors suggested that this lipid-specific effect on the conformation of the Phe508 mutant form of NBD-1 may have pathophysiological significance for cystic fibrosis and postulated that the trafficking defect for mutant CFTR might be based on aberrant interactions with PC since changes in cellular content of PC appear to have an drastic effect on mutant CFTR maturation and trafficking *in vivo* (Eidelman et al., 2002). The quantitative replacement of a large fraction of the PC with phospholipid analogues whose headgroups correspond to an analogue supplement (either 2-aminobutanol, methylethanolamine or 3-aminopropanol) in the media resulted not only in an increase in the total amount of mutant CFTR, but also in an effect on maturation leading to an increase in higher molecular weight forms of CFTR in a dose-dependent manner. In contrast cells expressing wild-type CFTR responded to the various supplements with either little change or reductions in CFTR level, indicating that choline headgroup replacement differentially affects the maturation and stability of wild-type and mutant CFTR.

#### 8.2.2.3 Lipid Involvement in Biogenesis of CD1 Molecules

Almost a decade ago, immunology faced a paradigm shift when it became apparent that specialized CD1 – restricted T lymphocytes recognize not only peptides but also lipid antigens (Gumperz, 2006; Joyce, 2001). Antigenic lipid – CD1 complexes can present foreign glycolipid antigens to a special subpopulation of T cells, the CD1-restricted invariant natural killer T (NKT) lymphocytes (Brutkiewicz et al., 2003; Parekh et al., 2005). Human CD1d molecules can present several types of foreign lipid and glycolipid components of the cell wall of pathogenic bacteria, mycobacteria and protozoa lipids. Exogenous lipids of foreign origin with large oligosaccharide headgroups or long alkyl chains require internalization via the endocytic system to the lysosomes prior to binding to CD1d molecules (Porcelli, 2001; Roberts et al., 2002).

*In vitro* CD1d molecules appear to bind a wide variety of lipids that vary in the chemical properties (Joyce, 2001). The crystallographic structure of CD1d protein shows a narrow hydrophobic ligand-binding cleft specifically designed to bind lipids rather than peptides (Koch et al., 2005). The lipid ligand of CD1d as initially isolated from *in vivo* sources is highly restricted to phosphatidylinositol (PI) and the glycosaminylated PI (GPI) of GPI-linked proteins (De Silva et al., 2002; Joyce et al., 1998). However, repeated attempts to activate NKT cells by the addition of purified PI or GPI have failed (De Silva et al., 2002; Molano et al., 2000). The fact that NKT cells do not recognize the CD1d-PI/GPI complex implies that they are not the ligand recognized by NKT cells (De Silva et al., 2002). The apparent exclusive association of PI with CD1d *in vivo* occurs during assembly in the ER membrane (De Silva et al., 2002). This paradox can be explained as follows. Loading of the lipid-binding site exclusively with PI/GPI occurs in the ER and is maintained through the normal secretory pathway to the plasma membrane. Upon endocytosis and passage through the lysosomes, CD1d protein exchanges lipids loaded in the ER for lipids of foreign origin that have also been internalized via the endocytic system. CD1d protein with its newly acquired lipid ligand is then recycled to the plasma membrane where the antigenic lipid-CD1d complex is recognized by NKT cells. This temporal association of ER lipids with CD1 protein during biogenesis can satisfy all structural and physiological requirements of a molecular chaperone by: first, protecting and preserving the integrity of the large hydrophobic lipid antigen-binding groove from collapse during trafficking; second, occupying the site to prevent premature lipid antigen loading; third, occupying the binding site with easily dissociated ligand that later might be exchanged for a lipid antigen in the lysosomes (De Silva et al., 2002; Joyce, 2001; Park et al., 2004). Thus, assembly of CD1 with cellular lipids in the ER is an evolutionarily conserved feature of a chaperone-like role for lipids involved in biogenesis of CD1.

#### 8.2.2.4 Lipid Involvement in Diabetes

Because of its importance to human health, insulin has been one of the most studied biological molecules. However, important details of the folding and assembly, storage and release of insulin remain unresolved. Insulin is synthesized as a single stranded immature precursor (preproinsulin) that contains an N-terminal signal sequence. After processing of the signal sequence, disulfide bonds between the A and B chains and within the A chain are formed followed by folding into mature monomeric reduced proinsulin form. Proinsulin exits from the Golgi network as a zinc associated hexamer and is stored in secretory granules where it is converted by proteolysis to the less soluble hexameric insulin crystalline form (Dodson and Steiner, 1998). The hexameric complex gradually dissolves to deliver the monomeric, bioactive form of insulin into the bloodstream stimulated by elevated plasma glucose levels. The crystalline forms of insulin can be either amorphic (irreversible misfolded form of insulin) or native (which can be monomerized in a bioactive form). Several lines of evidence suggest that sulfatides promote the productive folding of reduced proinsulin into its native crystalline form, indicating that a sulfatide (3'-sulfo-galactosylceramide) possesses molecular chaperone-like activity. Sulfatides are acidic glycosphingolipids that present primarily within the islets of Langerhans or ß cells of the pancreas where insulin is produced and in the nervous system.

Insulin-dependent diabetes mellitus is an autoimmune disease and high titers of auto-antibodies against both insulin and sulfatide were found in patients with insulin-dependent diabetes (Andersson et al., 2002; Buschard et al., 2005). Sulfatide and insulin are present in the same cellular compartments and share the same intracellular trafficking pathways (Buschard et al., 2005; Fredman et al., 2000). The inhibition of sulfatide synthesis with chloroquine and fumonisine B1 leads to inhibition of insulin granule formation *in vivo* (Fredman et al., 2000). Sulfatide binds directly to insulin, and sulfatide, but not its galactosylceramide precursor, was able to compete with monoclonal antibodies directed against a subdomain of the insulin molecule (Osterbye et al., 2001). Sulfatide specifically promotes *in vitro* refolding of proinsulin into its zinc-dependent hexameric form while without a sulfatide or in the presence of galactosylceramide, proinsulin dimers and hexamers were nearly absent (Osterbye et al., 2001). Sulfatide and not galactosylceramide specifically mediates the conversion of insulin hexamers to the biological active monomers at neutral pH, the pH at the ß-cell surface. The monomerization process is a dynamic equilibrium between insulin hexamers, dimers, and monomers. It appears that sulfatide is responsible for pushing the equilibrium toward the insulin monomer. Thus sulfatide has a dual role in supporting productive folding of the proinsulin form and promoting insulin monomerization. This is the first description of a lipid molecule acting as a molecular chaperone on the unfolded precursor of a protein (proinsulin) and demonstrating a functional interaction with mature protein (insulin) later in storage and release of the bioactive form.

Since sulfatide is the only glycolipid that so far has been associated with insulin-dependent diabetes mellitus, the therapeutic effect of sulfatide on the development of diabetes in the non-obese diabetic mouse was tested (Buschard et al., 2001). Diabetes was prevented in these mice by administration of sulfatide or its precursor, galactosylceramide. The mice were treated with either sulfatide, galactosylceramide (which is converted to sulfatide in the Golgi) or GM1 (a negatively charged glycosphingolipid lacking sulfate) or phosphate buffered saline control. Among all the control-treated mice 93% became diabetic. However sulfatide or galactosylceramide reduced the diabetes incidence by 50%. In contrast 86% of GM1-treated mice became diabetic. Therefore either administered or newly synthesized sulfatide is able to prevent diabetes in non-obese diabetic mice.

The composition and organization of membrane lipid species are altered in several human diseases (Alemany et al., 2007; Vigh et al., 2005). For example, in insulin-dependent diabetes mellitus massive changes occur in lipid composition of the rat myocardium causing a 46% increase in PI levels and a 22% decrease in PE content, which was prevented by insulin treatment after induction of the diabetic state (Han et al., 2000). The pathology of diabetes is not easily explained simply by a defect in glucose uptake by cells. Such drastic changes in membrane lipid composition could result in many unrecognized effects on the assembly, organization and function of membrane proteins.

Rearrangement of membrane lipid microdomains was recently recognized as crucial for proper compartmentalization and localization of insulin signaling. Insulin resistance due to the inhibition of insulin signaling can be caused by elimination of insulin receptors from caveolae microdomains induced by an accumulation of ganglioside GM3 (Kabayama et al., 2007), which triggers dissociation of the insulin receptor and caveolin-1 complex. The pharmacological inhibition of GM3 biosynthesis by a specific glucosylceramide synthase inhibitor resulted in almost complete recovery of insulin signaling (Kabayama et al., 2005), which may provide a new approach to treat insulin resistance. Such interference with glycosphingolipid biosynthesis not only enhanced insulin sensitivity (Aerts et al., 2007) but also improved glucose tolerance (Zhao et al., 2007).

## 8.3 Lipid-Dependent Membrane Protein Topogenesis

### *8.3.1 Membrane Protein Topological Organization*

A fundamental architectural principle of the structure of polytopic membrane proteins is membrane topology, i.e. the number of TM segments and their orientation relative to the membrane bilayer. Numerous algorithms exist that predict TM segments of membrane proteins based on the length of hydrophobic domains along the amino acid sequence (Elofsson and von Heijne, 2007). Such *in silico* approaches are about 85% accurate in predicting the alternating orientation of TMs connected by extramembrane domains. However, absolute orientation with respect to the membrane bilayer is less predictable unless the sidedness of the N-terminal and/or C-terminal extramembrane domains is known. Because these algorithms rely on cumulative short-range interactions and cannot incorporate the effects of long-range interactions such as stabilizing a charged hydrophilic domain within the membrane bilayer by salt bridging to another membrane domain, the organization of a significant number of membrane proteins cannot be accurately predicted. There has been extensive investigation aimed at understanding the features of the amino acid sequence that determine the insertion and orientation of membrane protein, but the role of the membrane lipid composition as a putative topological determinant has been largely ignored. However, membrane lipids manifest a diverse array of hydrophilic and hydrophobic surfaces and positive and negative charges that can influence the folding and orientation of integral membrane proteins. Therefore, *in silico* predictions of membrane protein topology is only a starting point providing a framework within which topology can be experimentally determined. Since protein-lipid interactions have now been established as a determinant of membrane protein topogenesis, the molecular basis for some diseases resulting from mutations in membrane proteins may be a result of topological misorientation of proteins.

### 8.3.2 Properties of Lipids That Determine Protein Topogenesis

Although topogenic signals encoded in the protein sequence of membrane proteins are primary determinants of final protein organization in the membrane, the topological organization of several twelve TM spanning secondary transporters of *E. coli* is dramatically influenced by the membrane lipid composition thus making the properties of the lipid bilayer a factor in determining protein topology. Development of viable *E. coli* strains with altered phospholipid composition (Dowhan et al., 2004) and advanced methods for determining protein topology, in particular the substituted cysteine accessibility method as applied to TM determination (Bogdanov et al., 2005), revealed that membrane lipid composition is a critical determinant of topology. The topologies of the N-terminal six-TM helical bundle of LacY (Bogdanov et al., 2002) (Fig. 8.2) and the N-terminal two-TM hairpin of phenylalanine permease (Zhang et al., 2003) and $\gamma$-aminobutyrate permease (Zhang et al., 2005b) are inverted with respect to the membrane bilayer and the remainder of each protein when assembled in membranes lacking PE, the major phospholipid of this organism, and containing only the anionic phospholipids PG and CL.

What specific structural, chemical or phase-forming features of lipid molecules are important for proper TM topogenesis? Structural features, individual chemical properties, and the collective physical properties of lipids in association with each other must be considered when assessing protein-lipid interactions. For instance PE is a zwitterionic glycerophosphate-based diacyl lipid with no net charge (Fig. 8.1) that favors the formation of a typical membrane bilayer structure when both fatty acids are saturated, but favors non-bilayer structures with increasing unsaturation of its fatty acids and with increasing temperature. However, the charge properties of PE are somewhat dampened by the formation of an internal charge paired ring between the phosphate and amine moieties, which cannot be formed by the zwitterionic phospholipid PC due to the trimethylated amine. PC on the other hand is nearly always bilayer forming and like PE capable of countering the high negative charge density contributed to the bilayer surface by net negatively charged phospholipids such as PS, CL, phosphatidic acid, PG or PI. The first three can form non-bilayer structures in the presence of divalent cations while all contribute to a net negative membrane surface. Finally, monoglucosyldiacylglycerol (MGlcDG) has no charge character (Fig. 8.1) and can be either bilayer or non-bilayer forming depending on temperature and fatty acid content. Diglucosyldiacylglycerol (DGlcDG) is similar to MGlcDG but only forms bilayer structures. The hydrophilic headgroup of these glycolipids are vastly different in their structure from that of phospholipids and although not charged can still participate in hydrogen bonding and dilution of overall membrane surface charge as does PE.

Strikingly, the replacement of PE *in vivo* by the foreign lipids MGlcDG (Xie et al., 2006) or DGlcDG (Bogdanov, Xie and Dowhan, unpublished) by introducing the *Acholeplasma laidlawii* MGlcDG and DGlcDG synthases, respectively, into an *E. coli* mutant lacking PE restored wild type TM topology of the N-terminal helical bundle of LacY. While replacement of PE by MGlcDAG restored the kinetic parameters of uphill transport of LacY, substitution of PE by DGlcDG did not. Similar to the effects of DGlcDG, LacY reconstituted into proteoliposomes containing PC and PG appears to display native topology but failed to show uphill transport function (Wang et al., 2002) while PG alone mimicked the aberrant orientation of LacY in PE-lacking cells. Therefore, TM topology is sensitive to the charge density on the membrane surface *i.e.* neutral MGlcDG and DGlcDG and zwitterionic PE and PC all dilute the negative charge of the membrane surface created by PG and CL and at the same time support native LacY topology. Moreover, the fact that charged but neutral PC and PE and the uncharged glycolipids support native topology strongly suggests that the net charge character of the membrane surface is a more important topological determinant than the structures of the headgroups. Finally, the phase-forming physical properties of the lipids appear not to be a factor in determining topology since MGlcDG and PE tend to be non-bilayer forming while DGlcDG and PC tend to be bilayer forming. Lack of structural specificity further suggests that binding of lipids to specific protein sites is less likely a factor in determining orientation than the properties of the bilayer matrix within which LacY is assembled. Since the predetermined molar ratio of charged/uncharged lipids is the most important lipid determinant of the TM topology, it is tempting to speculate that during the course of evolution both proteins and lipids co-evolved together in the context of the lipid environment of membrane systems in which both are mutually dependent on each other. Although it is surprising that *E. coli* is tolerant to such major changes in membrane lipid composition, these results emphasize the importance of similar charge properties of lipids in determining membrane protein topology rather than a strict structural requirement, as observed for proteins.

Phenylalanine permease and γ-aminobutyrate permease topology is also lipid dependent and secondary transporters for proline, melibiose, tryptophan and lysine are also defective in uphill transport in *E. coli* cells lacking PE (Bogdanov and Dowhan, unpublished) suggesting that function and possibly topology of a broad spectrum of transporters is dependent on membrane lipid composition. The orientation of OEP7, an outer envelope protein of spinach chloroplasts, was inverted with respect to its native orientation when reconstituted in liposomes made of a total lipid extract of chloroplasts containing mainly the phospholipids PC and PG (Schleiff et al., 2001). However when the ratio of these two lipids was adjusted to mimic the high PC content of the chloroplast outer membrane, then native topology was achieved. These results strongly support a general dependence of protein topology on lipid composition across species.

### 8.3.3 Lipid-Triggered TM Molecular Switch

The utilization of strains with tightly regulated inducible promoters controlling the expression of phospholipid biosynthetic enzymes not only allows for controlling steady state membrane lipid composition but dynamic switching of lipid composition. Use of a strain in which PE content can be controlled in a temporal manner revealed surprising topological dynamics of proteins after stable membrane insertion. PE is required to maintain correct relative orientations (Bogdanov et al., 2002) of the N- and C-terminal halves of LacY each independently folded (Nagamori et al., 2003) into a compact bundle of two six TM α-helices connected by the long hydrophilic cytoplasmic domain C6. Reintroduction of PE after assembly of LacY *in vivo* after membrane insertion and folding of LacY triggers a conformational change (see Fig. 8.2) resulting in a lipid-dependent restoration of uphill transport function and a near complete restoration of the wild type topological orientation (Bogdanov et al., 2002). Five of the six TMs regain native topological organization (Bogdanov et al., 2008). TMII adopts a "U"-shaped mini-loop configuration partially inserted into the cytoplasmic side of the membrane and allows the domains flanking TMII (i.e. P1 and C2) to remain on the same cytoplasmic side of the membrane with TMIII and the adjacent P3 domain adopting a proper topology in the cells with restored PE levels. TMVII, which is exposed to the periplasm in PE-lacking cells, re-inserts into the membrane after introduction of PE into cells. This result clearly demonstrates that changing the lipid composition of the membrane can induce large topology inversions of TMs in complex polytopic proteins after stable assembly.

Why can some proteins or protein domains undergo large TM movement but others cannot? What structural elements enable topological transitions dependent on lipid composition? In order for the N- and C-terminal six TM helical bundles to respond to the lipid environment independent of each other, either during initial assembly or during a change in lipid environment, there must exist flexible hinge regions between the independently folding domains. TMVII flanked by domains C6 and P7 (Fig. 8.2) was found to behave as a required molecular hinge by exiting the membrane to the periplasm in PE-lacking cells to allow sufficient flexibility so that the two halves of LacY could respond differentially to membrane lipid composition. TMVII displays low hydrophobicity due to two Asp residues that are normally salt bridged to neighboring TMs in the crystal structure. The low hydrophobicity of TMVII allows thermodynamically stable solvent exposure in PE-lacking cells. Increasing the hydrophobicity of TMVII by electrostatic neutralization of the Asp240 to an Ile within this hinge prevented TMVII from being released into the periplasm in PE-deficient cells and simultaneously blocked the inversion of the N-terminal bundle of the protein (Bogdanov et al., 2008). Finally, TMVII inserts back into the membrane upon reorganization of LacY after synthesis of PE (Fig. 8.2). Retention of aberrant orientation of TMI after introduction of PE was possible by a secondary hinge region where TMII assumed a mini-loop organization, which does not span the membrane bilayer, most likely facilitated by its high Gly content and a "kinked" structure (Abramson et al., 2004). Therefore, TM switching appears to rely on the intrinsic structural flexibility provided by TMVII as a mobile molecular hinge, which is necessary and sufficient for TM rearrangement in response to changes in lipid environment.

The inverted topology of the N-terminal two-TM hairpin of phenylalanine permease of *E. coli* once established in PE-lacking cells can also be changed in a reversible manner in response to alterations in PE levels (Zhang et al., 2003). An abnormally long TMIII of phenylalanine permease, which forms a "U"-shaped mini-loop in PE lacking cells, appears to provide the molecular hinge in this case. Therefore, TMs on either side of a flexible hinge region can organize independently of each other in response to lipid environment, whereas those proteins without such a hinge region (LacY with a mutation in TMVII) either cannot assume different topologies or cannot fold and are degraded. These results clearly demonstrated that the lipid composition is a determinant of TM orientation and challenges the dogma that once TM orientation is established during assembly it is static and not subject to change.

The results also lead to an interesting general conclusion about the dynamic structure of polytopic membrane proteins. Several other membrane proteins have highly flexible domains containing apparent hinge regions, which allow TM movement associated either with their biogenesis or function (Kanki et al., 2002; Lu et al., 2000; Moss et al., 1998; Zhang, 2001). The human P-glycoprotein is localized to mammalian cytoplasmic membranes and is an ATP-binding cassette transporter responsible for multidrug resistance. Like LacY (Abramson et al., 2004; Guan and Kaback, 2006) it is a highly flexible protein that undergoes large conformational changes during its catalytic cycle (Zhang, 2001) or biogenesis (Moss et al., 1998). In its native host, the protein exhibits twelve TMs with both the N- and C-terminus exposed to the cytoplasm. When expressed in *E. coli*, the N-terminal half of the protein assumes the same topology as in the native host. However, TM7 no longer spans the membrane and TMs 8-12 assume an inverted orientation. Therefore the whole C-terminal half, which includes the nucleotide-binding domain, is misoriented within bacterial membranes (Linton and Higgins, 2002).

### 8.3.4 Lipid-Protein Interactions that Determine Topology

What features of the lipid bilayer and what features of the amino sequence of integral membrane proteins determine orientation in the membrane? The initial topological decision appears to be made by the translocon which provides the permissive environment required for concurrent membrane insertion of TMs and orientation of flanking regions to the extramembrane space according to the positive-inside and/or charge difference rule (von Heijne, 1989). The fate of TMs after clearance of the translocon must follow thermodynamically driven routes involving direct interaction of the TMs and associated extramembrane domains with the surrounding lipid bilayer (Hessa et al., 2005; White and von Heijne, 2005). TMs passively partition into the bilayer based on their affinity for the hydrophobic lipid core of the membrane while flanking aromatic and charged residues position themselves near and within the aqueous-membrane interface, respectively (McKenzie et al., 2006; White and von Heijne, 2005). Therefore, early on TM interactions and folding events are impacted by the properties of the surrounding lipids, which further decode the topogenic signals within the nascent chain sequences to influence TM orientation and final folding events.

The final TM topology of polytopic membrane proteins is dictated primarily by the encoded amino acid sequence. Charged residues flanking the hydrophobic TMs are major determinants of the gross topology of polytopic membrane proteins and can in most cases be described by the statistically derived and experimentally confirmed positive inside rule (von Heijne, 1986; von Heijne, 1989), which states that loops retained on the cytoplasmic side of the membrane are enriched in positively charged residues compared to loops translocated across the membrane. However, it is not clear how positively charged residues exert their effect on topology, why they are retained in the cytosol, and what cellular factors govern their topological disposition. Although the positive inside rule discounts the importance of negatively charged residues, negatively charged residues can be topologically active if they are present in high numbers (Nilsson and von Heijne, 1990), flank a marginally hydrophobic TM (Delgado-Partin and Dalbey, 1998) or lie within a window of six residues from the end of a highly hydrophobic TM (Rutz et al., 1999). Several negative residues are required to translocate a cytoplasmic domain with even a single positive residue (Nilsson and von Heijne, 1990). Since the orientation of a membrane protein can be reversed either by the addition or removal of a single positively charged residue (Gafvelin and von Heijne, 1994) or by introduction of negatively charged residues near the ends of TMs (Rutz et al., 1999), the relative topological power of the charged residues needs further clarification. Recent studies on lipid-dependent topological organization of LacY strongly indicate that lipid-protein charge interactions modulate the topological signal potential of charged residues in extramembrane domains of proteins.

A distinguishing feature for the three permeases thus far established to be topologically responsive to membrane lipid composition is the presence of both conserved positively and negatively residues within cytoplasmic domains that are sensitive to lipid composition suggesting that protein-lipid interactions may be a determinant of final topology. The conservation of negatively charged residues within the N-terminal TM helical bundle of LacY across the sugar permease family and the low level of acidic residues within the cytoplasmic loops of the C-terminal five-TM helical bundle of LacY was postulated to be critical for these negative residues in lipid-dependent topogenesis (Bogdanov et al., 2008). The cytoplasmic extramembrane domains of LacY strictly follow the positive inside rule. It was shown previously that extramembrane domains containing only positively charged residues were more stabilized facing the cytoplasm as the anionic phospholipid content (PG and CL) was increase *in vivo* (van Klompenburg et al., 1997), which raised questions of how the presence of acidic residues in the cytoplasmic domains of the N-terminal bundle of LacY would be responsible for topological inversion in cells lacking PE and containing only anionic phospholipids. Remarkably, conversion of any one of the six acidic residues distributed among domains C2, C4, and C6 (see Fig. 8.2) to a neutral amino acid (increase of net charge by plus one) prevented topological inversion of the whole N-terminal bundle in PE-lacking cells (Bogdanov et al., 2008). However, in order to induce topological inversion in normal PE-containing cells, net charge had to be changed from plus two to minus two in all three cytoplasmic domains of the N-terminal bundle, i.e. a change from net plus six to net minus six for the extramembrane surface of this large domain. In all cases increasing the hydrophobicity of TMVII prevented topological inversion.

These results provide several new insights into how final topology of a polytopic membrane protein is determined and the factors that make protein domains sensitive to membrane lipid composition. Protein-lipid charge interactions must be considered as an important topological determinant particularly for extramembrane domains containing a mixture of negatively and positively charged residues. For such domains the net positive charge of the protein domains or the negative charge density of the membrane surface effects TM topological orientation in a complementary manner. Final topology is determined by cooperative short-range and long-range protein-lipid and protein-protein interactions that occur well after the nascent polypeptide exits the translocon. The N-terminal bundle behaves as a single TM unit in response to topogenic signals within the protein and to its environment. Increasing the positive charge within either the C4 or C6 domain or the hydrophobicity of TMVII affects the topology of all upstream sequences with the former changes being dependent on protein-lipid interactions. These interactions occur during late folding events most likely independent of interactions between the protein and the translocon. Therefore, candidate proteins that may be subject to lipid-dependent topological changes would be those containing TM bundles with cytoplasmic domains containing a mixture of negatively and positively charged amino acids connected by a flexible hinge region (an abnormally hydrophilic TM or an unusually long TM) to TM domains attached to primarily positively charged cytoplasmic domains.

The fact that the topology of fully assembled LacY can be changed by a change in lipid environment establishes that protein topology remains dynamic both during and after assembly. Although such lipid induced topological changes are unlikely in *E. coli*, this proof of principle observation has important implication for membrane proteins in eukaryotic cells. The lipid composition is significantly different between the plasma membrane and internal organelles. Thus during intracellular protein trafficking a membrane protein is exposed to different lipid environments that could affect topological organization post membrane insertion to either activate a latent activity or inactivate a protein. Similarly, local changes in lipid composition could result in large topological changes affecting function and stability. As shown in the LacY model system, single amino acid changes can result in vastly different topological responses to the lipid environment, which could be the molecular basis for some membrane protein related pathologies that may be due to mutations resulting in changes in lipid-dependent topogenic signals.

### 8.3.5 Lipids and Topological Disorders

Experimental findings in prokaryotes and eukaryotes suggest a conserved mechanism for establishing TM orientation within membrane proteins. TM segments of polytopic membrane proteins once membrane-inserted are generally considered stably oriented due to the assumed large free energy barrier to topological reorientation of adjacent extramembrane domains. Therefore, a topological "mistake" such as a segment being trapped on the "wrong" side of the membrane might be difficult to correct (Sanders and Myers, 2004). Moreover, TM domain insertion into the ER membrane proceeds simultaneously with glycosylation of extramembrane domains, which provides accuracy of topogenesis and contributes to topological stability by trapping a domain on one side of the membrane. Furthermore, in many cases topological mistakes made in the ER result in misfolded proteins that would be subject to rapid degradation by the proteosome (Sanders and Myers, 2004). Therefore, the general assumption is that TM topology is not easily perturbed by single point mutations and topological errors should be rare in disease-related misfolding (Sanders and Myers, 2004).

However, proteins in mammalian cells manage to escape from quality control in the ER to adopt alternative or dual topology in different intracellular membrane compartments. There are an increasing number of examples of proteins that are expressed in different topological forms with different functions. For example, ductin was found in two different orientations in cellular membranes, one of which serves as the subunit of the vacuolar $H^+$-ATPase and the other serves as a component of the microsomal connexin channel of gap junctions (Dunlop et al., 1995). Several members of the cytochrome P450 superfamily and NADPH cytochrome P450 reductase are also expressed on both the cell surface and in the ER membrane in different topological orientations (Levy, 1996; Zhu et al., 1999). Several other membrane proteins are expressed in more than one topological form within one membrane type, such as the prion protein (Hegde et al., 1998) that can result in neurodenegeration and P-glycoprotein (Moss et al., 1998) that can result in failure of cells to pump drugs out. Several viral envelope proteins adopt at least two different topological forms with respect to the membrane with only one isoform involved in cell-host fusion (Lambert and Prange, 2001; McGinnes et al., 2003).

The plasma membrane ATP-dependent P-glycoprotein is a broad spectrum multidrug antiporter of the plasma membrane, and its expression in many cancer cell lines causes multidrug resistance, which may be responsible for the failure of cancer chemotherapy. Discrepancy exists between two experimentally determined orientations of the P-glycoprotein depending on the mammalian system used to express and probe orientation; as noted earlier, the orientation of the protein expressed in *E. coli* is also different than in its native host. Different topologies were observed in cell-free protein synthesis system supplemented with ER membranes from dog pancreas (Skach et al., 1993; Zhang and Ling, 1991), *Xenopus* oocytes (Skach et al., 1993), Chinese hamster ovary cells (Zhang, 1996), and in *E. coli* (Beja and Bibi, 1995; Linton and Higgins, 2002) and human cell line (Loo and Clarke, 1995) expression systems. The major difference between these topologies is the membrane sidedness loop of C8 (between TM8 and TM9). TM8 in the C-terminal half does not contain an efficient stop-transfer signal (Zhang, 1996) and loop C8 linking TM 8 and TM9 contains a hidden glycosylation site (Zhang and Ling, 1991), which if glycosylated would stabilize topology in this region of the protein. The cytoplasmic location of the loop C8 was supported by both epitope mapping and cysteine scanning assays using human cell lines (Georges et al., 1993; Loo and Clarke, 1995) while extracellular location was demonstrated in dog pancreatic ER microsomes (Sahin-Toth et al., 1996; Skach et al., 1993) and further confirmed by limited proteolysis in Chinese hamster ovary cells (Zhang, 1996). The protein adopts a mixed topology with respect to this domain in *Xenopus* oocytes (Moss et al., 1998) also determined by a protease susceptibility assay. The glycosylation site within domain C8 is normally either sterically hidden by the membrane associated translation machinery (Zhang and Ling, 1991) or temporarily shielded by cytoplasmic protein factors (Zhang et al., 1995). Failure to glycosylate the protein in the ER could make the topology of this region of the protein unstable and sensitive to differences in membrane environment along the normal secretory pathway. Therefore, mutations that eliminate a functional glycosylation sites or introduce a new topogenic signals could result in an alternate topology for a protein at its final location in the cell.

In contrast to conformational disorders, experimental evidence of lipid effects on topology of protein membranes is lacking. However the existence of variations in lipid composition between different intracellular compartments may have relevance to membrane protein misorientation. Is topological fixed during co-translationally membrane insertion in the ER or does further remodeling of topology occur with changes in lipid composition as proteins move through different organelles to their final destination? The topogenesis of polytopic membrane proteins specifically routed to membrane domains having distinctive lipid compositions, such as lipid rafts, may be also evolutionarily "tailored" to the lipid composition of these domains (Sanders and Myers, 2004). Mis-targeting of proteins to such domains or lack of proper topogenic signals for raft proteins could be pathologically significant. Therefore, changes in membrane lipid composition either locally or during intracellular movement of proteins along the organelle-based secretory pathway can be potentially added to ligand binding (Ikeda et al., 2005), substrate binding (Gouffi et al., 2004) and membrane depolarization (Jakes et al., 1998) as modes for inducing such changes.

It is also generally assumed that alteration of protein topology requires extensive sequence changes (Sanders and Myers, 2004). Indeed most mutations result in subtle local conformational changes (Milenkovic et al., 2007). However, data are accumulating that even a single amino acid substitution within a membrane-flanking domain can profoundly alter TM topology that in turn would affect proper trafficking and processing due to aberrant localization of the targeting motif of the precursor protein. While a single mutation of either one of the N-terminal Lys or Arg residues to Ala in the lung-specific surfactant protein C precursor produced mixed orientation, double mutation resulted in complete reversal of orientation, thereby directing the targeting motif to the lumen of the ER instead of the cytosol (Mulugeta and Beers, 2003). Amazingly while the double mutant was retained in the ER, single mutants produced a mixed pattern of both ER (double mutant-like) and vesicular (wild type-like) expression, demonstrating that proper trafficking and processing of lung-specific surfactant protein C requires cytosolic localization of the targeting motif of the precursor protein. This study provides a likely precedent for a mechanism in disorders associated with misorientations of integral membrane proteins. For the growing number of lung associated pathological disorders associated with protein mutations in the membrane-flanking region, topological alteration should be considered as one significant contributing factor.

Hereditary spherocytosis is a common human inherited hemolytic anemia caused by mutations in erythrocyte anion exchanger type 1 known also as Band 3 protein. The substitution of the highly conserved Met663 to positively charged Lys located in the extracellular boundary of TM8 was recently described in a patient with the disease (Lima et al., 2005). This novel Band 3 Tambau mutant is retained in a pre-medial Golgi compartment likely due misorientation.

The vitelliform muscular dystrophy type 2 (*VMD2*) gene mutated in Best muscular dystrophy encodes a four TM protein termed bestrophin-1. The vast majority of known disease-associated alterations cluster near or within predicted TMs. Three out of eighteen extramembrane-associated mutations resulted in severe effects in ER membrane insertion. Four out of twelve TM-associated mutations showed an altered glycosylation pattern demonstrating that substitution of hydrophobic residues by positively charged amino acids exerts severe effects on TM properties and therefore membrane protein topology. These facts suggest that defective membrane integration or misorientation of bestrophin-1 may represent a potential disease mechanism for a subset of Best muscular dystrophy-related mutations (Milenkovic et al., 2007).

The three N-terminal TM helices of the Glu/Asp transporter are encoded by exons 2, 3 and 4, respectively. The loss of exon 3 results in the three instead of two TMs and inverts the whole topology of the protein (Huggett et al., 2000). Moreover, this splice variant encodes a functional transporter with inverted orientation within the plasma membrane. In some neurological disorders the release of glutamate due to anoxia was found to be largely due to an inverse operation of the transporter.

Pro-apoptotic proteins Bax (recruited from cytoplasm upon induction of apoptosis) and Bak have been reported to form a supermolecular pore in the outer mitochondrial membrane, which is large enough to release of cytochrome *c* and other proteins to the cytoplasm to initiate the apoptotic cascade (Kim et al., 2004). Upon induction of apoptosis Bax translocates and inserts into the outer mitochondrial membrane such that α-helices 5, 6 and 9 insert into the bilayer (Annis et al., 2005). The molecular mechanism by which anti-apoptotic Bcl-2 antagonizes the action of the proapoptotic proteins is still not completely understood. However unlike Bax, Bcl-2 constitutively resides in membranes as a monotopic protein anchored to the membrane via helix 9. This form of Bcl-2 is most likely inactive in preventing Bax oligomerization. However, during apoptosis preexisting, membrane-bound Bcl-2 appears to change membrane topology from a tail-anchored to multispanning form (Kim et al., 2004) in which cytoplasmic helices 5 and 6 become TMs. Bcl-2 most likely prevents productive oligomerization of membrane-bound Bax only when they are both multispanning transmembrane proteins (Dlugosz et al., 2006). Topologically changed Bcl-2 continues to inhibit apoptosis until the concentration of membrane-embedded Bax exceeds that of Bcl-2 (Dlugosz et al., 2006; Leber et al., 2007) allowing excess Bax to form a pore. Therefore, Bcl-2 membrane topology is not fixed during or immediately after biosynthesis and upon induction of apoptosis undergoes a major TM rearrangement.

## 8.4 Lipids in Organization of Protein Complexes

### *8.4.1 Lipids as Integral Components of Protein Complexes*

In addition to providing the amphipathic bilayer matrix within which membrane proteins reside, phospholipids are also specifically integrated between and within the subunits of oligomeric protein complexes. Phospholipids are structurally and functionally important components in the energy-transducing multimeric complexes of the bacterial cytoplasmic membrane and the inner mitochondrial membrane. For instances CL (Fig. 8.1) is essential for optimum activity of inner mitochondrial membrane proteins including NADH dehydrogenase, the cytochrome $bc_1$ complex, ATP synthase, cytochrome *c* oxidase, and the ATP/ADP translocase (for reviews and references see (Mileykovskaya et al., 2005; Schlame et al., 2000)). CL is also specifically integrated into the structure of *E. coli* succinate dehydrogenase and formate dehydrogenase-N (Jormakka et al., 2002; Yankovskaya et al., 2003).

The *Saccharomyces cerevisiae* ubiquinol:cytochrome *c* oxidoreductase or cytochrome $bc_1$ complex (Complex III), which is highly homologous to the mammalian complex, is a component of the inner mitochondrial membrane. The *b* subunit, which is encoded by mtDNA, and the $c_1$ and Rieske iron-sulfur protein subunits, which are encoded by nuclear DNA, make up the catalytic core of the complex. An additional 7 non-identical and non-catalytic nuclear encoded subunits, three heme groups, and two quinones make up the remainder of the complex (see (Hunte et al., 2008) for review). Complex III exists as a dimer in which fourteen phospholipid molecules have been identified in the 2.3-Å-crystal structure (Fig. 8.3A) (Hunte, 2005). These specifically localized phospholipid molecules are four CL, two PI, six PE, and two PC molecules. Six of these phospholipids localize within the oligomeric structure of the complex. Two PE molecules (one per monomer as seen in the front center of Fig. 8.3A) are at the interface between the two monomers making contact with the *b* subunits of both monomers. The two PE molecules are near two of the CL molecules (one per monomer as seen in the front center of Fig. 8.3A). Each PI molecule is intercalated between the three catalytic subunits of each monomer. PI acyl chains engulf the transmembrane helix of the Rieske subunit near the point of movement of its extrinsic domain and may dissipate torsion forces during the catalytic cycle of the complex. The remaining phospholipids lie at the surface of the complex as immobilized annular lipids in direct contact with the TMs of Complex III. CL and PE on the right front side and left rear side of Fig. 8.3A are annular lipids. These surfaces containing CL and PE form a cavity as shown in Fig. 8.3B that was proposed to interface with Complex IV in the formation of a supercomplex as described later. Proton uptake sites associated with quinone reduction lie near this CL binding site. Several mutations in this site either reduce electron transfer activity or reduce stability of associated subunits (Hunte, 2005; Lange et al., 2001; Palsdottir and Hunte, 2004).

The crystal structure of bovine cytochrome *c* oxidase (Complex IV) homodimer has been determined to a resolution of 1.8 Å (Shinzawa-Itoh et al., 2007). This integral membrane protein complex composed of thirteen different subunits per monomer is responsible for the reduction of molecular oxygen to water during aerobic respiration, with concomitant proton pumping across the mitochondrial inner membrane. A combination of high resolution X-ray structure analysis of the integral lipids in bovine Complex IV with mass spectroscopy analysis of their chain lengths and the positions of the unsaturated bonds of the

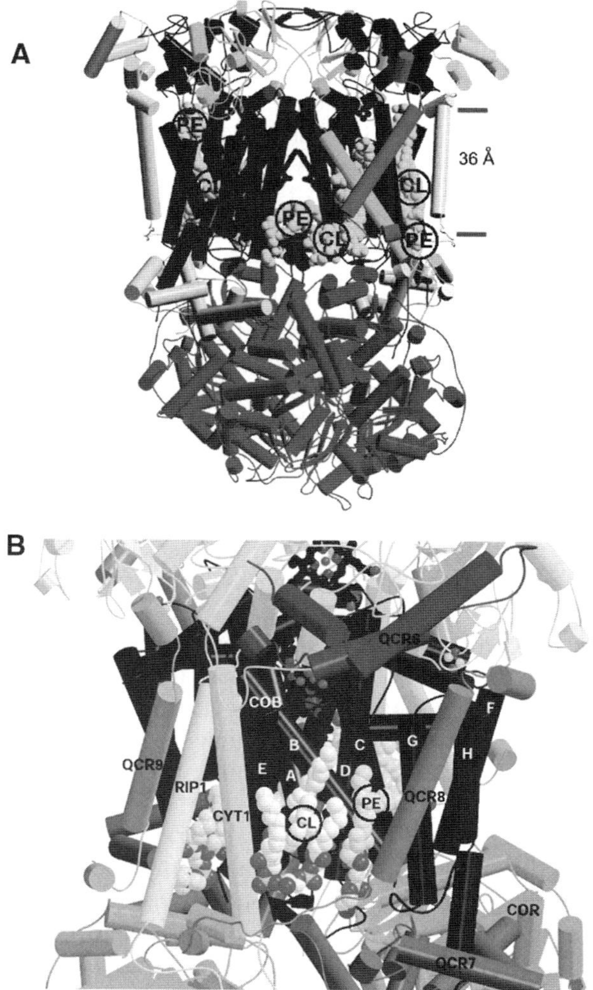


**Fig. 8.3** Crystal structure of Complex III dimer and model of putative Complex III surface that interacts with Complex IV from *S. cerevisiae*. (A) Dimer of Complex III based on the crystal structure (figure adapted from (Hunte, 2005)) with the interface between monomers in the center. α-Helices of different subunits within the dimer are shown as rods of different shading connected by non-helical domains. Note the positions of CL and PE (all circled) in the center front (also on back center but not shown) between the dimers and on the front (*right*) and back (*left*) sides of the diagram. The two bars on the right show the 36-Å width of the membrane. (B) The putative interface of Complex III monomer (figure adapted from (Pfeiffer et al., 2003)) with Complex IV with the various subunit domains labeled. This view, with a cavity containing PE and CL (both *circled*), corresponds to the right and left sides of the view shown in (A) and is positioned within the membrane bilayer. Bottom of both diagrams faces the mitochondrial matrix

hydrophobic tails provides understanding of structural and functional roles played by these lipids in the Complex IV (Shinzawa-Itoh et al., 2007). Thirteen lipids, including two CLs, one PC, three PEs, four PGs and three triacylglycerols were resolved in Complex IV. One CL, two PEs and one PG participate in the stabilization of the Complex IV dimer. However, the contacts made by PE and PG molecules between monomers are much weaker than those of CL, which interacts with subunits III and VIa within one monomer and bridges to the other monomer at subunits I and II. The four acyl chains of CL interact through van der Waals contacts with hydrophobic amino acid residues belonging to both monomers, and the two phosphate groups interact with both monomers via hydrogen bonds. Thus the dimer state of the bovine Complex IV is primarily stabilized by CL and subunit VIa. The bovine heart and liver Complex IV isozymes differ in that the N-terminal domains of their respective subunits VIa are not the same. The phosphorylated Thr11 residue of bovine heart Complex IV, which is proposed to stabilize the conformation of the N-terminal domain, is missing in liver isozyme. Therefore, the stability of the dimeric form of Complex IV may be different between the two isozymes and *in vivo* dimerization might respond differently to metabolic signals (Schmidt et al., 1997; Taanman and Capaldi, 1993). A third CL (not resolved in crystal structure of bovine heart cytochrome *c* oxidase) was found by photolabeling experiments with arylazido-containing CL analogues (Sedlak et al., 2006). This CL is located between subunits VIIa and VIIc near the entrance to the putative proton pumping channel, which contains a conserved aspartate. The authors suggested that this CL molecule could potentially function as a proton antenna to facilitate proton entry into the channel thus explaining the CL requirement for full enzymatic activity.

The high content of PG in Complex IV is remarkable in light of the low level (less than 1%) of this phospholipid in the inner membrane of mitochondria and its apparent absence from other mitochondrial complexes. Analysis of the X-ray structure of Complex IV revealed palmitate as one of the acyl chains of the two PG molecules of each dimer. These acyl chains lie near the putative $O_2$ transfer pathway in subunit III of each monomer of the Complex IV dimer. The other acyl chain of these PG molecules is vaccenate (*cis*-$\Delta^{11}$-octadecenoate), which appears to play a critical role in the specific binding of PG to this site. Only vaccenate-containing PG is found in bovine Complex IV in spite of the abundance of oleate (*cis*-$\Delta^{9}$-octadecenoate) in mitochondrial phospholipids. The X-ray structure demonstrates that only *cis*-vaccenate derivatives of PG will fit into the hydrophobic grooves of subunit III near the $O_2$ transfer pathway. All these data together suggest a unique role for PG bound to subunit III in the $O_2$ transfer process (Shinzawa-Itoh et al., 2007).

Cytochrome *c* oxidase is a highly conserved enzyme; the three core subunits (I, II and III) encoded by the mitochondrial genome in eukaryotes has high amino acid sequence homology with prokaryotic and eukaryotic species. Detailed analysis of evolutionary conservation of lipid-binding sites in cytochrome *c* oxidase (Qin et al., 2007) shows that lipid binding sites are specific and selective for both headgroups and alkyl tails. Strikingly, the overlay of bovine and bacterial crystal structures shows identical positions containing fatty acid chains of phospholipid or triacylglycerol in both enzymes. The headgroups of lipids interact with positively charged (Arg, Lys or His), aromatic (Trp or Tyr), or other polar residues (Thr, Ser, Gln or Asn) of cytochrome *c* oxidase at the membrane interface. The hydrophobic tails of these lipids are inserted into shallow grooves on the protein surface and are stabilized by precise fits through van der Waals contacts with hydrophobic residues (Qin et al., 2007). A similar analysis of the Complex III dimer crystal structure shows very similar protein-lipid interactions (Palsdottir and Hunte, 2004). Therefore, these protein-lipid associations are not the result of random or weak interactions but are extensive over whole lipid molecules and represent an integral part of the overall structure of these complexes.

## *8.4.2 Pathological Effects of Reduced CL Levels*

Therefore, any pathological state in which the above lipids, and in particular CL and PG, are significantly reduced (ischemia, hypothyroidism, aging, and heart failure; see (Chicco and Sparagna, 2007) for detailed review) would be expected to affect electron transfer efficiency and mitochondrial energy production. Mitochondrial dysfunction and diseases associated with inhibition of catalytic activity due to the loss of CL have been described for both Complexes III and IV. Decrease in Complex III activity coupled with a decrease in the content of CL was demonstrated for mitochondria isolated from rat heart subjected to ischemia and reperfusion presumably due to the oxidative damage of specifically heart mitochondrial CL. Remarkably, Complex III activity of mitochondria was restored to pre-ischemia levels after fusion of the mitochondria with CL-containing liposomes, while PC-, PE-, and oxidized CL-containing liposomes failed to restore the activity. It was suggested that the loss of Complex III activity results from oxidation of the high content of tetralinoleate-containing CL in heart mitochondria by oxygen free radicals produced upon reperfusion after ischemia (Petrosillo et al., 2003, 2005).

A similar connection between a decrease in CL levels and the age-linked decline of rat heart mitochondrial cytochrome *c* oxidase activity was demonstrated. Treatment of heart mitochondria from aged rats with CL-liposomes restored their lower cytochrome *c* oxidase activity to the level of young control rats (Paradies et al., 1997b). Again, no restoration of activity was found after treatment with other phospholipids or with peroxidized CL. CL level in heart mitochondria was shown to be regulated by thyroid hormone (Mutter et al., 2000). A decrease in cytochrome *c* oxidase activity in heart mitochondria isolated from hypothyroid rats can be also completely restored to the level of control rats by exogenously added CL but not by other phospholipids (Paradies et al., 1997a)

Decreased Complex I activity in fatty liver mitochondria isolated from rats fed with a choline-deficient diet to model in animals nonalcoholic fatty liver disease could also be completely restored to the level of control livers by exogenously added CL (Petrosillo et al., 2007). Under conditions of a choline-deficient diet the mitochondrial content of CL decreased due to reactive oxygen species-induced CL oxidation. Although no high-resolution crystal structure of the entire Complex I is available, these findings strongly suggest the presence of functionally important CL molecules in the complex.

### *8.4.3 Lipid Involvement in Supercomplex Formation*

Kinetic and structural analysis of the mammalian mitochondrial respiratory chain suggests that the individual Complexes I-IV (NADH:ubiquinone oxidoreductase, succinate:ubiquinone oxidoreductase, $bc_1$ complex (ubiqunol:cytochrome *c* oxidoreductase), and cytochrome *c* oxidase, respectively) exist in mitochondria under physiological conditions in equilibrium with supercomplexes or "respirasomes" composed of all of the above individual complexes (see (Mileykovskaya et al., 2005) for references). Depending on metabolic conditions electron transfer in the respiratory chain would occur (1) via substrate channeling of the small carriers, (ubiquinone, between Complexes I (or II) and III or cytochrome *c* between Complexes III and IV) due to supercomplex formation or (2) via random diffusion of these small carriers between individual complexes independently imbedded in the lipid bilayer (Genova et al., 2005). In *S. cerevisiae* mitochondria this equilibrium appears to be shifted to supercomplex organization of Complexes III and IV (Mileykovskaya et al., 2005), which may also contain Complex II as well as two peripheral NADH dehydrogenases (Boumans et al., 1998); *S. cerevisiae* lack Complex I and utilize the peripheral NADH dehydrogenases. $F_1F_0$-ATP synthase (Complex V) uses the electrochemical proton gradient generated in respiration to produce ATP.

The 3D structure for a respiratory supercomplex $I_1III_2IV_1$ (Fig. 8.4) from bovine heart mitochondria (Schafer et al., 2007) and a 2D model of *S. cerevisiae* supercomplex $IV_1III_2IV_1$ (Heinemeyer et al., 2007) have been recently proposed based on projection maps generated from single particle analysis of negatively stained samples visualized by electron microscopy. In both structures Complex III exists as a dimer, and Complex IV exists as a monomer. This is in agreement with numerous results demonstrating that the Complex III dimer represents one structural and functional unit (Hunte et al., 2008; Xia et al., 2007). In *S. cerevisiae* two Complex IV monomers are bound to opposite ends of the dimer of Complex III. In the bovine supercomplex the Complex III dimer interacts with Complex I and Complex IV monomers. The Complex IV face interacting with Complex III in the bovine supercomplex is the face forming the dimer with itself in the X-ray structure (Schafer et al., 2007). In contrast, in the 2D model of the *S. cerevisiae* supercomplex, an analogous side of the

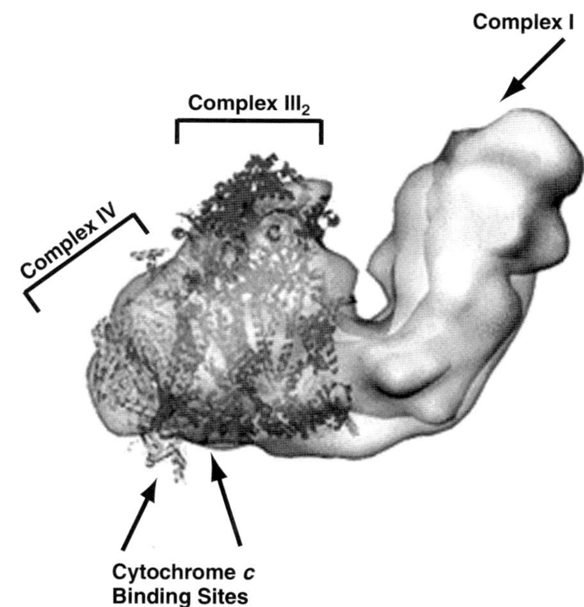


**Fig. 8.4** 3D structure of bovine heart supercomplex I $_1$III$_2$IV$_1$. The globular 3D structure of the supercomplex is overlayed with the crystal structures of the Complex III dimer and Complex IV monomer as indicated. The face of Complex III that interacts with Complex IV is shown in Fig. 8.3B. The close proximity of the cytochrome *c* binding sites on Complexes III and IV is indicated. The organization of the supercomplex and cytochrome *c* interaction shown in Fig. 8.5 is based on this structure. Figure adapted from (Schafer et al., 2007)

Complex IV was proposed to be oriented away from Complex III (Heinemeyer et al., 2007). It was suggested that in mammals the dimerization of Complex IV under different metabolic conditions might play a functional role, competing with supercomplex formation using the same contact interface (Schafer et al., 2007). In both structures the putative binding site for a mobile carrier (for ubiquinone or cytochrome *c* in bovine or only cytochrome *c* in *S. cerevisiae*) of each individual complex is in close proximity with the corresponding binding site of the neighboring complex (Fig. 8.4) supporting a substrate channeling mechanism in the supercomplex.

Phospholipids, in particular CL, play an important role in the association of these individual complexes into a functional supercomplex. For *S. cerevisiae* Complexes III and IV behave kinetically in intact mitochondria or display on gel electrophoresis after extraction from mitochondria with mild detergents as an associated supercomplex. However, in mutants of *S. cerevisiae* lacking CL (null mutants in the *CRDL* gene) but containing highly elevated amounts of PG (a precursor to CL), these complexes behave kinetically (Zhang et al., 2005a) and move during gel electrophoresis (Zhang et al., 2002) as individual complexes. Importantly, use of a mutant strain in which CL content of yeast mitochondria could be exogenously regulated *in vivo* (Zhang et al., 2002) demonstrated a direct correlation between levels of CL and the ratio of supercomplex to individual complexes in the mutant membrane. CL-lacking mutant cells, which in glucose media (anaerobic growth conditions) did not exhibit any alterations in their phenotype, grew considerably slower and to a lower final density on non-fermentable carbon sources (require oxidative phosphorylation) than wild type cells. Cells with intermediate CL content displayed intermediate growth phenotypes demonstrating the direct dependence of the efficiency of the energetic system on the level of CL (Zhang et al., 2002).

Kinetic experiments for the first time provided evidences for the role of CL in the organization of respiratory chain supercomplexes in intact mitochondria not disrupted with detergent. The classical kinetic approach used was based on the assumption that if the diffusion rate of a mobile electron carrier (in this case cytochrome *c*) is faster than the rates of its reduction and oxidation during steady-state respiration, the mobile carrier behaves kinetically as a homogeneous pool (Boumans et al., 1998). In this case the overall respiration rate should show a hyperbolic relation to the rate of either the reduction or oxidation of the carrier (pool behavior) indicative of a mobile carrier freely diffusing between individual respiratory complexes (i.e., III and IV). However, the relation approaches linearity (non-pool behavior) as an increasing amount of the mobile carrier is restricted due to substrate channeling between associated respiratory complexes organization into a supercomplex ($IV_1III_2IV_1$). A putative pool behavior of cytochrome *c* was studied by titration of the reduction rate of cytochrome *c* with the Complex III-specific inhibitor antimycin A (Boumans et al., 1998). In wild type yeast mitochondria containing normal levels of CL, cytochrome *c* did not show pool behavior consistent with the respiratory chain functioning as one unit in supercomplex organization with substrate channeling of cytochrome *c* between complexes III and IV (Boumans et al., 1998; Zhang et al., 2005a). However, in the case of the CL-lacking mutant cytochrome *c* exhibited pool behavior (Zhang et al., 2005a) consistent with the absence of respiratory chain supercomplex organization as previously shown in disrupted mitochondria (Zhang et al., 2002). These results further demonstrate an important role for CL in organization of not only the individual respiratory complexes but in higher order organization of these complexes into a "respirasome".

High concentrations of detergents at low ionic strength easily dissociate yeast Complexes III and IV from the supercomplex (Schagger and Pfeiffer, 2000) suggesting their association is through hydrophobic interactions. A cavity in the Complex III formed by membrane-embedded TM helices of cytochrome $c_1$ and cytochrome *b* with a lid on top of this cavity formed by subunits Qcr8 and Qcr6p has been suggested as a possible site of the interaction between Complexes III and IV (see Fig. 8.3B for the *S. cerevisiae* Complex III). CL together with PE fills this cavity and was proposed to act as a flexible amphipathic linkage between the above complexes (Pfeiffer et al., 2003). The structures of the supercomplexes from both bovine and *S. cerevisiae* are consistent with and support this lipid filled cavity of Complex III as the interface with Complex IV.

### *8.4.4 Disruption of Higher Complex Formation in Barth Syndrome and Apoptosis*

The higher order organization of electron transfer complexes in multicellular organisms into respirasomes dependent on CL is supported by studies of mitochondria from patients with Barth syndrome. The human *TAZ* gene encoding Tafazzin, a phospholipid acyltranferase that is involved in remodeling of CL to its mature highly unsaturated fatty acid composition, provides structural uniformity within the CL pool (Xu et al., 2006). Mutations in the *TAZ* gene are associated with Barth syndrome, an X-linked genetic disorder characterized by cardiomyopathy, skeletal myopathy, neutropenia, and growth retardation (see (Hauff and Hatch, 2006; Schlame and Ren, 2006; Xu et al., 2006) for reviews). Mitochondria from Barth syndrome patients show a lower CL content, but more striking is the polydispersity in the normally very narrow acyl chain composition of CL. Lymphoblasts from patients with Barth syndrome show decreased stability of the $I_1III_2IV_1$ supercomplex, where Complex IV is more dissociated from the supercomplex, suggesting that reduction in the mature species of CL results in unstable respiratory chain supercomplexes (McKenzie et al., 2006). The interaction between Complexes I and III was also affected resulting in decreased levels of the $I_1III_2$ supercomplex (McKenzie et al., 2006) . Interestingly, *S. cerevisiae* contain an orthologue of the *TAZ* gene and disruption of the *TAZ1* gene in *S. cerevisiae* affects the assembly and stability of Complex IV as well as its incorporation into the supercomplex (Brandner et al., 2005; Li et al., 2007).

The above abnormalities in molecular organization may be the molecular basis for or the result of abnormal architecture of mitochondrial cristae, which was observed using EM tomography in lymphoblast cells isolated from patients with Barth syndrome (Acehan et al., 2007). Although the authors pointed out that the absence of Tafazzin itself may cause a disruption of normal membrane adhesion due to localization of Tafazzin both in the inner and outer membrane surfaces encasing the inter-membrane space (Claypool et al., 2006), a direct relationship between cristae abnormality and the deficiency in CL was suggested. The inner membrane of mitochondria is organized in two morphologically distinct domains, the inner boundary membrane and the cristae membrane (Fig. 8.5A), which are connected by narrow tubular structures termed cristae junctions. The tomograms showed adhesion of two neighboring cristae membranes resulting in the disruption of the normal connection to the mitochondrial inner boundary membrane by the cristae junction structure, obliteration of the intra-cristae space, and stacking of the collapsed cristae. Cristae membranes appear to be enriched in Complexes I–V over the inner boundary membrane. However, there is normally dynamic redistribution of proteins between the inner boundary membrane and cristae membrane, which changes in accordance with changes in the physiological state of the cell (Vogel et al., 2006). Structural abnormalities in Barth syndrome mitochondria appear to

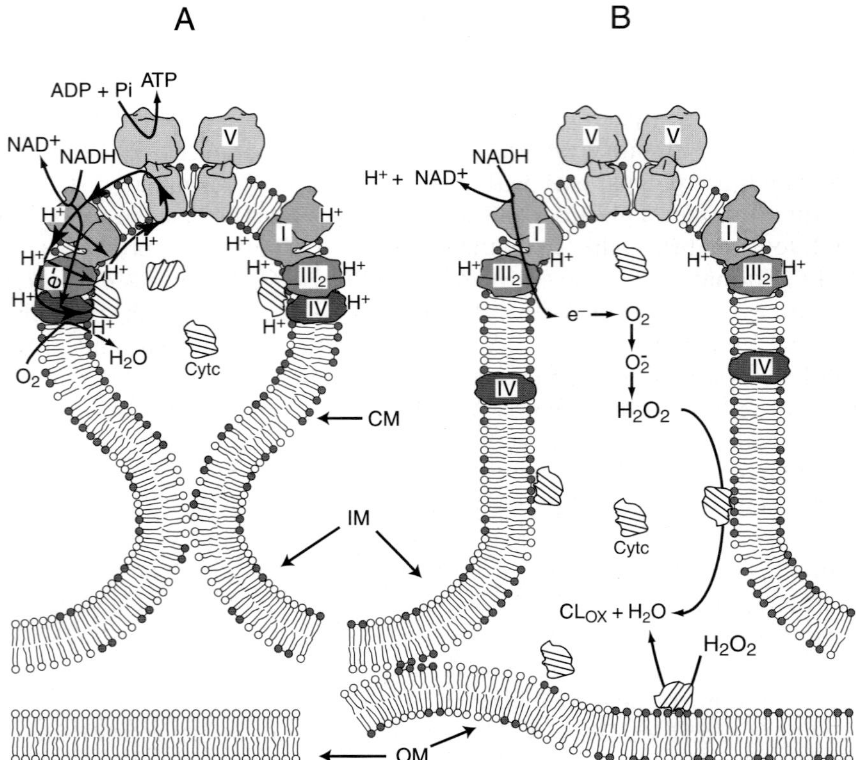


**Fig. 8.5** Model for involvement of CL in the energy coupling processes or apoptosis in mitochondria. (A) Under normal conditions cristae membranes (CM) are separated from the remaining mitochondrial inner boundary membranes (IM) by the cristae junctions. CM are composed of microdomains enriched in CL (depicted in the bilayer as dimer of gray headgroups) and Complexes I, III, IV and V ($F_1F_0$ATP synthase) organized in $I_1III_2IV_1$ (see Fig. 2.4) and $V_2$ supercomplexes. The proton gradient (positive outward) generated by electron transfer from NADH to molecular oxygen via the supercomplex $I_1III_2IV_1$ (with channeling of cytochrome *c* (Cytc) between Complex III and IV) is utilized to synthesize ATP. CL participates in the trapping of protons promoting localized protonic coupling between the respiratory supercomplex and Complex V located in close proximity to each other (figure adapted from (Papa et al., 2006)). (B) During early stages of apoptosis cristae junctions open allowing more equal distribution of CL between CM and IM with significant appearance of CL in the outer membrane (OM) via the contacts sites between the latter two membranes. Reduction of CL in the CM favors delocalization of the proton gradient and partial dissociation of Complex IV from the supercomplex. Binding of cytochrome *c* to CL disrupts electron flow between Complexes III and IV and induces increased production of reactive oxygen species by Complex III. Interaction of cytochrome *c* with CL converts the protein into a peroxidase, which catalyzes CL peroxidation ($CL_{ox}$), which further reduces CL levels (Basova et al., 2007)

block this dynamic. One possible explanation for the structural changes is that the abnormal spectrum of CL species may induce abnormalities in protein organization and aggregation. Large mitochondria with onion-like cristae structures were also present in lymphoblast cells isolated from patients with Barth syndrome. Similar onion-like organization of the inner membrane was demonstrated in *S. cerevisiae* mutants lacking small subunits of ATP synthase responsible for dimerization and oligomerization of Complex V (Paumard et al., 2002). Based on a decrease in membrane potential coupled with reduced oxidation of substrates by the respiratory chain in the mutant, disruption of microdomains organized both from the oligomers of Complex V and the respiratory chain supercomplexes was suggested (Bornhovd et al., 2006). Thus the change in the content of the mature species of CL may result in misassembly of respiratory chain supercomplexes, which in turn would result in distortion of microdomains formed by supercomplexes and changes in mitochondrial cristae morphology producing dysfunctional mitochondria.

CL also plays a direct structural and functional role in the regulation of the early steps of the apoptosis, which is dependent on the submitochondrial localization of CL. In the cristae membrane CL participates in organization of the respiratory chain complexes into supercomplexes as described above, which together with Complex V represent the main components of the cristae (Fig. 8.5A). Complex V was suggested to form oligomeric supercomplexes that are required for cristae formation (Dudkina et al., 2006; Giraud et al., 2002) and to create a scaffold to also keep respiratory supercomplexes in the cristae. Such an organization could trap supercomplexes within the cristae membrane microdomains and prevent them from diffusing to the inner boundary membrane via the cristae junctions thereby regulating dynamic redistribution of the components between sub-compartments of the inner membrane (Vogel et al., 2006). CL has been proposed to act as a proton sink at the membrane surface due to the abnormally high ionization constant (above 7.5) for one of its two phosphate groups (Kates et al., 1993). The ability of CL, enriched in the cristae, to trap protons and prevent escape of pumped protons to the bulk water phase would increase protonic coupling between respiratory supercomplexes and Complex V in cristae membrane microdomains (Haines and Dencher, 2002). Prevention of a delocalized proton electrochemical gradient would inhibit oxygen superoxide formation by the respiratory chain and the generation of other free radicals. In higher eukaryotes this would suppress induction of the mitochondrial pathway of apoptosis (Papa et al., 2006). During apoptosis CL levels increase in the outer mitochondrial membrane (Fig. 8.5B), which was proposed to be caused by redistribution of CL between the inner and outer leaflets of the inner membrane followed by movement of CL to the outer membrane via membrane contact sites between the two membranes (see (Lucken-Ardjomande and Martinou, 2005; Ott et al., 2007) for reviews). However, if compartmentalization of the inner membrane is taken into account, the redistribution of CL between cristae membranes and inner boundary membranes, due to disruption of cristae junctions during apoptosis (see below), might also play a critical role during induction of apoptosis. The reduction in cristae membrane CL might prevent the correct assembly of respiratory complexes and supercomplex formation (Fig. 8.5B).

CL can compete with Complex III for binding with cytochrome *c* in a dose dependent manner, and CL-bound cytochrome *c* does not effectively interact with respiratory supercomplexes (Basova et al., 2007). Thus, the increase in CL concentration in the outer leaflet of the inner boundary membrane would redistribute cytochrome *c* to the inner boundary membrane from the cristae making it unavailable for interaction with the supercomplexes (Fig. 8.5B). Interestingly, in mouse liver mitochondria under conditions where cells undergo apoptosis, cristae are remodeled and cristae junctions widen (Frezza et al., 2006; Scorrano et al., 2002). Cristae opening might increase the pool of cytochrome *c* available for binding with CL on the outer leaflet of the inner boundary membrane or the inner leaflet of the outer membrane. Interaction with CL causes structural changes in cytochrome *c* accompanied by a negative shift in its red-ox potential and conversion of the protein into a peroxidase, which catalyzes $H_2O_2$-dependent CL peroxidation (Basova et al., 2007). Reducing the amount of cytochrome *c* available as an electron acceptor from Complex III strongly promotes the production of superoxide by Complex III, which in turn would elevate $H_2O_2$ levels. CL peroxidation, which parallels the loss of CL during apoptosis, in turn might result in membrane detachment of cytochrome *c* and finally in its release into the cytosol through the permeabilized outer membrane (Ott et al., 2007) further activating downstream signals associated with apoptosis. Such a sequence of events may explain the paradox during progression of apoptosis where both an increase in outer membrane CL and reduction in total mitochondrial CL occur.

The exact mechanisms driving apoptotic redistribution of CL are still poorly understood although mitochondrial translocation and activity of tBid, a truncated form of the cytosolic protein Bid, seem to be associated with the apoptotic migration of CL (Lucken-Ardjomande and Martinou, 2005). *In vitro* experiments demonstrated that mitochondrial creatine kinase, which is localized to mitochondrial membrane contact sites, promotes segregation and clustering of CL in liposomes, suggesting that this phenomenon might occur at contact sites between the inner and outer mitochondrial membranes, creating nucleation sites for CL-domains capable of binding tBid and cytochrome *c* (Epand et al., 2007). Irrespective of the mechanisms leading to induction of mitochondrial-based apoptosis, it is clear that CL plays a central role in this process.

## 8.5 Concluding Remarks

Lipids have multiple functions in cells ranging from defining the bilayer permeability barrier of cell membranes and organelles to providing the matrix within which membrane proteins fold and function to being integral components of multisubunit protein complexes and supercomplexes. Diseases or physiological states that severely compromise membrane barrier function would be lethal which accounts for the lack of genetic disorders resulting in the lack of or extensive redistribution among major lipid classes. However, a large number of disorders directly involve lipids or most likely have a lipid involvement in the underlying molecular basis for the disease. Membrane proteins associated with some diseases express themselves as protein folding defects rather than defects in protein function and result in defects in trafficking and organelle targeting of proteins. Some membrane proteins may misfold in their proper location either spontaneously or in response to local changes in lipid composition. Mutation of structurally important residues could lead to a different organization of the mutant protein within the changing lipid environment as the protein moves through the same or altered organelle trafficking route. Studies on bacterial permeases clearly demonstrate that alterations in protein-lipid interactions can have dramatic effects during initial assembly and after stable assembly on the structure and function of membrane proteins. Therefore, lipid involvement in the underlying molecular mechanisms of protein folding disorders is certainly a distinct possibility.

Biochemical and genetic evidence has established a role for lipids as essential structural and functional components, equal to specific amino acid residues, in multicomponent molecular machines. In this regard CL is of particular clinical importance because of all major lipids CL levels are most dramatically affected by a number of pathological and physiological states. CL and changes in its levels are associated with diseases such as Barth syndrome and changes in cellular states brought about by ischemia/reperfusion, apoptosis, aging and oxidative damage. There are direct correlations between these changes in CL levels and the organization and function of mitochondrial respiratory components.

Therefore, defining the molecular basis for cellular dysfunction resulting from disrupted membrane associated processes must include a more extensive investigation of the role lipids play in these processes before a full understanding of normal function can be attained. Expansion of such basic information will uncover possible therapeutic treatments to modulate the seriousness of the genetic defects or pathophysiological states by affecting membrane lipid composition or distribution of individual lipid species thus opening new avenues for pharmacological intervention.

**Acknowledgments** The authors thank Philip Heacock for assistance in preparation of the figures. Work of the authors presented in this review was supported by grants GM20478 and GM56389 from the US National Institutes of Health and funds from the John S. Dunn, Sr., Research Foundation awarded to W.D.

## References

Abramson, J., Iwata, S. and Kaback, H.R., Lactose permease as a paradigm for membrane transport proteins (Review), Mol Membr Biol 21 (2004) 227–236.

Acehan, D., Xu, Y., Stokes, D.L. and Schlame, M., Comparison of lymphoblast mitochondria from normal subjects and patients with Barth syndrome using electron microscopic tomography, Lab Invest 87 (2007) 40–48.

Aerts, J.M., Ottenhoff, R., Powlson, A.S., Grefhorst, A., van Eijk, M., Dubbelhuis, P.F., Aten, J., Kuipers, F., Serlie, M.J., Wennekes, T., Sethi, J.K., O'Rahilly, S. and Overkleeft, H.S., Pharmacological inhibition of glucosylceramide synthase enhances insulin sensitivity, Diabetes 56 (2007) 1341–1349.

Alemany, R., Perona, J.S., Sanchez-Dominguez, J.M., Montero, E., Canizares, J., Bressani, R., Escriba, P.V. and Ruiz-Gutierrez, V., G protein-coupled receptor systems and their lipid environment in health disorders during aging, Biochim Biophys Acta 1768 (2007) 964–975.

Andersson, K., Buschard, K., Fredman, P., Kaas, A., Lidstrom, A.M., Madsbad, S., Mortensen, H. and Jan-Eric, M., Patients with insulin-dependent diabetes but not those with non-insulin-dependent diabetes have anti-sulfatide antibodies as determined with a new ELISA assay, Autoimmunity 35 (2002) 463–468.

Annis, M.G., Soucie, E.L., Dlugosz, P.J., Cruz-Aguado, J.A., Penn, L.Z., Leber, B. and Andrews, D.W., Bax forms multispanning monomers that oligomerize to permeabilize membranes during apoptosis, Embo J 24 (2005) 2096–2103.

Aridor, M. and Hannan, L.A., Traffic jam: a compendium of human diseases that affect intracellular transport processes, Traffic 1 (2000) 836–851.

Basova, L.V., Kurnikov, I.V., Wang, L., Ritov, V.B., Belikova, N.A., Vlasova, II, Pacheco, A.A., Winnica, D.E., Peterson, J., Bayir, H., Waldeck, D.H. and Kagan, V.E., Cardiolipin switch in mitochondria: shutting off the reduction of cytochrome c and turning on the peroxidase activity, Biochemistry 46 (2007) 3423–3434.

Beja, O. and Bibi, E., Multidrug resistance protein (Mdr)-alkaline phosphatase hybrids in Escherichia coli suggest a major revision in the topology of the C-terminal half of Mdr, J Biol Chem 270 (1995) 12351–12354.

Bogdanov, M. and Dowhan, W., Phospholipid-assisted protein folding: phosphatidylethanolamine is required at a late step of the conformational maturation of the polytopic membrane protein lactose permease, Embo J 17 (1998) 5255–5264.

Bogdanov, M. and Dowhan, W., Lipid-assisted protein folding, J Biol Chem 274 (1999) 36827–36830.

Bogdanov, M., Heacock, P.N. and Dowhan, W., A polytopic membrane protein displays a reversible topology dependent on membrane lipid composition, Embo J 21 (2002) 2107–2116.

Bogdanov, M., Sun, J., Kaback, H.R. and Dowhan, W., A phospholipid acts as a chaperone in assembly of a membrane transport protein, J Biol Chem 271 (1996) 11615–11618.

Bogdanov, M., Umeda, M. and Dowhan, W., Phospholipid-assisted refolding of an integral membrane protein. Minimum structural features for phosphatidylethanolamine to act as a molecular chaperone, J Biol Chem 274 (1999) 12339–12345.

Bogdanov, M., Xie, J., Heacock, P.N. and Dowhan, W., To flip or not to flip: protein-lipid charge interactions are a determinant of final membrane protein topology, J Cell Biol (2008), editorially accepted.

Bogdanov, M., Zhang, W., Xie, J. and Dowhan, W., Transmembrane protein topology mapping by the substituted cysteine accessibility method (SCAM(TM)): application to lipid-specific membrane protein topogenesis, Methods 36 (2005) 148–171.

Bornhovd, C., Vogel, F., Neupert, W. and Reichert, A.S., Mitochondrial membrane potential is dependent on the oligomeric state of F1F0-ATP synthase supracomplexes, J Biol Chem 281 (2006) 13990–13998.

Boumans, H., Grivell, L.A. and Berden, J.A., The respiratory chain in yeast behaves as a single functional unit, J Biol Chem 273 (1998) 4872–4877.

Brandner, K., Mick, D.U., Frazier, A.E., Taylor, R.D., Meisinger, C. and Rehling, P., Tazl, an outer mitochondrial membrane protein, affects stability and assembly of inner membrane protein complexes: implications for Barth Syndrome, Mol Biol Cell 16 (2005) 5202–5214.

Brutkiewicz, R.R., Lin, Y., Cho, S., Hwang, Y.K., Sriram, V. and Roberts, T.J., CD1d-mediated antigen presentation to natural killer T (NKT) cells, Crit Rev Immunol 23 (2003) 403–419.

Buschard, K., Blomqvist, M., Osterbye, T. and Fredman, P., Involvement of sulfatide in beta cells and type 1 and type 2 diabetes, Diabetologia 48 (2005) 1957–1962.

Buschard, K., Hanspers, K., Fredman, P. and Reich, E.P., Treatment with sulfatide or its precursor, galactosylceramide, prevents diabetes in NOD mice, Autoimmunity 34 (2001) 9–17.

Cheng, S.H., Gregory, R.J., Marshall, J., Paul, S., Souza, D.W., White, G.A., O'Riordan, C.R. and Smith, A.E., Defective intracellular transport and processing of CFTR is the molecular basis of most cystic fibrosis, Cell 63 (1990) 827–834.

Chicco, A.J. and Sparagna, G.C., Role of cardiolipin alterations in mitochondrial dysfunction and disease, Am J Physiol Cell Physiol 292 (2007) C33–C44.

Choo-Smith, L.P. and Surewicz, W.K., The interaction between Alzheimer amyloid beta(1-40) peptide and ganglioside GM1-containing membranes, FEBS Lett 402 (1997) 95–98.

Claypool, S.M., McCaffery, J.M. and Koehler, C.M., Mitochondrial mislocalization and altered assembly of a cluster of Barth syndrome mutant tafazzins, J Cell Biol 174 (2006) 379–390.

Crabbe, M.J., Cataract as a conformational disease – the Maillard reaction, alpha-crystallin and chemotherapy, Cell Mol Biol 44 (1998) 1047–1050.

De Silva, A.D., Park, J.J., Matsuki, N., Stanic, A.K., Brutkiewicz, R.R., Medof, M.E. and Joyce, S., Lipid protein interactions: the assembly of CD1d1 with cellular phospholipids occurs in the endoplasmic reticulum, J Immunol 168 (2002) 723–733.

Debnath, D., Bhattacharya, S. and Chakrabarti, A., Phospholipid assisted folding of a denatured heme protein: effect of phosphatidylethanolamine, Biochem Biophys Res Commun 301 (2003) 979–984.

Delgado-Partin, V.M. and Dalbey, R.E., The proton motive force, acting on acidic residues, promotes translocation of amino-terminal domains of membrane proteins when the hydrophobicity of the translocation signal is low, J Biol Chem 273 (1998) 9927–9934.

Dlugosz, P.J., Billen, L.P., Annis, M.G., Zhu, W., Zhang, Z., Lin, J., Leber, B. and Andrews, D.W., Bcl-2 changes conformation to inhibit Bax oligomerization, EMBO J 25 (2006) 2287–2296.

Dodson, G. and Steiner, D., The role of assembly in insulin's biosynthesis, Curr Opin Struct Biol 8 (1998) 189–194.

Dowhan, W., Molecular basis for membrane phospholipid diversity: Why are there so many lipids? Annu Rev Biochem 66 (1997) 99–232.

Dowhan, W., Bogdanov, M. and Mileykovskaya, E., Functional roles of lipids in membranes. In: Vance, D.E. and Vance, J.E. (Eds.), Biochemistry of Lipids. Lipoproteins and Membranes, 5th Edition, Amsterdam, Elsevier Press, (2008) pp. 1–37.

Dowhan, W., Mileykovskaya, E. and Bogdanov, M., Diversity and versatility of lipid-protein interactions revealed by molecular genetic approaches, Biochim Biophys Acta 1666 (2004) 19–39.

Drews, J., What's in a number? Nat Rev Drug Discov 5 (2006) 975.

Dudkina, N.V., Sunderhaus, S., Braun, H.P. and Boekema, E.J., Characterization of dimeric ATP synthase and cristae membrane ultrastructure from Saccharomyces and Polytomella mitochondria, FEBS Lett 580 (2006) 3427–3432.

Dunlop, J., Jones, P.C. and Finbow, M.E., Membrane insertion and assembly of ductin: a polytopic channel with dual orientations, Embo J 14 (1995) 3609–3616.

Eidelman, O., BarNoy, S., Razin, M., Zhang, J., McPhie, P., Lee, G., Huang, Z., Sorscher, E.J. and Pollard, H.B., Role for phospholipid interactions in the trafficking defect of Delta F508-CFTR, Biochemistry 41 (2002) 11161–11170.

Ellis, R.J., Do molecular chaperones have to be proteins? Biochem Biophys Res Commun 238 (1997) 687–692.

Elofsson, A. and von Heijne, G., Membrane protein structure: prediction versus reality, Annu Rev Biochem 76 (2007) 125–140.

Epand, R.F., Tokarska-Schlattner, M., Schlattner, U., Wallimann, T. and Epand, R.M., Cardiolipin clusters and membrane domain formation induced by mitochondrial proteins, J Mol Biol 365 (2007) 968–980.

Fadiel, A., Eichenbaum, K.D., Hamza, A., Tan, O., Lee, H.H. and Naftolin, F., Modern pathology: protein mis-folding and mis-processing in complex disease, Curr Protein Pept Sci 8 (2007) 29–37.

Fredman, P., Mansson, J.E., Rynmark, B.M., Josefsen, K., Ekblond, A., Halldner, L., Osterbye, T., Horn, T. and Buschard, K., The glycosphingolipid sulfatide in the islets of Langerhans in rat pancreas is processed through recycling: possible involvement in insulin trafficking, Glycobiology 10 (2000) 39–50.

Frezza, C., Cipolat, S., Martins de Brito, O., Micaroni, M., Beznoussenko, G.V., Rudka, T., Bartoli, D., Polishuck, R.S., Danial, N.N., De Strooper, B. and Scorrano, L., OPA1 controls apoptotic cristae remodeling independently from mitochondrial fusion, Cell 126 (2006) 177–189.

Gafvelin, G. and von Heijne, G., Topological "frustration" in multispanning E. coli inner membrane proteins, Cell 77 (1994) 401–412.

Gelman, M.S. and Kopito, R.R., Cystic fibrosis: premature degradation of mutant proteins as a molecular disease mechanism, Methods Mol Biol 232 (2003) 27–37.

Genova, M.L., Bianchi, C. and Lenaz, G., Supercomplex organization of the mitochondrial respiratory chain and the role of the Coenzyme Q pool: pathophysiological implications, Biofactors 25 (2005) 5–20.

Georges, E., Tsuruo, T. and Ling, V., Topology of P-glycoprotein as determined by epitope mapping of MRK-16 monoclonal antibody, J Biol Chem 268 (1993) 1792–1798.

Giraud, M.F., Paumard, P., Soubannier, V., Vaillier, J., Arselin, G., Salin, B., Schaeffer, J., Brethes, D., di Rago, J.P. and Velours, J., Is there a relationship between the supramolecular organization of the mitochondrial ATP synthase and the formation of cristae?, Biochim Biophys Acta 1555 (2002) 174–180.

Gouffi, K., Gerard, F., Santini, C.L. and Wu, L.F., Dual topology of the Escherichia coli TatA protein, J Biol Chem 279 (2004) 11608–11615.

Guan, L. and Kaback, H.R., Lessons from lactose permease, Annu Rev Biophys Biomol Struct 35 (2006) 67–91.

Gumperz, J.E., The ins and outs of CD1 molecules: bringing lipids under immunological surveillance, Traffic 7 (2006) 2–13.

Haines, T.H. and Dencher, N.A., Cardiolipin: a proton trap for oxidative phosphorylation, FEBS Lett 528 (2002) 35–39.

Han, X., Abendschein, D.R., Kelley, J.G. and Gross, R.W., Diabetes-induced changes in specific lipid molecular species in rat myocardium, Biochem J 352 (2000) 79–89.

Harrison, R.S., Sharpe, P.C., Singh, Y. and Fairlie, D.P., Amyloid peptides and proteins in review, Rev Physiol Biochem Pharmacol 159 (2007) 1–77.

Hauff, K.D. and Hatch, G.M., Cardiolipin metabolism and Barth Syndrome, Prog Lipid Res 45 (2006) 91–101.

Hayden, M.R., Tyagi, S.C., Kerklo, M.M. and Nicolls, M.R., Type 2 diabetes mellitus as a conformational disease, JOP 6 (2005) 287–302.

Hegde, R.S., Mastrianni, J.A., Scott, M.R., DeFea, K.A., Tremblay, P., Torchia, M., DeArmond, S.J., Prusiner, S.B. and Lingappa, V.R., A transmembrane form of the prion protein in neurodegenerative disease, Science 279 (1998) 827–834.

Heinemeyer, J., Braun, H.P., Boekema, E.J. and Kouril, R., A structural model of the cytochrome C reductase/oxidase supercomplex from yeast mitochondria, J Biol Chem 282 (2007) 12240–12248.

Hessa, T., White, S.H. and von Heijne, G., Membrane insertion of a potassium-channel voltage sensor, Science 307 (2005) 1427.

Huggett, J., Vaughan-Thomas, A. and Mason, D., The open reading frame of the Na(+)-dependent glutamate transporter GLAST-1 is expressed in bone and a splice variant of this molecule is expressed in bone and brain, FEBS Lett 485 (2000) 13–18.

Hunte, C., Specific protein-lipid interactions in membrane proteins, Biochem Soc Trans 33 (2005) 938–942.

Hunte, C., Solmaz, S., Palsdottir, H. and Wenz, T., A Structural Perspective on Mechanism and Function of the Cytochrome bc (1) Complex, Results Probl Cell Differ 45 (2008) 253–278.

Ikeda, M., Kida, Y., Ikushiro, S. and Sakaguchi, M., Manipulation of membrane protein topology on the endoplasmic reticulum by a specific ligand in living cells, J. Biochem. 138 (2005) 631–637.

Jakes, K.S., Kienker, P.K., Slatin, S.L. and Finkelstein, A., Translocation of inserted foreign epitopes by a channel-forming protein, Proc Natl Acad Sci USA 95 (1998) 4321–4326.

Jormakka, M., Tornroth, S., Byrne, B. and Iwata, S., Molecular basis of proton motive force generation: structure of formate dehydrogenase-N, Science 295 (2002) 1863–1868.

Joyce, S., CD1d and natural T cells: how their properties jump-start the immune system, Cell Mol Life Sci 58 (2001) 442–469.

Joyce, S., Woods, A.S., Yewdell, J.W., Bennink, J.R., De Silva, A.D., Boesteanu, A., Balk, S.P., Cotter, R.J. and Brutkiewicz, R.R., Natural ligand of mouse CD1d1: cellular glycosylphosphatidylinositol, Science 279 (1998) 1541–1544.

Kabayama, K., Sato, T., Kitamura, F., Uemura, S., Kang, B.W., Igarashi, Y. and Inokuchi, J., TNFalpha-induced insulin resistance in adipocytes as a membrane microdomain disorder: involvement of ganglioside GM3, Glycobiology 15 (2005) 21–29.

Kabayama, K., Sato, T., Saito, K., Loberto, N., Prinetti, A., Sonnino, S., Kinjo, M., Igarashi, Y. and Inokuchi, J., Dissociation of the insulin receptor and caveolin-1 complex by ganglioside GM3 in the state of insulin resistance, Proc Natl Acad Sci USA 104 (2007) 13678–13683.

Kanki, T., Sakaguchi, M., Kitamura, A., Sato, T., Mihara, K. and Hamasaki, N., The tenth membrane region of band 3 is initially exposed to the luminal side of the endoplasmic reticulum and then integrated into a partially folded band 3 intermediate, Biochemistry 41 (2002) 13973–13981.

Kates, M., Syz, J.Y., Gosser, D. and Haines, T.H., pH-dissociation characteristics of cardiolipin and its 2'-deoxy analogue, Lipids 28 (1993) 877–882.

Kim, P.K., Annis, M.G., Dlugosz, P.J., Leber, B. and Andrews, D.W., During apoptosis bcl-2 changes membrane topology at both the endoplasmic reticulum and mitochondria, Mol Cell 14 (2004) 523–529.

Koch, M., Stronge, V.S., Shepherd, D., Gadola, S.D., Mathew, B., Ritter, G., Fersht, A.R., Besra, G.S., Schmidt, R.R., Jones, E.Y. and Cerundolo, V., The crystal structure of human CD1d with and without alpha-galactosylceramide, Nat Immunol 6 (2005) 819–826.

Lambert, C. and Prange, R., Dual topology of the hepatitis B virus large envelope protein: determinants influencing post-translational pre-S translocation, J Biol Chem 276 (2001) 22265–22272.

Lange, C., Nett, J.H., Trumpower, B.L. and Hunte, C., Specific roles of protein-phospholipid interactions in the yeast cytochrome bc1 complex structure, Embo J 20 (2001) 6591–6600.

Leber, B., Lin, J. and Andrews, D.W., Embedded together: the life and death consequences of interaction of the Bcl-2 family with membranes, Apoptosis 12 (2007) 897–911.

Levy, D., Membrane proteins which exhibit multiple topological orientations, Essays Biochem 31 (1996) 49–60.

Li, G., Chen, S., Thompson, M.N. and Greenberg, M.L., New insights into the regulation of cardiolipin biosynthesis in yeast: implications for Barth syndrome, Biochim Biophys Acta 1771 (2007) 432–441.

Lima, P.R., Baratti, M.O., Chiattone, M.L., Costa, F.F. and Saad, S.T., Band 3Tambau: a de novo mutation in the AE1 gene associated with hereditary spherocytosis. Implications for anion exchange and insertion into the red blood cell membrane, Eur J Haematol 74 (2005) 396–401.

Lin, J.C. and Liu, H.L., Protein conformational diseases: from mechanisms to drug designs, Curr Drug Discov Technol 3 (2006) 145–153.

Linton, K.J. and Higgins, C.F., P-glycoprotein misfolds in *Escherichia coli*: evidence against alternating-topology models of the transport cycle, Mol Membr Biol 19 (2002) 51–58.

Loo, T.W. and Clarke, D.M., Membrane topology of a cysteine-less mutant of human P-glycoprotein, J Biol Chem 270 (1995) 843–848.

Lu, Y., Turnbull, I.R., Bragin, A., Carveth, K., Verkman, A.S. and Skach, W.R., Reorientation of aquaporin-1 topology during maturation in the endoplasmic reticulum, Mol Biol Cell 11 (2000) 2973–2985.

Lucken-Ardjomande, S. and Martinou, J.C., Newcomers in the process of mitochondrial permeabilization, J Cell Sci 118 (2005) 473–483.

McGinnes, L.W., Reitter, J.N., Gravel, K. and Morrison, T.G., Evidence for mixed membrane topology of the newcastle disease virus fusion protein, J Virol 77 (2003) 1951–1963.

McKenzie, M., Lazarou, M., Thorburn, D.R. and Ryan, M.T., Mitochondrial respiratory chain supercomplexes are destabilized in Barth Syndrome patients, J Mol Biol 361 (2006) 462–469.

Mendoza, J.L. and Thomas, P.J., Building an understanding of cystic fibrosis on the foundation of ABC transporter structures, J Bioenerg Biomembr 39 (2007) 499–505.

Milenkovic, V.M., Rivera, A., Horling, F. and Weber, B.H., Insertion and topology of normal and mutant bestrophin-1 in the endoplasmic reticulum membrane, J Biol Chem 282 (2007) 1313–1321.

Mileykovskaya, E., Zhang, M. and Dowhan, W., Cardiolipin in energy transducing membranes, Biochemistry (Mosc) 70 (2005) 154–158.

Molano, A., Park, S.H., Chiu, Y.H., Nosseir, S., Bendelac, A. and Tsuji, M., Cutting edge: the IgG response to the circumsporozoite protein is MHC class II-dependent and CD1d-independent: exploring the role of GPIs in NK T cell activation and antimalarial responses, J Immunol 164 (2000) 5005–5009.

Monaco, S., Zanusso, G., Mazzucco, S. and Rizzuto, N., Cerebral amyloidoses: molecular pathways and therapeutic challenges, Curr Med Chem 13 (2006) 1903–1913.

Morillas, M., Swietnicki, W., Gambetti, P. and Surewicz, W.K., Membrane environment alters the conformational structure of the recombinant human prion protein, J Biol Chem 274 (1999) 36859–36865.

Moss, K., Helm, A., Lu, Y., Bragin, A. and Skach, W.R., Coupled translocation events generate topological heterogeneity at the endoplasmic reticulum membrane, Mol Biol Cell 9 (1998) 2681–2697.

Mulugeta, S. and Beers, M.F., Processing of surfactant protein C requires a type II transmembrane topology directed by juxtamembrane positively charged residues, J Biol Chem 278 (2003) 47979–47986.

Mutter, T., Dolinsky, V.W., Ma, B.J., Taylor, W.A. and Hatch, G.M., Thyroxine regulation of monolysocardiolipin acyltransferase activity in rat heart, Biochem J 346 Pt 2 (2000) 403–406.

Nagamori, S., Vazquez-Ibar, J.L., Weinglass, A.B. and Kaback, H.R., In vitro synthesis of lactose permease to probe the mechanism of membrane insertion and folding, J Biol Chem 278 (2003) 14820–14826.

Nilsson, I. and von Heijne, G., Fine-tuning the topology of a polytopic membrane protein: role of positively and negatively charged amino acids, Cell 62 (1990) 1135–1141.

Osterbye, T., Jorgensen, K.H., Fredman, P., Tranum-Jensen, J., Kaas, A., Brange, J., Whittingham, J.L. and Buschard, K., Sulfatide promotes the folding of proinsulin, preserves insulin crystals, and mediates its monomerization, Glycobiology 11 (2001) 473–479.

Ott, M., Zhivotovsky, B. and Orrenius, S., Role of cardiolipin in cytochrome c release from mitochondria, Cell Death Differ 14 (2007) 1243–1247.

Palsdottir, H. and Hunte, C., Lipids in membrane protein structures, Biochim Biophys Acta 1666 (2004) 2–18.

Papa, S., Lorusso, M. and Di Paola, M., Cooperativity and flexibility of the protonmotive activity of mitochondrial respiratory chain, Biochim Biophys Acta 1757 (2006) 428–436.

Paradies, G., Petrosillo, G. and Ruggiero, F.M., Cardiolipin-dependent decrease of cytochrome c oxidase activity in heart mitochondria from hypothyroid rats, Biochim Biophys Acta 1319 (1997a) 5–8.

Paradies, G., Ruggiero, F.M., Petrosillo, G. and Quagliariello, E., Age-dependent decline in the cytochrome c oxidase activity in rat heart mitochondria: role of cardiolipin, FEBS Lett 406 (1997b) 136–138.

Parekh, V.V., Wilson, M.T. and Van Kaer, L., iNKT-cell responses to glycolipids, Crit Rev Immunol 25 (2005) 183–213.

Park, J.J., Kang, S.J., De Silva, A.D., Stanic, A.K., Casorati, G., Hachey, D.L., Cresswell, P. and Joyce, S., Lipid-protein interactions: biosynthetic assembly of CD1 with lipids in the endoplasmic reticulum is evolutionarily conserved, Proc Natl Acad Sci USA 101 (2004) 1022–1026.

Paumard, P., Vaillier, J., Coulary, B., Schaeffer, J., Soubannier, V., Mueller, D.M., Brethes, D., di Rago, J.P. and Velours, J., The ATP synthase is involved in generating mitochondrial cristae morphology, Embo J 21 (2002) 221–230.

Petrosillo, G., Di Venosa, N., Ruggiero, F.M., Pistolese, M., D'Agostino, D., Tiravanti, E., Fiore, T. and Paradies, G., Mitochondrial dysfunction associated with cardiac ischemia/reperfusion can be attenuated by oxygen tension control. Role of oxygen-free radicals and cardiolipin, Biochim Biophys Acta 1710 (2005) 78–86.

Petrosillo, G., Portincasa, P., Grattagliano, I., Casanova, G., Matera, M., Ruggiero, F.M., Ferri, D. and Paradies, G., Mitochondrial dysfunction in rat with nonalcoholic fatty liver Involvement of complex I, reactive oxygen species and cardiolipin, Biochim Biophys Acta 1767 (2007) 1260–1267.

Petrosillo, G., Ruggiero, F.M., Di Venosa, N. and Paradies, G., Decreased complex III activity in mitochondria isolated from rat heart subjected to ischemia and reperfusion: role of reactive oxygen species and cardiolipin, Faseb J 17 (2003) 714–716.

Pfeiffer, K., Gohil, V., Stuart, R.A., Hunte, C., Brandt, U., Greenberg, M.L. and Schagger, H., Cardiolipin stabilizes respiratory chain supercomplexes, J Biol Chem 278 (2003) 52873–52880.

Porcelli, S.A., Cutting glycolipids down to size, Nat Immunol 2 (2001) 191–192.

Qin, L., Sharpe, M.A., Garavito, R.M. and Ferguson-Miller, S., Conserved lipid-binding sites in membrane proteins: a focus on cytochrome c oxidase, Curr Opin Struct Biol 17 (2007) 444–450.

Qu, B.H., Strickland, E.H. and Thomas, P.J., Localization and suppression of a kinetic defect in cystic fibrosis transmembrane conductance regulator folding, J Biol Chem 272 (1997) 15739–15744.

Roberts, T.J., Sriram, V., Spence, P.M., Gui, M., Hayakawa, K., Bacik, I., Bennink, J.R., Yewdell, J.W. and Brutkiewicz, R.R., Recycling CD1d1 molecules present endogenous antigens processed in an endocytic compartment to NKT cells, J Immunol 168 (2002) 5409–5414.

Rutz, C., Rosenthal, W. and Schulein, R., A single negatively charged residue affects the orientation of a membrane protein in the inner membrane of Escherichia coli only when it is located adjacent to a transmembrane domain, J Biol Chem 274 (1999) 33757–33763.

Sahin-Toth, M., Kaback, H.R. and Friedlander, M., Association between the amino- and carboxyl-terminal halves of lactose permease is specific and mediated by multiple transmembrane domains, Biochemistry 35 (1996) 2016–2021.

Sambamurti, K., Suram, A., Venugopal, C., Prakasam, A., Zhou, Y., Lahiri, D.K. and Greig, N.H., A partial failure of membrane protein turnover may cause Alzheimer's disease: a new hypothesis, Curr Alzheimer Res 3 (2006) 81–90.

Sanders, C.R. and Myers, J.K., Disease-related misassembly of membrane proteins, Annu Rev Biophys Biomol Struct 33 (2004) 25–51.

Sanghera, N. and Pinheiro, T.J., Binding of prion protein to lipid membranes and implications for prion conversion, J Mol Biol 315 (2002) 1241–1256.

Schafer, E., Dencher, N.A., Vonck, J. and Parcej, D.N., Three-dimensional structure of the respiratory chain supercomplex I1III2IV1 from bovine heart mitochondria, Biochemistry 46 (2007) 12579–12585.

Schagger, H. and Pfeiffer, K., Supercomplexes in the respiratory chains of yeast and mammalian mitochondria, Embo J 19 (2000) 1777–1783.

Schlame, M. and Ren, M., Barth syndrome, a human disorder of cardiolipin metabolism, FEBS Lett 580 (2006) 5450–5455.

Schlame, M., Rua, D. and Greenberg, M.L., The biosynthesis and functional role of cardiolipin, Prog Lipid Res 39 (2000) 257–288.

Schleiff, E., Tien, R., Salomon, M. and Soll, J., Lipid composition of outer leaflet of chloroplast outer envelope determines topology of OEP7, Molecular Biology of the Cell 12 (2001) 4090–4102.

Schmelzer, K., Fahy, E., Subramaniam, S. and Dennis, E.A., The lipid maps initiative in lipidomics, Methods Enzymol 432 (2007) 171–183.

Schmidt, T.R., Jaradat, S.A., Goodman, M., Lomax, M.I. and Grossman, L.I., Molecular evolution of cytochrome c oxidase: rate variation among subunit VIa isoforms, Mol Biol Evol 14 (1997) 595–601.

Scorrano, L., Ashiya, M., Buttle, K., Weiler, S., Oakes, S.A., Mannella, C.A. and Korsmeyer, S.J., A distinct pathway remodels mitochondrial cristae and mobilizes cytochrome c during apoptosis, Dev Cell 2 (2002) 55–67.

Sedlak, E., Panda, M., Dale, M.P., Weintraub, S.T. and Robinson, N.C., Photolabeling of cardiolipin binding subunits within bovine heart cytochrome c oxidase, Biochemistry 45 (2006) 746–754.

Shinzawa-Itoh, K., Aoyama, H., Muramoto, K., Terada, H., Kurauchi, T., Tadehara, Y., Yamasaki, A., Sugimura, T., Kurono, S., Tsujimoto, K., Mizushima, T., Yamashita, E., Tsukihara, T. and Yoshikawa, S., Structures and physiological roles of 13 integral lipids of bovine heart cytochrome c oxidase, Embo J 26 (2007) 1713–1725.

Skach, W.R., Calayag, M.C. and Lingappa, V.R., Evidence for an alternate model of human P-glycoprotein structure and biogenesis, J Biol Chem 268 (1993) 6903–6908.

Sun, J., Wu, J., Carrasco, N. and Kaback, H.R., Identification of the epitope for monoclonal antibody 4B1 which uncouples lactose and proton translocation in the lactose permease of Escherichia coli, Biochemistry 35 (1996) 990–998.

Taanman, J.W. and Capaldi, R.A., Subunit VIa of yeast cytochrome c oxidase is not necessary for assembly of the enzyme complex but modulates the enzyme activity. Isolation and characterization of the nuclear-coded gene, J Biol Chem 268 (1993) 18754–18761.

van Klompenburg, W., Nilsson, I., von Heijne, G. and de Kruijff, B., Anionic phospholipids are determinants of membrane protein topology, Embo J 16 (1997) 4261–4266.

van Meer, G., Leeflang, B.R., Liebisch, G., Schmitz, G. and Goni, F.M., The European lipidomics initiative: enabling technologies, Methods Enzymol 432 (2007) 213–232.

Vigh, L., Escriba, P.V., Sonnleitner, A., Sonnleitner, M., Piotto, S., Maresca, B., Horvath, I. and Harwood, J.L., The significance of lipid composition for membrane activity: new concepts and ways of assessing function, Prog Lipid Res 44 (2005) 303–344.

Vogel, F., Bornhovd, C., Neupert, W. and Reichert, A.S., Dynamic subcompartmentalization of the mitochondrial inner membrane, J Cell Biol 175 (2006) 237–247.

von Heijne, G., The distribution of positively charged residues in bacterial inner membrane proteins correlates with the trans-membrane topology, Embo J 5 (1986) 3021–3027.

von Heijne, G., Control of topology and mode of assembly of a polytopic membrane protein by positively charged residues, Nature 341 (1989) 456–458.
Wang, X., Bogdanov, M. and Dowhan, W., Topology of polytopic membrane protein subdomains is dictated by membrane phospholipid composition, Embo J 21 (2002) 5673–5681.
White, S.H. and von Heijne, G., Do protein-lipid interactions determine the recognition of transmembrane helices at the ER translocon?, Biochem Soc Trans 33 (2005) 1012–1015.
Xia, D., Esser, L., Yu, L. and Yu, C.A., Structural basis for the mechanism of electron bifurcation at the quinol oxidation site of the cytochrome bc1 complex, Photosynth Res 92 (2007) 17–34.
Xie, J., Bogdanov, M., Heacock, P. and Dowhan, W., Phosphatidylethanolamine and monoglucosyldiacylglycerol are interchangeable in supporting topogenesis and function of the polytopic membrane protein lactose permease, J Biol Chem 281 (2006) 19172–19178.
Xu, Y., Malhotra, A., Ren, M. and Schlame, M., The enzymatic function of tafazzin, J Biol Chem 281 (2006) 39217–39224.
Yanagisawa, K., GM1 ganglioside and the seeding of amyloid in Alzheimer's disease: endogenous seed for Alzheimer amyloid, Neuroscientist 11 (2005) 250–260.
Yanagisawa, K., Odaka, A., Suzuki, N. and Ihara, Y., GM1 ganglioside-bound amyloid beta-protein (A beta): a possible form of preamyloid in Alzheimer's disease, Nat Med 1 (1995) 1062–1066.
Yankovskaya, V., Horsefield, R., Tornroth, S., Luna-Chavez, C., Miyoshi, H., Leger, C., Byrne, B., Cecchini, G. and Iwata, S., Architecture of succinate dehydrogenase and reactive oxygen species generation, Science 299 (2003) 700–704.
Zhang, J.T., Sequence requirements for membrane assembly of polytopic membrane proteins: molecular dissection of the membrane insertion process and topogenesis of the human MDR3 P-glycoprotein, Mol Biol Cell 7 (1996) 1709–1721.
Zhang, J.T., The multi-structural feature of the multidrug resistance gene product P-glycoprotein: implications for its mechanism of action (hypothesis), Mol Membr Biol 18 (2001) 145–152.
Zhang, J.T., Lee, C.H., Duthie, M. and Ling, V., Topological determinants of internal transmembrane segments in P-glycoprotein sequences, J Biol Chem 270 (1995) 1742–1746.
Zhang, J.T. and Ling, V., Study of membrane orientation and glycosylated extracellular loops of mouse P-glycoprotein by in vitro translation, J Biol Chem 266 (1991) 18224–18232.
Zhang, M., Mileykovskaya, E. and Dowhan, W., Gluing the respiratory chain together. Cardiolipin is required for supercomplex formation in the inner mitochondrial membrane, J Biol Chem 277 (2002) 43553–43556.
Zhang, M., Mileykovskaya, E. and Dowhan, W., Cardiolipin is essential for organization of complexes III and IV into a supercomplex in intact yeast mitochondria, J Biol Chem 280 (2005a) 29403–29408.
Zhang, W., Bogdanov, M., Pi, J., Pittard, A.J. and Dowhan, W., Reversible topological organization within a polytopic membrane protein is governed by a change in membrane phospholipid composition, J Biol Chem 278 (2003) 50128–50135.
Zhang, W., Campbell, H.A., King, S.C. and Dowhan, W., Phospholipids as determinants of membrane protein topology. Phosphatidylethanolamine is required for the proper topological organization of the gamma-aminobutyric acid permease (GabP) of Escherichia coli, J Biol Chem 280 (2005b) 26032–26038.
Zhao, H., Przybylska, M., Wu, I.H., Zhang, J., Siegel, C., Komarnitsky, S., Yew, N.S. and Cheng, S.H., Inhibiting glycosphingolipid synthesis improves glycemic control and insulin sensitivity in animal models of type 2 diabetes, Diabetes 56 (2007) 1210–1218.
Zhu, Q., von Dippe, P., Xing, W. and Levy, D., Membrane topology and cell surface targeting of microsomal epoxide hydrolase. Evidence for multiple topological orientations, J Biol Chem 274 (1999) 27898–27904.

# Chapter 9
# Altered Lipid Metabolism in Brain Injury and Disorders

**Rao Muralikrishna Adibhatla and J.F. Hatcher**

**Abstract** Deregulated lipid metabolism may be of particular importance for CNS injuries and disorders, as this organ has the highest lipid concentration next to adipose tissue. Atherosclerosis (a risk factor for ischemic stroke) results from accumulation of LDL-derived lipids in the arterial wall. Pro-inflammatory cytokines (TNF-$\alpha$ and IL-1), secretory phospholipase $A_2$ IIA and lipoprotein-$PLA_2$ are implicated in vascular inflammation. These inflammatory responses promote atherosclerotic plaques, formation and release of the blood clot that can induce ischemic stroke. TNF-$\alpha$ and IL-1 alter lipid metabolism and stimulate production of eicosanoids, ceramide, and reactive oxygen species that potentiate CNS injuries and certain neurological disorders. Cholesterol is an important regulator of lipid organization and the precursor for neurosteroid biosynthesis. Low levels of neurosteroids were related to poor outcome in many brain pathologies. Apolipoprotein E is the principal cholesterol carrier protein in the brain, and the gene encoding the variant Apolipoprotein E4 is a significant risk factor for Alzheimer's disease. Parkinson's disease is to some degree caused by lipid peroxidation due to phospholipases activation. Niemann-Pick diseases A and B are due to acidic sphingomyelinase deficiency, resulting in sphingomyelin accumulation, while Niemann-Pick disease C is due to mutations in either the *NPC1* or *NPC2* genes, resulting in defective cholesterol transport and cholesterol accumulation. Multiple sclerosis is an autoimmune inflammatory demyelinating condition of the CNS. Inhibiting phospholipase $A_2$ attenuated the onset and progression of experimental autoimmune encephalomyelitis. The endocannabinoid system is hypoactive in Huntington's disease. Ethyl-eicosapetaenoate showed promise in clinical trials. Amyotrophic lateral sclerosis causes loss of motorneurons. Cyclooxygenase-2 inhibition reduced spinal neurodegeneration in amyotrophic lateral sclerosis transgenic mice. Eicosapentaenoic acid supplementation provided improvement in schizophrenia

---

R.M. Adibhatla
Department of Neurological Surgery, Cardiovascular Research Center, Neuroscience Training Program, University of Wisconsin School of Medicine and Public Health, Madison, WI., William S. Middleton Veterans Affairs Hospital, Madison, WI53792, USA

P.J. Quinn, X. Wang (eds.), *Lipids in Health and Disease*, 

patients, while the combination of (eicosapentaenoic acid + docosahexaenoic acid) provided benefit in bipolar disorders. The ketogenic diet where >90% of calories are derived from fat is an effective treatment for epilepsy. Understanding cytokine-induced changes in lipid metabolism will promote novel concepts and steer towards bench-to-bedside transition for therapies.

**Keywords** Atherosclerosis · cholesterol · inflammation · neurodegenerative diseases · stroke

**Abbreviations** ArAc: Arachidonic acid; CDP-choline: Cytidine-5'-diphosphocholine; CNS: Central nervous system; COX/LOX: Cyclooxygenase/lipoxygenase; DHA: Docosahexaenoic acid; HNE: 4-Hydroxynonenal; IL-1ß: Interleukin 1ß; Lp-PLA$_2$: Lipoprotein-PLA$_2$; MMP: Matrix metalloprotease; OxPC: Oxidized phosphatidylcholine; PAF: Platelet activating factor; PC: Phosphatidylcholine; PE: phosphatidylethanolamine; PI: phosphatidylinositol; PLA$_2$: Phospholipase A$_2$; sPLA$_2$: Secretory PLA$_2$ or inflammatory PLA$_2$; PS: Phosphatidylserine; PUFA: Polyunsaturated fatty acid; ROS: Reactive oxygen species; SM: Sphingomyelin; SMase: Sphingomyelinase; TNF-$\alpha$: Tumor necrosis factor-$\alpha$.

## 9.1 Introduction

### *9.1.1 The Biological Membrane Structure and Function*

Cellular membranes are composed of glycerophospholipids [phosphatidylcholine (PC), phosphatidylethanolamine (PE), phosphatidylserine (PS) and phosphatidylinositol (PI)], sphingolipids [sphingomyelin (SM), ceramide and gangliosides], cholesterol and cholesterol esters, acylglycerols, and fatty acids. The phospholipid bilayer and associated lipids provide not only a permeability barrier but also a structured environment that is essential for the proper functioning of membrane-bound proteins (Maxfield and Tabas, 2005). Cholesterol is one of the most important regulators of lipid organization as its structure allows it to fill interstitial spaces between hydrophobic fatty acid chains of phospholipids. The neutral lipids such as PC and SM predominantly reside on the outer or exofacial leaflet, whereas anionic phospholipids PS (exclusively inner leaflet), PE, and PI reside on the inner or cytofacial leaflet of the biological membrane. The transbilayer distribution of cholesterol between the leaflets determines membrane fluidity and can alter the membrane function. Cardiolipin is a phospholipid that is exclusively limited to the mitochondrial membrane and is essential for proper assembly and functioning of the mitochondrial respiratory chain and oxidative phosphorylation.

In addition to their role as structural components of the cell membrane, phospholipids serve as precursors for various second messengers such as arachidonic acid (ArAc), docosahexaenoic acid (DHA), ceramide, 1,2-diacylglycerol, phosphatidic acid, and lyso-phosphatidic acid. Lipids comprise a large number of chemically distinct molecules arising from combinations of fatty acids with various backbone structures. Overall, mammalian cells may contain 1,000–2,000 lipid species. Lipid metabolism may be of particular importance for the CNS, as this organ has the highest concentration of lipids next to adipose tissue.

### *9.1.2 Lipids and the Central Nervous System (CNS)*

Neurodegenerative diseases, mental disorders, stroke and CNS traumas are problems of vast clinical importance. The crucial role of lipids in tissue physiology and cell signaling is demonstrated by the many neurological disorders, including bipolar disorders and schizophrenia, and neurodegenerative diseases such as Alzheimer's, Parkinson's, Niemann-Pick and Huntington diseases, that involve deregulated lipid metabolism (Fig. 9.1) (Adibhatla and

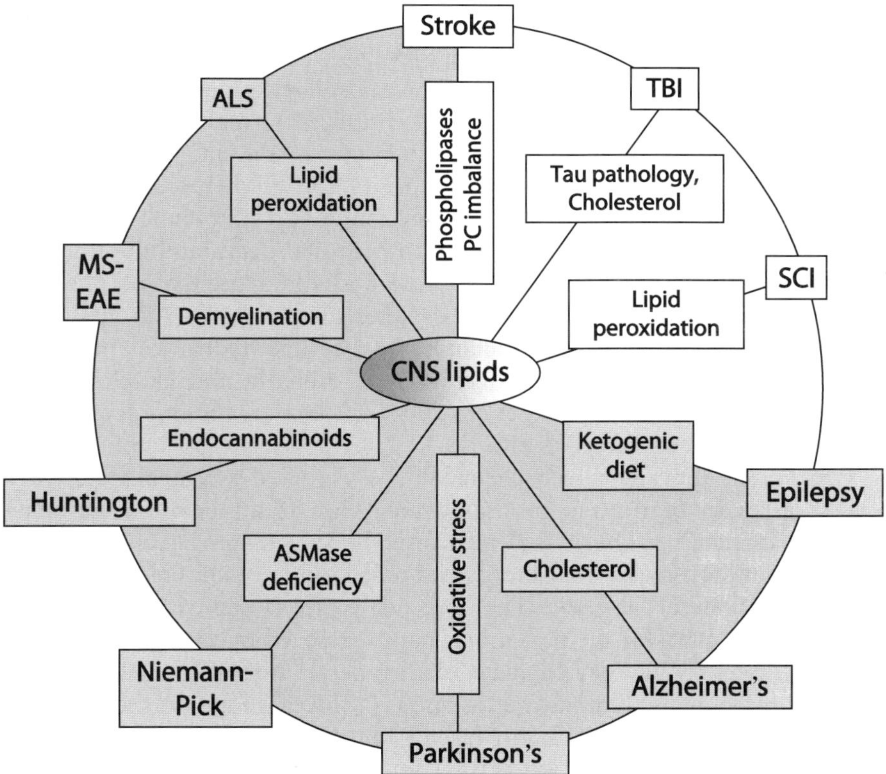


**Fig. 9.1** Lipid systems affected in CNS disorders (shaded) and injuries (clear). Neurosteroid synthesis in the brain is affected in various brain disorders and injuries and treatment with neurosteroids (pregnenolone, dehydroepiandrosterone and allopregnanolone) showed positive trend in these brain pathologies (Fig. 9.3)

Hatcher, 2007, and references cited therein). Altered lipid metabolism is also believed to be a key event which contributes to CNS injuries such as stroke (Adibhatla and Hatcher, 2008, and references cited therein; Adibhatla et al., 2006a).

## 9.2 Stroke, Traumatic Brain and Spinal Cord Injuries

### *9.2.1 Stroke or "Brain Attack": A Problem of Vast Clinical Importance*

Stroke generally refers to a local interruption of blood flow to the brain and is the leading cause of long-term disability, third leading cause of death (Young et al., 2007). Approximately 12% of strokes are hemorrhagic (rupture of a cerebral blood vessel), whereas the remaining 88% are ischemic and result from occlusion of a cerebral artery (either thrombolic or embolic). Blockage of a cerebral artery results in interruption of the blood flow and supply of nutrients, glucose and oxygen to the brain. The energy needs of the brain are supplied by metabolism of glucose and oxygen for the phosphorylation of ADP to ATP. Most of the ATP generated in the brain is utilized to maintain intracellular homeostasis and transmembrane ion gradients of sodium, potassium, and calcium. Energy failure results in collapse of ion gradients, and excessive release of neurotransmitters such as dopamine and glutamate (Adibhatla and Hatcher, 2006), ultimately leading to neuronal death and development of an infarction. Excess glutamate release and stimulation of its receptors results in activation of phospholipases/sphingomyelinases (Adibhatla and Hatcher, 2006; Adibhatla et al., 2006a), phospholipid hydrolysis and release of second messengers ArAc and ceramide (Adibhatla and Hatcher, 2006; Adibhatla et al., 2006b; Mehta et al., 2007). Ultimately these processes lead to apoptotic or necrotic cell death.

Focal cerebral ischemia or "ischemic stroke" is caused by a local blockage of a cerebral artery that results in loss of blood flow to a portion of the brain. Stroke is characterized by an ischemic core (infarct) surrounded by a "penumbra" (peri-infarct) region that has partial reduction in blood flow due to presence of collateral arteries. The ischemic core is generally considered unsalvageable, whereas the penumbra may be rescued by timely intervention and is a target for the development of therapeutic treatment. Local arterial blockage can be caused by either a thrombus (a clot that forms at the site of the arterial occlusion) or an embolus (a clot that forms peripherally, dislodges into the arterial circulation and is transported to the brain). Atherosclerosis, discussed in the next section, is the main risk factor for development of these embolisms (Fig. 9.2). Inflammation poses as one of the high risk factors for stroke for its role in the initiation, progression and maturation of atherosclerosis.

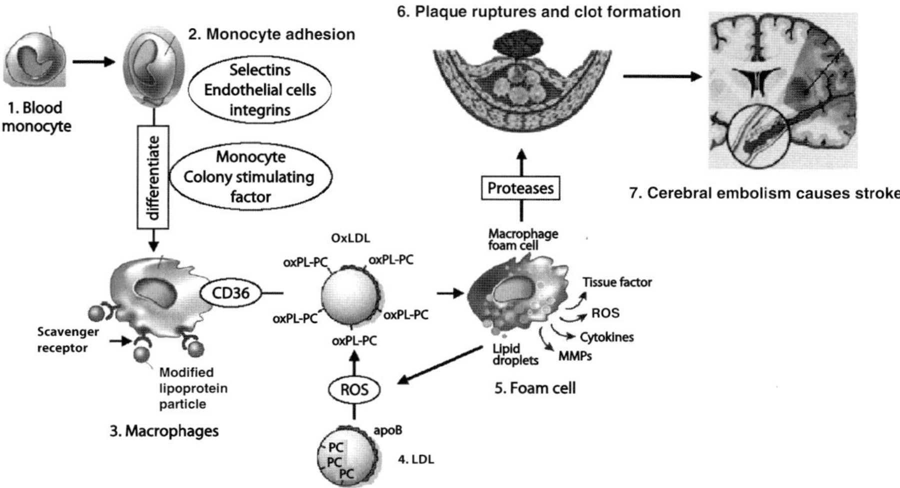


**Fig. 9.2** Atherosclerosis, a major risk factor for ischemic stroke (Adibhatla and Hatcher 2008a). Under inflammatory conditions (OxLDL, homocysteine, cigarette smoke, shear stress and infectious agents such as *Chlamydia pneumoniae*) endothelia cells of the artery express adhesion molecules that allow monocytes (1) to adhere to endothelia (2). Chemoattractants such as monocyte chemoattractant protein-1 (MCP-1) draw the monocytes through the endothelium into the arterial intima. Once resident in the intima, monocytes differentiate into macrophages (3) in response to locally produced agents such as monocyte colony stimulating factor. LDL (4) under oxidative stress gets oxidized to OxLDL. The macrophages increase expression of scavenging receptors such as CD36, SR-A and SR-B. These scavenger receptors then internalize specifically oxidized LDL (OxLDL, specifically OxPC) particles such that cholesteryl esters accumulate in cytoplasmic droplets, resulting in lipid-loaded macrophages (foam cells, 5). Foam cells produce ROS, which further propagate LDL oxidation, and secrete cytokines and matrix metalloproteinases (MMPs). The MMPs contribute to degradation of the fibrous cap surrounding the plaque, resulting in its rupture and formation of a blood clot (6). If the blood clot dislodges from the plaque, arterial blood flow can carry it to the brain, where it lodges in a cerebral artery (embolism) and causes an ischemic stroke (7)

#### 9.2.1.1 Atherosclerosis is a Risk Factor for Stroke

Atherosclerosis is believed to be predominantly an inflammatory condition produced as a response to injury (Elkind, 2006). Atherosclerosis is defined by the accumulation in the arterial intima of mainly low-density lipoprotein (LDL)-derived lipids along with apolipoprotein B-100 (apoB100). LDL is the major carrier of cholesterol in the circulation and is composed of one apoB-100 together with phosphatidylcholine (PC), sphingomyelin (SM) and unesterified cholesterol (500:200:400 molecules respectively) constituting a surface film surrounding a core of cholesteryl esters and triacylglycerols.

The traditional view of atherosclerosis has been simply the deposition and accumulation of cholesterol, other lipids, and cellular debris within the wall of medium to large arteries, resulting in plaque formation and disturbance of blood flow (Fig. 9.2). The role of cholesterol in atherosclerosis is well established and has been elegantly reviewed (Maxfield and Tabas, 2005). It is now believed that a complex endothelial injury and dysfunction induced by a variety of factors such as homocysteine, toxins (smoking), mechanical forces (shear stress), infections agents (*Chlamydia pneumoniae*) and oxidized LDL results in an inflammatory response that is instrumental in the formation and rupture of plaques, one of the greatest risk factors for ischemic stroke (Adibhatla and Hatcher 2008a) (Emsley and Tyrrell, 2002; Hansson and Libby, 2006).

Two critical events involved in atherogenesis involve accumulation and oxidation of LDL in the arterial intima and recruitment of monocytes to the developing lesion. After diffusion through the endothelial cell junctions into the arterial intima, LDL can be retained through interaction of apoB100 and matrix proteoglycans. LDL accumulates in the arterial intima when its rate of influx exceeds the rate of efflux. While the exact mechanisms governing LDL accumulation remain to be elucidated (Nicolo et al., 2007), evidence indicates that LDL uptake and retention are increased at plaque sites, which may involve degradation or binding to cellular and matrix components. Once in the arterial intima, LDL can be oxidized to OxLDL through oxidation of polyunsaturated fatty acids (PUFA) of LDL lipids, particularly PC of LDL to form OxPC.

A second critical event in atherosclerosis is an inflammatory response that triggers expression of adhesion molecules (selectins and integrins) in the arterial endothelium, stimulating adhesion of monocytes to the endothelium. Monocytes penetrate into the arterial intima, differentiate into macrophages and eventually become foam cells by binding and endocytosing OxLDL through CD36 scavenging receptors (Fig. 9.2). Studies showed that oxidized phospholipids bearing the PC headgroup as a ligand on OxLDL mediate uptake by macrophage scavenging receptors such as CD36 (Boullier et al., 2005). The macrophage foam cells generate ROS, produce tumor necrosis factor-α (TNF-α) and interleukin-1 (IL-1), and matrix metalloproteinase 9 (MMP-9) that promote atherosclerosis, degrade the fibrous cap, and eventually lead to plaque rupture.

Increased levels of TNF-α and IL-1 up-regulate expression of adhesion molecules and promote further monocyte recruitment into developing atherosclerotic lesions. Macrophage MMP-9 degrades extracellular matrix components including the fibrous cap of atheromatous plaques. Rupture of the fibrous cap exposes the blood to the inner components of the plaque, particularly tissue factor released from apoptotic macrophages. Tissue factor binds to activated coagulation factor VII and triggers the coagulation cascade, resulting in formation of a blood clot. Destabilization of this clot results in release of an embolus into the blood stream, which can be transported to the brain, where it can lodge in a cerebral artery and induce an ischemic stroke (Fig. 9.2) (Elkind, 2006; Emsley and Tyrrell, 2002; Hansson and Libby, 2006; Stoll and Bendszus, 2006).

#### 9.2.1.2 Lipoprotein-PLA$_2$ (Lp-PLA$_2$), also Known as Platelet Activating Factor (PAF) Acetylhydrolase

Lp-PLA$_2$, a 45 kDa protein, is a member of PLA$_2$ family classified as group VIIA PLA$_2$ and is also known as plasma PAF acetylhydrolase (Adibhatla and

Hatcher, 2008, and references cited therein). This enzyme is found in blood circulation in most animals, and in humans is associated with apoB-100 of LDL and is also found in atherosclerotic plaques (Lavi et al., 2007). Higher levels of Lp-PLA$_2$ are also associated with coronary heart disease, stroke and dementia (Lavi et al., 2007; Oei et al., 2005). Lp-PLA$_2$ is produced and secreted by cells of monocyte-macrophage series, T-lymphocytes and mast cells. In addition to PAF acetylhydrolase activity, Lp-PLA$_2$ also hydrolyzes oxidized PC of LDL to generate oxidized fatty acids and lyso-phosphatidylcholine (lyso-PC) (Zalewski et al., 2006). Local coronary lyso-PC formation is also associated with endothelial dysfunction and supports the role of this enzyme in vascular inflammation and atherosclerosis in humans. Lp-PLA$_2$ also has an anti-inflammatory function arising from hydrolysis of PAF, which is known to activate platelets, monocytes and macrophages.

#### 9.2.1.3 Atherosclerosis and Group IIA Secretory PLA$_2$ (Inflammatory PLA$_2$)

Group IIA phospholipase A$_2$ (secretory PLA$_2$, also known as inflammatory PLA$_2$) has been found in human atherosclerotic lesions. sPLA$_2$ IIA is implicated in chronic inflammatory conditions such as arthritis and may also contribute to atherosclerosis. sPLA$_2$ IIA is a pro-atherogenic factor and has been suggested to regulate collagen deposition in the plaque and fibrotic cap development (Ghesquiere et al., 2005, and references cited therein). sPLA$_2$ is one of the enzymes responsible for the release of lyso-PC *via* its catalytic action and these two play a crucial role in the development of atherosclerosis (Kougias et al., 2006). Non-catalytic (non-enzymatic) atherogenic effects of sPLA$_2$ II are thought to involve binding to a muscular-type (M-type) sPLA$_2$ receptor.

#### 9.2.1.4 Sphingomyelinase (SMase): A Link Between Atherosclerosis and Ceramide

LDL possesses SMase activity, which may be intrinsic to apoB-100. SMase hydrolyzes SM to release ceramide, which is elevated in atherosclerotic plaques as well as in LDL isolated from these lesions. Ceramide is believed to play an important role in aggregation of LDL within the arterial wall, a critical step in the initiation of atherosclerosis (Kinnunen and Holopainen, 2002).

### *9.2.2 Reactive Oxygen Species (ROS), Lipid Metabolism and Stroke*

The study of ROS and oxidative stress is difficult due to the transient nature of ROS, the number of complex ongoing processes, direct and reverse interactions between these processes, and the capacity of ROS to alter a large number of cellular components.

Oxidative stress results when production of ROS such as superoxide anion radicals, hydrogen peroxide, and hydroxyl radicals exceeds a biological system's ability to detoxify these reactive intermediates. Superoxide anion radicals can combine with reactive nitrogen species such as nitric oxide to generate the strong pro-oxidant peroxynitrite. ROS are produced by a number of cellular oxidative metabolic processes including oxidative phosphorylation by the mitochondrial respiratory chain, xanthine oxidase, NAD(P)H oxidases, monoamine oxidases, and metabolism of ArAc by lipoxygenases (LOX) (Adibhatla and Hatcher, 2006). It has been generally accepted that ArAc metabolism by cyclooxygenases (COX) also generates ROS, but recent literature shows that COX-2 does not directly produce ROS but does form carbon-centered radicals on ArAc (Kunz et al., 2007; Simmons et al., 2004). Most ROS are produced at low levels and any damage they cause to cells is constantly repaired. Low levels of ROS are used in *redox* (reduction/oxidation) cell signaling and may be important in prevention of aging by induction of mitochondrial *hormesis* (*hormesis:* a beneficial response to low dose exposure to toxins). Generation of ROS is also used by the immune system to destroy invading pathogens. Although there are intracellular defenses against ROS, increased production of ROS or loss of antioxidant defenses leads to progressive cell damage and decline in physiological function. Over-production of these free radicals can damage all components of the cell, including proteins, carbohydrates, nucleic acids, and lipids, leading to progressive decline in physiological function and ultimately cell death.

Beyond the initial damage to membranes, reaction of these radicals with double bonds of fatty acids in lipids produces peroxides that give rise to α,ß-unsaturated aldehydes including malondialdehyde (MDA), 4-hydroxynonenal (HNE) and acrolein. These aldehydes covalently bind to proteins through reaction with thiol groups and alter their function. We have previously shown that $CA_1$ hippocampal neurons were HNE positive by immunohistochemistry after transient cerebral ischemia (Adibhatla and Hatcher, 2006). Recently, elevated levels of an acrolein-protein conjugate were demonstrated in plasma of stroke patients (Adibhatla and Hatcher, 2007, and references cited therein).

The brain is believed to be particularly vulnerable to oxidative stress as it contains high concentrations of PUFA that are susceptible to lipid peroxidation, consumes relatively large amounts of oxygen for energy production, and has lower antioxidant defenses compared to other organs. Of all the brain cells, neurons are particularly vulnerable to oxidative insults due to low levels of reduced glutathione (Dringen, 2000). In addition to atherosclerosis, oxidative stress has been shown to be a component of many neurodegenerative disorders such as Parkinson's disease, Alzheimer's disease, Multiple Sclerosis, and Amyotrophic Lateral Sclerosis (Adibhatla and Hatcher 2008b).

Cessation of blood flow to the brain leads to energy loss and necrotic cell death. This initiates the immune response, activating inflammatory cells including microglia/macrophages and generating ROS. ROS can further stimulate release of cytokines that cause up-regulation of adhesion molecules, mobilization and activation of leukocytes, platelets and endothelium. These activated inflammatory cells also release cytokines, MMPs, nitric oxide and additional ROS in a feed-back fashion (Wang et al., 2007b). ArAc metabolites, synthesized by and liberated from astrocytes, microglial cells and macrophages, are intimately involved in the inflammatory process by enhancing vascular permeability and modulating inflammatory cell activities and ROS generation. The role of ROS in activation of various signaling pathways such as p38, JNK, p53, ERK1/2, Akt, NF-κB (Crack and Taylor, 2005), MMPs (Liu and Rosenberg, 2005) and stroke injury (Adibhatla and Hatcher, 2006, 2008b; Margaill et al., 2005) have been recently reviewed.

#### 9.2.2.1 Cytokines and Stroke

There are substantial data from both animal models and clinical studies that cytokines including TNF-α, IL-1 and IL-6 are up-regulated after stroke. Although the roles of cytokines in stroke pathology remain controversial, the majority of studies support their deleterious effects, at least in the early phase of stroke injury. Whether cytokines mediate pro-survival or pro-apoptotic signaling appears to depend on their concentration, the target cell, the activating signal, and the timing and sequence of action (Adibhatla and Hatcher, 2008). A number of studies have demonstrated that TNF-α and IL-1 modulate phospholipid and sphingolipid metabolism by up-regulating phospholipases and SMases and down-regulating enzymes of phospholipid/sphingolipid synthesis, although most of these studies were conducted in cell lines not of CNS origin. The release of ArAc during ischemia may be one of the initial events that up-regulate cytokine expression. Ceramide released by SMases triggers the MAP kinase cascade and can up-regulate cytokine expression through activation of NF-κB. Thus phospholipid metabolism and cytokine expression (the inflammatory response) may function through a feedback mechanism. The integration of cytokine biology and lipid metabolism in stroke is less explored and was recently reviewed (Adibhatla and Hatcher, 2008).

A systemic inflammatory response involving up-regulation of TNF-α and IL-1 is believed to be instrumental in the formation and destabilization of plaques, one of the risk factors for ischemic stroke (Emsley and Tyrrell, 2002; Hansson and Libby, 2006). There is considerable clinical data indicating that this systemic inflammation is associated with unfavorable outcome in stroke patients (McColl et al., 2007). However, this inter-relationship of systemic inflammation with stroke pathology has not been well studied.

#### 9.2.2.2 Inflammation and Resolution

A critical aspect of the inflammatory response is the ability to stop the inflammation, referred to as the resolution phase, an active process involving expression of anti-inflammatory agents. Activation of $PLA_2$s release ArAc, eicosapentaenoic acid, and DHA. ArAc is metabolized to eicosanoids (prostaglandins, leukotrienes, and thromboxanes) through the COX/LOX pathways, a major pathway mediating inflammation, but is also metabolized to anti-inflammatory lipoxins through the LOX pathway. Chemical mediators such as aspirin can acetylate COX-2; prostaglandin synthesis is inhibited and metabolism is shifted by acetylated COX-2/LOX pathway to generate pro-resolution lipoxins. Eicosapentaenoic acid and DHA, ω-3 fatty acids, are metabolized to resolvins and protectins such as neuroprotectin D1 that have important roles in resolution of inflammation (Serhan, 2007). For further reading refer to JX Kang's chapter on "Modulation of Cytokines by ω-3 Fatty Acid".

#### 9.2.2.3 Oxidized PC (OxPC) is an Inflammatory Marker

We have previously shown that PC loss, either due to activation of phospholipases or inhibition of its synthesis *via* CTP:phosphocholine cytidylyltransferase (CCT), may be a significant factor contributing to stroke injury that was attenuated by treatment with CDP-choline (a phase III clinical trail drug for stroke treatment) (Adibhatla and Hatcher, 2005, 2006, 2007; Adibhatla et al., 2006b). Another contributing factor for PC loss could be conversion to OxPC. Peroxidation of fatty acids in phospholipids results in an oxidized phospholipid. Scission of the peroxidized fatty acid results in formation of a phospholipid such as OxPC (Kadl et al., 2004) with a fatty acid containing an aldehyde residue, and the aldehyde cleavage fragments MDA, HNE, or acrolein discussed above. These reactive phospholipid aldehydes exhibit cytotoxicity by binding to lysine residues of cellular proteins. OxPC itself also changes the membrane properties, resulting in alterations in ion transport and membrane protein function. The presence of OxPC on the apoptotic cell surface has been characterized by EO6 monoclonal antibodies that exclusively bind to OxPC (Qin et al., 2007; Adibhatla and Hatcher 2008b). In addition to OxPC, EO6 antibodies also recognize OxPC bound to lysine residues of proteins. OxPC on apoptotic cells may enhance pro-inflammatory signals and also serve as a marker of inflammation and apoptosis (Bratton and Henson, 2005; Chang et al., 2004; Kadl et al., 2004). The presence of OxPC has been demonstrated in multiple sclerosis brain using EO6 monoclonal antibodies (Qin et al., 2007). Formation of OxPC species were also shown after permanent focal ischemia in mice (Gao et al., 2006) and transient focal cerebral ischemia in rat (Adibhatla and Hatcher 2008b). For additional details, please refer to chapters on "Role of Oxidized Phospholipids in Inflammation" by Norbert Leitinger and "Mediation of Apoptosis by Oxidized Phospholipids" by Albin Hermetter.

### *9.2.3 Cholesterol is the Precursor for Neurosteroid Synthesis*

The vast majority of cholesterol in the brain is derived from *de novo* synthesis as virtually no cholesterol is transported from the plasma; cholesterol is synthesized in the neurons, glia (astrocytes), oligodendrocytes. In the adult brain, the predominant synthesis is by astrocytes; cholesterol is then secreted *via* transport molecules such as ATP binding cassette protein (ABCA1), taken up by lipoprotein receptors on neurons and internalized to the endosome/lysosome (E/L) system (Fig. 9.3). Cholesterol will be transported to mitochondria by Niemann-Pick C1 (NPC1) protein where the neurosteroids such as dehydroepiandrosterone (DHEA) and allopregnanolone are synthesized *via* the rate-limiting intermediate, pregnenolone. These neurosteroids act on nuclear and NMDA/ $GABA_A$ receptors to promote neurogenesis and modulate neurotransmission.

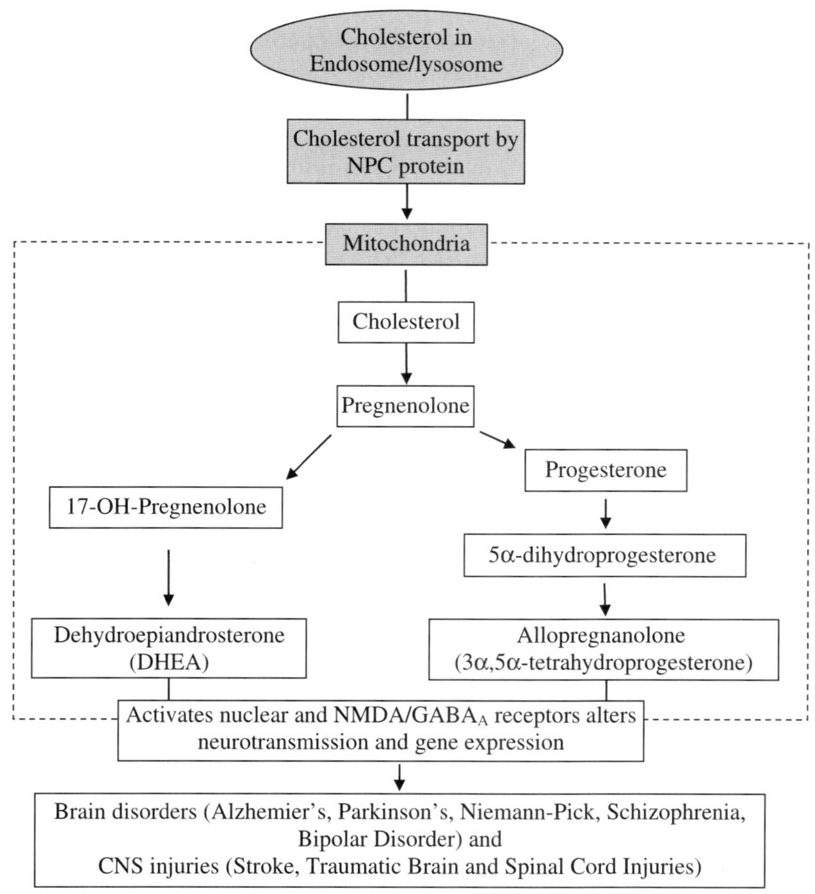


**Fig. 9.3** Synthesis of neurosteroids in the brain and their effects on various brain disorders and injuries. Niemann-Pick C protein transports cholesterol from endosome/lysosome system to mitochondria where the neurosteroid synthesis occurs. DHEA and allopregnanolone may provide beneficial effects based on the following studies: NPC (Griffin et al., 2004), AD, Schizophrenia, bipolar disorders, epilepsy (Marx et al., 2006a, 2006b), PD (Wojtal et al., 2006), stroke (Sayeed et al., 2006), TBI (Djebaili et al., 2005) and SCI (Lapchak et al., 2000). Studies showed both DHEA and allopregnanolone stimulated neurogenesis in stroke. (Marx et al., 2006b) and increased neuroprogenitor cells in AD models (Wang et al., 2007a)

While both provided benefit, allopregnanolone was more effective than progesterone in reducing infarction after stroke (Sayeed et al., 2006). DHEA and allopregnanolone also stimulated neurogenesis (Karishma and Herbert, 2002; Suzuki et al., 2004; Wang et al., 2005).

## *9.2.4 Traumatic Brain Injury (TBI)*

TBI is associated with significant neuropsychological deficits, primarily in the domains of attention, executive functioning and memory. In TBI, the initial traumatic event is shearing, laceration, and/or contusion of brain tissue resulting from a physical impact. Secondary injury after the initial trauma results from ischemia, alterations in ion and neuromodular levels, oxidative stress caused by ROS, edema and axonal swelling (Rigg and Zafonte, 2006). The neurosteroids such as DHEA and allopregnanolone reduced cell death, astrogliosis and functional deficits in rats after TBI (Djebaili et al., 2005). Corticosteroids have been proposed as therapies to reduce secondary injuries following TBI. Corticosteroids inhibit the $PLA_2$/COX/LOX pathways, thus limiting ArAc release and metabolism, down-regulating pro-inflammatory cytokines and attenuating inflammatory responses. However, large scale clinical trials of corticosteroids and lazaroids (21-aminosteroids) for treatment of TBI have either failed to demonstrate efficacy or found increased risk of mortality (Rigg and Zafonte, 2006).

### 9.2.4.1 TBI and ApoE

ApoE is an important mediator of cholesterol and lipid transport in the brain and is encoded by the polymorphic gene APOE. While it may be reasonable to relate effects of ApoE to cholesterol transport, the mechanism whereby ApoE elicits these effects has not been elucidated. ApoE has been shown to reduce glial activation and CNS inflammatory response. This action is isoform-specific with the ApoE4 isoform being less effective at down-regulating inflammatory cytokines (Lynch et al., 2005). A small peptide, apoE(133-149) was created from the receptor binding region that retains the ability of the native protein in down-regulating inflammatory responses. Administration of apoE(133-149) was shown to significantly improve histological and functional outcome after experimental TBI (Lynch et al., 2005).

While the APOE $\epsilon 4$ allele was first implicated as a significant risk factor for Alzheimer's Disease (see later section on Alzheimer's Disease), a number of clinical studies have indicated that performance on neuropsychological tasks is worse in TBI patients with the APOE $\epsilon 4$ allele than those without it (Ariza et al., 2006). While other clinical studies did not find an association of APOE $\epsilon 4$ allele with poorer outcome after TBI, these differences could be due to the severity of the TBI (no association in some studies with predominantly mild TBI), the neurological evaluation methods that assess the involvement of different brain regions, and the evaluation time post injury (up to 25 years after TBI). In other studies, the presence of the APOE ε4 allele has been associated with poorer outcome after cardiopulmonary resuscitation and intracerebral hemorrhage, but not after ischemic stroke (Smith et al., 2006).

Both human postmortem and experimental studies have shown Aß deposition and tau pathology after TBI (Jellinger, 2004). Statins have shown benefit in experimental TBI (Mahmood et al., 2007), but it is unknown if statin treatment affected Aß levels. The development of AD-like neuropathological and biochemical changes after severe TBI suggested that TBI may be a risk factor for subsequent development of dementia. Epidemiological studies have provided discrepant findings, thus the relationship between TBI and dementia remains a topic for further investigation.

#### 9.2.4.2 Spinal Cord Injury (SCI)

Similar to TBI, SCI is the result of an initial physical trauma followed by a secondary degenerative process. The majority of SCIs result from contusive, compressive, or stretch injury rather than physical transection of the spinal cord. The initial event after SCI is depolarization and opening of voltage-dependent ion channels, and consequent massive release of neurotransmitters including glutamate. This leads to accumulation of intracellular calcium, initiating a number of damaging events: mitochondrial dysfunction, activation of nitric oxide synthase (NOS) and $PLA_2$ (Adibhatla and Hatcher 2008b). $PLA_2$ releases ArAc into the COX/LOX pathways to generate eicosanoids. One consequence of mitochondrial dysfunction, activation of NOS, and COX/LOX activity is generation of free radicals (comprised of different species including reactive nitrogen species, ROS and other radicals) and subsequent lipid peroxidation, which is considered a major pathway of secondary injury in SCI (Hall and Springer, 2004).

The glucocorticoid steroids dexamethasone and methylprednisolone have been extensively used in clinical treatment of SCI. In animal studies, it was demonstrated that high dose methylprednisolone inhibited post-traumatic lipid peroxidation in spinal cord tissue. Beneficial effects secondary to inhibition of lipid peroxidation included preservation of ion homeostasis, mitochondrial energy metabolism, and attenuation of delayed glutamate release (Hall and Springer, 2004). It is believed that inhibition of lipid peroxidation is the principle neuroprotective mechanism of high-dose methylprednisolone and that glucocorticoid receptor-mediated anti-inflammatory effects have only a minor role (Hall and Springer, 2004).

Another agent that has undergone Phase III clinical trials for SCI is GM1 ganglioside. Since high-dose methylprednisolone had become widely accepted for treatment of SCI, GM1 was administered only after the completion of the 24 hr methylprednisolone dosing protocol. The results indicated that GM1 did not provide greater functional improvement compared to methylprednisolone alone (Hall and Springer, 2004). The neurosteroid, DHEA sulfate offered neuroprotection in a spinal cord ischemia model, which was believed to be mediated through $GABA_A$ receptors (Lapchak et al., 2000).

## 9.3 Lipids in CNS Disorders

### *9.3.1 Alzheimer's Disease (AD)*

AD is a progressive brain disorder affecting regions that control memory and cognitive functions, gradually destroying a person's memory and ability to learn, reason, communicate and carry out daily activities. One of the hallmarks of AD is overproduction of amyloid ß-peptide (Aß), resulting in the formation of plaques. A two-step cleavage of the neuronal membrane protein amyloid precursor protein (APP) (Ehehalt et al., 2003) results in two products, Aß40 and Aß42. Strong evidence for the role of Aß in the pathogenesis of AD was provided by the observation that mutations in APP or the enzymes that cleave it lead to over-production of Aß42 and rapid progression of the disease (Mandavilli, 2006). The second hallmark of AD is formation of neurofibrillary tangles due to hyperphosphorylation of tau protein.

Transgenic mouse models of AD require two or more mutations to reproduce all the physical features of AD (Aß plaques and tau tangles), however these models have not provided any leads to the relationship between Aß plaques and tau tangles. Most mouse models of AD are considered limited or incomplete: several models develop amyloid deposition but fail to develop neurofibrillary tangles that are an essential hallmark of AD. Neuritic atrophy is found in some transgenics, but of nearly one dozen mouse models, only one has reported loss of neurons that is characteristic of AD (Herrup et al., 2004). Some tau models develop severe memory deficits associated with AD but express little amyloid protein (Mandavilli, 2006). These shortfalls limit the usefulness of these models for developing therapeutic strategies for AD (Schwab et al., 2004).

There is growing evidence that cholesterol is of particular importance in development and progression of AD. ApoE is one of the major apolipoproteins in plasma and the principal cholesterol carrier protein in the brain. Identification of the gene encoding the variant ApoE4 (APOE $\epsilon$4 allele) as a significant risk factor for AD provided evidence for a role of cholesterol in the pathogenesis of AD (Puglielli, 2007). Elevated cholesterol levels increase Aß in cellular and animal models, and drugs such as statins and BM15.766 that inhibit cholesterol synthesis lower Aß levels (Hartmann et al., 2007; Puglielli, 2007; Whitfield, 2006). Statins inhibit 3-hydroxy-3-methylglutaryl-coenzyme A (HMG-CoA) reductase that initiates cholesterol and isoprenoid lipid synthesis while BM15.766 inhibits cholesterol synthesis at the ultimate step (Hartmann et al., 2007). Prospective trials evaluating statin therapy, however, did not demonstrate improvement in cognitive function in AD patients (Caballero and Nahata, 2004).

Cholesterol is needed to make the cellular membrane micro-domains referred to as lipid rafts. APP, ß-secretase, γ-secretase complex, and neutral sphingomyelinase (NSMase) are present in the lipid rafts that are rich in cholesterol and SM. Genetic mutations in APP or presenilins (part of the γ-secretase complex) increase production of Aß42. Recent studies suggest that Aß40 inhibits HMG-CoA reductase while Aß42 activates NSMase and increases ceramide production, which can accelerate the neurodegenerative process (Grimm et al., 2005; Mattson et al., 2005). It is unclear what regulates the cleavage of APP to Aß42 *vs* Aß40.

Cholesterol is the precursor for biosynthesis of neurosteroids. In AD brains, a general trend was observed towards decreased levels of all steroids, with significantly lower amounts of pregnenolone and dehydroepiandrosterone. These lower levels correlated with increased amounts of ß-amyloid peptides and phosphorylated tau proteins. It is not known whether these neurosteroid deficiencies contribute to or result from AD pathology, but since many neurosteroids have neuroprotective actions, their lower levels may contribute to Aß neurotoxicity (Wang et al., 2007a). Studies have indicated that allopregnanolone is the most effective neurosteroid for neuroprotection. Allopregnanolone also stimulated neurogenesis by increasing expression of genes that promote mitosis and inhibiting expression of those that repress cell proliferation, and may be a promising therapy for promoting cellular regeneration in AD and other neurodegenerative disorders (Wang et al., 2007a).

#### 9.3.1.1 AD, Oxidative Stress and Lipid Peroxidation

mRNA expression of pro-inflammatory $sPLA_2$ IIA was up-regulated in AD brains compared to non-dementia elderly brains. $sPLA_2$ IIA immunoreactive astrocytes in AD hippocampus were associated with Aß plaques (Moses et al., 2006; Adibhatla and Hatcher 2008b). $sPLA_2$ could contribute to lipid peroxidation through the release of ArAc. Studies demonstrating increased lipid peroxidation in AD support a role for oxidative damage in this disorder (Williams et al., 2006). HNE and acrolein levels were increased in the brain tissue from patients with mild cognitive disorder and early AD, indicating that lipid peroxidation occurs early in the pathogenesis of AD (Williams et al., 2006). Acrolein, by far the strongest electrophile among all α,ß-unsaturated aldehydes, reacts with DNA bases to form cyclic adducts, the major exocyclic adduct being acrolein-deoxyguanosine, which was elevated in brain tissue from AD patients (Liu et al., 2005). ROS may also play a role in amyloid deposition in AD as oxidizing conditions cause protein cross-linking and aggregation of Aß peptides, and also contribute to tau protein aggregation (Mariani et al., 2005). Aß aggregation stimulates ROS production, which may lead to cyclic or self-propagating oxidative damage. The DHA metabolite neuroprotectin D1 promoted neuronal survival and anti-apoptotic pathways that attenuated Aß42 neurotoxicity (Lukiw et al., 2005). AD and mild cognitive disorder subjects also showed lower levels of antioxidant defense systems.

A number of questions remain unanswered regarding development of therapies for AD. Are the familiar and sporadic forms of AD distinct? Are the animal models good enough to provide clear answers? Mouse models are based on familial AD (the rare form of the disease) and may not model the common sporadic form. Furthermore, most models do not exhibit the extent of neurodegeneration seen in AD patients (Mandavilli, 2006). The triple transgenic mouse (APP/PS1/tau) studies raise the possibility that a multi-targeted approach (i.e. simultaneously targeting Aß and tau) may provide the most significant clinical benefit for the treatment of AD (Oddo et al., 2004).

### *9.3.2 Parkinson's Disease (PD)*

PD is characterized by selective degeneration of dopaminergic neurons of the substantia nigra, resulting in bradykinesis, tremor and rigidity. Free radical generation and lipid peroxidation play a significant role in PD. One of the factors responsible for this is believed to be phospholipases activation in substantia nigra, supported by the fact that $cPLA_2$ deficient mice are resistant to 1-methyl-4-phenyl-1,2,3,6-tetrahydropyridine (MPTP) induced neurotoxicity (Farooqui et al., 2006; Adibhatla and Hatcher 2008b). The MPTP metabolite $MPP^+$ is taken up by nigrostriatal neurons where it inhibits mitochondrial oxidative phosphorylation and causes neuronal death. MPTP neurotoxicity has been used as an animal model for PD. Although MPTP produces virtually all the symptoms of PD, strictly it is not PD.

#### 9.3.2.1 Treatment of PD

The dopamine prodrug levodopa remains the treatment option for PD, however, long-term levodopa therapy leads to dyskinesia. Alternatives for early PD therapy include monoamine oxidase B inhibitors, dopamine agonists, catechol-*O*-methyltransferase (COMT) inhibitors, and amantadine (Hauser and Zesiewicz, 2007). The mechanism of action of amantadine remains unknown; however, it has been suggested to have anticholinergic properties in addition to acting as an NMDA receptor antagonist to increase dopamine release and inhibit its reuptake. COMT inhibitors are used in combination with levodopa, and act peripherally to increase the pool of available levodopa, optimize its transport to CNS, and decrease the side effects of levodopa by allowing lower doses (Bonifacio et al., 2007). CDP-choline increases tyrosine hydroxylase activity and has been used in combination with levodopa for PD treatment. CDP-choline showed functional improvements and allowed levodopa to be administered at lower doses, deceasing its side effects (Adibhatla and Hatcher, 2005). The neurosteroid pregnenolone enhances neuronal dopamine release and may provide a therapeutic option for PD.

#### 9.3.2.2 PD, Oxidative Stress and Lipid Peroxidation

In PD, the accelerated metabolism of dopamine by monoamine-oxidase-B may result in excessive ROS formation. A role for oxidative stress in PD was demonstrated by marked increases in 8-hydroxy-2'-deoxyguanosine, a hydroxyl radical-damaged guanine nucleotide. Several markers of lipid peroxidation were also found to be significantly increased in PD brain regions (Mariani et al., 2005).

#### 9.3.2.3 PD and α-Synuclein

PD is associated with the presence of Lewy bodies containing insoluble aggregates of α-synuclein in association with other proteins. Recently the presence of PUFA was linked to the appearance of soluble oligomers of α-synuclein that ultimately promote the formation of insoluble α-synuclein aggregates (Welch and Yuan, 2003). DHA was shown to stimulate oligomerization of α-synuclein, and DHA levels were elevated in PD brains (Sharon et al., 2003), suggesting that DHA could have a role in formation of α-synuclein aggregates. On the other hand, DHA reduced levodopa-induced dyskinesias in MPTP-treated monkeys (Samadi et al., 2006). This discrepancy for the role of DHA in PD warrants further investigation.

### *9.3.3 Niemann-Pick Diseases (NPD)*

NPD are genetic pediatric neurodegenerative conditions characterized by specific disorders in lipid metabolism and are categorized as types A, B and C. NP Types A and B (NPA and NPB) are caused by deficiencies in acidic sphingomyelinase (ASMase) (Schuchman, 2007, and references cited therein). NPA, the most common type, is caused by a nearly complete lack of ASMase and is characterized by jaundice, an enlarged liver, and profound brain damage. Since ASMase is localized to the lysosomes, NPA results in accumulation of SM and is classified as a lysosomal storage disorder (Futerman and van Meer, 2004). People with NPB have approximately 10% of the normal level of ASMase, generally have little or no neurologic involvement. Tricyclodecan-9-yl potassium xanthate (D609), a widely known PC-phospholipase C inhibitor, also inhibits SM synthase (Larsen et al., 2007) and may help prevent accumulation of SM in NPB, a possibility not yet tested.

NP type C (NPC) is always fatal but it differs from NPA and NPB at the biochemical and genetic level. NPC is caused by mutations in either the *NPC1* or *NPC2* genes (Vance, 2006). While the precise functions of the NPC1 and NPC2 proteins are not clear, these are involved in transport of lipids, particularly cholesterol, from the late endosomes/lysosomes (Maxfield and Tabas, 2005). Deficiencies in these proteins result in lysosomal accumulation of cholesterol and other lipids. In NPC, cholesterol accumulates in all tissues except the brain, where cholesterol levels decrease with age (Vance, 2006; Xie et al., 2000). Since 70–80% of cholesterol in the brain is contained in myelin, the extensive demyelination that occurs in NPC probably accounts for the net loss of cholesterol in the brain, which would likely mask accumulation of cholesterol in neurons or astrocytes. In a mouse model of NPC, significant neuronal accumulation of cholesterol was shown by post-natal day 9 when only mild signs of neurodegeneration were detectable (Reid et al., 2004).

Currently there are no treatments for NPC. Ezetimibe (Garcia-Calvo et al., 2005; Temel et al., 2007) is a novel cholesterol-lowering drug that acts at the brush border of the small intestine where it inhibits the absorption of cholesterol from the diet. Specifically, it appears to bind to a critical mediator of cholesterol absorption, the Niemann-Pick C1-Like 1 (NPC1L1) protein on the gastrointestinal tract epithelial cells as well as in hepatocytes. Despite the accumulation of cholesterol in late endosomes/lysosomes, neither cholesterol lowering agents nor dietary measures slowed the progression of the disease. These studies suggest that cholesterol accumulation *per se* is not the major contributor to the pathogenesis of NPC, but that disrupted cholesterol transport within the cell to the endoplasmic reticulum and mitochondria for cholesterol esterification and synthesis of neurosteroids may be the critical factor. Neurosteroids and enzymes involved in steroid synthesis were significantly reduced in NPC1-deficient mice. Administration of the neurosteroid allopregnanolone alleviated progression of the disease (Vance, 2006), suggesting that neurosteroid therapy might be a treatment option for NPC (Burns and Duff, 2004; Griffin et al., 2004).

In addition to accumulating cholesterol, NPC1-deficient cells also accumulate gangliosides and other glycosphingolipids, and neuropathological abnormalities in NPC disease closely resemble those seen in primary gangliosidoses. Treatment of NPC mice with N-butyldeoxynojirimycin, an inhibitor of glycosphingolipid synthesis, increased the average life span, and reduced ganglioside accumulation and neuropathological changes (Vance, 2006).

### *9.3.4 Multiple Sclerosis (MS)*

MS is an inflammatory demyelinating autoimmune disease affecting the CNS. In MS, the immune system attacks the myelin sheath of nerve cell fibers in the brain and spinal cord. MS is predominantly a T-lymphocyte mediated disorder, and cytokines may therefore have a key role in the pathogenesis of the disease. MS is the only neurological disorder where therapeutic manipulation of the cytokine system influences development of the disease (Adibhatla and Hatcher, 2007). Thiobarbituric acid reactive substances and $F_2$-isoprostane levels were shown to be elevated in CSF of MS patients, and HNE was associated with MS lesions, indicative that lipid peroxidation also occurs in MS (Carlson and Rose, 2006).

#### 9.3.4.1 Experimental Autoimmune Encephalomyelitis (EAE)

EAE is the immune response to immunization with myelin antigens (Marusic et al., 2005) and is an animal model for MS. Recent studies demonstrated a key role for $cPLA_2$ in EAE (Marusic et al., 2005, and references cited therein; Adibhatla and Hatcher 2008b). $cPLA_2$, which can be induced by TNF-$\alpha$ (Kronke and Adam-Klages, 2002), was highly expressed in EAE lesions. Blocking $cPLA_2$ showed a remarkable decrease in both the onset and progression of the disease (Marusic et al., 2005), indicating that $cPLA_2$ has a significant role in both the induction and effector phases of EAE. A second study showed that $cPLA_2$ null mice were resistant to EAE (Marusic et al., 2005). It should be noted that these studies were conducted using C57BL/6 or SV127 mouse strains which are naturally deficient in inflammatory $PLA_2$/$sPLA_2$ IIA (Adibhatla and Hatcher, 2006). Treatment of EAE rats with $sPLA_2$ inhibitor CHEC-9 (CHEA-SAAQC) significantly attenuated $sPLA_2$ activity, EAE symptoms, and ED-1 positive microglia/macrophages (Cunningham et al., 2006; Adibhatla and Hatcher 2008b). Recently, MS patients were shown to have elevated $sPLA_2$ activity. These studies suggest that both $cPLA_2$ as well as $sPLA_2$ inhibition may be treatment options for MS.

### *9.3.5 Huntington's Disease (HD)*

HD is a rare inherited neurological disorder characterized by abnormal body movements and lack of coordination; cognition may also be affected. HD is caused by a trinucleotide repeat expansion in the Huntingtin (*Htt*) gene. The normal gene has fewer than 36 repeats, whereas the mutated *Htt* gene has 40 or more CAG repeats. Since CAG is the codon for glutamine, HD is one of the polyglutamine disorders.

Endocannabinoids, endogenous agonists of cannabinoid receptors, are comprised of amides, esters and ethers of long chain PUFA. *N*-arachidonoylethanolamine (AEA, anandamide) and 2-arachidonylglycerol (2-AG) are well characterized lipid mediators of the endocannabinoid system (Maccarrone et al., 2007). The endocannabinoid system has been found to have an important neuroprotective role in CNS injury and neurodegenerative diseases; an extensive review of these studies was recently published (Pacher et al., 2006).

Endocannabinoids act as retrograde messengers, and upon release from postsynaptic neurons, regulate further neurotransmitter release by activating presynaptic cannabinoid receptors (Degroot and Nomikos, 2007). The endocannabinoid system is *hypoactive* in HD (Maccarrone et al., 2007), which may underlie the neurotransmission abnormalities of HD and may be the cause of the clinical manifestations of the disease. Inhibition of fatty acid amide hydrolase, monoacylglycerol lipase or the endocannabinoid membrane transporter can enhance endocannabinoid levels and counteract neurochemical deficits and the *hyperkinetic* effects of HD (Maccarrone et al., 2007). Dietary supplementation with essential fatty acids protected against motor deficits in a transgenic mouse model of HD (Clifford et al., 2002). Ethyl-eicosapentaenoate (Ethyl-EPA, LAX-101 or Miraxion) showed promise in clinical trials and its action is presumed to be through JNK pathway (Puri et al., 2005). HD is associated with up-regulated transglutaminase activity in selectively vulnerable brain regions and transglutaminase-catalyzed cross-links co-localize with hunttingtin (htt) protein aggregates (Muma, 2007). Combination therapy using minocycline and coenzyme $Q_{10}$ ($CoQ_{10}$) in R6/2 transgenic HD mouse model also provided synergistic benefit (minocycline attenuated microglia proliferation and $CoQ_{10}$ reduced htt protein aggregation) (Stack et al., 2006).

#### 9.3.5.1 HD and Lipid Peroxidation

In HD, one study reported no increase in 8-hydroxy-2'-deoxyguanosine or other markers of DNA oxidation, and no change in lipid peroxidation. In contrast, other studies have shown increases in the lipid peroxidation markers $F_2$-isoprostane and MDA in HD (Mariani et al., 2005).

### *9.3.6 Amyotrophic Lateral Sclerosis (ALS)*

ALS is an adult-onset neurodegenerative disease characterized by progressive loss of spinal cord and cortical motorneurons and is usually fatal within 2–5 years of diagnosis. Approximately 10% of ALS cases are familial (inherited), with the remaining 90% of cases being sporadic in origin (Kunst, 2004). Of the familial cases, approximately 20% (i.e., 2% of all ALS) are due to mutations in the gene for the cytosolic copper-zinc superoxide dismutase (SOD1), which detoxifies superoxide anion radicals to hydrogen peroxide. Expression of a mutant SOD1 protein, with or without residual SOD activity, is necessary to cause ALS phenotype, suggesting a dominant negative mechanism (Kunst, 2004). There is strong evidence that the toxicity of mutant SOD1 in ALS is not due to loss of activity, but to the gain of one or more toxic functions that are independent of SOD activity (Nirmalananthan and Greensmith, 2005, and references cited therein). It is believed that mutant SOD1 stimulates oxidative stress and induces mitochondrial dysfunction, excitotoxicity, inflammation, and protein aggregation.

The role of glutamate-mediated excitotoxicity in ALS was supported by the efficacy of riluzole, (a benzothiazole derivative that acts by reducing glutamate excitotoxicity) in slowing the progression of ALS (Kunst, 2004). Riluzole is the only available treatment for ALS, but is now known to have limited therapeutic benefits with minimal effects on survival (Nirmalananthan and Greensmith, 2005). Several inflammatory markers such as caspase 1, COX-2 and TNF-$\alpha$ are increased in spinal cord tissue in transgenic mouse models of ALS. Inhibition of COX-2 reduced spinal neurodegeneration and prolonged the survival of ALS transgenic mice (Minghetti, 2004). The role of COX-2 in ALS as well as the presence of TNF-$\alpha$ suggests that $cPLA_2$ and/or $sPLA_2$ may also be up-regulated in ALS to provide ArAc to the COX pathway. TNF-$\alpha$ induces $cPLA_2$, $sPLA_2$ as well as COX-2. This suggests that anti-TNF-α therapy could attenuate the progression of ALS, an option that has not been utilized yet.

#### 9.3.6.1 ALS and Lipid Peroxidation

Evidence of increased oxidative DNA damage in ALS was indicated by elevated levels of 8-hydroxy-2'-deoxyguanosine in plasma, urine and CSF (Mariani et al., 2005). Several studies have shown increased lipid peroxidation and DNA damage in transgenic mice expressing mutant SOD1 and in neural tissue or sera from ALS patients (Agar and Durham, 2003; Simpson et al., 2004).

### *9.3.7 Schizophrenia and Bipolar Disorders*

Schizophrenia is marked by disturbances in thinking, emotional reactions, social behavior, with delusions and hallucinations. Drugs that block dopamine receptors alleviate symptoms of schizophrenia, indicative of excess dopaminergic function, while agents that block glutamate receptors induce some of the symptoms of schizophrenia in otherwise normal persons (Horrobin, 2002).

Recent theories on the neurological deficits of schizophrenia have focused on abnormalities in phospholipid metabolism, particularly increased activity of $PLA_2$ enzymes and reduced activity of the system which incorporates PUFA into phospholipids (a simultaneous increase in phospholipid hydrolysis and decrease in synthesis) (Berger et al., 2006; Horrobin, 2002). Neither abnormality alone produces schizophrenia but the presence of both does. These abnormalities lead to changes in membrane structure and thus the function of membrane-bound proteins, availability of cell signaling molecules, and the behavior of neurotransmitter systems. This hypothesis is supported by animal studies demonstrating that application of $PLA_2$ into the brain produces alterations in the dopamine system (Horrobin, 2002). Also, since phospholipid metabolism has a crucial role in neuronal and synaptic growth and remodeling, it is plausible that defects in this system result in failure of normal neurodevelopment in schizophrenia. There is also evidence that schizophrenia is associated with alterations in lipid transport proteins and membrane phospholipid composition (increase in PS and decrease in PC and PE) (Berger et al., 2006). Genome studies have found that several genes involved in myelination have decreased expression levels in schizophrenia (Berger et al., 2006, and reference cited therein).

A number of reports indicate that at least a portion of schizophrenic patients have reduced levels of PUFA, particularly ArAc and DHA, in red cell phospholipids, with low levels particularly associated with negative symptoms (Horrobin, 2002). ArAc, DHA and EPA are important for monoaminergic neurotransmission, brain development, and synaptic functioning (Berger et al., 2006). This suggests that supplementation with essential fatty acids could alleviate symptoms of schizophrenia. In preliminary studies, however, DHA essentially had no effect and ArAc appeared to worsen symptoms in some schizophrenia patients. Unexpectedly, EPA provided significant improvement, comparable in magnitude to that produced by new atypical antipsychotic drugs, without any of the side effects characteristic of drug treatment. The combination of EPA and DHA was also beneficial in bipolar disorder (Horrobin, 2002).

Alterations in glutamatergic and GABAergic neurotransmitter systems have been implicated in various psychiatric disorders including schizophrenia and bipolar disorder. A number of neurosteroids exhibit the capacity to modulate excitatory and inhibitory neurotransmitter systems in the brain: allopregnanolone is a potent modulator of $GABA_A$ receptors and demonstrates anticonvulsant actions in seizure paradigms. Pregnenalone sulfate and DHEA modulate $GABA_A$ and NMDA receptors. In schizophrenia and bipolar disorder, pregnenolone and DHEA levels were elevated compared to control subjects, while allopregnanolone levels decreased in schizophrenia, suggesting alterations in pregnenolone metabolism. Thus neurosteroids may be modulators of the pathophysiology of schizophrenia and bipolar disorder, and relevant to treatment of these disorders (Marx et al., 2006a).

### *9.3.8 Epilepsy*

Epilepsy is a neurological disorder characterized by recurrent spontaneous seizures due to an imbalance between cerebral excitability and inhibition, with a tendency towards uncontrolled excitability (Papandreou et al., 2006). Recurrent severe seizures can lead to death of brain cells. Phenytoin (Dilantin, Phenytek) is a widely used anti-seizure medicine (LaRoche, 2007). The primary site of action appears to be the motor cortex where spread of seizure activity is inhibited, possibly by promoting sodium efflux from neurons. Phenytoin tends to stabilize the threshold against hyper-excitability caused by excessive stimulation. The current status of new (second generation) anti-epileptic drugs has been recently reviewed (Bialer et al., 2007).

The ketogenic diet is an established and effective non-pharmacological symptomatic treatment for epilepsy that has been in clinical use for more than 80 years (Bough and Rho, 2007). The ketogenic diet is a high fat, low protein and low carbohydrate diet in which >90% of calories are derived from fat and dietary availability of glucose is minimal. The hallmark feature of the ketogenic diet is production of ketone bodies (ß-hydroxybutyrate, acetoacetate, and acetone) in the liver with a concomitant rise in plasma levels. Since glucose (the preferred source of energy, particularly in the brain) is severely restricted, the ketone bodies are used as the energy source in extrahepatic tissues (Gasior et al., 2006). Despite its many years of use, there is still considerable debate over how the ketogenic diet works; several hypotheses have been advanced, but none are widely accepted. One hypothesis, which arrived out of the Epilepsy and Brain Mapping's research, suggests that ketosis, dehydration and acidosis each appear to play a role, and that there are alterations in (1) acid-base balance;

(2) water and electrolyte distribution; (3) lipid concentration (4) brain energy reserve or (5) a central action of ketones on the brain.

## 9.4 Summary and Perspective

Historically, the lipid field has been less prominent in neuroscience. Recent advances have demonstrated that lipids have broad information carrying functions in the CNS as both ligands and substrates for proteins. Lipids alter the geometric properties of membranes and control protein traffic, and provide messenger molecules that mediate communication between cells, suggesting that advances in our understanding of lipid metabolism could have far reaching implications in other genomic, proteomic and metabolomic fields (Feng and Prestwich, 2006; Piomelli et al., 2007). Lipidomic analyses together with RNA silencing (Aagaard and Rossi, 2007) may provide a powerful tool to elucidate the specific roles of lipid intermediates in cell signaling. A deeper knowledge of the complexity of lipid signaling will elevate our understanding of the role of lipid metabolism in various CNS disorders, opening new opportunities for drug development and therapies for neurological diseases.

**Acknowledgments** This work was supported by grants from NIH/NINDS (NS42008), American Heart Association Greater Midwest Affiliate Grant-in-Aid (0655757Z), UW-School of Medicine and Public Health Research Committee (161-PRJ13MX), UW-School of Medicine and Public Health and UW-Graduate school, and laboratory resources provided by William S. Middleton VA Hospital (to RMA).

## References

Aagaard, L. and Rossi, J. J. RNAi therapeutics: Principles, prospects and challenges. *Adv Drug Deliv Rev* 59 (2007) 75–86.

Adibhatla, R. M. and Hatcher, J. F. Cytidine 5'-diphosphocholine (CDP-choline) in stroke and other CNS disorders. *Neurochem Res* 30 (2005) 15–23.

Adibhatla, R. M. and Hatcher, J. F. Phospholipase $A_2$, reactive oxygen species, and lipid peroxidation in cerebral ischemia. *Free Radic Biol Med* 40 (2006) 376–387.

Adibhatla, R. M. and Hatcher, J. F. Role of lipids in brain injury and diseases. *Future Lipidol* 2 (2007) 403–422.

Adibhatla, R. M. and Hatcher, J. F. Integration of cytokine biology and lipid metabolism in stroke. *Front Biosci* 13 (2008) 1250–1270.

Adibhatla, R. M. and Hatcher, J. F. An odyssey of plaque to stroke: a lipid perspective. www.athero.org (2008a).

Adibhatla, R. M. and Hatcher, J. F. Phospholipase $A_2$, reactive oxygen species, and lipid peroxidation in CNS pathologies. *BMB Reports* (in press, 2008b).

Adibhatla, R. M., Hatcher, J. F. and Dempsey, R. J. Lipids and lipidomics in brain injury and diseases. *AAPS J* 8 (2006a) E314–E321.

Adibhatla, R. M., Hatcher, J. F., Larsen, E. C., Chen, X., Sun, D. and Tsao, F. CDP-choline significantly restores the phosphatidylcholine levels by differentially affecting phospholipase $A_2$ and CTP-phosphocholine cytidylyltransferase after stroke. *J Biol Chem* 281 (2006b) 6718–6725.

Agar, J. and Durham, H. Relevance of oxidative injury in the pathogenesis of motor neuron diseases. *Amyotroph Lateral Scler Other Motor Neuron Disord* 4 (2003) 232–242.

Ariza, M., Pueyo, R., Matarin, M. del M., Junque, C., Mataro, M., Clemente, I., Moral, P., Poca, M. A., Garnacho, A. and Sahuquillo, J. Influence of APOE polymorphism on cognitive and behavioural outcome in moderate and severe traumatic brain injury. *J Neurol Neurosurg Psychiatry* 77 (2006) 1191–1193.

Berger, G. E., Smesny, S. and Amminger, G. P. Bioactive lipids in schizophrenia. *Int Rev Psychiatry* 18 (2006) 85–98.

Bialer, M., Johannessen, S. I., Kupferberg, H. J., Levy, R. H., Perucca, E. and Tomson, T. Progress report on new antiepileptic drugs: A summary of the Eigth Eilat Conference (EILAT VIII). *Epilepsy Res* 73 (2007) 1–52.

Bonifacio, M. J., Palma, P. N., Almeida, L. and Soares-da-Silva, P. S. Catechol-O-methyltransferase and its inhibitors in Parkinson's disease. *CNS Drug Rev* 13 (2007) 352–379.

Bough, K. J. and Rho, J. M. Anticonvulsant mechanisms of the ketogenic diet. *Epilepsia* 48 (2007) 43–58.

Boullier, A., Friedman, P., Harkewicz, R., Hartvigsen, K., Green, S. R., Almazan, F., Dennis, E. A., Steinberg, D., Witztum, J. L. and Quehenberger, O. Phosphocholine as a pattern recognition ligand for CD36. *J Lipid Res* 46 (2005) 969–976.

Bratton, D. L. and Henson, P. M. Autoimmunity and apoptosis: Refusing to go quietly. *Nat Med* 11 (2005) 26–27.

Burns, M. P. and Duff, K. Brain on steroids resists neurodegeneration. *Nat Med* 10 (2004) 675–676.

Caballero, J. and Nahata, M. Do statins slow down Alzheimer's disease? A review. *J Clin Pharm Ther* 29 (2004) 209–213.

Carlson, N. G. and Rose, J. W. Antioxidants in multiple sclerosis: Do they have a role in therapy? *CNS Drugs* 20 (2006) 433–441.

Chang, M.-K., Binder, C. J., Miller, Y. I., Subbanagounder, G., Silverman, G. J., Berliner, J. A. and Witztum, J. L. Apoptotic cells with oxidation-specific epitopes are immunogenic and proinflammatory. *J Exp Med* 200 (2004) 1359–1370.

Clifford, J. J., Drago, J., Natoli, A. L., Wong, J. Y. F., Kinsella, A., Waddington, J. L. and Vaddadi, K. S. Essential fatty acids given from conception prevent topographies of motor deficit in a transgenic model of Huntington's disease. *Neuroscience* 109 (2002) 81–88.

Crack, P. J. and Taylor, J. M. Reactive oxygen species and the modulation of stroke. *Free Radic Biol Med* 38 (2005) 1433–1444.

Cunningham, T. J., Yao, L., Oetinger, M., Cort, L., Blankenhorn, E. P. and Greensteinm, J. I. Secreted phospholipase $A_2$ activity in experimental autoimmune encephalomyelitis and multiple sclerosis. *J Neuroinflammation* (2006) 3 doi: 10.1186/742-2094-3-26.

Degroot, A. and Nomikos, G. G. In vivo neurochemical effects induced by changes in endocannabinoid neurotransmission. *Curr Opin Pharmacol* 7 (2007) 62–68.

Djebaili, M., Guo, Q., Pettus, E. H., Hoffman, S. W. and Stein, D. G. The neurosteroids progesterone and allopregnanolone reduce cell death, gliosis, and functional deficits after traumatic brain injury in rats. *J Neurotrauma* 22 (2005) 106–118.

Dringen, R. Metabolism and functions of glutathione in brain. *Prog Neurobiol* 62 (2000) 649–671.

Ehehalt, R., Keller, P., Haass, C., Thiele, C. and Simons, K. Amyloidogenic processing of the Alzheimer β-amyloid precursor protein depends on lipid rafts. *J Cell Biol* 160 (2003) 113–123.

Elkind, M. S. Inflammation, atherosclerosis, and stroke. *Neurologist* 12 (2006) 140–148.

Emsley, H. C. A. and Tyrrell, P. J. Inflammation and infection in clinical stroke. *J Cereb Blood Flow Metab* 22 (2002) 1399–1419.

Farooqui, A. A., Ong, W.-Y. and Horrocks, L. A. Inhibitors of brain phospholipase $A_2$ activity: Their neuropharmacological effects and therapeutic importance for the treatment of neurologic disorders. *Pharmacol Rev* 58 (2006) 591–620.

Feng, L. and Prestwich, G. D. (2006) Functional Lipidomics. CRC Press, Taylor & Francis Group, Boca Raton

Futerman, A. H. and van Meer, G. The cell biology of lysosomal storage disorders. *Nat Rev Mol Cell Biol* 5 (2004) 554–565.

Gao, S., Zhang, R. L., Greenberg, M. E., Sun, M., Chen, X., Levison, B. S., Salomon, R. G. and Hazen, S. L. Phospholipid hydroxyalkenals, a subset of recently discovered endogenous CD36 ligands, spontaneously generate novel furan-containing phospholipids lacking CD36 binding activity *in vivo*. *J Biol Chem* 281 (2006) 31298–31308.

Garcia-Calvo, M., Lisnock, J., Bull, H. G., Hawes, B. E., Burnett, D. A., Braun, M. P., Crona, J. H., Davis, H. R., Jr., Dean, D. C., Detmers, P. A., Graziano, M. P., Hughes, M., MacIntyre, D. E., Ogawa, A., O'Neill, K. A., Iyer, S. P. N., Shevell, D. E., Smith, M. M., Tang, Y. S., Makarewicz, A. M., Ujjainwalla, F., Altmann, S. W., Chapman, K. T. and Thornberry, N. A. The target of ezetimibe is Niemann-Pick C1-Like 1 (NPC1L1). *Proc Natl Acad Sci USA* 102 (2005) 8132–8137.

Gasior, M., Rogawski, M. A. and Hartman, A. L. Neuroprotective and disease-modifying effects of the ketogenic diet. *Behav Pharmacol* 17 (2006) 431–439.

Ghesquiere, S. A. I., Gijbels, M. J. J., Anthonsen, M. W., van Gorp, P. J. J., van der Made, I., Johansen, B., Hofker, M. H. and de Winther, M. P. J. Macrophage-specific overexpression of group IIa $sPLA_2$ increases atherosclerosis and enhances collagen deposition. *J Lipid Res* 46 (2005) 201–210.

Griffin, L. D., Gong, W., Verot, L. and Mellon, S. H. Niemann-Pick type C disease involves disrupted neurosteroidogenesis and responds to allopregnanolone. *Nature Med* 10 (2004) 704–711.

Grimm, M. O. W., Grimm, H. S., Patzold, A. J., Zinser, E. G., Halonen, R., Duering, M., Tschape, J.-A., Strooper, B. D., Muller, U., Shen, J. and Hartmann, T. Regulation of cholesterol and sphingomyelin metabolism by amyloid-β and presenilin. *Nat Cell Biol* 7 (2005) 1118–1123.

Hall, E. D. and Springer, J. E. Neuroprotection and acute spinal cord injury: A reappraisal. *NeuroRx* 1 (2004) 80–100.

Hansson, G. K. and Libby, P. The immune response in atherosclerosis: A double-edged sword. *Nat Rev Immunol* 6 (2006) 508–519.

Hartmann, T., Kuchenbecker, J. and Grimm, M. O. W. Alzheimer's disease: The lipid connection. *J Neurochem* 103 (2007) 159–170.

Hauser, R. A. and Zesiewicz, T. A. Advances in the pharmacologic management of early Parkinson disease. *Neurologist* 13 (2007) 126–132.

Herrup, K., Neve, R., Ackerman, S. L. and Copani, A. Divide and die: Cell cycle events as triggers of nerve cell death. *J Neurosci* 24 (2004) 9232–9239.

Horrobin, D. (2002). The lipid hypothesis of schizophrenia. In E. R. Skinner (eds.), *Brain Lipids and Disorders in Biological Psychiatry*, pp. 39–52. Amsterdam: Elsevier Science.

Jellinger, K. A. Head injury and dementia. *Curr Opin Neurol* 17 (2004) 719–723.

Kadl, A., Bochkov, V. N., Huber, J. and Leitinger, N. Apoptotic cells as sources for biologically active oxidized phospholipids. *Antioxid Redox Signal* 6 (2004) 311–320.

Karishma, K. K. and Herbert, J. Dehydroepiandrosterone (DHEA) stimulates neurogenesis in the hippocampus of the rat, promotes survival of newly formed neurons and prevents corticosterone-induced suppression. *Eur J Neurosci* 16 (2002) 445–453.

Kinnunen, P. K. J. and Holopainen, J. M. Sphingomyelinase activity of LDL – A link between atherosclerosis, ceramide, and apoptosis? *Trends Cardiovasc Med* 12 (2002) 37–42.

Kougias, P., Chai, H., Lin, P. H., Lumsden, A. B., Yao, Q. and Chen, C. Lysophosphatidylcholine and secretory phospholipase $A_2$ in vascular disease: Mediators of endothelial dysfunction and atherosclerosis. *Med Sci Monit* 12 (2006) RA5–16.

Kronke, M. and Adam-Klages, S. Role of caspases in TNF-mediated regulation of $cPLA_2$. *FEBS Lett* 531 (2002) 18–22.

Kunst, C. B. Complex genetics of amyotrophic lateral sclerosis. *Am J Hum Genet* 75 (2004) 933–947.

Kunz, A., Anrather, J., Zhou, P., Orio, M. and Iadecola, C. Cyclooxygenase-2 does not contribute to postischemic production of reactive oxygen species. *J Cereb Blood Flow Metab* 27 (2007) 545–551.

Lapchak, P. A., Chapman, D. F., Nunez, S. Y. and Zivin, J. A. Dehydroepiandrosterone sulfate is neuroprotective in a reversible spinal cord ischemia model: Possible involvement of $GABA_A$ receptors. *Stroke* 31 (2000) 1953–1956.

LaRoche, S. M. A new look at the second-generation antiepileptic drugs: A decade of experience. *Neurologist* 13 (2007) 133–139.

Larsen, E. C., Hatcher, J. F. and Adibhatla, R. M. Effect of tricyclodecan-9-yl potassium xanthate (D609) on phospholipid metabolism and cell death during oxygen-glucose deprivation in PC12 cells. *Neuroscience* 146 (2007) 946–961.

Lavi, S., McConnell, J. P., Rihal, C. S., Prasad, A., Mathew, V., Lerman, L. O. and Lerman, A. Local production of lipoprotein-associated phospholipase $A_2$ and lyso-phosphatidylcholine in the coronary circulation: Association with early coronary atherosclerosis and endothelial dysfunction in humans. *Circulation* 115 (2007) 2715–2721.

Liu, K. J. and Rosenberg, G. A. Matrix metalloproteinases and free radicals in cerebral ischemia. *Free Radic Biol Med* 39 (2005) 71–80.

Liu, X., Lovell, M. A. and Lynn, B. C. Development of a method for quantification of acrolein-deoxyguanosine adducts in DNA using isotope dilution-capillary LC/MS/MS and its application to human brain tissue. *Anal Chem* 77 (2005) 5982–5989.

Lukiw, W. J., Cui, J.-G., Marcheselli, V. L., Bodker, M., Botkjaer, A., Gotlinger, K., Serhan, C. N. and Bazan, N. G. A role for docosahexaenoic acid-derived neuroprotectin D1 in neural cell survival and Alzheimer disease. *J Clin Invest* 115 (2005) 2774–2783.

Lynch, J. R., Wang, H., Mace, B., Leinenweber, S., Warner, D. S., Bennett, E. R., Vitek, M. P., McKenna, S. and Laskowitz, D. T. A novel therapeutic derived from apolipoprotein E reduces brain inflammation and improves outcome after closed head injury. *Exp Neurol* 192 (2005) 109–116.

Maccarrone, M., Battista, N. and Centonze, D. The endocannabinoid pathway in Huntington's disease: A comparison with other neurodegenerative diseases. *Prog Neurobiol* 81 (2007) 349–379.

Mahmood, A., Lu, D., Qu, C., Goussev, A. and Chopp, M. Treatment of traumatic brain injury with a combination therapy of marrow stromal cells and atorvastatin in rats. *Neurosurgery* 60 (2007) 546–554.

Mandavilli, A. The amyloid code. *Nat Med* 12 (2006) 747–751.

Margaill, I., Plotkine, M. and Lerouet, D. Antioxidant strategies in the treatment of stroke. *Free Radic Biol Med* 39 (2005) 429–443.

Mariani, E., Polidori, M. C., Cherubini, A. and Mecocci, P. Oxidative stress in brain aging, neurodegenerative and vascular diseases: An overview. *J Chromatog B* 827 (2005) 65–75.

Marusic, S., Leach, M. W., Pelker, J. W., Azoitei, M. L., Uozumi, N., Cui, J., Shen, M. W. H., DeClercq, C. M., Miyashiro, J. S., Carito, B. A., Thakker, P., Simmons, D. L., Leonard, J. P., Shimizu, T. and Clark, J. D. Cytosolic phospholipase $A_2\alpha$-deficient mice are resistant to experimental autoimmune encephalomyelitis. *J Exp Med* 202 (2005) 841–851.

Marx, C. E., Stevens, R. D., Shampine, L. J., Uzunova, V., Trost, W. T., Butterfield, M. I., Massing, M. W., Hamer, R. M., Morrow, A. L. and Lieberman, J. A. Neuroactive steroids are altered in schizophrenia and bipolar disorder: Relevance to pathophysiology and therapeutics. *Neuropsychopharmacology* 31 (2006a) 1249–1263.

Marx, C. E., Trost, W. T., Shampine, L. J., Stevens, R. D., Hulette, C. M., Steffens, D. C., Ervin, J. F., Butterfield, M. I., Blazer, D. G., Massing, M. W. and Lieberman, J. A. The neurosteroid allopregnanolone is reduced in prefrontal cortex in Alzheimer's disease. *Biol Psych* 60 (2006b) 1287–1294.

Mattson, M. P., Cutler, R. G. and Jo, D. G. Alzheimer peptides perturb lipid-regulating enzymes. *Nat Cell Biol* 7 (2005) 1045–1047.

Maxfield, F. R. and Tabas, I. Role of cholesterol and lipid organization in disease. *Nature* 438 (2005) 612–621.

McColl, B. W., Rothwell, N. J. and Allan, S. M. Systemic inflammatory stimulus potentiates the acute phase and CXC chemokine responses to experimental stroke and exacerbates brain damage via interleukin-1- and neutrophil-dependent mechanisms. *J Neurosci* 27 (2007) 4403–4412.

Mehta, S. L., Manhas, N. and Raghubir, R. Molecular targets in cerebral ischemia for developing novel therapeutics. *Brain Res Rev* 54 (2007) 34–66.

Minghetti, L. Cyclooxygenase-2 (COX-2) in inflammatory and degenerative brain diseases. *J Neuropath Exp Neurol* 63 (2004) 901–910.

Moses, G. S., Jensen, M. D., Lue, L. F., Walker, D. G., Sun, A. Y., Simonyi, A. and Sun, G. Y. Secretory $PLA_2$-IIA: A new inflammatory factor for Alzheimer's disease. *J Neuroinflammation* 3 (2006) doi:10.1186/742-2094-3-28.

Muma, N. A. Transglutaminase is linked to neurodegenerative diseases. *J Neuropathol Exp Neurol* 66 (2007) 258–263.

Nicolo, D., Varadhachary, A. S. and Monestier, M. Atherosclerosis, antiphospholipid syndrome, and antiphospholipid antibodies. *Front Biosci* 12 (2007) 2171–2182.

Nirmalananthan, N. and Greensmith, L. Amyotrophic lateral sclerosis: Recent advances and future therapies. *Curr Opin Neurol* 18 (2005) 712–719.

Oddo, S., Billings, L., Kesslak, J. P., Cribbs, D. H. and LaFerla, F. M. Aβ immunotherapy leads to clearance of early, but not late, hyperphosphorylated tau aggregates via the proteasome. *Neuron* 43 (2004) 321–332.

Oei, H.-H. S., van der Meer, I. M., Hofman, A., Koudstaal, P. J., Stijnen, T., Breteler, M. M. B. and Witteman, J. C. M. Lipoprotein-associated phospholipase $A_2$ activity is associated with risk of coronary heart disease and ischemic stroke: The Rotterdam study. *Circulation* 111 (2005) 570–575.

Pacher, P., Batkai, S. and Kunos, G. ,The endocannabinoid system as an emerging target of pharmacotherapy. *Pharmacol Rev* 58 (2006) 389–462.

Papandreou, D., Pavlou, E., Kalimeri, E. and Mavromichalis, I. The ketogenic diet in children with epilepsy. *Br J Nutr* 95 (2006) 5–13.

Piomelli, D., Astarita, G. and Rapaka, R. A neuroscientist's guide to lipidomics. *Nat Rev Neurosci* 8 (2007) 743–754.

Puglielli, L. Aging of the brain, neurotrophin signaling, and Alzheimer's disease: Is IGF1-R the common culprit? *Neurobiol Aging* 29 (2008) 795–811 doi:10.1016/j.neurobiolaging.2007.01.010.

Puri, B. K., Leavitt, B. R., Hayden, M. R., Ross, C. A., Rosenblatt, A., Greenamyre, J. T., Hersch, S., Vaddadi, K. S., Sword, A., Horrobin, D. F., Manku, M. and Murck, H. Ethyl-EPA in Huntington disease: A double-blind, randomized, placebo-controlled trial. *Neurology* 65 (2005) 286–292.

Qin, J., Goswami, R., Balabanov, R. and Dawson, G. Oxidized phosphatidylcholine is a marker for neuroinflammation in multiple sclerosis brain. *J Neurosci Res* 85 (2007) 977–984.

Reid, P. C., Sakashita, N., Sugii, S., Ohno-Iwashita, Y., Shimada, Y., Hickey, W. F. and Chang, T.-Y. A novel cholesterol stain reveals early neuronal cholesterol accumulation in the Niemann-Pick type C1 mouse brain. *J Lipid Res* 45 (2004) 582–591.

Rigg, J. L. and Zafonte, R. D. Corticosteroids in TBI: Is the story closed? *J Head Trauma Rehab* 21 (2006) 285–288.

Samadi, P., Grégoire, L., Rouillard, C., Bédard, P. J., Di Paolo, T. and Lévesque, D. Docosahexaenoic acid reduces levodopa-induced dyskinesias in 1-methyl-4-phenyl-1,2,3,6-tetrahydropyridine monkeys. *Ann Neurol* 59 (2006) 282–288.

Sayeed, I., Guo, Q., Hoffman, S. W. and Stein, D. G. Allopregnanolone, a progesterone metabolite, is more effective than progesterone in reducing cortical infarct volume after transient middle cerebral artery occlusion. *Ann Emerg Med* 47 (2006) 381–389.

Schuchman, E. The pathogenesis and treatment of acid sphingomyelinase-deficient Niemann–Pick disease. *J Inher Metabol Dis* 30 (2007) 654–663.

Schwab, C., Hosokawa, M. and McGeer, P. L. Transgenic mice overexpressing amyloid beta protein are an incomplete model of Alzheimer disease. *Exp Neurol* 188 (2004) 52–64.

Serhan, C. N. Resolution phases of inflammation: Novel endogenous anti-inflammatory and proresolving lipid mediators and pathways. *Ann Rev Immunol* 25 (2007) 101–137.

Sharon, R., Bar-Joseph, I., Frosch, M. P., Walsh, D. M., Hamilton, J. A. and Selkoe, D. J. The formation of highly soluble oligomers of α-synuclein is regulated by fatty acids and enhanced in Parkinson's disease. *Neuron* 37 (2003) 583–595.

Simmons, D. L., Botting, R. M. and Hla, T. Cyclooxygenase isozymes: The biology of prostaglandin synthesis and inhibition. *Pharmacol Rev* 56 (2004) 387–437.

Simpson, E. P., Henry, Y. K., Henkel, J. S., Smith, R. G. and Appel, S. H. Increased lipid peroxidation in sera of ALS patients: A potential biomarker of disease burden. *Neurology* 62 (2004) 1758–1765.

Smith, C., Graham, D. I., Murray, L. S., Stewart, J. and Nicoll, J. A. R. Association of *APOE* e4 and cerebrovascular pathology in traumatic brain injury. *J Neurol Neurosurg Psychiatry* 77 (2006) 363–366.

Stack, E. C., Smith, K. M., Ryu, H., Cormier, K., Chen, M., Hagerty, S. W., Del Signore, S. J., Cudkowicz, M. E., Friedlander, R. M. and Ferrante, R. J. Combination therapy using minocycline and coenzyme Q10 in R6/2 transgenic Huntington's disease mice. *Biochim Biophys Acta* 1762 (2006) 373–380.

Stoll, G. and Bendszus, M. Inflammation and atherosclerosis: Novel insights into plaque formation and destabilization. *Stroke* 37 (2006) 1923–1932.

Suzuki, M., Wright, L. S., Marwah, P., Lardy, H. A. and Svendsen, C. N. Mitotic and neurogenic effects of dehydroepiandrosterone (DHEA) on human neural stem cell cultures derived from the fetal cortex. *Proc Natl Acad Sci USA* 101 (2004) 3202–3207.

Temel, R. E., Tang, W., Ma, Y., Rudel, L. L., Willingham, M. C., Ioannou, Y. A., Davies, J. P., Nilsson, L.-M. and Yu, L. Hepatic Niemann-Pick C1-like 1 regulates biliary cholesterol concentration and is a target of ezetimibe. *J Clin Invest* 117 (2007) 1968–1978.

Vance, J. E. Lipid imbalance in the neurological disorder, Niemann-Pick C disease. *FEBS Lett* 580 (2006) 5518–5524.

Wang, J. M., Johnston, P. B., Ball, B. G. and Brinton, R. D. The neurosteroid allopregnanolone promotes proliferation of rodent and human neural progenitor cells and regulates cell-cycle gene and protein expression. *J Neurosci* 25 (2005) 4706–4718.

Wang, J. M., Liu, L., Irwin, R. W., Chen, S. and Brinton, R. D. Regenerative potential of allopregnanolone. *Brain Res Rev* 57 (2007a) 398–409. In press: doi:10.1016/j.brainresrev.2007.08.010.

Wang, Q., Tang, X. N. and Yenari, M. A. The inflammatory response in stroke. *J Neuroimmunol* 184 (2007b) 53–68.

Welch, K. and Yuan, J. α-Synuclein oligomerization: A role for lipids? *Trends Neurosci* 26 (2003) 517–519.

Whitfield, J. F. Can statins put the brakes on Alzheimer's disease? *Expert Opin Investig Drugs* 15 (2006) 1479–1485.

Williams, T. I., Lynn, B. C., Markesbery, W. R. and Lovell, M. A. Increased levels of 4-hydroxynonenal and acrolein, neurotoxic markers of lipid peroxidation, in the brain in mild cognitive impairment and early Alzheimer's disease. *Neurobiol Aging* 27 (2006) 1094–1099.

Wojtal, K., Trojnar, M. K. and Czuczwar, S. J. Endogenous neuroprotective factors: Neurosteroids. *Pharmacol Reports* 58 (2006) 335–340.

Xie, C., Burns, D. K., Turley, S. D. and Dietschy, J. M. Cholesterol is sequestered in the brains of mice with Niemann-Pick type C disease but turnover is increased. *J Neuropathol Exp Neurol* 59 (2000) 1106–1117.

Young, A. R., Ali, C., Duretete, A. and Vivien, D. Neuroprotection and stroke: Time for a compromise. *J Neurochem* 103 (2007) 1302–1309.

Zalewski, A., Nelson, J. J., Hegg, L. and Macphee, C. Lp-PLA$_2$: A new kid on the block. *Clin Chem* 52 (2006) 1645–1650.

# Chapter 10
# Lysophospholipid Activation of G Protein-Coupled Receptors

**Tetsuji Mutoh and Jerold Chun**

**Abstract** One of the major lipid biology discoveries in last decade was the broad range of physiological activities of lysophospholipids that have been attributed to the actions of lysophospholipid receptors. The most well characterized lysophospholipids are lysophosphatidic acid (LPA) and sphingosine 1-phosphate (S1P). Documented cellular effects of these lipid mediators include growth-factor-like effects on cells, such as proliferation, survival, migration, adhesion, and differentiation. The mechanisms for these actions are attributed to a growing family of 7-transmembrane, G protein-coupled receptors (GPCRs). Their pathophysiological actions include immune modulation, neuropathic pain modulation, platelet aggregation, wound healing, vasopressor activity, and angiogenesis. Here we provide a brief introduction to receptor-mediated lysophospholipid signaling and physiology, and then discuss potential therapeutic roles in human diseases.

**Keywords** Sphingosine 1-phosphate (S1P · autoimmune diseases · transplantation · cancer · cardiovascular diseases

## 10.1 Introduction

In addition to being integral for cell membranes and essential sources of energy, lipids also have a major function as signaling mediators. Lysophospholipids (LPs), are simple lipid molecules with a wide range of important signaling effects on many different organ systems. For example, LPs can act as extracellular signaling molecules that affect cardiovascular function, immune responses, pain transmission, embryo implantation, osteogenesis, the circulatory system, and brain development. A lysophospholipid is a 3-carbon backbone phospholipid derived from glycerophospholipids or sphingolipids that contain a single chain

T. Mutoh
Department of Molecular Biology, Helen L. Dorris Child and Adolescent Neuropsychiatric Disorder Institute, The Scripps Research Institute, 10550 North Torrey Pines Rd., ICND-118, La Jolla, CA 92037, USA

P.J. Quinn and X. Wang (eds.), *Lipids in Health and Disease*, and phosphate headgroup in the first position. Examples of LPs include LPA (lysophosphatidic acid), S1P (sphingosine 1-phosphate), LPC (lysophosphatidylcholine), SPC (sphingosylphosphorylcholine), LPS (lysophosphatidylserine), and LPE (lysophosphatydilethanolamine).

Despite the high concentration of LPC in blood (several hundred μM) (Croset et al., 2000), the physiological function of LPC remains largely unknown. On the other hand, LPs with relatively low concentrations (low μM range (Aoki, 2004; Okajima, 2002)) such as LPA, S1P, and LPS have documented functions *in vivo*. In particular the LPs, LPA and S1P, as well as their signaling cascades, have been extensively studied. Because lysophospholipids have a chemical makeup that allows them to enter the lipid bilayer, it was previously thought that the effect of LPs and their mechanisms of action were largely non-specific. However, this initial view changed with the identification of specific LP receptors that were essential to the physiological functions of LPs. Cloning and functional characterization of the lysophospholipid receptors represented a significant advance towards understanding this class of lipid signals. Today, ten *bona fide* lysophospholipid receptors have been reported, 5 for LPA ($LPA_{1\text{-}5}$) and 5 for S1P ($S1P_{1\text{-}5}$), with a number of additional putative lysophospholipid G protein-coupled receptors (GPCRs) existing in the literature (Anliker and Chun, 2004; Ishii et al., 2004; Lee et al., 2006; Rivera and Chun, 2007). Recently, a specific receptor for LPS was identified, however the associated signaling cascade(s) for this receptor is not fully understood (Sugo et al., 2006). Many of the LP receptors are necessary for normal embryonic development and have roles in normal adult physiologies as well as disease processes. Furthermore, the GPCRs for specific LPs are intriguing since they are attractive targets for drug discovery.

In this chapter, we will discuss the normal physiological functions of LPA and S1P mediated by their cognate receptors. In addition, we will discuss diseases associated with these bioactive LP molecules. Although LPA belongs to the glycerophospholipid family and S1P belongs to the sphingolipid group, the amino acid sequence of their receptors is generally conserved to a significant extent and they have overlapping but distinct biological functions. Originally known in the 1900s as a lipid metabolite, LPA was reported to have physiologically active properties that functioned to control blood pressure (Sen et al., 1968; Tokumura et al., 1978). S1P was originally identified as a mitogen capable of inducing intracellular calcium mobilization via proposed intracellular mechanisms (Zhang et al., 1991). Continuing research in the field of lipid biology revealed the importance of these two LP signaling molecules *in vivo*. For instance, the phosphorylated metabolite of FTY720, FTY720-P, is an S1P analog that was discovered to be a novel immunomodulator by inducing lymphopenia via S1P receptors (Mandala et al., 2002). While a number of recent reviews have covered many facets of this rapidly growing field, the purpose of this chapter is to provide a basic overview of LP signaling and discuss how LPs are relevant to both normal physiological functions and the pathology of human diseases.

## 10.2 Biochemistry of LP Signaling

### *10.2.1 Receptor Mediated Signaling Pathways*

Initially, LPs were shown to be precursors and metabolites in the *de novo* biosynthesis of phospholipids. However, other bioactive properties were subsequently discovered. For instance, LPA was shown to function as an anti-hypertensive agent (Sen et al., 1968; Tokumura et al., 1978). LPA was also discovered to act as a cell growth and motility factor present in serum, and the signaling cascades mediated by LPA were shown to involve G proteins (van Corven et al., 1989), suggesting the involvement of GPCRs, although other GPCR-independent mechanisms were also possible in the absence of identified receptors. The first LP receptor was cloned from mouse brain cDNA by degenerate PCR with primers designed against GPCRs (Hecht et al., 1996). This receptor, originally designated VZG-1, and now called $LPA_1$, was the first LP receptor discovered. Within several years of this initial report, several members of an orphan GPCR receptor family, called "endothelial differentiation genes (Edg)," were identified as GPCRs for both lysophosphatidic acid (LPA) and sphingosine 1-phosphate (S1P) (An et al., 1997; Ishii et al., 2004; Lee et al., 1998; van Brocklyn et al., 2000). All of these LP receptors are GPCRs capable of interacting with a number of heterotrimeric G proteins. The current nomenclature reflects the receptor's cognate ligand and chronological order of the relevant receptor's identification (Chun et al., 2002; Ishii et al., 2004; Table 10.1). LP receptor genes are distributed throughout the genome and are organized in a somewhat similar fashion.

The coding regions for each of the *lpa* genes in the genomes of human and mice, with the exception of $LPA_4$, are divided between two exons, while the coding region of each *s1p* gene is contained within a single exon, with only non-coding exon(s) upstream (Contos and Chun, 2001; Contos et al., 2000b, 2002). Several structural characteristics are shared between LPA and S1P receptors, including an extracellular N-terminus, seven $\alpha$-helical THs (transmembrane

**Table 10.1** Nomenclature of lysophospholipid receptors

| IUPHAR Nomenclature | Chromosomal location (Human) | Natural Agonist Ligand | Previous Names |
|---|---|---|---|
| LPA1 | 9q32 | LPA | Edg–2, LPA1, VZG–1, REC1.3 |
| LPA2 | 19p12 | LPA | Edg–4, LPA2 |
| LPA3 | 1p22.3–p31.1 | LPA | Edg–7, LPA3 |
| LPA4 | Xq13–q21.1 | LPA | GPR23, P2Y9 |
| LPA5 | 12p 13.31 | LPA | GPR92 |
| S1P1 | 1p21 | S1P > SPC | Edg–1, LPB1 |
| S1P2 | 19p 13.2 | S1P > SPC | Edg–5, LPB2, AGR16, H218 |
| S1P3 | 9q22.1–q22.2 | S1P > SPC | Edg–3, LPB3 |
| S1P4 | 19p 13.3 | S1P > SPC | Edg–6, LPB4, LPC1 |
| S1P5 | 19p 13.2 | S1P > SPC | Edg–8, LPB5, NRG–1 |

helices), and an intracellular C-terminus (Pierce et al., 2002). Studies of LPA and S1P receptor ligand binding mechanisms suggest that several specific amino acid residues are responsible for ligand interaction, e.g., Arginine 120 in TH3 is thought to be required for ligand binding and Glutamine 121 for ligand specificity recognition (Holdsworth et al., 2004; Parrill, 2005; Parrill et al., 2000), although no formal structural data have been reported for this family of GPCRs.

Lysophospholipid receptors each have a heterogeneous spatiotemporal gene expression pattern and multiple receptors may be expressed by the same cell. These data have been derived by examining mRNA combined with functional assays. Notably, no antibodies or antisera have been clearly proven for use in immunohistochemical studies of native proteins, although many can identify overexpressed proteins in cell lines. For example, $LPA_1$ was initially called "ventricular zone gene-1 (Vzg-1)" because of its enrichment in the neural progenitor zone of the embryonic cerebral cortex, the so called "ventricular zone." In adult mice, $LPA_1$ is widely expressed with high mRNA levels in brain, lung, heart, and other organs. $LPA_1$ and $S1P_1$ expression patterns are generally similar but differ in detail in both embryonic and adult tissues. For example, $S1P_1$ is expressed in the ventricular zone throughout the embryonic telencephalon, however, $LPA_1$ gene expression is limited to the neocortical ventricular zone as stated above (Anliker and Chun, 2004; Contos et al., 2000b; Hecht et al., 1996; McGiffert et al., 2002). Most cell types express multiple LPA and S1P receptors, and each receptor can activate multiple types of downstream molecules as mentioned below. LP signaling in each cell and tissue can vary depending upon the composition and expression level of the receptor family members and their downstream molecules. In addition, ligand availability, concentration, and half-life are also likely to influence cellular responses mediated by LP receptors. Their desensitization is probably mediated by known mechanisms in other systems of phosphorylation of GPCRs by kinases and or an uncoupling from G proteins by arrestins, followed by receptor internalization and degradation (Lefkowitz and Shenoy, 2005).

LPA and S1P receptors couple to heterotrimeric G proteins, which consist of a $G_\alpha$ and the associated $G_{\beta\gamma}$ subunits. The heterotrimeric G proteins are thought to be bound to the inner surface of the cell membrane. One receptor may couple to several different types of $G_\alpha$ protein subunits to form a complex signaling network (Fig. 10.1). $LPA_{1,2,4,5}$ and $S1P_{2\text{-}5}$ all signal via $G_{\alpha 12/13}$ to activate RhoA, a member of the family of Rho GTPases. $LPA_{1\text{-}5}$ and $S1P_{2,3}$ couple to $G_{\alpha q/11}$ to activate phospholipase C (PLC). $LPA_{1\text{-}4}$ and $S1P_{1\text{-}5}$also couple with $G_{\alpha i}$ to activate PLC, Ras, Phosphoinositide-3 Kinase (PI3K), and to inhibit adenylyl cyclase (AC), but $LPA_4$ can also couple to $G_{\alpha s}$ to activate AC. When a ligand binds to the receptor, it exchanges GDP for GTP on the $G_\alpha$ subunit, and then $G_\alpha$-GTP and $G_{\beta\gamma}$ can activate the effector molecule complex for each signaling cascade (Etienne-Manneville and Hall, 2002; Neves et al., 2002).

Furthermore, several reports suggest that LP receptor signaling can involve trans effects via receptor tyrosine kinases, as seen in the synergistic interaction between S1P and platelet derived growth factor (PDGF), as well as signaling via

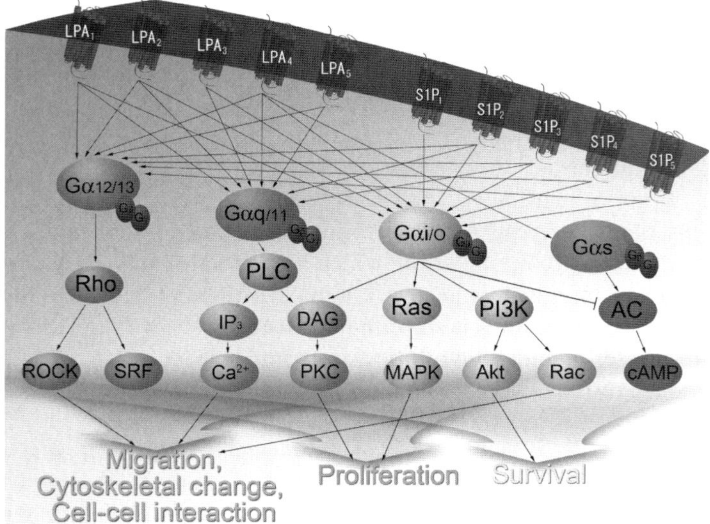


**Fig. 10.1** The network of LPA and S1P signaling through G protein-coupled receptors. Each LPA and S1P receptor couples to their specific class of G proteins. Ligand binding activates or inhibits downstream second messenger molecules, and the most prominent cellular effects are illustrated. *Rock*, Rho-associated kinase; *SRF*, serum response factor; *IP3*, inositol 1,4,5-trisphosphate; PLC, phospholipase C; DAG, diacylglycerol; *PKC*, protein kinase C; *MAPK*, mitogen-activated protein kinase; *PI3K*, phosphoinositol 3-kinase; *DAG*, diacylglycerol

other kinases such as p42/p44 MAPK or Akt activation in chemotaxis, and Erk1/2 mediated anti-apoptotic effects (Hobson et al., 2001; Pyne et al., 2007; Wong et al., 2007). In addition, PDGF, VEGF and TNF-$\alpha$ can stimulate sphingosine kinases (e.g., SPHK1) and increase S1P levels in an autocrine fashion. This has important implications for vascular maturation (Spiegel and Milstien, 2003). Interestingly, a recent report suggested a "criss-cross" transactivation between estrogen-S1P-EGFR pathways (Sukocheva et al., 2006), and that other interactions amongst GPCRs and receptor tyrosine kinases/other kinases are likely.

### *10.2.2 Variable Cellular Responses via LPA and S1P*

Before the discovery of specific LP receptors, there were multiple hypotheses proposed to explain the physiological signaling response mechanisms provoked by LPs. For instance, it was thought that LPs could act as calcium chelators, ionophores, membrane disruptors, second messengers, or act via intracellular receptors (reviewed in (Chun, 1999; Fukushima et al., 2001). Heterologous expression of cloned receptors was performed to prove that extracellular receptors mediated the LP signaling pathway. Two cell lines, RH7777 (hepatoma) and B103 (neuroblastoma) were identified, which lack endogenous responses to LPA and/or S1P and were useful in these mechanistic studies (Fukushima et al., 1998). Even though several non-GPCR mediated signaling pathways have been reported (Hooks et al., 2001; McIntyre et al., 2003), it is now clear that the dominant mechanism by which extracellular LPs function, at least in vertebrates, is through the actions of specific cell surface receptors (Chun and Rosen, 2006). In terms of pathology, there are numerous functional studies using cancer cells that show cell growth, apoptosis, invasion, cell migration, and extra-cellular matrix reorganization, which are mediated by LPs. Concomitant alterations in cell migration and invasion may further contribute to the growth of metastatic cancer. Studies with primary cells have shown that LPs influence immune responses including cytokine and chemokine secretion, platelet aggregation, smooth muscle contraction, and neurite retraction (Fig. 10.2).

#### 10.2.2.1 Cell Survival and Growth

LPA and S1P can signal via $G_i$, $G_{q/11}$, $G_{12/13}$, and Gs. In many well-documented studies using cultured cell lines, LPs function as survival factors (Ishii et al., 2004). LPA and S1P largely couple to the $G_i$ pathway that regulates PI3K and Akt, but other signals generated through different G protein pathways can also be initiated. The signaling cascades that are activated via the $G_i$ and Ras/MAPK pathway or $G_q$ and phospholipases generate second messengers that facilitate cell growth. While it is generally known that the activation of PLC, $Ca^{2+}$, or PKC signaling pathways are insufficient to promote cell proliferation, additional signaling pathways activated by LP receptors provide complementary proliferative stimuli. Signaling through the $G_{12/13}$ mediated Rho pathway also promotes cell proliferation. $G_i$-mediated signaling contributes to cell survival through PI3K/Akt (Radeff-Huang et al., 2004; Weiner et al., 2001).

#### 10.2.2.2 Cell Migration

LPs also affect cell migration of diverse normal and transformed cell types (Mills and Moolenaar, 2003). Depending on the combination of receptors expressed and downstream molecules, S1P signaling can also promote and inhibit cell migration (Okamoto et al., 2000; Sugimoto et al., 2003). $S1P_1$ is crucial for angiogenesis and lymphocyte trafficking, which is based on its ability to stimulate cell migration (Chun and Rosen, 2006). On the other hand, $S1P_2$ inhibits Rac and abolishes membrane ruffling and cell migration. This inhibition can be antagonized by concurrent $G_i$ mediated Rac activation (Sugimoto et al., 2003). Many studies have shown that $G_i$ and/or $G_{12/13}$ mediated pathways can control cell motility via changes in cytoskeletal organization (van Leeuwen et al., 2003). The actin cytoskeleton is regulated by the Rho-GTPase family: RhoA, Cdc42, and Rac. $G_i$ activates the PI3K-Rac pathway via RhoA (Neves et al., 2002). It is notable that $S1P_1$ is unique in that it only couples to $G_i$, which appears to utilize Rac signaling to promote migration via this particular receptor.

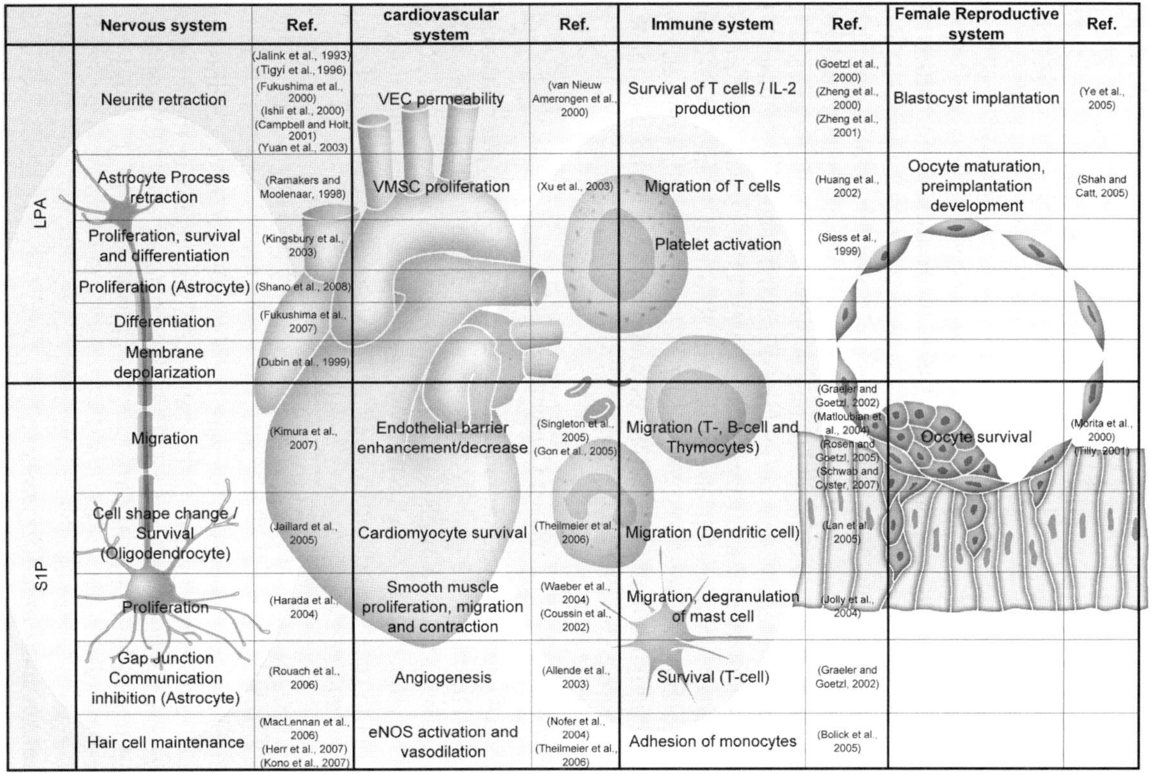

| | Nervous system | Ref. | cardiovascular system | Ref. | Immune system | Ref. | Female Reproductive system | Ref. |
|---|---|---|---|---|---|---|---|---|
| LPA | Neurite retraction | (Jalink et al., 1993) (Tigyi et al., 1996) (Fukushima et al., 2000) (Ishii et al., 2000) (Campbell and Holt, 2001) (Yuan et al., 2003) | VEC permeability | (van Nieuw Amerongen et al., 2000) | Survival of T cells / IL-2 production | (Goetzl et al., 2000) (Zheng et al., 2000) (Zheng et al., 2001) | Blastocyst implantation | (Ye et al., 2005) |
| | Astrocyte Process retraction | (Ramakers and Moolenaar, 1998) | VMSC proliferation | (Xu et al., 2003) | Migration of T cells | (Huang et al., 2002) | Oocyte maturation, preimplantation development | (Shah and Catt, 2005) |
| | Proliferation, survival and differentiation | (Kingsbury et al., 2003) | | | Platelet activation | (Siess et al., 1999) | | |
| | Proliferation (Astrocyte) | (Shano et al., 2008) | | | | | | |
| | Differentiation | (Fukushima et al., 2007) | | | | | | |
| | Membrane depolarization | (Dubin et al., 1999) | | | | | | |
| S1P | Migration | (Kimura et al., 2007) | Endothelial barrier enhancement/decrease | (Singleton et al., 2005) (Gon et al., 2005) | Migration (T-, B-cell and Thymocytes) | (Graeler and Goetzl, 2002) (Matloubian et al., 2004) (Rosen and Goetzl, 2005) (Schwab and Cyster, 2007) | Oocyte survival | (Morita et al., 2000) (Tilly, 2001) |
| | Cell shape change / Survival (Oligodendrocyte) | (Jaillard et al., 2005) | Cardiomyocyte survival | (Theilmeier et al., 2006) | Migration (Dendritic cell) | (Lan et al., 2005) | | |
| | Proliferation | (Harada et al., 2004) | Smooth muscle proliferation, migration and contraction | (Waeber et al., 2004) (Coussin et al., 2002) | Migration, degranulation of mast cell | (Jolly et al., 2004) | | |
| | Gap Junction Communication inhibition (Astrocyte) | (Rouach et al., 2006) | Angiogenesis | (Allende et al., 2003) | Survival (T-cell) | (Graeler and Goetzl, 2002) | | |
| | Hair cell maintenance | (MacLennan et al., 2006) (Herr et al., 2007) (Kono et al., 2007) | eNOS activation and vasodilation | (Nofer et al., 2004) (Theilmeier et al., 2006) | Adhesion of monocytes | (Bolick et al., 2005) | | |

**Fig. 10.2** Cellular response to LPA and S1P in different organ systems. Examples of receptor-mediated cellular responses to LPA and S1P. Cytoskeletal reorganization, migration, survival, and proliferation operate in many biological systems particularly within the nervous system, the cardiovascular system, the immune system, and the female reproductive system. *IL-2*, interleukin-2; *VEC*, vascular endothelial cells; *VSMCs*, vascular smooth muscle cells; *eNOS*, endothelial nitric oxide synthase

#### 10.2.2.3 Cell Shape Change

As mentioned above, LP signaling regulates F-Actin through $G_i$ and/or $G_{12/13}$. In this context, LPs affect cell shape changes not only by influencing motility, but also neurite retraction, growth cone collapse, repulsive growth cone turning, neuroblast and glial cell rounding, and smooth muscle cell contraction (Ishii et al., 2004; Moolenaar et al., 2004). For assessing the LP effect on cell shape and cytoskeletal changes, cell rounding assays, stress fiber formation assays, membrane ruffling assays with F-Actin staining, and real-time assays are utilized.

#### 10.2.2.4 Cell Adhesion and Aggregation

The activation of $G_{12/13}$ induces Rho and Rho kinase-mediated formation of actin stress fibers, focal adhesions, and cell contraction. Both LPA and S1P also mediate physiological wound healing processes and potentially atherogenic and thrombogenic processes (Siess, 2002). For example, LPA and mildly oxidized LDL (mox-LDL) promote monocyte binding to endothelial cells by increasing the cell surface expression of E-selectin and the vascular cell adhesion molecule-1 (VCAM-1) on endothelial cells (Rizza et al., 1999). LPA can also induce N-cadherin mediated Schwann cell clustering and Rho-ROCK mediated focal adhesion formation (Weiner et al., 2001). S1P can stimulate adhesion activity, but it can also inhibit cell adhesion via PI3K and nitric oxide synthase (eNOS) activation (Kimura et al., 2006).

#### 10.2.2.5 Inhibition of GAP Junction Communication

LPA inhibits GAP junction communication by connexin 43 phosphorylation in rat liver cells (Hill et al., 1994). MAPK and arachidonic acid cascades may transduce the signal, but the mechanism is still unclear (De Vuyst et al., 2007). Recently, it was reported that high S1P levels negatively affect gap junctions in astrocytes. In this case, the inhibitory effect of LPA is mediated through $G_i$ and Rho GTPases (Rouach et al., 2006).

#### 10.2.2.6 Transcription Regulation

LPA and S1P have been shown to activate NF-kappaB and induce expression of multiple effector genes. In endothelial cells, LPA also increases the levels of various adhesion molecule mRNAs and secreted factors, such as E-selectin, VCAM, and ICAM, as mentioned above (Li et al., 2005; Xia et al., 1998).

### *10.2.3 Metabolism and Enzymes*

To understand the dynamics of LPs *in vivo*, it is necessary to review the enzymes involved in LPA production and degradation. After the identification of the receptors, the identification of enzymes responsible for LP synthesis and degradation has accelerated our understanding of lipid biology (Fig. 10.3).

#### 10.2.3.1 Synthetic and Degradating Enzymes

As signaling mediators *in vivo*, the production and degradation of lysophospholipids should be tightly controlled. The metabolism of LPA has been partially characterized and involves a number of convergent biosynthetic pathways and enzymes of varied specificity (Meyer zu Heringdorf and Jakobs,

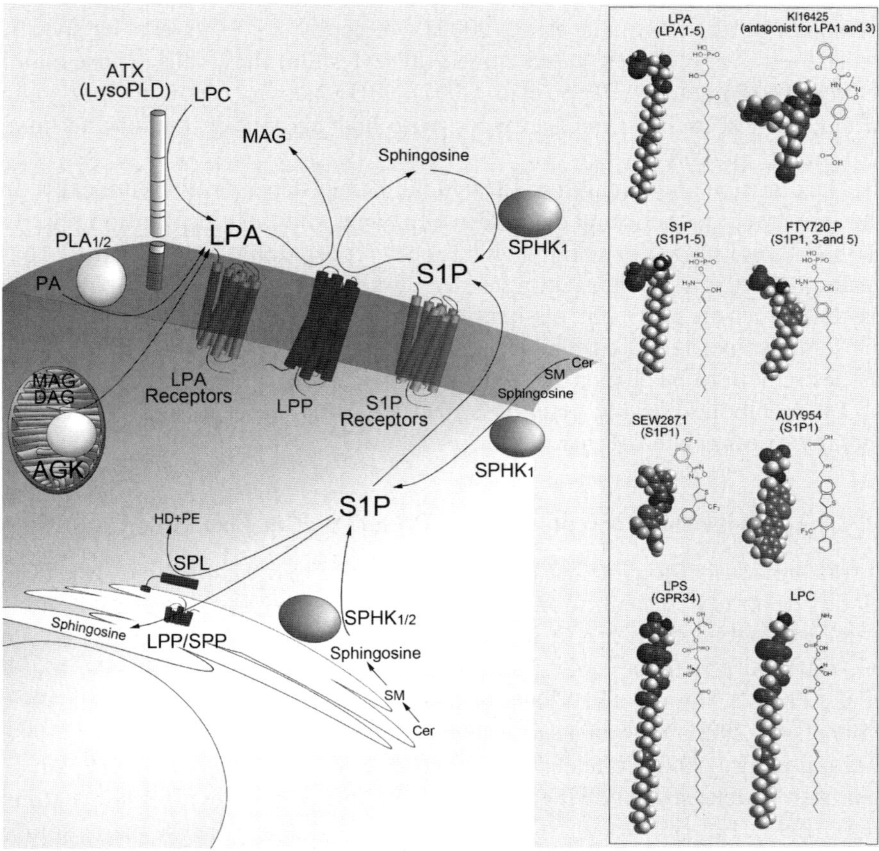


**Fig. 10.3** LPA and S1P metabolic pathways. Schematic representation of LPA and S1P metabolism. LPA is produced by ATX ("autotaxin," a lysophospholipase D or lysoPLD), phospholipase A1 and A2 ($PLA_{1/2}$), and acylglycerol kinase (AGK). PA is generated and transported from the inner leaflet of the plasma membrane, then subsequently converted to LPA by $PLA_1$ or $PLA_2$. According to a recent report, acylglycerol kinase phosphorylates monoacylglycerol (MAG) and DAG can produce LPA in mitochondria (Bektas et al., 2005). S1P is formed from sphingosine by sphingosine kinase 1 and 2 (SPHK1, SPHK2). Lipid phosphate phosphatases (LPPs) inactivate both LPA and S1P through dephosphorylation. Sphingosine phosphate phosphatase (SPP) specifically dephosphorylates S1P. S1P is also inactivated by S1P lyase (SPL) that produces irreversible cleavage. The space-filling molecular models and structures of LPA, S1P, and some major analogs are shown in the right box. High affinity LP receptors are indicated in parentheses under the name of each ligand

2007). To date, lysophospholipase D (lysoPLD), autotoxin (ATX), phospholipase A1 (PLA1), phospholipase A2 (PLA2), and acylglycerol kinase (AGK) are enzymes reported to be involved in LP synthesis (Chun and Rosen, 2006). There are multiple pathways responsible for LPA production (Meyer zu Heringdorf and Jakobs, 2007).

S1P metabolism involves a number of specific and highly conserved enzymes (Saba and Hla, 2004). Two sphingosine kinase isoforms, sphingosine kinase 1 (SPHK1) and sphingosine kinase 2 (SPHK 2), produce S1P from sphingosine (Kohama et al., 1998; Liu et al., 2000a). Recently it was shown by specific genetic removal of SPHK 1 and 2 in erythrocytes that these cells are the major source of S1P in blood (Pappu et al., 2007). SPHK activity is not only present in blood, but also in most mouse tissues (with high activity in thymus and lung) (Fukuda et al., 2003).

The duration and strength of LP signaling likely depends, at least locally, on the activity of synthetic and degradative enzymes and their localization relative to the LP receptors. For example, it has been reported that S1P lyase (SPL) has an important role in maintaining a steep gradient of S1P between blood and tissues, which in part controls lymphocyte localization (Schwab and Cyster, 2007; Schwab et al., 2005). The local distribution and potential LP gradients in tissues remain to be elucidated. Pharmacological and molecular manipulation of LP metabolic enzyme activity is also an intriguing approach for cancer therapy or other clinical treatments (see below).

#### 10.2.3.2 Endogenous Concentration of LPA and S1P *In Vivo*

It was previously thought that the major source of lysophospholipids was from blood. Reported concentrations of LPA and S1P vary in the literature, however most publications report around 1000 nM (200–5000 nM) in blood, and 0.2–100 nmol/g in tissues under basal, normal conditions (Aoki, 2004; Berdyshev et al., 2005; Bielawski et al., 2006; Das and Hajra, 1989; Eichholtz et al., 1993; Min et al., 2002; Murata et al., 2000; Okajima, 2002; Olivera et al., 1994; Yatomi et al., 1997). Platelets contain large amounts of LPA and S1P, which can both be released following platelet activation (Benton et al., 1982; Yatomi et al., 1997). For this reason, it was believed that platelets are the major source of S1P. However, it had been shown that erythrocytes can synthesize S1P by enzymatic pathways (Stoffel et al., 1970), and it is now clear that SPHKs present in erythrocytes are responsible for S1P in blood (Pappu et al., 2007). In addition, it has been shown that erythrocytes are able to import and store S1P that can be actively released upon stimulation (Hanel et al., 2007). It was also believed that LPA in blood is mainly derived from activated platelets, however it was recently reported that an LPA producing enzyme, lysophospholipase D(lysoPLD)in plasma, may also contribute to the total amount of LPA found in the blood (Aoki, 2004). The plasma LysoPLD activity was measured directly and half of this activity is attributed to autotaxin (ATX), one of the major LysoPLDs (Tanaka et al., 2006). In addition to the aforementioned pathway, there might be local LP synthesis in specific tissues. For example, the highest expression of ATX is found in the floor plate of the developing embryo and in the choroids plexus and osteoblasts throughout development (Bachner et al., 1999). Also, neurons are a potential source of LPA in the developing brain as nanomolar concentrations of LPA are found in conditioned medium from embryonic brain primary cultures (Fukushima et al., 2000).

### *10.2.4 Binding Proteins*

Lysophospholipids are usually bound to lipoproteins *in vivo*. Serum LPA binds to albumin, gelsolin, and other proteins (Moolenaar et al., 2004). S1P binds mainly to HDL and albumin (Levkau et al., 2004; Nofer et al., 2004; Okajima, 2002; Sato et al., 2007; Theilmeier et al., 2006). Such lipoproteins stabilize LPs in the hydrophilic environment and possibly protect them from rapid degradation. The stabilization effect of the lipoprotein is currently being studied (Moumtzi et al., 2007).

## 10.3 Physiology of LPs

As mentioned above, the LPA and S1P receptors are widely expressed throughout the body, however each receptor's expression is temporally and spatially distinct albeit often overlapping. To study the physiological function of each lipid receptor signal, targeted gene mutations in mice have been utilized to remove LP receptor genes or related enzymes. To date, null mutations for the LP receptor genes $LPA_1$-, $LPA_2$-, $LPA_3$, $S1P_1$-, $S1P_2$-, $S1P_3$, and $S1P_5$, and the LP producing enzymes ATX, SPHK1, SPHK2 have been reported.

### *10.3.1 LPA Receptor Mutant Mice*

Deletion of $LPA_1$ in mice causes a reduction in litter size primarily reflecting a 50% perinatal mortality rate. The observed mortality is due to poor suckling behavior that appears to result from an olfactory defect. However, they have a grossly normal cerebral cortex (Contos et al., 2000a). A smaller body size, shorter snouts, and cranial hematomas are characteristic features of surviving $LPA_1$ null mutant mice. Mild anatomical defects in the cerebral cortex and defective behavior in pre-pulse inhibition assays have also been reported in distinct null mutants or genetic variants (Estivill-Torrus et al., 2007; Harrison et al., 2003). $LPA_1$-null mice also show an increased number of apoptotic Schwann cells in the sciatic nerve (Contos et al., 2000a). This is consistent with the fact that $LPA_1$-null Schwann cells exhibit a reduced response to LPA (Weiner et al., 2001). Indeed, $LPA_1$-null mice do not show injury-induced dorsal root demyelination and neuropathic pain after peripheral nerve injury (Inoue et al., 2004).

The $LPA_2$-null mutation produces viable animals that are also grossly normal. The $LPA_1$ and $LPA_2$ double knockout animals show only a slight exacerbation of the hematoma defect that is seen in the $LPA_1$-null. However, primary fibroblasts and the embryonic cortex show vastly reduced responses to LPA (Contos et al., 2002; Kingsbury et al., 2003).

The functional loss of $LPA_3$ causes severe reproductive defects (Ye et al., 2005). The $LPA_3$-null litter sizes were less than 50% of wild type and heterozygote controls, and show delayed embryo implantation as well as spacing defects in the uterus. These phenotypes are attributable to the maternal genotype regardless of the male or embryo genotypes. The cyclooxygenase 2 (COX2) expression and prostaglandin levels are also reduced in the $LPA_3$-null uterus. This study demonstrated that $LPA_3$ is an indispensable upstream regulator of prostaglandin-mediated on-time implantation and embryo spacing.

### 10.3.2 S1P Receptor Mutant Mice

$S1P_1$-null mice have severe defects in vascular maturation, and die *in utero* because of hemorrhaging between E12.5 to E14.5 (Liu et al., 2000b). Because of the embryonic lethality, studies with $S1P_1$ conditionally deleted with the Cre-loxP system were used to analyze defects in specific cell and tissue types. Endothelial cell-specific deletion of $S1P_1$ showed that the vascular abnormality observed in $S1P_1$-null mice was due to a maturation defect in vascular endothelial cells (Allende et al., 2003). Also, T cell-specific deletion showed that $S1P_1$ was crucial for mature T cell egress from the thymus to the periphery (Allende et al., 2004a). To study lymphocyte egress using constitutive $S1P_1$ null mutant lymphocytes, hematopoietic precursors from $S1P_1$-null embryos were transferred to irradiated wild type adult mice and allowed to repopulate the lymphoid compartments. These elegant experiments showed that $S1P_1$ was intrinsically required for appropriate lymphocyte egress (Matloubian et al., 2004).

Interestingly, $S1P_2$-null mice show a degenerative and progressive loss of hearing and balance (Herr et al., 2007; Kono et al., 2007; MacLennan et al., 2006). $S1P_2$ is indispensable for maintenance of vestibular and cochlear hair cells *in vivo*. $S1P_2$-null mutants in the C57Bl/6 background have also been reported to show electrophysiological defects and develop seizures (MacLennan et al., 2001). In zebrafish, a single point mutation in the $S1P_2$-related *mil* gene leads to abnormal heart development (Kupperman et al., 2000), however, this defect in not recapitulated in $S1P_2$ knockout mice (Ishii et al., 2002). $S1P_3$-null mutant mice are grossly normal, but lack some of the S1P-mediated responses. For example, they show a loss of the vasodilation response to FTY720 (Tolle et al., 2005) with MEFs from $S1P_3$-nulls showing a marked decrease in PLC activation (Ishii et al., 2001). The knockout studies also revealed some functional redundancy in that mice lacking multiple receptors have new or exacerbated phenotypes. For example, mice lacking both $S1P_2$ and $S1P_3$ receptors have remarkably reduced litter sizes owing to an increase in perinatal lethality (Ishii et al., 2002). In addition, $S1P_{1,2,3}$ triple knockouts show severe defects in vascular development, to a greater extent than any single or double mutant, and are embryonically lethal at E10.5-11.5 (Kono et al., 2004).

The $S1P_5$-null mouse was recently reported. These mice do not have any apparent behavioral deficits or evident myelin deficiencies and their oligodendrocytes do not show defects in S1P induced process retraction and cell survival (Jaillard et al., 2005). This is surprising since $S1P_5$ expression is highly restricted and is present at significant levels only in oligodendrocytes and some hematopoetic cells. Further analyses of this mutant may reveal unrecognized phenotypes.

### *10.3.3 Others*

The genetic study of lipid metabolic enzymes has yielded complementary data. One of the most well characterized LPA producing enzymes is LysoPLD, originally known for its nucleotide phosphodiesterase activity as a protein called Autotoxin (ATX). ATX mutants have been generated by three different groups (Ferry et al., 2007; Tanaka et al., 2006; van Meeteren et al., 2006). Heterozygous deletion of ATX results in mice that are grossly normal, but have LPA plasma levels half of those in normal mice. Homozygotes are lethal at E9.5 due to severe defects in blood vessel development and neural tube formation (Tanaka et al., 2006; van Meeteren et al., 2006).

Similar to S1P receptor mutants, individual loss of either SPHK1 or SPHK2 does not produce an abnormal phenotype. However, SPHK 1 and 2 double mutant embryos lose detectable SPHK activity. As a consequence, the double mutants are lethal prior to E13.5 with severe vascular and neural tube defects (Allende et al., 2004b; Mizugishi et al., 2005).

Some of the LP kinase and lyase mutant mice have also been reported and their phenotypes are consistent with receptor mutant mice (Escalante-Alcalde et al., 2003; Schmahl et al., 2007). However, a number of enzymes are involved in lipid metabolic pathways so the existence of functional redundancy is therefore conceivable. Ongoing multiple and conditional gene targeting studies are helping to elucidate these pathways.

## 10.4 Possible Relevance of LPs to Human Diseases

### *10.4.1 Possible Clinical Applications*

#### 10.4.1.1 Immunity/Transplantation

Both LPA and S1P have been shown to act as immunomodulators in the regulation of T-cells, B-cells, and macrophages. These immune cells are likely regulated by combinations of LP receptors. LPA and S1P acting through $LPA_{1,2}$ and $S1P_{2,3}$ respectively may also serve as survival factors for T-cells by suppressing Bax (Goetzl et al., 1999). LPA induces migration and suppression of IL-2 production in unstimulated T-cells via $LPA_2$, however, once the T-cell is stimulated, LPA inhibits cell migration but activates IL-2 production via $LPA_1$ (Zheng et al., 2000, 2001). The expression pattern of LP receptors can also be changed during cell activation (Graler and Goetzl, 2002; Rosen et al., 2003; Zheng et al., 2000). According to recent models, S1P stimulates migration of inactive T-cells via $S1P_1$ and $S1P_4$. Upon activation, T-cells temporarily suppress receptor expression and lose the S1P mediated migration response. For retention in lymphoid organs, terminally differentiated effector T-cells then again upregulate $S1P_1$ to egress from lymph nodes (Graeler and Goetzl, 2002; Matloubian et al., 2004; Schwab and Cyster, 2007). The proper S1P gradient between plasma and lymph node is also important for lymphocyte migration (Schwab et al., 2005).

Studies with FTY720 have been of great importance in demonstrating the role of S1P signaling in immunomodulation. The phosphorylated metabolite of FTY720 (FTY720-P) is being evaluated as a clinically relevant immunosuppressant for organ transplantation. Conventional immunosuppressants like cyclosporine (cyclophilin inhibitor) and FK506 (calcineurin inhibitor) inhibit IL-2 dependent T-cell activation. The unique feature of FTY720-P is that it suppresses the immune system by inhibiting lymphocyte egress from lymphoid organs and acts as an S1P receptor modulator (Brinkmann, 2007). Thus, application of FTY720 with conventional immunosuppressants is expected to reduce the risk of conventional drug side effects like kidney toxicity from cyclosporine (Tedesco-Silva et al., 2005).

#### 10.4.1.2 Asthma

S1P levels are dramatically upregulated in the airways of asthmatic patients following allergen exposure. Cross-linking of IgE receptors on mast cells activates SPHK1 and increases S1P levels. Activation of $S1P_2$, and to a lesser extent $S1P_1$, promotes degranulation and chemotaxis of mast cells (Jolly et al., 2002, 2004). Also, airway smooth muscle cells (SMC) express $S1P_{1\text{-}4}$, and they could modulate the SMC contraction and proliferation via the $G_{12/13}$ and $G_{i/o}$ pathways (Jolly et al., 2002). This potential therapeutic modality has been demonstrated *in vivo* with the observation that FTY720 administration can reduce the Th1 or Th2 cell-mediated lung-inflammatory responses (Sawicka et al., 2003).

#### 10.4.1.3 Autoimmune Diseases

Since FTY720 does not generally impair lymphocyte proliferation and function, it could provide a new strategy for immunosuppression, which would be useful in transplantation, multiple sclerosis (MS), or autoimmune diabetes, leaving crucial functions of the immune system intact (Gardell et al., 2006; Rivera and Chun, 2007).

#### 10.4.1.4 Cancer

LP signaling has relevance to cancer. One of the better characterized cancer links is ovarian cancer. LPA elevation in the ascites of patients was reported to elicit growth factor-like activity (Mills et al., 1988), although there is controversy over the generality of this initial report. It has also been shown that LP receptors and the enzymes involved in LPA and S1P metabolisms are highly expressed in multiple cancer types, e.g. ovarian cancer and glioblastoma (Murph et al., 2006). S1P has both positive and negative effects on cancer cell growth (Hong et al., 1999). FTY720 has anti-tumor effects *in vitro* and *in vivo*, and this may be due to not only the effect on tumor cells, but also the inhibition of angiogenesis directly or indirectly (Azuma et al., 2002; Ho et al., 2005; LaMontagne et al., 2006).

#### 10.4.1.5 Cardiovascular

Both LPA and S1P have vaso-regulatory functions, such as regulation of heart rate, blood pressure, platelet aggregation, and smooth muscle contraction (Karliner, 2004; Siess et al., 2000). Atherosclerosis is a type of accelerated vasculitis that reduces blood flow leading to heart attacks and strokes (Siess, 2002). It is well known that HDL level correlates with a reduced risk of cardiovascular disease, such as atherosclerosis (Choi et al., 2006), and it has been recently shown that it is the S1P content of HDL that mediates many of its effects. For example, HDL induces vasodilation and myocardial perfusion by activation of $S1P_3$ (Levkau et al., 2004; Nofer et al., 2004). Furthermore, in an *in vivo* mouse study, HDL and S1P reduce the infarction size about by 20 and 40% and also inhibit inflammation caused by the recruitment of polymorphonuclear leukocytes and cardiomyocyte apoptosis via the $S1P_3$ receptor eNOS/NO pathway (Theilmeier et al., 2006).

#### 10.4.1.6 Hearing Loss

As mentioned above, $S1P_2$-null mice lose hearing and have balance defects (Herr et al., 2007; Kono et al., 2007; MacLennan et al., 2006). It may be possible to prevent the degeneration of hair cells with a selective S1P signaling modulator. These studies are ongoing and may offer novel treatment modalities for the prevention of age-related and ototoxic hearing loss.

#### 10.4.1.7 Wound Healing (CNS)

LPA and S1P in blood may enter the brain during central nervous system (CNS) injury. An experimentally caused brain hemorrhage provides an influx of 1–10 μM of LPA in the cerebrospinal fluid (Tigyi et al., 1995). In cerebral infarction, platelet aggregation can release micromolar concentrations of LPA and could also lead to increased LPA levels in CSF (Eichholtz et al., 1993). Indeed, intracranial injection of LPA or S1P causes astrogliosis *in vivo* (Sorensen et al., 2003). Reactive astrogliosis is a prominent component of CNS injury, and this would benefit from further study of LP signaling modulator applications.

#### 10.4.1.8 Pain

In animal models, nerve injury to the dorsal root results in the development of behavioral allodynia and hyperalgesia paralleled by demyelination. Intrathecal injection of LPA, but not S1P, initiates behavioral, morphological, and biochemical symptoms of neuropathic pain via an $LPA_1$-mediated Rho/Rho-kinase pathway (Inoue et al., 2004). LPA signaling modulation may be relevant for some forms of neuropathic pain, an area of significant, unmet medical need (Dworkin et al., 2007).

#### 10.4.1.9 Female Reproduction

Recent studies show that $LPA_3$ has a crucial role in blastocyst implantation through COX-2, which generates prostaglandins (PGs) $E_2$ and $I_2$ (Hama et al., 2006, 2007; Shah and Catt, 2005; Ye et al., 2005). S1P can also act to prevent intrinsic, chemical, and irradiation-induced oocyte apoptosis. S1P pretreatment improves the rate of successful pregnancy in irradiated mice (Morita et al., 2000; Tilly, 2001). Thus, controlling LP signaling could be a valuable therapeutic option in human infertility.

### *10.4.2 Pharmacology (Agonists and Antagonists)*

About 40% of drugs on the market in the United States target GPCRs. Furthermore, over 2% of genes in the human genome are estimated to encode GPCRs (over 1000) (Tyndall and Sandilya, 2005). Screening efforts are underway to identify chemicals that agonize and antagonize LP signaling (Chun and Rosen, 2006; Delgado et al., 2007; Herr and Chun, 2007). A computational approach is also being performed to design drugs and assess receptor specificity based on the structure of ligand binding pockets and amino acid residues required for ligand binding (as mentioned in Section 10.2.1.) (Holdsworth et al., 2004; Parrill et al., 2000).

#### 10.4.2.1 LPA Pharmacological Tools

Several LPA receptor agonists or antagonists have been reported, although most show modest selectivity and lack *in vivo* validation, which should be considered in any experimental usage, particularly for *in vivo* studies.

**Agonists:** N-acyl ethanolamide phosphate (NAEPA) is an LPA analog which has an ethanol amine backbone (Lynch et al., 1997). A screening of a 2-Oleoyl LPA derivative which had a pyran ring to stabilize the head group was performed and one $LPA_1$-selective agonist, two $LPA_3$-selective LPA agonists, and one $LPA_3$-selective antagonist were identified with this scheme (Tamaruya et al., 2004).

**Antagonists:** VPC-12449 is an $LPA_1$- and $LPA_3$-selective compound that can protect against $LPA_3$-mediated renal ischemia-reperfusion injury in a mouse model (Okusa et al., 2003). A natural lipid metabolite, diacylglycerol pyrophosphate (DGPP), was shown to act as an $LPA_1$ and $LPA_3$ specific antagonist (Fischer et al., 2001). Ki16425 is an $LPA_1$ and $LPA_3$ selective antagonist with little resemblance to LPA (Ohta et al., 2003). This compound can inhibit breast cancer cell proliferation and bone metastasis in mice (Boucharaba et al., 2006).

#### 10.4.2.2 S1P Pharmacological Tools

There are also several S1P receptor agonists and antagonists which have different receptor selectivities.

**Agonists:** AAL-(R) is non-selective S1P receptor agonist which has structural and functional similarities to FTY720 (Brinkmann et al., 2002; Rosen et al., 2003). Another agonist, KRP-203 prevents allograft rejection, but does not affect $S1P_3$ signaling (Fujishiro et al., 2006). KRP-203 ($S1P_1 > S1P_3$) is currently in Phase I clinical trials for the treatment of multiple sclerosis (Novartis). SEW2871 and AUY954 are $S1P_1$ specific agonists and have been shown to function to prevent appropriate lymphocyte egress and inhibit allograft rejection, respectively (Pan et al., 2006; Sanna et al., 2004). Receptor-selectivity is expected to show greater efficacy with minimal undesirable side effects.

**Antagonist**: JTE-013 is an $S1P_2$ specific antagonist (Yokoo et al., 2004). A recently reported $S1P_1$ specific antagonist called W146 can induce loss of capillary integrity (Sanna et al., 2006). In addition, there are a number of agonists/antagonists that have been described with varying affinities for the different receptor subtypes (Clemens et al., 2003, 2004; Davis et al., 2005; Im et al., 2001).

**FTY720 (FTY720-P):** FTY720 is perhaps the best characterized S1P receptor agonist and deserves special consideration. FTY720 is currently in Phase III clinical trials for the treatment of multiple sclerosis (Novartis). FTY720 was initially isolated from the fungi *Ascomycetes* in 1995 and identified as an immunosuppressive agent (Adachi et al., 1995). FTY720 administration significantly increases the survival rate of canine kidney allograft recipients (Suzuki et al., 1996). Recent reports have shown that FTY720 is phosphorylated by SPHK2 but not SPHK1 (Allende et al., 2004b; Kharel et al., 2005), and it inhibits T and B-cell egress from lymph nodes by modulating S1P signaling (Schwab and Cyster, 2007; Zemann et al., 2006). An increase in S1P levels, as seen by S1P lyase inhibition, also inhibits lymphocyte egress from lymph nodes (Schwab et al., 2005). Other $S1P_1$ antagonists, AAL-(R) and SEW2871, can inhibit thymocyte egress *in vivo* (Rosen et al., 2003), and an $S1P_1$ partially selective agonist, KRP-203, sequesters circulating lymphocytes into peripheral lymphoid organs (Shimizu et al., 2005). These data suggest that FTY720 and related agonists mimic high dose S1P exposure. FTY720 can also induce polyubiquitination and proteasomal degradation of $S1P_1$ (Oo et al., 2007), that could remove receptors from further agonism. In addition, an $S1P_1$ deficiency arrests thymocyte development at the CD69 positive stage, and prevents lymphocyte egress, as mentioned above (Allende et al., 2004a; Matloubian et al., 2004). These data suggest that FTY720 can result in degradation of the S1P receptor, acting as a functional antagonist of S1P signaling. However, some controversies concerning the immunosuppression mechanism by which FTY720 operates still remain (Chun, 2007), and this is still an active research area.

FTY720 represents the first generation of LP receptor modulators that may have therapeutic value. Other data concerning the efficacy of FTY720 administration for the treatment of type I diabetes, uveoretinitis, thyroiditis, myocarditis, systemic lupus erythematosus, rheumatoid arthritis, and multiple sclerosis in animal models have been reported (Fujino et al., 2003; Hozumi et al., 1999; Kurose et al., 2000; Matsuura et al., 2000; Okazaki et al., 2002; Suzuki et al., 1998; Webb et al., 2004). The potential to treat medically important diseases through LP receptor modulation represents an attractive and technically tractable approach that is being actively assessed.

**Acknowledgments** This work was supported by grants from the National Institutes of Health (MH051699, NS048478, HD050685, DA019674) to Jerold Chun.

## References

Adachi, K., Kohara, T., Nakao, N., Arita, M., Chiba, K., Mishina, T., Sasaki, S., Fujita, T. Design, synthesis, and structure-activity relationships of 2-substituted-2-amino-1,3-propanediols: Discovery of a novel immunosuppressant, FTY720. Bioorganic & Medicinal Chem Lett 5 (1995) 853–856.

Allende, M. L., Dreier, J. L., Mandala, S., Proia, R. L. Expression of the sphingosine 1-phosphate receptor, S1P1, on T-cells controls thymic emigration. J Biol Chem 279 (2004a) 15396–15401.

Allende, M. L., Sasaki, T., Kawai, H., Olivera, A., Mi, Y., van Echten-Deckert, G., Hajdu, R., Rosenbach, M., Keohane, C. A., Mandala, S., Spiegel, S., Proia, R. L. Mice deficient in sphingosine kinase 1 are rendered lymphopenic by FTY720. J Biol Chem 279 (2004b) 52487–52492.

Allende, M. L., Yamashita, T., Proia, R. L. G-protein-coupled receptor S1P1 acts within endothelial cells to regulate vascular maturation. Blood 102 (2003) 3665–3667.

An, S., Bleu, T., Huang, W., Hallmark, O. G., Coughlin, S. R., Goetzl, E. J. Identification of cDNAs encoding two G protein-coupled receptors for lysosphingolipids. FEBS Lett 417 (1997) 279–282.

Anliker, B., Chun, J. Cell surface receptors in lysophospholipid signaling. Semin Cell Dev Biol 15 (2004) 457–465.

Anliker, B., Chun, J. Lysophospholipid G protein-coupled receptors. J Biol Chem 279 (2004) 20555–20558.

Aoki, J. Mechanisms of lysophosphatidic acid production. Semin Cell Dev Biol 15 (2004) 477–489.

Azuma, H., Takahara, S., Ichimaru, N., Wang, J. D., Itoh, Y., Otsuki, Y., Morimoto, J., Fukui, R., Hoshiga, M., Ishihara, T., Nonomura, N., Suzuki, S., Okuyama, A., Katsuoka, Y. Marked prevention of tumor growth and metastasis by a novel immunosuppressive agent, FTY720, in mouse breast cancer models. Cancer Res 62 (2002) 1410–1419.

Bachner, D., Ahrens, M., Betat, N., Schroder, D., Gross, G. Developmental expression analysis of murine autotaxin (ATX). Mech Dev 84 (1999) 121–125.

Bektas, M., Payne, S. G., Liu, H., Goparaju, S., Milstien, S., Spiegel, S. A novel acylglycerol kinase that produces lysophosphatidic acid modulates cross talk with EGFR in prostate cancer cells. J Cell Biol 169 (2005) 801–811.

Benton, A. M., Gerrard, J. M., Michiel, T., Kindom, S. E. Are lysophosphatidic acids or phosphatidic acids involved in stimulus activation coupling in platelets? Blood 60 (1982) 642–649.

Berdyshev, E. V., Gorshkova, I. A., Garcia, J. G., Natarajan, V., Hubbard, W. C. Quantitative analysis of sphingoid base-1-phosphates as bisacetylated derivatives by liquid chromatography-tandem mass spectrometry. Anal Biochem 339 (2005) 129–136.

Bielawski, J., Szulc, Z. M., Hannun, Y. A., Bielawska, A. Simultaneous quantitative analysis of bioactive sphingolipids by high-performance liquid chromatography-tandem mass spectrometry. Methods 39 (2006) 82–91.

Bolick, D. T., Srinivasan, S., Kim, K. W., Hatley, M. E., Clemens, J. J., Whetzel, A., Ferger, N., Macdonald, T. L., Davis, M. D., Tsao, P. S., Lynch, K. R., Hedrick, C. C. Sphingosine-1-phosphate prevents tumor necrosis factor-{alpha}-mediated monocyte adhesion to aortic endothelium in mice. Arterioscler Thromb Vasc Biol 25 (2005) 976–981.

Boucharaba, A., Serre, C. M., Guglielmi, J., Bordet, J. C., Clezardin, P., Peyruchaud, O. The type 1 lysophosphatidic acid receptor is a target for therapy in bone metastases. Proc Natl Acad Sci USA 103 (2006) 9643–9648.

Brinkmann, V. Sphingosine 1-phosphate receptors in health and disease: mechanistic insights from gene deletion studies and reverse pharmacology. Pharmacol Ther 115 (2007) 84–105.

Brinkmann, V., Davis, M. D., Heise, C. E., Albert, R., Cottens, S., Hof, R., Bruns, C., Prieschl, E., Baumruker, T., Hiestand, P., Foster, C.A., Zollinger, M., Lynch, K.R. The immune modulator FTY720 targets sphingosine 1-phosphate receptors. J Biol Chem 277 (2002) 21453–21457.

Campbell, D. S., Holt, C. E. Chemotropic responses of retinal growth cones mediated by rapid local protein synthesis and degradation. Neuron 32 (2001) 1013–1026.

Choi, B. G., Vilahur, G., Viles-Gonzalez, J. F., Badimon, J. J. The role of high-density lipoprotein cholesterol in atherothrombosis. Mt Sinai J Med 73 (2006) 690–701.

Chun, J. The first cloned and identified lysophospholipid (LP) receptor gene, vzg-1: implications for related receptors and the nervous system. Adv Exp Med Biol 469 (1999) 357–362.

Chun, J. Immunology. The sources of a lipid conundrum. Science 316 (2007) 208–210.

Chun, J., Goetzl, E. J., Hla, T., Igarashi, Y., Lynch, K. R., Moolenaar, W., Pyne, S., Tigyi, G. International Union of Pharmacology. XXXIV. Lysophospholipid receptor nomenclature. Pharmacol Rev 54 (2002) 265–269.

Chun, J., Rosen, H. Lysophospholipid receptors as potential drug targets in tissue transplantation and autoimmune diseases. Curr Pharm Des 12 (2006) 161–171.

Clemens, J. J., Davis, M. D., Lynch, K. R., Macdonald, T.L. Synthesis of para-alkyl aryl amide analogues of sphingosine-1-phosphate: discovery of potent S1P receptor agonists. Bioorg Med Chem Lett 13 (2003) 3401–3404.

Clemens, J. J., Davis, M. D., Lynch, K. R., Macdonald, T. L. Synthesis of benzimidazole based analogues of sphingosine-1-phosphate: discovery of potent, subtype-selective S1P4 receptor agonists. Bioorg Med Chem Lett 14 (2004) 4903–4906.

Contos, J. J., Chun, J. The mouse lp(A3)/Edg7 lysophosphatidic acid receptor gene: genomic structure, chromosomal localization, and expression pattern. Gene 267 (2001) 243–253.

Contos, J. J., Fukushima, N., Weiner, J. A., Kaushal, D., Chun, J. Requirement for the lpA1 lysophosphatidic acid receptor gene in normal suckling behavior. Proc Natl Acad Sci USA 97 (2000a) 13384–13389.

Contos, J. J., Ishii, I., Chun, J. Lysophosphatidic acid receptors. Mol Pharmacol 58 (2000b) 1188–1196.

Contos, J. J., Ishii, I., Fukushima, N., Kingsbury, M. A., Ye, X., Kawamura, S., Brown, J. H., Chun, J. Characterization of lpa(2) (Edg4) and lpa(1)/lpa(2) (Edg2/Edg4) lysophosphatidic acid receptor knockout mice: signaling deficits without obvious phenotypic abnormality attributable to lpa(2). Mol Cell Biol 22 (2002) 6921–6929.

Coussin, F., Scott, R. H., Wise, A., Nixon, G. F. Comparison of sphingosine 1-phosphate-induced intracellular signaling pathways in vascular smooth muscles: differential role in vasoconstriction. Circ Res 91 (2002) 151–157.

Croset, M., Brossard, N., Polette, A., Lagarde, M. Characterization of plasma unsaturated lysophosphatidylcholines in human and rat. Biochem J 345 Pt 1 (2000) 61–67.

Das, A. K., Hajra, A. K. Quantification, characterization and fatty acid composition of lysophosphatidic acid in different rat tissues. Lipids 24 (1989) 329–333.

Davis, M. D., Clemens, J. J., Macdonald, T. L., Lynch, K. R. Sphingosine 1-phosphate analogs as receptor antagonists. J Biol Chem 280 (2005) 9833–9841.

De Vuyst, E., Decrock, E., De Bock, M., Yamasaki, H., Naus, C. C., Evans, W. H., Leybaert, L. Connexin hemichannels and gap junction channels are differentially influenced by lipopolysaccharide and basic fibroblast growth factor. Mol Biol Cell 18 (2007) 34–46.

Delgado, A., Casas, J., Llebaria, A., Abad, J. L., Fabrias, G. Chemical tools to investigate sphingolipid metabolism and functions. ChemMedChem 2 (2007) 580–606.

Dubin, A. E., Bahnson, T., Weiner, J. A., Fukushima, N., Chun, J. Lysophosphatidic acid stimulates neurotransmitter-like conductance changes that precede GABA and L-glutamate in early, presumptive cortical neuroblasts. J Neurosci 19 (1999) 1371–1381.

Dworkin, R. H., O'Connor, A. B., Backonja, M., Farrar, J. T., Finnerup, N. B., Jensen, T. S., Kalso, E. A., Loeser, J. D., Miaskowski, C., Nurmikko, T. J., Portenoy, R. K., Rice, A. S., Stacey, B. R., Treede, R. D., Turk, D. C., Wallace, M. S. Pharmacologic management of neuropathic pain: evidence-based recommendations. Pain 132 (2007) 237–251.

Eichholtz, T., Jalink, K., Fahrenfort, I., Moolenaar, W. H. The bioactive phospholipid lysophosphatidic acid is released from activated platelets. Biochem J 291 ( Pt 3) (1993) 677–680.

Escalante-Alcalde, D., Hernandez, L., Le Stunff, H., Maeda, R., Lee, H.S., Jr-Gang-Cheng, Sciorra, V. A., Daar, I., Spiegel, S., Morris, A. J., Stewart, C. L. The lipid phosphatase LPP3 regulates extra-embryonic vasculogenesis and axis patterning. Development 130 (2003) 4623–4637.

Estivill-Torrus, G., Llebrez-Zayas, P., Matas-Rico, E., Santin, L., Pedraza, C., De Diego, I., Del Arco, I., Fernandez-Llebrez, P., Chun, J., De Fonseca, F. R. Absence of LPA1 Signaling results in defective cortical development. Cereb Cortex 18 (2008) 938–950.

Etienne-Manneville, S., Hall, A. Rho GTPases in cell biology. Nature 420 (2002) 629–635.

Ferry, G., Giganti, A., Coge, F., Bertaux, F., Thiam, K., Boutin, J. A. Functional invalidation of the autotaxin gene by a single amino acid mutation in mouse is lethal. FEBS Lett 581 (2007) 3572–3578.

Fischer, D. J., Nusser, N., Virag, T., Yokoyama, K., Wang, D., Baker, D. L., Bautista, D., Parrill, A. L., Tigyi, G. Short-chain phosphatidates are subtype-selective antagonists of lysophosphatidic acid receptors. Mol Pharmacol 60 (2001) 776–784.

Fujino, M., Funeshima, N., Kitazawa, Y., Kimura, H., Amemiya, H., Suzuki, S., Li, X. K. Amelioration of experimental autoimmune encephalomyelitis in Lewis rats by FTY720 treatment. J Pharmacol Exp Ther 305 (2003) 70–77.

Fujishiro, J., Kudou, S., Iwai, S., Takahashi, M., Hakamata, Y., Kinoshita, M., Iwanami, S., Izawa, S., Yasue, T., Hashizume, K., Murakami, T., Kobayashi, E. Use of sphingosine-1-phosphate 1 receptor agonist, KRP-203, in combination with a subtherapeutic dose of cyclosporine A for rat renal transplantation. Transplantation 82 (2006) 804–812.

Fukuda, Y., Kihara, A., Igarashi, Y. Distribution of sphingosine kinase activity in mouse tissues: contribution of SPHK1. Biochem Biophys Res Commun 309 (2003) 155–160.

Fukushima, N., Ishii, I., Contos, J. J., Weiner, J. A., Chun, J. Lysophospholipid receptors. Annu Rev Pharmacol Toxicol 41 (2001) 507–534.

Fukushima, N., Kimura, Y., Chun, J. A single receptor encoded by vzg-1/lpA1/edg-2 couples to G proteins and mediates multiple cellular responses to lysophosphatidic acid. Proc Natl Acad Sci USA 95 (1998) 6151–6156.

Fukushima, N., Shano, S., Moriyama, R., Chun, J. Lysophosphatidic acid stimulates neuronal differentiation of cortical neuroblasts through the LPA1-G(i/o) pathway. Neurochem Int 50 (2007) 302–307.

Fukushima, N., Weiner, J. A., Chun, J. Lysophosphatidic acid (LPA) is a novel extracellular regulator of cortical neuroblast morphology. Dev Biol 228 (2000) 6–18.

Gardell, S. E., Dubin, A. E., Chun, J. Emerging medicinal roles for lysophospholipid signaling. Trends Mol Med 12 (2006) 65–75.

Goetzl, E. J., Kong, Y., Mei, B. Lysophosphatidic acid and sphingosine 1-phosphate protection of T cells from apoptosis in association with suppression of Bax. J Immunol 162 (1999) 2049–2056.

Goetzl, E. J., Kong, Y., Voice, J. K. Cutting edge: differential constitutive expression of functional receptors for lysophosphatidic acid by human blood lymphocytes. J Immunol 164 (2000) 4996–4999.

Graeler, M., Goetzl, E. J. Activation-regulated expression and chemotactic function of sphingosine 1-phosphate receptors in mouse splenic T cells, Faseb J 16 (2002) 1874–1878.

Graler, M. H., Goetzl, E. J. Lysophospholipids and their G protein-coupled receptors in inflammation and immunity. Biochim Biophys Acta 1582 (2002) 168–174.

Hama, K., Aoki, J., Bandoh, K., Inoue, A., Endo, T., Amano, T., Suzuki, H., Arai, H. Lysophosphatidic receptor, LPA3, is positively and negatively regulated by progesterone and estrogen in the mouse uterus. Life Sci 79 (2006) 1736–1740.

Hama, K., Aoki, J., Inoue, A., Endo, T., Amano, T., Motoki, R., Kanai, M., Ye, X., Chun, J., Matsuki, N., Suzuki, H., Shibasaki, M., Arai, H. Embryo spacing and implantation timing are differentially regulated by LPA3-mediated lysophosphatidic acid signaling in mice. Biol Reprod 77 (2007) 954–959.

Hanel, P., Andreani, P., Graler, M. H. Erythrocytes store and release sphingosine 1-phosphate in blood. Faseb J 21 (2007) 1202–1209.

Harada, J., Foley, M., Moskowitz, M. A., Waeber, C. Sphingosine-1-phosphate induces proliferation and morphological changes of neural progenitor cells. J Neurochem 88 (2004) 1026–1039.

Harrison, S. M., Reavill, C., Brown, G., Brown, J. T., Cluderay, J. E., Crook, B., Davies, C. H., Dawson, L. A., Grau, E., Heidbreder, C., Hemmati, P., Hervieu, G., Howarth, A., Hughes, Z. A., Hunter, A. J., Latcham, J., Pickering, S., Pugh, P., Rogers, D. C., Shilliam, C. S., Maycox, P. R. LPA1 receptor-deficient mice have phenotypic changes observed in psychiatric disease. Mol Cell Neurosci 24 (2003) 1170–1179.

Hecht, J. H., Weiner, J. A., Post, S. R., Chun, J. Ventricular zone gene-1 (vzg-1) encodes a lysophosphatidic acid receptor expressed in neurogenic regions of the developing cerebral cortex. J Cell Biol 135 (1996) 1071–1083.

Herr, D. R., Chun, J. Effects of LPA and S1P on the nervous system and implications for their involvement in disease. Curr Drug Targets 8 (2007) 155–167.

Herr, D. R., Grillet, N., Schwander, M., Rivera, R., Muller, U., Chun, J. Sphingosine 1-phosphate (S1P) signaling is required for maintenance of hair cells mainly via activation of S1P2. J Neurosci 27 (2007) 1474–1478.

Hill, C. S., Oh, S. Y., Schmidt, S. A., Clark, K. J., Murray, A. W. Lysophosphatidic acid inhibits gap-junctional communication and stimulates phosphorylation of connexin-43 in WB cells: possible involvement of the mitogen-activated protein kinase cascade. Biochem J 303 (Pt 2) (1994) 475–479.

Ho, J. W., Man, K., Sun, C. K., Lee, T. K., Poon, R. T., Fan, S. T. Effects of a novel immunomodulating agent, FTY720, on tumor growth and angiogenesis in hepatocellular carcinoma. Mol Cancer Ther 4 (2005) 1430–1438.

Hobson, J. P., Rosenfeldt, H. M., Barak, L. S., Olivera, A., Poulton, S., Caron, M. G., Milstien, S., Spiegel, S. Role of the sphingosine-1-phosphate receptor EDG-1 in PDGF-induced cell motility. Science 291 (2001) 1800–1803.

Holdsworth, G., Osborne, D. A., Pham, T. T., Fells, J. I., Hutchinson, G., Milligan, G., Parrill, A. L. A single amino acid determines preference between phospholipids and reveals length restriction for activation of the S1P4 receptor. BMC Biochem 5 (2004) 12.

Hong, G., Baudhuin, L. M., Xu, Y. Sphingosine-1-phosphate modulates growth and adhesion of ovarian cancer cells. FEBS Lett 460 (1999) 513–518.

Hooks, S. B., Santos, W. L., Im, D. S., Heise, C. E., Macdonald, T. L., Lynch, K. R. Lysophosphatidic acid-induced mitogenesis is regulated by lipid phosphate phosphatases and is Edg-receptor independent. J Biol Chem 276 (2001) 4611–4621.

Hozumi, Y., Kobayashi, E., Miyata, M., Fujimura, A. Immunotherapy for experimental rat autoimmune thyroiditis using a novel immunosuppressant, FTY720. Life Sci 65 (1999) 1739–1745.

Huang, M. C., Graeler, M., Shankar, G., Spencer, J., Goetzl, E. J. Lysophospholipid mediators of immunity and neoplasia. Biochim Biophys Acta 1582 (2002) 161–167.

Im, D. S., Clemens, J., Macdonald, T. L., Lynch, K. R. Characterization of the human and mouse sphingosine 1-phosphate receptor, S1P5 (Edg-8): structure-activity relationship of sphingosine1-phosphate receptors. Biochemistry 40 (2001) 14053–14060.

Im, D. S., Heise, C. E., Ancellin, N., O'Dowd, B. F., Shei, G. J., Heavens, R. P., Rigby, M. R., Hla, T., Mandala, S., McAllister, G., George, S. R., Lynch, K. R. Characterization of a novel sphingosine 1-phosphate receptor, Edg-8. J Biol Chem 275 (2000) 14281–14286.

Inoue, M., Rashid, M. H., Fujita, R., Contos, J. J., Chun, J., Ueda, H. Initiation of neuropathic pain requires lysophosphatidic acid receptor signaling. Nat Med 10 (2004) 712–718.

Ishii, I., Contos, J. J., Fukushima, N., Chun, J. Functional comparisons of the lysophosphatidic acid receptors, LP(A1)/VZG-1/EDG-2, LP(A2)/EDG-4, and LP(A3)/EDG-7 in neuronal cell lines using a retrovirus expression system. Mol Pharmacol 58 (2000) 895–902.

Ishii, I., Friedman, B., Ye, X., Kawamura, S., McGiffert, C., Contos, J. J., Kingsbury, M. A., Zhang, G., Brown, J. H., Chun, J. Selective loss of sphingosine 1-phosphate signaling with no obvious phenotypic abnormality in mice lacking its G protein-coupled receptor, LP(B3)/EDG-3. J Biol Chem 276 (2001) 33697–33704.

Ishii, I., Fukushima, N., Ye, X., Chun, J. Lysophospholipid receptors: signaling and biology. Annu Rev Biochem 73 (2004) 321–354.

Ishii, I., Ye, X., Friedman, B., Kawamura, S., Contos, J. J., Kingsbury, M. A., Yang, A. H., Zhang, G., Brown, J. H., Chun, J. Marked perinatal lethality and cellular signaling deficits in mice null for the two sphingosine 1-phosphate (S1P) receptors, S1P(2)/LP(B2)/EDG-5 and S1P(3)/LP(B3)/EDG-3. J Biol Chem 277 (2002) 25152–25159.

Jaillard, C., Harrison, S., Stankoff, B., Aigrot, M. S., Calver, A. R., Duddy, G., Walsh, F. S., Pangalos, M. N., Arimura, N., Kaibuchi, K., Zalc, B., Lubetzki, C. Edg8/S1P5: an oligodendroglial receptor with dual function on process retraction and cell survival. J Neurosci 25 (2005) 1459–1469.

Jolly, P. S., Bektas, M., Olivera, A., Gonzalez-Espinosa, C., Proia, R. L., Rivera, J., Milstien, S., Spiegel, S. Transactivation of sphingosine-1-phosphate receptors by FcepsilonRI triggering is required for normal mast cell degranulation and chemotaxis. J Exp Med 199 (2004) 959–970.

Jolly, P. S., Rosenfeldt, H. M., Milstien, S., Spiegel, S. The roles of sphingosine-1-phosphate in asthma. Mol Immunol 38 (2002) 1239–1245.

Karliner, J. S. Mechanisms of cardioprotection by lysophospholipids. J Cell Biochem 92 (2004) 1095–1103.

Kharel, Y., Lee, S., Snyder, A. H., Sheasley-O'Neill S. L., Morris, M. A., Setiady, Y., Zhu, R., Zigler, M. A., Burcin, T. L., Ley, K., Tung, K. S., Engelhard, V. H., Macdonald, T. L., Pearson-White, S., Lynch, K. R. Sphingosine kinase 2 is required for modulation of lymphocyte traffic by FTY720. J Biol Chem 280 (2005) 36865–36872.

Kimura, A., Ohmori, T., Ohkawa, R., Madoiwa, S., Mimuro, J., Murakami, T., Kobayashi, E., Hoshino, Y., Yatomi, Y., Sakata, Y. Essential roles of sphingosine 1-phosphate/S1P1 receptor axis in the migration of neural stem cells toward a site of spinal cord injury. Stem Cells 25 (2007) 115–124.

Kimura, T., Tomura, H., Mogi, C., Kuwabara, A., Ishiwara, M., Shibasawa, K., Sato, K., Ohwada, S., Im, D. S., Kurose, H., Ishizuka, T., Murakami, M., Okajima, F. Sphingosine 1-phosphate receptors mediate stimulatory and inhibitory signalings for expression of adhesion molecules in endothelial cells. Cell Signal 18 (2006) 841–850.

Kingsbury, M. A., Rehen, S. K., Contos, J. J., Higgins, C. M., Chun, J. Non-proliferative effects of lysophosphatidic acid enhance cortical growth and folding. Nat Neurosci 6 (2003) 1292–1299.

Kohama, T., Olivera, A., Edsall, L., Nagiec, M. M., Dickson, R., Spiegel, S. Molecular cloning and functional characterization of murine sphingosine kinase. J Biol Chem 273 (1998) 23722–23728.

Kono, M., Belyantseva, I. A., Skoura, A., Frolenkov, G. I., Starost, M. F., Dreier, J. L., Lidington, D., Bolz, S. S., Friedman, T. B., Hla, T., Proia, R. L. Deafness and stria vascularis defects in S1P2 receptor-null mice. J Biol Chem 282 (2007) 10690–10696.

Kono, M., Mi, Y., Liu, Y., Sasaki, T., Allende, M. L., Wu, Y. P., Yamashita, T., Proia, R. L. The sphingosine-1-phosphate receptors S1P1, S1P2, and S1P3 function coordinately during embryonic angiogenesis. J Biol Chem 279 (2004) 29367–29373.

Kupperman, E., An, S., Osborne, N., Waldron, S., Stainier, D. Y. A sphingosine-1-phosphate receptor regulates cell migration during vertebrate heart development. Nature 406 (2000) 192–195.

Kurose, S., Ikeda, E., Tokiwa, M., Hikita, N., Mochizuki, M. Effects of FTY720, a novel immunosuppressant, on experimental autoimmune uveoretinitis in rats. Exp Eye Res 70 (2000) 7–15.

LaMontagne, K., Littlewood-Evans, A., Schnell, C., O'Reilly, T., Wyder, L., Sanchez, T., Probst, B., Butler, J., Wood, A., Liau, G., Billy, E., Theuer, A., Hla, T., Wood, J. Antagonism of sphingosine-1-phosphate receptors by FTY720 inhibits angiogenesis and tumor vascularization. Cancer Res 66 (2006) 221–231.

Lan, Y. Y., De Creus, A., Colvin, B. L., Abe, M., Brinkmann, V., Coates, P. T., Thomson, A. W. The sphingosine-1-phosphate receptor agonist FTY720 modulates dendritic cell trafficking in vivo. Am J Transplant 5 (2005) 2649–2659.

Lee, C. W., Rivera, R., Gardell, S., Dubin, A. E., Chun, J. GPR92 as a new G12/13- and Gq-coupled lysophosphatidic acid receptor that increases cAMP, LPA5. J Biol Chem 281 (2006) 23589–23597.

Lee, M. J., van Brocklyn, J. R., Thangada, S., Liu, C. H., Hand, A. R., Menzeleev, R., Spiegel, S., Hla, T. Sphingosine-1-phosphate as a ligand for the G protein-coupled receptor EDG-1. ,Science 279 (1998) 1552–1555.

Lefkowitz, R. J., Shenoy, S. K. Transduction of receptor signals by beta-arrestins. Science 308 (2005) 512–517.

Levkau, B., Hermann, S., Theilmeier, G., van der Giet, M., Chun, J., Schober, O., Schafers, M. High-density lipoprotein stimulates myocardial perfusion in vivo. Circulation 110 (2004) 3355–3359.

Li, H., Ye, X., Mahanivong, C., Bian, D., Chun, J., Huang, S. Signaling mechanisms responsible for lysophosphatidic acid-induced urokinase plasminogen activator expression in ovarian cancer cells. J Biol Chem 280 (2005) 10564–10571.

Liu, H., Sugiura, M., Nava, V. E., Edsall, L. C., Kono, K., Poulton, S., Milstien, S., Kohama, T., Spiegel, S. Molecular cloning and functional characterization of a novel mammalian sphingosine kinase type 2 isoform. J Biol Chem 275 (2000a) 19513–19520.

Liu, Y., Wada, R., Yamashita, T., Mi, Y., Deng, C. X., Hobson, J. P., Rosenfeldt, H. M., Nava, V. E., Chae, S. S., Lee, M. J., Liu, C. H., Hla, T., Spiegel, S., Proia, R. L. Edg-1, the G protein-coupled receptor for sphingosine-1-phosphate, is essential for vascular maturation. J Clin Invest 106 (2000b) 951–961.

Lynch, K. R., Hopper, D. W., Carlisle, S. J., Catalano, J. G., Zhang, M., MacDonald, T. L. Structure/activity relationships in lysophosphatidic acid: the 2-hydroxyl moiety. Mol Pharmacol 52 (1997) 75–81.

MacLennan, A. J., Benner, S. J., Andringa, A., Chaves, A. H., Rosing, J. L., Vesey, R., Karpman, A. M., Cronier, S. A., Lee, N., Erway, L. C., Miller, M. L. The S1P2 sphingosine 1-phosphate receptor is essential for auditory and vestibular function. Hear Res 220 (2006) 38–48.

MacLennan, A. J., Carney, P. R., Zhu, W. J., Chaves, A. H., Garcia, J., Grimes, J. R., Anderson, K. J., Roper, S. N., Lee, N. An essential role for the H218/AGR16/Edg-5/LP(B2) sphingosine 1-phosphate receptor in neuronal excitability. Eur J Neurosci 14 (2001) 203–209.

Mandala, S., Hajdu, R., Bergstrom, J., Quackenbush, E., Xie, J., Milligan, J., Thornton, R., Shei, G. J., Card, D., Keohane, C., Rosenbach, M., Hale, J., Lynch, C. L., Rupprecht, K., Parsons, W., Rosen, H. Alteration of lymphocyte trafficking by sphingosine-1-phosphate receptor agonists. Science 296 (2002) 346–349.

Matloubian, M., Lo, C. G., Cinamon, G., Lesneski, M. J., Xu, Y., Brinkmann, V., Allende, M. L., Proia, R. L., Cyster, J. G. Lymphocyte egress from thymus and peripheral lymphoid organs is dependent on S1P receptor 1. Nature 427 (2004) 355–360.

Matsuura, M., Imayoshi, T., Chiba, K., Okumoto, T. Effect of FTY720, a novel immunosuppressant, on adjuvant-induced arthritis in rats. Inflamm Res 49 (2000) 404–410.

McGiffert, C., Contos, J. J., Friedman, B., Chun, J. Embryonic brain expression analysis of lysophospholipid receptor genes suggests roles for s1p(1) in neurogenesis and s1p(1-3) in angiogenesis. FEBS Lett 531 (2002) 103–108.

McIntyre, T. M., Pontsler, A. V., Silva, A. R., St Hilaire, A., Xu, Y., Hinshaw, J. C., Zimmerman, G. A., Hama, K., Aoki, J., Arai, H., Prestwich, G. D. Identification of an intracellular receptor for lysophosphatidic acid (LPA): LPA is a transcellular PPAR-gamma agonist. Proc Natl Acad Sci USA 100 (2003) 131–136.

Meyer zu Heringdorf, D., Jakobs, K. H. Lysophospholipid receptors: signalling, pharmacology and regulation by lysophospholipid metabolism. Biochim Biophys Acta 1768 (2007) 923–940.

Mills, G. B., May, C., McGill, M., Roifman, C. M., Mellors, A. A putative new growth factor in ascitic fluid from ovarian cancer patients: identification, characterization, and mechanism of action. Cancer Res 48 (1988) 1066–1071.

Mills, G. B., Moolenaar, W. H. The emerging role of lysophosphatidic acid in cancer. Nat Rev Cancer 3 (2003) 582–591.

Min, J. K., Yoo, H. S., Lee, E. Y., Lee, W J., Lee, Y.M. Simultaneous quantitative analysis of sphingoid base 1-phosphates in biological samples by o-phthalaldehyde precolumn derivatization after dephosphorylation with alkaline phosphatase. Anal Biochem 303 (2002) 167–175.

Mizugishi, K., Yamashita, T., Olivera, A., Miller, G. F., Spiegel, S., Proia, R. L. Essential role for sphingosine kinases in neural and vascular development. Mol Cell Biol 25 (2005) 11113–11121.

Moolenaar, W. H., van Meeteren, L. A., Giepmans, B. N. The ins and outs of lysophosphatidic acid signaling. Bioessays 26 (2004) 870–881.

Morita, Y., Perez, G. I., Paris, F., Miranda, S. R., Ehleiter, D., Haimovitz-Friedman, A., Fuks, Z., Xie, Z., Reed, J. C., Schuchman, E. H., Kolesnick, R. N., Tilly, J. L. Oocyte apoptosis is suppressed by disruption of the acid sphingomyelinase gene or by sphingosine-1-phosphate therapy. Nat Med 6 (2000) 1109–1114.

Moumtzi, A., Trenker, M., Flicker, K., Zenzmaier, E., Saf, R., Hermetter, A. Import and fate of fluorescent analogs of oxidized phospholipids in vascular smooth muscle cells. J Lipid Res 48 (2007) 565–582.

Murata, N., Sato, K., Kon, J., Tomura, H., Okajima, F. Quantitative measurement of sphingosine 1-phosphate by radioreceptor-binding assay. Anal Biochem 282 (2000) 115–120.

Murph, M., Tanaka, T., Liu, S., Mills, G. B. Of spiders and crabs: the emergence of lysophospholipids and their metabolic pathways as targets for therapy in cancer. Clin Cancer Res 12 (2006) 6598–5602.

Neves, S. R., Ram, P. T., Iyengar, R. G protein pathways. Science 296 (2002) 1636–1639.

Nofer, J. R., van der Giet, M., Tolle, M., Wolinska, I., von Wnuck Lipinski, K., Baba, H. A., Tietge, U. J., Godecke, A., Ishii, I., Kleuser, B., Schafers, M., Fobker, M., Zidek, W., Assmann, G., Chun, J., Levkau, B. HDL induces NO-dependent vasorelaxation via the lysophospholipid receptor S1P3. J Clin Invest 113 (2004) 569–581.

Ohta, H., Sato, K., Murata, N., Damirin, A., Malchinkhuu, E., Kon, J., Kimura, T., Tobo, M., Yamazaki, Y., Watanabe, T., Yagi, M., Sato, M., Suzuki, R., Murooka, H., Sakai, T., Nishitoba, T., Im, D. S., Nochi, H., Tamoto, K., Tomura, H., Okajima, F. Ki16425, a subtype-selective antagonist for EDG-family lysophosphatidic acid receptors. Mol Pharmacol 64 (2003) 994–1005.

Okajima, F. Plasma lipoproteins behave as carriers of extracellular sphingosine 1-phosphate: is this an atherogenic mediator or an anti-atherogenic mediator? Biochim Biophys Acta 1582 (2002) 132–137.

Okamoto, H., Takuwa, N., Yokomizo, T., Sugimoto, N., Sakurada, S., Shigematsu, H., Takuwa, Y. Inhibitory regulation of Rac activation, membrane ruffling, and cell migration by the G protein-coupled sphingosine-1-phosphate receptor EDG5 but not EDG1 or EDG3. Mol Cell Biol 20 (2000) 9247–9261.

Okazaki, H., Hirata, D., Kamimura, T., Sato, H., Iwamoto, M., Yoshio, T., Masuyama, J., Fujimura, A., Kobayashi, E., Kano, S., Minota, S. Effects of FTY720 in MRL-lpr/lpr mice: therapeutic potential in systemic lupus erythematosus. J Rheumatol 29 (2002) 707–716.

Okusa, M. D., Ye, H., Huang, L., Sigismund, L., Macdonald, T., Lynch, K.R. Selective blockade of lysophosphatidic acid LPA3 receptors reduces murine renal ischemia-reperfusion injury. Am J Physiol Renal Physiol 285 (2003) F565–F574.

Olivera, A., Rosenthal, J., Spiegel, S. Sphingosine kinase from Swiss 3T3 fibroblasts: a convenient assay for the measurement of intracellular levels of free sphingoid bases. Anal Biochem 223 (1994) 306–312.

Oo, M. L., Thangada, S., Wu, M. T., Liu, C. H., Macdonald, T. L., Lynch, K. R., Lin, C. Y., Hla, T. Immunosuppressive and anti-angiogenic sphingosine 1-phosphate receptor-1 agonists induce ubiquitinylation and proteasomal degradation of the receptor. J Biol Chem 282 (2007) 9082–9089.

Pan, S., Mi, Y., Pally, C., Beerli, C., Chen, A., Guerini, D., Hinterding, K., Nuesslein-Hildesheim, B., Tuntland, T., Lefebvre, S., Liu, Y., Gao, W., Chu, A., Brinkmann, V., Bruns, C., Streiff, M., Cannet, C., Cooke, N., Gray, N. A monoselective sphingosine-1-phosphate receptor-1 agonist prevents allograft rejection in a stringent rat heart transplantation model. Chem Biol 13 (2006) 1227–1234.

Pappu, R., Schwab, S. R., Cornelissen, I., Pereira, J. P., Regard, J. B., Xu, Y., Camerer, E., Zheng, Y. W., Huang, Y., Cyster, J. G., Coughlin, S.R. Promotion of lymphocyte egress into blood and lymph by distinct sources of sphingosine-1-phosphate. Science 316 (2007) 295–298.

Parrill, A. L. Structural characteristics of lysophosphatidic acid biological targets. Biochem Soc Trans 33 (2005) 1366–1369.

Parrill, A. L., Wang, D., Bautista, D. L., van Brocklyn, J. R., Lorincz, Z., Fischer, D. J., Baker, D. L., Liliom, K., Spiegel, S., Tigyi, G. Identification of Edg1 receptor residues that recognize sphingosine 1-phosphate. J Biol Chem 275 (2000) 39379–39384.

Pierce, K. L., Premont, R. T., Lefkowitz, R J. Seven-transmembrane receptors. Nat Rev Mol Cell Biol 3 (2002) 639–650.

Pyne, N. J., Waters, C. M., Long, J. S., Moughal, N. A., Tigyi, G., Pyne, S. Receptor tyrosine kinase-G-protein coupled receptor complex signaling in mammalian cells. Adv Enzyme Regul 47 (2007) 271–280.

Radeff-Huang, J., Seasholtz, T .M., Matteo, R. G., Brown, J. H. G protein mediated signaling pathways in lysophospholipid induced cell proliferation and survival. J Cell Biochem 92 (2004) 949–966.

Ramakers, G. J., Moolenaar, W. H. Regulation of astrocyte morphology by RhoA and lysophosphatidic acid. Exp Cell Res 245 (1998) 252–262.

Rivera, R., Chun, J. Potential therapeutic roles of lysophospholipid signaling in autoimmune-related diseases. Future Lipidol 2 (2007) 535–545.

Rizza, C., Leitinger, N., Yue, J., Fischer, D. J., Wang, D. A., Shih, P. T., Lee, H., Tigyi, G., Berliner, J. A. Lysophosphatidic acid as a regulator of endothelial/leukocyte interaction. Lab Invest 79 (1999) 1227–1235.

Rosen, H., Alfonso, C., Surh, C. D., McHeyzer-Williams, M. G. Rapid induction of medullary thymocyte phenotypic maturation and egress inhibition by nanomolar sphingosine 1-phosphate receptor agonist. Proc Natl Acad Sci USA 100 (2003) 10907–10912.

Rosen, H., Goetzl, E. J. Sphingosine 1-phosphate and its receptors: an autocrine and paracrine network. Nat Rev Immunol 5 (2005) 560–570.

Rouach, N., Pebay, A., Meme, W., Cordier, J., Ezan, P., Etienne, E., Giaume, C., Tence, M. S1P inhibits gap junctions in astrocytes: involvement of Gi and Rho GTPase/ROCK. Eur J Neurosci 23 (2006) 1453–1464.

Saba, J. D., Hla, T. Point-counterpoint of sphingosine 1-phosphate metabolism. Circ Res 94 (2004) 724–734.

Sanna, M. G., Liao, J., Jo, E., Alfonso, C., Ahn, M. Y., Peterson, M. S., Webb, B., Lefebvre, S., Chun, J., Gray, N., Rosen, H. Sphingosine 1-phosphate (S1P) receptor subtypes S1P1 and S1P3, respectively, regulate lymphocyte recirculation and heart rate. J Biol Chem 279 (2004) 13839–13848.

Sanna, M. G., Wang, S. K., Gonzalez-Cabrera, P. J., Don, A., Marsolais, D., Matheu, M. P., Wei, S. H., Parker, I., Jo, E., Cheng, W. C., Cahalan, M. D., Wong, C. H., Rosen, H. Enhancement of capillary leakage and restoration of lymphocyte egress by a chiral S1P1 antagonist in vivo. Nat Chem Biol 2 (2006) 434–441.

Sato, K., Malchinkhuu, E., Horiuchi, Y., Mogi, C., Tomura, H., Tosaka, M., Yoshimoto, Y., Kuwabara, A., Okajima, F. HDL-like lipoproteins in cerebrospinal fluid affect neural cell activity through lipoprotein-associated sphingosine 1-phosphate. Biochem Biophys Res Commun 359 (2007) 649–654.

Sawicka, E., Zuany-Amorim, C., Manlius, C., Trifilieff, A., Brinkmann, V., Kemeny, D. M., Walker, C. Inhibition of Th1- and Th2-mediated airway inflammation by the sphingosine 1-phosphate receptor agonist FTY720. J Immunol 171 (2003) 6206–6214.

Schmahl, J., Raymond, C. S., Soriano, P. PDGF signaling specificity is mediated through multiple immediate early genes. Nat Genet 39 (2007) 52–60.

Schwab, S. R., Cyster, J. G. Finding a way out: lymphocyte egress from lymphoid organs. Nat Immunol 8 (2007) 1295–1301.

Schwab, S. R., Pereira, J. P., Matloubian, M., Xu, Y., Huang, Y., Cyster, J. G. Lymphocyte sequestration through S1P lyase inhibition and disruption of S1P gradients. Science 309 (2005) 1735–1739.

Sen, S., Smeby, R. R., Bumpus, F. M. Antihypertensive effect of an isolated phospholipid. Am J Physiol 214 (1968) 337–341.

Shah, B. H., Catt, K. J. Roles of LPA3 and COX-2 in implantation. Trends Endocrinol Metab 16 (2005) 397–399.

Shano, S., Moriyama, R., Chun, J., Fukushima, N. Lysophosphatidic acid stimulates astrocyte proliferation through LPA(1). Neurochem Int 52 (2008) 216–220.

Shimizu, H., Takahashi, M., Kaneko, T., Murakami, T., Hakamata, Y., Kudou, S., Kishi, T., Fukuchi, K., Iwanami, S., Kuriyama, K., Yasue, T., Enosawa, S., Matsumoto, K., Takeyoshi, I., Morishita, Y., Kobayashi, E. KRP-203, a novel synthetic immunosuppressant, prolongs graft survival and attenuates chronic rejection in rat skin and heart allografts. Circulation 111 (2005) 222–229.

Siess, W. Athero- and thrombogenic actions of lysophosphatidic acid and sphingosine-1-phosphate. Biochim Biophys Acta 1582 (2002) 204–215.

Siess, W., Essler, M., Brandl, R. Lysophosphatidic acid and sphingosine 1-phosphate: two lipid villains provoking cardiovascular diseases? IUBMB Life 49 (2000) 67–71.

Siess, W., Zangl, K. J., Essler, M., Bauer, M., Brandl, R., Corrinth, C., Bittman, R., Tigyi, G., Aepfelbacher, M. Lysophosphatidic acid mediates the rapid activation of platelets and endothelial cells by mildly oxidized low density lipoprotein and accumulates in human atherosclerotic lesions. Proc Natl Acad Sci USA 96 (1999) 6931–6936.

Singleton, P. A., Dudek, S. M., Chiang, E. T., Garcia, J. G. Regulation of sphingosine 1-phosphate-induced endothelial cytoskeletal rearrangement and barrier enhancement by S1P1 receptor, PI3 kinase, Tiam1/Rac1, and alpha-actinin. Faseb J 19 (2005) 1646–1656.

Sorensen, S. D., Nicole, O., Peavy, R. D., Montoya, L. M., Lee, C. J., Murphy, T. J., Traynelis, S. F., Hepler, J. R. Common signaling pathways link activation of murine PAR-1, LPA, and S1P receptors to proliferation of astrocytes. Mol Pharmacol 64 (2003) 1199–1209.

Spiegel, S., Milstien, S. Sphingosine-1-phosphate: an enigmatic signalling lipid. Nat Rev Mol Cell Biol 4 (2003) 397–407.

Stoffel, W., Assmann, G., Binczek, E. Metabolism of sphingosine bases. 13. Enzymatic synthesis of 1-phosphate esters of 4t-sphingenine (sphingosine), sphinganine (dihydrosphingosine), 4-hydroxysphinganine (phytosphingosine) and 3-dehydrosphinganine by erythrocytes. Hoppe Seylers Z Physiol Chem 351 (1970) 635–642.

Sugimoto, N., Takuwa, N., Okamoto, H., Sakurada, S., Takuwa, Y. Inhibitory and stimulatory regulation of Rac and cell motility by the G12/13-Rho and Gi pathways integrated downstream of a single G protein-coupled sphingosine-1-phosphate receptor isoform. Mol Cell Biol 23 (2003) 1534–1545.

Sugo, T., Tachimoto, H., Chikatsu, T., Murakami, Y., Kikukawa, Y., Sato, S., Kikuchi, K., Nagi, T., Harada, M., Ogi, K., Ebisawa, M., Mori, M. Identification of a lysophosphatidylserine receptor on mast cells. Biochem Biophys Res Commun 341 (2006) 1078–1087.

Sukocheva, O., Wadham, C., Holmes, A., Albanese, N., Verrier, E., Feng, F., Bernal, A., Derian, C. K., Ullrich, A., Vadas, M. A., Xia, P. Estrogen transactivates EGFR via the sphingosine 1-phosphate receptor Edg-3: the role of sphingosine kinase-1. J Cell Biol 173 (2006) 301–310.

Suzuki, K., Yan, H., Li, X. K., Amemiya, H., Suzuki, S., Hiromitsu, K. Prevention of experimentally induced autoimmune type I diabetes in rats by the new immunosuppressive reagent FTY720. Transplant Proc 30 (1998) 1044–1045.

Suzuki, S., Enosawa, S., Kakefuda, T., Shinomiya, T., Amari, M., Naoe, S., Hoshino, Y., Chiba, K. A novel immunosuppressant, FTY720, with a unique mechanism of action, induces long-term graft acceptance in rat and dog allotransplantation. Transplantation 61 (1996) 200–205.

Tamaruya, Y., Suzuki, M., Kamura, G., Kanai, M., Hama, K., Shimizu, K., Aoki, J., Arai, H., Shibasaki, M. Identifying specific conformations by using a carbohydrate scaffold: discovery of subtype-selective LPA-receptor agonists and an antagonist. Angew Chem Int Ed Engl 43 (2004) 2834–2837.

Tanaka, M., Okudaira, S., Kishi, Y., Ohkawa, R., Iseki, S., Ota, M., Noji, S., Yatomi, Y., Aoki, J., Arai, H. Autotaxin stabilizes blood vessels and is required for embryonic vasculature by producing lysophosphatidic acid. J Biol Chem 281 (2006) 25822–25830.

Tedesco-Silva, H., Mourad, G., Kahan, B. D., Boira, J. G., Weimar, W., Mulgaonkar, S., Nashan, B., Madsen, S., Charpentier, B., Pellet, P. and Vanrenterghem, Y. FTY720, a novel immunomodulator: efficacy and safety results from the first phase 2A study in de novo renal transplantation. Transplantation 79 (2005) 1553–1560.

Theilmeier, G., Schmidt, C., Herrmann, J., Keul, P., Schafers, M., Herrgott, I., Mersmann, J., Larmann, J., Hermann, S., Stypmann, J., Schober, O., Hildebrand, R., Schulz, R., Heusch, G., Haude, M., von Wnuck Lipinski, K., Herzog, C., Schmitz, M., Erbel, R., Chun, J., Levkau, B. High-density lipoproteins and their constituent, sphingosine-1-phosphate, directly protect the heart against ischemia/reperfusion injury in vivo via the S1P3 lysophospholipid receptor. Circulation 114 (2006) 1403–1409.

Tigyi, G., Hong, L., Yakubu, M., Parfenova, H., Shibata, M. and Leffler, C. W. Lysophosphatidic acid alters cerebrovascular reactivity in piglets. Am J Physiol 268 (1995) H2048–H2055.

Tilly, J. L. Commuting the death sentence: how oocytes strive to survive. Nat Rev Mol Cell Biol 2 (2001) 838–848.

Tokumura, A., Fukuzawa, K., Tsukatani, H. Effects of synthetic and natural lysophosphatidic acids on the arterial blood pressure of different animal species. Lipids 13 (1978) 572–574.

Tolle, M., Levkau, B., Keul, P., Brinkmann, V., Giebing, G., Schonfelder, G., Schafers, M., von Wnuck Lipinski, K., Jankowski, J., Jankowski, V., Chun, J., Zidek, W., Van der Giet, M. Immunomodulator FTY720 Induces eNOS-dependent arterial vasodilatation via the lysophospholipid receptor S1P3. Circ Res 96 (2005) 913–920.

Tyndall, J. D., Sandilya, R. GPCR agonists and antagonists in the clinic. Med Chem 1 (2005) 405–421.

van Brocklyn, J. R., Graler, M. H., Bernhardt, G., Hobson, J. P., Lipp, M., Spiegel, S. Sphingosine-1-phosphate is a ligand for the G protein-coupled receptor EDG-6. Blood 95 (2000) 2624–2629.

van Corven, E. J., Groenink, A., Jalink, K., Eichholtz, T., Moolenaar, W. H. Lysophosphatidate-induced cell proliferation: identification and dissection of signaling pathways mediated by G proteins. Cell 59 (1989) 45–54.

van Leeuwen, F. N., Giepmans, B. N., van Meeteren, L. A., Moolenaar, W. H. Lysophosphatidic acid: mitogen and motility factor. Biochem Soc Trans 31 (2003) 1209–1212.

van Meeteren, L. A., Ruurs, P., Stortelers, C., Bouwman, P., van Rooijen, M. A., Pradere, J. P., Pettit, T. R., Wakelam, M. J., Saulnier-Blache, J. S., Mummery, C. L., Moolenaar, W. H., Jonkers, J. Autotaxin, a secreted lysophospholipase D, is essential for blood vessel formation during development. Mol Cell Biol 26 (2006) 5015–5022.

van Nieuw Amerongen, G. P., Vermeer, M. A., van Hinsbergh, V. W. Role of RhoA and Rho kinase in lysophosphatidic acid-induced endothelial barrier dysfunction. Arterioscler Thromb Vasc Biol 20 (2000) E127–E133.

Waeber, C., Blondeau, N., Salomone, S. Vascular sphingosine-1-phosphate S1P1 and S1P3 receptors. Drug News Perspect 17 (2004) 365–382.

Webb, M., Tham, C. S., Lin, F. F., Lariosa-Willingham, K., Yu, N., Hale, J., Mandala, S., Chun, J., Rao, T. S. Sphingosine 1-phosphate receptor agonists attenuate relapsing-remitting experimental autoimmune encephalitis in SJL mice. J Neuroimmunol 153 (2004) 108–121.

Weiner, J. A., Fukushima, N., Contos, J. J., Scherer, S. S., Chun, J. Regulation of Schwann cell morphology and adhesion by receptor-mediated lysophosphatidic acid signaling. J Neurosci 21 (2001) 7069–7078.

Wong, R. C., Tellis, I., Jamshidi, P., Pera, M., Pebay, A. Anti-apoptotic effect of sphingosine-1-phosphate and platelet-derived growth factor in human embryonic stem cells. Stem Cells Dev (2007).

Xia, P., Gamble, J. R., Rye, K. A., Wang, L., Hii, C. S., Cockerill, P., Khew-Goodall, Y., Bert, A. G., Barter, P. J., Vadas, M.A. Tumor necrosis factor-alpha induces adhesion molecule expression through the sphingosine kinase pathway. Proc Natl Acad Sci USA 95 (1998) 14196–14201.

Xu, Y. J., Aziz, O. A., Bhugra, P., Arneja, A. S., Mendis, M. R., Dhalla, N. S. Potential role of lysophosphatidic acid in hypertension and atherosclerosis. Can J Cardiol 19 (2003) 1525–1536.

Yatomi, Y., Welch, R. J., Igarashi, Y. Distribution of sphingosine 1-phosphate, a bioactive sphingolipid, in rat tissues. FEBS Lett 404 (1997) 173–174.

Ye, X., Hama, K., Contos, J. J., Anliker, B., Inoue, A., Skinner, M. K., Suzuki, H., Amano, T., Kennedy, G., Arai, H., Aoki, J., Chun, J. LPA3-mediated lysophosphatidic acid signalling in embryo implantation and spacing. Nature 435 (2005) 104–108.

Yokoo, E., Yatomi, Y., Takafuta, T., Osada, M., Okamoto, Y., Ozaki, Y. Sphingosine 1-phosphate inhibits migration of RBL-2H3 cells via S1P2: cross-talk between platelets and mast cells. J Biochem (Tokyo) 135 (2004) 673–681.

Yuan, X. B., Jin, M., Xu, X., Song, Y. Q., Wu, C. P., Poo, M. M., Duan, S. Signalling and crosstalk of Rho GTPases in mediating axon guidance. Nat Cell Biol 5 (2003) 38–45.

Zemann, B., Kinzel, B., Muller, M., Reuschel, R., Mechtcheriakova, D., Urtz, N., Bornancin, F., Baumruker, T., Billich, A. Sphingosine kinase type 2 is essential for lymphopenia induced by the immunomodulatory drug FTY720. Blood 107 (2006) 1454–1458.

Zhang, H., Desai, N. N., ,Olivera, A., Seki, T., Brooker, G., Spiegel, S. Sphingosine-1-phosphate, a novel lipid, involved in cellular proliferation. J Cell Biol 114 (1991) 155–167.

Zheng, Y., Kong, Y., Goetzl, E. J. Lysophosphatidic acid receptor-selective effects on Jurkat T cell migration through a Matrigel model basement membrane. J Immunol 166 (2001) 2317–2322.

Zheng, Y., Voice, J. K., Kong, Y., Go etzl, E. J. Altered expression and functional profile of lysophosphatidic acid receptors in mitogen-activated human blood T lymphocytes. Faseb J 14 (2000) 2387–2389.

# Chapter 11
# Phospholipid-Mediated Signaling and Heart Disease

**Paramjit S. Tappia and Tushi Singal**

**Abstract** Cardiac hypertrophy, congestive heart failure, diabetic cardiomyopathy and myocardial ischemia-reperfusion injury are associated with a disturbance in cardiac sarcolemmal membrane phospholipid homeostasis. The contribution of the different phospholipases and their related signaling mechanisms to altered function of the diseased myocardium is not completely understood. Resolution of this issue is essential for both the understanding of the pathophysiology of heart disease and for determining if components of the phospholipid signaling pathways could serve as appropriate therapeutic targets. This review provides an outline of the role of phospholipase $A_2$, C and D and subsequent signal transduction mechanisms in different cardiac pathologies with a discussion of their potential as targets for drug development for the prevention/treatment of heart disease.

**Keywords** Cardiac hypertrophy · congestive heart failure · diabetic cardiomyopathy · ischemia-reperfusion · phospholipases

## 11.1 Introduction

Phospholipases play an important role in cellular metabolism, including the biosynthesis and degradation of membrane lipids. This leads to the production of many types of lipidic second messengers that mediate changes in the function of important intracellular proteins as well as the signaling of nuclear transcription factors and subsequent gene expression (Lamers et al., 1992; Dhalla et al., 2006; Tappia et al., 2006). The cardiac sarcolemmal (SL) membrane phospholipids serve as substrates for 3 major phospholipase families, phospholipase $A_2$ ($PLA_2$), phospholipase C (PLC) and phospholipase D (PLD), which produce important lipid signaling molecules (Buckland and Wilton, 2000) and have been

P.S. Tappia
Institute of Cardiovascular Sciences, St. Boniface General Hospital Research Centre & Department of Human Anatomy & Cell Science, Faculty of Medicine, University of Manitoba, Winnipeg, Canada

P.J. Quinn, X. Wang (eds.), *Lipids in Health and Disease*,

localized to the SL membrane. These phospholipase enzymes are defined and named by the site of action on the phospholipid structure and thus differ in their catalytic and regulatory properties. It is pointed out that other cellular membranes such as mitochondrial, sarcoplasmic reticulum and nuclear are also important sites for the localization and action of phospholipases and thus may also play an important role in the regulation of heart function (Panagia et al., 1991; Williams et al., 1997; Cocco et al., 2006; McHowat and Creer, 2004); however, this review will largely focus on the significance of the phospholipases localized to the cardiac SL membrane. The $PLA_2$ isozymes represent a large family of distinct enzymes, each of which demonstrates unique characteristics (Diaz and Arm, 2003). At least four different $PLA_2$ isozymes exist in mammalian cells (Six and Dennis, 2000; Tanaka et al., 2000; Ohto et al., 2005). Secretory $PLA_2$ ($sPLA_2$), also called group II $PLA_2$, requires millimolar $Ca^{2+}$ concentrations for activity, and is secreted into the extracellular space. The cytosolic $PLA_2$ ($cPLA_2$) also known as group IV $PLA_2$, requires increases in intracellular $Ca^{2+}$ for phosphorylation of the enzyme and translocation to intracellular membrane, but does not require $Ca^{2+}$ for its catalytic activity. The $Ca^{2+}$-independent $PLA_2$ ($iPLA_2$) or group VI $PLA_2$, does not require $Ca^{2+}$ for activity. Another isoform of $PLA_2$ is platelet activating factor acetylhydrolase, also known as lipoprotein-associated $PLA_2$ (Lp- $PLA_2$), is a calcium-independent enzyme that cleaves oxidized and polar phospholipids (Allison et al., 2007). The expression of $PLA_2$ is regulated at the transcriptional level by mediators such as cytokines and growth factors, including interferon-$\gamma$, macrophage stimulating factor, tumor necrosis factor and epidermal growth factor (Hirabayashi et al., 2004). The $PLA_2$ enzyme activity is also enhanced by phosphorylation, a process mediated by mitogen activated protein kinases, as well as indirect activation by protein kinase C (PKC) and G-protein coupled receptors (GPCR) (Stahelin et al., 2003). To date, $sPLA_2$, $cPLA_2$ and $iPLA_2$ isozymes have been localized to the cardiac SL membrane whereas Lp-$PLA_2$ is bound predominantly to low density lipoprotein cholesterol (Allison et al., 2007).

The cardiac SL membrane associated PLC isozymes play a central role in activating intracellular signal transduction pathways, especially during early key events in the regulation of various cell functions (Rhee, 2001). A number of different agonists including norepinephrine (NE) and angiotensin II (ANG II), which are released by ischemic myocardial cells, bind to their respective receptors on the cell surface resulting in G-protein (Gq subfamily) activation, which can lead to subsequent stimulation of PLC (Rhee, 2001). The activation of PLC results in the hydrolysis of phosphatidylinositol 4,5 bisphosphate ($PIP_2$) to produce diacylglycerol (DAG) and inositol 1,4,5-trisphosphate ($IP_3$). While $IP_3$ may serve to enhance the sarcoplasmic reticulum (SR) $Ca^{2+}$ release, DAG functions as a potent activator of most PKC isozymes, which in turn phosphorylate several cardiac proteins and stimulate $Ca^{2+}$-influx (Malhotra et al., 2001b; Kamp and Hell, 2000). Recently, the phenylephrine-induced transient increase in intracellular $Ca^{2+}$-concentration caused by $Ca^{2+}$-release from SR has been demonstrated to result in a transient suppression of L-type $Ca^{2+}$ current ($I_{Ca\text{-}L}$) in rat ventricular cardiomyocytes (Zhang et al., 2005). Interestingly, this response is followed by a potentiation of $I_{Ca\text{-}L}$ that may involve PKC. In addition, this biphasic modulation of $I_{Ca\text{-}L}$ by phenylephrine can be blocked by prazosin, indicating that these responses were mediated by the GPCR, $\alpha_1$-adrenoceptor, and thus implicating a role for PLC. The PLC family consists of 6 subfamilies: PLC β, γ, δ, ε, ζ and η (Rhee, 2001; Rebecchi and Pentyala, 2000; Song et al., 2001; Saunders et al., 2002; Wing et al., 2003; Hwang et al., 2005) and are activated by different mechanisms (Lee et al., 1994; Rhee and Bae, 1997; Katan, 1998; Yagisawa et al., 1998; Fukami, 2002; Lopez et al., 2001; Yin et al., 2003). PLC $\beta_1$, $\delta_1$, $\gamma_1$ and two forms of ε are the predominant forms expressed in the heart (Tappia et al., 1999; Lopez et al., 2001). ANG II, $\alpha_1$-adrenergic agonists and endothelin-1 are relevant stimulants of PLC β isozymes via the α subunits of the heterotrimeric Gq subfamily; PLC β has also been shown to be activated by Gβγ dimer (Lee et al., 1994). A non-tyrosine kinase activation of PLC γ isozymes has been reported (Rhee and Bae, 1997), furthermore activation of PLC γ isozymes independent of tyrosine kinase has also been reported (Sekiya et al., 1999). The receptor initiated events for the activation of PLC δ isozymes are considered to be mediated via transglutaminase II, $G_h$, a new class of GTP binding protein (Im et al., 1997; Park et al., 2001). Although the PLC δ-$G_h$ pathway may be an important player in the signaling pathway that regulates calcium homeostasis and modulates physiological processes. PLC ε isozymes are activated by Ras, Rho and Rap 2B as well as by $G\alpha_{12}$ (Song et al., 2001). The activation of PLC ζ and η is far less characterized.

The PLD isozymes hydrolyze phosphatidylcholine (PC) to produce PA, which is considered to be an important lipid signaling molecule. PA can be dephosphorylated to DAG by the action of phosphatidate phosphohydrolase (PAP). Thus, both PLD and PAP can modulate the levels of both PA and PLD-derived DAG in the heart. Different agents such as norepinephrine (NE), endothelin-1 and angiotensin 11 (ANG II) have been shown to increase the formation of PA in the cardiomyocytes (Sadoshima and Izumo, 1993; Ye et al., 1994). The importance of PA in heart function is evident from its ability to stimulate SL and SR $Ca^{2+}$-related transport systems (Dhalla et al., 1997; Xu et al., 1996b) and to increase the intracellular $Ca^{2+}$ concentration in adult cardiomyocytes as well as augment cardiac contractile activity in the normal heart (Xu et al., 1996b). On the other hand, the *in vivo* significance of the PLD-derived DAG remains to be defined (Lamers et al., 1995; Martin et al., 1997; Hodgkin et al., 1998). Two mammalian PLD isozymes, PLD1 and PLD2, have been cloned and have ~50% identity and have been shown to be differentially regulated (Colley et al., 1997; Frohman and Morris, 1999). PLD1, which exhibits low basal activity, requires $PIP_2$ for its activity, and is activated by PKC and Rho small G-proteins family members (Yamazaki et al., 1999). PLD1 is localized to perinuclear regions such as endoplasmic reticulum, golgi apparatus, and endosomes. PLD2 is constitutively active and is the major SL PLD isozyme

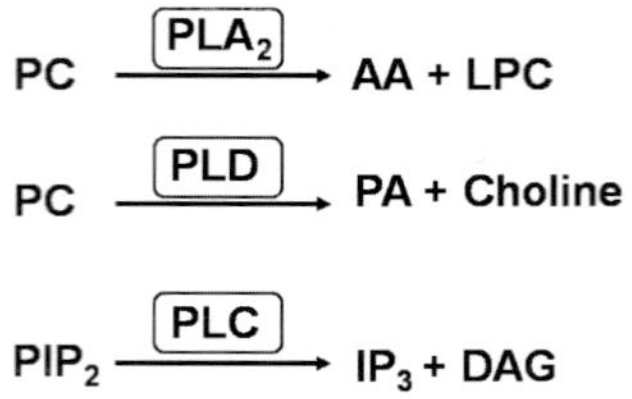

$$PC \xrightarrow{PLA_2} AA + LPC$$

$$PC \xrightarrow{PLD} PA + Choline$$

$$PIP_2 \xrightarrow{PLC} IP_3 + DAG$$

**Fig. 11.1** Products of phospholipase hydrolytic activity on phospholipid substrate. PC, phosphatidylcholine; AA, arachidonic acid; LPC, lysophosphatidic acid; PA, phosphatidic acid: $PIP_2$, phosphatidylinositol 4,5-bisphosphate; DAG, diacylglycerol; $IP_3$, inositol 1,4,5-trisphosphate; $PLA_2$, Phospholipase $A_2$; PLC, Phospholipase C; PLD, Phospholipase D

in the myocardium (Park et al., 2000). PLD2 also requires $PIP_2$ for its activity, but unlike PLD1, it is activated by unsaturated fatty acids such as AA and oleate (Dai et al., 1995; Liu et al., 1998) and is insensitive to the PLD1 activating factors (Colley et al., 1997). Figure 11.1 illustrates the hydrolysis of membrane phospholipids by 3 major phospholipases resulting in the production of intracellular second messengers that may lead to altered signaling in the heart.

## 11.2 Cardiac Hypertrophy is Associated with Activation of Different Phospholipases

The American Heart Association Statistics Committee, 2008 has reported heart disease to be the leading cause of death in the western world. Cardiovascular disease claims one life every 33 seconds (Rosamond et al., 2008). In the year 2007, 15.8 million Americans suffer from cardiovascular disease and 0.45 million died of it and 1234 Americans die everyday (National Centre for Health Statistics). Pathological or reactive cardiac growth is triggered by autocrine and paracrine neurohormonal factors released during biomechanical stress that signal through the Gq/PLC pathway, leading to an increase in cytosolic $Ca^{2+}$ and activation of PKC (Dorn and Force, 2005). In this regard, in stroke prone spontaneously hypertensive rats, the development of cardiac hypertrophy have been suggested to involve an increase in PLC signaling pathway (Kawaguchi et al., 1993). PLC ε-deficient mouse strain had decreased cardiac function and decreased contractile response to acute isoproterenol administration (Wang et al., 2005). These studies suggested that loss of PLC ε signaling sensitizes the heart to development of hypertrophy during chronic cardiac stress. Studies with the cardiomyopathic hamster (BIO 14.6) have shown that cardiac hypertrophy is due to increase in PLC activity as a consequence of enhanced responsiveness to ANG II (Sakata, 1993); in fact, ANG II can initiate cardiac hypertrophy and upregulate signal molecules in the Gαq/11-mediated signal transduction pathway, such as PLC $\beta_3$, and ERK1/2, at both tissue and cellular levels (Bai et al., 2004). Cardiac hypertrophy due to volume overload induced by arteriovenous

(AV) shunt has been linked to PLC $\beta_1$ and $\gamma_1$ activation (Dent et al., 2004a). Furthermore, increases in specific PLC isozyme mRNA levels has also been observed in atrial and right ventricular hypertrophy due to volume overload (Dent et al., 2006). It is interesting to note that pressure overload induced cardiac hypertrophy in guinea pigs, due to a ligature around the descending thoracic aorta, is associated with an increase in PLC $\beta_1$ activity (Jalili et al., 1999). Fas receptor activation is an important component in hypertrophy induced by pressure- and volume-overload and recently it has been reported that Fas-mediated hypertrophy is dependent on the $IP_3$ pathway, which is functionally inter-connected to the PI3K/AKT/GSK3β pathway; both pathways act in concert to cause NFAT nuclear translocation and subsequent hypertrophy (Barac et al., 2005). Recent evidence suggests that ventricular pressure overload hypertrophy led to an upregulation of PLD isozymes both in rat and human heart which may be due to potentiation of PLD activation by α-adrenoceptor and PKC stimulation (Peivandi et al., 2005). PA produced by the activation of PLD also stimulates SL PLC activity (Dhalla et al., 1997; Tappia et al., 2001b). Since DAG, formed due to the activation of PLC, is considered to play a crucial role in regulating the activity of PKC, the positive feedback effect of PA on this pathway may be essential for maintaining the sustained elevation in the activity of PKC during the development of cardiac hypertrophy.

Stimulation of signaling pathways via Gqα provokes cardiac hypertrophy in cultured cardiomyocytes and transgenic mouse models (D'Angelo et al., 1997; Sakata et al., 1998). ANG II receptor, type 1 overexpression has been reported to induce cardiac hypertrophy (Paradis et al., 2000). In this regard, it is pointed out that an essential downstream effector for Gqα is PLC β (Rhee, 2001). The first transgenic murine cardiac hypertrophy model to support a Gqα mechanism of hypertrophy was over expression of the constitutively activated Gq coupled to $\alpha_{iA}$ adrenergic receptor (Milano et al., 1994). In these hearts a chronic activation of PLC resulted in hypertrophy and an increase in the hypertrophic marker gene atrial natriuretic factor (ANF). In isolated adult left ventricular cardiomyocytes we have reported that the NE-induced increases in ANF gene expression and protein synthesis that can be attenuated by a PLC inhibitor, U73122, as well as by an $\alpha_1$-adrenoceptor blocker, prazosin (Singal et al., 2004). The receptor of most growth factors are transmembrane tyrosine kinases and transduce their signal via PLC γ, thus implicating this family of PLC isozymes in cardiac hypertrophy (Hefti et al., 1997). We have recently established that PLC activities regulate their own isozyme gene expression in a PKC-ERK1/2-dependent pathway, which may represent a cycle of events associated with the cardiomyocyte hypertrophic response to NE (Singal et al., 2006). The development of hypertrophy in cultured rat neonatal cardiomyocytes induced by endothelin-1 has been reported to be due to activation of PLC β isozymes (Lamers et al., 1995). In addition, recent studies in neonatal rat cardiomyocytes stimulated with different hypertrophic stimuli, have shown an increased mRNA expression and protein level of PLC β isozymes (Schnabel

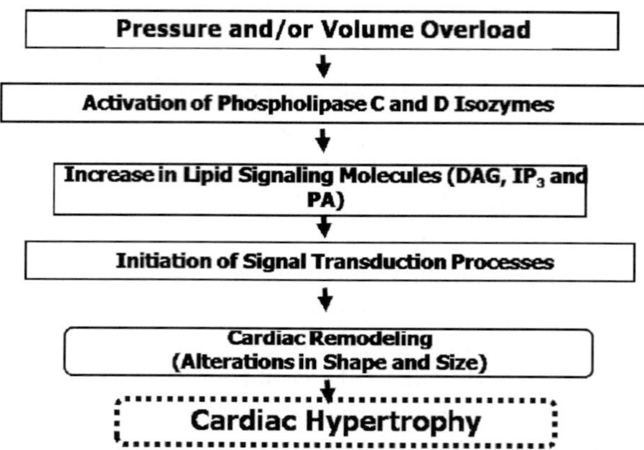


**Fig. 11.2** Hemodynamic overload leading to cardiac hypertrophy through activation of Phospholipase C and D. DAG, diacylglycerol; $IP_3$, inositol 1,4,5-trisphosphate; PA, phosphatidic acid

et al., 2000). In brief, the above evidences suggest that activation of phospholipases play a key role in mediating cardiac hypertrophy as shown in Fig. 11.2.

## 11.3 Differential Changes in Phospholipases in Congestive Heart Failure

In 2004, 1 in 8 death certificates (284,365 deaths) in the US indicates heart failure as the cause of death with total death rate for heart failure were 52 per 1000. Total death rate was 63.2 for white males, 43.5 for white females and 78.8 for black males and 58.7 for black females. The estimated direct and indirect cost of heart failure in the US for 2008 is 34.8 billion (Heart disease and stroke statistics-2008 update, American Heart Association, Rosamond et al., 2008). Congestive heart failure (CHF) is invariably preceded by cardiac hypertrophy as an adaptive mechanism due to a wide variety of neurohumoral changes; however, the mechanisms for the transition of cardiac hypertrophy to heart failure are poorly understood. Although the initial outcome is a compensatory growth of the heart, prolonged development of hypertrophy leads to CHF. We have reported that abnormalities in protein abundance and activity and in the cellular localization of myocardial $PLA_2$ isozymes occur at the overt stage of CHF after myocardial infarction (MI) (McHowat et al., 2001). Since PLD-derived PA influences intracellular $Ca^{2+}$ concentration and contractile performance of the cardiomyocytes, changes in $iPLA_2$ activity may contribute to abnormal contractile performance of the failing heart via an impaired interaction of $PLA_2$ with PLD pathway.

We have examined the changes in activities and SL protein abundance of PLC isozymes in CHF due to MI (Tappia et al., 1999, 2001b; Ju et al., 1998).

While profound decreases in SL PLC $\gamma_1$ and $\delta_1$ activities and protein levels occur in CHF, treatment of animals with the ACE inhibitor imidapril partially corrected these changes in PLC isozymes forms as well as cardiac function (Tappia et al., 1999). This would suggest that RAS may be involved in mediating alterations in PLC isozymes and that PLC isozymes could serve as novel targets for the treatment of CHF. In addition, PLC isozymes could be another target for the mechanisms of action of ACE inhibitors. It is pointed out that an upregulation of the Gq$\alpha$/PLC-$\beta$ pathway in the viable, border, and scar tissues in the post-MI hearts is also seen, which may play an important role in cardiac fibrosis and scar remodeling (Ju et al., 1998).

Differential changes in SL PLC isozyme activities and their SL abundance in the failing heart of the cardiomyopathic hamster (UM X7.1) have also been detected (Ziegelhoffer et al., 2001). Although a decrease in PLC $\delta_1$ isozyme activity and an increase in PLC $\beta_1$ and $\gamma_1$ isozyme activities were detected, an elevation of $IP_3$ levels were seen; while the relevance of the changes of PLC isozymes remains to be defined, the increase in $IP_3$ may contribute to intracellular calcium overload in the failing cardiomyocytes of cardiomyopathic hamster. Volume overload induced cardiac hypertrophy invariably precedes CHF and in this model we have also reported decreases in PLC isozymes and SL level of $PIP_2$ (Dent et al., 2004b). It is interesting to note that in the cardiomyopathic hamster and in CHF due myocardial infarction and volume overload the decrease in SL PLC $\delta_1$ activity correlates to its reduced SL protein abundance (Dent et al., 2004a; Ziegelhoffer et al., 2001). The $NH_2$-terminal part of the PH domain of PLC $\delta_1$ has a high affinity for the polar head group of $PIP_2$ that confers a unique capacity of PLC $\delta_1$ to associate with the SL membrane (Tall et al., 1997). The reported decrease in the SL $PIP_2$ content may be a mechanism, which reduces the attachment of PLC $\delta_1$ to the SL membrane. It should be noted that in addition to $PIP_2$ serving as a substrate for PLC and as a membrane attachment site, a number of diverse biochemical events are also regulated by $PIP_2$ and are affected by the altered concentration of this lipid in the membrane. Of note, the decreased number of $PIP_2$ molecules could compromise the contractile performance of the heart by directly causing a depression of the inward rectifier $K^+$ channels (Huang et al., 1998), as well as depression of the SL $Na^+$-$Ca^{2+}$ exchanger and $Ca^{2+}$-pump activities (Hilgemann and Ball, 1996). Therefore, a diminished amount of $PIP_2$ inside the SL membrane may be critical for cardiac dysfunction during CHF.

Although in CHF due to MI an increase in PLD2 gene expression as well as SL PLD2 protein abundance and activity has been reported, an increase in PLD2 protein and activity has also been detected in scar tissue that may be involved in scar remodeling (Dent et al., 2004b). In addition, a greater increase in SL phosphatidate phosphohydrolase (PAP) type 2 activity was observed; the net effect of PLD-derived PA formation and PAP-mediated dephosphorylation of PA was a decrease in the SL PA level and an impairment of the bioprocesses mediated by SL PA as well as a defective interaction between SL PLD and PLC signaling pathways as evidenced by a loss of PA-induced increase in $Ca^{2+}$

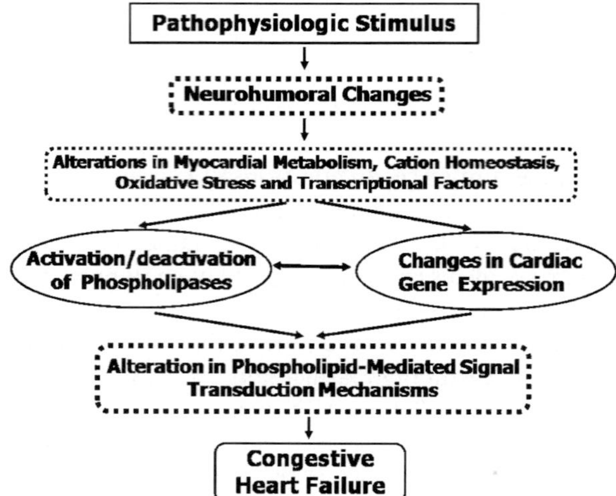


**Fig. 11.3** Mechanisms contributing to altered phospholipid-mediated signal transduction in congestive heart failure

transients due to a diminished stimulation of PLC activities in failing cardiomyocytes (Tappia et al., 2001b, 2003). This signal transduction pathway could constitute an important target for pharmacological interventions as treatment of CHF animals with imidapril, an angiotensin converting enzyme (ACE) inhibitor, normalized PA levels in the cardiac SL membrane (Yu et al., 2002). Figure 11.3 depicts the possible mechanisms that lead to alterations in the different phospholipase isozyme activities and subsequent changes in phospholipid-mediated signal transduction mechanisms leading to congestive heart failure. In summary, it is evident that phospholipid-signaling systems may have an important role in cardiac hypertrophy as well as contractile dysfunction in CHF and may be potential targets for drug development.

## 11.4 Phospholipid-Mediated Signaling in Diabetic Cardiomyopathy

The World Health Organization estimates that by 2015, the number of overweight people globally will increase to 2.3 billion, and more than 700 million people will be obese. Mayo clinic data indicate that the prevalence of diabetes increased 3.8% every year. The total prevalence of diabetes mellitus in US is expected to more than double from 2005 to 2050 in all age, sex and race groups. Data from the National Diabetes Information Clearing House, states at least 65% of people with diabetes die of some form of heart disease or stroke. Among women with CHD, diabetes is the strongest risk factor for heart failure in US (American Heart Association, 2008, Rosamond et al., 2008). In 2002, the direct and indirect cost attributable to diabetes was \$132 billion. A disproportionately high prevalence of diabetes in African and Mexican Americans when compared to Caucasians has been observed. Cardiovascular disease is responsible for 80% of deaths among diabetic patients much of which has been attributed to coronary artery disease (Hayat et al., 2004). In fact, the incidence of heart disease is greater in the diabetic population than the non-diabetic population (Regan, 1983). However, the presence of a primary cardiomyopathy in diabetes has been long identified (Dhalla et al., 1985; Fein et al., 1980; Fein and Sonnenblick, 1985; Penpargkul et al., 1980; Regan et al., 1974). Diabetic cardiomyopathy is a cardiac disease that arises as a result of the diabetic state, independent of vascular or valvular pathology. It manifests initially as asymptomatic diastolic dysfunction, which progresses to symptomatic heart failure. The compliance of the heart wall is decreased and contractile function is impaired. The pathophysiology is incompletely understood, but appears to be initiated both by hyperglycemia and changes in cardiac metabolism. These changes induce oxidative stress and activate a number of secondary messenger pathways, leading to cardiac hypertrophy, fibrosis and cell death (Sharma and McNeill, 2006). A number of animal studies have found that this cardiomyopathic condition is associated with defects in the capacity of cardiomyocytes to regulate intracellular ionic homeostasis in a normal manner (Allo et al., 1991) resulting in abnormal $Ca^{2+}$ transients and contractile activity (Lagadic-Gossmann et al., 1996; Ganguly et al., 1983; Heyliger et al., 1987; Horackova and Murphy, 1988). Furthermore, alterations in contractile proteins and intracellular ions impair excitation-contraction coupling, while decreased autonomic responsiveness and autonomic neuropathy impair its regulation. Extensive structural abnormalities also occur, which have deleterious mechanical and functional consequences (Sharma and McNeill, 2006). The overpresentation of diabetic patients with heart failure trials such as SOLVD (Studies Of Left Ventricular Dysfunction); 26% (Shindler et al., 1996), ATLAS (Assessment Trial of Lisinopril And Survival); 19% (Ryden et al., 2000) and V-HeFT II (Vasodilator-Heart Failure Trial II); 20% (Cohn et al., 1991) attests to the prevalence of this condition in the diabetic population.

Diabetic cardiomyopathy is a result of maladaptive changes in energy homeostasis. Although diabetes-induced changes in $PLA_2$ activities have been measured in several tissues, very little is known about the status of $PLA_2$ isozymes in diabetic myocardium. An increased membrane-associated $iPLA_2$ activity however, has been observed in the hearts of rats with streptozotocin (STZ)-induced diabetes (McHowat et al., 2000), which may be due to the diabetes-induced increase in $iPLA_2$ mRNA level in rat myocardium (Su et al., 2005). This increase in $iPLA_2$ activity was accompanied by an increase in LPC production. These investigators also demonstrated that the diabetes-induced changes in $iPLA_2$ activity and LPC production were reversed by insulin treatment of diabetic animals and concluded that diabetes-induced changes in membrane phospholipid content and phospholipid hydrolysis may contribute to some of the alterations in myocardial function that are observed in diabetic patients. In addition, the molecular species of the major phospholipid classes in SL membrane of STZ-diabetic rats have also been examined (Vecchini et al., 2000). The relative content of plasmalogens increased in all the phospholipid classes of diabetic SL membrane. PC and phosphatidylethanolamine were mostly enriched with molecular species containing linoleic acid and deprived of the molecular species containing AA. The molecular species of phosphatidylserine containing either AA or docosahexaenoic acid were less abundant in membranes from diabetic rats than in membranes from controls. Insulin treatment of diabetic rats restored the species profile of phosphatidylethanolamine and overcorrected the changes in molecular species of PC. These investigators concluded that the high SL level of plasmalogens and the abnormal molecular species of glycerophospholipids may be critical for the membrane dysfunction and defective contractility of the diabetic heart.

While ANG II and PKC have been implicated in cardiac dysfunction during diabetes (Malhotra et al., 2001a; Liu et al., 1999), virtually nothing is known about the status of PLC in the diabetic heart. In acute diabetes (3 days, after the induction) the enhanced inotropic response to methoxamine, an $\alpha_1$-adrenoceptor agonist, was ascribed to an increased PLC activity (Wald et al., 1988). We have earlier reported that the total cardiac SL PLC activities are significantly decreased in STZ-induced chronic diabetic rats under *in vitro* assay conditions (Tappia et al., 2001a). In isolated cardiomyocytes we also observed a reduced concentration of basal as well as PA-induced $IP_3$ generation in diabetic rats (Tappia et al., 2004b), suggesting that decreased basal PLC activities *in vivo* may exist in diabetic cardiomyopathy. We have recently reported that the decrease in the total SL PLC in diabetes is associated with a decrease in SL PLC $\beta_3$ activity, which immunofluorescence in frozen diabetic left ventricular tissue sections revealed to be due to a decrease in PLC $\beta_3$ protein abundance; a 2-week insulin treatment of 6 wk diabetic animals partially normalized these parameters (Tappia et al., 2004a). The functional significance of the defective total PLC activities, including PLC $\beta_3$ activity, and diminished levels of $IP_3$, is that it may constitute a mechanism for the reported reduced force of contraction in response to $\alpha_1$-adrenergic stimulation of the isolated papillary muscle (Heyliger et al., 1982), however an enhanced inotropic response to $\alpha_1$-adrenergic stimulation in the isolated working heart from diabetic rats has also been reported (Heijnis et al., 1992). On the other hand, a reduced production of PLC-derived DAG would affect several cellular processes (Puceat and Vassort et al., 1996). While abnormalities in other signaling pathways occur during diabetes, in particular, the β-adrenoceptor induced increases in contractions and $[Ca^{2+}]_i$ transients which are markedly diminished (Tamada et al., 1998; Ha et al., 1999), it can be suggested that an impairment of PLC signaling mechanisms may also significantly contribute to a defective cardiac contractile performance during diabetes. Furthermore, it is pointed out that depressed activities of other PLC isozymes in diabetic cardiomyopathy have also been observed (Tong et al., 1998; Tappia et al., 2000).

Although the positive inotropic effect of PA on the isolated perfused heart of STZ and alloxan-induced diabetic rats has been demonstrated (Xu et al., 1996a), the effects of exogenous PA on $Ca^{2+}$ transients and contractile activity have also been reported in cardiomyocytes isolated from chronic STZ-induced diabetic rats. The PA induced contractility was correlated to an attenuated PA-induced $IP_3$ generation in diabetic rat cardiomyocytes (Tappia et al., 2004b). Insulin treatment of the diabetic animals resulted in a partial recovery of PA responses and it was suggested that a defect in the PA-PLC signaling pathway in diabetic rat cardiomyocytes may contribute to the depressed cardiac contractile performance during diabetes, similar to the defect in CHF (Tappia et al., 2003). A number of possible mechanisms can be proposed to explain the depressed PLC activities in diabetic cardiomyopathy. As already indicated PA is a potent stimulator of PLC. A decrease in SL PA formation due to an impaired PLD activity has been reported (Williams et al., 1998), which could result in an attenuated stimulation of PLC. In this regard, it is interesting to note that the marked reduction of AA content of PC in SL membrane of diabetic heart could represent a mechanism of a defective PLD activity. An increase in total myocardial DAG level has been reported in STZ-induced diabetic rats and in spontaneous autoimmune diabetic BB rats (Okumura et al., 1988; Inoguchi et al., 1992). Increase in membrane DAG content has been shown to destabilize the membrane and structural transitions (Das and Rand, 1984; Das and Rand, 1986) and this may have an inhibitory effect on PLC activity. It is pointed out that oxidative stress has been shown to occur during diabetic cardiomyopathy (Dhalla et al., 1998). Since SL PLC is inhibited by oxidants through reversible modification of the associated thiol groups (Meij et al., 1994), the depressed PLC activities seen in diabetes could also in part be explained by the oxidant-induced alteration of thiol groups. *In vitro* studies have demonstrated that LPC inhibits both SL PI 4 kinase and PI 4-P 5 kinase activities (Liu et al., 1997). This is of particular relevance as LPC accumulates in SL during diabetic cardiomyopathy, suggestive of a diminished synthesis of $PIP_2$ substrate for PLC. Furthermore, oxidants have also been shown to inhibit both SL PI 4 kinase and PI 4-P 5 kinase activities (Mesaeli et al., 2000), as well as SL total PLC activity (Meij et al., 1994). Since substrate availability determines hydrolytic activity of PLC, such mechanisms could additionally contribute to a decrease in PLC activity. In addition, the decrease SL $PIP_2$ level may also contribute to the depressed cardiac contractility independent of the effects on PLC activities (Huang et al., 1998). It is also interesting to note that the hexosamine pathway has been suggested to inhibit phenylephrine-induced inotropy of the diabetic heart (Pang et al., 2004), which may be related to defective PLC activities. In summary, on the basis of the limited information available in the literature, it can be suggested that diabetes-induced changes in the membrane composition as well as phospholipid-mediated signaling systems may contribute to the depressed contractility of the diabetic myocardium (Fig. 11.4).

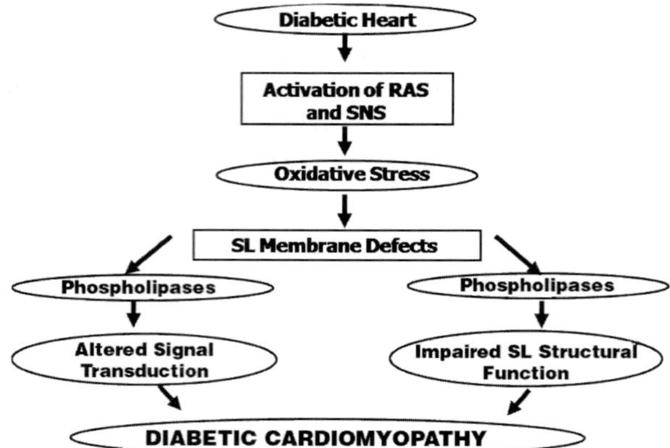


**Fig. 11.4** Diabetes-induced generation of free radicals leading to altered membrane composition and sarcolemmal phospholipid-mediated signaling systems. RAS, renin angiotensin system; SNS, sympathetic nervous system; SL, sarcolemma

## 11.5 Myocardial Ischemia-Reperfusion and Alterations in Phospholipid-Mediated Signal Transduction

Approximately 12 million visited a physician's office for ischemic heart disease (IHD) in the USA 2001 (National Hospital Ambulatory Medical Care Survey: 2001). Presently, it is estimated that 18.5 million people in the USA suffer from IHD. Of particular concern are the growing disparities in IHD mortality among African Americans relative to Hispanics and non-Hispanic whites (Karter et al., 1998). Myocardial ischemia is known to produce dramatic changes in cardiac function, metabolism and ultrastructure (Jennings and Reimer, 1991; Hearse and Bolli, 1992) as well as proteolysis (Yoshida et al., 1995), DNA fragmentation (Scarabelli et al., 2001) and oxidative stress (Dhalla et al., 1999; Dhalla et al., 2000) (Fig. 11.5); however, the cellular and molecular events leading to contractile dysfunction and derangement of cardiac structure are not clearly understood. Although re-institution of coronary flow to the ischemic heart is considered beneficial for the recovery of cardiac pump function, reperfusion after a certain period of ischemia has been shown to further aggravate the myocardial abnormalities (Bolli and Marban, 1999; Dhalla et al., 1999, 2000; Piper et al., 2003; Marczin et al., 2003; Kim et al., 2003). Ischemia-reperfusion (I-R) injury is known to occur during, clinical procedures such as coronary bypass surgery, angioplasty, thrombolytic therapy and cardiac transplantation (Dhalla et al., 1999). The cardiac pump failure and changes in cardiac cell ultrastructure due to I-R or hypoxia-reoxygenation involve a variety of complex pathophysiological abnormalities and our current information on these aspects is largely based on the beneficial effects of a number of drug interventions for the treatment of IHD. For example, the beneficial effects of $Ca^{2+}$ antagonists (Urquhart et al., 1985; Cavero and Spedding, 1983) and $Na^{+}$-$H^{+}$ exchange

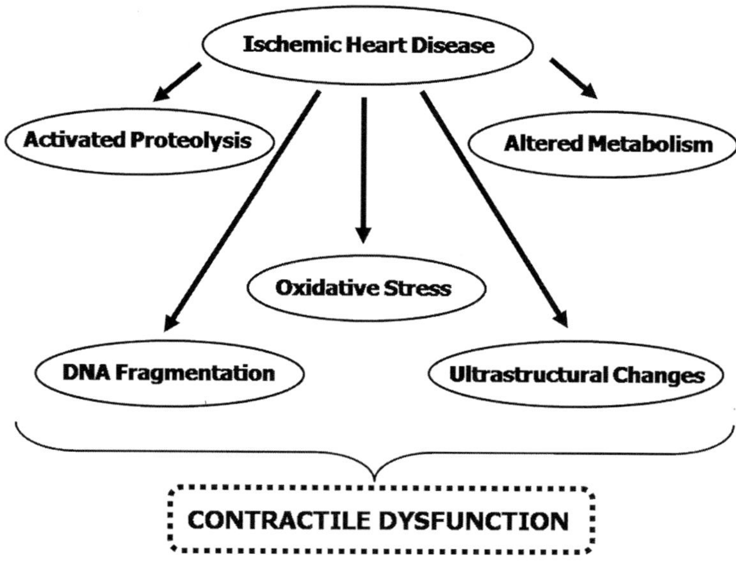


**Fig. 11.5** Cardiac and other defects associated with ischemic heart disease

inhibitors (Avkiran and Snabaitiset al, 1999; Hartmann et al., 1999; Haist et al., 2003; Xiao et al., 2003) have supported a role of intracellular $Ca^{2+}$-overload (Dhalla et al., 2001; Miyamae et al., 1996; Jeremy et al., 1992), whereas those of antioxidants (Dhalla et al., 1999, 2000) suggest the involvement of oxidative stress in the pathophysiology of IHD. Indeed, the role of reactive oxygen species (ROS) in the genesis of myocardial cell damage and subsequent contractile dysfunction is established (Kloner et al., 1983; Ungvari et al., 2005). Since intracellular $Ca^{2+}$ overload is considered to play a crucial role in the I-R injury and cardiac dysfunction (Dhalla et al., 1978, 1982), it is possible that several mechanisms, which are involved in the regulation of $Ca^{2+}$ movements in the myocardial cell, are altered by ROS, including phospholipases.

The activation of $PLA_2$ is known to lead to an accumulation of AA within the membrane phospholipid pool of the ischemic myocardium. Some metabolites of AA produce detrimental effects in the heart in ischemia and proinflammatory effects in reperfusion, whereas others have been recently shown to reduce I-R injury in the heart (Gross et al., 2005). Choline released into the heart perfusate is found to be a useful indicator of phospholipid degradation caused by I-R (Bruhl et al., 2004). Choline glycerophospholipid, in particular PC and plasmenylcholines, are degraded by $PLA_2$ to lysophospholipids and lyosophosphatidylcholine (LPC), which is a known arrhythmogenic agent. Recent studies have revealed that LPC produces mechanical and metabolic derangements in working rat hearts, and $Ca^{2+}$ overload in isolated cardiomyocytes. Thus, LPC possesses an ischemia like effect on the heart. LPC accumulated in the myocardium activates $PLA_2$, establishing a vicious cycle of exacerbated LPC production (Hashizume et al., 1997).

Although I-R induced changes in PLC activities have been reported to be associated with cardiac dysfunction due to I-R (Anderson et al., 1995; Otani et al., 1988; Moraru et al., 1995; Kurz et al., 1999; Mouton et al., 1991; Schwertz and Halverson, 1992; Munakata et al., 2002), in these studies no attempt was made to identify specific PLC isozyme changes. We have been the first to report that while cardiac ischemia is associated with an activation of SL PLC $\beta_1$ and decreased SL PLC $\gamma_1$ and $\delta_1$ activities, reperfusion of the ischemic heart results in activation of SL PLC $\gamma_1$ and $\delta_1$ isozymes, whereas PLC $\beta_1$ activity progressively declines (Asemu et al., 2003, 2004). Although exposure of SL membranes and isolated cardiomyocytes to oxidants induces changes in PLC and components of the phosphoinositide pathway (Meij et al., 1994; Mesaeli et al., 2000), the effects of oxidants on specific PLC isozymes has not been completely examined. In this regard, we are also the first to have reported that treatment of cardiomyocytes with $H_2O_2$ results in an activation of PLC $\gamma_1$ (Mangat et al., 2006). It was suggested that PLC $\gamma_1$ might play a role in cardiomyocyte survival during oxidative stress via PKC $\varepsilon$ and phosphorylation of Bcl-2. Furthermore, blockade of PLC activities with U73122 results in an augmentation of the $H_2O_2$ induced cardiomyocyte apoptosis, while no effect on $H_2O_2$ induced necrotic cell death (Asemu et al., 2003). These data suggest that PLC-mediated signaling transduction may initiate anti-apoptotic signals in cardiomyocytes during oxidative stress. Cardiac I-R is also associated with an increase in PLC $\delta_1$. Given that PLC isozymes are dependent on $Ca^{2+}$, they activate $Ca^{2+}$-transporting systems and that PLC $\delta$ isozymes are considered $Ca^{2+}$-amplifiers (Rebecchi and Pentyala, 2000), it is conceivable that activated PLC $\delta_1$ may contribute to a self-perpetuating cycle that exacerbates cardiomyocyte $Ca^{2+}$-overload and subsequent cardiac dysfunction during I-R. Although it appears that $Ca^{2+}$ may be involved in the activation of this PLC isozyme (Asemu et al., 2004), the role of oxidants cannot be excluded. However, the mechanisms responsible for and the significance of the changes in specific SL PLC isozyme activities, protein contents and gene expression with respect to $Ca^{2+}$-homeostasis and cardiac dysfunction in I-R have not been investigated.

The I-R-induced changes are not limited to $PLA_2$ and PLC. Some studies have examined the redox regulation of the cardiac PLD activities in the setting of I-R or with isolated cardiac SL and SR membrane preparations. A recent study has suggested that a direct Rho A-PLD1 interaction stimulates PLD1 activity, which mediates the cardioprotective effect of adenosine A3 receptor, establishing an important antiischemic role of PLD (Mozzicato et al., 2004). While some investigators have reported that the activation of PLD is associated with an improvement of post-ischemic functional recovery and attenuation of cellular injury (Tosaki et al., 1997), other studies, as well as work from our laboratory have found variable changes in the PLD activity in the ischemic heart (Bruhl et al., 2003; Kurz et al., 2004; Asemu et al., 2005). Our studies have revealed that PLD2 activity is increased in early reperfusion of the 30 min ischemic heart, whereas in prolonged reperfusion PLD2 activity is significantly depressed (Asemu et al., 2005). While the activation of PLD2 may represent a cell survival response, evoked by low concentrations of oxidants for a relatively short period of time, the depressed activity may be in response to cardiomyocyte damage due to a more prolonged exposure to high concentrations of oxidants and may be linked to the poor functional recovery of the heart due to oxidative stress following I-R. In this regard, it is known that I-R leads to depletion of GSH content (Ferrari et al., 1991; Ozer et al., 2005), which is the major intracellular non-protein sulfydryl and plays an important role in the maintenance of cellular proteins and lipids in their functional state and acts primarily to protect these important structures against the threat of oxidation (Wu et al., 2004; Hurd et al., 2005). In view of the observed depression of PLD2 activity in prolonged reperfusion (Asemu et al., 2005), it is conceivable that this may be due to oxidation of PLD2. The increase in the SL PLD2 activity in early reperfusion may be due to post-translational modifications as a result of the formation of ROS. On the other hand, we have reported that a $Ca^{2+}$-independent $PLA_2$ and subsequent mobilization of the unsaturated fatty acid has been shown to modulate the activity of PLD in heart SL (Liu et al., 1998). Interestingly, the $PLA_2$ is also activated by $H_2O_2$ (Sapirstein et al., 1996), which could provide a mechanism of an indirect regulation of the SL PLD2 activity by $H_2O_2$. It is pointed out that the basal myocardial PLD activity has been reported to be decreased by the tyrosine kinase inhibitor, genistein, and increased by vanadate, a tyrosine phosphatase inhibitor (Lindmar and Loffelholz, 1998). In view of the fact that tyrosine kinases are activated in response to oxidants (Snabaitis et al., 2002; Purdom and Chen, 2005), it is conceivable that SL PLD isozyme activities could also be indirectly regulated by oxidant-induced activation of tyrosine kinase activities. Since a significant degree of PLD activity is also localized in myocardial SR membranes (Panagia et al., 1991), it is possible that it may take part in the regulation of $Ca^{2+}$-movements (Asemu et al., 2005). Thus, we also observed a decrease in the SR PLD2 activity after 5 min of reperfusion, which was suggested to be a reversible oxidation because the SR PLD2 activity was recovered after 30 min reperfusion.

While both SL and SR PLD activities, *in vitro*, have been reported to be inhibited by oxidants such as $H_2O_2$ and HOCI, through reversible modification of associated thiol groups (Dai et al., 1992, 1995), some of the inconsistencies between the observations in the isolated perfused heart and oxidant effects on SR and SL preparations could be explained on the basis that the functional thiol groups of the SL PLD2 in the isolated perfused heart are not as readily accessible by oxidants as these are in the isolated SL preparation. Such responses may also be due to differences in the sensitivity of the SR and SL PLD to different concentrations of oxidant molecules as well as ROS. Nonetheless, the elucidation of the detailed mechanism of PLD activation by thiol modulating agents will be of importance for clearly understanding the oxidant-induced signal transduction pathways and PLD regulation under different types of oxidative stress.

A comment about redox signaling must also be made. Although it is well known that I-R results in cardiomyocyte death by apoptosis as well as necrosis (Olivetti et al., 1997), it has also been shown that ROS produced during I-R can induce a number of anti-apoptotic genes and transcription factors (Nishio et al., 1998).

Thus, it appears that cardiomyocyte death induced by I-R is a net effect of the redox regulated cell survival signals and ROS triggered cell death mechanisms. The first unequivocal evidence for the role of ROS as a second messenger for cell survival was observed with the production of ROS during the agonist-induced activation of NFκB which regulates the inducible expression of a number of genes such as Bcl-2 and pro-apoptotic factors including Bax and p53, in the I-R myocardium (Bromme and Holtz, 1996) via a signal transduction pathway that could involve PLD. Subsequently, myocardial adaptation to ischemia due to ischemic preconditioning (IP), where a brief period of ischemia prior to a prolonged period improves myocardial function and diminishes the infarct size, was also observed to be associated with the generation of ROS (Tritto and Ambrosio, 2001; Dhalla et al., 1998). It was indicated that PLD may play an important role in IP and may be related to ROS generation during IP (Tritto and Ambrosio, 2001; Dhalla et al., 1998). In fact, the redox signaling which is considered to protect the heart during IP involves the phosphorylation of tyrosine kinases and activation of multiple kinases (Moraru et al., 1992; Cohen et al., 1996), including PKC isozymes (Eskildsen-Helmond et al., 1996). Such a cascade of signal transduction for all survival has been linked to the activation of PLD (Moraru et al., 1992; Cohen et al., 1996; Trifan et al., 1996). It should be noted that agonists of PLD simulate the effects of IP, whereas the inhibition of PLD blocks the beneficial effects of IP (Ozer et al., 2005; Hashizume et al., 1997). In summary, the two extremes of stress imposed on the heart under I-R change the redox potential of the cardiomyocyte and affect redox-sensitive molecules involved in phospholipid-mediated signal transduction mechanisms. The early activation of specific PLC and PLD isozymes may represent initiation of redox signaling and cell survival pathways, whereas inactivation of specific PLC and PLD isozymes may contribute to impaired cardiac contractile function (Fig. 11.6).

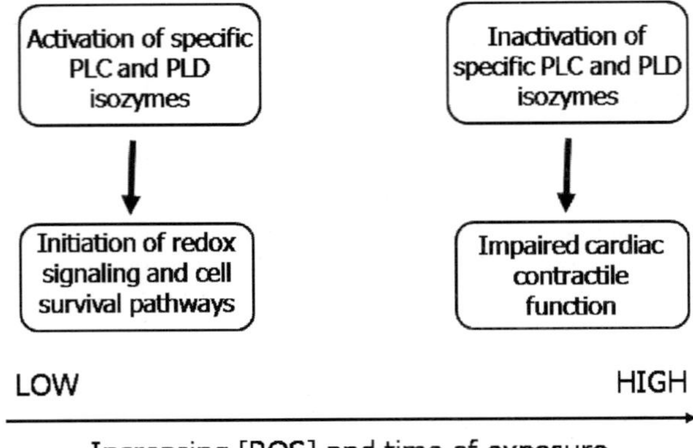


**Fig. 11.6** Role of Phospholipase C and D in the cardiac response to different concentrations and time of exposure to reactive oxygen species. ROS, reactive oxygen species; PLC, Phospholipase C; PLD, Phospholipase D

## 11.6 Concluding Remarks

Several lines of evidence have revealed that phospholipid-mediated signal transduction mechanisms are impaired in the diseased myocardium. However, the contribution of these signaling systems with respect to other myocardial signaling systems, to cardiac function and the myocardial genetic machinery needs to be defined. The precise mechanisms of regulation of the cardiac phospholipase activities also remain to be completely understood. Although the cellular and molecular mechanisms responsible for changes in phospholipid signaling systems are being investigated, it could emerge that specific phospholipase isozymes might constitute additional therapeutic targets for drug discovery for the treatment of heart disease, however with the increasing complexity of phospholipid-mediated signal transduction mechanisms, the achievement of this outcome will be a major challenge.

**Acknowledgments** The work reported in this article was supported by grants from the Canadian Institutes of Health Research, Heart and Stroke Foundation of Manitoba, Manitoba Health Research Council and the St. Boniface Hospital Research Foundation. TS is a recipient of a Manitoba Health Research Council PhD Studentship Award.

## References

Allison M.A., Denenberg J.O., Nelson J.J., Natarajan L. Criqui M.H. The association between lipoprotein-associated phospholipase $A_2$ and cardiovascular disease and total mortality in vascular medicine patients. J. Vasc. Surg. 46 (2007) 500–506.

Allo S.N., Lincoln T.M., Wilson G.L., Green F.J. Watanabe A.M., Schaffer S.W. Non-insulin-dependent diabetes-induced defects in cardiac cellular calcium regulation. Am. J. Physiol. Cell. Physiol. 260 (1991) C1165–C1171.

Anderson K.E., Dart A.M., Woodcock E.A. Inositol phosphate release and metabolism during myocardial ischemia and reperfusion in rat heart. Circ. Res. 76 (1995) 261–268.

Asemu G., Dent M.R., Singal T., Dhalla N.S., Tappia P.S. Differential changes in phospholipase D and phosphatidate phosphohydrolase activities in ischemia-reperfusion of rat heart. Arch. Biochem. Biophys. 436 (2005) 136–144.

Asemu G., Dhalla N.S., Tappia P.S. Inhibition of PLC improves postischemic recovery in isolated rat heart. Am. J. Physiol. Heart Circ. Physiol. 287 (2004) H2598–H2605.

Asemu G., Tappia P.S., Dhalla N.S. Identification of the changes in phospholipase C isozymes in ischemic-reperfused rat heart. Arch. Biochem. Biophys. 411 (2003) 174–182.

Avkiran M., Snabaitis A.K. Regulation of cardiac sarcolemmal $Na^+/H^+$ exchanger activity: potential pathophysiological significance of endogenous mediators and oxidant stress. J. Thromb. Thrombolysis. 8 (1999) 25–32.

Bai H., Wu L.L., Xing D.Q., Liu J., Zhao Y.L. Angiotensin II induced upregulation of $G\alpha q/11$, phospholipase C $\beta 3$ and extracellular signal-regulated kinase 1/2 via angiotensin II type 1 receptor. Chin. Med. J. (Engl). 117 (2004) 88–93.

Barac Y.D., Zeevi-Levin N., Yaniv G., Reiter I., Milman F., Shilkrut M., Coleman R., Abassi Z., Binah O. The 1,4,5-inositol trisphosphate pathway is a key component in Fas-mediated hypertrophy in neonatal rat ventricular myocytes. Cardiovasc. Res. 68 (2005) 75–86.

Bolli R., Marban E. Molecular and cellular mechanisms of myocardial stunning. Physiol. Rev. 79 (1999) 609–634.

Bromme H.J., Holtz J. Apoptosis in the heart: when and why? Mol. Cell Biochem. 163–164 (1996) 261–275.

Bruhl A., Faldum A., Loffelholz K. Degradation of phosphatidylethanol counteracts the apparent phospholipase D-mediated formation in heart and other organs. Biochim. Biophys. Acta. 1633 (2003) 84–89.

Bruhl A., Hafner G., Loffelholz K. Release of choline in the isolated heart, an indicator of ischemic phospholipid degradation and its protection by ischemic preconditioning: no evidence for a role of phospholipase D. Life Sci. 75 (2004) 1609–1620.

Buckland, A.G., Wilton, D.C. Anionic phospholipids, interfacial binding and the regulation of cell functions. Biochim. Biophys. Acta. 1483 (2000) 199–216.

Cavero I., Spedding M. "Calcium antagonists": a class of drugs with a bright future. Part I. Cellular calcium homeostasis and calcium as a coupling messenger. Life Sci. 33 (1983) 2571–2581.

Cocco L., Faenza I., Fiume R., Maria Billi A., Gilmour R.S., Manzoli F.A. Phosphoinositide-specific phospholipase C (PI-PLC) $\beta_1$ and nuclear lipid-dependent signaling. Biochim. Biophys. Acta 1761 (2006) 509–521.

Cohen M.V., Liu Y., Liu G.S., Wang P., Weinbrenner C., Cordis G.A., Das D.K., Downey J.M. Phospholipase D plays a role in ischemic preconditioning in rabbit heart. Circulation 94 (1996) 1713–1718.

Cohn J.N., Johnson G., Ziesche S., Cobb F., Francis G., Tristani F., Smith R., Dunkman W.B., Loeb H., Wong M., et al. A comparison of enalapril with hydralazine-isosorbide dinitrate in the treatment of chronic congestive heart failure. N. Engl. J. Med. 325 (1991) 303–310.

Colley W.C., Sung T.C., Roll R., Jenco J., Hammond S.M., Altshuller Y., Bar-Sagi D., Morris A.J., Frohman M.A. Phospholipase D2, a distinct phospholipase D isoform with novel regulatory properties that provokes cytoskeletal reorganization. Curr. Biol. 7 (1997) 191–201.

Dai J., Meij J.T., Dhalla V., Panagia V. Involvement of thiol groups in the impairment of cardiac sarcoplasmic reticular phospholipase D activity by oxidants. J. Lipid Mediat. Cell Signal. 11 (1995) 107–118.

Dai J., Meij J.T.A., Padua R.R., Panagia V. Depression of cardiac sarcolemmal phospholipase D activity by oxidant-induced thiol modification. Circ. Res. 71 (1992) 970–977.

D'Angelo D.D., Sakata Y., Lorenz J.N., Boivin G.P., Walsh R.A., Liggett S.B., Dorn G.W. 2nd. Transgenic G$\alpha$q overexpression induces cardiac contractile failure in mice. Proc. Natl. Acad. Sci. USA 94 (1997) 8121–8126.

Das S., Rand R.P. Diacylglycerol causes major structural transitions in phospholipid bilayer membranes. Biochem. Biophys. Res. Commun. 124 (1984) 491–496.

Das S., Rand R.P. Modification by diacylglycerol of the structure and interaction of various phospholipid bilayer membranes. Biochemistry. 25 (1986) 2882–2889.

Dent M.R., Aroutiounova N., Dhalla N.S., Tappia P.S. Losartan attenuates Phospholipase C isozyme gene expression in hypertrophied hearts due to volume overload. J. Cell. Mol. Med. 10 (2006) 470–479.

Dent M.R., Dhalla N.S., Tappia P.S. Phospholipase C gene expression, protein content, and activities in cardiac hypertrophy and heart failure due to volume overload. Am. J. Physiol. Heart Circ. Physiol. 287 (2004a) H719–H727.

Dent M.R., Singal T., Dhalla N.S., Tappia P.S. Expression of phospholipase D isozymes in scar and viable tissue in congestive heart failure due to myocardial infarction. J. Cell. Mol. Med. 8 (2004b) 526–536.

Dhalla N.S., Dent M.R., Tappia P.S., Sethi R., Barta J., Goyal R.K. Subcellular remodeling as a viable target for the treatment of congestive heart failure. J. Cardiovasc. Pharmacol. Ther. 11 (2006) 31–45.

Dhalla N.S., Liu X., Panagia V., Takeda N. Subcellular remodeling and heart dysfunction in chronic diabetes. Cardiovasc. Res. 40 (1998) 239–247.

Dhalla N.S., Xu Y.J., Sheu S.S., Tappia P.S., Panagia V. Phosphatidic acid: a potential signal transducer for cardiac hypertrophy. J. Mol. Cell. Cardiol. 29 (1997) 2865–2871.

Dhalla N.S., Das P.K., Sharma G.P. Subcellular basis of cardiac contractile failure. J. Mol. Cell Cardiol. 10 (1978) 363–385.

Dhalla N.S., Elmoselhi A.B., Hata T., Makino N. Status of myocardial antioxidants in ischemia-reperfusion injury. Cardiovasc. Res. 47 (2000) 446–456.

Dhalla N.S., Golfman L., Takeda S., Takeda N., Nagano M. Evidence for the role of oxidative stress in acute ischemic heart disease: a brief review. Can. J. Cardiol. 15 (1999) 587–593.

Dhalla N.S., Pierce G.N., Innes I.R., Beamish R.E. Pathogenesis of cardiac dysfunction in diabetes mellitus. Can. J. Cardiol. 1 (1985) 263–281.

Dhalla N.S., Pierce G.N., Panagia V., Singal P.K., and Beamish R.E. Calcium movements in relation to heart function. Basic Res. Cardiol. 77 (1982) 117–139.

Dhalla N.S., Temsah R.M., Netticadan T., Sandhu M.S. Calcium overload in ischemia/reperfusion injury. In: Heart Physiology and Pathophysiology, edited by Sperelakis N, Kurachi Y, Terzic A, and Cohen M. Academic Press, San Diego, (2001) 949–965.

Diaz B.L., Arm J.P. Phospholipase $A_2$. Prostaglandins Leukot. Essent. Fatty Acids. 69 (2003) 87–97.

Dorn G.W. 2nd, Force T. Protein kinase cascades in the regulation of cardiac hypertrophy. J. Clin. Invest. 115 (2005) 527–537.

Eskildsen-Helmond Y.E., Gho B.C., Bezstarosti K., Dekkers D.H., Soei L.K., Van Heugten H.A., Verdouw P.D., Lamers J.M. Exploration of the possible roles of phospholipase D and protein kinase C in the mechanism of ischemic preconditioning in the myocardium. Ann. NY Acad. Sci. 793 (1996) 210–225.

Fein F.S., Sonnenblick E.H. Diabetic cardiomyopathy. Prog. Cardiovasc. Dis. 27 (1985) 255–270.

Fein F.S., Kornstein L.B., Strobeck J.E., Capasso J.M., Sonnenblick E.H. Altered myocardial mechanics in diabetic rats. Circ. Res. 47 (1980) 922–933.

Ferrari R., Ceconi C., Curello S., Cargnoni A., Alfieri O., Pardini A., Marzollo P., Visioli O. Oxygen free radicals and myocardial damage: protective role of thiol-containing agents. Am. J. Med. 91 (1991) 95S–105S.

Frohman M.A., Morris A.J. Phospholipase D structure and regulation. Chem. Phys. Lipids. 98 (1999) 127–140.

Fukami K. Structure, regulation, and function of phospholipase C isozymes. J. Biochem. (Tokyo). 131 (2002) 293–299.

Ganguly P.K., Pierce G.N., Dhalla K.S., Dhalla N.S. Defective sarcoplasmic reticular calcium transport in diabetic cardiomyopathy. Am. J. Physiol. 244 (1983) E528–E535.

Gross G.J., Falck J.R., Gross E.R., Isbell M., Moore J., Nithipatikom K. Cytochrome P450 and arachidonic acid metabolites: role in myocardial ischemia/reperfusion injury revisited. Cardiovasc. Res. 68 (2005) 18–25.

Ha T., Kotsanas G., Wendt I. Intracellular $Ca^{2+}$ and adrenergic responsiveness of cardiac myocytes in streptozotocin-induced diabetes. Clin. Exp. Pharmacol. Physiol. 26 (1999) 347–353.

Haist J.V., Hirst C.N., Karmazyn M. Effective protection by NHE-1 inhibition in ischemic and reperfused heart under preconditioning blockade. Am. J. Physiol. Heart Circ. Physiol. 284 (2003) H798–H803.

Hartmann M., Decking U.K. Blocking $Na^+$-$H^+$ exchange by cariporide reduces $Na^+$-overload in ischemia and is cardioprotective. J. Mol. Cell. Cardiol. 31 (1999) 1985–1995.

Hashizume H., Hoque A.N., Magishi K., Hara A., Abiko Y. A new approach to the development of anti-ischemic drugs. Substances that counteract the deleterious effect of lysophosphatidylcholine on the heart. Jpn. Heart J. 38 (1997) 11–25.

Hayat S.A., Patel B., Khattar R.S., Malik R.A. Diabetic cardiomyopathy: mechanisms, diagnosis and treatment. Clin. Sci. 107 (2004) 539–557.

Hearse D.J., Bolli R. Reperfusion induced injury: manifestations, mechanisms, and clinical relevance. Cardiovasc. Res. 26 (1992) 101–108.

Hefti M.A., Harder B.A., Eppenberger H.M., Schaub M.C. Signaling pathways in cardiac myocyte hypertrophy. J. Mol. Cell. Cardiol. 29 (1997) 2873–2892.

Heijnis J.B., van Zwieten P.A. Enhanced inotropic responsiveness to $\alpha_1$-adrenoceptor stimulation in isolated working hearts from diabetic rats. J. Cardiovasc. Pharmacol. 20 (1992) 559–562.

Heyliger C.E., Pierce G.N., Singal P.K., Beamish R.E., Dhalla N.S. Cardiac α- and β-adrenergic receptor alterations in diabetic cardiomyopathy. Basic Res. Cardiol. 77 (1982) 610–618.

Heyliger C.E., Prakash A., McNeill J.H. Alterations in cardiac sarcolemmal $Ca^{2+}$ pump activity during diabetes mellitus. Am. J. Physiol. 252 (1987) H540–H544.

Hilgemann D.W., Ball R. Regulation of cardiac $Na^+$, $Ca^{2+}$ exchange and $K_{ATP}$ potassium channels by $PIP_2$. Science 273 (1996) 956–959.

Hirabayashi T., Murayama T., Shimizu T. Regulatory mechanism and physiological role of cytosolic phospholipase $A_2$. Biol. Pharm. Bull. 27 (2004) 1168–1173.

Hodgkin M.N., Pettitt T.R., Martin A., Michell R.H., Pemberton A.J., Wakelam M.J. Diacylglycerols and phosphatidates: which molecular species are intracellular messengers? Trends. Biochem. Sci. 23 (1998) 200–204.

Horackova M., Murphy M.G. Effects of chronic diabetes mellitus on the electrical and contractile activities, $^{45}Ca^{2+}$ transport, fatty acid profiles and ultrastructure of isolated rat ventricular myocytes. Pflug. Arch. 411 (1988) 564–572.

Huang C.L., Feng S., Hilgemann D.W. Direct activation of inward rectifier potassium channels by $PIP_2$ and its stabilization by Gβγ. Nature 391 (1998) 803–806.

Hurd T.R, Costa N.J., Dahm C.C., Beer S.M., Brown S.E., Filipovska A., Murphy M.P. Glutathionylation of mitochondrial proteins. Antioxid. Redox. Signal. 7 (2005) 999–1010.

Hwang J.I., Oh Y.S., Shin K.J., Kim H., Ryu S.H., Suh P.G. Molecular cloning and characterization of a novel phospholipase C, PLC-η. Biochem J. 389 (2005) 181–186.

Im H-J., Russell M.A. Feng J-F. Transglutaminase II: a new class of GTP-binding protein with new biological functions. Cell Signal. 9 (1997) 477–482.

Inoguchi T., Battan R., Handler E., Sportsman J.R., Heath W., King G.L. Preferential elevation of protein kinase C isoform βII and diacylglycerol levels in the aorta and heart of diabetic rats: differential reversibility to glycemic control by islet cell transplantation. Proc. Natl. Acad. Sci. USA 89 (1992) 11059–11063.

Jalili T., Takeishi Y., Song G., Ball N.A., Howles G., Walsh R.A. PKC translocation without changes in Gαq and PLC-β protein abundance in cardiac hypertrophy and failure. Am. J. Physiol. 277 (1999) H2298–H2304.

Jennings R.B., Reimer K.A. The cell biology of acute myocardial ischemia. Annu. Rev. Med. 42 (1991) 225–246.

Jeremy R.W., Koretsune Y., Marban E., Becker L.C. Relation between glycolysis and calcium homeostasis in postischemic myocardium. Circ. Res. 70 (1992) 1180–1190.

Ju H., Zhao S., Tappia P.S., Panagia V., Dixon I.M. Expression of Gqα and PLC-β in scar and border tissue in heart failure due to myocardial infarction. Circulation. 97 (1998) 892–899.

Kamp T.J., Hell J.W. Regulation of cardiac L-type calcium channels by protein kinase A and protein kinase C. Circ. Res. 87 (2000) 1095–1102.

Karter A.J., Gazzaniga J.M., Cohen R.D., Casper M.L., Davis B.D., Kaplan G.A. Ischemic heart disease and stroke mortality in African-American and non-Hispanic white men and women, 1985–1991. West. J. Med. 169 (1998) 139–145.

Katan M. Families of phosphoinositide-specific phospholipase C: structure and function. Biochim. Biophys. Acta. 1436 (1998) 5–17.

Kawaguchi H., Sano H., Iizuka K., Okada H., Kudo T., Kageyama K., Muramoto S., Murakami T., Okamoto H., Mochizuki N., Kitabatake A. Phosphatidylinositol metabolism in hypertrophic rat heart. Circ. Res. 72 (1993) 966–972.

Kim S.J., Depre C., Vatner S.F. Novel mechanisms mediating stunned myocardium. Heart Fail. Rev. 8 (2003) 143–153.

Kloner R.A., Ellis S.G., Lange R., Braunwald E. Studies of experimental coronary artery reperfusion. Effects on infarct size, myocardial function, biochemistry, ultrastructure and microvascular damage. Circulation 68 (1983) 18–15.

Kurz T., Kemken D., Mier K., Weber I., Richardt G. Human cardiac phospholipase D activity is tightly controlled by phosphatidylinositol 4,5-bisphosphate. J. Mol. Cell. Cardiol. 36 (2004) 225–232.

Kurz T., Schneider I., Tolg R., Richardt G. $\alpha_1$-adrenergic receptor-mediated increase in the mass of phosphatidic acid and 1,2-diacylglycerol in ischemic rat heart. Cardiovasc. Res. 42 (1999) 48–56.

Lagadic-Gossmann D., Buckler K.J., Le Prigent K., Feuvray D. Altered $Ca^{2+}$ handling in ventricular myocytes isolated from diabetic rats. Am. J. Physiol. 270 (1996) H1529–H1537.

Lamers J.M., Eskildsen-Helmond Y.E., Resink A.M., de Jonge H.W., Bezstarosti K., Sharma H.S., van Heugten H.A. Endothelin-1-induced phospholipase C-β and D and protein kinase C isoenzyme signaling leading to hypertrophy in rat cardiomyocytes. J. Cardiovasc. Pharmacol. 26 (1995) S100–S103.

Lamers J.M.J., Dekkers D.H., Bezstarosti K., Meij J.T., van Heugten H.A. Occurrence and functions of the phosphatidylinositol cycle in the myocardium. Mol. Cell. Biochem. 116 (1992) 59–67.

Lee C.W., Lee K.H., Lee S.B., Park D. Rhee S.G. Regulation of phospholipase C-$\beta_4$ by ribonucleotides and the α subunit of Gq. J. Biol. Chem. 269 (1994) 25335–25338.

Lindmar R., Loffelholz K. Phospholipase D in rat myocardium: formation of lipid messengers and synergistic activation by G-protein and protein kinase C. Biochem. Pharmacol. 56 (1998) 799–805.

Liu S.Y., Tappia P.S., Dai J., Williams S.A., Panagia V. Phospholipase $A_2$-mediated activation of phospholipase D in rat heart sarcolemma. J. Mol. Cell. Cardiol. 30 (1998) 1203–1214.

Liu S.Y., Yu C.H., Hays J.A., Panagia V., Dhalla N.S. Modification of heart sarcolemmal phosphoinositide pathway by lysophosphatidylcholine. Biochim. Biophys. Acta. 1349 (1997) 264–274.

Liu X., Wang J., Takeda N., Binaglia L., Panagia V., Dhalla N.S. Changes in cardiac protein kinase C activities and isozymes in streptozotocin-induced diabetes. Am. J. Physiol. 277 (1999) E798–E804.

Lopez I., Mak E.C., Ding J., Hamm H.E., Lomasney J.W. A novel bifunctional phopspholipase C that is regulated by $G\alpha_{12}$ and stimulates the Ras/mitogen-activated protein kinase pathway. J. Biol. Chem. 276 (2001) 2758–2765.

Malhotra A., Kang B.P., Cheung S., Opawumi D., Meggs L.G. Angiotensin II promotes glucose-induced activation of cardiac protein kinase C isozymes and phosphorylation of troponin I. Diabetes 50 (2001a) 1918–1926.

Malhotra A., Kang B.P., Opawumi D., Belizaire W., Meggs L.G. Molecular biology of protein kinase C signaling in cardiac myocytes. Mol. Cell Biochem. 25 (2001b) 97–107.

Mangat R., Singal T., Dhalla N.S., Tappia P.S. Inhibition of phospholipase C $\gamma_1$ augments the decrease in cardiomyocyte viability by $H_2O_2$. Am J. Physiol. Heart Circ. Physiol. 291 (2006) H854–H860.

Marczin N., El-Habashi N., Hoare G.S., Bundy R.E., Yacoub M. Antioxidants in myocardial ischemia-reperfusion injury: therapeutic potential and basic mechanisms. Arch. Biochem. Biophys. 420 (2003) 222–236.

Martin A., Saqib K.M., Hodgkin M.N., Brown F.D., Pettit T.R., Armstrong S., Wakelam M.J. Role and regulation of phospholipase D signalling. Biochem. Soc. Trans. 25 (1997) 1157–1160.

McHowat J., Creer M.H., Hicks K.K., Jones J.H., McCrory R., Kennedy R.H. Induction of Ca-independent $PLA_2$ and conservation of plasmalogen polyunsaturated fatty acids in diabetic heart. Am. J. Physiol. Endocrinol. Metab. 279 (2000) E25–E32.

McHowat J., Creer M.H. Catalytic features, regulation and function of myocardial phospholipase $A_2$. Curr. Med. Chem. Cardiovasc. Hematol. Agents 2 (2004) 209–218.

McHowat J., Tappia P.S., Liu S., McCrory R., Panagia V. Redistribution and abnormal activity of phospholipase $A_2$ isoenzymes in postinfarct congestive heart failure. Am. J. Physiol. Cell. Physiol. 280 (2001) C573–C580.

Meij J.T., Suzuki S., Panagia V., Dhalla N.S. Oxidative stress modifies the activity of cardiac sarcolemmal phospholipase C. Biochim. Biophys. Acta. 1199 (1994) 6–12.

Mesaeli N., Tappia P.S., Suzuki S., Dhalla N.S., Panagia V. Oxidants depress the synthesis of phosphatidylinositol 4,5-bisphosphate in heart sarcolemma. Arch. Biochem. Biophys. 382 (2000) 48–56.

Milano C.A., Dolber P.C., Rockman H.A., Bond R.A., Venable M.E., Allen L.F., Lefkowitz R.J. Myocardial expression of a constitutively active $\alpha_{1B}$-adrenergic receptor in transgenic mice induces cardiac hypertrophy. Proc. Natl. Acad. Sci. USA 91 (1994) 10109–10113.

Miyamae M., Camacho S.A., Weiner M.W., Figueredo V.M. Attenuation of postischemic reperfusion injury is related to prevention of $[Ca^{2+}]_m$ overload in rat hearts. Am J Physiol. Heart Circ. Physiol. 271 (1996) H2145–H2153.

Moraru II, Jones R.M., Popescu L.M., Engelman R.M., Das D.K. Prazosin reduces myocardial ischemia/reperfusion-induced $Ca^{2+}$ overloading in rat heart by inhibiting phosphoinositide signaling. Biochim. Biophys. Acta. 1268 (1995) 1–8.

Moraru II, Popescu L.M., Maulik N., Liu X., Das D.K. Phospholipase D signaling in ischemic heart. Biochim. Biophys. Acta 1139 (1992) 148–154.

Mouton R., Huisamen B., Lochner A. The effect of ischaemia and reperfusion on sarcolemmal inositol phospholipid and cytosolic inositol phosphate metabolism in the isolated perfused rat heart. Mol. Cell. Biochem. 105 (1991) 127–135.

Mozzicato S., Joshi B.V., Jacobson K.A., Liang B.T. Role of direct RhoA-phospholipase D1 interaction in mediating adenosine-induced protection from cardiac ischemia. FASEB. J. 18 (2004) 406–408.

Munakata M., Stamm C., Friehs I., Zurakowski D., Cowan D.B., Cao-Danh H., McGowan F.X. Jr, del Nido P.J. Protective effects of protein kinase C during myocardial ischemia require activation of phosphatidyl-inositol specific phospholipase C. Ann. Thorac. Surg. 73 (2002) 1236–1245.

Nishio Y., Kashiwagi A., Taki H., Shinozaki K., Maeno Y., Kojima H., Maegawa H., Haneda M., Hidaka H., Yasuda H., Horiike K., Kikkawa R. Altered activities of transcription factors and their related gene expression in cardiac tissues of diabetic rats. Diabetes 47(8): (1998) 1318–1325.

Ohto T., Uozumi N., Hirabayashi T., Shimizu T. Identification of novel cytosolic phospholipase $A_2$s, murine cPLA$_2\delta$, $\varepsilon$, and $\zeta$, which form a gene cluster with cPLA$_2\beta$. J. Biol. Chem. 280 (2005); 24576–24583.

Okumura K., Akiyama N., Hashimoto H., Ogawa K., Satake T. Alteration of 1,2-diacylglycerol content in myocardium from diabetic rats. Diabetes 37(9) (1988) 1168–1172.

Olivetti G., Abbi R., Quaini F., Kajstura J., Cheng W., Nitahara J.A., Quaini E., Di Loreto C., Beltrami C.A., Krajewski S., Reed J.C., Anversa P. Apoptosis in the failing human heart. N. Engl. J. Med. 336(16) (1997) 1131–1141.

Otani H., Prasad M.R., Engelman R.M., Otani H., Cordis G.A., Das D.K. Enhanced phosphodiesteratic breakdown and turnover of phosphoinositides during reperfusion of ischemic rat heart. Circ Res. 63 (1988) 930–936.

Ozer M.K., Parlakpinar H., Cigremis Y., Ucar M., Vardi N., Acet A. Ischemia-reperfusion leads to depletion of glutathione content and augmentation of malondialdehyde production in the rat heart from overproduction of oxidants: can caffeic acid phenethyl ester (CAPE) protect the heart? Mol. Cell. Biochem. 273 (2005) 169–175.

Panagia V., Ou C., Taira Y., Dai J., Dhalla N.S. Phospholipase D actvity in subcellular membranes of rat ventricular myocardium. Biochim. Biophys Acta 1064 (1991) 242–250.

Pang Y., Bounelis P., Chatham J.C., Marchase R.B. Hexosamine pathway is responsible for inhibition by diabetes of phenylephrine-induced inotropy. Diabetes 53 (2004) 1074–1081.

Paradis P., Dali-Youcef N., Paradis F.W., Thibault G., Nemer M. Overexpression of angiotensin II type I receptor in cardiomyocytes induces cardiac hypertrophy and remodeling. Proc. Natl. Acad. Sci. USA 97 (2000) 931–936.

Park H., Park E.S., Lee H.S., Yun H.Y., Kwon N.S. Baek K.J. Distinct characteristic of Gαh (transglutaminase II) by compartment: GTPase and transglutaminase activities. Biochem. Biophys. Res. Commun. 284 (2001) 496–500.

Park J.B., Kim J.H., Kim Y., Ha S.H., Yoo J.S., Du G., Frohman M.A., Suh P.G., Ryu S.H. Cardiac phospholipase D2 localizes to sarcolemmal membranes and is inhibited by α-actinin in an ADP-ribosylation factor-reversible manner. J. Biol. Chem. 275 (2000) 21295–21301.

Peivandi A.A., Huhn A., Lehr H.A., Jin S., Troost J., Salha S., Weismüller T., Löffelholz K. Upregulation of phospholipase D expression and activation in ventricular pressure-overload hypertrophy. J. Pharmacol. Sci. 98 (2005) 244–254.

Penpargkul S., Schaible T., Yipintsoi T., Scheuer J. The effects of diabetes on performance and metabolism of rat hearts. Circ. Res. 47 (1980) 911–924.

Piper H.M., Meuter K., Schafer C. Cellular mechanisms of ischemia-reperfusion injury. Ann. Thorac. Surg. 75 (2003) S644–S648.

Puceat M., Vassort G. Signalling by protein kinase C isoforms in the heart. Mol. Cell Biochem. 157 (1996) 65–72.

Purdom S., Chen Q.M. Epidermal Growth factor receptor-dependent and -independent pathways in hydrogen peroxide-induced mitogen-activated protein kinase activation in cardiomyocytes and heart fibroblasts. J. Pharmacol. Exp. Ther. 312 (2005) 1179–1186.

Rebecchi M.J., Pentyala S.N. Structure, function, and control of phosphoinositide-specific phospholipase C. Physiol. Rev. 80 (2000) 1291–1335.

Regan T.J. Congestive heart failure in the diabetic. Annu. Rev. Med. 34 (1983) 161–168.

Regan T.J., Ettinger P.O., Khan M.I., Jesrani M.V., Lyons M.M., Oldewurtel H.A., Weber M. Altered myocardial function and metabolism in chronic diabetes mellitus without ischemia in dogs. Circ. Res. 35 (1974) 222–237.

Rhee S.G., Bae Y.S. Regulation of phosphoinositide-specific phospholipase C isozymes. J. Biol. Chem. 272 (1997) 15045–15048.

Rhee S.G. Regulation of phosphoinositide-specific phospholipase C. Annu. Rev. Biochem. 70 (2001) 281–312.

Rosamond W., Flegal K., Furie K., Go A., Greenlund K., Haase N., Hailpern S.M., Ho M., Howard V., Kissela B., et al. American Heart Association Statistics Committee and Stroke Statistics Subcommittee. Heart disease and stroke statistics-(2008) update: A report from the American Heart Association Statistics Committee and Stroke Statistics Subcommittee. Circulation. 117 (2008) e25–e146.

Ryden L., Armstrong P.W., Cleland J.G., Horowitz J.D., Massie B.M., Packer M., Poole-Wilson P.A. Efficacy and safety of high-dose lisinopril in chronic heart failure patients at high cardiovascular risk, including those with diabetes mellitus. Results from the ATLAS trial. Eur. Heart J. 21 (2000) 1967–1978.

Sadoshima J., Izumo S. Signal transduction pathways of angiotensin II–induced c-fos gene expression in cardiac myocytes in vitro. Roles of phospholipid-derived second messengers. Circ. Res. 73 (1993) 424–438.

Sakata Y., Hoit B.D., Liggett S.B., Walsh R.A., Dorn G.W. 2nd. Decompensation of pressure-overload hypertrophy in Gαq-overexpressing mice. Circulation. 97 (1998) 1488–1495.

Sakata Y. Tissue factors contributing to cardiac hypertrophy in cardiomyopathic hamsters (BIO14.6): involvement of transforming growth factor-β1 and tissue renin-angiotensin system in the progression of cardiac hypertrophy. Hokkaido Igaku Zasshi. 68 (1993) 18–28.

Sapirstein A., Spech R.A., Witzgall R., Bonventre J.V. Cytosolic phospholipase $A_2$ ($PLA_2$), but not secretory $PLA_2$, potentiates hydrogen peroxide cytotoxicity in kidney epithelial cells. J. Biol. Chem. 271 (1996) 21505–21513.

Saunders C.M., Larman M.G., Parrington J., Cox L.J., Royse J., Blayney L.M., Swann K., Lai F.A. PLC ζ: a sperm-specific trigger of $Ca^{2+}$ oscillations in eggs and embryo development. Development 129 (2002) 3533–3544.

Scarabelli T., Stephanou A., Rayment N., Pasini E., Comini L., Curello S., Ferrari R., Knight R. Apoptosis of endothelial cells precedes myocytes cell apoptosis in ischemia/reperfusion injury. Circulation 104 (2001) 253–256.

Schnabel P., Mies F., Nohr T., Geisler M., Bohm M. Differential regulation of phospholipase C- β isozymes in cardiomyocyte hypertrophy. Biochem. Biophys. Res. Commun. 275 (2000) 1–6.

Schwertz D.W., Halverson J. Changes in phosphoinositide-specific phospholipase C and phospholipase $A_2$ activity in ischemic and reperfused rat heart. Basic Res. Cardiol. 87 (1992) 113–127.

Sekiya F., Bae Y.S., Rhee S.G. Regulation of phospholipase C isoenzymes: activation of phospholipase C-γ in the absence of tyrosine phosphorylation. Chem. Phys. Lipids 98 (1999) 3–11.

Sharma V., and McNeill J.H. Diabetic cardiomyopathy: where are we 40 years later? Can. J. Cardiol. 22 (2006) 305–308.

Shindler D.M., Kostis J.B., Yusuf S., Quinones M.A., Pitt B., Stewart D., Pinkett T., Ghali J.K., Wilson A.C. Diabetes mellitus, a predictor of morbidity and mortality in the Studies of Left Ventricular Dysfunction (SOLVD) Trials and Registry. Am. J. Cardiol. 77 (1996) 1017–1020.

Singal T., Dhalla N.S., Tappia P.S. Norepinephrine-induced changes in gene expression of phospholipase C in cardiomyocytes. J. Mol. Cell. Cardiol. 41 (2006) 126–137.

Singal T., Dhalla N.S., Tappia P.S. Phospholipase C may be involved in norepinephrine-induced cardiac hypertrophy. Biochem. Biophys. Res. Commun. 320 (2004) 1015–1019.

Six D.A., Dennis E.A. The expanding superfamily of phospholipase $A_2$ enzymes: classification and characterization. Biochim. Biophys. Acta. 1488 (2000) 1–19.

Snabaitis A.K., Hearse D.J., Avkiran M. Regulation of sarcolemmal $Na^+/H^+$ exchange by hydrogen peroxide in adult rat ventricular myocytes. Cardiovasc. Res. 53 (2002) 470–480.

Song C., Hu C.D., Masago M., Kariyai K., Yamawaki-Kataoka Y., Shibatohge M., Wu D., Satoh T., Kataoka T. Regulation of a novel human phospholipase C, PLCε, through membrane targeting by Ras. J. Biol. Chem. 276 (2001) 2752–2757.

Stahelin R.V., Rafter J.D., Das S., Cho W. The molecular basis of differential subcellular localization of C2 domains of protein kinase C-α and group IVa cytosolic phospholipase $A_2$. J. Biol. Chem. 278 (2003) 12452–12460.

Su X., Han X., Mancuso D.J., Abendschein D.R., Gross R.W. Accumulation of long-chain acylcarnitine and 3-hydroxy acylcarnitine molecular species in diabetic myocardium: identification of alterations in mitochondrial fatty acid processing in diabetic myocardium by shotgun lipidomics. Biochemistry 44 (2005) 5234–5245.

Tall E., Dorman G., Garcia P., Runnels L., Shah S., Chen J., Profit A., Gui Q.M., Chaudhary A., Prestwich G.D., Rebecchi M.J. Phosphoinositide binding specificity among phospholipase C isozymes as determined by photo-cross-linking to novel substrate and product analogs. Biochemistry 36 (1997) 7239–7248.

Tamada A., Hattori Y., Houzen H., Yamada Y., Sakuma I., Kitabatake A., Kanno M. Effects of β-adrenoceptor stimulation on contractility, $[Ca^{2+}]i$, and $Ca^{2+}$ current in diabetic rat cardiomyocytes. Am. J. Physiol. 274 (1998) H1849–H1857.

Tanaka H., Takeya R., Sumimoto H. A novel intracellular membrane-bound calcium-independent phospholipase $A_2$. Biochem. Biophys. Res. Commun. 272 (2000) 320–326.

Tappia P.S., Asemu G., Aroutiounova N., Dhalla N.S. Defective sarcolemmal phospholipase C signaling in diabetic cardiomyopathy. Mol. Cell. Biochem. 261 (2004a) 193–199.

Tappia P.S., Bibeau M., Sahi N., Pierce G.N., Dixon I.M.C., Panagia V. Defective sarcolemmal Gqα/PLCβ$_1$ signaling in diabetic cardiomyopathy. J. Mol. Cell. Cardiol. 32 (2000) A46.

Tappia P.S., Dent M.R., Dhalla N.S. Oxidative stress and redox regulation of phospholipase D in myocardial disease. Free Radic. Biol. Med. 41 (2006) 349–361.

Tappia P.S., Liu S.Y., Shatadal S., Takeda N., Dhalla N.S., Panagia V. Changes in sarcolemmal PLC isoenzymes in postinfarct congestive heart failure: partial correction by imidapril. Am. J. Physiol. Heart Circ. Physiol. 277 (1999) H40–H49.

Tappia P.S., Liu S.Y., Tong Y., Ssenyange S., Panagia V. Reduction of phosphatidyliositol-4,5-bisphosphate mass in heart sarcolemma during diabetic cardiomyopathy. Adv. Exp. Med. Biol. 498 (2001a) 183–190.

Tappia P.S., Maddaford T.G., Hurtado C., Dibrov E., Austria J.A., Sahi N., Panagia V., Pierce G.N. Defective phosphatidic acid-phospholipase C signaling in diabetic cardiomyopathy. Biochem. Biophys. Res. Commun. 316 (2004b) 280–289.

Tappia P.S., Maddaford T.G., Hurtado C., Panagia V., Pierce G.N. Depressed phosphatidic acid-induced contractile activity of failing cardiomyocytes. Biochem. Biophys. Res. Commun. 300 (2003) 457–463.

Tappia P.S., Yu C.H., Di Nardo P., Pasricha A.K., Dhalla N.S., Panagia V. Depressed responsiveness of phospholipase C isoenzymes to phosphatidic acid in congestive heart failure. J. Mol. Cell. Cardiol. 33 (2001b) 431–440.

Tong Y., Liu S-Y., Tappia P.S., Panagia V. Sarcolemmal PLC $\gamma_1$ is hypoactive but hyperresponsive to phosphatidic acid in diabetic cardiomyopathy (Abstract). J. Mol. Cell. Cardiol. 30 (1998) A257.

Tosaki A., Maulik N., Cordis G., Trifan O.C., Popescu L.M., Das D.K. Ischemic preconditioning triggers phospholipase D signaling in rat heart. Am. J. Physiol. 273 (1997) H1860–H1866.

Trifan O.C., Popescu L.M., Tosaki A., Cordis G., Das D.K. Ischemic preconditioning involves phospholipase D. Ann. N Y. Acad. Sci. 793 (1996) 485–488.

Tritto I., Ambrosio G. Role of oxidants in the signaling pathway of preconditioning. Antioxid. Redox. Signal. 3 (2001) 3–10.

Ungvari Z., Gupte S.A., Recchia F.A., Batkai S., Pacher P. Role of oxidative-nitrosative stress and downstream pathways in various forms of cardiomyopathy and heart failure. Curr. Vasc. Pharmacol. 3 (2005) 221–229.

Urquhart J., Epstein S.E., Patterson R.E. Comparative effects of calcium-channel blocking agents on left ventricular function during acute ischemia in dogs with and without congestive heart failure. Am. J. Cardiol. 55 (1985) 10B–16B.

Vecchini A., Del Rosso F., Binaglia L., Dhalla N.S., Panagia V. Molecular defects in sarcolemmal glycerophospholipid subclasses in diabetic cardiomyopathy. J. Mol. Cell. Cardiol. 32 (2000) 1061–1074.

Wald M., Borda E.S., Sterin-Borda L. α-adrenergic supersensitivity and decreased number of α-adrenoceptors in heart from acute diabetic rats. Can. J. Physiol. Pharmacol. 66 (1988) 1154–1160.

Wang H., Oestreich E.A., Maekawa N., Bullard T.A., Vikstrom K.L., Dirksen R.T., Kelley G.G., Blaxall B.C., Smrcka A.V. Phospholipase C ε modulates β-adrenergic receptor-dependent cardiac contraction and inhibits cardiac hypertrophy. Circ. Res. 97 (2005) 1305–1313.

Williams S.A., Tappia P.S., Yu C.H., Bibeau M., Panagia V. Impairment of the sarcolemmal phospholipase D-phosphatidate phosphohydrolase pathway in diabetic cardiomyopathy. J. Mol. Cell. Cardiol. 30 (1998) 109–118.

Williams S.A., Tappia P.S., Yu C.H., Binaglia L., Panagia V., Dhalla N.S. Subcellular alterations in cardiac phospholipase D activity in chronic diabetes. Protaglandins Leukot. Essent. Fatty Acids 57 (1997) 95–99.

Wing M.R., Bourdon D.M., Harden T.K. PLC-ε: a shared effector protein in Ras-, Rho-, and Gα βγ-mediated signaling. Mol. Interv. 3 (2003) 273–280.

Wu G., Fang Y.Z., Yang S., Lupton J.R., Turner N.D. Glutathione metabolism and its implications for health. J. Nutr. 134 (2004) 489–492.

Xiao X.H., Allen D.G. The cardioprotective effects of $Na^+/H^+$ exchange inhibition and mitochondrial $K_{ATP}$ channel activation are additive in the isolated rat heart. Pflugers Arch. 447 (2003) 272–279.

Xu Y.J., Botsford M.W., Panagia V., Dhalla N.S. Responses of heart function and intracellular free $Ca^{2+}$ to phosphatidic acid in chronic diabetes. Can. J. Cardiol. 12 (1996a) 1092–1098.

Xu Y.J., Panagia V., Shao Q., Wang X., Dhalla N.S. Phosphatidic acid increases intracellular free $Ca^{2+}$ and cardiac contractile force. Am. J. Physiol. 271 (1996b) H651–H659.

Yagisawa H., Sakuma K., Paterson H.E., Cheung R., Allen V., Hirata H., Watanabe Y., Hirata M., Williams R.L., Katan M. Replacements of single basic amino acids in the pleckstrin homology domain of phospholipase C-$\delta_1$ alter the ligand binding, phospholipase activity and interaction with the plasma membrane. J. Biol. Chem. 273 (1998) 417–424.

Yamazaki M., Zhang Y., Watanabe H., Yokozeki T., Ohno S., Kaibuchi K., Shibata H., Mukai H., Ono Y., Frohman M.A., Kanaho Y. Interaction of the small G protein RhoA with the C terminus of human phospholipase D1. J. Biol. Chem. 274 (1999) 6035–6038.

Ye H., Wolf R.A., Kurz T., Corr P.B. Phosphatidic acid increases in response to noradrenaline and endothelin-1 in adult rabbit ventricular myocytes. Cardiovasc. Res. 28 (1994) 1828–1834.

Yin G., Yan C., Berk B.C. Angiotensin II signaling pathways mediated by tyrosine kinases. Int. J. Biochem. Cell Biol. 35 (2003) 780–783.

Yoshida K., Inui M., Harada L., Saido T.C., Sorimachi Y., Ishihara T., Kawashima S., Sobue K. Reperfusion of rat heart after brief ischemia induces proteolysis of calspectine (nonerythroid spectrin or fodrin) by calpain. Circ. Res. 77 (1995) 603–610.

Yu C.H., Panagia V., Tappia P.S., Liu S.Y., Takeda N., Dhalla N.S. Alterations of sarcolemmal phospholipase D and phosphatidate phosphohydrolase in congestive heart failure. Biochim. Biophys. Acta 1584 (2002) 65–72.

Zhang S., Lin J., Hirano Y., Hiraoka M. Modulation of ICa-L by $\alpha_1$-adrenergic stimulation in rat ventricular myocytes. Can. J. Physiol. Pharmacol. 83 (2005) 1015–1024.

Ziegelhoffer A., Tappia P.S., Mesaeli N., Sahi N., Dhalla N.S., Panagia V. Low level of sarcolemmal phosphatidylinositol 4,5-bisphosphate in cardiomyopathic hamster (UM-X7.1) heart. Cardiovasc. Res. 49 (2001) 118–126.

# Chapter 12
# The Role of Phospholipid Oxidation Products in Inflammatory and Autoimmune Diseases

## Evidence from Animal Models and in Humans

**Norbert Leitinger**

**Abstract** Since the discovery of oxidized phospholipids (OxPL) and their implication as modulators of inflammation in cardiovascular disease, roles for these lipid oxidation products have been suggested in many other disease settings. Lipid oxidation products accumulate in inflamed and oxidatively damaged tissue, where they are derived from oxidative modification of lipoproteins, but also from membranes of cells undergoing apoptosis. Thus, increased oxidative stress as well as decreased clearance of apoptotic cells has been implied to contribute to accumulation of OxPL in chronically inflamed tissues.

A central role for OxPL in disease states associated with dyslipedemia, including atherosclerosis, diabetes and its complications, metabolic syndrome, and renal insufficiency, as well as general prothrombotic states, has been proposed. In addition, in organs which are constantly exposed to oxidative stress, including lung, skin, and eyes, increased levels of OxPL are suggested to contribute to inflammatory conditions. Moreover, accumulation of OxPL causes general immunmodulation and may lead to autoimmune diseases. Evidence is accumulating that OxPL play a role in lupus erythematosus, anti-phospholipid syndrome, and rheumatoid arthritis. Last but not least, a role for OxPL in neurological disorders including multiple sclerosis (MS), Alzheimer's and Parkinson's disease has been suggested.

This chapter will summarize recent findings obtained in animal models and from studies in humans that indicate that formation of OxPL represents a general mechanism that may play a major role in chronic inflammatory and autoimmune diseases.

**Keywords** Oxidized phospholipids · chronic inflammation · autoimmune disease

N. Leitinger
Department of Pharmacology and Robert M. Berne Cardiovascular Research Center, University of Virginia, Charlottesville, VA, USA

P.J. Quinn, X. Wang (eds.), *Lipids in Health and Disease*,

**Abbreviations** AMD: age-related macular degeneration; apoB: apolipoprotein B; β2/GPI: beta-2-glycoprotein; BMP: bone-morphogenic protein; DC: dendritic cell; EC: endothelial cell; HDL: high-density lipoprotein; IL: interleukin; KLF: kruppel-like factor; LDL: low-density lipoprotein; Lp-PLA$_2$: lipoprotein-associated phospholipase A$_2$; NADPH-oxidase: nicotinamide adenine dinucleotide phosphate-oxidase; NAFLD: nonalcoholic fatty liver disease; NASH: nonalcoholic steatohepatitis; PAPC: 1-palmitoyl-2-arachidonoyl-*sn*-3-glycero-phosphorylcholine; POVPC: 1-palmitoyl-2-oxovaleroyl-*sn*-glycero-3-phosphorylcholine; PGPC: 1-palmitoyl-2-glutaroyl-*sn*-glycero-3-phosphorylcholine; OxPAPC: oxidized PAPC, OxPL: oxidized phospholipids; PAF: platelet-activating factor; PAF-AH: PAF-acetylhydrolase; PON: paraoxonase; PTH: parathyroid hormone; SLE: systemic lupus erythematosus; TNF: tumor necrosis factor; TLR: toll-like receptor; UV: ultraviolet.

## 12.1 Introduction

### *12.1.1 Formation of OxPL in Tissues Exposed to Oxidative Stress*

Non-enzymatic oxidative modification of phospholipids in chronically inflamed tissue is mediated by free radicals which are produced by enzymes including NADPH oxidase and myeloperoxidase (Zhang et al. 2002). While ozone may play an additional role in lipid oxidation in the lung, singlet oxygen produced by ultraviolet (UV) light is the mechanism particularly relevant for skin and eye pathologies.

Polyunsaturated fatty acids and especially arachidonic acid are highly susceptible to lipid peroxidation, which leads to the generation of lipid hydroperoxides, which then undergo carbon-carbon bond cleavage giving rise to the formation of short chain, unesterified aldehydes and aldehydes still esterified to the parent lipid, termed core-aldehydes (Esterbauer et al. 1987). Considerable progress has been made in recent years in dissecting the molecular structures of OxPL, which consequently allowed for the experimental use of defined compounds rather than complex lipoproteins and lipid mixtures.

Oxidation of phospholipids, such as 1-palmitoyl-2-arachidonoyl-*sn*-3-glycero-phosphorylcholine (PAPC), yields a series of oxidation products (OxPAPC) some of which have been structurally identified and shown to accumulate in atherosclerotic lesions (Watson et al. 1997). Examples for structures derived from oxidation of PAPC include 1-palmitoyl-2-oxovaleroyl-*sn*-glycero-3--phosphorylcholine (POVPC) and 1-palmitoyl-2-glutaroyl-*sn*-glycero-3--phosphorylcholine (PGPC) (Watson et al. 1997), as well as a series of CD36 binding motifs, which contain an oxidatively truncated *sn*-2 acyl group with a terminal gamma- hydroxy(or oxo)-α,β-unsaturated carbonyl (Podrez et al. 2002a). The molecular properties of identified OxPL are described in detail in a recent review (Fruhwirth et al. 2007).

The exact mechanisms how OxPL induce an inflammatory response remain a matter of speculation. Since OxPL are structurally similar to some bacterial components, it is possible that these modified lipids are recognized by cells as so called "danger signals" and consequently induce an initially protective immune response involving inflammatory reactions. Oxidized moieties of phospholipids are exposed on membranes (Greenberg et al. 2008), and thereby can be recognized by "pattern recognition receptors" including scavenger receptors such as CD36 (Hazen and Chisolm 2002). Due to their structural similarity to platelet activating factor (PAF), fragmented alkyl-OxPL ("PAF-like lipids") serve as specific agonists for the PAF receptor (PAF-R). It was shown that PAF-like lipids derived from oxidized LDL act on cells that express the PAF-R at low concentrations (Marathe et al. 1999, 2002; Smiley et al. 1991; Zimmerman et al. 2002).

Initially, the formation of lipid oxidation products has been considered solely detrimental, since it was demonstrated that these toxic products propagate inflammation and tissue damage. Only recently it became evident that certain OxPL also exert antiiflammatory, tissue protective effects (Bochkov 2007). This is the case when cells and tissues respond towards these oxidatively modified stress signals with upregulation of protective genes. In addition, OxPL derived from arachidonic acid-containing PL potently inhibit effects of lipopolysaccharide (LPS) (Bochkov et al. 2002a). Therefore, development of pharmacological agents based on structures of OxPL has been implied, and various treatment scenarios such as in lung disease and in sepsis have been suggested.

Together, oxidative modification of phospholipids represents a common underlying mechanism in many diseases where tissue damage is involved. The formation of OxPL seems to be a general feature in chronic inflammatory settings that often lead to debilitating states in many patients. Among those diseases are cardiovascular disease, diabetes and its complications including eye and kidney diseases, but also general immune modulation that could play an important role in sepsis and autoimmune diseases. Furthermore, evidence is accumulating that OxPL may play a role in neurological disorders including Alzheimer's and Parkinson's disease (Fig. 12.1).

### *12.1.2 Catabolism of OxPL*

OxPL have to be promptly removed *in vivo* in order to diminish adverse biological effects that may arise through their uncontrolled accumulation. One of the candidate enzymes responsible for the hydrolysis of OxPL is PAF-acetylhydrolase (PAF-AH), also termed lipoprotein associated phospholipase $A_2$ (Lp-PLA2) (Dada et al. 2002; Zalewski et al. 2006; Khuseyinova et al. 2005). PAF-AHs were originally identified as enzymes that hydrolyze the acetyl group at the *sn*-2 position of PAF (1-*O*-alkyl-2-acetyl-*sn*-glycero-3-phosphocholine). Oxidation of the *sn*-2 acyl group of a phospholipid results in aldehydic or

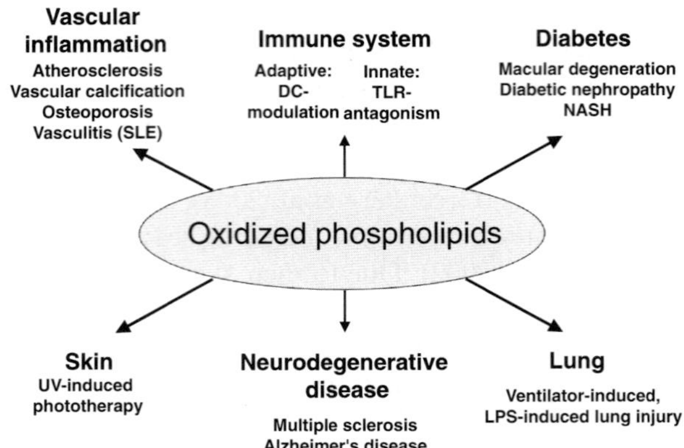


**Fig. 12.1** Oxidized phospholipids play a role in pathologies that have been associated with oxidative stress. DC, dendritic cell; LPS, lipopolysaccharide; NASH, nonalcoholic steatohepatitis; SLE, systemic lupus erythematosus; TLR, toll-like receptor; UV, ultraviolet

carboxylic groups (as in POVPC or PGPC, respectively), formation of which increases the substrate affinity for plasma PAF-AH. Moreover, oxidatively modified, non-fragmented phospholipids such as esterified F2-isoprostanes and phospholipid hydroperoxides are good substrates for PAF-AH. Consequently, PAF-AH catabolizes PAF as well as OxPL in oxidized LDL (OxLDL). It is assumed that hydrolysis OxPL by PAF-AH limits the activity of these proatherogenic lipids. Indeed, overexpression of PAF-AH in cells or tissues was shown to suppress oxidative stress-induced cell death. Adenovirus-mediated overexpression of PAF-AH in rabbits resulted in reduced foam cell formation (Turunen et al. 2004). Moreover, recombinant plasma PAF-AH was effective in treating acute pancreatitis, asthma and anaphylactic shock in animal models (Arai et al. 2002). Many recent studies have correlated PAF-AH (Lp-PLA2) activity with severity of vascular disease and atherosclerosis (Tsimikas et al. 2007). The role of PAF-AH in atherogenesis is not quite clear yet; the degradation of OxPL may be protective, however, accumulation of the enzymatic products lyso-PC and fatty acids may contribute to disease progression.

Paraoxonase (PON) is an enzyme associated with HDL and was shown to degrade OxPL (Mackness et al. 1998b, 2006b; Mackness and Durrington 1995b; Marathe et al. 2003). Beneficial effects of degradation of OxPL by PON are demonstrated by studies where overexpression of PON protected against atherosclerosis, while PON deficiency led to increased lesion formation (Ng et al. 2005; Shih et al. 1998, 2000; Tward et al. 2002; Watson et al. 1995). A recent study shows that PON overexpression inhibits intimal hyperplasia (Miyoshi et al. 2007).

Gene polymorphisms in both PAF-AH and PON have been correlated with the incidence of atherosclerosis in humans (Agachan et al. 2004; Ahmed et al.

2001; Bilge et al. 2007; Cao et al. 1998; Dessi et al. 1999; Fortunato et al. 2003; Arai et al. 2002; Kakafika et al. 2003; Yamada et al. 2000).

## 12.2 OxPL in Vascular Disease

### *12.2.1 Atherosclerosis*

Atherosclerosis is characterized by chronic inflammation of large arteries with clinical consequences including myocardial infarction and stroke. Although the knowledge about the mechanisms underlying atherosclerosis and its complications has dramatically increased, the question about the initiating factors of atherogenesis remains unsolved.

There is extensive evidence that accumulation and subsequent oxidative modification of LDL particles in the subendothelial space play a key role in development and progression of atherosclerosis (Lusis 2000; Berliner et al. 1995; Leitinger 2005). Phospholipid oxidation products are found at high concentrations within fatty streak lesions of cholesterol fed rabbits, mice, and in human atherosclerotic lesions (Watson et al. 1997; Berliner et al. 2001; Subbanagounder et al. 2000; Subbanagounder et al. 2000; Huber et al. 2002). Antibodies against OxPL are present in the serum of apoE-deficient mice and the presence of antibodies against OxPL in patients with atherosclerosis, diabetes, hypertension and other chronic inflammatory diseases further underlines the importance and potential functional relevance of these molecules (Binder et al. 2005).

OxPL are thought to play an essential role in various settings and stages of atherosclerotic lesion initiation and progression. OxPL were shown to activate endothelial cells to specifically bind monocytes, an initiating step in the development of atherosclerotic lesions (Huber et al. 2006; Berliner et al. 1990; Leitinger et al. 1999). Moreover, OxPL may compromise the barrier properties of the vascular endothelium, which may facilitate transmigration of LDL and inflammatory cells into the subendothelial layers. In this context, it was demonstrated that OxPAPC alters the expression, phosphorylation, and localization of tight junction proteins, such as occludin (DeMaio et al. 2006). On the other hand, endothelial barrier function was shown to be enhanced by certain OxPL via activation of Cdc42 and Rac (Birukov et al. 2004). Recently, it was shown that OxPL potently affect connexin expression and function in endothelial (EC) and smooth muscle cells (SMC) of the vascular wall (Isakson et al. 2006) which may have significant effects on the progression of atherosclerotic lesions (Chadjichristos and Kwak 2007; Kwak et al. 2003).

OxPL also potently activate vascular SMC and macrophages, resulting in changes of gene expression profiles and phenotypes of the respective cell type. Several papers have reported that OxPL, including POVPC, stimulated increased cell division and possibly differentiation of SMC (Heery et al. 1995;

Chatterjee et al. 2004), while others showed activation of apoptotic signaling pathways (Loidl et al. 2003). Recently, it was shown that POVPC caused phenotypic switching in SMC by suppressing the expression of multiple KLF4-dependent differentiation markers and induction of proinflammatory gene expression (Pidkovka et al. 2007).

Finally, OxPL are implicated to control end stage disease consequences including plaque rupture. There is evidence that OxPAPC activates alkaline phosphatase in calcifying vascular cells, suggesting a role for OxPL in vascular calcification (Parhami et al. 1997). We have recently shown that OxPL may increase the propensity of atherosclerotic lesions towards rupture by upregulation of matrix metalloproteinases and by inducing angiogenesis (Bochkov et al. 2006), both of which may contribute to destabilization of advanced plaques.

Further evidence for a role of OxPL in atherogenesis *in vivo* comes from studies showing that knocking out or inhibiting putative receptors for OxPL (including the PAF receptor, CD36, and toll-like receptors (TLRs) 2 and 4) leads to a decrease in experimental atherogenesis (Berliner and Watson 2005; Subbanagounder et al. 1999; Febbraio et al. 2000; Mullick et al. 2005; Tobias and Curtiss 2005). Moreover, when OxPAPC was directly applied to murine carotids *in vivo* using a pluronic gel, a pattern of inflammatory genes was upregulated, similar to that seen in experimental atherosclerosis (Furnkranz et al. 2005). Taken together, these studies clearly demonstrate that OxPL contribute to atherogenesis through their effects on vascular wall cells influencing several functions that are important for the initiation, progression and end stage events including plaque rupture.

### *12.2.2 Plasma Levels of OxPL as Cardiovascular Risk Markers*

There is extensive evidence that changes in plasma OxPL/apoB ratios, measured using the murine monoclonal antibody E06 (Tsimikas 2006b; Tsimikas and Witztum 2001) may reflect the extent of atherosclerotic disease burden (Tsimikas et al. 2005, 2006). It was shown that OxPL/apoB levels are increased in patients with coronary, carotid or femoral artery disease, acute coronary syndromes and after percutaneous coronary intervention (Tsimikas et al. 2006).

Moreover, *Tsimikas et al.* examined whether plasma OxPL/apoB levels may indicate the vessel wall content of OxPL during atherosclerosis progression or regression. They used cynomolgus monkeys and New Zealand White rabbits and OxPL content was measured in plasma and immunohistochemically in aortic plaques at baseline, after a high-fat/high-cholesterol diet and after reversion to normal chow. Immunostaining revealed that during atherosclerosis progression OxPL co-localized with apoB-100, whereas during regression OxPL virtually disappeared. These data suggest that changes in plasma levels of OxPL/apoB ratios reflect changes in OxPL content in atherosclerotic plaques during dietary-induced atherosclerosis progression and regression (Tsimikas et al. 2007).

In humans, OxPL were shown to circulate on apoB-containing lipoproteins, primarily on Lp(a) (Tsimikas et al. 2005, 2006a). Circulating levels of OxPL were strongly associated with angiographically documented coronary artery disease, and data from these studies suggest that the atherogenicity of Lp(a) may be mediated in part by associated OxPL (Tsimikas et al. 2005). Data obtained from the Bruneck study demonstrate that the OxPL/apoB levels predict 10 year CVD event rates independently of traditional risk factors such as hsCRP (Kiechl et al. 2007). Increasing Lp-PLA2 activity was shown to further amplify the risk of CVD mediated by oxPL/apoB (Tsimikas et al. 2006). Thus, OxPL/apoB and Lp-PLA2 levels can be used to predict symptomatic cardiovascular disease and even new cardiovascular events.

### *12.2.3 OxPL as Link Between Atherosclerosis and SLE*

Associations of several autoimmune diseases with atherosclerosis have been observed and a role for LDL oxidation especially in systemic lupus erythematosus (SLE) has been suggested (Frostegard et al. 2005; Hayem et al. 2001; Svenungsson et al. 2001). OxLDL forms immune complexes with β2GPI, which can be detected in the plasma of patients. OxLDL/β2GPI complexes have been demonstrated in patients with syphilis, infectious endocarditis, diabetes mellitus, antiphospholipid syndrome and chronic nephritis, indicating that oxidation of LDL and the formation of complexes with β2GPI is not restricted to SLE. It is hypothesized that these autoantibodies accelerate the development of atherosclerosis in autoimmune patients.

Further evidence for OxPL as a link between atherosclerosis and lupus comes from a recently described new mouse model (Feng et al. 2007). These authors created double knockout apoE2/2Fas2/2 mice, which spontaneously develop lupus-like disease, increased atherosclerotic lesions, accompanied by decreased bone density. Interestingly, apoE2/2Fas2/2 mice had decreased serum OxPL on apoB-100-containing particles but an increase in serum IgG antibodies to OxPL. Serum IgG antibodies to OxPL correlated positively with glomerular tuft areas and aortic lesion areas (Feng et al. 2007). It was presviously shown that IgG autoantibodies to OxPL correlated with aortic lesion areas in apoE-deficient mice, and therefore were considered atherogenic, while IgM antibodies that recognize phosphorylcholine in OxPL and OxLDL seem to be protective against atherosclerosis (Binder et al. 2002). These data provide evidence that IgG autoantibodies to OxPLs and immune complexes are important contributors to atherosclerosis and autoimmune diseases including SLE and glomerulonephritis, likely triggering shared pathways that promote the pathogenesis of these diseases.

### 12.2.4 OxPL as Link Between Atherosclerosis and Osteoporosis

Atherosclerosis has been associated with reduced bone mineral density and fracture risk and effects of high fat diet on bone formation have been reviewed (Parhami et al. 2001). It was demonstrated that OxPAPC inhibits spontaneous osteogenic differentiation of marrow stromal cells and mineralization of calvarial preosteoblasts, suggesting that OxPL may account for the clinical link between atherosclerosis and osteoporosis (Huang et al. 2007). Moreover, OxPAPC attenuated induction of osteogenic markers alkaline phosphatase and osteocalcin, as well as expression of PTH receptor by BMP-2. It was further shown that OxPAPC affected osteogenic signaling by inhibiting PTH signaling. Since anabolic agents that promote bone formation are increasingly used as treatment for osteoporosis, these data also suggest that OxPL may interfere with anabolic therapies for osteoporosis (Huang et al. 2007).

In the apoE2/2Fas2/2 mouse it was shown that serum levels of IgG autoantibodies to OxPL, which positively correlated with aortic lesion areas and glomerular tuft areas, correlated negatively with bone density, indicating that IgG anti-OxPL may also be involved in the process of osteoporosis. Interestingly, osteoporosis has been observed more frequently in SLE patients with cardiovascular disease, which is associated with increased OxLDL and autoantibodies to OxLDL (Svenungsson et al. 2001). The precise mechanism of osteopenia in association with atherosclerosis and the potential role of OxPL and their autoantibodies in bone loss require further investigation.

### 12.2.5 Platelets and Thrombosis

Activation of the endothelium leads to a prothrombotic phenotype. We have shown that OxPL stimulate production of tissue factor in cultured human EC (Bochkov et al. 2002b), an effect that could be inhibited by genestein (Holzer et al. 2007). The prothrombotic phenotype in EC is further enhanced by inhibition of the anticoagulant activity of tissue factor pathway inhibitor (Hiraishi et al. 2002) and downregulation of the anticoagulant glycoprotein thrombomodulin (Ishii et al. 2003). Collectively, these findings demonstrate that OxPL promote a transition from an anticoagulant endothelium to the procoagulant state such as seen in hyperlipidemia.

Hyperlipidemia is also associated with enhanced platelet reactivity. Activated platelets adhere to the endothelium and release vasoactive mediators which induce vasoconstriction and endothelial dysfunction. While preincubation of platelets with OxPAPC did not induce platelet aggregation, it resulted in increased surface expression of CD62p and CD41. Moreover, binding of an anti-CD36 Ab was inhibited, indicating that platelet CD36 is a main receptor responsible for binding of oxidized lipoproteins (Hartwich et al. 2002). Podrez et al. demonstrated that the prothrombotic platelet phenotype in hyperlipidemia was mediated by CD36. Genetic deletion of *Cd36* was protective against enhanced platelet reactivity, and OxPL were able to bind and activate platelets via CD36. Thus, interactions of platelet CD36 with specific endogenous OxPL play a crucial role in the well-known clinical associations between dyslipidemia, oxidant stress and a prothrombotic phenotype (Podrez et al. 2007). Direct platelet-activating activity by OxPL may also be induced by OxPL that structurally mimic PAF ("PAF-like lipids") (Marathe et al. 1999, 2002; Smiley et al. 1991; Zimmerman et al. 2002).

Furthermore, OxPL were shown to modulate the activity of the platelet prothrombinase complex, a major contributor to overall thrombin formation. Platelet-dependent thrombin generation was induced by ethanolamine phospholipids (PE) present in oxidized LDL. It was shown that oxidation products of unsaturated diacyl-PE were mainly responsible for the increased prothrombinase activity and synthetic aldehyde-PE adducts largely reproduced the stimulation of the thrombin generation (Zieseniss et al. 2001). These data suggest that oxidized PE contribute to the prothrombotic phenotype by increasing prothrombinase activity in platelets.

Recently, it was shown that protein C inhibitor (PCI) interacted with different phospholipids and their oxidized forms, thereby increasing its activity (Malleier et al. 2007). The protein C system represents a major anticoagulant pathway. PCI, a member of the serpin (serine protease inhibitor) family of protease inhibitors, was originally described as an inhibitor of the anticoagulant serine protease activated protein C. These results suggest that phospholipids are important endogenous cofactors of PCI and stimulation of PCI activity by OxPL could therefore increase the risk for thrombotic events.

## 12.3 Diabetes and Associated Diseases

### *12.3.1 Type II Diabetes*

Oxidative tissue damage, as evidenced by increased production of oxidized lipids has been reported in type II diabetes (Sampson et al. 2002). Moreover, the degradation of OxPL was shown to be defective in type II diabetic patients, possibly due to decreased PON activity on the HDL from these patients. In this study the authors compared HDL from three different populations, controls, patients with type II diabetes and patients with cardiovascular disease. The data demonstrated that the ability of HDL from patients with cardiovascular disease but also from type II diabetic patients without cardiovascular disease to degrade OxPAPC *in vitro* was significantly reduced (Mastorikou et al. 2006). This was accompanied by significantly higher levels of circulating plasma oxidized LDL. The authors conclude that increased OxPL levels could contribute to the increased susceptibility of type II diabetic patients to develop cardiovascular disease.

Furthermore, serum PON 55 and 192 polymorphsms and PON activity have been implicated to correlate with incidence of non-insulin dependent DM and also in type II diabetes complicated by retinopathy (Mackness et al. 1998a, 2005, 2006a; Mackness and Durrington 1995a; Sampson et al. 2005). Although it remains to be shown whether OxPL contribute to adipose tissue inflammation and insulin resistance, these results imply a role for OxPL in the development of complications associated with type II diabetes.

### *12.3.2 Eye Disease*

One of the complications associated with type II diabetes is eye disease and several studies have also shown an association of macular degeneration and atherosclerosis (Mullins et al. 2000; Friedman 2000). In that context, it was suggested that PON gene polymorphisms and plasma OxLDL levels may be risk factors for age-related macular degeneration (Ikeda et al. 2001).

Oxidative modification of phospholipids has been implicated in various pathologies of the eye. Lipid oxidation was postulated as the underlying pathologic trigger in age-related macular degeneration (AMD), cataractogenesis (reviewed in (Huang et al. 2006)) and the association between lens opacification and lipid oxidation has been demonstrated. Oxidation of phospholipids in the eye may be initiated by singlet oxygen produced by UV light, which acts on lens epidermal cells and causes oxidative membrane damage. Due to the lack of cell turnover in the lens, the prolonged exposure to oxidative stress results in compositional changes greater than those reported in any other organ (Huang et al. 2006).

A study using human donor eyes showed that OxPL were present in the photoreceptors and retinal pigment epithelium of the normal human macular area, and their levels increased with age. Using the monoclonal antibody DLH3 against OxPL (Itabe et al. 1994), the authors demonstrate that eyes with AMD showed more intense immunoreactivity than age-matched normal eyes (Suzuki et al. 2007). These findings suggest that oxidative stress is involved in the pathogenesis of AMD, possibly by oxidizing phospholipids in the photoreceptors. Thus, it is conceivable that controlling oxidation of phospholipids may be a potential treatment for AMD.

Studies by Hoppe et al. show that OxLDL as well as oxidized lipid-protein complexes inhibit processing of photoreceptor outer segments and alter phagosome maturation in retinal pigment epithelium (Hoppe et al. 2001, 2004a, b; Sun et al. 2006). Clearance of shed photoreceptor outer segments by the retinal pigment epithelium, a tissue with one of the highest turnover rates in the body, is critical to the maintenance and normal function of the retina. It was shown that OxPL may serve as endogenous ligands mediating the uptake of photoreceptor outer segments via the scavenger receptor CD36. OxPL, that contained a CD36 recognition motif, were formed *in vivo* in the retinas of dark-adapted rats following physiological light exposure. These studies suggest that intense light exposure initiates oxidative modification of PL in photoreceptor outer segments, necessary for CD36-mediated phagocytosis under oxidant stress conditions (Sun et al. 2006).

CD36 was also shown to be essential for inhibiting angiogenesis when activated by thrombospondin-1. Consequently, CD36 and thrombospondin-1 were shown to be involved in mediating antiangiogenic signals in ischemic proliferative retinopathy (Mwaikambo et al. 2006). These authors report that expression of CD36 in macrophages and microvascular endothelial cells after corneal injury suppresses corneal angiogenesis. They further show that POVPC inhibited neovessel formation, which could be blocked by an antibody recognizing CD36. POVPC also diminished VEGF-A expression and induced regression of newly grown vessels. These data indicate that certain OxPL species may negatively affect corneal neovascularisation.

### *12.3.3 Kidney*

Diabetic nephropathy consequently results in chronic renal insufficiency and need for dialysis. End-stage renal failure is accompanied by increased oxidative stress and patients are at increased risk to develop cardiovascular disease. Oxidatively modified LDL was shown to be localized in kidneys (Exner et al. 1996) and a role for OxPL in kidney disease was suggested.

Evidence for a role of OxPL in kidney disease comes from studies in mouse models where infusion of an apoE-derived peptide (D-4F) lowered the presence of glomerular and tubulo-interstitial OxPL resulting in reduced renal inflammation (Buga et al. 2008). Evidence that OxPL could be involved in kidney disease in humans comes from a recent study demonstrating that OxPL serve as biomarkers in patients with end-stage renal failure (Bossola et al. 2007). When the effect of hemodyalysis on OxPL levels was analysed, the results demonstrated a significant reduction in OxPL/apoB following dialysis, despite the pro-oxidant effects of the procedure.

IgG antibodies to OxPL may also contribute to renal complications, especially in glomerulonephritis, where anti-dsDNA antibodies were shown to play a role. Moreover, autoantibodies to other oxidation-specific epitopes, including malondialdehyde, have been suggested to participate in a model of human membrane nephropathy (Dominguez et al. 2000). In the apoE2/2Fas2/2 mouse, increased IgG deposition in glomeruli with increased IgG anti-POVPC and anti-PGPC suggested that IgG anti-OxPL might act similarly to IgG anti-dsDNA to induce nephritis (Feng et al. 2007).

### *12.3.4 Liver*

Nonalcoholic fatty liver disease (NAFLD) is a growing hepatological problem in Western countries. NAFLD may be limited to the fatty liver alone, or it may progress to nonalcoholic steatohepatitis (NASH) (Alisi and Nobili 2007). Although the etiology of NASH is unknown, it is frequently associated with obesity, type II diabetes mellitus, and hyperlipidemia. The pathogenesis of NASH is not yet fully understood, but it is widely accepted that insulin resistance is a culprit leading to hepatic lipid accumulation followed by oxidative stress, inflammation and finally necrosis. These modifications enhance lipid peroxidation, hepatocyte injury resulting in inflammation and fibrosis, and may evolve into cirrhosis and hepatocellular carcinoma.

OxPL have been found in fatty liver, and their levels were increased especially in NAFLD (Ikura et al. 2006), implicating a role for OxPL in disease progression. Evidence for OxPL as important contributors to liver injury comes from a study demonstrating that PAF-AH (II)-deficient mice were no longer protected against oxidative stress-induced hepatic injury (Kono et al. 2008). Cells derived from PAF-AH (II)-deficient mice were more sensitive to oxidative stress than those derived from wild-type mice and treatment of PAF-AH (II)-deficient mice with carbon tetrachloride demonstrated a delay in hepatic injury recovery, which correlated with increased production of OxPL. These results indicated that PAF-AH (II) metabolizes OxPL thereby protecting liver tissue from oxidative stress-induced injury.

NAFLD has become one of the critical chronic liver diseases worldwide. There is a great urgency to clarify the pathogenesis of NAFLD/NASH to establish reasonable treatment strategies (Alisi and Nobili 2007). Further understanding of the role of OxPL in NAFLD/NASH may lead to development of such strategies.

## 12.4 Role of OxPL in Immune Response

### *12.4.1 Innate Immune System*

In chronically inflamed tissue, the local immune response is determined by environmental factors, such as accumulating lipid oxidation products. Phospholipid oxidation products resemble danger signals, which accumulate under conditions of increased oxidative stress and cell death, and as such were shown to act as endogenous regulators of the innate immune response (Binder et al. 2002, 2003; Bochkov et al. 2002a). Upon oxidative modification, phospholipids are recognized by certain pattern recognition receptors (PRRs), indicating the formation of structural motifs, which are sensed by the innate immune system as danger signals or “altered-self”. PRRs that have been implicated in the recognition of OxPL include scavenger receptors such as CD36, TLRs, CD14, LPS-binding protein and C-reactive protein (Miller et al. 2003a; Walton et al. 2003b; Chang et al. 2002; Hazen and Chisolm 2002; Podrez et al. 2002b; Boullier et al. 2005; Miller et al. 2003b). The fact that the oxidation process renders phospholipids "visible" to the innate immune system bears important implications for the pathogenesis of both chronic inflammatory and autoimmune diseases (Kronke and Leitinger 2006).

Further evidence for recognition of endogenous phospholipid oxidation products by PRRs comes from our studies demonstrating that OxPAPC competes with LPS for the binding to PRRs such as LPS-binding protein and CD14. In this case oxidized phospholipids act as endogenous "LPS receptor-antagonists" and block LPS-induced signaling events including the activation of the pro-inflammatory transcription factor NFκB (Bochkov et al. 2002a). This protective role of phospholipid oxidation products during an acute LPS-induced inflammation may represent an important feedback mechanism whereby accumulating oxidized phospholipids limit further tissue-damage by blocking the innate immune response.

Based on findings that OxPL inhibit LPS-induced inflammation, possible therapeutic effects of OxPAPC in rodents with acute necrotizing pancreatitis were investigated. In this study, treatment with OxPAPC decreased the severity of experimental pancreatitis in mice and rats and the protective effect of OxPAPC was mediated, at least in part, through blocking the LPS signaling pathway (Li, Wang, and Wu 2007).

### 12.4.2 Endotoxemia and Sepsis

We and others have shown previously that OxPL effectively inhibit the activation of TLR4 (Bochkov et al. 2002a; Walton et al. 2003a; Eligini et al. 2002), the mechanism of which may involve inhibition of interaction of LPS with LBP, CD14. The biological significance of this finding is underlined by studies demonstrating that the exogenous administration of OxPAPC could prevent mortality of mice exposed to high doses of LPS (Bochkov et al. 2002a). Thus, the formation of OxPL seems to inhibit LPS-induced inflammation *in vivo* and may be protective in settings of endotoxic shock.

Based on these findings, Knapp et al. investigated the effects of OxPL during *E. coli*-induced abdominal sepsis *in vivo*. In contrast to the protective effects observed in endotoxemia, administration of OxPAPC rendered mice highly susceptible to *E. coli* peritonitis, as indicated by an accelerated mortality and enhanced bacterial outgrowth and dissemination. The mechanism by which OxPAPC impaired the immune response was by diminishing phagocytosing capacity of neutrophils and macrophages. These data suggest that OxPL arising as a byproduct of the respiratory burst during acute inflammation may contribute to mortality during Gram-negative sepsis *via* impairment of the phagocytic properties of professional phagocytes involved in innate immunity (Knapp et al. 2007).

Although OxPL generated at sites of inflammation might be able to prevent overwhelming inflammation in settings of sterile inflammation, impairment of the innate immune response to bacterial infections by OxPL is detrimental for the outcome of host defense reactions.

## *12.4.3 Adaptive Immune System*

### 12.4.3.1 Oxidative Modification of Phospholipids Provides Epitopes for the Adaptive Immune System

The patterns that are generated during oxidation of phospholipids are also recognized by the humoral part of the adaptive immune system. Both IgG and IgM antibodies directed against OxLDL are present in the plasma of humans and animals and their titers have been shown to correlate with atherosclerosis progression (Horkko et al. 2000; Tsimikas et al. 2001; Cyrus et al. 2001), as well as in several autoimmune disorders such as systemic lupus erythematosus and rheumatoid arthritis (Amara et al. 1995; Cvetkovic et al. 2002; Hayem et al. 2001; Wu and Rao 1999; Tsimikas et al. 2005). Detailed studies in ApoE-deficient mice, which show increased levels of OxLDL, led to the cloning of a set of abundant monoclonal IgM antibodies directed against OxLDL, which includes the prototypic EO6 antibody that specifically binds to OxPL on the surface of OxLDL and apoptotic cells (Palinski et al. 1996; Horkko et al. 1999; Chang et al. 1999).

### 12.4.3.2 OxPL as Regulators of Dendritic Cell (DC) Maturation

Recent *in vitro* and *in vivo* evidence highlights the impact OxPL exert on the maturation process of DC, where these molecules mediate both stimulatory and inhibitory effects, thereby influencing decisive steps of the adaptive immune response. Studies performed by our lab have shown that phospholipid oxidation products such as OxPAPC inhibit basic steps of the classical DC maturation process. For instance, OxPAPC blocks LPS-induced expression of costimulatory molecules including CD40 and CD83 and inhibits secretion of pro-inflammatory cytokines such as IL-12 and TNF. In parallel with the inhibitory influence on DC maturation, OxPAPC treatment of DCs dampened DC-induced T-cell proliferation as well interferon-γ secretion by T-cells (Bluml et al. 2005). This reduction in interferon-γ levels in T-cells, together with the inhibition of IL-12 expression in DCs, indicates a specific block of the Th1-orientated immune response by OxPAPC.

Important evidence for a role of OxPL as modulators of DC function *in vivo* comes from studies performed in ApoE-deficient mice, which have elevated levels of OxLDL and suffer from severe dyslipidemia and accelerated atherosclerosis. ApoE-deficient mice show a severely compromised immune response, resulting in a depressed delayed-type hypersensitivity reaction (Laskowitz et al. 2000; Ludewig et al. 2001). It was shown that these mice exhibit an altered DC function, indicated by impaired migratory capabilities of skin DCs. Elevated levels of PAF and PAF-like OxPL caused these immunological alterations in ApoE-deficient mice. Consequently, the treatment with recombinant PAF-AH restored normal immunological parameters in these mice (Angeli et al. 2004).

Using ApoE-deficient mice on a high-fat/cholesterol diet, Shamshiev et al. showed that dyslipidemia inhibited TLR-induced production of proinflammatory cytokines, including IL-12, IL-6, and TNF, as well as up-regulation of costimulatory molecules by CD8α(–) DCs, but not by CD8α(+) DCs, *in vivo*. Decreased DC activation profoundly influenced Th cell responses, leading to impaired Th1 and enhanced Th2 responses. As a consequence of this immune modulation, host resistance to Leishmania major was compromised (Shamshiev et al. 2007).

These results show that a dyslipidemic microenvironment can directly interfere with DC responses to pathogens and skew the development of T cell-mediated immunity. It will be important to specify the molecular structures within these phospholipid oxidation products that are responsible for their effects on DC function. This will eventually allow designing low molecular substances, which mimic their immunomodulatory effects.

## 12.5 Lung

OxPL are constantly formed and accumulate in murine lung tissue (Nakamura et al. 1998) and in the lung circulation as a result of increased oxidative stress that accompanies pathological conditions such as acute lung injury, lung inflammation, acute respiratory distress syndrome (ARDS), ventilator-induced lung injury (VILI), systemic inflammatory response syndrome (SIRS) and sepsis. Under these conditions, lung vascular barrier function is largely compromised.

Interestingly, studies by Birukov et al. and other reports strongly suggest barrier-protective effects of certain OxPL on human pulmonary endothelium (Birukova et al. 2007). It was shown that OxPL produce a sustained increase in transendothelial electrical resistance of human pulmonary EC and restore barrier disruption induced by edemagenic agonist thrombin *in vitro* (Birukov et al. 2004). Subsequently it has been shown that intravenously injected OxPAPC, but not unoxidized PAPC, protects rats from lung inflammation and injury induced by intratracheal application of LPS (Nonas et al. 2006). Measurements of endothelial transmonolayer electrical resistance and immunofluorescent analysis of monolayer integrity demonstrated that OxPAPC markedly attenuated LPS-induced tissue inflammation, barrier disruption, and cytokine production (Nonas et al. 2006). These studies demonstrate protective effects of OxPL on LPS-induced lung dysfunction *in vivo* and *in vitro*. Evidence for protective effects of OxPL in the lung being independent of their LPS-antagonizing properties comes from a recent study that demonstrates that OxPL are also protective in ventilator-induced lung injury (Nonas et al. 2008). Thus, OxPL, which are constantly present in the lung (Uhlson et al. 2002), have a potential to preserve function of lung during life-threatening systemic inflammation as well as in settings of non-infectious lung injury.

## 12.6 Skin

Long wave ultraviolet (UV) irradiation causes oxidizing stress to the skin that provokes synthesis of antioxidant stress response genes. UVA1 (340–400 nm) is frequently used in clinical UV therapy for inflammatory skin diseases such as psoriasis (Legat et al. 2004a, b), morphea (Gruss et al. 2001), atopic dermatitis, scleroderma and graft versus host disease (Mang and Krutmann 2005). Especially cells in the dermis respond to UVA1 irradiation by inducing synthesis of protective stress response genes that have anti-inflammatory effects. The generation of singlet oxygen by UVA1 (340–390 nm) leads to subsequent oxidation of intracellular membrane lipids. Strong evidence suggests that oxidation of membrane lipids by UVA1 is involved in response gene induction (Baier et al. 2007; Basu-Modak and Tyrrell 1993; Basu-Modak et al. 1996). UVA irradiation also perturbs the activity of the immune system (Furio et al. 2005). However, underlying mechanisms and structures of responsible lipid oxidation products are poorly understood.

Lipid mediators that are formed upon UV irradiation are classically regarded as inducers and propagators of chronic inflammatory reactions, but recently their role is being redefined since data have accumulated that demonstrate their potential to induce anti-inflammatory genes, to prevent expression of pro-inflammatory genes, and to interfere with the function of professional antigen presenting cells (Nonas et al. 2006; Gruber et al. 2007; Bochkov and Leitinger 2003). Phospholipid mediators that arise upon UV irradiation are fragmented PAF-like lipids, that can elicit immuno-suppressive effects in the skin (Walterscheid et al. 2002), as well as long chain acyl-OxPL, that were shown to induce HO-1 expression (Kronke et al. 2003), which, besides being a major antioxidant response gene, itself has immunomodulatory effects (Allanson and Reeve 2004). Investigating the formation and biological properties of OxPL in skin cells after UVA-1 irradiation, we found that the common membrane phospholipid PAPC is oxidized by UVA1 *in vitro* and in living dermal fibroblasts, the major UVA1 responsive dermal cell type. We have shown that UVA1-OxPL regulate synthesis of stress response enzymes in cultured skin cells and in cells of the immune system (Gruber et al. 2007; Ishikawa et al. 1997).

These results show that certain OxPL species may mediate anti-inflammatory and immunomodulatory effects in the skin, particularly after UV irradiation.

## 12.7 Neurological Disorders

It has been long known that oxidative modification of membrane lipids accompanies neurological tissue damage, which may ultimately lead to multiple sclerosis (MS), Parkinson's and Alzheimer's disease. The formation of phospholipid hydroperoxides was demonstrated in rat brain synaptosomes, a process that could be inhibited by alpha-tocopherol (Shi et al. 1999). Moreover, oxidative modification in membrane phospholipids of the central nervous system was detected after chemotherapy (Miketova et al. 2005). Recently, the same family of oxidized choline glycerophospholipids that previously has been identified in atheroma to serve as endogenous ligands for the scavenger receptor CD36, has been found in the brain. A subset of these OxPL possessing *sn*-2 esterified fatty acyl hydroxyalkenal groups, can undergo intramolecular cyclization and dehydration to form a terminal furyl moiety (oxPC-furan). Generation of oxPC-furans was demonstrated in brain tissues following cerebral ischemia (Gao et al. 2006).

OxPL, including the aldehyde-containing POVPC, were detected in the brain of MS patients. MS is an inflammatory neurodegenerative autoimmune disease, which involves formation of plaques of demyelination in the brain, eventually resulting in axonal degeneration. In brains from MS patients, E06-positive areas were present in MS plaques, which also showed evidence of OxPL-modified proteins, while E06 reactivity was largely absent from control tissue. Moreover, spinal cords from mice trated to develop experimental allergic encephalomyelitis also showed strong immunoreactivity for OxPL (Qin et al. 2007). These authors conclude that the formation of OxPL could play a role in the progression of MS.

Investigation of phospholipid catabolic and anabolic enzymes revealed lower activity of $PLA_2$ and phosphoethanolamine- and phosphocholine-cytidylyltransferases in autopsied substantia nigra of patients with idiopathic Parkinson's disease. The decreased rate of phospholipid turnover in the pigmented neurons of the substantia nigra might result in reduced ability to repair oxidative membrane damage and thus accumulation of OxPL in Parkinson's disease (Ross et al. 2001).

Lesions of Alzheimer's disease, evident as dense plaques composed of fibrillar amyloid beta-proteins, likely develop when these proteins are first induced to form beta-sheet secondary structures. It was demonstrated that membranes containing oxidatively damaged phospholipids accumulated amyloid beta-protein significantly faster than membranes containing unoxidized phospholipids. The protein on oxidized membranes more readily changed conformation to a beta-sheet, indicating that oxidatively damaged phospholipid membranes promote beta-sheet formation by amyloid beta-proteins. These data suggest a possible role for lipid peroxidation in the pathogenesis of Alzheimer's Disease (Koppaka et al. 2003; Koppaka and Axelsen 2000).

## 12.8 Concluding Remarks

The formation of biologically active phospholipid oxidation products represents a general concept that occurs in all inflamed and stressed tissues. The relative abundance of individual structures, as well as presence of recognition systems for these "danger signals" determines cell and tissue response, which span from inflammation-propagating responses to anti-inflammatory, tissue-protective effects. Anti-inflammatory, tissue protective effects of certain structures derived from oxidative modification of phospholipids could potentially be used to develop pharmacological agents with immune-modulatory activity. Treatment strategies based on structures derived from OxPL may evolve from studies showing beneficial effects of OxPL administration in several organs and tissues, including lung, pancreas, skin but also settings of systemic inflammation including endotoxemia and sepsis. On the other hand, it has been suggested that for some drugs, the mechanism of action and possible side effects are closely linked to oxidative damage of cell membranes. A recent study shows that several antipsychotic and antineoplastic agents interact with OxPL, which may lead to modulation of their function (Mattila et al. 2007). Definitely more knowledge about pharmacodynamic properties, receptors and signaling pathways that are induced by OxPL needs to be gained in order to devise such strategies.

**Acknowledgments** I would like to thank Alexandra Kadl, M.D., for stimulating discussions and carefully reading this manuscript. I apologize to those whose work has not been cited due to space limitations. This work was supported by the NIH and AHA.

## References

Agachan, B. et al. "Paraoxonase 55 and 192 polymorphism and its relationship to serum paraoxonase activity and serum lipids in Turkish patients with non-insulin dependent diabetes mellitus." *Cell Biochem. Funct.* 22 (2004) 163–8.

Ahmed, Z. et al. "Apolipoprotein A-I promotes the formation of phosphatidylcholine core aldehydes that are hydrolyzed by paraoxonase (PON-1) during high density lipoprotein oxidation with a peroxynitrite donor." *J. Biol. Chem.* 276 (2001) 24473–81.

Alisi, A. and V. Nobili "Molecular pathogenesis of nonalcoholic steatohepatitis: today and tomorrow." *Am. J. Pathol.* 171 (2007) 712–3.

Allanson, M. and V.E. Reeve "Immunoprotective UVA (320–400 nm) irradiation upregulates heme oxygenase-1 in the dermis and epidermis of hairless mouse skin." *J. Invest Dermatol.* 122 (2004) 1030–6.

Amara, A. et al. "Autoantibodies to malondialdehyde-modified epitope in connective tissue diseases and vasculitides." *Clin. Exp. Immunol* 101 (1995) 233–8.
Angeli, V. et al. "Dyslipidemia associated with atherosclerotic disease systemically alters dendritic cell mobilization." *Immunity* 21 (2004) 561–74.
Arai, H. et al. "Platelet-activating factor acetylhydrolase (PAF-AH)." *J. Biochem. (Tokyo)* 131 (2002) 635–40.
Baier, J. et al. "Direct detection of singlet oxygen generated by UVA irradiation in human cells and skin." *J. Invest Dermatol.* 127 (2007) 1498–506.
Basu-Modak, S., P. Luscher, and R.M. Tyrrell "Lipid metabolite involvement in the activation of the human heme oxygenase-1 gene." *Free Radic. Biol. Med.* 20 (1996) 887–97.
Basu-Modak, S. and R.M. Tyrrell "Singlet oxygen: a primary effector in the ultraviolet A/near-visible light induction of the human heme oxygenase gene." *Cancer Res.* 53 (1993) 4505–10.
Berliner, J.A. et al. "Atherosclerosis: basic mechanisms. Oxidation, inflammation, and genetics." *Circulation* 91 (1995) 2488–96.
Berliner, J.A. et al. "Evidence for a role of phospholipid oxidation products in atherogenesis." *Trends Cardiovasc. Med.* 11 (2001) 142–7.
Berliner, J.A. et al. "Minimally modified low density lipoprotein stimulates monocyte endothelial interactions." *J. Clin. Invest* 85 (1990) 1260–6.
Berliner, J.A. and A.D. Watson "A role for oxidized phospholipids in atherosclerosis." *N. Engl. J. Med.* 353 (2005) 9–11.
Bilge, I. et al. "Is paraoxonase 192 gene polymorphism a risk factor for membranoproliferative glomerulonephritis in children?" *Cell Biochem. Funct.* 25 (2007) 159–65.
Binder, C.J. et al. "Innate and acquired immunity in atherogenesis." *Nat. Med.* 8 (2002): 1218–26.
Binder, C.J. et al. "Pneumococcal vaccination decreases atherosclerotic lesion formation: molecular mimicry between Streptococcus pneumoniae and oxidized LDL." *Nat. Med.* 9 (2003) 736–43.
Binder, C.J. et al. "The role of natural antibodies in atherogenesis." *J. Lipid Res.* 46 (2005) 1353–63.
Birukov, K.G. et al. "Epoxycyclopentenone-containing oxidized phospholipids restore endothelial barrier function via Cdc42 and Rac." *Circ. Res.* 95 (2004) 892–901.
Birukova, A.A. et al. "Polar head groups are important for barrier-protective effects of oxidized phospholipids on pulmonary endothelium." *Am. J. Physiol Lung Cell Mol. Physiol* 292 (2007) L924–L935.
Bluml, S. et al. "Oxidized phospholipids negatively regulate dendritic cell maturation induced by TLRs and CD40." *J. Immunol.* 175 (2005) 501–8.
Bochkov, V.N. "Inflammatory profile of oxidized phospholipids." *Thromb. Haemost.* 97 (2007) 348–54.
Bochkov, V.N. et al. "Protective role of phospholipid oxidation products in endotoxin-induced tissue damage." *Nature* 419 (2002a) 77–81.
Bochkov, V.N. and N. Leitinger "Anti-inflammatory properties of lipid oxidation products." *J. Mol. Med.* 81 (2003) 613–26.
Bochkov, V.N. et al. "Oxidized phospholipids stimulate tissue factor expression in human endothelial cells via activation of ERK/EGR-1 and Ca(+ +)/NFAT." *Blood* 99 (2002b) 199–206.
Bochkov, V.N. et al. "Oxidized phospholipids stimulate angiogenesis via autocrine mechanisms, implicating a novel role for lipid oxidation in the evolution of atherosclerotic lesions." *Circ. Res.* 99 (2006) 900–8.
Bossola, M. et al. "Oxidized low-density lipoprotein biomarkers in patients with end-stage renal failure: Acute effects of hemodialysis." *Blood Purif.* 25 (2007) 457–65.
Boullier, A. et al. "Phosphocholine as a pattern recognition ligand for CD36." *J. Lipid Res.* 46 (2005) 969–76.

Buga, G.M. et al. "D-4F reduces EO6 immunoreactivity, SREBP-1c mRNA levels, and renal inflammation in LDL receptor-null mice fed a Western diet." *J. Lipid Res.* 49 (2008) 192–205.

Cao, H. et al. "Lack of association between carotid intima-media thickness and paraoxonase gene polymorphism in non-insulin dependent diabetes mellitus." *Atherosclerosis* 138 (1998) 361–6.

Chadjichristos, C.E. and B.R. Kwak "Connexins: new genes in atherosclerosis." *Ann. Med.* 39 (2007) 402–11.

Chang, M.K. et al. "Monoclonal antibodies against oxidized low-density lipoprotein bind to apoptotic cells and inhibit their phagocytosis by elicited macrophages: evidence that oxidation-specific epitopes mediate macrophage recognition." *Proc. Natl. Acad. Sci. USA* 96 (1999) 6353–8.

Chang, M.K. et al. "C-reactive protein binds to both oxidized LDL and apoptotic cells through recognition of a common ligand: Phosphorylcholine of oxidized phospholipids." *Proc. Natl. Acad. Sci. USA* 99 (2002) 13043–8.

Chatterjee, S. et al. "Identification of a biologically active component in minimally oxidized low density lipoprotein (MM-LDL) responsible for aortic smooth muscle cell proliferation." *Glycoconj. J* 20 (2004) 331–8.

Cvetkovic, J.T. et al. "Increased levels of autoantibodies against copper-oxidized low density lipoprotein, malondialdehyde-modified low density lipoprotein and cardiolipin in patients with rheumatoid arthritis." *Rheumatology (Oxford)* 41 (2002) 988–95.

Cyrus, T. et al. "Absence of 12/15-lipoxygenase expression decreases lipid peroxidation and atherogenesis in apolipoprotein e-deficient mice." *Circulation* 103 (2001) 2277–82.

Dada, N., N.W. Kim, and R.L. Wolfert "Lp-PLA2: an emerging biomarker of coronary heart disease." *Expert. Rev. Mol. Diagn.* 2 (2002) 17–22.

DeMaio, L. et al. "Oxidized phospholipids mediate occludin expression and phosphorylation in vascular endothelial cells." *Am. J. Physiol Heart Circ. Physiol* 290 (2006) H674–H683.

Dessi, M. et al. "Influence of the human paraoxonase polymorphism (PON1 192) on the carotid-wall thickening in a healthy population." *Coron. Artery Dis.* 10 (1999) 595–9.

Dominguez, J.H. et al. "Studies of renal injury III: Lipid-induced nephropathy in type II diabetes." *Kidney Int.* 57 (2000) 92–104.

Eligini, S. et al. "Oxidized phospholipids inhibit cyclooxygenase-2 in human macrophages via nuclear factor-kappaB/IkappaB- and ERK2-dependent mechanisms." *Cardiovasc. Res.* 55 (2002) 406–15.

Esterbauer, H. et al. "Autoxidation of human low density lipoprotein: loss of polyunsaturated fatty acids and vitamin E and generation of aldehydes." *J. Lipid. Res.* 28 (1987) 495–509.

Exner, M. et al. "Lipoproteins accumulate in immune deposits and are modified by lipid peroxidation in passive Heymann nephritis." *Am. J. Pathol.* 149 (1996) 1313–20.

Febbraio, M. et al. "Targeted disruption of the class B scavenger receptor CD36 protects against atherosclerotic lesion development in mice." *J. Clin. Invest* 105 (2000) 1049–56.

Feng, X. et al. "ApoE-/-Fas-/- C57BL/6 mice: a novel murine model simultaneously exhibits lupus nephritis, atherosclerosis, and osteopenia." *J. Lipid Res.* 48 (2007) 794–805.

Fortunato, G. et al. "A paraoxonase gene polymorphism, PON 1 (55), as an independent risk factor for increased carotid intima-media thickness in middle-aged women." *Atherosclerosis* 167 (2003) 141–8.

Friedman, E. "The role of the atherosclerotic process in the pathogenesis of age-related macular degeneration." *Am. J. Ophthalmol.* 130 (2000) 658–63.

Frostegard, J. et al. "Lipid peroxidation is enhanced in patients with systemic lupus erythematosus and is associated with arterial and renal disease manifestations." *Arthritis Rheum.* 52 (2005) 192–200.

Fruhwirth, G.O., A. Loidl, and A. Hermetter "Oxidized phospholipids: From molecular properties to disease." *Biochim. Biophys. Acta* 1772 (2007) 718–36.

Furio, L. et al. "UVA radiation impairs phenotypic and functional maturation of human dermal dendritic cells." *J. Invest Dermatol.* 125 (2005) 1032–8.

Furnkranz, A. et al. "Oxidized phospholipids trigger atherogenic inflammation in murine arteries." *Arteriosclerosis, Thrombosis, and Vascular Biology* 25 (2005) 633–8.

Gao, S. et al. "Phospholipid hydroxyalkenals, a subset of recently discovered endogenous CD36 ligands, spontaneously generate novel furan-containing phospholipids lacking CD36 binding activity in vivo." *J. Biol. Chem.* 281 (2006) 31298–308.

Greenberg, M.E. et al. "The lipid whisker model of the structure of oxidized cell membranes." *J. Biol. Chem.* 283 (2008) 2385–96.

Gruber, F. et al. "Photooxidation generates biologically active phospholipids that induce heme oxygenase-1 in skin cells." *J Biol. Chem.* 282 (2007) 16934–41.

Gruss, C.J. et al. "Effects of low dose ultraviolet A-1 phototherapy on morphea." *Photodermatol. Photoimmunol. Photomed.* 17 (2001) 149–55.

Hartwich, J. et al. "Regulation of platelet adhesion by oxidized lipoproteins and oxidized phospholipids." *Platelets.* 13(May 2002) 141–51.

Hayem, G. et al. "Anti-oxidized low-density-lipoprotein (OxLDL) antibodies in systemic lupus erythematosus with and without antiphospholipid syndrome." *Lupus* 10 (2001) 346–51.

Hazen, S.L. and G.M. Chisolm "Oxidized phosphatidylcholines: pattern recognition ligands for multiple pathways of the innate immune response." *Proc. Natl. Acad. Sci. USA* 99 (2002) 12515–7.

Heery, J.M. et al. "Oxidatively modified LDL contains phospholipids with platelet-activating factor-like activity and stimulates the growth of smooth muscle cells." *J. Clin. Invest.* 96 (1995) 2322–30.

Hiraishi, S., S. Horie, and Y. Seyama "Oxidation products of phospholipid-containing delta-9 fatty acids specifically impair the activity of tissue factor pathway inhibitor." *Biochem. Biophys. Res. Commun.* 298 (2002) 468–73.

Holzer, G. et al. "The dietary soy flavonoid genistein abrogates tissue factor induction in endothelial cells induced by the atherogenic oxidized phospholipid oxPAPC." *Thromb. Res.* 120 (2007) 71–9.

Hoppe, G. et al. "Oxidized low density lipoprotein-induced inhibition of processing of photoreceptor outer segments by RPE." *Invest Ophthalmol. Vis. Sci.* 42 (2001) 2714–20.

Hoppe, G. et al. "Products of lipid peroxidation induce missorting of the principal lysosomal protease in retinal pigment epithelium." *Biochim. Biophys. Acta* 1689 (2004a) 33–41.

Hoppe, G. et al. "Accumulation of oxidized lipid-protein complexes alters phagosome maturation in retinal pigment epithelium." *Cell Mol. Life Sci.* 61 (2004b) 1664–74.

Horkko, S. et al. "Immunological responses to oxidized LDL." *Free Radic. Biol. Med.* 28 (2000) 1771–9.

Horkko, S. et al. "Monoclonal autoantibodies specific for oxidized phospholipids or oxidized phospholipid-protein adducts inhibit macrophage uptake of oxidized low-density lipoproteins." *J. Clin. Invest.* 103 (1999) 117–28.

Huang, L. et al. "Oxidation-induced changes in human lens epithelial cells. 1. Phospholipids." *Free Radic. Biol. Med.* 41 (2006) 1425–32.

Huang, M.S. et al. "Atherogenic phospholipids attenuate osteogenic signaling by BMP-2 and parathyroid hormone in osteoblasts." *J. Biol. Chem.* 282 (2007) 21237–43.

Huber, J. et al. "Specific monocyte adhesion to endothelial cells induced by oxidized phospholipids involves activation of cPLA2 and lipoxygenase." *J. Lipid Res.* 47 (2006) 1054–62.

Huber, J. et al. "Oxidized membrane vesicles and blebs from apoptotic cells contain biologically active oxidized phospholipids that induce monocyte-endothelial interactions." *Arterioscler Thromb VAsc Biol* 22 (2002) 101–7.

Ikeda, T. et al. "Paraoxonase gene polymorphisms and plasma oxidized low-density lipoprotein level as possible risk factors for exudative age-related macular degeneration." *Am. J. Ophthalmol.* 132 (2001) 191–5.

Ikura, Y. et al. "Localization of oxidized phosphatidylcholine in nonalcoholic fatty liver disease: impact on disease progression." *Hepatology* 43 (2006) 506–14.

Isakson, B.E. et al. "Oxidized phospholipids alter vascular connexin expression, phosphorylation, and heterocellular communication." *Arterioscler. Thromb. Vasc. Biol.* 26 (2006) 2216–21.

Ishii, H. et al. "Oxidized phospholipids in oxidized low-density lipoprotein downregulate thrombomodulin transcription in vascular endothelial cells through a decrease in the binding of RAR{beta}-RXR{alpha} heterodimers and Sp1 and 3 to their binding sequences in the TM promoter." *Blood* 101 (2003) 4767–74.

Ishikawa, K. et al. "Induction of heme oxygenase-1 inhibits the monocyte transmigration induced by mildly oxidized LDL." *J. Clin. Invest.* 100 (1997) 1209–16.

Itabe, H. et al. "A monoclonal antibody against oxidized lipoprotein recognizes foam cells in atherosclerotic lesions. Complex formation of oxidized phosphatidylcholines and polypeptides." *J. Biol. Chem.* 269 (1994) 15274–9.

Kakafika, A.I. et al. "The PON1 M55L gene polymorphism is associated with reduced HDL-associated PAF-AH activity." *J. Lipid Res.* 44 (2003) 1919–26.

Khuseyinova, N. et al. "Association between Lp-PLA2 and coronary artery disease: focus on its relationship with lipoproteins and markers of inflammation and hemostasis." *Atherosclerosis* 182 (2005) 181–8.

Kiechl, S. et al. "Oxidized phospholipids, lipoprotein(a), lipoprotein-associated phospholipase A2 activity, and 10-year cardiovascular outcomes: prospective results from the Bruneck study." *Arterioscler. Thromb. Vasc. Biol.* 27 (2007) 1788–95.

Knapp, S. et al. "Oxidized phospholipids inhibit phagocytosis and impair outcome in gram-negative sepsis in vivo." *J. Immunol.* 178 (2007) 993–1001.

Kono, N. et al. "Protection against oxidative stress-induced hepatic injury by intracellular type II platelet-activating factor acetylhydrolase by metabolism of oxidized phospholipids in vivo." *J. Biol. Chem.* 283 (2008) 1628–36.

Koppaka, V. and P.H. Axelsen "Accelerated accumulation of amyloid beta proteins on oxidatively damaged lipid membranes." *Biochemistry* 39 (2000) 10011–6.

Koppaka, V. et al. "Early synergy between Abeta42 and oxidatively damaged membranes in promoting amyloid fibril formation by Abeta40." *J. Biol. Chem.* 278 (2003) 36277–84.

Kronke, G. et al. "Oxidized phospholipids induce expression of human heme oxygenase-1 involving activation of cAMP-responsive element-binding protein." *J. Biol. Chem.* 278 (2003) 51006–14.

Kronke, G. and N. Leitinger "Oxidized phospholipids at the interface of innate and adaptive immunity." *Future Lipidol.* 1 (2006) 623–30.

Kwak, B.R. et al. "Reduced connexin43 expression inhibits atherosclerotic lesion formation in low-density lipoprotein receptor-deficient mice." *Circulation* 107 (2003) 1033–9.

Laskowitz, D.T. et al. "Altered immune responses in apolipoprotein E-deficient mice." *J. Lipid Res.* 41 (2000) 613–20.

Legat, F.J. et al. "Reduction of treatment frequency and UVA dose does not substantially compromise the antipsoriatic effect of oral psoralen-UVA." *J. Am. Acad. Dermatol.* 51 (2004a) 746–54.

Legat, F.J. et al. "The role of calcitonin gene-related peptide in cutaneous immunosuppression induced by repeated subinflammatory ultraviolet irradiation exposure." *Exp. Dermatol.* 13 (2004b) 242–50.

Leitinger, N. "Oxidized phospholipids as triggers of inflammation in atherosclerosis." *Mol. Nutr. Food Res.* 49 (2005) 1063–71.

Leitinger, N. et al. "Structurally similar oxidized phospholipids differentially regulate endothelial binding of monocytes and neutrophils." *Proc. Natl. Acad. Sci. USA* 96 (1999) 12010–5.

Li, L., X.P. Wang, and K. Wu "The therapeutic effect of oxidized 1-palmitoyl-2-arachidonoyl-sn-glycero-3-phosphorylcholine in rodents with acute necrotizing pancreatitis and its mechanism." *Pancreas* 35 (2007) e27–e36.

Loidl, A. et al. "Oxidized phospholipids in minimally modified low density lipoprotein induce apoptotic signaling via activation of acid sphingomyelinase in arterial smooth muscle cells." *J Biol Chem.* 278 (2003) 32921–8.

Ludewig, B. et al. "Dendritic cells in autoimmune diseases." *Curr. Opin. Immunol.* 13 (2001) 657–62.

Lusis, A.J. "Atherosclerosis." *Nature* 407 (2000) 233–41.

Mackness, B. et al. "High C-reactive protein and low paraoxonase1 in diabetes as risk factors for coronary heart disease." *Atherosclerosis* 186 (2006a) 396–401.

Mackness, B. et al. "Serum paraoxonase (PON1) 55 and 192 polymorphism and paraoxonase activity and concentration in non-insulin dependent diabetes mellitus." *Atherosclerosis* 139 (1998a) 341–9.

Mackness, B., P. McElduff, and M.I. Mackness "The paraoxonase-2-310 polymorphism is associated with the presence of microvascular complications in diabetes mellitus." *J. Intern. Med.* 258 (2005) 363–8.

Mackness, B. et al. "Human paraoxonase-1 overexpression inhibits atherosclerosis in a mouse model of metabolic syndrome." *Arterioscler. Thromb. Vasc. Biol.* 26 (2006b) 1545–50.

Mackness, M.I. and P.N. Durrington "HDL, its enzymes and its potential to influence lipid peroxidation." *Atherosclerosis* 115 (1995b) 243–53.

Mackness, M.I. and P.N. Durrington "Paraoxonase: another factor in NIDDM cardiovascular disease." *Lancet* 346 (1995a) 856.

Mackness, M.I. et al. "Paraoxonase and coronary heart disease." *Curr. Opin. Lipidol.* 9 (1998b) 319–24.

Malleier, J.M. et al. "Regulation of protein C inhibitor (PCI) activity by specific oxidized and negatively charged phospholipids." *Blood* 109 (2007) 4769–76.

Mang, R. and J. Krutmann "UVA-1 Phototherapy." *Photodermatol. Photoimmunol. Photomed.* 21 (2005) 103–8.

Marathe, G.K. et al. "Inflammatory platelet-activating factor-like phospholipids in oxidized low density lipoproteins are fragmented alkyl phosphatidylcholines." *J. Biol. Chem.* 274 (1999) 28395–404.

Marathe, G.K., G.A. Zimmerman, and T.M. McIntyre "Platelet-activating factor acetylhydrolase, and not paraoxonase-1, is the oxidized phospholipid hydrolase of high density lipoprotein particles." *J. Biol. Chem.* 278 (2003) 3937–47.

Marathe, G.K. et al. "Activation of vascular cells by PAF-like lipids in oxidized LDL." *Vascular Pharmacol.* 38 (2002) 193–200.

Mastorikou, M., M. Mackness, and B. Mackness "Defective metabolism of oxidized phospholipid by HDL from people with type 2 diabetes." *Diabetes* 55 (2006) 3099–103.

Mattila, J.P., K. Sabatini, and P.K. Kinnunen "Oxidized phospholipids as potential novel drug targets." *Biophys. J.* 93 (2007) 3105–12.

Miketova, P. et al. "Oxidative changes in cerebral spinal fluid phosphatidylcholine during treatment for acute lymphoblastic leukemia." *Biol. Res. Nurs.* 6 (2005) 187–95.

Miller, Y.I. et al. "Oxidized low density lipoprotein and innate immune receptors." *Curr. Opin. Lipidol.* 14 (2003a) 437–45.

Miller, Y.I. et al. "Minimally modified LDL binds to CD14, induces macrophage spreading via TLR4/MD-2, and inhibits phagocytosis of apoptotic cells." *J. Biol. Chem.* 278 (2003b) 1561–8.

Miyoshi, M. et al. "Gene delivery of paraoxonase-1 inhibits neointimal hyperplasia after arterial balloon-injury in rabbits fed a high-fat diet." *Hypertens. Res.* 30 (2007) 85–91.

Mullick, A.E., P.S. Tobias, and L.K. Curtiss "Modulation of atherosclerosis in mice by Toll-like receptor 2." *J. Clin. Invest* 115 (2005) 3149–56.

Mullins, R.F. et al. "Drusen associated with aging and age-related macular degeneration contain proteins common to extracellular deposits associated with atherosclerosis, elastosis, amyloidosis, and dense deposit disease." *FASEB J.* 14 (2000) 835–46.

Mwaikambo, B.R. et al. "Activation of CD36 inhibits and induces regression of inflammatory corneal neovascularization." *Invest Ophthalmol. Vis. Sci.* 47 (2006) 4356–64.

Nakamura, T., P.M. Henson, and R.C. Murphy "Occurrence of oxidized metabolites of arachidonic acid esterified to phospholipids in murine lung tissue." *Anal. Biochem.* 262 (1998) 23–32.

Ng, C.J. et al. "The paraoxonase gene family and atherosclerosis." *Free Radic. Biol. Med.* 38 (2005) 153–63.

Nonas, S. et al. "Oxidized phospholipids reduce ventilator-induced vascular leak and inflammation in vivo." *Critical Care,* 12(1): R27, 2008.

Nonas, S. et al. "Oxidized phospholipids reduce vascular leak and inflammation in rat model of acute lung injury." *Am. J Respir. Crit Care Med.* 173 (2006) 1130–8.

Palinski, W. et al. "Cloning of monoclonal autoantibodies to epitopes of oxidized lipoproteins from apolipoprotein E-deficient mice. Demonstration of epitopes of oxidized low density lipoprotein in human plasma." *J. Clin. Invest.* 98 (1996) 800–14.

Parhami, F. et al. "Lipid oxidation products have opposite effects on calcifying vascular cell and bone cell differentiation. A possible explanation for the paradox of arterial calcification in osteoporotic patients." *Arterioscler. Thromb. Vasc. Biol.* 17 (1997) 680–7.

Parhami, F. et al. "Atherogenic high-fat diet reduces bone mineralization in mice." *J. Bone Miner. Res.* 16 (2001) 182–8.

Pidkovka, N.A. et al. "Oxidized phospholipids induce phenotypic switching of vascular smooth muscle cells in vivo and in vitro." *Circ. Res.* 101 (2007) 792–801.

Podrez, E.A. et al. "Platelet CD36 links hyperlipidemia, oxidant stress and a prothrombotic phenotype." *Nat. Med.* 13 (2007) 1086–95.

Podrez, E.A. et al. "A novel family of atherogenic oxidized phospholipids promotes macrophage foam cell formation via the scavenger receptor CD36 and is enriched in atherosclerotic lesions 1." *J. Biol. Chem.* 277 (2002a) 38517–23.

Podrez, E.A. et al. "Identification of a novel family of oxidized phospholipids that serve as ligands for the macrophage scavenger receptor CD36." *J. Biol. Chem.* 277 (2002b) 38503–16.

Qin, J. et al. "Oxidized phosphatidylcholine is a marker for neuroinflammation in multiple sclerosis brain." *J. Neurosci. Res.* 85 (2007) 977–84.

Ross, B.M. et al. "Elevated activity of phospholipid biosynthetic enzymes in substantia nigra of patients with Parkinson's disease." *Neuroscience* 102 (2001) 899–904.

Sampson, M.J. et al. "Paraoxonase-1 (PON-1) genotype and activity and in vivo oxidized plasma low-density lipoprotein in Type II diabetes." *Clin. Sci. (Lond)* 109 (2005) 189–97.

Sampson, M.J. et al. "Plasma F2 isoprostanes: direct evidence of increased free radical damage during acute hyperglycemia in type 2 diabetes." *Diabetes Care* 25 (2002) 537–41.

Shamshiev, A.T. et al. "Dyslipidemia inhibits Toll-like receptor-induced activation of CD8alpha-negative dendritic cells and protective Th1 type immunity." *J. Exp. Med.* 204 (2007) 441–52.

Shi, H. et al. "Formation of phospholipid hydroperoxides and its inhibition by alpha-tocopherol in rat brain synaptosomes induced by peroxynitrite." *Biochem. Biophys. Res. Commun.* 257 (1999) 651–6.

Shih, D.M. et al. "Mice lacking serum paraoxonase are susceptible to organophosphate toxicity and atherosclerosis." *Nature* 394 (1998) 284–7.

Shih, D.M. et al. "Combined serum paraoxonase knockout/apolipoprotein E knockout mice exhibit increased lipoprotein oxidation and atherosclerosis." *J. Biol. Chem.* 275 (2000) 17527–35.

Smiley, P.L. et al. "Oxidatively fragmented phosphatidylcholines activate human neutrophils through the receptor for platelet-activating factor." *J. Biol. Chem.* 266 (1991) 11104–10.

Subbanagounder, G. et al. "Determinants of bioactivity of oxidized phospholipids. Specific oxidized fatty acyl groups at the sn-2 position." *Arterioscler. Thromb. Vasc. Biol.* 20 (2000) 2248–54.

Subbanagounder, G. et al. "Evidence that phospholipid oxidation products and/or platelet-activating factor play an important role in early atherogenesis : in vitro and In vivo inhibition by WEB 2086." *Circ. Res.* 85 (1999) 311–8.

Subbanagounder, G., A.D. Watson, and J.A. Berliner "Bioactive products of phospholipid oxidation: isolation, identification, measurement and activities." *Free Radic. Biol. Med.* 28 (2000) 1751–61.

Sun, M. et al. "Light-induced oxidation of photoreceptor outer segment phospholipids generates ligands for CD36-mediated phagocytosis by retinal pigment epithelium: a potential mechanism for modulating outer segment phagocytosis under oxidant stress conditions." *J. Biol. Chem.* 281 (2006) 4222–30.

Suzuki, M. et al. "Oxidized phospholipids in the macula increase with age and in eyes with age-related macular degeneration." *Mol. Vis.* 13(2007): 772–8.

Svenungsson, E. et al. "Risk factors for cardiovascular disease in systemic lupus erythematosus." *Circulation* 104 (2001) 1887–93.

Tobias, P. and L.K. Curtiss "Thematic review series: The immune system and atherogenesis. Paying the price for pathogen protection: toll receptors in atherogenesis." *J. Lipid Res.* 46 (2005) 404–11.

Tsimikas, S. "Oxidized low-density lipoprotein biomarkers in atherosclerosis." *Curr. Atheroscler. Rep.* 8 (2006b) 55–61.

Tsimikas, S. "Oxidative biomarkers in the diagnosis and prognosis of cardiovascular disease." *Am. J. Cardiol.* 98 (2006a) 9P–17P.

Tsimikas, S. et al. "Increased plasma oxidized phospholipid:apolipoprotein B-100 ratio with concomitant depletion of oxidized phospholipids from atherosclerotic lesions after dietary lipid-lowering: a potential biomarker of early atherosclerosis regression." *Arterioscler. Thromb. Vasc. Biol.* 27 (2007) 175–81.

Tsimikas, S. et al. "Oxidized phospholipids, Lp(a) lipoprotein, and coronary artery disease." *N. Engl. J Med.* 353 (2005) 46–57.

Tsimikas, S. et al. "Oxidized phospholipids predict the presence and progression of carotid and femoral atherosclerosis and symptomatic cardiovascular disease: five-year prospective results from the Bruneck study." *J. Am. Coll. Cardiol.* 47 (2006) 2219–28.

Tsimikas, S., W. Palinski, and J.L. Witztum "Circulating autoantibodies to oxidized LDL correlate with arterial accumulation and depletion of oxidized LDL in LDL receptor-deficient mice." *Arterioscler. Thromb. Vasc. Biol.* 21 (2001) 95–100.

Tsimikas, S., L.D. Tsironis, and A.D. Tselepis "New insights into the role of lipoprotein(a)-associated lipoprotein-associated phospholipase A2 in atherosclerosis and cardiovascular disease." *Arterioscler. Thromb. Vasc. Biol.* 27 (2007) 2094–9.

Tsimikas, S. and J.L. Witztum "Measuring circulating oxidized low-density lipoprotein to evaluate coronary risk." *Circulation* 103 (2001) 1930–2.

Turunen, P. et al. "Adenovirus-mediated gene transfer of Lp-PLA2 reduces LDL degradation and foam cell formation in vitro." *J. Lipid Res.* 45 (2004) 1633–9.

Tward, A. et al. "Decreased atherosclerotic lesion formation in human serum paraoxonase transgenic mice." *Circulation* 106 (2002) 484–90.

Uhlson, C. et al. "Oxidized phospholipids derived from ozone-treated lung surfactant extract reduce macrophage and epithelial cell viability." *Chem. Res. Toxicol.* 15 (2002) 896–906.

Walterscheid, J.P., S.E. Ullrich, and D.X. Nghiem "Platelet-activating factor, a molecular sensor for cellular damage, activates systemic immune suppression." *J. Exp. Med.* 195 (2002) 171–9.

Walton, K.A. et al. "Specific phospholipid oxidation products inhibit ligand activation of toll-like receptors 4 and 2." *Arterioscler. Thromb. Vasc. Biol.* 23 (2003a) 1197–203.

Walton, K.A. et al. "Receptors involved in the oxidized 1-palmitoyl-2-arachidonoyl-sn-glycero-3-phosphorylcholine-mediated synthesis of interleukin-8. A role for Toll-like receptor 4 and a glycosylphosphatidylinositol-anchored protein." *J. Biol. Chem.* 278 (2003b) 29661–6.

Watson, A.D. et al. "Protective effect of high density lipoprotein associated paraoxonase. Inhibition of the biological activity of minimally oxidized low density lipoprotein." *J. Clin. Invest.* 96 (1995) 2882–91.

Watson, A.D. et al. "Structural identification by mass spectrometry of oxidized phospholipids in minimally oxidized low density lipoprotein that induce monocyte/endothelial interactions and evidence for their presence in vivo." *J. Biol. Chem.* 272 (1997) 13597–607.

Wu, G.S. and N.A. Rao "Activation of NADPH oxidase by docosahexaenoic acid hydroperoxide and its inhibition by a novel retinal pigment epithelial protein." *Invest Ophthalmol. Vis. Sci.* 40 (1999) 831–9.

Yamada, Y. et al. "Correlations between plasma platelet-activating factor acetylhydrolase (PAF-AH) activity and PAF-AH genotype, age, and atherosclerosis in a Japanese population." *Atherosclerosis* 150 (2000) 209–16.

Zalewski, A. et al. "Lp-PLA2: a new kid on the block." *Clin. Chem.* 52 (2006) 1645–50.

Zhang, R. et al. "Myeloperoxidase functions as a major enzymatic catalyst for initiation of lipid peroxidation at sites of inflammation." *J Biol Chem.* 277 (2002) 46116–22.

Zieseniss, S. et al. "Modified phosphatidylethanolamine as the active component of oxidized low density lipoprotein promoting platelet prothrombinase activity." *J. Biol. Chem.* 276 (2001) 19828–35.

Zimmerman, G.A. et al. "The platelet-activating factor signaling system and its regulators in syndromes of inflammation and thrombosis." *Crit Care Med.* 30 (2002) S294–S301.

# Chapter 13
# Mediation of Apoptosis by Oxidized Phospholipids

**Gilbert O. Fruhwirth and Albin Hermetter**

**Abstract** Free radical-mediated oxidation of (poly)unsaturated glycerophospholipids in membranes and lipoproteins leads to the formation of a plethora of products. Some of these oxidized phospholipids, especially the truncated forms, induce apoptosis depending on their chemical structure, concentration and cell type. Depending on the phospholipid and the cell type, two pathways have so far been identified for the intracellular transmission of the apoptotic signals. One pathway involves activation of acid sphingomyelinase, which gives rise to the formation of ceramide and is followed by phosphorylation of pro-apoptotic mitogen-activated protein kinases. Alternatively, oxidized phospholipids act directly on mitochondria leading to efflux of pro-apoptotic effectors in endothelial cells. During the execution of the apoptotic program additional oxidized phospholipids are generated. The apoptotic cascade itself leads to oxidation and exposure of e.g. membrane phosphatidylserine. Oxidized phospholipids on the outer leaflet of the plasma membrane can form surface lipid patterns that specifically bind to phagocytic cells, e.g. macrophages.

In this manuscript we review the recent literature reporting on apoptosis inducing glycerophospholipids. In addition, we describe the cellular processes that lead to phospholipid oxidation as part of the apoptotic mode of cell death and are likely to enhance the recognition of apoptotic cells by phagocytic macrophages.

**Keywords** Phagocytosis · signaling · sphingomyelinase · truncated phospholipids

A. Hermetter
Institute of Biochemistry, Graz University of Technology, Petersgasse 12/2, A-8010 Graz, Austria

P.J. Quinn, X. Wang (eds.), *Lipids in Health and Disease*,

## 13.1 Structure, Formation, and Reactivity of Oxidized Phospholipids

Glycerophospholipids usually contain saturated and (poly)unsaturated fatty acids at their *sn*-1 and *sn*-2 positions, respectively. The latter are prone to oxidative modification, which leads to many different reaction products, depending on chain length and degree of unsaturation. *In vivo*, enzyme-catalyzed as well as non-enzymatic reactions contribute to this process. Endogenous lipid oxidizing enzymes include lipoxygenases, myeloperoxidase, and amongst others NADPH oxidase which generates reactive oxygen species (ROS) and is involved in innate immune defense and cell growth. However, since a considerable amount of the oxidized lipids detected in biological samples (tissues and fluids) exist in both enantiomeric forms, some compounds are likely to be formed by non-enzymatic radical-induced reactions.

Phosphatidylcholine (PC) is the main phospholipid in all mammalian cells (40–50%) and lipoproteins. As a consequence, most oxidized phospholipids detected in mammalian tissues contain a choline head group. In addition to oxidized PC (oxPC), oxidized phosphatidylserine (oxPS) and phosphatidylethanolamine (oxPE) have been detected. The latter was found in the retina, which is a tissue containing high amounts of phosphatidylethanolamine (Gugiu et al., 2006), while OxPS was reported to be present on the surface of apoptotic cells (Kagan et al., 2002; Matsura et al., 2005).

The phospholipids of eukaryotic cells may contain acyl chains or alk(en)yl residues linked to the *sn*-1 position of glycerol. This position is usually esterified with saturated fatty acids. Plasmalogens are 1-O-alkenyl-2-acylglycerophospholipids containing hydrocarbon chains that are attached to position *sn*-1 of the glycerol backbone by a vinyl ether bond which is susceptible to modification by ROS. In contrast, the *sn*-2 hydroxy group of glycerol in phospholipids is esterified with mostly (poly)unsaturated fatty acids. These residues are highly oxidizable and, as a consequence, most of the oxidized phospholipids are modified at the *sn*-2 position of glycerol. Polyunsaturated fatty acids (PUFAs) of mammalian glycerophospholipids not only comprise essential fatty acids like arachidonic acid (AA) and linoleic acid (LA), but also all long-chain fatty acids generated endogenously from AA or LA including e.g. docosahexaenoic acid or eicosapentaenoic acid.

Oxidation of AA in 1-palmitoyl-2-arachidonoylphosphatidylcholine (PAPC) leads to the formation of a plethora of different reaction products. The first step in lipid peroxidation is initiated by hydrogen abstraction followed by rearrangement of double bonds and addition of triplet oxygen (Min, 1997; Spiteller, 2005) leading to highly reactive peroxyl radicals. These radicals can eventually undergo a large variety of consecutive reactions including further hydrogen abstraction and fragmentation. Alternatively, other oxidized phospholipid species are formed via the isoprostane pathways. The result of all these reactions are several classes of different phospholipid oxidation products

A

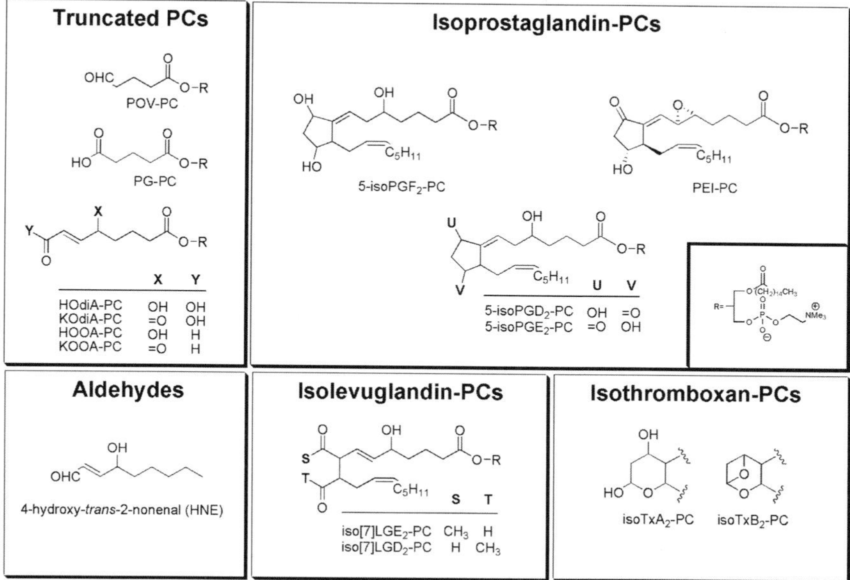


B

| L-PC series | | A-PC series | |
|---|---|---|---|
| Abbrevation | *sn*-2 position esterified to | Abbrevation | *sn*-2 position esterified to |
| ON-PC | 9-oxononanoic acid | OV-PC | 5-oxovaleric acid |
| Az-PC | Azelaic acid (nonanedioic acid) | G-PC | Glutaric acid |
| HDdiA-PC | 9-hydroxy-10-dodecenedioic acid | HOdiA-PC | 5-hydroxy-8-oxo-6-octenedioic acid |
| HODA-PC | 9-hydroxy-12-oxo-10-dodecenoic acid | HOOA-PC | 5-hydroxy-8-oxo-6-octenoic acid |
| KDdiA-PC | 9-keto-10-dodecendioic acid | KOdiA-PC | 5-keto-6-octendioic acid |
| KODA-PC | 9-keto-12-oxo-10-dodecenoic acid | KOOA-PC | 5-keto-8-oxo-6-octenoic acid |

**Fig. 13.1** Chemical structures of oxidized phospholipids. (A) Free radical-induced oxidation of 1-palmitoyl-2-arachidonoyl-*sn*-glycero-3-phosphocholine (PA-PC) leads to a plethora of different oxidation products such as peroxidized phospholipids (not shown), truncated phospholipids, isoprostanes, isolevuglandins, and isothromboxanes.
(B) As described for the oxidation of PA-PC, oxidation of 1-hexadecanoyl-2-octadecadi-9',12'-enoyl-sn-glycero-3-phosphocholine (PL-PC) leads to several products. Most of them are truncated lipids. The analogs of the PL-PC series are compared with those of the PA-PC series shown in (A).
Other abbreviations used: lyso-PC: 1-acyl-*sn*-glycero-3-phosphocholine (1-acyl group not specified); PEI-PC: 1-hexadecanoyl-2-(5,6-epoxyisoprostane E2 oyl)-*sn*-glycero-3-phosphocholine

including phospholipids containing peroxy fatty acids, truncated phospholipids and low molecular weight aldehydes (e.g. hydroxynonenal or malondialdehyde), isoprostane phospholipids, isolevuglandin phospholipids, and isothromboxan phospholipids (Fig. 13.1). Many phospholipid oxidation products contain chemically reactive groups including aldehydo or keto carbonyls, epoxides or double bonds in conjugation to carbonyl groups (Fig. 13.1). As a consequence, these compounds can undergo chemical reactions with a large variety of biological molecules containing nucleophilic functional groups. Modification of biomolecules by oxidized phospholipids may give rise to the formation of stable compounds (e.g. protein adducts with isolevuglandins) or substances that still show considerable reactivity (e.g. Michael adducts or Schiff bases formed with $\gamma$-hydroxyalkenals). A comprehensive review describing the various reaction pathways involved in the formation of the most important oxidized phospholipids has been published recently (Fruhwirth et al., 2007) and for further details the reader is referred to this publication.

Moreover, the stability of oxidized phospholipids in biological systems is limited. The oxidized acyl chains of glycerophospholipids can be released by intracellular and plasma phospholipases including platelet activating factor-acetyl hydrolase (PAF-AH, Lp-$PLA_2$) leading to the formation of lyso-phospholipids and the corresponding free oxidized (short-chain) carboxylic acids (Stafforini et al., 2006). The latter compounds exhibit basically the same reactivities in free form as compared to their chemical properties in the phospholipid-bound state. However, the target molecules of both types of lipids are likely to be different since their polarities and thus, their partition within aqueous and lipid compartments are different, too. The free oxidized carboxylic acids are more polar and therefore they can be expected to preferentially react with hydrophilic molecules or hydrophilic domains (e.g. of proteins). Some of these reactions can be considered relevant for detoxification. However, most of them are leading to modification of important functional molecules and thus, negatively influence their cellular function.

## 13.2 Oxidized Phospholipids in Apoptosis and Apoptotic Signaling

Evidence is growing that oxidized phospholipids play a key role in the development of several chronic diseases including atherosclerosis being one of the main health risks in the Western civilized world. Increasing evidence also suggests that apoptosis is a major event in the pathophysiology of atherosclerosis (Martinet and Kockx, 2001). In primary and secondary atherosclerotic lesions, apoptosis may help to reduce lesion size (Pollman et al., 1998; Wang et al., 1999), whereas at later stages apoptosis may contribute to the formation of unstable plaques (Kockx, 1998; Libby et al., 1997). On the other hand, apoptotic cells are an additional source of oxidized phospholipids and may actively contribute to inflammation (Huber et al., 2002). Oxidized phospholipids have been shown to accumulate in atherosclerotic lesions (Berliner et al., 2001) and show a great variety of biological effects *in vivo* and *in vitro*. They were identified as the essential molecular components that are responsible for the pathophysiological actions of oxidized low density lipoprotein (oxLDL) on vascular cells. Many studies have been performed investigating the effects of oxidized LDL and its constituent oxidized phospholipids on cell proliferation and cell death (Baird et al., 2005; Salvayre et al., 2002; Yaraei et al., 2005). In addition, several studies have been performed using oxidized PA-PC which is generated from the polyunsatured parent compound PA-PC by oxidation *in vitro* (Fruhwirth et al., 2007). It is a lipid mixture that contains several oxidized phospholipids species. The complexity of the cellular effects induced by oxidized phospholipids became clear only recently when a microarray study revealed that more than 1000 genes in endothelial cells are affected by the different products of PA-PC oxidation (Gargalovic et al., 2006). This is just one more example showing that there is a need to perform studies with chemically defined compounds rather than complex mixtures to gain a better understanding of the biological functions of oxidized phospholipids on a molecular basis. The following chapters will describe how oxidized phospholipids are formed as a result of pro-apoptotic processes and how these compounds trigger apoptosis.

### *13.2.1 Apoptotic Cells Generate Oxidized Phospholipids*

A fundamental feature of cell architecture is the asymmetric distribution of lipids across bilayer membranes. For instance, the aminophospholipid phosphatidylserine (PS) preferentially localizes to the inner leaflet of the plasma membrane which contains more than 65% of this lipid class (Allan, 1996; Daleke, 2003). As a consequence of apoptosis, lipid asymmetry is gradually lost leading to exposure of phosphatidylserine on the outer leaflet of the plasma membrane which is a recognition signal for macrophages to phagocytose the apoptotic cell (Savill and Fadok, 2000). In general, aminophospholipid translocases (Bevers et al., 1999) and a nonspecific phospholipid scramblase (Daleke, 2003; Frasch et al., 2000) facilitate directed and random transbilayer movement of PS, respectively. During apoptosis, oxPS contributes to externalisation of native and oxidized PS species, which both serve as a signal for recognition by macrophages (Borisenko et al., 2004; Fadok et al., 1992; Matsura et al., 2005; Tyurina et al., 2004b). The oxidation of phosphatidylserine in apoptotic cells is mediated by the peroxidase activity of cytochrome *c* which is released into the cytosol during programmed cell death. The specific PS oxidation occurs within PS-cytochrome *c* complexes directly at the inner side of the plasma membrane (Tyurina et al., 2004a). Kagan et al speculate that oxPS improve the recognition pattern for phagocytosis and increase the probability that the apoptotic cell is taken up by a macrophage (Borisenko et al., 2003; Kagan et al., 2003; Tyurina et al., 2004b). Several macrophage receptors are likely to be involved in the recognition of phosphatidylserine and/or its oxidized form by the phagocytic cell. These proteins belong to the scavenger-receptor superfamily (Platt et al., 1998), integrins (Fadok et al., 1998), complement receptors (Mevorach et al., 1998), the lectin-like oxidized LDL receptor-1 (Murphy et al., 2006), and the phosphatidylserine receptor (Arur et al., 2003; Fadok et al., 2000; Li et al., 2003; Wang et al., 2003). Scavenger receptor B (CD36) was shown to play an essential role in macrophage-mediated clearance of apoptotic cells *in vivo* (Greenberg et al., 2006). A comparison of wildtype with CD36 knock-out mice led to the observation that macrophage recognition of apoptotic cells via CD36 occured almost exclusively through interaction with plasma-membrane-associated oxPS while non-oxidized PS hardly showed any effect. Identification of lipid species via LC/ESI-MS-MS revealed that the membranes of apoptotic cells contain various truncated phospholipids originating from oxidation of phosphatidylserine species containing unsaturated fatty acids at the *sn*-2 position. The following lipids were identified: OV-PS, G-PS, KOOA-PS, HOOA-PS, KOdiA-PS, HOdiA-PS (from 2-arachidonoyl-PS) and ON-PS, Az-PS, KODA-PS, HODA-PS, KDdiA-PS, HDdiA-PS (from 2-linoleoyl-PS) (for detailed structures and abbreviations see Fig. 13.1). It has been shown previously that the choline phospholipid analogs of all these species which were found in atherosclerotic lesions *in vivo* are also ligands of CD36 (Podrez et al., 2002a, b). OxPC was also detected on the surface of apoptotic cells using the monoclonal antibody EO6 which recognizes POV-PC but not PA-PC in protein-bound form (Chang et al., 1999). EO6 effectively inhibits phagocytosis of apoptotic cells by macrophages, suggesting that, in addition to oxPS, the presence of oxidatively modified PC at the cell surface is also important for phagocytosis of apoptotic cells. Thus, externalization of oxidized phospholipids is mandatory for clearance of apoptotic cells by macrophages (Kagan et al., 2000). *In vitro* competition experiments showed that cultured cells expressing CD36 preferentially interacted with vesicles of oxPS as compared to oxPC, revealing a primary role for oxPS-CD36 interaction in macrophage phagcytosis of apoptotic cells.

Sabatini et al observed that the truncated oxidized phospholipids, PAz-PC and PON-PC lead to phase separation in dipalmitoylphosphatidylcholine (DPPC) monolayers which was associated with upfolding of the truncated polar fatty acid chain towards the lipid-water interface and as a result monolayer expansion (Sabatini et al., 2006). Li et al obtained similar results in a more physiologically relevant bilayer model using two-dimensional nuclear Overhauser effect spectroscopy. The respective vesicles contained 20 mol% KOdiA-PC and 80 mol% dimyristoylphosphatidylcholine (DMPC). It was found that the oxidized *sn*-2 fatty acid of KOdiA-PC localizes in close proximity to its polar head group on the bilayer surface (Li et al., 2007). Further experiments with lipids containing different polar headgroups and truncated *sn*-2 fatty acids revealed two interesting effects. Firstly, the oxovaleroyl and the glutaroyl moieties of oxidized PC show the same behaviour as KOdiA-PC. Secondly, the polar headgroup has no influence on the protrusion of the

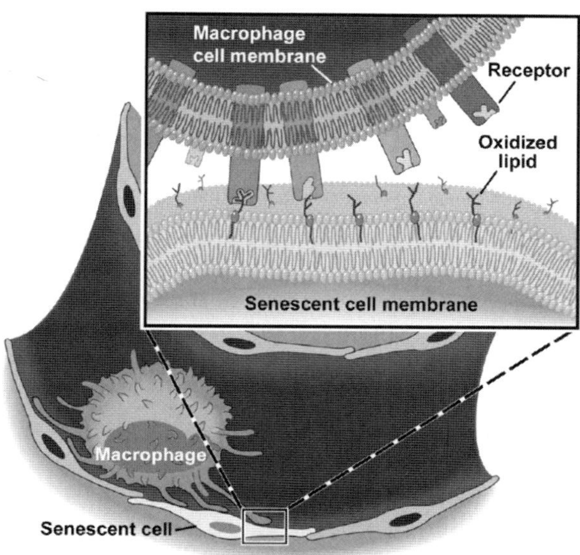


**Fig. 13.2** The lipid "whisker" model. Cell membranes of senescent or apoptotic cells contain oxidized phospholipids with *sn*-2 oxidized fatty acid acyl chains protruding into the extracellular space. This conformation renders them accessible to interact with scavenger receptors and other pattern recognition receptors on the surface of probing macrophages of the innate immune system (Greenberg et al., 2008; Permission for use of this figure has been requested from The Journal of Biological Chemistry.)

oxidized fatty acid moiety (Greenberg et al., 2008). Taken together, these results led to the development of the so-called "lipid whisker model" which is based on the assumption that various truncated phospholipids form a molecular pattern on the surface of apoptotic or senescent cells for recognition and subsequent phagocytosis by macrophages (Fig. 13.2).

### *13.2.2 Oxidized Phospholipids as Inducers of Apoptosis*

It has already been emphasized that apoptotic cells can be both, the source of oxidized phospholipids as well as the initiators of apoptosis. Oxidation of (poly)-unsaturated phospholipids leads to a large variety of oxidation products (see Chapter 1 and (Fruhwirth et al., 2007)). These compounds may have very different chemical structures. It can be anticipated that they exert very different physical effects in membranes and undergo specific interactions with their target proteins and lipids. As a consequence, the great variety of oxidized phospholipids should be reflected by a corresponding diversity of biological effects. In many studies investigating cell survival and/or cell death, the effects of oxidized lipid mixtures or oxidized LDL have been studied so far. Information on the apoptotic effects of defined oxidized phospholipids is still scarce. It has already been shown earlier that the fatty acid oxidation products of free AA, namely hydroperoxyeicosatetraenoic acids, hydroxynonenal and malondialdehyde induce apoptosis of vascular smooth muscle cells (Kalyankrishna et al., 2002). The following chapters will report on the apoptotic effects of pure oxidized phospholipids.

#### 13.2.2.1 Truncated Oxidized Phospholipids Induce Apoptosis in Vascular Smooth Muscle Cells

The short-chain oxidized phospholipids PG-PC and POV-PC are mediators of apoptosis in rat vascular smooth muscle cells (rVSMCs). Both oxidized phospholipids are very cytotoxic under low serum conditions (Fruhwirth et al., 2006). Under the influence of PG-PC and POV-PC, the cells show morphological changes typical for apoptosis. Both oxidized phospholipids lead to DNA laddering, which is a hallmark of apoptotic cell death (Collins et al., 1997; Compton, 1992). This time-dependent effect was only observed under low serum conditions. The sizes of the DNA fragments were estimated to be multiples of about 180 base pairs, which is typical for this mode of cell death (Majno and Joris, 1995). Exposure of phosphatidylserine on the outer leaflet of the plasma membrane was observed as another apoptotic marker (Koopman et al., 1994). PG-PC and POV-PC induced only apoptosis in VSMCs, whereas no evidence was found for necrosis under the chosen experimental conditions. Transmission electron microscopy showed a reduction of the nuclear volume (pyknosis) five hours after exposure to these lipids. Typical apoptotic phenotypes like cell shrinkage as well as condensation of cytoplasmic organelles were also evident at this time point, while marked convolution of the nuclear envelope followed by partial nuclear fragmentation (karyorhexis) and chromatin condensation became apparent later (14 hrs). The aldehydo lipid POV-PC showed much higher activity than PG-PC. Membrane blebbing started earlier and the formation of apoptotic vesicles was more pronounced.

Apoptosis of VSMCs as induced by POV-PC and PG-PC is the result of a distinct signaling pathway (Fig. 13.3). Both oxidized phospholipids rapidly activate acid sphingomyelinase in a time-dependent manner characterstic for a signaling event (Loidl et al., 2003), whereas neutral sphingomyelinase activity is persistently increased. As a consequence, the lipid messenger ceramide is formed, which mediates numerous cellular phenomena including apoptosis, proliferation, cytokine release and differentiation (Gomez-Munoz, 2006; Taha et al., 2006). Downstream of acid sphingomyelinase activation and ceramide formation, the mitogen-activated protein kinases (MAPK) c-Jun N-terminal kinase (JNK) and p38 MAPK are phosphorylated in a time-dependent manner. The maximum of activation was detected 10 min after stimulation by PG-PC or POV-PC (Loidl et al., 2003). The “sphingomyelinase-MAPK pathway” as described above has already been identified as a

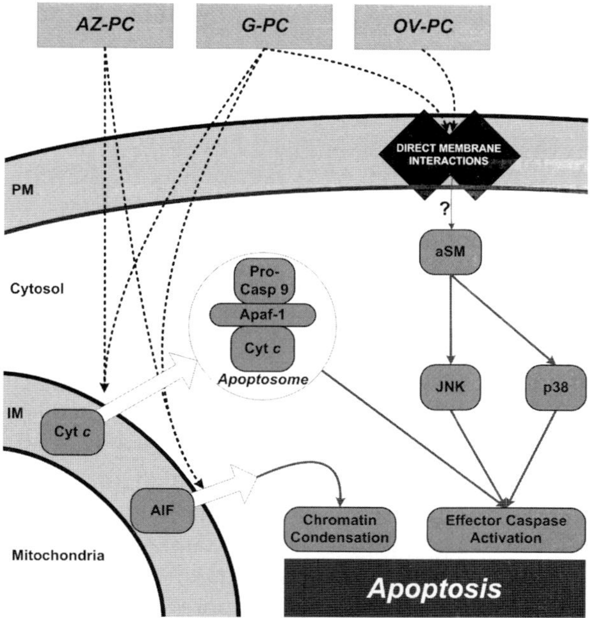


**Fig. 13.3** Signaling pathways involved in oxidized phospholipid-induced apoptosis in mammalian smooth muscle and endothelial cells. AIF, apoptosis inducing factor; Apaf-1, apoptosis activating factor 1; aSM, acid sphingomyelinase; Az-PC azelaoyl acid ester of lyso-PC; Cyt *c*, cytochrome *c*; G-PC, glutaric acid ester of lyso-PC; IM, intermembrane space of the mitochondria; JNK, c-Jun N-terminal kinase; OV-PC, 5-oxovaleric acid ester of lyso-PC; p38, p38 mitogen-activated protein kinase; PM, plasma membrane

general signaling mechanism for many phenomena associated with stress response and apoptosis (Johnson and Lapadat, 2002). If transcription of acid sphingomyelinase is specifically inhibited by NB6 (Deigner et al., 2001) prior to stimulation with oxidized phospholipids, p38 MAPK and JNK are no longer activated, indicating a causal relationship between sphingomyelinase activation and MAPK signaling. This is in line with the observation that the apoptotic executor caspase 3 is not activated by the oxidized phospholipids, if the cells were desensitized by NB6 treatment. Minimally modified LDL in which mainly the lipids were oxidized activated the same apoptotic signaling cascade as induced by PG-PC and POV-PC. The time-dependent signaling of the oxidized lipoprotein even perfectly matched the sphingomyelinase-MAPK activation observed with POV-PC and PG-PC. Obviously, the oxidized phospholipids are largely responsible for the effects of this lipoprotein on apoptotic signaling in VSMC (Loidl et al., 2003). Taken together, POV-PC and PG-PC trigger a sequence of apoptotic events by activation of the "sphingomyelinase-MAPK" pathway, subsequent effector caspase activation, phosphatidylserine exposure, and late morphological changes of the cellular structure. Finally, it is interesting to mention that at least the truncated phospholipids did not activate signaling components of cell proliferation and survival. AKT-kinase or NF-κB activities remained unaffected if cells were exposed to POV-PC and PG-PC.

#### 13.2.2.2 Truncated Oxidized Phospholipids Induce Apoptosis in Human Endothelial Cells

Very recently, the receptor-independent cytotoxic effects of synthetic 1-O-hexadecyl-2-azelaoyl-sn-glycero-3-phosphocholine (HAz-PC) were studied. This compound is an ether analogue of 1-hexadecanoyl-2-azelaoyl-sn-glycero-3-phosphocholine (PAz-PC) and induces apoptosis in promyelocytic HL60 cells and human umbelial vein endothelial cells (HUVEC) (Chen et al., 2007). Az-PCs induced changes of cell morphology typical for apoptosis, triggered phosphatidylserine exposure on the outer leaflet of the plasma membrane, stimulated the release of mitochondrial cytochrome *c* and apoptosis-inducing factor (AIF) and activated caspase 9 (Fig. 13.3). The requirement for the apoptosomal caspase 9 activity and the marked reduction of the apoptotic effect by overexpression of the apoptosis suppressor protein Bcl-$X_L$ (a mitochondrial ion channel) supported the assumption that the mitochondria were the predominant targets of HAz-PC action. In addition, the same authors established a correlation between early mitochondrial permeabilization (cyochrome *c* and AIF release) and later phosphatdylserine exposure at the plasma membrane. In addition to AzPCs, Chen et al. compared the effects of different oxidized phospholipids on swelling of isolated mitochondria from Sprague-Dawley rat liver (Chen et al., 2007). In a light scattering assay, the ether phopsholipid HAz-PC was found to be more effective than its acyl derivative PAz-PC ($\sim$50%). The oxidized diacylphospholipid PG-PC containing a shorter and thus more polar acyl residue was more active than PAz-PC. However, POV-PC, lyso-phosphatidylcholine (lyso-PC) and PAF hardly induced mitochondrial swelling in this *in vitro* assay.

#### 13.2.2.3 Influence of Phospholipases on Apoptosis Induced by Truncated Phospholipids

The above described pro-apoptotic effects were observed when cells were exposed to chemically defined phospholipids. However, it has to be taken into account that the respective 2-acyl-glycerophospholipids may be subject to hydrolysis by phospholipases inside and outside cells. As a consequence, the apoptotic oxidized phospholipids are degraded leading to harmless compounds or lipids of even higher toxicity. We found that POV-PCand PG-PC were inactivated in serum (Fruhwirth et al., 2006) and that this effect was due to hydrolysis by serum phospholipases since the main degradation product was lysophosphatidylcholine (Björkerud and Björkerud, 1996; Matsuzawa et al., 1997). This data is in agreement with the results of a more recent study (Chen et al., 2007) reporting that overexpression of PAF-AH in HL60 cells abolished the pro-apoptotic effect of HAz-PC as determined from phosphatidylserine exposure. The pro-apoptotic effects of POV-PC, PG-PC and Az-PCs (0–50 μM) in rVSMCs and HUVECs seem to be specific for this lipid class, since they can neither be induced by unmodified phospholipids nor by the degradation products of the oxidized phospholipids, namely lysophosphatidylcholine and the free oxidized fatty acids released from the glycerol backbone.

#### 13.2.2.4 Lyso-Phospholipids can Induce Apoptosis or Necrosis Depending on the Cell Type

Lysophosphatidylcholine is a frequent product of oxidized phospholipid hydrolysis and shows structural similarities to its diacyl counterparts containing a short acyl chain in *sn*-2 position. Therefore, its cellular activities deserve particular attention, especially in the context of its cytotoxicity. The effects of phospholipid oxidation products and lyso-PC depend not only on their concentration but also on the cell type. Lyso-PC containing a long acyl chain in *sn*-1 position (e.g. C16:0, C18:0) is an amphiphilic phospholipid that is generated by phospholipase-catalyzed hydrolysis of phosphatidylcholine or extensive oxidation leading to loss of the entire *sn*-2 acyl chain. Its critical micellar concentration (CMC) is around 50 μM. It is easily taken up into lipid membranes and increases their "fluidities". Above the CMC it forms micelles that destroy membrane integrity also by removal of proteins as shown in erythrocytes (Bierbaum et al., 1979; Colles and Chisholm, 2000). Lyso-PC exerts apoptotic effects in rVSMCs at concentrations below its CMC and induces necrotic cell death at concentrations above its CMC (Hsieh et al., 2000). According to another recent study lyso-PC above its CMC induces rather necrosis than apoptosis in mouse aortic endothelial cells (Zhou et al., 2006). After preincubation with lyso-PC, these cells showed features typical for necrosis such as total disruption of the plasma membrane integrity and depletion of intracellular ATP. In addition, the same authors showed that lyso-PC led to a prolonged increase in intracellular calcium and ROS levels. Whereas the elevated calcium level is persistent in necrosis, it is only transient in apoptosis. However, in human endothelial cells the situation is different. Lyso-PC above its CMC (75 μM) induces apoptosis in human aortic endothelial cells (HAECs) (Matsubara and Hasegawa, 2005). This is also in line with the observation that lyso-PC induces apoptosis, but not necrosis, in HUVECs even at concentrations far beyond its CMC (300 μM) while it shows hardly any effect at concentrations below 100 μM (Heermeier et al., 2001). Another study confirmed the results obtained with HUVECs (Tsutsumi et al., 2006). Unfortunately, both studies were based solely on a single assay measuring phosphatidylserine exposure. Other independent methods for the assessment of apoptotic and necrotic cell death were not performed. Takahashi et al.

reported that lyso-PC-induced apoptosis was mediated at least in part by p38 MAPK followed by caspase 3 activation in HUVECs. The lyso-PC concentration in these experiments was above its CMC (75 μM) (Takahashi et al., 2002). Taken together, these results show that lyso-PC may induce different modes of cell death depending on the cell line. Apoptosis prevails in human endothelial cell lines and necrosis is found in rodent endothelial and smooth muscle cell lines. It has been suggested a long time ago that the susceptibility of a cell towards lyso-PC is determined by its amount and distribution in the cellular membranes of various species (Bierbaum et al., 1979). Obviously, there must be specific differences in membrane architecture between human and rodent cells that are responsible for the individual susceptibilities towards the cytotoxic effects of lysophospholipids. Currently these differences are still far from clear and await further investigation.

## 13.3 Future Perspectives

In recent years most studies on oxidized phospholipids have been performed using mixtures of oxidation products that are generated from their polyunsaturated parent compounds like PA-PC or oxidized lipoprotein particles (e.g. oxLDL). These preparations contain a large variety of different substances differing in structure, polarity and hydrophobicity as well as bioavailability. In many of these oxidized lipid preparations, neither the type nor the content of the individual oxidized components was known. Therefore, the respective compounds contribute to the apparent biological activities of oxidized lipid mixtures to a different and unpredictable extent. Thus, it will be desirable to concentrate on chemically defined lipid species in the future.

Furthermore, it is mandatory to specify and standardize the solubilization of oxidized phospholipids for incubation with cells or individual target molecules. Physiologically relevant systems are pure lipid micelles or vesicles depending on the chemical structure of the lipid, complexes with proteins (e.g. albumin) and plasma lipoproteins. Furthermore, it has to be taken into account that oxidized phospholipids exchange between lipid surfaces much faster than regular membrane phospholipids containing two long hydrophobic acyl chains (Li et al., 2007). Therefore, they do not only express their activity at the site of production. They may also transmit "signals" within a larger cross section of a tissue. This phenomenon is particularly relevant to the more polar oxidized phospholipids. Once these compounds have been generated within a cell or a lipoprotein they may interact with many different target proteins and lipids on the surface of and inside a cell leading to impaired functions of these target molecules. Recently, we showed that fluorescent analogs of oxidized phospholipids are easily taken up into cultured vascular cells and react with a rather limited fraction of proteins and lipids (Moumtzi et al., 2006). Identification of individual molecular targets as well as the specific signalling effects downstream of the individual targets will be a challenge for the future, which will help to understand the complex molecular network underlying the (patho)physiological effects of oxidized phospholipids.

A major focus of current biomedical studies is still on the role of these compounds in common lipid-associated disorders including atherosclerosis. Other research fields that are relevant to human health and disease such as cancer, autoimmune and infectious diseases can be expected to become also emerging topics in the context of oxidative stress and (phospho)lipid oxidation. The interdependence of cell growth, (programmed) cell death and cell differentiation is a key aspect in the context of (patho)physiological lipid activity. Thus, further research on the great variety of oxidized phospholipid species will undoubtedly add new and perhaps surprising facets to the understanding of the biological functions of oxidized lipid mediators.

**Acknowledgment** Work in the authors' laboratory was supported by grants from the Austrian Science Fund (SFB Lipotox, project F30-B06).

## References

Allan, D, *Mapping the lipid distribution in the membranes of BHK cells (mini-review)*, Mol. Membr. Biol. (1996) 13 81–84.

Arur, S, Uche, E, Rezaul, K, Fong, M, Scranton, V, Cowan, A E, Mohler, W, and Han, D K, *Annexin I is an endogenous ligand that mediates apoptotic cell engulfment*, Dev. Cell 4 (2003) 587–598.

Baird, S K, Reid, L, Hampton, M B, and Gieseg, S P, *OxLDL induced cell death is inhibited by the macrophage synthesised pterin, 7,8-dihydroneopterin, in U937 cells but not THP-1 cells*, Biochim. Biophys. Acta – Molecular Cell Research 1745 (2005) 361–369.

Berliner, J A, Subbanagounder, G, Leitinger, N, Watson, A D, and Vora, D, *Evidence for a role of phospholipid oxidation products in atherogenesis*, Trends Cardiovasc. Med. 11 (2001) 142–147.

Bevers, E M, Comfurius, P, Dekkers, D W C, and Zwaal, R F A, *Lipid translocation across the plasma membrane of mammalian cells*, Biochim. Biophys. Acta – Molecular and Cell Biology of Lipids 1439 (1999) 317–330.

Bierbaum, T J, Bouma, S R, and Huestis, W-H, *A mechanism of erythrocyte lysis by lysophosphatidylcholine*, Biochim. Biophys. Acta 555 (1979) 102–110.

Björkerud, B and Björkerud, S, *Contrary effects of lightly and strongly oxidized LDL with potent promotion of growth versus apoptosis on arterial smooth muscle cells, macrophages, and fibroblasts*, Arterioscler. Thromb. Vasc. Biol. 16 (1996) 416–424.

Borisenko, G G, Iverson, S L, Ahlberg, S, Kagan, V E, and Fadeel, B, *Milk fat globule epidermal growth factor 8 (MFG-E8) binds to oxidized phosphatidylserine: Implications for macrophage clearance of apoptotic cells*. Cell Death Differ 11 (2004) 943–945.

Borisenko, G G, Matsura, T, Liu, S X, Tyurin, V A, Jianfei, J, Serinkan, F B, and Kagan, V E, *Macrophage recognition of externalized phosphatidylserine and phagocytosis of apoptotic Jurkat cells–existence of a threshold*, Arch. Biochem. Biophys. 413 (2003) 41–52.

Chang, M K, Bergmark, C, Laurila, A, Horkko, S, Han, K H, Friedman, P, Dennis, E A, and Witztum, J L, *Monoclonal antibodies against oxidized low-density lipoprotein bind to apoptotic cells and inhibit their phagocytosis by elicited macrophages: Evidence that oxidation-specific epitopes mediate macrophage recognition*, PNAS 96 (1999) 6353–6358.

Chen, R, Yang, L, and McIntyre, T M, *Cytotoxic Phospholipid Oxidation Products: Cell Death from Mitochondrial Damage and the Intrinsic Caspase Cascade*, J. Biol. Chem. 282 (2007) 24842–24850.

Colles, S M and Chisholm, G M, *Lysophosphatidylcholine-induced cellular injury in cultured fibroblasts involves oxidative events*, J. Lipid Res. 41 (2000) 1188–1198.

Collins, J A, Schandl, C A, Young, K K, Vesely, J, and Willingham, M C, *Major DNA fragmentation is a late event in apoptosis*, J. Histochem. Cytochem. 45 (1997) 923–934.

Compton, M M, *A biochemical hallmark of apoptosis: Internucleosomal degradation of the genome*, Cancer Metastasis Rev. 11 (1992) 105–119.

Daleke, D L, *Regulation of transbilayer plasma membrane phospholipid asymmetry*, J. Lipid Res. 44 (2003) 233–242.

Deigner, H-P, Claus, R, Bonaterra, G A, Gehrke, C, Bibak, N, Blaess, M, Cantz, M, Metz, J, and Kinscherf, R, *Ceramide induces aSMase expression: Implications for oxLDL-induced apoptosis*, FASEB J. 15 (2001) 807–814.

Fadok, V A, Voelker, D R, Campbell, P A, Cohen, J J, Bratton, D L, and Henson, P M, *Exposure of phosphatidylserine on the surface of apoptotic lymphocytes triggers specific recognition and removal by macrophages*, J. Immunol. 148 (1992) 2207–2216.

Fadok, V A, Bratton, D L, Rose, D M, Pearson, A, Ezekewitz, R A, and Henson, P M, *A receptor for phosphatidylserine-specific clearance of apoptotic cells*, Nature 405 (2000) 85–90.

Fadok, V A, Warner, M L, Bratton, D L, and Henson, P M, *CD36 is required for phagocytosis of apoptotic cells by human macrophages that use either a phosphatidylserine receptor or the vitronectin receptor ({alpha}v{beta}3)*, J. Immunol. 161 (1998) 6250–6257.

Frasch, S C, Henson, P M, Kailey, J M, Richter, D A, Janes, M S, Fadok, V A, and Bratton, D L, *Regulation of phospholipid scramblase activity during apoptosis and cell activation by protein kinase Cdelta*, J. Biol. Chem. 275 (2000) 23065–23073.

Fruhwirth, G O, Loidl, A, and Hermetter, A, *Oxidized phospholipids: From molecular properties to disease*, Biochim. Biophys. Acta – Molecular Basis of Disease 1772 (2007) 718–736.

Fruhwirth, G O, Moumtzi, A, Loidl, A, Ingolic, E, and Hermetter, A, *The oxidized phospholipids POVPC and PGPC inhibit growth and induce apoptosis in vascular smooth muscle cells*, Biochim. Biophys. Acta – Molecular and Cell Biology of Lipids 1761 (2006) 1060–1069.

Gargalovic, P S, Imura, M, Zhang, B, Gharavi, N M, Clark, M J, Pagnon, J, Yang, W P, He, A, Truong, A, Patel, S, Nelson, S F, Horvath, S, Berliner, J A, Kirchgessner, T G, and Lusis, A J, *Identification of inflammatory gene modules based on variations of human endothelial cell responses to oxidized lipids*, PNAS 103 (2006) 12741–12746.

Gomez-Munoz, A, *Ceramide 1-phosphate/ceramide, a switch between life and death*, Biochim. Biophys. Acta 1758 (2006) 2049–2056.

Greenberg, M E, Li, X M, Gugiu, B G, Gu, X, Qin, J, Salomon, R G, and Hazen, S L, *The lipid whisker model of the structure of oxidized cell membranes*, J. Biol. Chem. 283 (2008) 2385–2396.

Greenberg, M E, Sun, M, Zhang, R, Febbraio, M, Silverstein, R, and Hazen, S L, *Oxidized phosphatidylserine-CD36 interactions play an essential role in macrophage-dependent phagocytosis of apoptotic cells*, J. Exp. Med. 203 (2006) 2613–2625.

Gugiu, B, Mesaros, C, Sun, M, Gu, X, Crabb, J W, and Salomon, R G, *Identification of oxidatively truncated ethanolamine phospholipids in retina and their generation from polyunsaturated phosphatidylethanolamines*, Chem. Res. Toxicol. 19 (2006) 262–271.

Heermeier, K, Leicht, W, Palmetshofer, A, Ullrich, M, Wanner, C, and Galle, J, *Oxidized LDL Suppresses NF-kappaB and Overcomes Protection from Apoptosis in Activated Endothelial Cells*, J. Am. Soc. Nephrol. 12 (2001) 456–463.

Hsieh, C C, Yen, M H, Liu, H W, and Lau, Y T, *Lysophosphatidylcholine induces apoptotic and non-apoptotic death in vascular smooth muscle cells: In comparison with oxidized LDL*, Atherosclerosis 151 (2000) 481–491.

Huber, J, Vales, A, Mitulovic, G, Blumer, M, Schmid, R, Witztum, J L, Binder, B R, and Leitinger, N, *Oxidized membrane vesicles and blebs from apoptotic cells contain biologically active oxidized phospholipids that induce monocyte-endothelial interactions*, Arterioscler. Thromb. Vasc. Biol. 22 (2002) 101–107.

Johnson, G L and Lapadat, R, *Mitogen-activated protein kinase pathways mediated by ERK, JNK, and p38 protein kinases*, Science 298 (2002) 1911–1912.

Kagan, V E, Borisenko, G G, Serinkan, B F, Tyurina, Y Y, Tyurin, V A, Jiang, J, Liu, S X, Shvedova, A A, Fabisiak, J P, Uthaisang, W, and Fadeel, B, *Appetizing rancidity of apoptotic cells for macrophages: Oxidation, externalization, and recognition of phosphatidylserine*, Am. J. Physiol. Lung Cell Mol. Physiol. 285 (2003) L1–L17.

Kagan, V E, Fabisiak, J P, Shvedova, A A, Tyurina, Y Y, Tyurin, V A, Schor, N F, and Kawai, K, *Oxidative signaling pathway for externalization of plasma membrane phosphatidylserine during apoptosis*, FEBS Lett. 477 (2000) 1–7.

Kagan, V E, Gleiss, B, Tyurina, Y Y, Tyurin, V A, Elenstrom-Magnusson, C, Liu, S X, Serinkan, F B, Arroyo, A, Chandra, J, Orrenius, S, and Fadeel, B, *A role for oxidative stress in apoptosis: Oxidation and externalization of phosphatidylserine is required for macrophage clearance of cells undergoing fas-mediated apoptosis*, J. Immunol. 169 (2002) 487–499.

Kalyankrishna, S, Parmentier, J H, and Malik, K U, *Arachidonic acid-derived oxidation products initiate apoptosis in vascular smooth muscle cells*, Prostaglandins Other Lipid Mediat. 70 (2002) 13–29.

Kockx, M M, *Apoptosis in atherosclerotic plaques. Quantitative and qualitative aspects*, Arterioscler. Thromb. Vasc. Biol. 18 (1998) 1519–1522.

Koopman, G, Reutelingsperger, C P, Kuijten, G A, Keehnen, R M, Pals, S T, and van Oers, M H, *Annexin V for flow cytometric detection of phosphatidylserine expression on B cells undergoing apoptosis*, Blood 84 (1994) 1415–1420.

Li, M O, Sarkisian, M R, Mehal, W Z, Rakic, P, and Flavell, R A, *Phosphatidylserine receptor is required for clearance of apoptotic cells*, Science 302 (2003) 1560–1563.

Li, X M, Salomon, R G, Qin, J, and Hazen, S L, *Conformation of an endogenous ligand in a membrane bilayer for the macrophage scavenger receptor CD36*, Biochemistry (Mosc). 46 (2007) 5009–5017.

Libby, P, Geng, YJ, Sukhova, G K, Simon, D I, and Lee, R T, *Molecular determinants of atherosclerotic plaque vulnerability*, Ann. N.Y. Acad. Sci. 811 (1997) 134–142.

Loidl, A, Sevcsik, E, Riesenhuber, G, Deigner, H P, and Hermetter, A, *Oxidized phospholipids in minimally modified low density lipoprotein induce apoptotic signaling via activation of acid sphingomyelinase in arterial smooth muscle cells*, J. Biol. Chem. 278 (2003) 32921–32928.

Majno, G and Joris, I, *Apoptosis, oncosis, and necrosis. An overview of cell death*, Am. J. Pathol. 146 (1995) 3–15.

Martinet, W and Kockx, M M, *Apoptosis in atherosclerosis: Focus on oxidized lipids and inflammation*, Curr. Opin. Lipidol. 12 (2001) 535–541.

Matsubara, M and Hasegawa, K, *Benidipine, a dihydropyridine-calcium channel blocker, prevents lysophosphatidylcholine-induced injury and reactive oxygen species production in human aortic endothelial cells*, Atherosclerosis 178 (2005) 57–66.

Matsura, T, Togawa, A, Kai, M, Nishida, T, Nakada, J, Ishibe, Y, Kojo, S, Yamamoto, Y, and Yamada, K, *The presence of oxidized phosphatidylserine on Fas-mediated apoptotic cell surface*, Biochim. Biophys. Acta – Molecular and Cell Biology of Lipids 1736 (2005) 181–188.

Matsuzawa, A, Hattori, K, Aoki, J, Arai, H, and Inoue, K, *Protection against oxidative stress-induced cell death by intracellular platelet-activating factor-acetylhydrolase II*, J. Biol. Chem. 272 (1997) 32315–32320.

Mevorach, D, Mascarenhas, J O, Gershov, D, and Elkon, K B, *Complement-dependent clearance of apoptotic cells by human macrophages*, J. Exp. Med. 188 (1998) 2313–2320.

Min, D B, Lipid oxidation of edible oil, in: Akoh, C C and Min, D B, *Food lipids – chemistry, nutrition, and biotechnology*, Marcel Dekker Inc., New York, Basel, Hong Kong, 1997, pp. 283–296.

Moumtzi, A, Trenker, M, Flicker, K, Zenzmaier, E, Saf, R, and Hermetter, A, *Import and fate of fluorescent analogs of oxidized phospholipids in vascular smooth muscle cells*, J.Lipid Res. 48 (2006) 565–582.

Murphy, J E, Tacon, D, Tedbury, P R, Hadden, J M, Knowling, S, Sawamura, T, peckham, M, Phillips, S E V, Walker, J H, and Ponnambalam, S, *LOX-1 scavenger receptor mediates calcium-dependent recognition of phosphatidylserine and apoptotic cells*, Biochem. J. 393 (2006) 107–115.

Platt, N, da Silva, R P, and Gordon, S, *Recognizing death: The phagocytosis of apoptotic cells*, Trends Cell Biol. 8 (1998) 365–372.

Podrez, E A, Poliakov, E, Shen, Z, Zhang, R, Deng, Y, Sun, M, Finton, P J, Shan, L, Febbraio, M, Hajjar, D P, Silverstein, R L, Hoff, H F, Salomon, R G, and Hazen, S L, *A novel family of atherogenic oxidized phospholipids promotes macrophage foam cell formation via the scavenger receptor CD36 and is enriched in atherosclerotic lesions*, J. Biol. Chem. 277 (2002a) 38517–38523.

Podrez, E A, Poliakov, E, Shen, Z, Zhang, R, Deng, Y, Sun, M, Finton, P J, Shan, L, Gugiu, B, Fox, P L, Hoff, H F, Salomon, R G, and Hazen, S L, *Identification of a novel family of oxidized phospholipids that serve as ligands for the macrophage scavenger receptor CD36*, J. Biol. Chem. 277 (2002b) 38503–38516.

Pollman, M J, Hall, J L, Mann, M J, Zhang, L, and Gibbons, G H, *Inhibition of neointimal cell Bcl-x expression induces apoptosis and regression of vascular disease*, Nat. Med. 4 (1998) 222–227.

Sabatini, K, Mattila, J P, Megli, F M, and Kinnunen, P K, *Characterization of two oxidatively modified phospholipids in mixed monolayers with DPPC*, Biophys. J. 90 (2006) 4488–4499.

Salvayre, R, Auge, N, Benoist, H, and Negre-Salvayre, A, *Oxidized low-density lipoprotein-induced apoptosis*, Biochim. Biophys. Acta – Molecular and Cell Biology of Lipids 1585 (2002) 213–221.

Savill, J and Fadok, V, *Corpse clearance defines the meaning of cell death*, Nature 407 (2000) 784–788.

Spiteller, G, *Is atherosclerosis a multifactorial disease or is it induced by a sequence of lipid peroxidation reactions?* Ann. N.Y. Acad. Sci. 1043 (2005) 355–366.

Stafforini, D M, Sheller, J R, Blackwell, T S, Sapirstein, A, Yull, F E, McIntyre, T M, Bonventre, J V, Prescott, S M, and Roberts, L J, *Release of free F2-isoprostanes from esterified phospholipids is catalyzed by intracellular and plasma platelet-activating factor acetylhydrolases*, J. Biol. Chem. 281 (2006) 4616–4623.

Taha, T A, Mullen, T D, and Obeid, L M, *A house divided: Ceramide, sphingosine, and sphingosine-1-phosphate in programmed cell death*, Biochim. Biophys. Acta 1758 (2006) 2027–2036.

Takahashi, M, Okazaki, H, Ogata, Y, Takeuchi, K, Ikeda, U, and Shimada, K, *Lysophosphatidylcholine induces apoptosis in human endothelial cells through a p38-mitogen-activated protein kinase-dependent mechanism*, Atherosclerosis 161 (2002) 387–394.

Tsutsumi, H, Kumagai, T, Naitoo, S, Ebina, K, and Yokota, K, *Synthetic peptide (P-21) derived from asp-hemolysin inhibits the induction of apoptosis on HUVECs by lysophosphatidylcholine*, Biol. Pharm. Bull. 29 (2006) 907–910.

Tyurina, Y Y, Kawai, K, Tyurin, V A, Liu, S X, Kagan, V E, and Fabisiak, J P, *The plasma membrane is the site of selective phosphatidylserine oxidation during apoptosis: Role of cytochrome c*, Antioxidants & Redox Signaling 6 (2004a) 209–225.

Tyurina, Y Y, Serinkan, F B, Tyurin, V A, Kini, V, Yalowich, J C, Schroit, A J, Fadeel, B, and Kagan, V E, *Lipid antioxidant, etoposide, inhibits phosphatidylserine externalization and macrophage clearance of apoptotic cells by preventing phosphatidylserine oxidation*, J. Biol. Chem. 279 (2004b) 6056–6064.

Wang, B Y, Ho, H K, Lin, P S, Schwarzacher, S P, Pollman, M J, Gibbons, G H, Tsao, P S, and Cooke, J P, *Regression of atherosclerosis:role of nitric oxide and apoptosis*, Circulation 99 (1999) 1236–1241.

Wang, X, Wu, Y C, Fadok, V A, Lee, M C, Gengyo-Ando, K, Cheng, L C, Ledwich, D, Hsu, P K, Chen, J Y, Chou, B K, Henson, P, Mitani, S, and Xue, D, *Cell corpse engulfment mediated by C. elegans phosphatidylserine receptor through CED-5 and CED-12*, Science 302 (2003) 1563–1566.

Yaraei, K, Campbell, L A, Zhu, X, Liles, W C, Kuo, C C, and Rosenfeld, M E, *Chlamydia pneumoniae Augments the oxidized low-density lipoprotein-induced death of mouse macrophages by a caspase-independent pathway*, Infect. Immun. 73 (2005) 4315–4322.

Zhou, L, Shi, M, Guo, Z, Brisbon, W, Hoover, R, and Yang, H, *Different cytotoxic injuries induced by lysophosphatidylcholine and 7-ketocholesterol in mouse endothelial cells*, Endothelium 13 (2006) 213–226.

# Part III
# Sphingolipids

# Chapter 14
# Regulation of Lipid Metabolism by Sphingolipids

**Tilla S. Worgall**

**Abstract** Sphingolipids, together with phospholipids and cholesterol are key components of membrane lipid bilayers, Sphingolipids and cholesterol contribute to specialized membrane domains called rafts. Sphingolipids have been recognized to exert a distinct role in the post-transcriptional regulation of the sterol-regulatory element binding proteins (SREBPs), key transcription factors of lipid synthesis. Sphingolipid synthesis is an obligate activator of SREBP. Inhibition of sphingolipid synthesis decreases SREBP on a post-transcriptional level. SREBPs regulate the transcription of key enzymes that synthesize cholesterol, phospholipids and fatty acids but not enzymes that synthesize sphingolipids. This observation suggests a distinct role for sphingolipids in the regulation of cellular lipid metabolism. Although exact mechanisms how sphingolipids regulate lipid metabolism are currently not known, this relationship has important implications with regard to cellular lipid homeostasis, composition of lipoproteins and development of atherosclerosis.

**Keywords** Ceramide · cholesterol · SREBP · serine-palmitoyl-transferase · sphingomyelin

**Abbreviations** ABCA-1: ATP-binding cassette transporter A1; CerS: ceramide synthase proteins; HDL: high-density lipoproteins; LCAT: lecithin-cholesterol acyl transferase; LCB: long-chain base; PDMP: 1-phenyl-2-decanoylamino-3-morpholino-1-propanol; SMS: sphingomyelin synthase; SPT: serine-palmitoyl-transferase; pSREBP: precursor (inactive) sterol regulatory element binding protein; mSREBP: mature (active) sterol regulatory element binding protein.

T.S. Worgall
Department of Pathology, Columbia University, 168 W 168 St, BB 457, New York, NY 10032, USA

P.J. Quinn, X. Wang (eds.), *Lipids in Health and Disease*,

## 14.1 The Interaction of Sphingomyelin, Cholesterol and Phospholipids Defines Membrane Structure and Functional Properties

All eukaryotic cells are contained within a lipid bilayer membrane that is composed of phospholipids, sphingolipids and cholesterol. Although all lipids in eukaryotic membranes are fluid, membranes are not uniform. Membrane lipids in the endoplasmic reticulum are generally unsaturated, a characteristic that promotes spontaneous transbilayer lipid translocation and budding of transport vesicles (Baumgart et al., 2003). In contrast, the plasma membrane contains more saturated lipids and segregates into membrane domains that differ in structure and function. Most of the cellular sphingomyelin is found in the outer leaflet of the bilayer membrane where it preferentially associates with cholesterol. Cholesterol and sphingomyelin together define membrane domains described as liquid ordered domains or lipid rafts. The presence of a saturated long chain sphingoid base and a saturated fatty acyl chain conveys rigidity to sphingomyelin rich membranes. In contrast, phospholipids typically contain one saturated and one unsaturated acyl chain and define membrane domains that are described as liquid disordered domains (Lange et al., 1989; Ohvo-Rekila et al., 2002; van Meer, 1989). The segregation of the plasma membrane into different fluid phases affects membrane structure, protein sorting and protein activity. Mechanisms how membrane lipid homeostasis is maintained are currently not fully understood.

### *14.1.1 Sphingolipid Synthesis Correlates with the Activation of SREBP – Inhibition of Sphingolipid Synthesis Decreases SREBP*

Sphingolipid homeostasis is achieved through a sophisticated system of end-product inhibition and feed-forward regulation. Sphingolipids are generated through de-novo and recycling pathways. Ceramide is the central product of synthesis. Ceramide forms the backbone of sphingomyelin and glucosylceramides or, is further metabolized to sphingosine or ceramide-1-phosphate. Plasma sphingomyelin concentration. Two different, recently identified sphingomyelin synthases (SMS), located at the Golgi and at the plasma membrane, contribute to the regulation of plasma sphingomyelin concentration. Ceramides have multiple roles ranging from lipid second messenger to the induction of apoptosis, cell growth and differentiation, and also function as feedback inhibitors within the sophisticated system that regulates sphingolipid synthesis (Hannun and Obeid, 2002). We and others made the observation that sphingolipid synthesis correlates with the regulation of the sterol-regulatory element binding proteins (SREBP), key transcription factor of lipid synthesis (Dobrosotskaya et al., 2002; Worgall et al., 2004a). SREBPs directly activate the expression of more than 30 genes dedicated to the synthesis and uptake of cholesterol, fatty acids, triglyceride, and phospholipids, as well as the NADPH cofactor required to synthesize these molecules (Horton et al., 2002). We observed that exogenous and endogenous ceramides decreases transcriptionally active SREBP and promote accumulation of the precursor form of SREBP in the endoplasmic reticulum (Worgall et al., 2004a). The demonstration that ceramide mediated decrease of SREBP correlates with the inhibition of sphingolipid synthesis and that inhibition of rate-limiting enzymes of sphingolipid synthesis equally decreases SREBP and SRE-mediated gene transcription suggested an important role of sphingolipid synthesis in the regulation of SREBP. The latter observation was made in cells that are mutated in serine-palmitoyl-CoA synthase, the rate limiting enzyme of sphingolipid de-novo synthesis. These cells, unable to produce sphingolipids de novo, fail to increase SREBP and SRE-mediated gene transcription in sterol depletion. The ability to increase SREBP is recovered when sphingolipids are supplied exogenously in the incubation medium. The observed regulation of SREBP by sphingolipids occurs independently of cellular cholesterol concentration and independent of the source of cellular ceramide. Increasing cellular ceramide by either exogenous short-chain ceramides, inhibition of ceramide catabolism or sphingomyelinase mediated sphingomyelin hydrolysis equally decreases SREBP and SRE-mediated gene transcription. Of note, sphingolipid mediated regulation of SREBP is an evolutionary conserved mechanism that is also found in *D. melanogaster*. SREBP is also found in *C. elegans* and *Giardia lamblia* (Dobrosotskaya et al., 2002; Kunte et al., 2006; McKay et al., 2003; Worgall et al., 2004b). In *G. lamblia*, SREBP regulates the transcription of the cyst wall

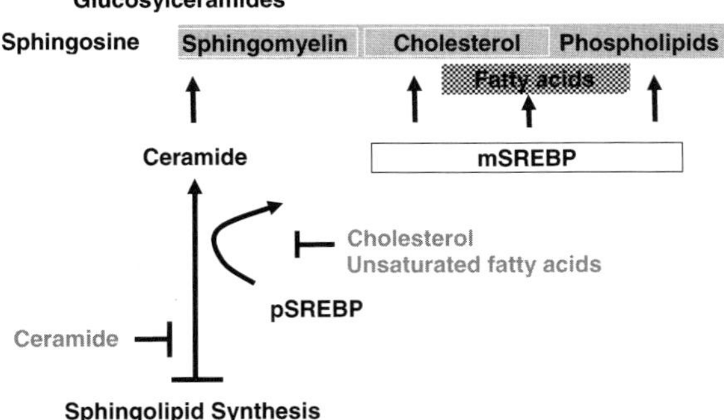


**Fig. 14.1** Sphingolipid synthesis generates ceramide, sphingomyelin, sphingosine and glucosylceramides and correlates with the regulation of the sterol-regulatory element binding proteins (SREBP). SREBP are membrane proteins of the endoplasmic reticulum. Post-transcriptional regulation involves the vesicular transfer of the inactive precursor SREBP (pSREBP) to the Golgi apparatus. Here, two distinct proteases cleave pSREBP and release the transcriptional active mature SREBP (mSREBP). mSREBP regulates a total of thirty-three genes of cholesterol, triglyceride, fatty acid and phospholipid metabolism. Unsaturated fatty acid and cholesterol decrease activation of mSREBP on a post-transcriptional level. Ceramide inhibits sphingolipid synthesis and decreases activation of mSREBP

proteins that ensure progression of the *G. lamblia* life-cycle from trophozoites to the encysted forms. *In D. melanogaster,* an organism that does not produce sterols, one SREBP protein regulates non-sterol lipid metabolism and is regulated by metabolites of the membrane lipid phosphatidylethanolamine. Together, the data suggest that sphingolipids can function as regulators of SREBP and that sphingolipids together with SREBP integrate signals that originate from the composition of membranes with transcriptional control of cellular lipid synthesis (Fig. 14.1).

### *14.1.2 SREBPs Activate the Synthesis of Proteins That Regulate Phospholipid, Fatty Acid and Cholesterol But Not Sphingolipid Synthesis*

There are three SREBP isoforms (Horton et al., 2002; Horton et al., 2003). SREBP-1a activates all SREBP regulated genes, i.e. genes that regulate phospholipid, cholesterol and fatty acid synthesis. In contrast, SREBP-1 c preferentially activates genes of fatty acid synthesis. SREBP-2 activates genes of cholesterol synthesis. According to current knowledge, proteins that synthesize sphingolipids are not transcriptionally regulated by SREBPs. The regulation of SREBPs occurs transcriptionally and post-transcriptionally. The inactive precursor form of SREBP is located in the endoplasmic reticulum bound to a sterol-sensing protein, SCAP. Cholesterol promotes the association of the Insig protein to the SREBP-SCAP complex. This association anchors SREBP in the endoplasmic reticulum and the transcriptionally active form of SREBP is not generated (Horton et al., 2002). When cellular cholesterol levels are low, SREBP and SCAP are trafficked by vesicular transport to the Golgi apparatus. SCAP is recycled to the endoplasmic reticulum and SREBP is cleaved within the Golgi. Two distinct proteolytic steps within the Golgi generate the transcriptionally active mature SREBP (mSREBP) (Sakai et al., 1996). mSREBP is a transcription factor of the basic helix-loop-helix-leucine zipper (bHLH) class of transcription factors that binds to the promoter region of genes that regulate cholesterol, fatty acid and phospholipid synthesis. SREBP-2 activates the LDL receptor and essentially all genes required for cholesterol synthesis (Horton et al., 2003). Genes regulated by SREBP-1 include the rate limiting enzymes of fatty acid synthesis, acetyl-CoA carboxylase and fatty acid synthase that generate palmitic acid, a saturated fatty acid. SREBP also regulates the generation of oleic acid from stearic acid by transcriptional activation of stearoyl-CoA desaturase, promotes the desaturation of essential fatty acids by increasing transcription of $\Delta$6-desaturase, as well as elongation of long-chain fatty acids through long-chain acyl elongases. Of note, SREBP-1 also regulates acyl-CoA synthetase and fatty acid CoA ligase, enzymes that regulate the incorporation of fatty acids into triglycerides and phospholipids.

### *14.1.3 Cholesterol and Polyunsaturated Fatty Acids Inhibit Activation of SREBP*

SREBP regulation is inhibited post-transcriptionally by two key end products of synthesis, cholesterol and polyunsaturated fatty acids. Experimentally, cholesterol inhibits the translocation of SREBP by promoting the interaction of SCAP with Insig. Unsaturated fatty acids and polyunsaturated fatty acids are the other known inhibitors of SREBP. However, the molecular mechanisms how polyunsaturated fatty acids inhibit SREBP are less clear. The ability of polyunsaturated fatty acids to inhibit SREBP increases with chain length and degree of unsaturation (Worgall et al., 1998). Another observation is that oleic acid and polyunsaturated fatty acids maintain the ability to decrease SREBP in the presence of cationic amphiphiles that induce a state of relative sterol-depletion by sequestering cholesterol in lysosomes. This observation strongly suggests that polyunsaturated fatty acids decrease SREBP by a cholesterol independent effect.

### *14.1.4 Fatty Acids Can Inhibit SREBP via Metabolites of Sphingomyelin*

One possible mechanism how fatty acids decrease SREBP is linked to cellular sphingolipid metabolism. Similar to the observation that sphingomyelinase decreases SRE-mediated gene by releasing cholesterol out of its association with sphingomyelin in the plasma membrane, fatty acids have been shown to mobilize cholesterol and sphingomyelin between model membranes (Johnson et al., 2003; Scheek et al., 1997). Furthermore, fatty acids were also shown to promote hydrolysis of sphingomyelin in cell culture (Johnson et al., 2003). Of note, ceramide and phosphocholine are products of sphingomyelin hydrolysis. Phosphocholine does not affect SREBP. However, ceramide alone and additively to polyunsaturated fatty acids, decreases SREBP and expression of SRE-mediated genes. The observation that ceramide decreases SREBP in the presence of cationic amphiphiles supports regulatory mechanisms that are independent of cholesterol-mediated regulation.

## 14.2 Sphingolipids Can Regulate Key Mechanisms of Cellular Cholesterol Efflux

Cholesterol efflux is one mechanism that maintains cellular cholesterol homeostasis. An important regulator of cellular cholesterol efflux is the ATP-binding cassette receptor A1 (ABCA-1). ABCA-1 is critical for the formation of plasma high-density lipoproteins (HDL) that are inversely related to cardiovascular disease risk (Schaefer et al., 1994). Two groups investigated the role of sphingolipids in the regulation of ABCA-1 (Ghering and Davidson, 2006; Glaros et al., 2005; Witting et al., 2003). One group demonstrates that ceramide increases expression of ABCA-1 and ABCA-1 mediated cellular cholesterol efflux (Witting et al., 2003). The other group demonstrates that the glucosylceramide synthase inhibitor, 1-phenyl-2-decanoylamino-3-morpholino-1-propanol (PDMP) increases transcription, expression and activity of ABCA-1 at the plasma membrane (Glaros et al., 2005). Possibly, both groups investigated a similar mechanism as PDMP also increases cellular ceramide concentrations (Shayman et al., 2000). Although it has been suggested that the overall ceramide shape and the amide bond are critical for the cholesterol efflux effect, exact mechanisms how ceramide or possibly glucosylceramide regulate ABCA-1 are currently unknown (Ghering and Davidson, 2006). Together the data suggest that ceramide, and possibly glucosylceramide, could have a specific role in the regulation of ABCA-1 that promotes cholesterol efflux.

### *14.2.1 Inhibition of Sphingolipid Synthesis Affects Plasma Lipoproteins and the Development of Atherosclerosis in Mouse Models*

In-vivo studies support the relevance of sphingolipid synthesis as a regulator of lipid metabolism. Two independent groups (Hojjati et al., 2005a,b; Park et al., 2004; Park et al., 2006) demonstrated that inhibition of SPT, the rate-limiting step in sphingolipid de-novo synthesis, significantly decreases atherosclerotic lesion in apolipoprotein E knock-out mice. Both groups investigated the effect of myriocin, a competitive inhibitor of SPT, on lipid parameters and development of atherosclerosis. The experimental approach differed as one group administered myriocin intraperitoneally and the other group evaluated an oral administration route. However, both groups demonstrate that inhibition of sphingolipid synthesis has beneficial effects on lipoprotein metabolism. This correlates with a decreased formation of atherosclerotic lesions and macrophage accumulation. In contrast to oral administration, intraperitoneal administration of myriocin did not affect plasma cholesterol concentrations but significantly decreased the sphingomyelin to phosphatidylcholine ratio, a desirable profile that is associated with decreased cardiovascular events (Jiang et al., 2000). A follow-up study examined molecular mechanisms for these effects (Park et al., 2006). Using different mouse models, the investigators demonstrate that inhibition of sphingolipid synthesis significantly decreases SREBP-1 on a transcriptional and post-transcriptional level and induces the expression of apolipoprotein A1 and lecithin-cholesterol acyl transferase (LCAT) proteins that promote increased plasma concentrations of anti-atherogenic HDL.

### 14.2.2 Overexpression of Sphingomyelin Synthase Induces Lipoprotein Patterns That Correlate With Increased Cardiovascular Risk

Epidemiological observations have demonstrated that absolute and relative plasma sphingomyelin levels are independent risk factors for coronary heart disease (Jiang et al., 2000; Nelson et al., 2006). Possible mechanisms for this effect are beginning to be understood. Hepatic over-expression of sphingomyelin synthase in wild-type mice shifts plasma cholesterol and sphingomyelin concentration to the pro-atherogenic non-HDL lipoproteins (Dong et al., 2006). This observation is of interest, because sphingomyelin concentration not only enhances the susceptibility of lipoproteins to aggregate but also affects lipoprotein kinetics (Jeong et al., 1998; Schissel et al., 1996). Increased sphingomyelin content in triglyceride rich particles decreases both affinity for and catalytic activity of lipoprotein lipase resulting in decreased lipolysis (Arimoto et al., 1998). Increased sphingomyelin in HDL particles decreases the activity of LCAT resulting in decreased generation of mature HDL and reverse cholesterol flux (Bolin and Jonas, 1996; Lee et al., 2006). Together, increased triglyceride and decreased plasma HDL concentration constitute a lipoprotein profile associated with increased cardiovascular risk. Therefore, the demonstration of a causal relationship between sphingomyelin synthase activity and lipoprotein sphingomyelin concentration could be of specific interest in the quest to identify novel targets for anti-atherogenic therapies.

### 14.2.3 Perturbed Lipid Metabolism and Atherosclerotic Lipoprotein Patterns in Sphingolipid Storage Disorders

Further support for the role of sphingolipids in the regulation of lipid metabolism and development of atherosclerosis is obtained from the study of sphingolipid storage diseases in which cholesterol homeostasis and intracellular distribution of cholesterol are perturbed secondary to accumulation of endogenous sphingolipids (McGovern et al., 2004; Puri et al., 2003; Schuchmann and Desnick, 1995). A summary of major findings is the subject of a review (Pagano et al., 2000). One example for this relationship is Niemann Pick Disease Type A and B, a disease is caused by mutations in the gene for acid sphingomyelinase that significantly decreases enzyme activity. In Niemann Pick Disease Type A and B cellular concentrations of sphingomyelin and cholesterol are severely elevated. Sphingomyelin constitutes up to 70% of phospholipids compared to normal concentrations of 5–20%. At the same time, cholesterol mass is increased up to ten fold (Schuchmann and Desnick, 1995). The lipid abnormalities are clinically significant. A cross sectional study of 10 children and adolescents with Niemann Pick Disease type A and 30 patients with Niemann Pick Disease Type B demonstrated significant atherogenic lipoprotein profiles specifically characterized by low plasma HDL-cholesterol (McGovern et al., 2004). In addition, evaluation of coronary artery calcium scores, an indicator for the development of atherosclerosis, were positive in 10 of 18 type B patients studied. The observation of increased sphingomyelin content in HDL that impairs maturation and results in low plasma HDL levels supports earlier described mechanisms of impaired LCAT activity (Arimoto et al., 1998; Lee et al., 2006)

## 14.3 Mechanisms of Regulation

The molecular mechanisms how sphingolipids regulate lipid metabolism are currently unknown. Experimental data support regulatory mechanisms that are based on the physical interaction of sphingolipids with cholesterol as well on the functional effects of sphingolipids. A good example for the physical interaction between sphingolipids and cholesterol is the model proposed for the effect of sphingolipids on cholesterol metabolism in sphingolipid storage disorders (Demel et al., 1977). According to this model, excess sphingolipids sequester cholesterol resulting in a reduction of regulatory cholesterol pools at the endoplasmic reticulum. This mechanism can explain the observed increase in SREBP in sphingolipid storage disease (Puri et al., 2003) as well as the decrease in SREBP activity in cells in which treatment with myriocin decreases sphingomyelin mass (Park et al., 2006).

However, other experimental observations do not quite fit this model that centers on the physical interaction of sphingomyelin with cholesterol and current data suggest at least one other level of regulation. For example, 25-hydroxy-cholesterol and ceramide both increase sphingomyelin mass but decreases SREBP and SRE-mediated gene transcription (Ridgway, 1995; Sakai et al., 1996).Ceramide dose-dependently decreases SREBP and HMG-CoA synthase activity by a mechanism independent of a cholesterol-mediated decrease of SREBP (Ridgway and Merriam, 1995).

The observation that ceramide, in contrast to cholesterol, causes an accumulation of inactive SREBP in the endoplasmic reticulum furthermore suggests a regulation that is different from cholesterol-mediated inhibition of SREBP. One possibility is that ceramide promotes the interaction of SCAP with Insig resulting in an anchoring of pSREBP in the endoplasmic reticulum. Another possibility is that ceramide decreases vesicular trafficking from the endoplasmic reticulum to the Golgi. This model is supported by the demonstration that sphingolipid synthesis facilitates vesicular transport and that ceramide inhibits coated vesicle formation (Abousalham et al., 2002; Rosenwald and Pagano, 1993; Zanolari et al., 2000). In contrast to a model that focuses on sphingomyelin as the main regulator of lipid homeostasis, the later observation implicate ceramide synthesis or ceramide itself with a role in the regulation of SREBP and lipid homeostasis.

### *14.3.1 Novel Proteins That Regulate Sphingolipid Synthesis Could Affect the Regulation of Lipid Metabolism*

Could sphingolipid synthesis regulate genes of lipid synthesis, independent of its effects on sphingomyelin mass? The majority of sphingolipid synthesis occurs through the recycling pathway, cells can however increase sphingolipid synthesis through the de-novo pathway, specifically when proliferation rates are high (Batheja et al., 2003). Do both pathways differ in their capacity to regulate lipid synthesis? Notable is the observation that increasing sphingolipid synthesis through the recycling of sphingosine promotes the synthesis of C16:0, C18:0, and C20:0 long-chain ceramides (Le Stunff et al., 2007). This observation could have relevance with regard to the synthesis of sphingolipids that confer different membrane properties. For example, increased mass of saturated ceramide species could affect the degree of fluidity and ability of the endoplasmic reticulum to generate transport vesicles. Molecular mechanisms that explain how sphingolipids affect lipid metabolism are currently unknown. However, much progress has been made in the understanding of the molecular machinery that regulates sphingolipid synthesis. A new subunit of SPT has been described, a class of ceramide synthases has been discovered and two genes that regulate sphingomyelin synthesis were identified in addition to the discovery of a protein that regulates the trafficking of ceramide destined for sphingomyelin synthesis at the Golgi.

Sphingolipid synthesis is tightly regulated by a sophisticated feed-back system that balances de-novo and recycling pathways (Fig. 14.2). The initial step in de-novo sphingolipid synthesis is the condensation of L-serine and palmitoyl-CoA. This reaction takes place on the cytosolic surface of the endoplasmic reticulum and is catalyzed by SPT. Serine, palmitoyl-CoA and select saturated fatty acids promote this reaction that generates 3-ketodihydrosphingosine that is then reduced to sphinganine (Merrill et al., 2005). Until recently, two subunits LCB1 and LCB2 that heterodimerize and constitute SPT were known. The description of a third subunit, LCB3 that can replace LCB2 and heterodimerizes with LCB1 is expected to reveal new insight into tissue and possibly substrate dependent regulation as this subunit is expressed in all human tissues, except at very low levels in fetal brain and not at all in peripheral blood cells and bone marrow (Hornemann et al., 2006). SPT activity is increased in cells that proliferate. One potential mechanism how sphingolipid synthesis increases lipid synthesis could be related to the activity and expression patterns of SPT.

Ceramide and dihydroceramide are generated by acylation of sphinganine with mostly saturated fatty acyl-CoAs, a reaction catalyzed by a recently discovered family of mammalian ceramide synthase (CerS) genes. (Pewzner-Jung et al., 2006; Venkataraman et al., 2002). The six known CerS proteins are specific for fatty acids of different chain length. How might ceramides containing different fatty acids impact upon cell physiology? One or more biophysical properties of the membrane lipid bilayer could be influenced by the fatty

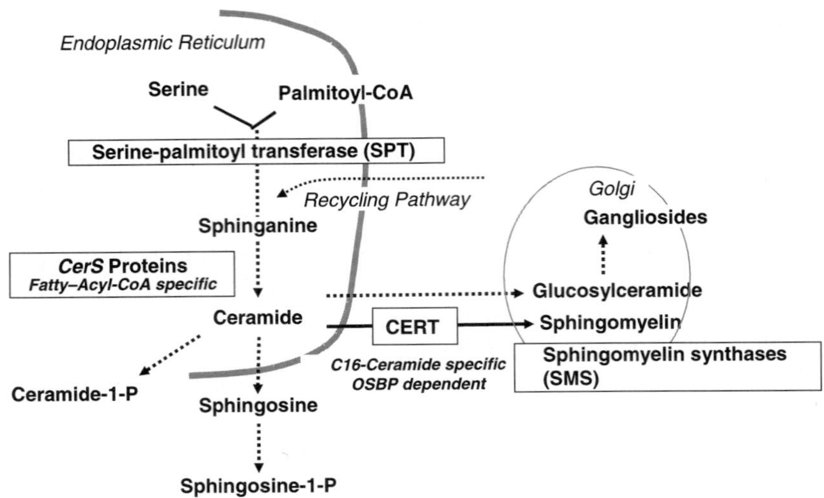


**Fig. 14.2** Sphingolipid Synthesis – Recently described genes and proteins of sphingolipid synthesis. Sphingolipid homeostasis is achieved through a sophisticated system of end-product inhibition and feed-forward regulation. Sphingolipids are generated through de-novo and recycling pathways. Two proteins, LCB2 or the recently described LCB3 heterodimerize with LCB1 and constitute serine-palmitoyl transferase (SPT), the rate-limiting enzyme of de-novo synthesis. The currently known six ceramide synthases (CerS proteins) are fatty-acyl-CoA specific (dihydro)ceramide synthases that acylate sphinganine. Ceramide is the central product of synthesis and forms the backbone of sphingomyelin, and glucosylceramides or is further metabolized to sphingosine or ceramide-1-phosphate. The CERT protein delivers ceramide to the Golgi for sphingomyelin synthesis, a reaction specific for C16-ceramide and described to necessitate the oxysterol binding protein (OSBP) for basal activity. Two different, recently identified sphingomyelin synthases (SMS), located at the Golgi and at the plasma membrane can regulate plasma sphingomyelin concentration

acid composition of ceramide (Sot et al., 2005). Alternatively, specific ceramides could directly interact with downstream components in signaling pathways or, unsaturated fatty acids, possibly by depleting a pool of acyl-CoA, could decrease sphingolipid synthesis which in turn would reduce generation of SREBP and related genes of lipid synthesis (Lahiri and Futerman, 2005; Spassieva et al., 2006).

The discovery of the CERT protein revealed the molecular mechanisms how ceramide is extracted from the ER membrane and delivered to the luminal side of the Golgi apparatus for sphingomyelin synthesis (Hanada, 2006; Hanada et al., 2003; Kawano et al., 2006). The activity of CERT is subject to phosphorylation. Protein kinase D mediated phosphorylation inactivates CERT and reduces the synthesis of sphingomyelin (Fugmann et al., 2007). Two observations support a coordinated regulation of CERT and sterol metabolism. One, the demonstration that oxysterol binding protein (OSBP) is required for basal CERT activity, and two, the presence of a START domain that is necessary for the transport of ceramide and shared by at least 15 other known mammalian proteins that regulate intracellular lipid transport (Soccio and Breslow, 2003).

Finally, significant progress has been made in the understanding of the molecular basis of sphingomyelin synthesis with the identification of two sphingomyelin synthases. Sphingomyelin synthase 1 is located in the Golgi apparatus and sphingomyelin synthase 2 is located at the plasma membrane. Both sphingomyelinases transfer the phosphorylcholine head group from phosphatidylcholine to ceramide to yield sphingomyelin and diacylglyerol (Huitema et al., 2004; Tafesse et al., 2006). Although a specific role in the regulation of SREBP is not known, diacylglyerol can recruit protein kinase D to the Golgi to initiate vesicular transport (Baron and Malhotra, 2002). The relevance of sphingomyelin synthases is supported by experimental studies that demonstrated development of an atherogenic lipoprotein profile when sphingomyelin synthase was over expressed in the liver (Dong et al., 2006).

## 14.4 Conclusions

Sphingolipids interact physically and functionally with cholesterol, fatty acids and phospholipids. Recently, much insight has been gained in the understanding of the role of sphingolipids in the regulation of genes that synthesize key proteins of cholesterol, phospholipid and fatty acid metabolism. In-vitro and in-vivo experimental data demonstrate that sphingolipid synthesis correlates with the activation of SREBPs, key transcription factors of lipid synthesis, and that inhibition of serine-palmitoyl transferase, SPT, the rate-limiting enzyme of sphingolipid de-novo synthesis decreases the activity of SREBPs. Metabolites of sphingolipid synthesis, i.e. ceramide, not only inhibit sphingolipid synthesis but also correlate with increased transcription and expression of genes and proteins that promote cholesterol efflux and increase plasma HDL-cholesterol. Experimentally, inhibition of the sphingolipid biosynthetic pathway correlates with generation of anti-atherogenic lipoprotein profiles and a decrease in atherosclerosis. In turn, increased sphingomyelin synthesis correlates with increased plasma sphingomyelin concentration of pro-atherogenic lipoproteins and an increased susceptibility to particle aggregation. Specific mechanisms how sphingolipids regulate lipid synthesis are currently unknown. One important regulatory mechanism is derived from the physical affinity of sphingomyelin to cholesterol. However, other mechanisms related to sphingolipid synthesis could also play a role. The continuing discovery of genes and proteins that regulate sphingolipid synthesis has revealed important insight into molecular mechanisms. Together, manipulation of the sphingolipid synthetic pathway could potentially be a promising therapeutic target for treatment of dyslipidemia and atherosclerosis.

**Acknowledgment** T. S. Worgall is supported by NIH Grant NIA KO8 AG025833 and Cystic Fibrosis Pilot and Feasibility Award 07010.

## References

Abousalham A, Hobman TC, Dewald J, Garbutt M, Brindley DN. Cell-permeable ceramides preferentially inhibit coated vesicle formation and exocytosis in Chinese hamster ovary compared with Madin- Darby canine kidney cells by preventing the membrane association of ADP- ribosylation factor. *Biochem J* 361(2002) 653–661.

Arimoto I, Saito H, Kawashima Y, Miyajima K, Handa T. Effects of sphingomyelin and cholesterol on lipoprotein lipase-mediated lipolysis in lipid emulsions. *J Lipid Res* 39(1998) 143–151.

Baron CL, Malhotra V. Role of diacylglycerol in PKD recruitment to the TGN and protein transport to the plasma membrane. *Science* 295(2002) 325–328.

Batheja AD, Uhlinger DJ, Carton JM, Ho G, D'Andrea MR. Characterization of serine palmitoyltransferase in normal human tissues. *J Histochem Cytochem* 51(2003) 687–696.

Baumgart T, Hess ST, Webb WW. Imaging coexisting fluid domains in biomembrane models coupling curvature and line tension. *Nature* 425(2003) 821–824.

Bolin DJ, Jonas A. Sphingomyelin inhibits the lecithin-cholesterol acyltransferase reaction with reconstituted high density lipoproteins by decreasing enzyme binding. *J Biol Chem* 271(1996) 19152–19158.

Demel RA, Jansen JW, van Dijck PW, van Deenen LL. The preferential interaction of cholesterol with different classes of phospholipids. *Biochim Biophys Acta* 465(1977) 1–10.

Dobrosotskaya IY, Seegmiller AC, Brown MS, Goldstein JL, Rawson RB. Regulation of SREBP processing and membrane lipid production by phospholipids in Drosophila. *Science* 296(2002) 879–883.

Dong J, Liu J, Lou B, Li Z, Ye X, Wu M, Jiang XC. Adenovirus-mediated overexpression of sphingomyelin synthases 1 and 2 increases the atherogenic potential in mice. *J Lipid Res* 47(2006) 1307–1314.

Fugmann T, Hausser A, Schoffler P, Schmid S, Pfizenmaier K, Olayioye MA. Regulation of secretory transport by protein kinase D-mediated phosphorylation of the ceramide transfer protein. *J Cell Biol* 178(2007) 15–22.

Ghering AB, Davidson WS. Ceramide structural features required to stimulate ABCA1-mediated cholesterol efflux to apolipoprotein A-I. *J Lipid Res* 47(2006) 2781–2788.

Glaros EN, Kim WS, Quinn CM, Wong J, Gelissen I, Jessup W, Garner B. Glycosphingolipid accumulation inhibits cholesterol efflux via the ABCA1/apolipoprotein A-I pathway: 1-phenyl-2-decanoylamino-3-morpholino-1-propanol is a novel cholesterol efflux accelerator. *J Biol Chem* 280(2005) 24515–24523.

Hanada K. Discovery of the molecular machinery CERT for endoplasmic reticulum-to-Golgi trafficking of ceramide. *Mol Cell Biochem* 286(2006) 23–31.

Hanada K, Kumagai K, Yasuda S, Miura Y, Kawano M, Fukasawa M, Nishijima M. Molecular machinery for non-vesicular trafficking of ceramide. *Nature* 426(2003) 803–809.

Hannun YA, Obeid LM. The Ceramide-centric universe of lipid-mediated cell regulation: stress encounters of the lipid kind. *J Biol Chem* 277(2002) 25847–25850.

Hojjati MR, Li Z, Jiang XC. Serine palmitoyl-CoA transferase (SPT) deficiency and sphingolipid levels in mice. *Biochim Biophys Acta* 1737(2005a) 44–51.

Hojjati MR, Li Z, Zhou H, Tang S, Huan C, Ooi E, Lu S, Jiang XC. Effect of myriocin on plasma sphingolipid metabolism and atherosclerosis in apoE-deficient mice. *J Biol Chem* 280(2005b) 10284–10289.

Hornemann T, Richard S, Rutti MF, Wei Y, von Eckardstein A. Cloning and initial characterization of a new subunit for mammalian serine-palmitoyltransferase. *J Biol Chem* 281(2006) 37275–37281.

Horton JD, Goldstein JL, Brown MS. SREBPs: activators of the complete program of cholesterol and fatty acid synthesis in the liver. *J Clin Invest* 109(2002) 1125–1131.

Horton JD, Shah NA, Warrington JA, Anderson NN, Park SW, Brown MS, Goldstein JL. Combined analysis of oligonucleotide microarray data from transgenic and knockout mice identifies direct SREBP target genes. *Proc Natl Acad Sci U S A* 100(2003) 12027–12032.

Huitema K, van den Dikkenberg J, Brouwers JF, Holthuis JC. Identification of a family of animal sphingomyelin synthases. *Embo J* 23(2004) 33–44.

Jeong T, Schissel SL, Tabas I, Pownall HJ, Tall AR, Jiang X. Increased sphingomyelin content of plasma lipoproteins in apolipoprotein E knockout mice reflects combined production and catabolic defects and enhances reactivity with mammalian sphingomyelinase. *J Clin Invest* 101(1998) 905–912.

Jiang XC, Paultre F, Pearson TA, Reed RG, Francis CK, Lin M, Berglund L, Tall A. Plasma sphingomyelin level as a risk factor for coronary artery disease. *Arterioscler Thromb Vasc Biol* 20(2000) 2614–2618.

Johnson RA, Hamilton JA, Worgall TS, Deckelbaum RJ. Free fatty acids modulate intermembrane trafficking of cholesterol by increasing lipid mobilities: novel 13C NMR analyses of free cholesterol partitioning. *Biochemistry* 42(2003) 1637–1645.

Kawano M, Kumagai K, Nishijima M, Hanada K. Efficient trafficking of ceramide from the endoplasmic reticulum to the Golgi apparatus requires a VAMP-associated protein-interacting FFAT motif of CERT. *J Biol Chem* 281(2006) 30279–30288.

Kunte AS, Matthews KA, Rawson RB. Fatty acid auxotrophy in Drosophila larvae lacking SREBP. *Cell Metab* 3(2006) 439–448.

Lahiri S, Futerman AH. LASS5 is a bona fide dihydroceramide synthase that selectively utilizes palmitoyl-CoA as acyl donor. *J Biol Chem* 280(2005) 33735–33738.

Lange Y, Swaisgood MH, Ramos BV, Steck TL. Plasma membranes contain half the phospholipid and 90% of the cholesterol and sphingomyelin in cultured human fibroblasts. *J Biol Chem* 264(1989) 3786–3793.

Le Stunff H, Giussani P, Maceyka M, Lepine S, Milstien S, Spiegel S. Recycling of sphingosine is regulated by the concerted actions of sphingosine-1-phosphate phosphohydrolase 1 and sphingosine kinase 2. *J Biol Chem* 282(2007) 34372–34380.

Lee CY, Lesimple A, Denis M, Vincent J, Larsen A, Mamer O, Krimbou L, Genest J, Marcil M. Increased sphingomyelin content impairs HDL biogenesis and maturation in human Niemann-Pick disease type B. *J Lipid Res* 47(2006) 622–632.

McGovern MM, Pohl-Worgall T, Deckelbaum RJ, Simpson W, Mendelson D, Desnick RJ, Schuchman EH, Wasserstein MP. Lipid abnormalities in children with types A and B Niemann Pick disease. *J Pediatr* 145(2004) 77–81.

McKay RM, McKay JP, Avery L, Graff JM. C elegans: a model for exploring the genetics of fat storage. *Dev Cell* 4(2003) 131–142.

Merrill AH, Jr., Sullards MC, Allegood JC, Kelly S, Wang E. Sphingolipidomics: high-throughput, structure-specific, and quantitative analysis of sphingolipids by liquid chromatography tandem mass spectrometry. *Methods* 36(2005) 207–224.

Nelson JC, Jiang XC, Tabas I, Tall A, Shea S. Plasma sphingomyelin and subclinical atherosclerosis: findings from the multi-ethnic study of atherosclerosis. *Am J Epidemiol* 163(2006) 903–912.

Ohvo-Rekila H, Ramstedt B, Leppimaki P, Slotte JP. Cholesterol interactions with phospholipids in membranes. *Prog Lipid Res* 41(2002) 66–97.

Pagano RE, Puri V, Dominguez M, Marks DL. Membrane traffic in sphingolipid storage diseases. *Traffic* 1(2000) 807–815.

Park TS, Panek RL, Mueller SB, Hanselman JC, Rosebury WS, Robertson AW, Kindt EK, Homan R, Karathanasis SK, Rekhter MD. Inhibition of sphingomyelin synthesis reduces atherogenesis in apolipoprotein E-knockout mice. *Circulation* 110(2004) 3465–3471.

Park TS, Panek RL, Rekhter MD, Mueller SB, Rosebury WS, Robertson A, Hanselman JC, Kindt E, Homan R, Karathanasis SK. Modulation of lipoprotein metabolism by inhibition of sphingomyelin synthesis in ApoE knockout mice. *Atherosclerosis* 189(2006) 264–272.

Pewzner-Jung Y, Ben-Dor S, Futerman AH. When do Lasses (longevity assurance genes) become CerS (ceramide synthases)? Insights into the regulation of ceramide synthesis. *J Biol Chem* 281(2006) 25001–25005.

Puri V, Jefferson JR, Singh RD, Wheatley CL, Marks DL, Pagano RE. Sphingolipid storage induces accumulation of intracellular cholesterol by stimulating SREBP-1 cleavage. *J Biol Chem* 278(2003) 20961–20970.

Ridgway ND. 25-Hydroxycholesterol stimulates sphingomyelin synthesis in Chinese hamster ovary cells. *J Lipid Res* 36(1995) 1345–1358.

Ridgway ND, Merriam DL. Metabolism of short-chain ceramide and dihydroceramide analogues in Chinese hamster ovary (CHO) cells. *Biochim Biophys Acta* 1256(1995) 57–70.

Rosenwald AG, Pagano RE. Inhibition of glycoprotein traffic through the secretory pathway by ceramide. *J Biol Chem* 268(1993) 4577–4579.

Sakai J, Duncan EA, Rawson RB, Hua X, Brown MS, Goldstein JL. Sterol-regulated release of SREBP-2 from cell membranes requires two sequential cleavages, one within a transmembrane segment. *Cell* 85(1996) 1037–1046.

Schaefer EJ, Lamon-Fava S, Ordovas JM, Cohn SD, Schaefer MM, Castelli WP, Wilson PW. Factors associated with low and elevated plasma high density lipoprotein cholesterol and apolipoprotein A-I levels in the Framingham Offspring Study. *J Lipid Res* 35(1994) 871–882.

Scheek S, Brown MS, Goldstein JL. Sphingomyelin depletion in cultured cells blocks proteolysis of sterol regulatory element binding proteins at site 1. *Proc Natl Acad Sci U S A* 94(1997) 11179–11183.

Schissel SL, Tweedie-Hardman J, Rapp JH, Graham G, Williams KJ, Tabas I. Rabbit aorta and human atherosclerotic lesions hydrolyze the sphingomyelin of retained low-density lipoprotein. Proposed role for arterial-wall sphingomyelinase in subendothelial retention and aggregation of atherogenic lipoproteins. *J Clin Invest* 98(1996) 1455–1464.

Schuchmann EH, Desnick RJ: Niemann-Pick Diseases Type A and B: Acid sphingomyelinase deficiencies, Vol. 2. New York: McGraw Hill; (1995) 2601–2624.

Shayman JA, Lee L, Abe A, Shu L. Inhibitors of glucosylceramide synthase. *Methods Enzymol* 311(2000) 373–387.

Soccio RE, Breslow JL. StAR-related lipid transfer (START) proteins: mediators of intracellular lipid metabolism. *J Biol Chem* 278(2003) 22183–22186.

Sot J, Aranda FJ, Collado MI, Goni FM, Alonso A. Different effects of long- and short-chain ceramides on the gel-fluid and lamellar-hexagonal transitions of phospholipids: a calorimetric, NMR, and x-ray diffraction study. *Biophys J* 88(2005) 3368–3380.

Spassieva S, Seo JG, Jiang JC, Bielawski J, Alvarez-Vasquez F, Jazwinski SM, Hannun YA, Obeid LM. Necessary role for the Lag1p motif in (dihydro)ceramide synthase activity. *J Biol Chem* 281(2006) 33931–33938.

Tafesse FG, Ternes P, Holthuis JC. The multigenic sphingomyelin synthase family. *J Biol Chem* 281(2006) 29421–29425.

van Meer G. Lipid traffic in animal cells. *Annu Rev Cell Biol* 5(1989) 247–275.

Venkataraman K, Riebeling C, Bodennec J, Riezman H, Allegood JC, Sullards MC, Merrill AH, Jr., Futerman AH. Upstream of growth and differentiation factor 1 (uog1), a mammalian homolog of the yeast longevity assurance gene 1 (LAG1), regulates N-stearoyl-sphinganine (C18-(dihydro)ceramide) synthesis in a fumonisin B1-independent manner in mammalian cells. *J Biol Chem* 277(2002) 35642–35649.

Witting SR, Maiorano JN, Davidson WS. Ceramide enhances cholesterol efflux to apolipoprotein A-I by increasing the cell surface presence of ATP-binding cassette transporter A1. *J Biol Chem* 278(2003) 40121–40127.

Worgall TS, Sturley SL, Seo T, Osborne TF, Deckelbaum RJ. Polyunsaturated fatty acids decrease expression of promoters with sterol regulatory elements by decreasing levels of mature sterol regulatory element-binding protein. *J Biol Chem* 273(1998) 25537–25540.

Worgall TS, Juliano RA, Seo T, Deckelbaum RJ. Ceramide synthesis correlates with the posttranscriptional regulation of the sterol-regulatory element-binding protein. *Arterioscler Thromb Vasc Biol* 24(2004a) 943–948.

Worgall TS, Davis-Hayman SR, Magana MM, Oelkers PM, Zapata F, Juliano RA, Osborne TF, Nash TE, Deckelbaum RJ. Sterol and fatty acid regulatory pathways in a Giardia lamblia-derived promoter: evidence for SREBP as an ancient transcription factor. *J Lipid Res* 45(2004b) 981–988.

Zanolari B, Friant S, Funato K, Sutterlin C, Stevenson BJ, Riezman H. Sphingoid base synthesis requirement for endocytosis in Saccharomyces cerevisiae. *Embo J* 19(2000) 2824–2833.

# Chapter 15
# Multiple Roles for Sphingolipids in Steroid Hormone Biosynthesis

**Natasha C. Lucki and Marion B. Sewer**

**Abstract** Steroid hormones are essential regulators of a vast number of physiological processes. The biosynthesis of these chemical messengers occurs in specialized steroidogenic tissues via a multi-step process that is catalyzed by members of the cytochrome P450 superfamily of monooxygenases and hydroxysteroid dehydrogenases. Though numerous signaling mediators, including cytokines and growth factors control steroidogenesis, trophic peptide hormones are the primary regulators of steroid hormone production. These peptide hormones activate a cAMP/cAMP-dependent kinase (PKA) signaling pathway, however, studies have shown that crosstalk between multiple signal transduction pathways and signaling molecules modulates optimal steroidogenic capacity. Sphingolipids such as ceramide, sphingosine, sphingosine-1-phosphate, sphingomyelin, and gangliosides have been shown to control the steroid hormone biosynthetic pathway at multiple levels, including regulating steroidogenic gene expression and activity as well as acting as second messengers in signaling cascades. In this review, we provide an overview of recent studies that have investigated the role of sphingolipids in adrenal, gonadal, and neural steroidogenesis.

**Keywords** Steroidogenesis · sphingolipids · CYP · sphingosine-1-phosphate · ceramide

## 15.1 Introduction

Steroid hormones like testosterone, progesterone, cortisol, aldosterone, and estradiol are important endocrine chemical messengers that are involved in a vast number of physiological processes including metabolism, inflammation, electrolyte and fluid balance, and secondary sex differentiation (Foster, 2004; Ghayee and Auchus, 2007; Newton and Holden, 2007; Williams-Ashman and

N.C. Lucki
School of Biology and Parker H, Petit Institute for Bioengineering and Bioscience, Georgia Institute of Technology, 310 Ferst Drive, Atlanta, GA, 30332-0230, USA

P.J. Quinn, X. Wang (eds.), *Lipids in Health and Disease*,

Reddi, 1971). Steroid hormone synthesis occurs in the gonads, adrenal gland, placenta, intestines (Abdallah et al., 2004; Ghayee and Auchus, 2007; Mueller et al., 2007; Payne and Hales, 2004), and has recently been characterized in brain and peripheral nervous system tissues (Mellon et al., 2004; Mellon, 2007; Tsutsui et al., 2000). Biosynthesis is catalyzed by the sequential activities of cytochrome P450 monooxygenases and hydroxysteroid dehydrogenases that convert cholesterol into the many steroid hormones. Selective expression of these steroidogenic enzymes assures the production of steroid hormones in a tissue-specific manner. Activation of steroidogenesis is initiated by the binding of trophic peptide hormones – adrenocorticotrophin (ACTH), leutenizing hormone (LH), follicle stimulating hormone (FSH) – derived from the anterior pituitary to cognate receptors in target tissues which activates a cAMP/cAMP-dependent protein kinase (PKA) signaling pathway. Activation of cAMP-dependent signaling leads to a rapid increase in cholesterol mobilization and a chronic induction of steroidogenic gene transcription (Fig. 15.1). Although this cAMP-dependent pathway is the main regulator of steroid hormone production, many other signaling systems involving many cytokines and sphingolipids have been reported to modulate steroidogenesis (Sewer et al., 2007).

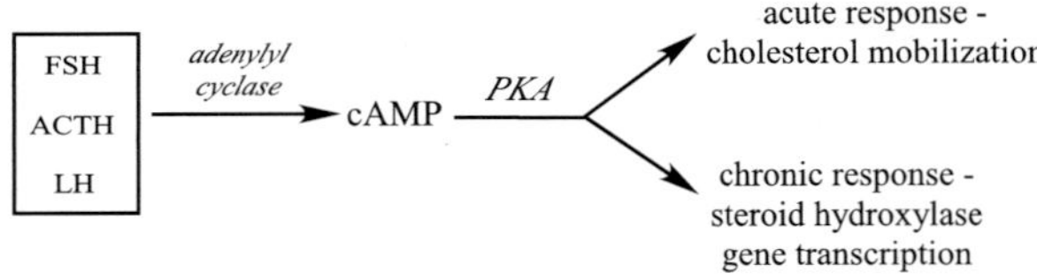


**Fig. 15.1** Temporal regulation of steroidogenesis by peptide hormones

Sphingolipids, a family of lipids with a common sphingoid base backbone, have recently been identified as important bioactive molecules involved in a variety of cellular processes, including steroidogenesis (Ozbay et al., 2004; Urs et al., 2007; Ledeen and Wu, 2006; Spiegel and Milstein, 2007, 2003b; Zheng et al., 2006). Sphingolipids such as ceramide (Cer), sphingosine (SPH), sphingosine-1-phosphate (S1P), sphingomyelin (SM), and gangliosides (GMs) have been shown to modulate the steroidogenic pathway at multiple levels including regulating steroidogenic gene expression and activity as well as acting as secondary messengers in signaling cascades. In this review, we will provide a summary of recent studies that have investigated the role of sphingolipids in steroid hormone biosynthesis, with special emphasis on the role of sphingolipids in adrenal, gonadal, and neural steroidogenesis.

## 15.2 Review of Steroidogenesis

Steroidogenesis is a highly regulated biological process that is required for physiological homeostasis. This process takes place in specialized steroidogenic tissues where the biosynthesis of steroid hormones occurs in a highly synchronized manner via the coordinated activity of a series of steroidogenic enzymes. The primary steroidogenic tissues involved in the *de novo* steroid hormone biosynthesis include the gonads (ovaries and testis), the adrenal glands, and the placenta. Cholesterol, the substrate for the synthesis of all steroid hormones, is differentially metabolized into steroid hormones via the concerted action of enzymes localized in these steroidogenic centers (Fig. 15.2).

Activation of steroidogenesis is initiated upon binding of peptide trophic hormones to their cognate G protein-coupled receptors (GPCRs) in the target tissues. As shown in Fig. 15.1, peptide hormones activate two temporally distinct phases of steroid production: a rapid acute response and a slower chronic phase. The acute phase of steroidogenesis involves activation of the steroidogenic acute regulatory protein (StAR) for rapid cholesterol mobilization from the outer to the inner mitochondrial membrane (Miller, 2007; Sewer and Waterman, 2001; Thomson, 1997). Because cholesterol transport to the inner mitochondrial membrane is the rate-limiting step in steroid hormone production, StAR is a key protein necessary for the acute response. In most steroidogenic tissues, StAR expression is mediated by cAMP; in addition, intracellular $Ca^{2+}$ may also play a role in StAR transcription in the adrenal cortex. In addition to StAR, the peripheral benzodiazepine receptor (PBR) and

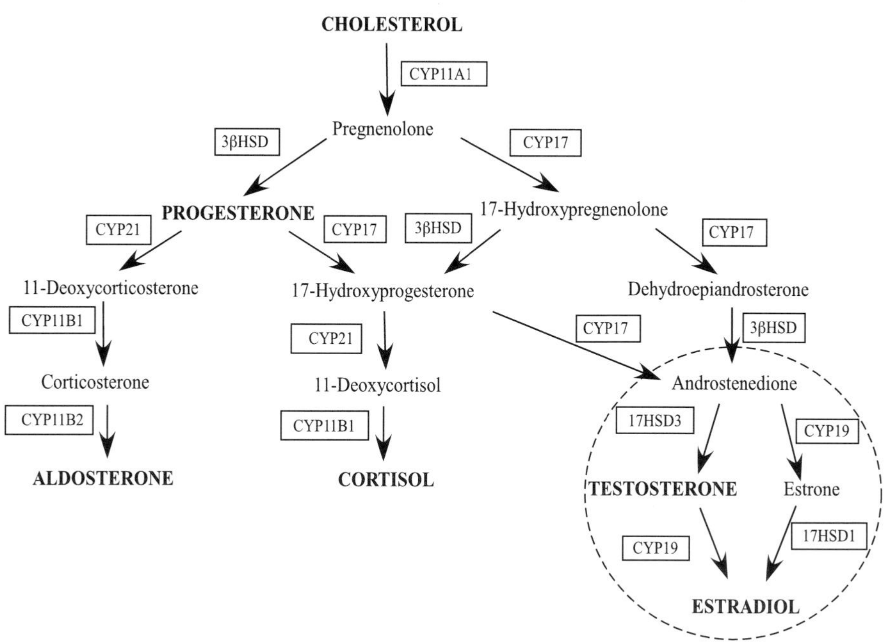


**Fig. 15.2** Steroid hormone biosynthetic pathways

hormone-sensitive lipase (HSL) are essential proteins for the intracellular transport and production of free cholesterol, respectively (Kraemer et al., 2004; Papadopoulos, 1993). In rodents, the scavenger receptor class B type I (SR-BI) plays a key role in the uptake of cholesterol esters from lipoproteins (Krieger, 1999). The de-esterification of cholesterol esters by HSL is essential for its transport into the mitochondria and therefore its utilization in steroidogenesis (Kraemer et al., 2004). Once cholesterol-laden vesicles are at the outer mitochondrial membrane, a large macromolecular complex containing StAR, PBR, and voltage-dependent anion channel (VDAC) (Hauet et al., 2002; Liu et al., 2003b, 2006; Miller, 2007) facilitate import of the substrate into the inner mitochondrial membrane.

The chronic phase of steroidogenesis involves the transcriptional activation of steroidogenic genes that are responsible for cholesterol metabolism. This occurs via activation of adenylyl cyclase and signaling through multiple signaling mechanisms, including a cAMP/PKA pathway, leading to the activation of varied downstream effectors that ultimately promote the binding of transcription factors to the promoters of steroidogenic genes (Arlt and Stewart, 2005; Bassett et al., 2004a; Bornstein et al., 2004; Condon et al., 2002; Jamnongjit and Hammes, 2006; Mendelson et al., 2005; Okamoto et al., 2004; Otis and Gallo-Payet, 2007; Sewer et al., 2007; Sewer and Waterman, 2003; Sirianni et al., 2003). One of the major transcription factors that regulates the expression of most steroidogenic genes in the adrenal gland and gonads of mammals is the nuclear receptor steroidogenic factor-1 (SF1/Ad4BP/NR5A1) (Lala et al., 1992; Morohashi et al., 1992; Parker et al., 2002). The ability of SF-1 to bind to target genes is regulated by post-translational modifications including phosphorylation and acetylation (Chen et al., 2004; Hammer et al., 1999; Ishihara and Morohashi, 2005). More recently, ligand binding has also been implicated in the regulation of SF1 activity (Ishihara and Morohashi, 2005; Krylova et al., 2005; Li et al., 2005, 2007; Urs et al., 2006).

### *15.2.1 Adrenal Steroidogenesis*

The adrenal gland is divided into two regions: cortex and medulla. The medulla, which comprises about 10% of the gland, is made up of neuroendocrine cells that synthesize catecholamines. The adrenal cortex, on the other hand, comprises most of the adrenal gland and is the site of adrenal steroid hormone biosynthesis. This region can be further subdivided into 3 distinct zones, each with a characteristic steroidogenic profile: (1) the zona glomerulosa is the outer cortical zone where the mineralcorticoid aldosterone is produced. (2) The middle zone, zona fasciculata, makes glucocorticoids, (3) while the inner zone, zona reticularis, is the site of androgen biosynthesis. Each of the three cortical zones expresses a unique profile of steroidogenic genes, thereby allowing for zone-specific cholesterol metabolism (Bassett et al., 2004b; Rainey, 1999).

Adrenocortical steroid hormones have a vast array of biological functions. Cortisol, the primary human glucocorticoid, regulates the inflammatory response (Newton and Holden, 2007), carbohydrate and lipid metabolism, and stress response (Kassel and Herrlich, 2007). Aldosterone regulates blood pressure by modulating fluid and electrolyte balance (Brizuela et al., 2006; Foster, 2004). In the adrenal cortex, dehydroepiandrosterone (DHEA), dehydroepiandrosterone sulfate (DHEA-S), and androstenedione are the androgens produced (Havelock et al., 2004; Rainey et al., 2002).

In the zona fasciculata and reticularis of the adrenal cortex, steroidogenesis is mainly regulated by the binding of ACTH to the melanocortin-2 receptor, whereas in the zona glomerulosa, angiotensin II (AngII) directs aldosterone production by binding to angiotensin receptors. As shown in Fig. 15.2, there are two classes of steroidogenic enzymes whose transcription is activated in the chronic phase of steroid hormone production: the cytochrome P450 heme-containing proteins (CYPs) and hydroxysteroid dehydrogenase (HSD) enzymes (Payne and Hales, 2004; Sewer and Waterman, 2003). Additionally, the expression of StAR (Caron et al., 1997; Clark and Combs, 1999; Clem et al., 2005; Reinhart et al., 1999), PBR (Besman et al., 1989), HSL (Kraemer et al., 2004), SR-BI (Azhar et al., 1998), and adrenodoxin (Brentano and Miller, 1992; Chen and Waterman, 1992), the iron-sulfur electron transfer protein, are also induced by trophic hormone stimulation. The P450 side chain cleavage enzyme (encoded by CYP11A1) is an inner mitochondria membrane-bound enzyme that catalyzes the first enzymatic reaction in the synthesis of all steroid hormones: cleavage of free cholesterol into pregnenolone. P450c17α is encoded by CYP17 and is localized in the endoplasmic reticulum (ER). P450c17α catalyzes the hydroxylation of progesterone and pregnenolone at the carbon-17 and the conversion of pregnenolone to DHEA in the zona reticularis and the conversion of progesterone into androstenedione in the zona fasciculata. The microsomal 3β-hydroxysteroid dehydrogenase (3βHSD) catalyzes the conversion of pregnenolone, 17α-hydroxypregnenolone, and DHEA into progesterone, 17α-hydroxyprogesterone, and androstenedione, respectively. P45021 hydroxylase, encoded by CYP21, is also microsomal and catalyzes the conversion of progesterone and 17α-hydroxyprogesterone into 11-deoxycorticosterone and 11-deoxycortisol, respectively. P45011-β-hydroxylase (CYP11B1) is localized at the inner mitochondrial membrane in the zona fasciculata and converts 11-deoxycorticosterone or 11-deoxycortisol into corticosterone or cortisol. In the zona glomerulosa, aldosterone synthase (encoded by CYP11B2) is expressed in the inner mitochondrial membrane and catalyzes the conversion of 11-deoxycorticosterone into aldosterone. As previously discussed, the zone-specific expression of CYP11B1 and CYP17 in the zona fasciculata and reticularis and CYP11B2 in the zona glomerulosa allow for the differential steroid hormone biosynthesis.

### 15.2.2 Gonadal Steroidogenesis

The primary sites of androgen and estrogen biosynthesis are the testis and ovaries. Analogous to adrenocortical steroidogenesis, tissue-specific steroidogenic gene expression accounts for differential production of sex hormones. Also, gonadal steroidogenesis occurs via the two above-mentioned temporally distinct phases (acute and chronic) that assures proper and controlled steroid hormone output. In the gonads, LH and FSH regulate acute and chronic steroidogenesis. LH and FSH activate an adenylyl-cyclase/cAMP-dependent pathway in the same manner as ACTH in the adrenal cortex (Jamnongjit and Hammes, 2006; Mendelson et al., 2005; Sewer and Waterman, 2003; Tajima et al., 2005). The steroid hormones produced in the gonads -testosterone, estradiol, and progesterone – function primarily in the control of secondary male and female sex characteristics and in embryogenesis.

As in the adrenal cortex, a specific pattern of steroidogenic enzymes expression allow for the production of the gonadal steroid hormones in a cell-specific manner. Some enzymes expressed in the adrenal are also equally expressed in the gonads: CYP11A1 and 3βHSD are expressed in the ovaries and testis and catalyzes the same reactions as in the adrenal cortex (Fig. 15.2). CYP17 is also expressed in the Leydig cells of the testis and theca cells of the ovary but, in contrast to the 17α-hydroxylase reaction that is prevalent in the adrenal cortex, the lyase reaction predominates, resulting in the convertion of pregnenolone into androstenedione. Moreover, additional gonadal-specific enzymes, such as aromatase (CYP19) and 17α-HSD types 1 and 3 direct the production of gonadal-specific hormones. Testosterone biosynthesis is terminated in the Leydig cells of the testis by the activity of 17αHSD3, which catalyses the conversion of androstenedione to testosterone. In ovarian granulosa cells, aromatase catalyzes the conversion of androstenedione or testosterone into estrone or estradiol, respectively, while 17αHSD1 converts estrone to estradiol (Fig. 15.2).

### 15.2.3 Neurosteroidogenesis

Adrenal and gonadal steroid hormones have long been known to regulate many important brain functions (Fuxe et al., 1981; McEwen, 1991). In addition, circulating steroids like progesterone, 11-deoxycorticosterone, and testosterone can be converted to neuroactive steroids within the brain (Mellon and Griffin, 2002). More recently, however, many laboratories have reported that nervous tissue is capable of expressing essential steroidogenic enzymes and therefore *de novo* synthesize steroid hormones, that are at least in part independent from classic steroidogenic tissues (Mellon and Griffin, 2002; Tsutsui et al., 2000). Moreover, the expression of StAR has also been detected in neural tissues (Lavaque et al., 2006; Sierra, 2004). Today, the term neurosteroid refers to both *de novo* synthesized steroids by nervous cells and circulating steroids that are subsequently converted to neuroactive forms within nervous tissues.

Neurosteroids appear to mainly function as neurotransmitters in a paracrine and autocrine fashion in the modulation of many brain functions including myelination, inhibition of neuronal toxicity and ischemia, behavioral aspects, and neuronal survival, growth, and differentiation (Griffin et al., 2004; Mellon et al., 2004; Mellon, 2007; Mukai et al., 2006). Such neurosteroids include progesterone, pregnenolone, allopregnanolone, DHEA, their sulfate esters, and 5α/5β-tetrahydroprogesterone, some of which were previously viewed as inactive metabolites or steroid precursors (Plassart-Schiess and Baulieu, 2001; Sakamoto et al., 2007).

Glial cells, oligodentrocytes and type I astrocytes, are considered the major neuronal steroidogenic cells (Tsutsui and Ukena, 1999). However, other cell types including Schwann cells, cerebellar Purkinje cells, and neurons have also been reported as capable of steroid hormone production (Mellon and Griffin, 2002; Plassart-Schiess and Baulieu, 2001). Purkinje cells, for example, are one of the major site of *de novo* progesterone and pregnenolone sulfate production (Tsutsui and Ukena, 1999).

The neurosteroidogenic pathway involves the same group of cytochrome P450 and HSDs enzymes as in classical steroidogenic tissues. Tissue-specific expression of a unique panel of steroidogenic enzymes also occurs in different brain regions and nerve cells (Mellon and Deschepper, 1993). However, some enzymes are expressed at higher levels in the brain than in other steroidogenic tissues and vary during development (Mellon et al., 2004). CYP11A1, 3β-HSD, and CYP17 are expressed in many brain regions including the cortex, cerebellum, and hypothalamus (Mellon and Griffin, 2002). These three brain regions also express CYP11B1 and CYP11B2 as well as StAR (reviewed in ref. (Mellon and Griffin, 2002)).

A significant amount of data relating the function of neurosteroids to nervous system development comes from the Niemann-Pick Type C-1 (NPC-1) knockout mouse model (Griffin et al., 2004; Mellon et al., 2004). This mouse has a mutation in the npc1 gene, which codes for a late endosomal membrane protein that is involved in the trafficking of cholesterol out of late endosomes for steroidogenesis (Blanchette-Mackie, 2000). Mutation of this gene leads to the accumulation of cholesterol and GMs in lysosomes and impaired neurosteroidogenesis (Mellon et al., 2004).

### 15.2.4 Other Steroidogenic Tissues

The liver, placenta, and small intestines have also been reported as steroidogenic centers where a selective set of cytochrome P450 and HSD enzymes are expressed. CYP17 is expressed in the rat liver and this expression fluctuates during development, indicating a pattern of expression exclusively tuned for the needs of this tissue (Vianello et al., 1997). Mueller et al., (Mueller et al., 2007) reported that murine intestinal epithelial cells are capable of producing glucocorticoids utilizing a differently regulated steroidogenic pathway than in adrenocortical cells. cAMP accumulation abrogates glucocorticoid production in this intestinal cells, illustrating a diversification of the classical steroidogenic pathway, which has likely evolved as a result of the different requirements of their environment. The placenta is an important steroidogenic tissue during pregnancy, which secretes human chorionic gonadotropin (hCG) hormone that stimulates progesterone production by the corpus luteum (Pepe and Albrecht, 1995). It has been reported that in addition to the corpus luteum, other non-gonadal tissues including the kidney, lungs, pancreas and liver express hCG receptors (Abdallah et al., 2004). Although the role of these receptors is unknown, the expression of steroidogenic genes in these tissues suggests that placental hCG may mediate additional pleiotropic effects in the developing fetus.

## 15.3 Review of Sphingolipids

Sphingolipids comprise a family of phospholipids and glycolipids that are characterized by the presence of a common sphingoid base (such as SPH) backbone. This family of lipids has a large structural diversity that allows for the existence of many structurally similar moieties yet with crucial differences in biochemical and biophysical properties. The role of this class of lipids in membrane structure is well-established (Goni and Alonso, 2006). Although the precise amount may vary considerably, as high as 30% of lipids that make up plasma membranes are sphingolipids, especially complex sphingolipids such as SM and glycosphingolipids (GSL) (Smith and Merrill, 2002). Furthermore, certain sphingolipids like SM, Cer, and glucosylceramides may aggregate to form higher order domains termed “lipid rafts”, which are believed to be important in cell signaling (Huwiler et al., 2000; Smith and Merrill, 2002; Tani et al., 2007). Significantly, a large body of data has established a role for sphingolipids as key bioactive molecules in a variety of biological processes (Brizuela et al., 2006; Budnik et al., 1999; Cuvillier et al., 1996; Degnan et al., 1996; Gomez-Munoz, 2006; Hannun, 1996; Kihara et al., 2007; Meroni et al., 2000; Rabano et al., 2003; Thon et al., 2005) (Table 15.1). These processes include cell differentiation, growth, apoptosis, cell-cell interaction, and mediation of signaling pathways and gene expression (Merrill et al., 1999; Zeidan and Hannun, 2007; Zheng et al., 2006). Roles for sphingolipids in vascular function (Lorenz et al., 2007), neurodegeneration (Tamboli et al., 2005), cancer [reviewed in ref. (Ogretmen, 2006)], autophagy (Lavieu et al., 2006, 2007), and insulin resistance (Adams et al., 2004; Turinsky et al., 1990) have also been reported, which illustrate the broad spectrum of cellular processes that can be directed or indirectly mediated by these bioactive lipid molecules. In

**Table 15.1** Summary of recent data obtained for selected bioactive sphingolipids in steroidogenesis as well as their multiple downstream molecular targets and upstream molecular inducers

| Sphingolipid | Role in steroidogenesis | Upstream inducers | Downstream targets |
|---|---|---|---|
| Cer | Suppress progesterone and testosterone production; attenuate StAR expression; regulate CYP17α activity; inhibit cAMP production; regulate 11β-HSD1 and CYP19arom activity. | TNF-α, Fas ligand, INF-γ, IL-1β, SMase | ERK1/2, SAPK, p38, c-Jun, caspases, CAPK, CAPP |
| S1P | Upregulate CYP17 and CYP19 expression; increase cortisol, estrogen, and aldosterone secretion; | SK1/2 activity. PKC, ERK2, and external S1PR ligands regulate SK activity. | ERK and PI3K/Akt pathways, PKC, PLD, PLC, S1PR ligand |
| SPH | Antagonist ligand | | SF1 |
| SM | Precursor of ceramide | Acid/neutral/alkaline ceramidase enzymatic activity | |
| GMs | Neurosteroidogenesis: degradation by allopregnanolone | allopregnanolone | |

[1]Abbreviations used: tumor necrosis factor-α (TNF-α), interferon-γ (INF-γ), interleukin-1β (IL-1β), sphingosine kinase (SK), protein kinase C (PKC), sphingomyelinase (SMase), extracellular-regulated mitogen-activated protein kinase (ERK), stress-activated protein kinase (SAPK), ceramide activated protein kinase (CAPK), ceramide activated protein phosphatase (CAPP), mitogen activated kinase (MEK), phosphatidylinositol-3-kinase (PI3K), phospholipase C (PLC), phospholipase D (PLD), S1PR (S1P receptor).

addition, a series of recent studies examining the role of sphingolipids, primarily Cer and S1P, in adrenal and gonadal steroidogenesis (Brizuela et al., 2007; Brizuela et al., 2006; Budnik et al., 1999; Li et al., 2001; Meroni et al., 2000; Ozbay et al., 2006; Rabano et al., 2003) add to a rapidly expanding list of roles for sphingolipids in cellular processes.

The simplest sphingolipids, SPH ((2S, 3R, 4E)-2-aminooctadec-4-ene-1,3-diol - Ser) and Cer (N-acylsphingosine) shown in Fig. 15.3, constitute the basic structure of higher order sphingolipids like SM, cerebrosides, GMs, and other GSLs (Goni and Alonso, 2006). Complex sphingolipids are synthesized by Cer metabolism. Cer can be produced by the degradation of SM or by *de novo* biosynthesis. *De novo* Cer biosynthesis occurs from the condensation of serine and palmitoyl-CoA by the sequential action of the enzymes serine palmitoyltransferase, ketodihydro-sphingosine reductase, and ceramide synthase (Hanada et al., 2003) (Fig. 15.3). SM hydrolysis is the major cellular source of Cer and it can be activated in response to a variety of extracellular signals, e.g. vitamin D3, cytokines, growth factors, and cytotoxic agents (Andrieu-Abadie

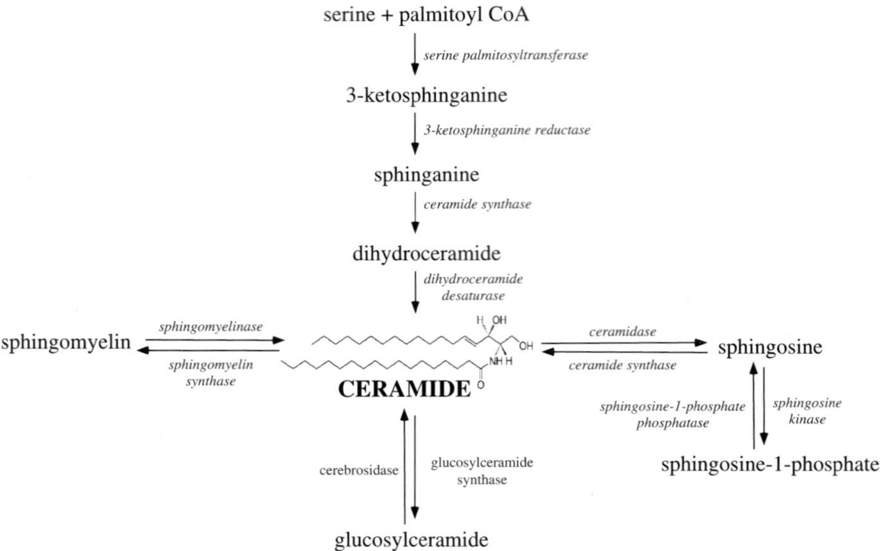


**Fig. 15.3** Overview of sphingolipid metabolic pathways

and Levade, 2002; Huwiler et al., 2000; Okazaki et al., 1989). Alternatively, SPH formed as a product of complex sphingolipid metabolism can be recycled back into Cer via the scavenger pathway (Zeidan and Hannun, 2007).

The amounts of different sphingolipid moieties within a cell is controlled by multiple enzymes including sphingomyelinases (SMases) that hydrolyze SM into Cer; ceramidases that convert Cer into SPH and a free fatty acid; ceramide synthase which catalyzes the *de novo* synthesis of Cer from SPH; sphingomyelinase synthase that catalyses the production of SM; and sphingosine kinases (SKs) and ceramide kinases (Cerk) that phosphorylate SPH and Cer, respectively (Fig. 15.3) (Huwiler et al., 2000; Maceyka et al., 2002; Maceyka et al., 2005; Pettus et al., 2003; Spiegel and Milstien, 2007). All of the above-mentioned enzymes play a central role in regulating the many bioactive sphingolipid types. Because the activity of these enzymes can be regulated by multiple signaling factors (Huwiler et al., 2000; Maceyka et al., 2002, 2005; Pettus et al., 2003; Spiegel and Milstien, 2007), the sphingolipid metabolic profile in any given cell at any given point in time is comprised of a unique set of bioactive sphingolipids. Therefore, the amounts and molecular species of the different sphingolipids are in a constant dynamic flux.

Of all known sphingolipid moieties, a large body of research has established physiological roles for SPH, Cer, S1P, and C1P (Adams et al., 2004; Andrieu-Abadie and Levade, 2002; Brizuela et al., 2006, 2007; Budnik et al., 1999; Hadizadeh et al., 2007; Kihara et al., 2007; Li et al., 2001; Meroni et al., 2000; Ozbay et al., 2006; Pettus et al., 2003; Rabano et al., 2003). Although these molecules are structurally similar and can be interconverted by one or two-step reactions, they have unique, and sometimes opposing, biological functions (reviewed in ref. (Kihara et al., 2007)). As shown in Fig. 15.3, the dynamic balance of these sphingolipid metabolites is maintained by enzymes such as ceramidases, SKs, and S1P phosphatases whose activity is regulated by a variety of intra- and extracellular signals including TNF-α, Fas ligand, and cytokines (Cai et al., 1997, 2007; Hannun, 1996; Ruvolo et al., 2002; Sawai et al., 1997; Spiegel and Milstien, 2003a).

SM, the most abundant sphingolipid in mammalian cells, has also been implicated as a bioactive sphingolipid (Degnan et al., 1996; Ding et al., 2007; Porn et al., 1991). SMase activity is triggered by diverse stimuli, which results in the cleavage of membrane SM into free Cer. This process, now called the "sphingomyelin cycle", is a receptor-mediated signaling system that plays an important role in regulating sphingolipids homeostasis (Andrieu-Abadie and Levade, 2002; Hannun, 1996; Ziulkoski et al., 2001). Cytokines such as TNF-α and interleukin 1β (IL-1β) have been found to activate SMase activity (Santana et al., 1996; Zeidan et al., 2006).

Cer, aside from the precursor of complex sphingolipids, is a second messenger molecule involved in a series of cellular events including differentiation, senescence, proliferation, mediation of stress response, cell cycle arrest, and apoptosis (Gomez-Munoz, 2006; Kolesnick, 2002; Pettus et al., 2002). Cytokines and fatty acids have been reported as mediators of intracellular ceramide production and subsequent activity (Lu et al., 2003; Osawa et al., 2005; Zeidan et al., 2006). Conversely, C1P, a major metabolite of Cer produced by Cerk-catalytic phosphorylation of Cer, has been identified as having antagonistic effects than Cer by being an inhibitor of apoptosis and a cell survival inducer (Gomez-Munoz et al., 2004). C1P has also been reported to play a role in inflammation and phagocytosis (Gomez-Munoz, 2006) and in arachidonic acid release (AA) and prostanoid production (Pettus et al., 2003). Since a role for AA in steroid hormone production stimulation has been found (Castilla et al., 2004), C1P may be a regulator of steroidogenesis.

Like Cer, SPH acts as a pro-apoptotic signal (Hung et al., 1999; Sakakura et al., 1998; Sweeney et al., 1998) as well as an inhibitor of protein kinase C (PKC) (Hannun et al., 1986), phospholipase D (PLD) (Natarajan et al., 1994), and calmodulin-dependent kinase (Jefferson and Schulman, 1988) in diverse cell types. SPH has also been shown to activate diacylglycerol (DAG) kinase (Yamada and Sakane, 1993). S1P is involved in cell survival and proliferation (Olivera et al., 1999; Olivera and Spiegel, 1993; Spiegel and Milstien, 2002). In contrast to Cer, which is membrane-bound, S1P can diffuse into the cytosol and be secreted into the extracellular space, where it can exert its signaling properties by binding to cell surface receptors (reviewed in ref. (Kihara et al., 2007)).

S1P is formed by phosphorylation of SPH by SKs. There are two isoforms of SK (1 and 2) whose function and subcellular location are distinct (Liu et al., 2003a; Maceyka et al., 2005; Okada et al., 2005). SK2 prevents apoptosis and leads to increase in S1P metabolism back to ceramide (Spiegel and Milstien, 2007). SK1, however, appears to be a critical regulator in the intracellular amounts of S1P and its precursors SPH and Cer. Overexpresion of SK1 leads to accumulation of S1P and tumorgenesis (Le Stunff et al., 2007; Spiegel and Milstien, 2007). Therefore, there is evidence to suggest that these two SK isoforms may have distinct physiological functions and effect on sphingolipid metabolism. External ligands including acetylcholamine, prosaposin, lysophosphatidic acid (LPA), formylmethionine peptide, and even S1P itself have been reported as agonists of SK activity via GPCR signaling (Maceyka et al., 2002). Downstream targets of GPCR signaling are still under investigation, but PKC and ERK2 activation have been shown to regulate SK1 activity (Spiegel and Milstien, 2007) and may be important in regulating the dynamic balance of intracellular S1P, SPH, and Cer.

Given the opposite functions of Cer/SPH and S1P in cell viability, it has been proposed that the balance of these lipid mediators is a key factor in determining cell fate toward proliferation or apoptosis (Cuvillier et al., 1996). As a result of such important role in cell viability, these sphingolipid metabolites are under extensive investigation as potential anti-cancer targets (Mimeault, 2002; Pettus et al., 2002; Thon et al., 2005). Nonetheless, the dynamic balance of these sphingolipid molecules will also affect other biological processes including steroidogenesis.

## 15.4 The Role of Sphingolipids in Steroid Hormone Production

As previously summarized, steroidogenesis involves the binding of the trophic hormones ACTH, LH, and FSH to their cognate receptors which lead to the subsequent activation of a series of cascade pathways (Fig. 15.1), primarily the cAMP-dependent/PKA pathway leading to the activation of many downstream targets. Calcium is also important for maximal steroidogenesis (Gallo-Payet and Payet, 1989). Additionally, cAMP-independent signaling systems play an integral role in the regulation of steroidogenesis (Bornstein et al., 2004; Okamoto et al., 2004; Otis and Gallo-Payet, 2007). Interleukins (IL-3, IL-6, IL-1β, TNF-α) (Budnik et al., 1999; Hedger, 1997; Weber et al., 1997), calmidazolium (Choi and Cooke, 1992), steroidogenic-inducing protein (SIP) (Stocco and Khan, 1992), and chloride ions (Choi and Cooke, 1990; Gallo-Payet et al., 1999; Ramnath et al., 1997) are a few of the many regulators of steroid hormone production.

In addition, recent data suggests that sphingolipids can also act as secondary modulators of steroid hormone production. In the adrenal cortex, for example, ACTH/cAMP rapidly activates sphingolipid metabolism by decreasing intracellular amounts of SM, Cer, and SPH while increasing S1P production via SK activation (Ozbay et al., 2004, 2006). Bioactive sphingolipids have been identified as modulators of steroidogenesis by acting at different levels of the steroidogenic signaling pathway (Table 15.1). Some points of regulation include: (1) regulating cytochrome P450 gene expression, (2) serving as ligands for the major steroidogenic transcription factor SF-1, and (3) participating in secondary signaling cascades and second messenger systems. Some of these key finding are described below.

### *15.4.1 Ceramide*

Current data suggests that the primary role of Cer in steroidogenesis is as a mediator in cytokine and growth factors signaling pathways, which untimely lead to a change in basal steroid hormone production (Arai et al., 2007; Budnik et al., 1999; Cai et al., 1997; Degnan et al., 1996; Meroni et al., 2000; Santana et al., 1995, 1996). TNF-α, Fas ligand, interferon-γ (INF-γ), and IL-1β modulate intracellular Cer concentrations via the activation of SMases (Cai et al., 1997; Hannun, 1996; Sawai et al., 1997). Of note, although some of the end point results of Cer accumulation in steroid production have been reported (Budnik et al., 1999; Cai et al., 1997; Degnan et al., 1996; Meroni et al., 2000; Santana et al., 1995, 1996), the precise molecular mechanism of action of Cer is still mostly unknown.

Cer has been shown to negatively regulate progesterone production in granulosa cells (Santana et al., 1996). The accumulation of intracellular Cer is a result of activation of SM hydrolysis by IL-1β. Similarly, TNF- α, which has many overlapping cellular effects with IL-1β, was shown to activate SMases and generate intracellular Cer in both MA-10 murine Leydig cells (Degnan et al., 1996) and Jeg-3 human choriocarcinoma cells (McClellan et al., 1997). Cer was also reported to suppress human choriogonadotropin (hCG)-stimulated testosterone production in rat Leydig cells (Meroni et al., 2000) and progesterone production in rat luteal cells in a dose-dependent manner (Li et al., 2001). Budnik et al. (Budnik et al., 1999) reported that regulation of progesterone biosynthesis by TNF- α/Cer signaling occurs via inhibition of StAR expression in MA-10 cells. Attenuation of StAR expression and reduced testosterone secretion by TNF-α and Cer was also seen in rat Leydig cells (Morales et al., 2003). It has been proposed that the apoptotic effect of TNF- α signaling is mediated via SM degradation into Cer (Mimeault, 2002; Thon et al., 2005). It is important to point out that the role of Cer in steroid hormone production is likely to be independent from the role of Cer in cell proliferation and/or apoptosis because in the majority of these studies, apoptosis was not a reported reason for the decrease in steroid hormone production (Son et al., 2004).

Cer can also modulate the activity and/or expression of steroidogenic enzymes. Meroni and colleagues have shown that Cer can regulate P450c17α enzymatic activity as well as inhibit cAMP production (Meroni et al., 2000). In another report, Cer was identified as a novel regulator of 11β-HSD1 in preadipocytes (Arai et al., 2007). Cer was observed to cause an increase in CCAAT/enhancer binding protein-β (C/EBPβ) recruitment to the 11β-HSD1 promoter, and in this way activate gene expression (Arai et al., 2007). Furthermore, Cer can inhibit hCG-induced aromatase activity and estradiol production in granulosa cells (Santana et al., 1995; Son et al., 2004).

Enzymes involved in Cer metabolism may also play a key role in steroidogenic regulation. Cortisol can activate the expression of the acid ceramidase gene (ASAH1) and promote Cer degradation (Lucki and Sewer, unpublished observations), thereby increasing the cellular concentrations of SPH, the antagonist for SF-1 (Urs et al., 2006). The regulation of ASAH1 by glucocorticoids may represent an intra-adrenal feedback mechanism that controls optimal steroid hormone output by repressing the ability of SF-1 to induce the transcription of steroidogenic genes.

### 15.4.2 Sphingosine

SF-1 was once classified as an orphan nuclear receptor because the identity of the endogenous ligand for the receptor was unknown. However, investigations of the endogenous receptor have identified SPH as a *bonafide* ligand (Urs et al., 2006) and recent crystallographic studies in bacterially expressed SF-1 confirmed that the ligand binding domain (LBD) of SF-1 is bound by phospholipids (Krylova et al., 2005; Li et al., 2005; Wang et al., 2005). SPH is an antagonist for SF-1 and its binding decreases CYP17 expression (Urs et al., 2006). cAMP stimulation reverses the antagonistic action of SPH possibly by displacing SPH from the SF-1 ligand binding pocket and promoting PA binding (Fig. 15.4). Notably, we have also demonstrated that phosphatidic acid (PA) is an endogenous activating ligand for SF1 in H295R adrenocortical cells (Li et al., 2007). The identification of SPH as a ligand for SF-1 adds yet another level of regulation for sphingolipids in controlling steroid hormone production. Moreover, since SF-1 Also regulates the expression of genes involved in endocrine development and sex differentiation, it is possible that SPH may also control these processes by antagonizing SF-1.

### 15.4.3 Sphingosine-1-Phosphate

As previously mentioned, S1P is an amphiphilic molecule that can exert its signaling functions both within the cell and extracellularly by binding to specific G protein-coupled receptors (S1PRs) (Kihara et al., 2007). There are 5 S1PRs and each couples to multiple heterotrimeric G protein leading to the activation of specific intracellular targets (An et al., 1997; Im et al., 2000; Lee et al., 1998; Van Brocklyn et al., 2000; Yamazaki et al., 2000). S1PR$_1$ couples to G$_i$ and

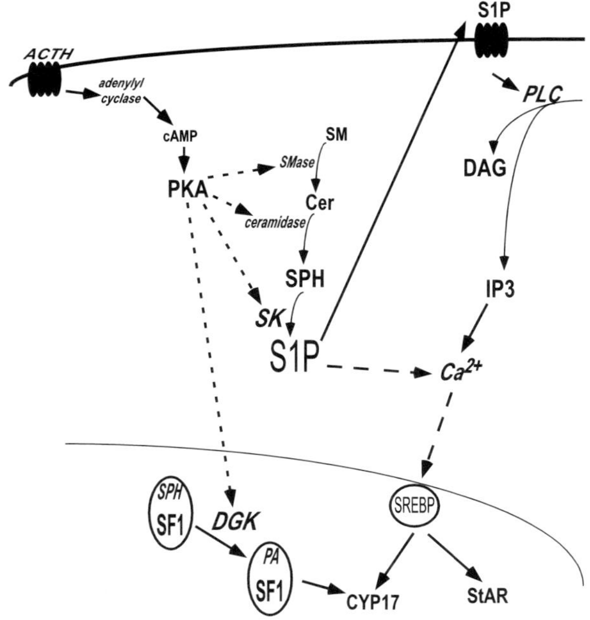


**Fig. 15.4** Relationship between sphingolipid metabolism and steroid hormone biosynthesis in the human adrenal cortex

activates the phospholipase C (PLC), phosphatidylinositol 3-kinase (PI3K), and extracellular regulated kinase (ERK) pathways. $S1PR_2$ and $S1PR_3$ couple to $G_i$, $G_{12/13}$, and $G_q$ and activate pathways such as PLC, PI3K, ERK, and Rho. $S1PR_4$ activates the PLC and ERK pathways, while $S1P_5$ has been shown to inhibit adenylyl cyclase and ERK (reviewed in ref. (Kihara et al., 2007). Binding of S1P to its cognate receptor seems to be the major mode of signaling and the PI3K/Akt and ERK pathways have been identified as downstream targets (An et al., 1997; Payne et al., 2004; Spiegel and Milstien, 2002, 2003b). Previous reports have found a link between PI3K and ERK activation and steroidogenesis, which supports a role for S1P in steroid hormone production (Meroni et al., 2002; Shah et al., 2005).

As depicted in Fig. 15.4, our laboratory has demonstrated that cAMP-mediated S1P accumulation lead to an increase in CYP17 expression by activating cleavage of the sterol regulatory element binding protein 1 (SREBP1) in H295R human adrenocortical cells (Ozbay et al., 2006). In the zona fasciculata of bovine adrenal cells, S1P was demonstrated to illicit cortisol secretion via binding to a S1PR and subsequent activation of PKC and PLD (Rabano et al., 2003). In addition, Brizuela et al. (Brizuela et al., 2006) reported that S1P stimulate aldosterone secretion in bovine glomerulosa adrenal cells. This aldosterone modulation is mediated by S1P binding to its cognate S1PR and activating the PLD/phosphatidate phosphohydrolase pathway with PKC and extraceullar $Ca^{2+}$ as possible components of the signaling cascade.

Interestingly, we have also found that SK1 is rapidly translocated to the nucleus of H295R human adrenocortical cells in response to ACTH/cAMP stimulation (Li et al., unpublished observations). Since SPH inhibits steroidogenesis by antagonizing SF-1, it is likely that nuclear import of SK1 facilitates conversion of SPH to S1P and acts to promote activated gene steroidogenic gene transcription. Given the identification of sphingolipids in the nucleus (Ledeen and Wu, 2006), future investigation into the significance of these molecules in nuclear function is likely to reveal novel roles for these bioactive lipids in controlling diverse cellular processes.

Akt and ERK 1/2 have also been identified as direct S1P targets (Brizuela et al., 2007). The finding that inhibiting PI3K and MEK prevents S1P-mediated aldosterone secretion led Brizuela et al. to propose a model for S1P in the regulation of aldosterone production in which stimulation of PI3K/Akt and MEK/ERK pathways activate PLD and ultimately result in aldosterone production. S1P was identified as an inducer of CYP19 expression and estrogen production in granulosa cells by mediating the production of prostaglandin-2 ($PGE_2$), which is an established activator of CYP19 (Cai et al., 1997). Given that S1P induces the expression of liver receptor homologue-1 (LRH-1) (Hadizadeh et al., 2007) and LRH-1 regulates CYP19 expression in breast cancer cells (Clyne et al., 2002), it is likely that estradiol production in both physiological and pathophysiological conditions is controlled by the amounts of S1P.

### *15.4.4 Sphingomyelin*

SM is present in the plasma membrane and can be hydrolyzed into Cer via the action of SMases. SM hydrolysis is mediated by a series of stimuli including TNF-α, IL-β1, and Fas ligands (Degnan et al., 1996; Hannun, 1996), which, as described above, activate ceramide production. Therefore, SM seems to be a target for a series of external signals that modulate, among other cellular processes, steroidogenesis. S1P and C1P have been reported to repress acid SMase and may perhaps be part of a negative feedback loop that regulates flux through the sphingolipid metabolic pathway (Gomez-Munoz et al., 2003, 2004).

In mouse Leydig testicular cells, SM degradation was shown to be correlated with cholesterol movement from the plasma membrane to the mitochondria and subsequent increase in steroid hormone secretion (Porn et al., 1991). Conversely, SMase activity was shown to inhibit Leydig cell function via degradation of SM and accumulation of the pro-apoptotic sphingolipid Cer (Degnan et al., 1996). In addition, our laboratory has found that lysoSM (sphingosylphosphorylcholine) is able to bind to SF-1 in H295R adrenocortical cells under basal conditions, and that cAMP treatment promotes dissociation of the sphingolipid from SF1 (Urs et al., 2006). Given that different sphingolipid species are in a dynamic balance at any given point within a cell, SM can play an important role in steroidogenesis regulation by serving as a precursor for Cer, lysoSM, and S1P.

### *15.4.5 Glycosphingolipids (Gangliosides)*

Thus far, the roles of bioactive sphingolipids in the modulation of steroidogenesis have been discussed. However, an equally important aspect of the relationship between sphingolipids and steroid hormones involves the regulation of sphingolipid metabolism by steroid hormones. This concept is exemplified in neurosteroidogenesis between the complex sphingolipids, GMs, and the steroid hormone allopregnanolone (Griffin et al., 2004; Mellon et al., 2004). Mellon et al. have demonstrated that allopregnanolone treatment reduces GM accumulation and ameliorates neurodegeneration in the NPC-1 mouse model (Mellon, 2007). Even though the precise molecular mechanism by which these molecules regulate NPC progression is not unclear, these findings indicate the intimate relationship between steroid hormone biosynthesis and sphingolipid metabolism and emphasize the complexity of the regulatory mechanisms that control steroidogenesis. Complex GMs have also been found to be essential for optimal testosterone production and spermatogenesis in mice (Takamiya et al., 1998). Testosterone has also been found to regulate GM levels in rat kidney (Anic and Mesaric, 1998).

## 15.5 Summary and Future Outlook

A significant amount of recent studies have pointed toward an important role for sphingolipids in the regulation of steroidogenesis. These bioactive lipids are key mediators that act at different levels of steroid hormone production including regulation of steroidogenic enzymes expression and activity as well as participating in regulatory signaling cascades. Although the studies discussed in this review significantly increased our understanding of the multiple mechanisms by which sphingolipids control steroidogenesis, it is evident that future studies are necessary to fully elucidate the roles of sphingolipids in steroid hormone biosynthesis. Technologies such as mass spectroscopy, metabolomic profiling, and proteomics are likely to be valuable tools in continued study of the relationship between these two classes of lipids. The identification and characterization of novel bioactive sphingolipids, quantification of flux through both the sphingolipid and steroidogenic metabolic pathways in response to various factors, and the examination of the role of steroid hormones as regulators of sphingolipid production are just a few research avenues that will provide more insight into the roles of sphingolipids in steroid hormone biosynthesis and endocrine function.

## References

Abdallah, M. A., Lei, Z. M., Li, X., Greenwold, N., Nakajima, S. T., Jauniaux, E., and Rao Ch, V. (2004). Human fetal nongonadal tissues contain human chorionic gonadotropin/luteinizing hormone receptors. *J Clin Endocrinol Metab* **89**, 952–956.

Adams, J. M., Pratipanawatr, T., Berria, R., Wang, E., DeFronzo, R. A., Sullards, M. C., and Mandarino, L. J. 2nd. (2004). Ceramide content is increased in skeletal muscle from obese insulin-resistant humans. *Diabetes* **53**, 25–31.

An, S., Bleu, T., Huang, W., Hallmark, O. G., Coughlin, S. R., and Goetzl, E. J. (1997). Identification of cDNAs encoding two G protein-coupled receptors for lysosphingolipids. *FEBS Lett* **417**, 279–282.

Andrieu-Abadie, N., and Levade, T. (2002). Sphingomyelin hydrolysis during apoptosis. *Biochim Biophys Acta* **1585**, 126–134.

Anic M., and Mesaric M. (1998). The influence of sex steroid hormones on ganglioside biosynthesis in rat kidney. *Biol Chem* **379**, 693–697.

Arai, N., Masuzaki, H., Tanaka, T., Ishii, T., Yasue, S., Kobayashi, N., Tomita, T., Noguchi, M., Kusakabe, T., Fujikura, J., Ebihara, K., Hirata, M., Hosoda, K., Hayashi, T., Sawai, H., Minokoshi, Y., and Nakao, K. (2007). Ceramide and Adenosine 5'-Monophosphate-Activated Protein Kinase Are Two Novel Regulators of 11{beta}-Hydroxysteroid Dehydrogenase Type 1 Expression and Activity in Cultured Preadipocytes. *Endocrinology* **148**, 5268–5277.

Arlt, W., and Stewart, P. M. (2005). Adrenal corticosteroid biosynthesis, metabolism, and action. *Endocrinol Metab Clin North Am* **34**, 293–313.

Azhar, S., Nomoto, A., Leers-Sucheta, S., and Reaven, E. (1998). Simultaneous induction of an HDL receptor protein (SR-BI) and the selective uptake of HDL-cholesteryl esters in a physiologically relevant steroidogenic cell model. *J Lipid Res* **39**, 1616–1628.

Bassett, M. H., White, P C., and Rainey, W. E. (2004a). Regulation of aldosterone synthase expression. *Mol Cell Endocrinol* **217**, 67–74.

Bassett, M. H., White, P. C., and Rainey, W. E. (2004b). A role for the NGFI-B family in adrenal zonation and adrenocortical disease. *Endocr Res* **30**, 567–574.

Besman, M. J., Yanagibashi, K., Lee, T. D., Kawamura, M., Hall, P. F., and Shively, J. E. (1989). Identification of des-(Gly-Ile)-endozepine as an effector of corticotropin-dependent adrenal steroidogenesis: stimulation of cholesterol delivery is mediated by the peripheral benzodiazepine receptor. *Proc Natl Acad Sci U S A* **86**, 4897–4901.

Blanchette-Mackie, E. J. (2000). Intracellular cholesterol trafficking: role of the NPC1 protein. *Biochim Biophys Acta* **1486**, 171–183.

Bornstein, S. R., Rutkowshi, H., and Vrezas, I. (2004). Cytokines and steroidogenesis. *Mol Cell Endocrinol* **215**, 135–141.

Brentano, S. T., and Miller, W. L. (1992). Regulation of human cytochrome P450scc and adrenodoxin messenger ribonucleic acids in JEG-3 cytotrophoblast cells. *Endocrinology* **131**, 3010–3018.

Brizuela, L., Rabano, M., Gangoiti, P., Narbona, N., Macarulla, J. M., Trueba, M., and Gomez-Munoz, A. (2007). Sphingosine-1-phosphate stimulates aldosterone secretion through a mechanism involving the PI3K/PKB and MEK/ERK 1/2 pathways. *J Lipid Res* **48**, 2264–2274.

Brizuela, L., Rabano, M., Pena, A., Gangoiti, P., Macarulla, J. M., Trueba, M., and Gomez-Munoz, A. (2006). Sphingosine 1-phosphate: a novel stimulator of aldosterone secretion. *J Lipid Res* **47**, 1238–1249.

Budnik, L. T., Jahner, D., and Mukhopadhyay, A. K. (1999). Inhibitory effects of TNF alpha on mouse tumor Leydig cells: possible role of ceramide in the mechanism of action. *Mol Cell Endocrinol* **150**, 39–46.

Cai, Z., Bettaieb, A., Mahdani, N. E., Legres, L. G., Stancou, R., Masliah, J., and Chouaib, S. (1997). Alteration of the sphingomyelin/ceramide pathway is associated with resistance of human breast carcinoma MCF7 cells to tumor necrosis factor-alpha-mediated cytotoxicity. *J Biol Chem* **272**, 6918–6926.

Cai, Z., Kwintkiewicz, J., Young, M., and Stocco, C. (2007). Prostaglandin E2 increases cyp19 expression in rat granulosa cells: implications of GATA-4. *Molecular and Cellular Endocrinology* **263**, 181–189.

Caron, K. M., Ikeda, Y., Soo, S. C., Stocco, D. M., Parker, K. L., and Clark, B. J. (1997). Characterization of the promoter region of the mouse gene encoding the steroidogenic acute regulatory protein. *Mol Endocrinol* **11**, 138–147.

Castilla, R., Maloberti, P., Castillo, F., Duarte, A., Cano, F., Maciel, F. C., Neuman, I., Mendez, C. F., Paz, C., and Podesta, E. J. (2004). Arachidonic acid regulation of steroid synthesis: new partners in the signaling pathway of steroidogenic hormones. *Endocr Res* **30**, 599–606.

Chen, J. Y., and Waterman, M. R. (1992). Two promoters in the bovine adrenodoxin gene and the role of associated, unique cAMP-responsive sequences. *Biochemistry* **31**, 2400–2407.

Chen, W. Y., Lee, W. C., Hsu, N. C., Huang, F., and Chung, B. C. (2004). SUMO modification of repression domains modulates function of nuclear receptor 5A1 (steroidogenic factor-1). *J Biol Chem* **279**, 38730–38735.

Choi, M. S., and Cooke, B. A. (1990). Evidence for two independent pathways in the stimulation of steroidogenesis by luteinizing hormone involving chloride channels and cyclic AMP. *FEBS Lett* **261**, 402–404.

Choi, M. S., and Cooke, B. A. (1992). Calmidazolium is a potent stimulator of steroidogenesis via mechanisms not involving cyclic AMP, calcium or protein synthesis. *Biochem J* **281 (Pt 1)**, 291–296.

Clark, B. J., and Combs, R. (1999). Angiotensin II and cyclic adenosine 3',5'-monophosphate induce human steroidogenic acute regulatory protein transcription through a common steroidogenic factor-1 element. *Endocrinology* **140**, 4390–4398.

Clem, B. F., Hudson, E. A., and Clark, B. J. (2005). Cyclic adenosine 3',5'-monophosphate (cAMP) enhances cAMP-responsive element binding (CREB) protein phosphorylation and phospho-CREB interaction with the mouse steroidogenic acute regulatory protein gene promoter. *Endocrinology* **3**, 1348–1356.

Clyne, C. D., Speed, C. J., Zhou, J., and Simpson, E. R. (2002). Liver receptor homologue-1 (LRH-1) regulates expression of aromatase in preadipocytes. *J Biol Chem* **277**, 20591–20597.

Condon, J. C., Pezzi, V., Drummond, B. M., Yin, S., and Rainey, W. E. (2002). Calmodulin-dependent kinase I regulates adrenal cell expression of aldosterone synthase. *Endocrinology* **143**, 3651–3657.

Cuvillier, O., Pirianov, G., Kleuser, B., Vanek, P. G., Coso, O. A., Gutkind, S., and Spiegel, S. (1996). Suppression of ceramide-mediated programmed cell death by sphingosine-1-phosphate. *Nature* **381**, 800–803.

Degnan, B. M., Bourdelat-Parks, B., Daniel, A., Salata, K., and Francis, G. L. (1996). Sphingomyelinase inhibits in vitro Leydig cell function. *Ann Clin Lab Sci* **26**, 234–242.

Ding, T., Li, Z., Hailemariam, T., Mukherjee, S., Maxfield, F. R., Wu, M., and Jiang, X. C. (2007). Sphingomyelin synthase (SMS) overexperssion and knockdown: Impact on cellular sphingomyelin and diacylglycerol metabolism, and cell apopotosis. *J Lipid Res* **49 (2)**, 376–385.

Foster, R. H. (2004). Reciprocal influences between the signalling pathways regulating proliferation and steroidogenesis in adrenal glomerulosa cells. *J Mol Endocrinol* **32**, 893–902.

Fuxe, K., Gustafsson, J. A., and Wetterberg, L. (1981). Steroid hormone regulation of the brain. *Pergamon Press, Oxford.*

Gallo-Payet, N., Cote, M., Chorvatova, A., Guillon, G., and Payet, M. D. (1999). Cyclic AMP-independent effects of ACTH on glomerulosa cells of the rat adrenal cortex. *J Steroid Biochem Mol Biol* **69**, 335–342.

Gallo-Payet, N., and Payet, M. D. (1989). Excitation-secretion coupling: involvement of potassium channels in ACTH-stimulated rat adrenocortical cells. *J Endocrinol* **120**, 409–421.

Ghayee, H. K., and Auchus, R. J. (2007). Basic concepts and recent developments in human steroid hormone biosynthesis. *Rev Endocr Metab Disord* **8 (4)**, 289–300.

Gomez-Munoz, A. (2006). Ceramide 1-phosphate/ceramide, a switch between life and death. *Biochim Biophys Acta* **1758**, 2049–2056.

Gomez-Munoz, A., Kong, J., Salh, B., and Steinbrecher, U. P. (2003). Sphingosine-1-phosphate inhibits acid sphingomyelinase and blocks apoptosis in macrophages. *FEBS Lett* **539**, 56–60.

Gomez-Munoz, A., Kong, J. Y., Salh, B., and Steinbrecher, U. P. (2004). Ceramide-1-phosphate blocks apoptosis through inhibition of acid sphingomyelinase in macrophages. *J Lipid Res* **45**, 99–105.

Goni, F. M., and Alonso, A. (2006). Biophysics of sphingolipids I. Membrane properties of sphingosine, ceramides and other simple sphingolipids. *Biochim Biophys Acta* **1758**, 1902–1921.

Griffin, L. D., Gong, W., Verot, L., and Mellon, S. H. (2004). Niemann-Pick type C disease involves disrupted neurosteroidogenesis and responds to allopregnanolone. *Nat Med* **10**, 704–711.

Hadizadeh, S., King, D., Shah, S., and Sewer, M. B. (2007). Sphingosine-1-phosphate regulates the expression of the liver receptor homologue-1. *Molecular and Cellular Endocrinology* ***In press, accepted manuscript, available online 5 December 2007, doi:10.1016/j.mce.2007.11.030.***

Hammer, G. D., Krylova, I., Zhang, Y., Darimont, B. D., Simpson, K., Weigel, N. L., and Ingraham, H. A. (1999). Phosphorylation of the nuclear receptor SF-1 modulates cofactor recruitment: integration of hormone signaling in reproduction and stress. *Mol Cell* **3**, 521–526.

Hanada, K., Kumagai, K., Yasuda, S., Miura, Y., Kawano, M., Fukasawa, M., and Nishijima, M. (2003). Molecular machinery for non-vesicular trafficking of ceramide. *Nature* **426**, 803–809.

Hannun, Y. A. (1996). Functions of ceramide in coordinating cellular responses to stress. *Science* **274**, 1855–1859.

Hannun, Y. A., Loomis, C. R., Merrill, A. H., Jr., and Bell, R. M. (1986). Sphingosine inhibition of protein kinase C activity and of phorbol dibutyrate binding in vitro and in human platelets. *J Biol Chem* **261**, 12604–12609.

Hauet, T., Liu, J., Li, H., Gazouli, M., Culty, M., and Papadopoulos, V. (2002). PBR, StAR, and PKA: partners in cholesterol transport in steroidogenic cells. *Endocr Res* **28**, 395–401.

Havelock, J. C., Auchus, R. J., and Rainey, W. E. (2004). The rise in adrenal androgen biosynthesis: adrenarche. *Semin Reprod Med* **22**, 337–347.

Hedger, M. P. (1997). Testicular leukocytes: what are they doing? *Rev Reprod* **2**, 38–47.

Hung, W. C., Chang, H. C., and Chuang, L. Y. (1999). Activation of caspase-3-like proteases in apoptosis induced by sphingosine and other long-chain bases in Hep3B hepatoma cells. *Biochem J* **338 (Pt 1)**, 161–166.

Huwiler, A., Kolter, T., Pfeilschifter, J., and Sandhoff, K. (2000). Physiology and pathophysiology of sphingolipid metabolism and signaling. *Biochim Biophys Acta* **1485**, 63–99.

Im, D. S., Heise, C. E., Ancellin, N., O'Dowd, B. F., Shei, G. J, Heavens, R. P., Rigby, M. R., Hla, T., Mandala, S., McAllister, G., George, S. R., and Lynch, K. R. (2000). Characterization of a Novel Sphingosine 1-Phosphate Receptor, Edg-8. *J Biol Chem* **275**, 14281–14286.

Ishihara, S. L., and Morohashi, K. (2005). A boundary for histone acetylation allows distinct expression patterns of the Ad4BP/SF-1 and GCNF loci in adrenal cortex cells. *Biochem Biophys Res Commun* **329**, 554–562.

Jamnongjit, M., and Hammes, S. R. (2006). Ovarian steroids: the good, the bad, and the signals that raise them. *Cell Cycle* **5**, 1178–1183.

Jefferson, A. B., and Schulman, H. (1988). Sphingosine inhibits calmodulin-dependent enzymes. *J Biol Chem* **263**, 15241–15244.

Kassel, O., and Herrlich, P. (2007). Crosstalk between the glucocorticoid receptor and other transcription factors: Molecular aspects. *Molecular and Cellular Endocrinology* **275**, 13–29.

Kihara, A., Mitsutake, S., Mizutani, Y., and Igarashi, Y. (2007). Metabolism and biological functions of two phosphorylated sphingolipids, sphingosine 1-phosphate and ceramide 1-phosphate. *Prog Lipid Res* **46**, 126–144.

Kolesnick, R. (2002). The therapeutic potential of modulating the ceramide/sphingomyelin pathway. *J Clin Invest* **110**, 3–8.

Kraemer, F. B., Shen, W. J., Harada, K., Patel, S., Osuga, J., Ishibashi, S., and Azhar, S. (2004). Hormone-sensitive lipase is required for high-density lipoprotein cholesteryl ester-supported adrenal steroidogenesis. *Mol Endocrinol* **18**, 549–557.

Krieger, M. (1999). Charting the fate of the "good cholesterol": identification and characterization of the high-density lipoprotein receptor SR-BI. *Annu Rev Biochem* **68**, 523–558.

Krylova, I. N., Sablin, E. P., Moore, J., Xu, R. X., Waitt, G. M., MacKay, J. A., Juzumiene, D., Bynum, J. M., Madauss, K., Montana, V., Lebedeva, L., Suzawa, M., Williams, J. D., Williams, S. P., Guy, R. K., Thornton, J. W., Fletterick, R. J., Willson, T. M., and Ingraham, H. A. (2005). Structural analyses reveal phosphatidyl inositols as ligands for the NR5 orphan receptors SF-1 and LRH-1. *Cell* **120**, 343–355.

Lala, D. S., Rice, D.A., and Parker, K. L. (1992). Steroidogenic factor I, a key regulator of steroidogenic enzyme expression, is the mouse homolog of fushi tarazu-factor I. *Mol Endocrinol* **6**, 1249–1258.

Lavaque, E., Sierra, A., Azcoitia, I., and Garcia-Segura, L. M. (2006). Steroidogenic acutre regulatory protein in the brain. *Neuroscience* **138**, 741–747.

Lavieu, G., Scarlatti, F., Sala, G., Carpentier, S., Levade, T., Ghidoni, R., Botti, J., and Codogno, P. (2006). Regulation of autophagy by sphingosine kinase 1 and its role in cell survival during nutrient starvation. *J Biol Chem* **281**, 8518–8527.

Lavieu, G., Scarlatti, F., Sala, G., Levade, T., Ghidoni, R., Botti, J., and Codogno, P. (2007). Is autophagy the key mechanism by which the sphingolipid rheostat controls the cell fate decision? *Autophagy* **3**, 45–47.

Le Stunff, H., Giussani, P., Maceyka, M., Lépine, S., Milstien, S., and Spiegel, S. (2007). Recycling of sphingosine is regulated by the concerted actions of sphingosine-1-phosphate phosphohydrolase 1 and sphingosine kinase 2. *J Biol Chem* **282**, 34372–34380.

Ledeen, R. W., and Wu, G. (2006). Sphingolipids of the nucleus and their roles in nuclear signaling. *Biochim Biophys Acta* **1761**, 588–598.

Lee, M. J., Thangada, S., Liu, C. H., Thompson, B. D., and Hla, T. (1998). Lysophosphatidic acid stimulates the G-protein-coupled receptor EDG-1 as a low affinity agonist. *J Biol Chem* **273**, 22105–22112.

Li, D., Urs, A. N., Allegood J., Leon, A., Merrill, A. H., Jr., and Sewer, M. B. (2007). Cyclic AMP-stimulated interaction between steroidogenic factor 1 and diacylglycerol kinase theta facilitates induction of CYP17. *Mol Cell Biol* **27**, 6669–6685.

Li, Q., Ni, J., Bian, S., Yao, L., Zhu, H., and Zhang, W. (2001). Inhibition of steroidogenesis and induction of apoptosis in rat luteal cells by cell-permeable ceramide in vitro. *Sheng Li Xue Bao* **53**, 142–146.

Li, Y., Choi, M., Cavey, G., Daugherty, J., Suino, K., Kovach, A., Bingham, N. C., Kliewer, S. A., and Xu, H. E. (2005). Crystallographic identification and functional characterization of phospholipids as ligands for the orphan nuclear receptor steroidogenic factor-1. *Mol Cell* **17**, 491–502.

Liu, H., Toman, R. E., Goparaju, S. K., Maceyka, M., Nava, V. E., Sankala, H., Payne, S. G., Bektas, M., Ishii, I., Chun, J., Milstien, S., and Spiegel, S. (2003a). Sphingosine kinase type 2 is a putative BH3-only protein that induces apoptosis. *J Biol Chem* **278**, 40330–40336.

Liu, J., Li, H., and Papadopoulos, V. (2003b). PAP7, a PBR/PKA-RIalpha-associated protein: a new element in the relay of the hormonal induction of steroidogenesis. *J Steroid Biochem Mol Biol* **85**, 576–586.

Liu, J., Rone, M. B., and Papadopoulos, V. (2006). Protein-protein interactions mediate mitochondrial cholesterol transport and steroid biosynthesis. *J Biol Chem* **281**, 38879–38893.

Lorenz, J. N., Arend, L. J., Robitz, R., Paul, R. J., and MacLennan, A. J. (2007). Vascular dysfunction in S1P2 sphingosine 1-phosphate receptor knockout mice. *Am J Physiol Regul Integr Comp Physiol* **292**, R440–446.

Lu, Z. H., Mu, Y. M., Wang, B. A., Li, X. L., Lu, J. M., Li, J. Y., Pan, C. Y., Yanase, T., and Nawata, H. (2003). Saturated free fatty acids, palmitic acid and stearic acid, induce apoptosis by stimulation of ceramide generation in rat testicular Leydig cell. *Biochem Biophys Res Commun* **303**, 1002–1007.

Maceyka, M., Payne, S. G., Milstien, S., and Spiegel, S. (2002). Sphingosine kinase, sphingosine-1-phosphate, and apoptosis. *Biochim Biophys Acta* **1585**, 193–201.

Maceyka, M., Sankala, H., Hait, N. C., Le Stunff, H., Liu, H., Toman, R., Collier, C., Zhang, M., Satin, L. S., Merrill, A. H., Jr., Milstien, S., and Spiegel, S. (2005). SphK1 and SphK2, sphingosine kinase isoenzymes with opposing functions in sphingolipid metabolism. *J Biol Chem* **280**, 37118–37129.

McClellan, D. R., Bourdelat-Parks, B., Salata, K., and Francis, G. L. (1997). Sphingomyelinase affects hormone production in Jeg-3 choriocarcinoma cells. *Cell Endocrinology Metabolism* **9,** 19–24.

McEwen, B. S. (1991). Steroid hormones are multifuctional messengers in the brain. *Trends in Endocrinology Metabolism* **2**, 62–67.

Mellon, S. H., Gong, W., and Griffin, L. D. (2004). Niemann pick type C disease as a model for defects in neurosteroidogenesis. *Endocr Res* **30**, 727–735.

Mellon, S. H. (2007). Neurosteroid regulation of central nervous system development. *Pharmacol Ther* **116**, 107–124.

Mellon, S. H., and Deschepper, C. F. (1993). Neurosteroid biosynthesis: genes for adrenal steroidogenic enzymes are expressed in the brain. *Brain Res* **629**, 283–292.

Mellon, S. H., and Griffin, L. D. (2002). Neurosteroids: biochemistry and clinical significance. *Trends Endocrinol Metab* **13**, 35–43.

Mendelson, C. R., Jiang, B., Shelton, J. M., Richardson, J. A., and Hinshelwood, M. M. (2005). Transcriptional regulation of aromatase in placenta and ovary. *J Steroid Biochem Mol Biol* **95**, 25–33.

Meroni, S. B., Pellizzari, E. H., Canepa, D. F., and Cigorraga, S. B. (2000). Possible involvement of ceramide in the regulation of rat Leydig cell function. *J Steroid Biochem Mol Biol* **75**, 307–313.

Meroni, S. B., Riera, M. F., Pellizzari, E. H., and Cigorraga, S. B. (2002). Regulation of rat Sertoli cell function by FSH: possible role of phosphatidylinositol 3-kinase/protein kinase B pathway. *J Endocrinol* **174**, 195–204.

Merrill, A. H., Jr., Nikolova-Karakashian, N., Schmelz, E. M., Morgan, E. T., and Stewart, J. (1999). Regulation of cytochrome P450 expression by sphingolipids. *Chemistry and Physics of Lipids* **102**, 131–139.

Miller, W. L. (2007). Mechanism of StAR's regulation of mitochondrial cholesterol import. *Mol Cell Endocrinol* **265–266**, 46–50.

Mimeault, M. (2002). New advances on structural and biological functions of ceramide in apoptotic/necrotic cell death and cancer. *FEBS Lett* **530**, 9–16.

Morales, V., Santana, P., Diaz, R., Tabraue, C., Gallardo, G., Blanco, F. L., Hernandez, I., Fanjul, L. F., and Ruiz de Galarreta, C. M. (2003). Intratesticular Delivery of Tumor

Necrosis Factor-{alpha} and Ceramide Directly Abrogates Steroidogenic Acute Regulatory Protein Expression and Leydig Cell Steroidogenesis in Adult Rats. *Endocrinology* **144**, 4763–4772.

Morohashi, K., Honda, S., Inomata, Y., Handa, H., and Omura, T. (1992). A common trans-acting factor, Ad4-binding protein, to the promoters of steroidogenic P-450s. *J Biol Chem* **267**, 17913–17919.

Mueller, M., Atanasov, A., Cima, I., Corazza, N., Schoonjans, K., and Brunner, T. (2007). Differential regulation of glucocorticoid synthesis in murine intestinal epithelial versus adrenocortical cell lines. *Endocrinology* **148**, 1445–1453.

Mukai, H., Tsurugizawa, T., Ogiue-Ikeda, M., Murakami, G., Hojo, Y., Ishii, H., Kimoto, T., and Kawato, S. (2006). Local neurosteroid production in the hippocampus: influence on synaptic plasticity of memory. *Neuroendocrinology* **84**, 255–263.

Natarajan, V., Jayaram, H. N., Scribner, W. M., and Garcia, J. G. (1994). Activation of endothelial cell phospholipase D by sphingosine and sphingosine-1-phosphate. *Am J Respir Cell Mol Biol* **11**, 221–229.

Newton, R., and Holden, N. S. (2007). Separating transrepression and transactivation: A distressing divorce for the glococorticoid receptor? *Molecular Pharmacology* **72**, 799–809.

Ogretmen, B. (2006). Sphingolipids in cancer: regulation of pathogenesis and therapy. *FEBS Lett* **580**, 5467–5476.

Okada, T., Ding, G., Sonoda, H., Kajimoto, T., Haga, Y., Khosrowbeygi, A., Gao, S., Miwa, N., Jahangeer, S., and Nakamura, S. (2005). Involvement of N-terminal-extended form of sphingosine kinase 2 in serum-dependent regulation of cell proliferation and apoptosis. *J Biol Chem* **280**, 36318–36325.

Okamoto, M., Takemori, H., and Katoh, Y. (2004). Salt-inducible kinase in steroidogenesis and adipogenesis. *Trends Endocrinol Metab* **15**, 21–26.

Okazaki, T., Bell, R. M., and Hannun, Y. A. (1989). Sphingomyelin turnover induced by vitamin D3 in HL-60 cells. Role in cell differentiation. *J Biol Chem* **264**, 19076–19080.

Olivera, A., Kohama, T., Edsall, L., Nava, V., Cuvillier, O., Poulton, S., and Spiegel, S. (1999). Sphingosine kinase expression increases intracellular sphingosine-1-phosphate and promotes cell growth and survival. *J Cell Biol* **147**, 545–558.

Olivera, A., and Spiegel, S. (1993). Sphingosine-1-phosphate as second messenger in cell proliferation induced by PDGF and FCS mitogens. *Nature* **365**, 557–560.

Osawa, Y., Uchinami, H., Bielawski, J., Schwabe, R. F., Hannun, Y. A., and Brenner, D. A. (2005). Roles for C16-ceramide and sphingosine 1-phosphate in regulating hepatocyte apoptosis in response to tumor necrosis factor-alpha. *J Biol Chem* **280**, 27879–27887.

Otis, M., and Gallo-Payet, N. (2007). Role of MAPKs in angiotensin II-induced steroidogenesis in rat glomerulosa cells. *Mol Cell Endocrinol* **265–266**, 126–130.

Ozbay, T., Merrill, A. H., Jr., and Sewer, M. B. (2004). ACTH regulates steroidogenic gene expression and cortisol biosynthesis in the human adrenal cortex via sphingolipid metabolism. *Endocr Res* **30**, 787–794.

Ozbay, T., Rowan, A., Leon, A., Patel, P., and Sewer, M. B. (2006). Cyclic adenosine 5'-monophosphate-dependent sphingosine-1-phosphate biosynthesis induces human CYP17 gene transcription by activating cleavage of sterol regulatory element binding protein 1. *Endocrinology* **147**, 1427–1437.

Papadopoulos, V. (1993). Peripheral-type benzodiazepine/diazepam binding inhibitor receptor: biological role in steroidogenic cell function. *Endocr Rev* **14**, 222–240.

Parker, K. L., Rice, D. A., Lala, D. S., Ikeda, Y., Luo, X., Wong, M., Bakke, M., Zhao, L., Frigeri, C., Hanley, N. A., Stallings, N., and Schimmer, B. P. (2002). Steroidogenic factor 1: an essential mediator of endocrine development. *Recent Prog Horm Res* **57**, 19–36.

Payne, A. H., and Hales, D. B. (2004). Overview of steroidogenic enzymes in the pathway from cholesterol to active steroid hormones. *Endocr Rev* **25**, 947–970.

Payne, S. G., Milstien, S., Barbour, S. E., and Spiegel, S. (2004). Modulation of adaptive immune responses by sphingosine-1-phosphate. *Semin Cell Dev Biol* **15**, 521–527.

Pepe, G. J., and Albrecht, E. D. (1995). Actions of placental and fetal adrenal steroid hormones in primate pregnancy. *Endocr Rev* **16**, 608–648.

Pettus, B. J., Bielawska, A., Spiegel, S., Roddy, P., Hannun, Y. A., and Chalfant, C. E. (2003). Ceramide kinase mediates cytokine- and calcium ionophore-induced arachidonic acid release. *J Biol Chem* **278**, 38206–38213.

Pettus, B. J., Chalfant, C. E., and Hannun, Y. A. (2002). Ceramide in apoptosis: an overview and current perspectives. *Biochim Biophys Acta* **1585**, 114–125.

Plassart-Schiess, E., and Baulieu, E. E. (2001). Neurosteroids: recent findings. *Brain Res Brain Res Rev* **37**, 133–140.

Porn, M. I., Tenhunen, J., and Slotte, J. P. (1991). Increased steroid hormone secretion in mouse Leydig tumor cells after induction of cholesterol translocation by sphingomyelin degradation. *Biochim Biophys Acta* **1093**, 7–12.

Rabano, M., Pena, A., Brizuela, L., Marino, A., Macarulla, J. M., Trueba, M., and Gomez-Munoz, A. (2003). Sphingosine-1-phosphate stimulates cortisol secretion. *FEBS Lett* **535**, 101–105.

Rainey, W. E. (1999). Adrenal zonation: clues from 11beta-hydroxylase and aldosterone synthase. *Mol Cell Endocrinol* **151**, 151–160.

Rainey, W. E., Carr, B. R., Sasano, H., Suzuki, T., and Mason, J. I. (2002). Dissecting human adrenal androgen production. *Trends Endocrinol Metab* **13**, 234–239.

Ramnath, H. I., Peterson, S., Michael, A. E., Stocco, D. M., and Cooke, B. A. (1997). Modulation of steroidogenesis by chloride ions in MA-10 mouse tumor Leydig cells: roles of calcium, protein synthesis, and the steroidogenic acute regulatory protein. *Endocrinology* **138**, 2308–2314.

Reinhart, A. J., Williams, S. C., Clark, B. J., and Stocco, D. M. (1999). SF-1 (steroidogenic factor-1) and C/EBP beta (CCAAT/enhancer binding protein-beta) cooperate to regulate the murine StAR (steroidogenic acute regulatory) promoter. *Mol Endocrinol* **13**, 729–741.

Ruvolo, P. P., Clark, W., Mumby, M., Gao, F., and May, W. S. (2002). A functional role for the B56 alpha-subunit of protein phosphatase 2A in ceramide-mediated regulation of Bcl2 phosphorylation status and function. *J Biol Chem* **277**, 22847–22852.

Sakakura, C., Sweeney, E. A., Shirahama, T., Hagiwara, A., Yamaguchi, T., Takahashi, T., Hakomori, S., and Igarashi, Y. (1998). Selectivity of sphingosine-induced apoptosis. Lack of activity of DL-erythyro-dihydrosphingosine. *Biochem Biophys Res Commun* **246**, 827–830.

Sakamoto, H., Ukena, K., Kawata, M., and Tsutsui, K. (2007). Expression, localization and possible actions of 25-Dx, a membrane-associated putative progesterone-binding protein, in the developing Purkinje cell of the cerebellum: A new insight into the biosynthesis, metabolism and multiple actions of progesterone as a neurosteroid. *Cerebellum* **Mar 2:1–8 [Epub ahead of print]**.

Santana, P., Llanes, L., Hernandez, I., Gallardo, G., Quintana, J., Gonzalez, J., Estevez, F., Ruiz de Galarreta, C., and Fanjul, L. F. (1995). Ceramide mediates tumor necrosis factor effects on P450-aromatase activity in cultured granulosa cells. *Endocrinology* **136**, 2345–2348.

Santana, P., Llanes, L., Hernandez, I., Gonzalez-Robayna, I., Tabraue, C., Gonzalez-Reyes, J., Quintana, J., Estevez, F., Ruiz de Galarreta, C. M., and Fanjul, L. F. (1996). Interleukin-1 beta stimulates sphingomyelin hydrolysis in cultured granulosa cells: evidence for a regulatory role of ceramide on progesterone and prostaglandin biosynthesis. *Endocrinology* **137**, 2480–2489.

Sawai, H., Okazaki, T., Takeda, Y., Tashima, M., Sawada, H., Okuma, M., Kishi, S., Umehara, H., and Domae, N. (1997). Ceramide-induced translocation of protein kinase C-delta and -epsilon to the cytosol. Implications in apoptosis. *J Biol Chem* **272**, 2452–2458.

Sewer, M. B., Dammer, E. B., and Jagarlapudi, S. (2007). Transcriptional Regulation of Adrenocortical Steroidogenic Gene Expression. *Drug Metab Rev* **39**, 371–388.

Sewer, M. B., and Waterman, M. R. (2001). Insights into the transcriptional regulation of steroidogenic enzymes and StAR. *Rev Endocr Metab Disord* **2**, 269–274.

Sewer, M. B., and Waterman, M. R. (2003). ACTH modulation of transcription factors responsible for steroid hydroxylase gene expression in the adrenal cortex. *Microsc Res Tech* **61**, 300–307.

Shah, B. H., Baukal, A. J., Shah, F. B., and Catt, K. J. (2005). Mechanisms of extracellularly regulated kinases 1/2 activation in adrenal glomerulosa cells by lysophosphatidic acid and epidermal growth factor. *Mol Endocrinol* **19**, 2535–2548.

Sierra, A. (2004). Neurosteroids: the StAR protein in the brain. *J Neuroendocrinol* **16**, 787–793.

Sirianni, R., Carr, B. R., Ando, S., and Rainey, W. E. (2003). Inhibition of Src tyrosine kinase stimulates adrenal androgen production. *J Mol Endocrinol* **30**, 287–299.

Smith, W. L., and Merrill, A. H., Jr. (2002). Sphingolipid metabolism and signaling minireview series. *J Biol Chem* **277**, 25841–25842.

Son, D. S., Arai, K. Y., Roby, K. F., and Terranova, P. F. (2004). Tumor necrosis factor alpha (TNF) increases granulosa cell proliferation: dependence on c-Jun and TNF receptor type 1. *Endocrinology* **145**, 1218–1226.

Spiegel, S., and Milstien, S. (2002). Sphingosine 1-phosphate, a key cell signaling molecule. *J Biol Chem* **277**, 25851–25854.

Spiegel, S., and Milstien, S. (2003a). Exogenous and intracellularly generated sphingosine 1-phosphate can regulate cellular processes by divergent pathways. *Biochem Soc Trans* **31**, 1216–1219.

Spiegel, S., and Milstien, S. (2003b). Sphingosine-1-phosphate: an enigmatic signalling lipid. *Nat Rev Mol Cell Biol* **4**, 397–407.

Spiegel, S., and Milstien, S. (2007). Functions of the multifaceted family of sphingosine kinases and some close relatives. *J Biol Chem* **282**, 2125–2129.

Stocco, D. M., and Khan, S. A. (1992). Effects of steroidogenesis inducing protein (SIP) on steroid production in MA-10 mouse Leydig tumor cells: utilization of a non-cAMP second messenger pathway. *Mol Cell Endocrinol* **84**, 185–194.

Sweeney, E. A., Inokuchi, J., and Igarashi, Y. (1998). Inhibition of sphingolipid induced apoptosis by caspase inhibitors indicates that sphingosine acts in an earlier part of the apoptotic pathway than ceramide. *FEBS Lett* **425**, 61–65.

Tajima, K., Yoshii, K., Fukuda, S., Orisaka, M., Miyamoto, K., Amsterdam, A., and Kotsuji, F. (2005). Luteinizing hormone-induced extracellular-signal regulated kinase activation differently modulates progesterone and androstenedione production in bovine theca cells. *Endocrinology* **146**, 2903–2910.

Takamiya, K., Yamamoto, A., Furukawa, Z. K., Zhao, J., Fukumoto, S., Yamashiro, S., Okada, M., Hraguchi, M., Shin, M., Kishikawa, M., Shiku, H., Aizawa, S., and Furukawa, K. (1998). Complex gangliosides are essential in spermatogenesis of mice: Possible roles in the transport of testosterone. *Proc Natl Acad Sci* **95**, 12147–12152.

Tamboli, I. Y., Prager, K., Barth, E., Heneka, M., Sandhoff, K., and Walter, J. (2005). Inhibition of glycosphingolipid biosynthesis reduces secretion of the beta-amyloid precursor protein and amyloid beta-peptide. *J Biol Chem* **280**, 28110–28117.

Tani, M., Ito, M., and Igarashi, Y. (2007). Ceramide/sphingosine/sphingosine 1-phosphate metabolism on the cell surface and in the extracellular space. *Cell Signal* **19**, 229–237.

Thomson, M. (1997). Molecular and cellular mechanisms used in the acute phase of stimulated steroidogenesis. *Hormone Metabolism Research* **30**, 16–28.

Thon, L., Mohlig, H., Mathieu, S., Lange, A., Bulanova, E., Winoto-Morbach, S., Schutze, S., Bulfone-Paus, S., and Adam, D. (2005). Ceramide mediates caspase-independent programmed cell death. *Faseb J* **19**, 1945–1956.

Tsutsui, K., and Ukena, K. (1999). Neurosteroids in the cerebellar Purkinje neuron and their actions (review). *Int J Mol Med* **4**, 49–56.

Tsutsui, K., Ukena, K., Usui, M., Sakamoto, H., and Takase, M. (2000). Novel brain function: biosynthesis and actions of neurosteroids in neurons. *Neurosci Res* **36**, 261–273.

Turinsky, J., O'Sullivan, D. M., and Bayly, B. P. (1990). 1,2-Diacylglycerol and ceramide levels in insulin-resistant tissues of the rat in vivo. *J Biol Chem* **265**, 16880–16885.

Urs, A. N., Dammer, E. B., and Sewer, M. B. (2006). Sphingosine regulates the transcription of CYP17 by binding to steroidogenic factor-1. *Endocrinology* **147**, 5249–5258.

Urs, A. N., Dammer, E., Kelly, S., Wang, E., Merrill, A.H., Jr., and Sewer, M.B. (2007). Steroidogenic Factor-1 is a sphingolipid binding protein. *Mol Cell Endocrinology* **265–266**, 174–178.

Van Brocklyn, J. R., Graler, M. H., Bernhardt, G., Hobson, J. P., Lipp, M., and Spiegel, S. (2000). Sphingosine-1-phosphate is a ligand for the G protein-coupled receptor EDG-6. *Blood* **95**, 2624–2629.

Vianello, S., Waterman, M. R., Dalla Valle, L., and Colombo, L. (1997). Developmentally regulated expression and activity of 17alpha-hydroxylase/C-17,20-lyase cytochrome P450 in rat liver. *Endocrinology* **138**, 3166–3174.

Wang, W., Zhang, C., Marimuthu, A., Krupka, H. I., Tabrizizad, M., Shelloe, R., Mehra, U., Eng, K., Nguyen, H., Settachatgul, C., Powell, B., Milburn, M. V., and West, B. L. (2005). The crystal structures of steroidogenic factor-1 and liver receptor homologue-1. *Proc Natl Acad Sci U S A* **102**, 7505–7510.

Weber, M. M., Michl, P., Auernhammer, C. J., and Engelhardt, D. (1997). Interleukin-3 and interleukin-6 stimulate cortisol secretion from adult human adrenocortical cells. *Endocrinology* **138**, 2207–2210.

Williams-Ashman, H. G., and Reddi, A. H. (1971). Actions of vertebrate sex hormones. *Annu Rev Physiol* **33**, 31–82.

Yamada, K., and Sakane, F. (1993). The different effects of sphingosine on diacylglycerol kinase isozymes in Jurkat cells, a human T-cell line. *Biochim Biophys Acta* **1169**, 211–216.

Yamazaki, Y., Kon, J., Sato, K., Tomura, H., Sato, M., Yoneya, T., Okazaki, H., Okajima, F., and Ohta, H. (2000). Edg-6 as a putative sphingosine 1-phosphate receptor coupling to Ca(2+) signaling pathway. *Biochem Biophys Res Commun* **268**, 583–589.

Zeidan, Y. H., and Hannun, Y. A. (2007). Translational aspects of sphingolipid metabolism. *Trends Mol Med* **13**, 327–336.

Zeidan, Y. H., Pettus, B. J., Elojeimy, S., Taha, T., Obeid, L. M., Kawamori, T., Norris, J. S., and Hannun, Y. A. (2006). Acid ceramidase but not acid sphingomyelinase is required for tumor necrosis factor-{alpha}-induced PGE2 production. *J Biol Chem* **281**, 24695–24703.

Zheng, W., Kollmeyer, J., Symolon, H., Momin, A., Munter, E., Wang, E., Kelly, S., Allegood, J. C., Liu, Y., Peng, Q., Ramaraju, H., Sullards, M. C., Cabot, M., and Merrill, A. H. J. (2006). Ceramides and other bioactive sphingolipid backbones in health and disease: lipidomic analysis, metabolism and roles in membrane structure, dynamics, signaling and autophagy. *Biochem Biophys Res Commun* **1758**, 1864–1884.

Ziulkoski, A. L., Zimmer, A. R., and Guma, F. C. (2001). De novo synthesis and recycling pathways of sphingomyelin in rat Sertoli cells. *Biochem Biophys Res Commun* **281**, 971–975.

# Chapter 16
# Roles of Bioactive Sphingolipids in Cancer Biology and Therapeutics

**Sahar A. Saddoughi, Pengfei Song and Besim Ogretmen**

**Abstract** In this chapter, roles of bioactive sphingolipids in the regulation of cancer pathogenesis and therapy will be reviewed. Sphingolipids have emerged as bioeffector molecules, which control various aspects of cell growth, proliferation, and anti-cancer therapeutics. Ceramide, the central molecule of sphingolipid metabolism, generally mediates anti-proliferative responses such as inhibition of cell growth, induction of apoptosis, and/or senescence. On the other hand, sphingosine 1-phosphate (S1P) plays opposing roles, and induces transformation, cancer cell growth, or angiogenesis. A network of metabolic enzymes regulates the generation of ceramide and S1P, and these enzymes serve as transducers of sphingolipid-mediated responses that are coupled to various exogenous or endogenous cellular signals. Consistent with their key roles in the regulation of cancer growth and therapy, attenuation of ceramide generation and/or increased S1P levels are implicated in the development of resistance to drug-induced apoptosis, and escape from cell death. These data strongly suggest that advances in the molecular and biochemical understanding of sphingolipid metabolism and function will lead to the development of novel therapeutic strategies against human cancers, which may also help overcome drug resistance.

**Keywords** Apoptosis · ceramide · drug resistance · cancer therapeutics · sphingolipids

## 16.1 Introduction

Sphingolipids are a family of membrane lipids that contribute to the regulation of the fluidity and the sub-domain structure of the lipid bilayers (Futerman and Hannun, 2004). In addition, bioactive sphingolipids such as ceramide,

B. Ogretmen
Department of Biochemistry and Molecular Biology, Hollings Cancer Center, Medical University of South Carolina, 173 Ashley Avenue, Charleston, SC 29425, USA

P.J. Quinn, X. Wang (eds.), *Lipids in Health and Disease*,

Ceramide-1-phosphate, sphingosine 1-phophate (S1P), sphingosine, and glucosylceramide (GlcCer), function as bioeffector molecules, which are involved in the regulation of various aspects of cancer pathogenesis and therapy, including apoptosis, cell proliferation, cell migration, senescence, or inflammation (Futerman and Hannun, 2004; Kok and Sietsma, 2004; Ogretmen and Hannun, 2004; Reynolds et al., 2004; Fox et al., 2006; Modrak et al., 2006).

### *16.1.1 Structure and Metabolism of Ceramide*

Ceramide is composed of sphingosine, which is an amide-linked to a fatty acyl chain, varying in length from $C_{14}$ to $C_{26}$ (Fig. 16.1). Ceramide then serves as the metabolic and structural precursor for complex sphingolipids, which are composed of hydrophilic head groups, such as sphingomyelin (SM), ceramide 1-phosphate (C1P), and GlcCer (Gouaze-Andersson and Cabot, 2006). The synthesis of GlcCer is the precursor for the generation of complex glycosphingolipids and gangliosides (Fig. 16.1) (Futerman and Hannun, 2004; Futerman and Riezman, 2005). Endogenous ceramide levels are regulated by complex and integrated metabolic pathways, and each of these pathways involves a number of specialized enzymes (Fig. 16.1) (Futerman and Hannun, 2004; Futerman and Riezman, 2005). In addition to the activation of sphingomyelinases (SMases) (Andrieu-Abadie and Levade, 2002; Clarke and Hannun, 2006; Clarke et al., 2006) that hydrolyze SM to yield ceramide, endogenous ceramide can be generated via the *de novo* pathway (Dolgachev et al., 2004; Reynolds et al., 2004). In the *de novo* pathway, serine and palmitoyl CoA condense to form 3-ketosphinganine by serine palmitoyl transferase (SPT) (Merrill et al., 1988; Nagiec et al., 1996), leading to the synthesis of dihydroceramide by dihydroceramide synthases (dhCerS1-6) (Bose et al., 1995; Venkataraman et al., 2002). Then, a double bond is inserted between carbons 4–5 in the sphingosine backbone of dihydroceramide to generate ceramide (Michel et al., 1997; Kraveka et al., 2007). Ceramide can then be utilized as a substrate by ceramidases (CDases) to liberate sphingosine (Park and Schuchman, 2006), which is phosphorylated to generate S1P. Ceramide is also metabolized by the functions of ceramide kinase (CK), or SM synthase (Sugiura et al., 2002; Van der Luit et al., 2007) (Fig. 16.1), which requires the transport of ceramide from the endoplasmic reticulum (ER) to the Golgi apparatus by ceramide transporter protein, CERT, via non-vesicular transport (Hanada et al., 2003; Kumagai et al., 2005; Rao et al., 2007; Kudo et al., 2008). Ceramide can also be converted into glucosylceramide (GlcCer) in the Golgi, however this process is CERT independent (D'Angelo et al., 2007). Importantly, non-vesicular transport of GlcCer from its site of synthesis (early Golgi) to distal Golgi compartments is carried out by FAPP2, four-phosphate adaptor protein, controlling the synthesis of glycosphingolipids,

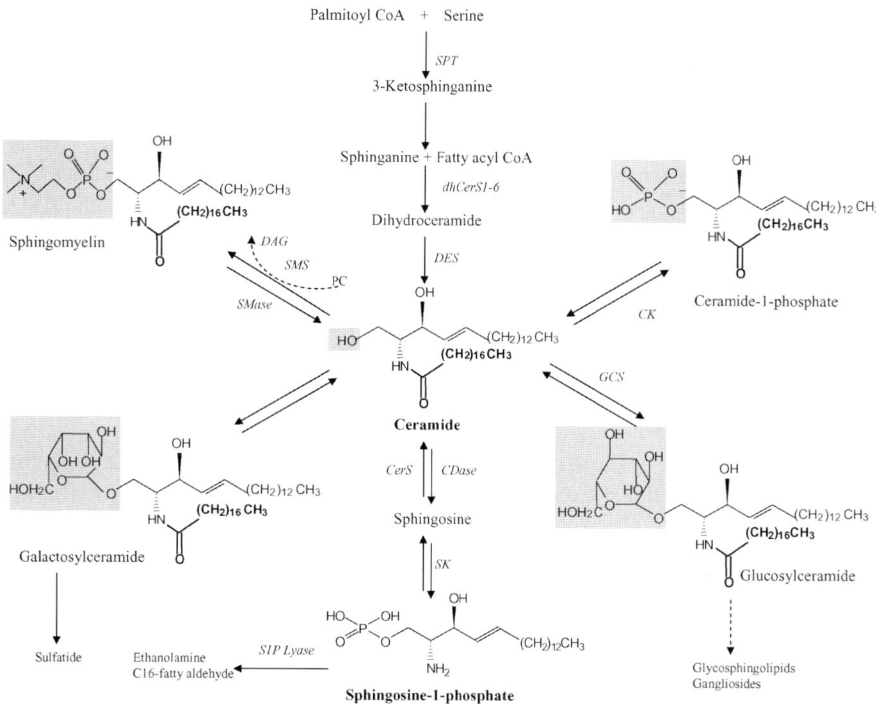


**Fig. 16.1** Sphingolipid metabolism. Sphingolipids are comprised of three main components: a sphingosine backbone, a fatty-acid chain, and a head group. Characteristics of sphingolipids change based on the head group, and recently, it was shown that the fatty-acid chain length could influence sphingolipid function. Ceramide is central in sphingolipid metabolism. Ceramide can be generated through several pathways, in particular it can be synthesized *de novo* from palmitoyl CoA and serine. Also, interestingly, ceramide, which is a pro-apoptotic sphingolipid, can proceed on to form S1P, a pro-survival sphingolipid. Thus, sphingolipid metabolic enzymes play a crucial role in determining the fate of cancer cells

which might essentially play crucial roles in determining the lipid composition of the plasma membrane (D'Angelo et al., 2007).

## *16.1.2 Ceramide Synthases and the* de novo *Generation of Ceramide*

Recently, biochemical and clinical studies indicate that different fatty-acid chain lengths of ceramide may have different functions within the cell, highlighting the importance of ceramide synthase (CerS) in sphingolipid metabolism (Pewzner-Jung et al., 2006). CerS, identified as the yeast longevity assurance gene 1 (LAG1), is known to regulate life-span/longevity in *Saccharomyces cerevisiae*, and its deletion prolongs the replicative life-span of yeast (Jazwinski and Conzelmann, 2002; Obeid and Hannun, 2003). Additionally, a LAG1 homologue, LAC1 was determined as a key component of CerS (Jazwinski and Conzelmann, 2002; Obeid and Hannun, 2003). The discovery of the mouse homologue of LAG1, also known as LASS1, or the upstream of growth and differentiation factor 1 (UOG1) (Lee, 1991; Venkataraman et al., 2002), demonstrated that it specifically regulates the synthesis of $C_{18}$-ceramide with a high degree of fatty-acid chain length specificity (Venkataraman et al., 2002; Riebeling et al., 2003; Mizutani et al., 2005; Kageyama-Yahara and Riezman, 2006). Further studies showed that there are six LASS proteins (LASS1–6), which were recently renamed as ceramide synthases 1–6 (CerS1–6) (Pewzner-Jung et al., 2006). CerS1–6 are associated with the ER membrane and contain a crucial TRAM-Lag1p-CLN8 (TLC) domain. The TLC domain constitutes the catalytic activity of CerS, and is required for the generation of ceramide (Schulz et al., 2006; Spassieva et al., 2006). All of the CerS proteins, except CerS1, contain a homeobox transcription factor HOX domain found at the N-terminus (Mesika et al., 2007) except for CerS1, which might be important for the enzymatic activity of CerS2–6. However, the physiological roles of the HOX domain of CerS2–5 are still unclear.

Importantly, as mentioned above, each CerS exerts specificity for the generation of endogenous ceramides with distinct fatty-acid chain lengths (Pewzner-Jung et al., 2006). For example, CerS1 specifically generates ceramide with an 18-carbon containing fatty-acid chain ($C_{18}$-ceramide), whereas CerS5–6 mainly generate $C_{16}$-ceramide, and to a lesser extent $C_{12}$- and $C_{14}$-ceramides (Fig. 16.2) (Pewzner-Jung et al., 2006; Cerantola et al., 2007). Indeed, CerS5 was shown to be the *bona fide* ceramide synthase for the generation of $C_{16}$-ceramide (Lahiri and Futerman, 2005), and CerS2 generates very long chain ceramides, particularly C24-ceramide (Laviad et al., 2007).

Interestingly, recent data (Koybasi et al., 2004; Karahatay et al., 2007) suggest while the levels of $C_{18}$-ceramide are generally lower, $C_{16}$-ceramide is significantly up-regulated in the majority (about 80%) of tumor tissues of head and neck squamous cell carcinoma (HNSCC) patients when compared to their adjacent normal tissues (Karahatay et al., 2007). Decreased $C_{18}$-ceramide and increased $C_{16}$-ceramide in HNSCC tumor tissues were associated with decreased and increased expression of CerS1 and 6, respectively. Remarkably, clinical analyses revealed that lower levels of $C_{18}$-ceramide in HNSCC tumor tissues are significantly associated with higher incidences of lymphovascular invasion and nodal metastasis in HNSCC patients, indicating the clinical relevance of LASS1/$C_{18}$-ceramide metabolism and signaling in HNSCC pathogenesis and progression (Karahatay et al., 2007). More importantly, these data were also supported by studies which demonstrated that defects in the LASS1-dependent generation of $C_{18}$-ceramide play important roles in HNSCC growth (Koybasi et al., 2004), and/or response to therapy in human HNSCC cells *in situ* and *in vivo* (Senkal et al., 2007). Recently, a role for LASS1 in the regulation of sensitivity to various chemotherapeutic agents has been further confirmed in an independent study using human cancer cell lines *in situ* (Min et al., 2007).

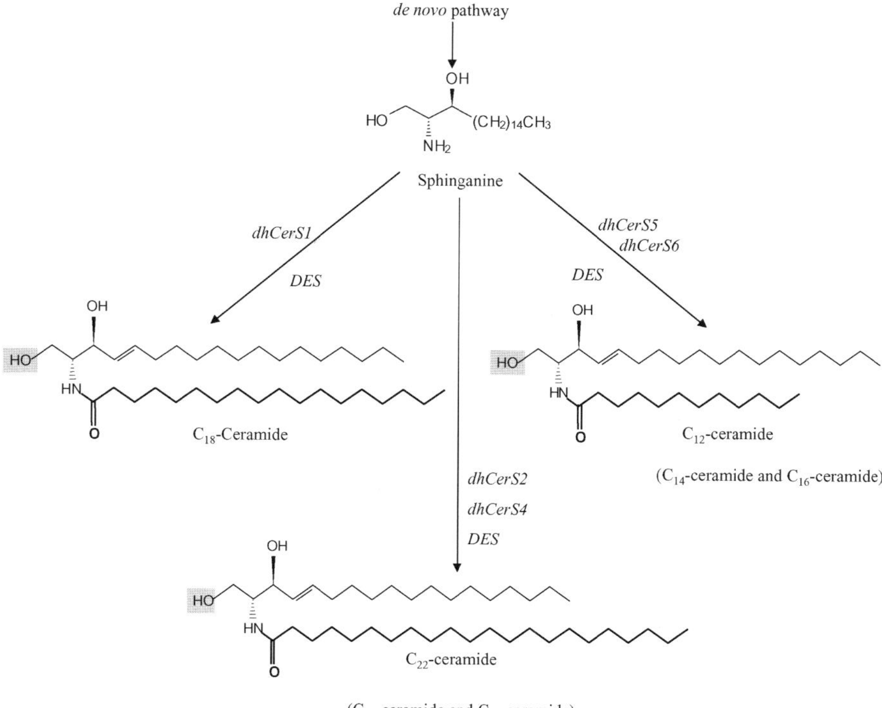


**Fig. 16.2** *De novo* generation of ceramide via the function of dhCerS. Recently identified dhCerS1–6 are responsible for the generation and determining the fatty-acid chain length of ceramide in the *de novo* pathway. For example, dhCerS1, previously known as LASS1, is responsible for generating dihydro-$C_{18}$-ceramide, whereas CersS2 and CerS4 synthesize dihydro-$C_{22}$-, $C_{24}$-, and $C_{26}$-ceramides. In addition, dihydro-$C_{12}$-, $C_{14}$- and $C_{16}$-ceramides are generated by CerS5 and CerS6. These dihydro-ceramides are then desaturated to form ceramides by DES, dihydroceramide desaturase

### *16.1.3 Down-Stream Targets of Ceramide Signaling in Cancer*

Ceramide mediates the regulation of growth arrest, senescence, and/or apoptosis (Ogretmen and Hannun, 2004). Some of these biological functions might be controlled through novel sphingolipid-protein interactions (Snook et al., 2006). Most frequently, these direct targets of ceramide constitute protein phosphatases and kinases that regulate important signaling pathways in cancer, such as Akt, protein kinase C (PKC), MAP kinases, or phospholipase D (Hannun and Obeid, 2002; Ogretmen and Hannun, 2004) (Fig. 16.3). The regulation of protein phosphatase-1 and -2 (PP1 and PP2A)-family enzymes, also referred to as ceramide activated protein phosphatases (CAPPs), by ceramide has been well documented previously (Dobrowsky and Hannun, 1992; Dobrowsky et al., 1993; Fishbein et al., 1993; Wolff et al., 1994).

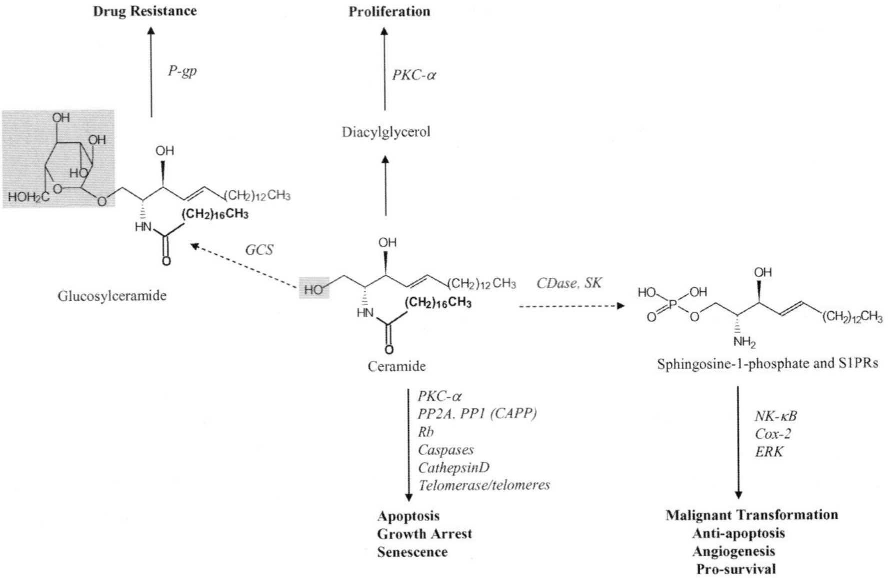

**Fig. 16.3** Ceramide and S1P signaling in cancer cells. Ceramide signaling mainly occurs through the regulation of its immediate and direct targets, leading to apoptosis, growth arrest, and senescence. Conversely, when ceramide is metabolized into S1P, cells will undergo cellular transformation, anti-apoptosis, or induction of angiogenesis. Conversion of ceramide to SM leads to the liberation of DAG from PC, and DAG is a known activator of PKC, which is involved in promoting cellular proliferation. Also, ceramide can be further metabolized into glucosylceramide, which leads to drug resistance. Phosphorylation of ceramide by CK to generate C1P also associated with induction of cell growth. Thus, these data support the hypothesis that while ceramide induces anti-proliferation, alterations in its generation and/or accumulation might result in pro-survival and anti-apoptosis

Upon ceramide-mediated activation of CAPPs, various down-stream targets, such as Bcl-2-family proteins, cyclin dependent kinases, Rb, and c-Myc oncoprotein, are regulated (Fig. 16.3) (Ogretmen and Hannun, 2004). Another important ceramide binding protein is cathepsin D, which is activated via ceramide interaction, leading to induction of apoptosis (Fig. 16.3) (Heinrich et al., 2004). Similarly, ceramide was shown to associate with protein kinase zeta (PKC-zeta) (Fig. 16.3), which seems to be important for its activation and the formation of a pro-apoptotic complex between PKC-zeta and prostate apoptotic response-4 (PAR-4) in differentiating stem cells (Wang et al., 2005). Further studies also indicated that the activation of PKC-zeta by ceramide is involved in the inactivation of Akt within the structured membrane microdomains, leading to growth arrest in vascular smooth muscle cells (Fox et al., 2007). Recently, one of the best characterized ceramide-binding proteins, CERT, was implicated in cancer biology, and the data showed that down-regulation of CERT results in increased sensitivity against chemotherapeutic agents (Swanton et al., 2007), suggesting that alterations of sphingolipid metabolism by CERT might confer a survival advantage to cancer cells. Thus, these data underscore the importance of identification of novel ceramide-protein interactions as immediate and direct mechanisms involved in the regulation of ceramide-mediated biological responses in various patho-physiological conditions, including cancer.

### 16.1.4 Sub-Cellular Functions of Ceramide

The biological roles of ceramide might also be controlled by its subcellular localization. For example, when ceramide is generated in the plasma membrane via the hydrolysis of SM by SMases, it activates pathways associated with growth inhibition, oxidative stress-mediated cell death, and lipid raft functions (Segui et al., 2001; Testai et al., 2004). Cell signaling pathways activated by mitochondrial ceramide proceeds through ceramide-activated protein phosphatases, PP1 and PP2A, involved in the regulation of Bcl-2 family proteins, cytochrome C release, and loss of mitochondrial membrane potential, leading to intrinsic cell death (Smyth et al., 1996; Thon et al., 2005). Ceramide can also be generated in the lysosomes by the function of acid SMase, which interacts with cathepsin D, leading to the cleavage of BID and subsequent apoptotic cell death in cancer cells (Heinrich et al., 2004). Lysosomal ceramides can also be metabolized by the function of acid CDase, which is over-expressed in majority of human cancer cells, and may result in resistance to apoptosis (Saad et al., 2007; Liu et al., 2008).

Ceramides generated in the ER might be topologically associated with the nucleus, since the nuclear membrane is a continuous structure of the ER membranes. In the nucleus, ceramide activates protein phosphatase-1, which then dephosphorylates serine/arginine-rich proteins (SR-proteins), which induce the alternative splicing of pro-apoptotic proteins Bcl-XS or caspase-9 (Chalfant et al., 2001, 2002). Another recently identified nuclear target of ceramide includes a pro-survival protein telomerase, which catalyzes the elongation/maintenance of telomeres at the end of chromosomes (Blackburn, 2005). Core telomerase contains two main subunits, telomerase reverse transcriptase (hTERT), and the RNA component (hTR), which acts as an intrinsic template (Blackburn, 2005). Ceramide mediates the repression of the hTERT promoter, and mediates the inhibition of telomerase activity (Ogretmen et al., 2001a). Specifically, in addition to inhibition of c-Myc-dependent activation of hTERT promoter (Ogretmen et al., 2001b), exogenous $C_6$-ceramide, or the $C_{18}$-ceramide generated by CerS1 induces the deacetylation of Sp3 transcription factor by histone deacetylase 1 (HDAC1) (Wooten and Ogretmen, 2005), and deacetylated Sp3 then helps the recruitment of HDAC1 to the promoter of hTERT, which causes local histone deacetylation and repression of the hTERT promoter in human lung cancer cells (Wooten-Blanks et al., 2007). Interestingly, overexpression of CerS6, which generates $C_{16}$-ceramide did not inhibit hTERT expression (Wooten-Blanks et al., 2007), supporting the novel view that endogenous ceramides with different fatty-acid chain lengths might have distinct biological roles, and targets.

## 16.2 Anti-Proliferative Roles of Ceramide

### *16.2.1 Ceramide and Apoptosis*

Apoptosis can be induced by various factors including chemotherapeutic agents, CD95, tumor necrosis factor-1, growth factor withdrawal, hypoxia, or DNA damage. Many of these mediators of apoptosis are regulators of ceramide generation, suggesting a role for ceramide in apoptosis (Pettus et al., 2002). There are a myriad of studies which indicate that the changes in endogenous levels of ceramide in response to these agents occur before triggering an apoptotic cascade (Dbaibo et al., 1997, Ogretmen and Hannun, 2004). Additionally, increasing the endogenous levels of ceramide with inhibitors of ceramide metabolism enzymes or by overexpression of ceramide-generating enzymes results in apoptosis and/or growth arrest (Abe et al., 1995; Bielawska et al., 1996). For example, in leukemia cells, expression of bacterial SMase, which generates ceramide from intracellular pools of SM, has been shown to cause a significant increase in ceramide levels and to induce apoptosis (Zhang et al., 1997). In addition, ionizing radiation activates acid SMase for ceramide generation. Importantly, human lymphoblasts and mice deficient in SMase are resistant to high doses of irradiation, suggesting an active role of ceramide in the regulation of apoptosis (Santana et al., 1996). On the other hand, inhibitors of the *de novo* pathway, such as fumonisin B1 (Plo et al., 1999) prevents apoptosis in response to these agents, further providing evidence for the role of ceramide generation in mediating apoptosis.

One of the mechanisms by which ceramide regulates apoptosis is via the induction of Fas capping, which involves the lateral segregation of cross-linked Fas ligand with its surface receptor at the SM-enriched plasma membrane of Jurkat T lymphocytes, necessary for its optimal function in cell killing (Cremesti et al., 2001). On the other hand, cells that are resistant to ceramide- and CD95/Fas-induced apoptosis have defective mitochondrial apoptosis (Raisova et al., 2000), indicating that perturbations of ceramide-CD95/Fas signaling can result in the development of resistance to cell death in human cancer cells.

In conclusion, these studies show that ceramide is closely associated with apoptosis, and that it plays an important role in how cells respond to various stress stimuli for induction of apoptosis in various cancer models.

### *16.2.2 Ceramide in Growth Inhibition and Differentiation*

One of the well characterized functions of ceramide is its capability to induce a G0/G1 cell cycle arrest, which can be linked to the activation of the retinoblastoma gene product (Rb) (Dbaibo et al., 1995). In addition, ceramide specifically inactivates the cyclin-dependent kinase cdk2, but not cdk4, through activation of a phosphatase (Lee et al., 2000). An important *in vivo* example of ceramide-mediated growth arrest was observed with the use of ceramide-coated balloon catheters, which caused growth arrest of vascular smooth muscle-cells (VSMC) after stretch injury *in vivo* (Charles et al., 2000). Mechanistically, this growth arrest of VSMC was linked to ceramide-induced Akt inhibition, which was mediated through PKC-zeta (Bourbon et al., 2002).

The idea that ceramide is a regulator of cell differentiation has been recognized since the discovery that vitamin D3-induced differentiation of HL-60 and U037 human leukemia cells resulted in a progressive increase in the hydrolysis of SM by neutral-SMase (N-SMase), resulting in the elevation of ceramide, which induced monocytic, but not neutrophilic or macrophage-type, differentiation of these cells (Okazaki et al., 1989). In neuronal cell lines, ceramide mimics nerve growth factor function, and induces differentiation in T9 glioma cells, Purkinje cells, and hippocampal neurons (Dobrowsky et al., 1994).

### *16.2.3 Ceramide and Senescence*

A breakthrough in understanding the relationship between ceramide and senescence came with the observation that ceramide increased significantly as human fibroblasts entered the senescent phase (Venable et al., 1995). This was supported by the fact that fibroblasts that were treated with ceramide recapitulated the morphologic and biochemical changes of senescence such as activation of Rb, regulation of cdk's, or inhibition of growth factor signaling (Venable et al., 1995). Mechanistically, these changes induced by ceramide occur through inhibition of phospholipase D, which leads to the reduction of diacylglycerol (DAG) generation, and results in the failure to translocate and activate PKC to the membrane, a critical response in transducing mitogenic stimuli (Venable et al., 1995). Additionally, the fact that yeast aging genes, *lac1* and *lag1* are known to be essential components of ceramide synthase (Guillas et al., 2001) provides a genetic link between ceramide and aging.

Senescence is also regulated by alterations in telomere length, which is maintained by telomerase, one of the down-stream targets of ceramide signaling. The telomerase regulation by ceramide involves two distinct mechanisms; the inactivation of c-Myc transcription factor via increased ubiquitin/proteasome function for its rapid proteolysis, which otherwise activates the hTERT promoter, and the recruitment of Sp3/HDAC1 repressor complex into the hTERT promoter for repression in the A549 human lung adenocarcinoma cell line (Ogretmen et al., 2001a,b; Wooten and Ogretmen, 2005; Wooten-Blanks et al., 2007). In addition, ceramide also mediates telomerase-independent rapid shortening of telomere length in A549 cells (Sundararaj et al., 2004). Ceramide-mediated telomere shortening was linked to the inhibition of an unexpected role of a nuclear form of glyceraldehydes 3-phosphate dehydrogenase (GAPDH) in telomere/binding and protection function (Sundararaj et al., 2004). Taken together, these results support that ceramide and sphingolipids play important roles in the regulation of senescence and aging.

## 16.3 S1P and S1P Receptor Signaling in Cancer Biology

S1P, a product of SK (Fig. 16.1), is considered to be a pro-survival lipid, because of its involvement in malignant transformation, cancer proliferation, inflammation, vasculorogenesis, and resistance to apoptotic cell death (Maceyka et al., 2002; Hla, 2004; Taha et al., 2006a). Increased generation of S1P triggers signaling pathways that mediate these pro-survival processes mainly by engaging with S1P receptors 1–5 (S1PR), members of the family of transmembrane hepta-helical, G-protein-coupled receptors, which are encoded by the endothelial differentiation genes (EDG), in paracrine or autocrine modes after being secreted from the cell (Fig. 16.3) (Rosen and Goetzl, 2005). Overexpression of SK1 results in malignant transformation and tumor formation in 3T3 fibroblasts (Xia et al., 2000). Increased S1P levels promote proliferation and survival in human glioma and breast cancer cells (Nava et al., 2002; Van Brocklyn et al., 2002; Sarkar et al., 2005). Additionally, in endothelial and smooth muscle cells, SK1/S1P/S1PR signaling also has several specific effects on the promotion of vascular development and angiogenesis (Argraves et al., 2004; Chae et al., 2004; Mizugishi et al., 2005). Importantly, in ovarian cancer patient samples, elevated S1P levels are found in the serum (Tilly and Kolesnick, 2002), and S1PR1 has been recently discovered to be a requirement for tumor angiogenesis *in vivo* (Rosen and Goetzl, 2005).

Interestingly, partial inhibition of SK1 expression results in apoptosis in MCF-7 human breast cancer cells (Taha et al., 2006b). An important role for SK1/S1P pathway has also been shown in a colon carcinogenesis model in rats, which was linked to the up-regulation of Cox-2 (Kawamori et al., 2006).

S1P generation can also be catalyzed by SK2, and overexpression of SK2, interestingly, mediates growth inhibitory effects, possibly via the pro-apoptotic functions of its BH3-like domain (Liu et al., 2003). However, recently, it was suggested that endogenous SK2, which is localized mainly to the nucleus, might act similar to SK1, providing a pro-survival characteristics to cancer cells. Specifically, inhibition of the expression of SK2 in the nucleus results in p21-mediated growth inhibition (Sankala et al., 2007). To address these ambiguous data regarding opposing roles of exogenous and endogenous SK2 in the regulation of cancer cell death or survival, it was suggested that overexpressed SK2 might not be effectively targeted to the nucleus, leading to the release of its BH3-like domain into the cytosol, and this might be a reasons why it acts as a pro-apoptotic molecule in studies in which it is exogenously introduced into the cells (Sankala et al., 2007). These interesting data further confirm the importance of subcellular localization of sphingolipid metabolism in the regulation of cancer cell growth and/or apoptosis (Birbes et al., 2001).

## 16.4 Sphingolipid Signaling in Cancer Therapy

Previous studies suggest that understanding the intrinsic mechanisms of action for ceramide and S1P can open doors to new therapies to battle cancer. It has been well established that increases in intracellular ceramide will promote apoptosis. Thus, finding ways to intrinsically elevate ceramide in cancer cells is desirous. Conversely, S1P has been shown to promote cancer pathogenesis, thus, suppression of its generation/accumulation could suppress tumor growth. Some of these therapeutic approaches are summarized in Table 16.1.

**Table 16.1** Sphingolipid analogues & inhibitors of ceramide metabolism

| Compound | Mode of Action | Cancer Type |
|---|---|---|
| B13 | Acid ceramidase inhibitor | Prostate, Colon and HNSCC |
| D-MAPP | Neutral/ Alkaline ceramidase inhibitor | Squamous cell carcinoma |
| Pyridinium ceramide | Mitochondrial targeting | HNSCC, lung, colon and breast |
| 4,6-diene-ceramide | Ceramide analogue | Breast |
| C16-serinol | Ceramide analogue | Neuroblastoma |
| PPMP, PPPP | GCS inhibitors | Solid tumors |
| Dimethylsphingosine | Sphingosine Kinase inhibitor | Leukemia, colon and breast |
| Anti-S1P monoclonal antibody | Binds S1P | Solid tumors |
| Pegylated lyposomes with ceramide | Improved delivery | Breast |
| Vincristine in sphingomyelin-liposomes | Improved delivery | Acute lymphoid leukemia |
| Safingol (L-t-dihydro-sphingosine) | Sphingosine kinase inhibitor | Solid tumors |
| FTY-720 | Myriocine analogue | Bladder, prostate, breast, lymphoma |

D-MAPP, D-*erythro*-2-(N-myristoylamino)-1-phenyl-1-propanol; HNSCC, Head and neck squamous cell carcinoma; PPMP, 1-phenyl-2-palmitoylamino-3-morpholino-1-propanol; GCS, glucosylceramide synthase.

### *16.4.1 Ceramide Metabolism in Cancer Therapeutics*

Increasing endogenous ceramide has been suggested as an effective method to regulate cancer cell growth. To this end, certain chemotherapeutic agents such as daunorubicin, camptothecin, fludarabine, etoposide, and gemcitabine increase ceramide generation through the *de novo* pathway, or via activation of SMases (Futerman and Hannun, 2004; Ogretmen and Hannun, 2004). Alternatively, targeting enzymes of ceramide clearance appears to elevate endogenous ceramide, leading to increased anti-proliferative responses in various cancer cells (Fox et al., 2006; Ogretmen, 2006). Moreover, combining the chemotherapeutic agent gemcitabine with SM synergistically inhibits pancreatic tumor growth *in vivo* (Modrak et al., 2004). The therapeutic effects of doxorubicin were also enhanced when used in combination with SM in various human cancer cell lines (Veldman et al., 2004). Mechanistically, SM was shown to increase the cellular uptake of doxorubicin via altering cell membrane permeability, leading to increased accumulation and bioavailability of the drug in these cells (Veldman et al., 2004). Also, $\Delta^9$-tetrahydrocannabinol exerts apoptosis in Jurkat cells via the CB(2) cannabinoid receptor by activation of ceramide generation *de novo*, which plays a role in the mitochondrial intrinsic pathway (Carracedo et al., 2006; Herrera et al., 2006).

Additionally, small molecule inhibitors of the sphingolipid pathway to induce the accumulation of ceramide have been used in some cancers. For example, B13, an inhibitor of acid CDase was used in a metastatic colon cancer mouse model and a prostate cancer xenograph model (Selzner et al., 2001; Samsel et al., 2004). In both cases, B13 caused the accumulation of ceramide and resulted in prevention of tumor growth. Another effective approach to increase ceramide accumulation in cancer cells has been to inhibit SM synthase , or acid CDase (Meng et al., 2004; Saad et al., 2007).

In addition, the use of ceramide analogues or mimetics could also promote apoptotic pathways in cancer cells. In fact, many studies report that exogenous treatment with ceramides induces cell death, and/or growth arrest (Szulc et al., 2006; Bielawska et al., 2008). These findings were supported with *in vivo* studies, in which treatment with recently developed exogenous ceramides (ceramidoids) inhibited cancer cell growth, and decreased tumor progression in HNSCC and other cancer models (Senkal et al., 2006; Szulc et al., 2006; Bielawska et al., 2008).

It should also be noted that treatment of cells with exogenous ceramides may result in the generation of endogenous long chain ceramides via the sphingosine recycling pathway, which can be blocked by FB1, and not by myriocin. This alternative pathway for the generation of endogenous ceramide seems to be important for the regulation of telomerase and c-Myc in A549 cells (Ogretmen et al., 2002; Sultan et al., 2006).

In a recent study, treatment of prostate and lung cancer cells with γ–tocopherol (γT), the main dietary form of vitamin E, inhibited cell proliferation and induced apoptosis (Jiang et al., 2004), which were concomitant with the accumulation of dihydroceramides. Importantly, fenretinide, which was initially thought to increase ceramide generation, has been reported by various, independent groups, to elevate dihydroceramides, possibly via the negative regulation of dihydroceramide desaturase (DES) (Schulz et al., 2006; Zheng et al., 2006; Kraveka et al., 2007). In fact, recent data revealed that accumulation of endogenous dihydroceramides via down-regulation of DES results in Rb-dependent growth arrest in human neuroblastoma cells (Kraveka et al., 2007). Although dihydroceramides are thought to be inert or biologically inactive molecules, these data suggest that they, too—at least when generated in cells—might be important in the regulation of cancer cell growth and/or survival.

### *16.4.2 Targeting the SK1/S1P Pathway as an Anti-Cancer Therapeutic*

Down-regulation of S1P biosynthesis provides another therapeutic modality for the treatment of cancers. On the other hand, exogenous S1P treatment exerts a protective role against cell death in normal (non-cancerous) human cells (Tilly and Kolesnick, 2002). Additionally, S1P has suppressive effects against chemotherapy-induced apoptosis in the ovary (Tilly and Kolesnick, 2002), and also blocked male germ cell apoptosis in the human testis (Suomalainen et al., 2005). Thus, one anti-cancer therapeutic strategy has been to use inhibitors of SK1. Preliminary studies *in situ* and in animal models indicate that SK1 inhibitors prevent cancer cell proliferation and tumor growth (French et al., 2006). There are also novel S1PR1 and S1PR3 antagonists (Davis et al., 2005), and inhibition of these receptors with these compounds may inhibit cancer cell growth. Another novel approach to cancer therapeutics is the use of a monoclonal antibody that binds S1P with high affinity and specificity (Visentin et al., 2006). The anti-S1P monoclonal antibody significantly reduced tumor progression in various murine xenograft and allograft models (Visentin et al., 2006). The anti-S1P monoclonal antibody also prevented S1P-induced cell proliferation, release of pro-angiogenic cytokines, and protection of tumor cells from apoptosis by S1P (Visentin et al., 2006). Thus, these data strongly suggest that S1P may present an important target for anti-cancer therapeutics.

Additional experimental evidence supporting the pro-survival roles of S1P/S1PR axis was obtained by employing the potent immunosuppressive agent FTY720, which is known to engage with S1PRs (Brinkmann et al., 2004; Chun and Rosen, 2006). FTY-20 is phosphorylated *in vivo* to FTY720-P possibly by SK2 (Billich et al., 2003; Paugh et al., 2003), which then induces sequestration of lymphocytes in lymph tissues by engaging S1PRs with high affinity and specificity (Brinkmann et al., 2004; Chun and Rosen, 2006). These data implicate that S1PRs play important roles in immunosuppression. More importantly, treatment with FTY720 inhibits angiogenesis and tumor vascularization, and mediates cell death, suggesting that it might be exploited as an anti-cancer therapeutic agent (LaMontagne et al., 2006).

Thus, these data support the hypothesis that inhibition of SK1 may enhance the treatment of cancer cells, whereas selective elevation of S1P in normal cells may provide protection against toxicity during therapy.

## 16.5 Sphingolipids in Drug Resistance and Chemoprevention

One of the main obstacles involved in cancer therapy is the development of drug resistance. Interestingly, a relationship between the changes in the sphingolipid metabolism and development of drug resistance in human cancer cells has been documented (Ogretmen and Hannun, 2001; Senchenkov et al., 2001; Radin, 2002; Hinrichs et al., 2005). Therefore, one possible approach to overcome this resistance could be through modulation of the sphingolipid pathway.

### *16.5.1 Role of Ceramide in Drug Resistance*

Recent studies indicate that one mechanism of resistance that cancer cells develop against chemotherapy is the alteration of ceramide accumulation. In fact, ceramide is highly metabolized into GlcCer due to an increase in glucosylceramide synthase (GCS) activity and/or expression in some cancer cells (Gouaze-Andersson and Cabot, 2006). This phenomenon has been implicated in development of drug resistance in various cancer cell types, especially in breast cancer cells (Fig. 16.3) (Senchenkov et al., 2001). Although the role of GCS in the development of drug-resistance has been challenged in some cancer models (Veldman et al., 2003; Norris-Cervetto et al., 2004), a mechanistic link between GCS and P-glycoprotein (P-gp), an ABC transporter implicated in drug resistance, has been recently revealed (Gouaze et al., 2005; Gouaze-Andersson et al., 2007). The data demonstrated that knockdown of GCS expression with small interfering RNA (siRNA) significantly inhibits the expression of MDR1, a gene that encodes for P-gp, and reverses drug resistance (Gouaze et al., 2005; Gouaze-Andersson et al., 2007). These data are consistent with an earlier study which showed that, SDZ PSC 833, an inhibitor of P-gp, inhibits GCS, and alters GlcCer levels (Goulding et al., 2000). Additionally, increased accumulation of GlcCer is found in cells overexpressing P-gp (Gouaze et al., 2004). Interestingly, several members of the ABC transporter family are implicated in the translocation of phospholipids and sphingolipids across the lipid bilayer, and P-gp has been proposed as a specific transporter for glucosylceramide that translocates this molecule across the Golgi to deliver it for the synthesis of neutral glycosphingolipids (De Rosa et al., 2004). Thus, P-gp and

GCS appear to function in the same pathway of ceramide/GlcCer metabolism, and this may provide an important link for the function of GCS in drug resistance.

These results also suggest that inhibitors of GCS may be useful in preventing chemotherapy resistance. For example, combinations of fenretinide (4-HPR), which is known to elevate ceramide (Wang et al., 2001) and (dihydro)ceramide levels (Schulz et al., 2006; Zheng et al., 2006; Kraveka et al., 2007), with the inhibitors of GCS or SK, such as PPMP (1-phenyl-2-palmitoylamino-3-morpholino-1-propanol) or safingol (L-threo-dihydrosphingosine), were reported to suppress the growth of various human cancer cells, synergistically (Maurer et al., 1999, 2000).

### *16.5.2 Sphingosine Kinase and S1P in Drug Resistance*

SK1/S1PR signaling protects cancer cells from chemotherapy-induced apoptosis; therefore, changes in sphingolipid metabolism play functional roles in conferring drug resistance. For example, in prostate adenocarcinoma SK-1 regulates drug-induced apoptosis and serves as a chemotherapy sensor both in culture and in animal models (Pchejetski et al., 2005). In parallel with these data, increasing the expression of SK-1 reduced the sensitivity of A-375 melanoma cells to Fas- and ceramide-mediated apoptosis that could be reversed by inhibition of SK-1 expression (Bektas et al., 2005). Also, high expression levels of SK1 and S1PRs were detected in camptothecin- (CPT) resistant PC3 prostate cancer cells (Akao et al., 2006). Specifically, inhibition of SK1 expression or S1PR signaling significantly inhibited PC3 cell growth, and treatment of these cells with CPT induced upregulation of SK1/S1PR signaling (Akao et al., 2006). These data are also supported by studies conducted using the model organism *Dictyostelium discoideum*, in which modulation of SK or S1P-lyase contribute to altered sensitivity to cisplatin (Alexander et al., 2006). The role of S1P-lyase in increased sensitivity to this drug in a p38-, and to lesser extent, c-Jun NH2-terminal kinase- (mitogen activated protein kinases) dependent manner, was confirmed in human A549 lung cancer and HEK293 cells (Min et al., 2005). The role of S1P-lyase in the regulation of apoptosis has been also demonstrated in various human cancer models previously (Oskouian et al., 2006).

Importantly, in a recent study, overexpression of SK1 is linked to the upregulation of Bcr-Abl, leading to alterations of the balance between pro-apoptotic $C_{18}$-ceramide and pro-survival S1P, leading to resistance to imatinib mesylate in K562 human CML cells (Baran et al., 2007). Importantly, down-regulation of SK1 significantly reversed resistance to drug-induced apoptosis in these cells (Baran et al., 2007).

Thus, taken together, these studies indicate that therapeutic strategies that induce tumor levels of ceramide, particularly $C_{18}$-ceramide in HNSCC, while decreasing S1P accumulation in the serum would be ideal for improving the therapeutic outcome of some cancers in the clinic (Fig. 16.4).

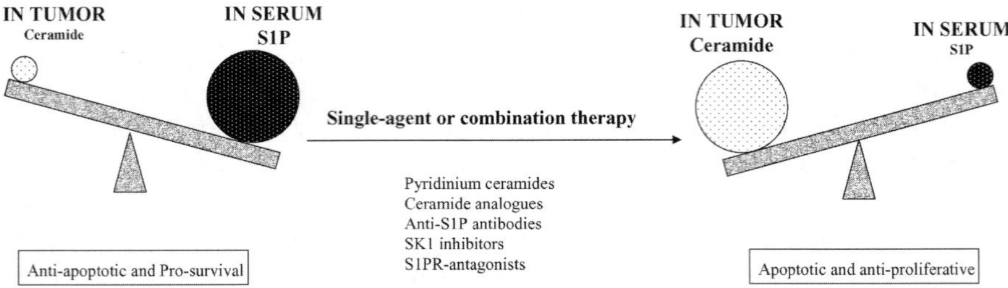


**Fig. 16.4** Roles of ceramide and S1P in anti-cancer therapeutics. The cellular balance between ceramide in tumor samples and S1P is believed to determine the fate of cancer cells. Often, within a tumor, there are altered levels/accumulation of ceramide (such as low $C_{18}$-ceramide levels in HNSCC) and high S1P levels, which are often secreted into the serum at relatively high concentrations. This causes the metabolic balance to favor S1P resulting in a pro-survival outcome. However, new therapies are being explored to shift this balance in favor of ceramide. Ultimately, these new treatments are aimed at increasing ceramide levels while inhibiting S1P generation, secretion, or signaling through S1PRs

### *16.5.3 Chemopreventive Roles of Sphingolipids*

The roles of sphingolipids in chemoprevention have also been reported previously (Borek and Merrill, 1993). Administration of SM in the diet prevents the formation of chemically induced colon cancer tumors and aberrant colonic crypts by decreasing the rate of cell proliferation and increasing apoptosis in mice (Schmelz et al., 1996). Diets supplemented with ceramide, SM, glucosylceramide, lactosylceramide, or ganglioside GD3 to C57B1/6 J(Min/+) mice with a truncated APC gene product, reduced the number of tumors in the intestine (Schmelz et al., 1996, 2000).

Moreover, resveratrol, a compound found at high levels in red wine, induces apoptosis via increased generation of *de novo*-generated ceramide in breast cancer cells *in situ* (Scarlatti et al., 2003, 2007; Minutolo et al., 2005).

In addition, the role of S1P in the induction of Cox2 expression, and in the production of prostaglandins (Pettus et al., 2003), coupled with findings of increased SK1 in colon cancer and colon carcinogenesis (Kawamori et al., 2006), raises the possibility of utilizing SK1 as a novel target for chemoprevention. Thus, there are strong leads which suggest that sphingolipids may play roles in chemoprevention.

## 16.6 Sphingolipid-Based Anti-Cancer Therapeutics

One common approach to promote apoptosis in cancer cells is the use of exogenous ceramide analogues or mimetics as therapeutic agents (Table 16.1). There have been tremendous recent improvements in the design and delivery of these ceramides. For example, varied chain pyridinium ceramides (Pyr-Cers)

have been synthesized with increased water solubility and cell-membrane permeability (Novgorodov et al., 2005; Rossi et al., 2005; Dindo et al., 2006; Senkal et al., 2006; Szulc et al., 2006; Dahm et al., 2008). The positive charge of the pyridinium ring in these structures allows targeting and accumulation of these ceramide analogues mainly into mitochondria, and to a lesser extent, to the nucleus of cancer cells. There are studies which suggest that cancer cells acquire, in general, a more negative charge in their sub-cellular structures (especially mitochondria) (Modica-Napolitano and Aprille, 2001), therefore, Pyr-Cer can preferentially target cancer cells with minimum toxicity to normal cells. In fact, data indicate that L-t-$C_6$-Pyr-Cer and D-e-$C_{16}$-Pyr-Cer preferentially accumulates in mitochondria-, and nuclei-enriched fractions in several human cancer cells *in vitro* (Novgorodov et al., 2005; Rossi et al., 2005; Dindo et al., 2006; Senkal et al., 2006; Dahm et al., 2008), and this was consistent with the higher accumulation of the compound in the HNSCC tumor site, compared to the liver and intestines *in vivo* (Senkal et al., 2006). The accumulation of Pyr-Cer in the mitochondria dramatically altered the structures and functions of mitochondria, resulting in a decrease of the mitochondrial membrane potential, release of mitochondrial cytochrome C, activation of caspase-3 and caspase-9 and causing apoptotic cell death (Novgorodov et al., 2005; Rossi et al., 2005; Dindo et al., 2006).

Other novel structural analogs of ceramide, such as $C_{16}$-serinol and (2S,3R)-(4E,6E)-2-octanoylamidooctadecadiene-1,3-diol (4,6-diene-ceramide) induced apoptosis in various human cancer cells (Bieberich et al., 2000; Struckhoff et al., 2004). Other ceramide analogs, 5R-OH-3E-$C_8$-ceramide, adamantyl-ceramide and benzene-$C_4$-ceramide selectively inhibited the growth of drug-resistant human breast cancer cell lines (SKBr3 and MCF-7/Adr) (Crawford et al., 2003).

In another approach, delivery of exogenous ceramide in pegylated liposomes, which are known to be more effective at crossing the cell membrane, increased growth inhibitory effects in human breast cancer cells, via enhanced accumulation of ceramide (Stover and Kester, 2003). The liposomal delivery of exogenous natural ceramide also resulted in the inhibition of phosphorylated Akt and stimulated the activity of caspase-3/7 more effectively than non-liposomal ceramide (Stover and Kester, 2003). *In vivo* therapeutic efficacy of the pegylated ceramide for the delivery of exogenous ceramide, which resulted in slower tumor growth in murine models of breast cancer, was also demonstrated (Stover et al., 2005). Encapsulated vincristine in SM-liposomes, also called sphingosomes has improved efficacy compared to the conventional drug in animal models of adult acute lymphocytic leukemia (ALL), and sphingosomal vincristine is now in clinical phase II trials for treatment of patients with recurrent and refractory adult ALL (Thomas et al., 2006). Polymeric nanoparticle delivery systems can also be used to improve the delivery and efficacy of ceramides for the treatment of cancer cells to overcome resistance (van Vlerken et al., 2007).

These preliminary approaches point to the feasibility of developing more active and perhaps more selective analogs of ceramide, which can be tested in clinical trials for the treatment of patients with cancer. They also raise the possibility that several enzymes of ceramide clearance (SM synthases, CDases, CK, and GCS) may serve as novel therapeutic targets. In fact, several independent studies showed that down-regulation of acid CDase, or inhibition of its activity, induces apoptosis, and inhibits tumor growth (Samsel et al., 2004; Morales et al., 2007; Holman et al., 2008). Similar data were also reported for the down-regulation of CK, which inhibited cellular proliferation and enhanced apoptosis induced by serum starvation in A549 human lung cancer cells (Mitra et al., 2007).

## 16.7 Summary and Future Perspectives

The emerging roles of bioactive sphingolipids in regulating various facets of cancer pathogenesis and therapeutics have been demonstrated in various models. Experimental evidence suggests that there is altered regulation of ceramide levels, coupled with alterations of expression and/or activity of enzymes of sphingolipid metabolism in several cancers, that is consistent with the potential tumor-suppressor functions of ceramide. On the other hand, the SK/S1P/S1PR axis is increasingly implicated in pro-survival, anti-apoptosis, neovascularization, and inflammation. In addition, *in situ* and *in vivo* studies indicate functions of sphingolipids in chemoprevention, especially in colon cancers. Therefore, ceramide analogues/mimetics, and inhibitors of SK or enzymes of ceramide clearance might be exploited for the development of novel strategies for anticancer therapeutics (Fig. 16.4 and Table 16.1).

These therapeutic efforts will be improved when the complexities of the networks of sphingolipid metabolism, subcellular compartmentalization, and signaling are better understood by using conventional biochemical and molecular biological techniques integrated with more sophisticated approaches such as lipidomics and bioinformatics. To date, most, if not all, of the enzymes involved in sphingolipid metabolism have been identified and cloned, providing the molecular tools and insights required to discover novel functions of sphingolipids in cancer pathogenesis and therapy. Thus, in the next few years, significant advances are expected in understanding the mechanistic function of bioactive sphingolipids in the regulation of cancer progression, metastasis, and therapeutics.

**Acknowledgments** We thank Drs. Z.M. Szulc and Jennifer Schnellmann for their critical review of this chapter. We also thank the members of the Ogretmen laboratory for their helpful discussions. We apologize to those investigators whose important work was not included in this chapter because of space limitations. The Ogretmen laboratory is supported by research grants from the National Institutes of Health. S.A. Saddoughi is sponsored by funding from Wachovia, and Abney Foundations.

## References

Abe, A., Radin, N.S., Shayman, J.A., Wotring, L.L., Zipkin, R.E., Sivakumar, R., Ruggieri, J.M., Carson, K.G., and Ganem, B. Structural and stereochemical studies of potent inhibitors of glucosylceramide synthase and tumor cell growth. *J Lipid Res*, **36,** 1995, 611–621.

Akao, Y., Banno, Y., Nakagawa, Y., Hasegawa, N., Kim, T.J., Murate, T., Igarashi, Y., and Nozawa, Y. High expression of sphingosine kinase 1 and S1P receptors in chemotherapy-resistant prostate cancer PC3 cells and their camptothecin-induced up-regulation. *Biochem Biophys Res Commun*, **342,** 2006, 1284–1290.

Alexander, S., Min, J., and Alexander, H. Dictyostelium discoideum to human cells: pharmacogenetic studies demonstrate a role for sphingolipids in chemoresistance. *Biochim Biophys Acta*, **1760,** 2006, 301–309.

Andrieu-Abadie, N. and Levade, T. Sphingomyelin hydrolysis during apoptosis. *Biochim Biophys Acta*, **1585,** 2002, 126–134.

Argraves, K.M., Wilkerson, B.A., Argraves, W.S., Fleming, P.A., Obeid, L.M., and Drake, C.J. Sphingosine-1-phosphate signaling promotes critical migratory events in vasculogenesis. *J Biol Chem*, **279,** 2004, 50580–50590.

Baran, Y., Salas, A., Senkal, C.E., Gunduz, U., Bielawski, J., Obeid, L.M., and Ogretmen, B. Alterations of ceramide/sphingosine 1-phosphate rheostat involved in the regulation of resistance to imatinib-induced apoptosis in K562 human chronic myeloid leukemia cells. *J Biol Chem*, **282,** 2007, 10922–10934.

Bektas, M., Jolly, P.S., Muller, C., Eberle, J., Spiegel, S., and Geilen, C.C. Sphingosine kinase activity counteracts ceramide-mediated cell death in human melanoma cells: role of Bcl-2 expression. *Oncogene*, **24,** 2005, 178–187.

Bieberich, E., Kawaguchi, T., and Yu, R.K. N-acylated serinol is a novel ceramide mimic inducing apoptosis in neuroblastoma cells. *J Biol Chem*, **275,** 2000, 177–181.

Bielawska, A., Greenberg, M.S., Perry, D., Jayadev, S., Shayman, J.A., McKay, C., and Hannun, Y.A. (1S,2R)-D-erythro-2-(N-myristoylamino)-1-phenyl-1-propanol as an inhibitor of ceramidase. *J Biol Chem*, **271,** 1996, 12646–12654.

Bielawska, A., Bielawski, J., Szulc, Z.M., Mayroo, N., Liu, X., Bai, A., Elojeimy, S., Rembiesa, B., Pierce, J., Norris, J.S., and Hannun, Y.A. Novel analogs of d-e-MAPP and B13. Part 2: Signature effects on bioactive sphingolipids. *Bioorg Med Chem*, **16,** 2008, 1032–1045.

Billich, A., Bornancin, F., Devay, P., Mechtcheriakova, D., Urtz, N., and Baumruker, T. Phosphorylation of the immunomodulatory drug FTY720 by sphingosine kinases. *J Biol Chem*, **278,** 2003, 47408–47415.

Birbes, H., El Bawab, S., Hannun, Y.A., and Obeid, L.M. Selective hydrolysis of a mitochondrial pool of sphingomyelin induces apoptosis. *Faseb J*, **15,** 2001, 2669–2679.

Blackburn, E.H. Telomeres and telomerase: their mechanisms of action and the effects of altering their functions. *FEBS Lett*, **579,** 2005, 859–862.

Borek, C. and Merrill, A.H., Jr. Sphingolipids inhibit multistage carcinogenesis and protein kinase C. *Basic Life Sci*, **61,** 1993, 367–371.

Bose, R., Verheij, M., Haimovitz-Friedman, A., Scotto, K., Fuks, Z., and Kolesnick, R. Ceramide synthase mediates daunorubicin-induced apoptosis: an alternative mechanism for generating death signals. *Cell*, **82,** 1995, 405–414.

Bourbon, N.A., Sandirasegarane, L., and Kester, M. Ceramide-induced inhibition of Akt is mediated through protein kinase Czeta: implications for growth arrest. *J Biol Chem*, **277,** 2002, 3286–3292.

Brinkmann, V., Cyster, J.G., and Hla, T. FTY720: sphingosine 1-phosphate receptor-1 in the control of lymphocyte egress and endothelial barrier function. *Am J Transplant*, **4,** 2004, 1019–1025.

Carracedo, A., Lorente, M., Egia, A., Blazquez, C., Garcia, S., Giroux, V., Malicet, C., Villuendas, R., Gironella, M., Gonzalez-Feria, L., Piris, M.A., Iovanna, J.L., Guzman, M., and Velasco, G. The stress-regulated protein p8 mediates cannabinoid-induced apoptosis of tumor cells. *Cancer Cell*, **9,** 2006, 301–312.

Cerantola, V., Vionnet, C., Aebischer, O.F., Jenny, T., Knudsen, J., and Conzelmann, A. Yeast sphingolipids do not need to contain very long chain fatty acids. *Biochem J*, **401,** 2007, 205–216.

Chae, S.S., Paik, J.H., Furneaux, H., and Hla, T. Requirement for sphingosine 1-phosphate receptor-1 in tumor angiogenesis demonstrated by in vivo RNA interference. *J Clin Invest*, **114,** 2004, 1082–1089.

Chalfant, C.E., Ogretmen, B., Galadari, S., Kroesen, B.J., Pettus, B.J., and Hannun, Y.A. FAS activation induces dephosphorylation of SR proteins; dependence on the de novo generation of ceramide and activation of protein phosphatase 1. *J Biol Chem*, **276,** 2001, 44848–44855.

Chalfant, C.E., Rathman, K., Pinkerman, R.L., Wood, R.E., Obeid, L.M., Ogretmen, B., and Hannun, Y.A. De novo ceramide regulates the alternative splicing of caspase 9 and Bcl-x in A549 lung adenocarcinoma cells. Dependence on protein phosphatase-1. *J Biol Chem*, **277,** 2002, 12587–12595.

Charles, R., Sandirasegarane, L., Yun, J., Bourbon, N., Wilson, R., Rothstein, R.P., Levison, S.W., and Kester, M. Ceramide-coated balloon catheters limit neointimal hyperplasia after stretch injury in carotid arteries. *Circ Res*, **87,** 2000, 282–288.

Chun, J. and Rosen, H. Lysophospholipid receptors as potential drug targets in tissue transplantation and autoimmune diseases. *Curr Pharm Des*, **12,** 2006, 161–171.

Clarke, C.J. and Hannun, Y.A. Neutral sphingomyelinases and nSMase2: bridging the gaps. *Biochim Biophys Acta*, **1758,** 2006, 1893–1901.

Clarke, C.J., Snook, C.F., Tani, M., Matmati, N., Marchesini, N., and Hannun, Y.A. The extended family of neutral sphingomyelinases. *Biochemistry*, **45,** 2006, 11247–11256.

Crawford, K.W., Bittman, R., Chun, J., Byun, H.S., and Bowen, W.D. Novel ceramide analogues display selective cytotoxicity in drug-resistant breast tumor cell lines compared to normal breast epithelial cells. *Cell Mol Biol (Noisy-le-grand)*, **49,** 2003, 1017–1023.

Cremesti, A., Paris, F., Grassme, H., Holler, N., Tschopp, J., Fuks, Z., Gulbins, E., and Kolesnick, R. Ceramide enables fas to cap and kill. *J Biol Chem*, **276,** 2001, 23954–23961.

D'Angelo, G., Polishchuk, E., Di Tullio, G., Santoro, M., Di Campli, A., Godi, A., West, G., Bielawski, J., Chuang, C.C., van der Spoel, A.C., Platt, F.M., Hannun, Y.A., Polishchuk, R., Mattjus, P., and De Matteis, M.A. Glycosphingolipid synthesis requires FAPP2 transfer of glucosylceramide. *Nature*, **449,** 2007, 62–67.

Dahm, F., Bielawska, A., Nocito, A., Georgiev, P., Szulc, Z.M., Bielawski, J., Jochum, W., Dindo, D., Hannun, Y.A., and Clavien, P.A. Mitochondrially targeted ceramide LCL-30 inhibits colorectal cancer in mice. *Br J Cancer*, **98,** 2008, 98–105.

Davis, M.D., Clemens, J.J., Macdonald, T.L., and Lynch, K.R. Sphingosine 1-phosphate analogs as receptor antagonists. *J Biol Chem*, **280,** 2005, 9833–9841.

Dbaibo, G.S., Pushkareva, M.Y., Jayadev, S., Schwarz, J.K., Horowitz, J.M., Obeid, L.M., and Hannun, Y.A. Retinoblastoma gene product as a downstream target for a ceramide-dependent pathway of growth arrest. *Proc Natl Acad Sci U S A*, **92,** 1995, 1347–1351.

Dbaibo, G.S., Perry, D.K., Gamard, C.J., Platt, R., Poirier, G.G., Obeid, L.M., and Hannun, Y.A. Cytokine response modifier A (CrmA) inhibits ceramide formation in response to tumor necrosis factor (TNF)-alpha: CrmA and Bcl-2 target distinct components in the apoptotic pathway. *J Exp Med*, **185,** 1997, 481–490.

De Rosa, M.F., Sillence, D., Ackerley, C., and Lingwood, C. Role of multiple drug resistance protein 1 in neutral but not acidic glycosphingolipid biosynthesis. *J Biol Chem*, **279,** 2004, 7867–7876.

Dindo, D., Dahm, F., Szulc, Z., Bielawska, A., Obeid, L.M., Hannun, Y.A., Graf, R., and Clavien, P.A. Cationic long-chain ceramide LCL-30 induces cell death by mitochondrial targeting in SW403 cells. *Mol Cancer Ther*, **5,** 2006, 1520–1529.

Dobrowsky, R.T. and Hannun, Y.A. Ceramide stimulates a cytosolic protein phosphatase. *J Biol Chem*, **267,** 1992, 5048–5051.

Dobrowsky, R.T., Kamibayashi, C., Mumby, M.C., and Hannun, Y.A. Ceramide activates heterotrimeric protein phosphatase 2A. *J Biol Chem*, **268,** 1993, 15523–15530.

Dobrowsky, R.T., Werner, M.H., Castellino, A.M., Chao, M.V., and Hannun, Y.A. Activation of the sphingomyelin cycle through the low-affinity neurotrophin receptor. *Science*, **265,** 1994, 1596–1599.

Dolgachev, V., Farooqui, M.S., Kulaeva, O.I., Tainsky, M.A., Nagy, B., Hanada, K., and Separovic, D. De novo ceramide accumulation due to inhibition of its conversion to complex sphingolipids in apoptotic photosensitized cells. *J Biol Chem*, **279,** 2004, 23238–23249.

Fishbein, J.D., Dobrowsky, R.T., Bielawska, A., Garrett, S., and Hannun, Y.A. Ceramide-mediated growth inhibition and CAPP are conserved in Saccharomyces cerevisiae. *J Biol Chem*, **268,** 1993, 9255–9261.

Fox, T.E., Finnegan, C.M., Blumenthal, R., and Kester, M. The clinical potential of sphingolipid-based therapeutics. *Cell Mol Life Sci*, **63,** 2006, 1017–1023.

Fox, T.E., Houck, K.L., O'Neill, S.M., Nagarajan, M., Stover, T.C., Pomianowski, P.T., Unal, O., Yun, J.K., Naides, S.J., and Kester, M. Ceramide recruits and activates protein kinase C zeta (PKC zeta) within structured membrane microdomains. *J Biol Chem*, **282,** 2007, 12450–12457.

French, K.J., Upson, J.J., Keller, S.N., Zhuang, Y., Yun, J.K., and Smith, C.D. Antitumor activity of sphingosine kinase inhibitors. *J Pharmacol Exp Ther*, **318,** 2006, 596–603.

Futerman, A.H. and Hannun, Y.A. The complex life of simple sphingolipids. *EMBO Rep*, **5,** 2004, 777–782.

Futerman, A.H. and Riezman, H. The ins and outs of sphingolipid synthesis. *Trends Cell Biol*, **15,** 2005, 312–318.

Gouaze-Andersson, V. and Cabot, M.C. Glycosphingolipids and drug resistance. *Biochim Biophys Acta*, **1758,** 2006, 2096–2103.

Gouaze-Andersson, V., Yu, J.Y., Kreitenberg, A.J., Bielawska, A., Giuliano, A.E., and Cabot, M.C. Ceramide and glucosylceramide upregulate expression of the multidrug resistance gene MDR1 in cancer cells. *Biochim Biophys Acta*, **1771,** 2007, 1407–1417.

Gouaze, V., Liu, Y.Y., Prickett, C.S., Yu, J.Y., Giuliano, A.E., and Cabot, M.C. Glucosylceramide synthase blockade down-regulates P-glycoprotein and resensitizes multidrug-resistant breast cancer cells to anticancer drugs. *Cancer Res*, **65,** 2005, 3861–3867.

Gouaze, V., Yu, J.Y., Bleicher, R.J., Han, T.Y., Liu, Y.Y., Wang, H., Gottesman, M.M., Bitterman, A., Giuliano, A.E., and Cabot, M.C. Overexpression of glucosylceramide synthase and P-glycoprotein in cancer cells selected for resistance to natural product chemotherapy. *Mol Cancer Ther*, **3,** 2004, 633–639.

Goulding, C.W., Giuliano, A.E., and Cabot, M.C. SDZ PSC 833 the drug resistance modulator activates cellular ceramide formation by a pathway independent of P-glycoprotein. *Cancer Lett*, **149,** 2000, 143–151.

Guillas, I., Kirchman, P.A., Chuard, R., Pfefferli, M., Jiang, J.C., Jazwinski, S.M., and Conzelmann, A. C26-CoA-dependent ceramide synthesis of Saccharomyces cerevisiae is operated by Lag1p and Lac1p. *Embo J*, **20,** 2001, 2655–2665.

Hanada, K., Kumagai, K., Yasuda, S., Miura, Y., Kawano, M., Fukasawa, M., and Nishijima, M. Molecular machinery for non-vesicular trafficking of ceramide. *Nature*, **426,** 2003, 803–809.

Hannun, Y.A. and Obeid, L.M. The Ceramide-centric universe of lipid-mediated cell regulation: stress encounters of the lipid kind. *J Biol Chem*, **277,** 2002, 25847–25850.

Heinrich, M., Neumeyer, J., Jakob, M., Hallas, C., Tchikov, V., Winoto-Morbach, S., Wickel, M., Schneider-Brachert, W., Trauzold, A., Hethke, A., and Schutze, S. Cathepsin D links TNF-induced acid sphingomyelinase to Bid-mediated caspase-9 and -3 activation. *Cell Death Differ*, **11,** 2004, 550–563.

Herrera, B., Carracedo, A., Diez-Zaera, M., Gomez del Pulgar, T., Guzman, M., and Velasco, G. The CB2 cannabinoid receptor signals apoptosis via ceramide-dependent activation of the mitochondrial intrinsic pathway. *Exp Cell Res*, **312,** 2006, 2121–2131.

Hinrichs, J.W., Klappe, K., and Kok, J.W. Rafts as missing link between multidrug resistance and sphingolipid metabolism. *J Membr Biol*, **203,** 2005, 57–64.

Hla, T. Physiological and pathological actions of sphingosine 1-phosphate. *Semin Cell Dev Biol*, **15,** 2004, 513–520.

Holman, D.H., Turner, L.S., El-Zawahry, A., Elojeimy, S., Liu, X., Bielawski, J., Szulc, Z.M., Norris, K., Zeidan, Y.H., Hannun, Y.A., Bielawska, A., and Norris, J.S. Lysosomotropic acid ceramidase inhibitor induces apoptosis in prostate cancer cells. *Cancer Chemother Pharmacol*, **61,** 2008, 231–242.

Jazwinski, S.M. and Conzelmann, A. LAG1 puts the focus on ceramide signaling. *Int J Biochem Cell Biol*, **34,** 2002, 1491–1495.

Jiang, Q., Wong, J., Fyrst, H., Saba, J.D., and Ames, B.N. gamma-Tocopherol or combinations of vitamin E forms induce cell death in human prostate cancer cells by interrupting sphingolipid synthesis. *Proc Natl Acad Sci U S A*, **101,** 2004, 17825–17830.

Kageyama-Yahara, N. and Riezman, H. Transmembrane topology of ceramide synthase in yeast. *Biochem J*, **398,** 2006, 585–593.

Karahatay, S., Thomas, K., Koybasi, S., Senkal, C.E., Elojeimy, S., Liu, X., Bielawski, J., Day, T.A., Gillespie, M.B., Sinha, D., Norris, J.S., Hannun, Y.A., and Ogretmen, B. Clinical relevance of ceramide metabolism in the pathogenesis of human head and neck squamous cell carcinoma (HNSCC): attenuation of C(18)-ceramide in HNSCC tumors correlates with lymphovascular invasion and nodal metastasis. *Cancer Lett*, **256,** 2007, 101–111.

Kawamori, T., Osta, W., Johnson, K.R., Pettus, B.J., Bielawski, J., Tanaka, T., Wargovich, M.J., Reddy, B.S., Hannun, Y.A., Obeid, L.M., and Zhou, D. Sphingosine kinase 1 is up-regulated in colon carcinogenesis. *Faseb J*, **20,** 2006, 386–388.

Kok, J.W. and Sietsma, H. Sphingolipid metabolism enzymes as targets for anticancer therapy. *Curr Drug Targets*, **5,** 2004, 375–382.

Koybasi, S., Senkal, C.E., Sundararaj, K., Spassieva, S., Bielawski, J., Osta, W., Day, T.A., Jiang, J.C., Jazwinski, S.M., Hannun, Y.A., Obeid, L.M., and Ogretmen, B. Defects in cell growth regulation by C18:0-ceramide and longevity assurance gene 1 in human head and neck squamous cell carcinomas. *J Biol Chem*, **279,** 2004, 44311–44319.

Kraveka, J.M., Li, L., Szulc, Z.M., Bielawski, J., Ogretmen, B., Hannun, Y.A., Obeid, L.M., and Bielawska, A. Involvement of dihydroceramide desaturase in cell cycle progression in human neuroblastoma cells. *J Biol Chem*, **282,** 2007, 16718–16728.

Kudo, N., Kumagai, K., Tomishige, N., Yamaji, T., Wakatsuki, S., Nishijima, M., Hanada, K., and Kato, R. Structural basis for specific lipid recognition by CERT responsible for nonvesicular trafficking of ceramide. *Proc Natl Acad Sci U S A*, **105,** 2008, 488–493.

Kumagai, K., Yasuda, S., Okemoto, K., Nishijima, M., Kobayashi, S., and Hanada, K. CERT mediates intermembrane transfer of various molecular species of ceramides. *J Biol Chem*, **280,** 2005, 6488–6495.

Lahiri, S. and Futerman, A.H. LASS5 is a bona fide dihydroceramide synthase that selectively utilizes palmitoyl-CoA as acyl donor. *J Biol Chem*, **280,** 2005, 33735–33738.

LaMontagne, K., Littlewood-Evans, A., Schnell, C., O'Reilly, T., Wyder, L., Sanchez, T., Probst, B., Butler, J., Wood, A., Liau, G., Billy, E., Theuer, A., Hla, T., and Wood, J. Antagonism of sphingosine-1-phosphate receptors by FTY720 inhibits angiogenesis and tumor vascularization. *Cancer Res*, **66,** 2006, 221–231.

Laviad, E.L., Albee, L., Pankova-Kholmyansky, I., Epstein, S., Park, H., Merrill, A.H., Jr., and Futerman, A.H. Characterization of ceramide synthase 2: Tissue distribution, substrate specificity and inhibition by sphingosine 1-phosphate. *J Biol Chem*, **283**, 2008, 5677–5684.

Lee, J.Y., Bielawska, A.E., and Obeid, L.M. Regulation of cyclin-dependent kinase 2 activity by ceramide. *Exp Cell Res*, **261,** 2000, 303–311.

Lee, S.J. Expression of growth/differentiation factor 1 in the nervous system: conservation of a bicistronic structure. *Proc Natl Acad Sci U S A*, **88,** 1991, 4250–4254.

Liu, H., Toman, R.E., Goparaju, S.K., Maceyka, M., Nava, V.E., Sankala, H., Payne, S.G., Bektas, M., Ishii, I., Chun, J., Milstien, S., and Spiegel, S. Sphingosine kinase type 2 is a putative BH3-only protein that induces apoptosis. *J Biol Chem*, **278,** 2003, 40330–40336.

Liu, X., Elojeimy, S., Turner, L.S., Mahdy, A.E., Zeidan, Y.H., Bielawska, A., Bielawski, J., Dong, J.Y., El-Zawahry, A.M., Guo, G.W., Hannun, Y.A., Holman, D.H., Rubinchik, S., Szulc, Z., Keane, T.E., Tavassoli, M., and Norris, J.S. Acid ceramidase inhibition: a novel target for cancer therapy. *Front Biosci*, **13,** 2008, 2293–2298.

Maceyka, M., Payne, S.G., Milstien, S., and Spiegel, S. Sphingosine kinase, sphingosine-1-phosphate, and apoptosis. *Biochim Biophys Acta*, **1585,** 2002, 193–201.

Maurer, B.J., Metelitsa, L.S., Seeger, R.C., Cabot, M.C., and Reynolds, C.P. Increase of ceramide and induction of mixed apoptosis/necrosis by N-(4-hydroxyphenyl)- retinamide in neuroblastoma cell lines. *J Natl Cancer Inst*, **91,** 1999, 1138–1146.

Maurer, B.J., Melton, L., Billups, C., Cabot, M.C., and Reynolds, C.P. Synergistic cytotoxicity in solid tumor cell lines between N-(4-hydroxyphenyl)retinamide and modulators of ceramide metabolism. *J Natl Cancer Inst*, **92,** 2000, 1897–1909.

Meng, A., Luberto, C., Meier, P., Bai, A., Yang, X., Hannun, Y.A., and Zhou, D. Sphingomyelin synthase as a potential target for D609-induced apoptosis in U937 human monocytic leukemia cells. *Exp Cell Res*, **292,** 2004, 385–392.

Merrill, A.H., Jr., Wang, E., and Mullins, R.E. Kinetics of long-chain (sphingoid) base biosynthesis in intact LM cells: effects of varying the extracellular concentrations of serine and fatty acid precursors of this pathway. *Biochemistry*, **27,** 1988, 340–345.

Mesika, A., Ben-Dor, S., Laviad, E.L., and Futerman, A.H. A new functional motif in Hox domain-containing ceramide synthases: identification of a novel region flanking the Hox and TLC domains essential for activity. *J Biol Chem*, **282,** 2007, 27366–27373.

Michel, C., van Echten-Deckert, G., Rother, J., Sandhoff, K., Wang, E., and Merrill, A.H., Jr. Characterization of ceramide synthesis. A dihydroceramide desaturase introduces the 4,5-trans-double bond of sphingosine at the level of dihydroceramide. *J Biol Chem*, **272,** 1997, 22432–22437.

Min, J., Van Veldhoven, P.P., Zhang, L., Hanigan, M.H., Alexander, H., and Alexander, S. Sphingosine-1-phosphate lyase regulates sensitivity of human cells to select chemotherapy drugs in a p38-dependent manner. *Mol Cancer Res*, **3,** 2005, 287–296.

Min, J., Mesika, A., Sivaguru, M., Van Veldhoven, P.P., Alexander, H., Futerman, A.H., and Alexander, S. (Dihydro)ceramide synthase 1 regulated sensitivity to cisplatin is associated with the activation of p38 mitogen-activated protein kinase and is abrogated by sphingosine kinase 1. *Mol Cancer Res*, **5,** 2007, 801–812.

Minutolo, F., Sala, G., Bagnacani, A., Bertini, S., Carboni, I., Placanica, G., Prota, G., Rapposelli, S., Sacchi, N., Macchia, M., and Ghidoni, R. Synthesis of a resveratrol analogue with high ceramide-mediated proapoptotic activity on human breast cancer cells. *J Med Chem*, **48,** 2005, 6783–6786.

Mitra, P., Maceyka, M., Payne, S.G., Lamour, N., Milstien, S., Chalfant, C.E., and Spiegel, S. Ceramide kinase regulates growth and survival of A549 human lung adenocarcinoma cells. *FEBS Lett*, **581,** 2007, 735–740.

Mizugishi, K., Yamashita, T., Olivera, A., Miller, G.F., Spiegel, S., and Proia, R.L. Essential role for sphingosine kinases in neural and vascular development. *Mol Cell Biol*, **25,** 2005, 11113–11121.

Mizutani, Y., Kihara, A., and Igarashi, Y. Mammalian Lass6 and its related family members regulate synthesis of specific ceramides. *Biochem J*, **390,** 2005, 263–271.

Modica-Napolitano, J.S. and Aprille, J.R. Delocalized lipophilic cations selectively target the mitochondria of carcinoma cells. *Adv Drug Deliv Rev*, **49,** 2001, 63–70.

Modrak, D.E., Gold, D.V., and Goldenberg, D.M. Sphingolipid targets in cancer therapy. *Mol Cancer Ther*, **5,** 2006, 200–208.

Modrak, D.E., Cardillo, T.M., Newsome, G.A., Goldenberg, D.M., and Gold, D.V. Synergistic interaction between sphingomyelin and gemcitabine potentiates ceramide-mediated apoptosis in pancreatic cancer. *Cancer Res*, **64,** 2004, 8405–8410.

Morales, A., Paris, R., Villanueva, A., Llacuna, L., Garcia-Ruiz, C., and Fernandez-Checa, J.C. Pharmacological inhibition or small interfering RNA targeting acid ceramidase sensitizes hepatoma cells to chemotherapy and reduces tumor growth in vivo. *Oncogene*, **26,** 2007, 905–916.

Nagiec, M.M., Lester, R.L., and Dickson, R.C. Sphingolipid synthesis: identification and characterization of mammalian cDNAs encoding the Lcb2 subunit of serine palmitoyltransferase. *Gene*, **177,** 1996, 237–241.

Nava, V.E., Hobson, J.P., Murthy, S., Milstien, S., and Spiegel, S. Sphingosine kinase type 1 promotes estrogen-dependent tumorigenesis of breast cancer MCF-7 cells. *Exp Cell Res*, **281,** 2002, 115–127.

Norris-Cervetto, E., Callaghan, R., Platt, F.M., Dwek, R.A., and Butters, T.D. Inhibition of glucosylceramide synthase does not reverse drug resistance in cancer cells. *J Biol Chem*, **279,** 2004, 40412–40418.

Novgorodov, S.A., Szulc, Z.M., Luberto, C., Jones, J.A., Bielawski, J., Bielawska, A., Hannun, Y.A., and Obeid, L.M. Positively charged ceramide is a potent inducer of mitochondrial permeabilization. *J Biol Chem*, **280,** 2005, 16096–16105.

Obeid, L.M. and Hannun, Y.A. Ceramide, stress, and a "LAG" in aging. *Sci Aging Knowledge Environ*, **2003,** 2003, pe27.

Ogretmen, B. Sphingolipids in cancer: regulation of pathogenesis and therapy. *FEBS Lett*, **580,** 2006, 5467–5476.

Ogretmen, B. and Hannun, Y.A. Biologically active sphingolipids in cancer pathogenesis and treatment. *Nat Rev Cancer*, **4,** 2004, 604–616.

Ogretmen, B. and Hannun, Y.A. Updates on functions of ceramide in chemotherapy-induced cell death and in multidrug resistance. *Drug Resist Updat*, **4,** 2001, 368–377.

Ogretmen, B., Kraveka, J.M., Schady, D., Usta, J., Hannun, Y.A., and Obeid, L.M. Molecular mechanisms of ceramide-mediated telomerase inhibition in the A549 human lung adenocarcinoma cell line. *J Biol Chem*, **276,** 2001a, 32506–32514.

Ogretmen, B., Schady, D., Usta, J., Wood, R., Kraveka, J.M., Luberto, C., Birbes, H., Hannun, Y.A., and Obeid, L.M. Role of ceramide in mediating the inhibition of telomerase activity in A549 human lung adenocarcinoma cells. *J Biol Chem*, **276,** 2001b, 24901–24910.

Ogretmen, B., Pettus, B.J., Rossi, M.J., Wood, R., Usta, J., Szulc, Z., Bielawska, A., Obeid, L.M., and Hannun, Y.A. Biochemical mechanisms of the generation of endogenous long chain ceramide in response to exogenous short chain ceramide in the A549 human lung adenocarcinoma cell line. Role for endogenous ceramide in mediating the action of exogenous ceramide. *J Biol Chem*, **277,** 2002, 12960–12969.

Okazaki, T., Bell, R.M., and Hannun, Y.A. Sphingomyelin turnover induced by vitamin D3 in HL-60 cells. Role in cell differentiation. *J Biol Chem*, **264,** 1989, 19076–19080.

Oskouian, B., Sooriyakumaran, P., Borowsky, A.D., Crans, A., Dillard-Telm, L., Tam, Y.Y., Bandhuvula, P., and Saba, J.D. Sphingosine-1-phosphate lyase potentiates apoptosis via p53- and p38-dependent pathways and is down-regulated in colon cancer. *Proc Natl Acad Sci U S A*, **103,** 2006, 17384–17389.

Park, J.H. and Schuchman, E.H. Acid ceramidase and human disease. *Biochim Biophys Acta*, **1758,** 2006, 2133–2138.

Paugh, S.W., Payne, S.G., Barbour, S.E., Milstien, S., and Spiegel, S. The immunosuppressant FTY720 is phosphorylated by sphingosine kinase type 2. *FEBS Lett*, **554,** 2003, 189–193.

Pchejetski, D., Golzio, M., Bonhoure, E., Calvet, C., Doumerc, N., Garcia, V., Mazerolles, C., Rischmann, P., Teissie, J., Malavaud, B., and Cuvillier, O. Sphingosine kinase-1 as a chemotherapy sensor in prostate adenocarcinoma cell and mouse models. *Cancer Res*, **65,** 2005, 11667–11675.

Pettus, B.J., Chalfant, C.E., and Hannun, Y.A. Ceramide in apoptosis: an overview and current perspectives. *Biochim Biophys Acta*, **1585,** 2002, 114–125.

Pettus, B.J., Bielawski, J., Porcelli, A.M., Reames, D.L., Johnson, K.R., Morrow, J., Chalfant, C.E., Obeid, L.M., and Hannun, Y.A. The sphingosine kinase 1/sphingosine-1-phosphate pathway mediates COX-2 induction and PGE2 production in response to TNF-alpha. *Faseb J*, **17,** 2003, 1411–1421.

Pewzner-Jung, Y., Ben-Dor, S., and Futerman, A.H. When do Lasses (longevity assurance genes) become CerS (ceramide synthases)?: Insights into the regulation of ceramide synthesis. *J Biol Chem*, **281,** 2006, 25001–25005.

Plo, I., Ghandour, S., Feutz, A.C., Clanet, M., Laurent, G., and Bettaieb, A. Involvement of de novo ceramide biosynthesis in lymphotoxin-induced oligodendrocyte death. *Neuroreport*, **10,** 1999, 2373–2376.

Radin, N.S. The development of aggressive cancer: a possible role for sphingolipids. *Cancer Invest*, **20,** 2002, 779–786.

Raisova, M., Bektas, M., Wieder, T., Daniel, P., Eberle, J., Orfanos, C.E., and Geilen, C.C. Resistance to CD95/Fas-induced and ceramide-mediated apoptosis of human melanoma cells is caused by a defective mitochondrial cytochrome c release. *FEBS Lett*, **473,** 2000, 27–32.

Rao, R.P., Yuan, C., Allegood, J.C., Rawat, S.S., Edwards, M.B., Wang, X., Merrill, A.H., Jr., Acharya, U., and Acharya, J.K. Ceramide transfer protein function is essential for normal oxidative stress response and lifespan. *Proc Natl Acad Sci U S A*, **104,** 2007, 11364–11369.

Reynolds, C.P., Maurer, B.J., and Kolesnick, R.N. Ceramide synthesis and metabolism as a target for cancer therapy. *Cancer Lett*, **206,** 2004, 169–180.

Riebeling, C., Allegood, J.C., Wang, E., Merrill, A.H., Jr., and Futerman, A.H. Two mammalian longevity assurance gene (LAG1) family members, trh1 and trh4, regulate dihydroceramide synthesis using different fatty acyl-CoA donors. *J Biol Chem*, **278,** 2003, 43452–43459.

Rosen, H. and Goetzl, E.J. Sphingosine 1-phosphate and its receptors: an autocrine and paracrine network. *Nat Rev Immunol*, **5,** 2005, 560–570.

Rossi, M.J., Sundararaj, K., Koybasi, S., Phillips, M.S., Szulc, Z.M., Bielawska, A., Day, T.A., Obeid, L.M., Hannun, Y.A., and Ogretmen, B. Inhibition of growth and telomerase activity by novel cationic ceramide analogs with high solubility in human head and neck squamous cell carcinoma cells. *Otolaryngol Head Neck Surg*, **132,** 2005, 55–62.

Saad, A.F., Meacham, W.D., Bai, A., Anelli, V., Elojeimy, S., Mahdy, A.E., Turner, L.S., Cheng, J., Bielawska, A., Bielawski, J., Keane, T.E., Obeid, L.M., Hannun, Y.A., Norris, J.S., and Liu, X. The functional effects of acid ceramidase overexpression in prostate cancer progression and resistance to chemotherapy. *Cancer Biol Ther*, **6,** 2007, 1455–1460.

Samsel, L., Zaidel, G., Drumgoole, H.M., Jelovac, D., Drachenberg, C., Rhee, J.G., Brodie, A.M., Bielawska, A., and Smyth, M.J. The ceramide analog, B13, induces apoptosis in prostate cancer cell lines and inhibits tumor growth in prostate cancer xenografts. *Prostate*, **58,** 2004, 382–393.

Sankala, H.M., Hait, N.C., Paugh, S.W., Shida, D., Lepine, S., Elmore, L.W., Dent, P., Milstien, S., and Spiegel, S. Involvement of sphingosine kinase 2 in p53-independent induction of p21 by the chemotherapeutic drug doxorubicin. *Cancer Res*, **67,** 2007, 10466–10474.

Santana, P., Pena, L.A., Haimovitz-Friedman, A., Martin, S., Green, D., McLoughlin, M., Cordon-Cardo, C., Schuchman, E.H., Fuks, Z., and Kolesnick, R. Acid sphingomyelinase-deficient human lymphoblasts and mice are defective in radiation-induced apoptosis. *Cell*, **86,** 1996, 189–199.

Sarkar, S., Maceyka, M., Hait, N.C., Paugh, S.W., Sankala, H., Milstien, S., and Spiegel, S. Sphingosine kinase 1 is required for migration, proliferation and survival of MCF-7 human breast cancer cells. *FEBS Lett*, **579,** 2005, 5313–5317.

Scarlatti, F., Sala, G., Somenzi, G., Signorelli, P., Sacchi, N., and Ghidoni, R. Resveratrol induces growth inhibition and apoptosis in metastatic breast cancer cells via de novo ceramide signaling. *Faseb J*, **17,** 2003, 2339–2341.

Scarlatti, F., Sala, G., Ricci, C., Maioli, C., Milani, F., Minella, M., Botturi, M., and Ghidoni, R. Resveratrol sensitization of DU145 prostate cancer cells to ionizing radiation is associated to ceramide increase. *Cancer Lett*, **253,** 2007, 124–130.

Schmelz, E.M., Sullards, M.C., Dillehay, D.L., and Merrill, A.H., Jr. Colonic cell proliferation and aberrant crypt foci formation are inhibited by dairy glycosphingolipids in 1, 2-dimethylhydrazine-treated CF1 mice. *J Nutr*, **130,** 2000, 522–527.

Schmelz, E.M., Dillehay, D.L., Webb, S.K., Reiter, A., Adams, J., and Merrill, A.H., Jr. Sphingomyelin consumption suppresses aberrant colonic crypt foci and increases the proportion of adenomas versus adenocarcinomas in CF1 mice treated with 1,2-dimethylhydrazine: implications for dietary sphingolipids and colon carcinogenesis. *Cancer Res*, **56,** 1996, 4936–4941.

Schulz, A., Mousallem, T., Venkataramani, M., Persaud-Sawin, D.A., Zucker, A., Luberto, C., Bielawska, A., Bielawski, J., Holthuis, J.C., Jazwinski, S.M., Kozhaya, L., Dbaibo, G.S., and Boustany, R.M. The CLN9 protein, a regulator of dihydroceramide synthase. *J Biol Chem*, **281,** 2006, 2784–2794.

Segui, B., Cuvillier, O., Adam-Klages, S., Garcia, V., Malagarie-Cazenave, S., Leveque, S., Caspar-Bauguil, S., Coudert, J., Salvayre, R., Kronke, M., and Levade, T. Involvement of FAN in TNF-induced apoptosis. *J Clin Invest*, **108,** 2001, 143–151.

Selzner, M., Bielawska, A., Morse, M.A., Rudiger, H.A., Sindram, D., Hannun, Y.A., and Clavien, P.A. Induction of apoptotic cell death and prevention of tumor growth by ceramide analogues in metastatic human colon cancer. *Cancer Res*, **61,** 2001, 1233–1240.

Senchenkov, A., Litvak, D.A., and Cabot, M.C. Targeting ceramide metabolism–a strategy for overcoming drug resistance. *J Natl Cancer Inst*, **93,** 2001, 347–357.

Senkal, C.E., Ponnusamy, S., Rossi, M.J., Bialewski, J., Sinha, D., Jiang, J.C., Jazwinski, S.M., Hannun, Y.A., and Ogretmen, B. Role of human longevity assurance gene 1 and C18-ceramide in chemotherapy-induced cell death in human head and neck squamous cell carcinomas. *Mol Cancer Ther*, **6,** 2007, 712–722.

Senkal, C.E., Ponnusamy, S., Rossi, M.J., Sundararaj, K., Szulc, Z., Bielawski, J., Bielawska, A., Meyer, M., Cobanoglu, B., Koybasi, S., Sinha, D., Day, T.A., Obeid, L.M., Hannun, Y.A., and Ogretmen, B. Potent antitumor activity of a novel cationic pyridinium-ceramide alone or in combination with gemcitabine against human head and neck squamous cell carcinomas in vitro and in vivo. *J Pharmacol Exp Ther*, **317,** 2006, 1188–1199.

Smyth, M.J., Perry, D.K., Zhang, J., Poirier, G.G., Hannun, Y.A., and Obeid, L.M. prICE: a downstream target for ceramide-induced apoptosis and for the inhibitory action of Bcl-2. *Biochem J*, **316 (Pt 1),** 1996, 25–28.

Snook, C.F., Jones, J.A., and Hannun, Y.A. Sphingolipid-binding proteins. *Biochim Biophys Acta*, **1761,** 2006, 927–946.

Spassieva, S., Seo, J.G., Jiang, J.C., Bielawski, J., Alvarez-Vasquez, F., Jazwinski, S.M., Hannun, Y.A., and Obeid, L.M. Necessary role for the Lag1p motif in (dihydro)ceramide synthase activity. *J Biol Chem*, **281,** 2006, 33931–33938.

Stover, T. and Kester, M. Liposomal delivery enhances short-chain ceramide-induced apoptosis of breast cancer cells. *J Pharmacol Exp Ther*, **307,** 2003, 468–475.

Stover, T.C., Sharma, A., Robertson, G.P., and Kester, M. Systemic delivery of liposomal short-chain ceramide limits solid tumor growth in murine models of breast adenocarcinoma. *Clin Cancer Res*, **11,** 2005, 3465–3474.

Struckhoff, A.P., Bittman, R., Burow, M.E., Clejan, S., Elliott, S., Hammond, T., Tang, Y., and Beckman, B.S. Novel ceramide analogs as potential chemotherapeutic agents in breast cancer. *J Pharmacol Exp Ther*, **309,** 2004, 523–532.

Sugiura, M., Kono, K., Liu, H., Shimizugawa, T., Minekura, H., Spiegel, S., and Kohama, T. Ceramide kinase, a novel lipid kinase. Molecular cloning and functional characterization. *J Biol Chem*, **277,** 2002, 23294–23300.

Sultan, I., Senkal, C.E., Ponnusamy, S., Bielawski, J., Szulc, Z., Bielawska, A., Hannun, Y.A., and Ogretmen, B. Regulation of the sphingosine-recycling pathway for ceramide generation by oxidative stress, and its role in controlling c-Myc/Max function. *Biochem J*, **393,** 2006, 513–521.

Sundararaj, K.P., Wood, R.E., Ponnusamy, S., Salas, A.M., Szulc, Z., Bielawska, A., Obeid, L.M., Hannun, Y.A., and Ogretmen, B. Rapid shortening of telomere length in response to ceramide involves the inhibition of telomere binding activity of nuclear glyceraldehyde-3-phosphate dehydrogenase. *J Biol Chem*, **279,** 2004, 6152–6162.

Suomalainen, L., Pentikainen, V., and Dunkel, L. Sphingosine-1-phosphate inhibits nuclear factor kappaB activation and germ cell apoptosis in the human testis independently of its receptors. *Am J Pathol*, **166,** 2005, 773–781.

Swanton, C., Marani, M., Pardo, O., Warne, P.H., Kelly, G., Sahai, E., Elustondo, F., Chang, J., Temple, J., Ahmed, A.A., Brenton, J.D., Downward, J., and Nicke, B. Regulators of mitotic arrest and ceramide metabolism are determinants of sensitivity to paclitaxel and other chemotherapeutic drugs. *Cancer Cell*, **11,** 2007, 498–512.

Szulc, Z.M., Bielawski, J., Gracz, H., Gustilo, M., Mayroo, N., Hannun, Y.A., Obeid, L.M., and Bielawska, A. Tailoring structure-function and targeting properties of ceramides by site-specific cationization. *Bioorg Med Chem*, **14,** 2006, 7083–7104.

Taha, T.A., Hannun, Y.A., and Obeid, L.M. Sphingosine kinase: biochemical and cellular regulation and role in disease. *J Biochem Mol Biol*, **39,** 2006a, 113–131.

Taha, T.A., Kitatani, K., El-Alwani, M., Bielawski, J., Hannun, Y.A., and Obeid, L.M. Loss of sphingosine kinase-1 activates the intrinsic pathway of programmed cell death: modulation of sphingolipid levels and the induction of apoptosis. *Faseb J*, **20,** 2006b, 482–484.

Testai, F.D., Landek, M.A., and Dawson, G. Regulation of sphingomyelinases in cells of the oligodendrocyte lineage. *J Neurosci Res*, **75,** 2004, 66–74.

Thomas, D.A., Sarris, A.H., Cortes, J., Faderl, S., O'Brien, S., Giles, F.J., Garcia-Manero, G., Rodriguez, M.A., Cabanillas, F., and Kantarjian, H. Phase II study of sphingosomal vincristine in patients with recurrent or refractory adult acute lymphocytic leukemia. *Cancer*, **106,** 2006, 120–127.

Thon, L., Mohlig, H., Mathieu, S., Lange, A., Bulanova, E., Winoto-Morbach, S., Schutze, S., Bulfone-Paus, S., and Adam, D. Ceramide mediates caspase-independent programmed cell death. *Faseb J*, **19,** 2005, 1945–1956.

Tilly, J.L. and Kolesnick, R.N. Sphingolipids, apoptosis, cancer treatments and the ovary: investigating a crime against female fertility. *Biochim Biophys Acta*, **1585,** 2002, 135–138.

Van Brocklyn, J., Letterle, C., Snyder, P., and Prior, T. Sphingosine-1-phosphate stimulates human glioma cell proliferation through Gi-coupled receptors: role of ERK MAP kinase and phosphatidylinositol 3-kinase beta. *Cancer Lett*, **181,** 2002, 195–204.

Van der Luit, A.H., Budde, M., Zerp, S., Caan, W., Klarenbeek, J.B., Verheij, M., and Van Blitterswijk, W.J. Resistance to alkyl-lysophospholipid-induced apoptosis due to down-regulated sphingomyelin synthase 1 expression with consequent sphingomyelin- and cholesterol-deficiency in lipid rafts. *Biochem J*, **401,** 2007, 541–549.

van Vlerken, L.E., Duan, Z., Seiden, M.V., and Amiji, M.M. Modulation of intracellular ceramide using polymeric nanoparticles to overcome multidrug resistance in cancer. *Cancer Res*, **67,** 2007, 4843–4850.

Veldman, R.J., Zerp, S., van Blitterswijk, W.J., and Verheij, M. N-hexanoyl-sphingomyelin potentiates in vitro doxorubicin cytotoxicity by enhancing its cellular influx. *Br J Cancer*, **90,** 2004, 917–925.

Veldman, R.J., Mita, A., Cuvillier, O., Garcia, V., Klappe, K., Medin, J.A., Campbell, J.D., Carpentier, S., Kok, J.W., and Levade, T. The absence of functional glucosylceramide synthase does not sensitize melanoma cells for anticancer drugs. *Faseb J*, **17,** 2003, 1144–1146.

Venable, M.E., Lee, J.Y., Smyth, M.J., Bielawska, A., and Obeid, L.M. Role of ceramide in cellular senescence. *J Biol Chem*, **270,** 1995, 30701–30708.

Venkataraman, K., Riebeling, C., Bodennec, J., Riezman, H., Allegood, J.C., Sullards, M.C., Merrill, A.H., Jr., and Futerman, A.H. Upstream of growth and differentiation factor 1 (uog1), a mammalian homolog of the yeast longevity assurance gene 1 (LAG1), regulates N-stearoyl-sphinganine (C18-(dihydro)ceramide) synthesis in a fumonisin B1-independent manner in mammalian cells. *J Biol Chem*, **277,** 2002, 35642–35649.

Visentin, B., Vekich, J.A., Sibbald, B.J., Cavalli, A.L., Moreno, K.M., Matteo, R.G., Garland, W.A., Lu, Y., Yu, S., Hall, H.S., Kundra, V., Mills, G.B., and Sabbadini, R.A. Validation of an anti-sphingosine-1-phosphate antibody as a potential therapeutic in reducing growth, invasion, and angiogenesis in multiple tumor lineages. *Cancer Cell*, **9,** 2006, 225–238.

Wang, G., Silva, J., Krishnamurthy, K., Tran, E., Condie, B.G., and Bieberich, E. Direct binding to ceramide activates protein kinase Czeta before the formation of a pro-apoptotic complex with PAR-4 in differentiating stem cells. *J Biol Chem*, **280,** 2005, 26415–26424.

Wang, H., Maurer, B.J., Reynolds, C.P., and Cabot, M.C. N-(4-hydroxyphenyl)retinamide elevates ceramide in neuroblastoma cell lines by coordinate activation of serine palmitoyltransferase and ceramide synthase. *Cancer Res*, **61,** 2001, 5102–5105.

Wolff, R.A., Dobrowsky, R.T., Bielawska, A., Obeid, L.M., and Hannun, Y.A. Role of ceramide-activated protein phosphatase in ceramide-mediated signal transduction. *J Biol Chem*, **269,** 1994, 19605–19609.

Wooten-Blanks, L.G., Song, P., Senkal, C.E., and Ogretmen, B. Mechanisms of ceramide-mediated repression of the human telomerase reverse transcriptase promoter via deacetylation of Sp3 by histone deacetylase 1. *Faseb J*, **21,** 2007, 3386–3397.

Wooten, L.G. and Ogretmen, B. Sp1/Sp3-dependent regulation of human telomerase reverse transcriptase promoter activity by the bioactive sphingolipid ceramide. *J Biol Chem*, **280,** 2005, 28867–28876.

Xia, P., Gamble, J.R., Wang, L., Pitson, S.M., Moretti, P.A., Wattenberg, B.W., D'Andrea, R.J., and Vadas, M.A. An oncogenic role of sphingosine kinase. *Curr Biol*, **10,** 2000, 1527–1530.

Zhang, P., Liu, B., Jenkins, G.M., Hannun, Y.A., and Obeid, L.M. Expression of neutral sphingomyelinase identifies a distinct pool of sphingomyelin involved in apoptosis. *J Biol Chem*, **272,** 1997, 9609–9612.

Zheng, W., Kollmeyer, J., Symolon, H., Momin, A., Munter, E., Wang, E., Kelly, S., Allegood, J.C., Liu, Y., Peng, Q., Ramaraju, H., Sullards, M.C., Cabot, M., and Merrill, A.H., Jr. Ceramides and other bioactive sphingolipid backbones in health and disease: lipidomic analysis, metabolism and roles in membrane structure, dynamics, signaling and autophagy. *Biochim Biophys Acta*, **1758,** 2006, 1864–1884.

# Chapter 17
# Glycosphingolipid Disorders of the Brain

**Stephanie D. Boomkamp and Terry D. Butters**

**Abstract** Glycosphingolipids, comprising a ceramide lipid backbone linked to one/more saccharides, are particularly abundant on the outer leaflet of the eukaryotic plasma membrane and play a role in a wide variety of essential cellular processes. Biosynthesis and subsequently degradation of these lipids is tightly regulated via the involvement of numerous enzymes, and failure of an enzyme to participate in the metabolism results in storage of the enzyme's substrate, giving rise to a lysosomal storage disease. The characteristics, severity and onset of the disease are dependent on the enzyme deficient and the residual activity. Most lysosomal storage disorders found thus far are caused by a defect in the catabolic activity of a hydrolase, causing progressive accumulation of its substrate, predominantly in the lysosome. Storage of gangliosides, sialic acid containing glycosphingolipids, mostly found in the central nervous system, is a hallmark of neuronopathic forms of the disease, that include GM1 and GM2 gangliosidoses, Gaucher type II and III and Niemann-Pick C. Models for these diseases have provided valuable insight into the disease pathology and potential treatment methods.

Treatment of these rare but severe disorders proves challenging due to restricted access of therapeutics through the blood-brain barrier. However, recent advances in enzyme replacement, bone marrow transplantation, gene transfer, substrate reduction and chaperon-mediated therapy provide great potential in treating these devastating disorders.

**Keywords** Glycosphingolipid · imino sugar · lysosomal storage disease · neurodegeneration · therapeutic strategies

**Abbreviations** CBE: conduritol B-epoxide; CSF: cerebrospinal fluide; CMT: chaperon-mediated therapy; CNS: central nervous system; ER: endoplasmic reticulum; ERAD: endoplasmic reticulum associated degradation; ERT:

S.D. Boomkamp
Glycobiology Institute, Department of Biochemistry, University of Oxford, Oxford, OX1 3QU, UK

P.J. Quinn, X. Wang (eds.), *Lipids in Health and Disease*,

enzyme replacement therapy; GalNAc: *N*-acetylgalactosamine; GalNAcT: *N*-acetylgalactosaminyltransferase; GA2: gangliotriglycosylceramide; Gb3: globotriaosylceramide; GlcCer: glucosylceramide; GlcNAc: *N*-acetylglucosamine; GM1a: II$^3$-α-N-acetylneuraminylgangliotetraglycosylceramide; GcGM1a: II$^3$-α-N-glycolylneuraminylgangliotetraglycosylceramide; GM2: II$^3$-α-*N*-acetylneuraminylgangliotriglycosylceramide; GM3: II$^3$-α-N-acetylneuraminyllactosylceramide; GM2AP: GM2 activator protein; GSL(s): glycosphingolipid(s); Hex: hexosaminidase; IL, interleukin; LacCer: lactosylceramide; LDL: low density lipoprotein; LSD(s): lysosomal storage disease(s); M-CSF: macrophage colony stimulating factor; MCB: membranous cytoplasmic body; MHC: major histocompatibility complex; MIP-1α: macrophage inflammatory protein-1α; NeuAc: sialic acid; *N*B-DGJ: *N*-butyldeoxygalactonojirimycin; *N*B-DNJ: *N*-butyldeoxynojirimycin; NPC: Niemann-Pick C; OS: oligosaccharide; SAP(s): sphingolipid activator protein(s); SRT: substrate reduction therapy; TGF-β1: transforming growth factor-β1; TNF-α: tumour necrosis factor-α; TNFR1: tumour necrosis factor receptor 1.

## 17.1 Introduction to Glycosphingolipid Metabolism and Disease

All eukaryotic cells contain a lipid outer membrane composed of glycerolipids, sphingolipids and sterols. All three lipids have been found to exhibit a wide range of combinatorial diversity and their biochemical and biophysical properties determine functionality.

The backbone of all sphingolipids from which the name is derived, is the sphingoid long-chain base. The most common of these lipids are sphinganine and sphingosine. Ceramide, the simplest sphingolipid and the common precursor for more complex lipids, consists of sphingosine to which a fatty acid is attached. Coupling of a glucose or galactose monosaccharide to ceramide is the first step in the formation of glycosphingolipids (GSLs). GSLs are the most structurally diverse sphingolipids and carry out an enormous range of essential cellular functions. In order to maintain the integrity of the cell, GSLs are continuously synthesised in the endoplasmic reticulum (ER) and the Golgi apparatus, and degraded in the lysosome. If there is a metabolic deficiency in the enzyme required for the degradation of a GSL, substrate accumulates to pathological levels, giving rise to a lysosomal storage disease (LSD). Therapy for treating these relatively rare but severe disorders proves challenging, especially when the nervous system is affected.

## 17.2 Metabolism

Glycosphingolipids are ubiquitous components of all eukaryotic plasma cell membranes and are particularly abundant at the cell surface. GSLs and their metabolites have been shown to act as intracellular signalling molecules that modulate numerous essential processes such as cell-cell interaction, proliferation, differentiation, cell death and stress response dependent on cell-type (Zeller and Marchase, 1992).

From our knowledge of human disease states, it can be speculated that GSLs are required for at least one stage of human embryogenesis, as there are no diseases resulting from mutations encoding enzymes involved in the initial steps in the GSL biosynthetic pathway. Knock-out mice that lack ceramide glucosyltransferase, the first enzyme required for GSL biosynthesis, die *in utero* due to widespread apoptosis (Yamashita et al., 1999). Besides playing a vital role in mammalian embryonic development and cellular differentiation, GSLs have also been found to play a crucial role in spermatogenesis (Sandhoff et al., 2005).

Other genetically engineered mouse models deficient in specific genes encoding GSL biosynthetic enzymes ß-1,4-*N*-acetylgalactosaminyltransferase and/or GD3 synthase, are viable despite the absence of complex gangliosides and do exhibit neurological abnormalities (Proia, 2003). Thus, gangliosides are essential for the stabilisation of the central nervous system (Yamashita et al., 2005).

The wide variety of cellular functions demonstrates that GSLs are not merely structural components of the plasma membrane and in order to maintain the integrity of the cell, the lipid bilayer is in a state of constant remodelling, and therefore GSLs are synthesised and degraded continuously.

Biosynthesis of glycolipids, and subsequently GSLs, commences in the ER with the formation of a long-chain aliphatic amino alcohol, sphingosine. This reaction is catalysed by the enzyme serine palmitoyl transferase via two steps, coupling of palmitoyl CoA and serine, forming 3-ketosphinganine, followed by subsequent acylation to form *N*-acyl sphingosine and ceramide when various fatty acids are linked to the 2-amino group of sphingosine (Fig. 17.1).

Hereafter, the *de novo* synthesised ceramide is translocated to the Golgi apparatus where stepwise glycosylation takes place. The mechanism of transport from the ER to the Golgi has been studied extensively but yet remains unclear, and vesicular membrane flow as well as non-vesicular transport has been described in the literature (van Meer and Lisman, 2002).

Coupling of a glucose residue to ceramide by glucosyltransferase gives rise to glucosylceramide (GlcCer), which can then be converted to lactosylceramide (LacCer), the common precursor for all GSLs.

The sequential addition of further monosaccharides and sialic acid residues by specific glycosyltransferases generates the manifold members of the ganglioside series. The newly synthesized GSLs are then transported to the plasma membrane via exocytotic membrane flow, where they carry out a wide variety of cell-type dependent functions.

It has been estimated that as much as 90% of cell GSLs are synthesised after endocytosis of their precursors (Gillard et al., 1998). This process can be disrupted via administration of *N*-alkylated imino sugars e.g. *N*-butyldeoxynojirimycin (*N*B-DNJ) or its galactose analogue, *N*-butyldeoxygalactonojirimycin (*N*B-DGJ), which inhibit the key enzyme involved in the GSL biosynthetic pathway, ceramide glucosyltransferase, and is discussed in further detail later (see Section 17.5.2).

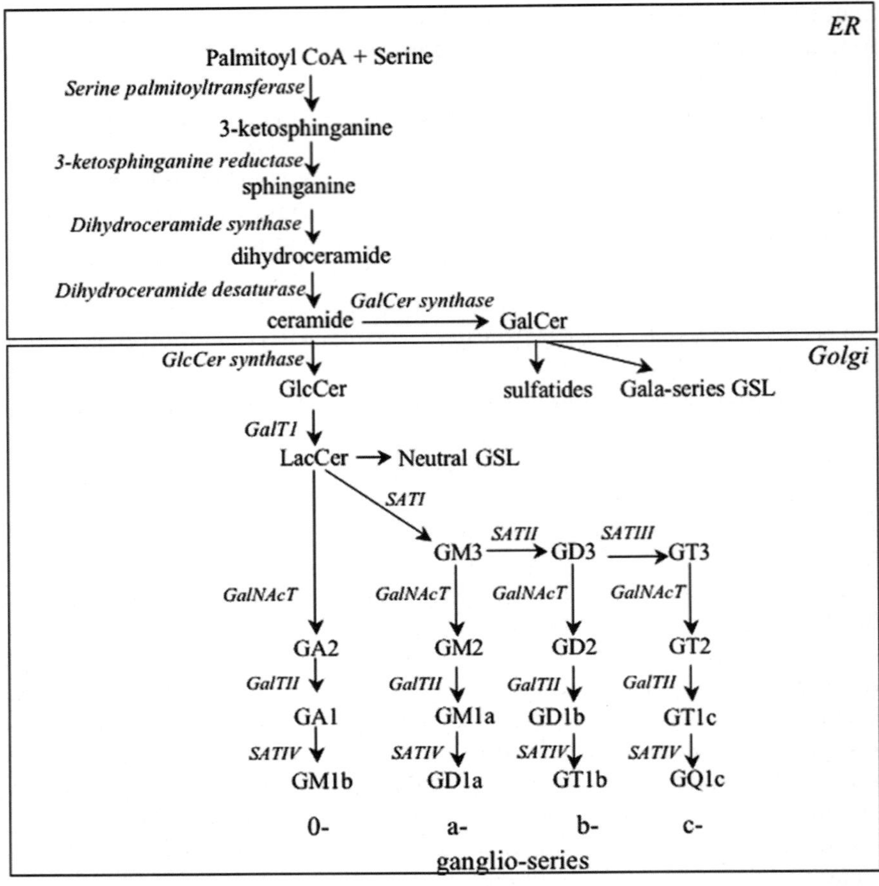


**Fig. 17.1** Synthesis of sphingolipids in the endoplasmic reticulum (ER) and the Golgi apparatus. GalNAcT: *N*-acetylgalactosamine transferase, GalT: galactosyltransferase, SAT: sialyl transferase

GSLs can be recycled in the Golgi and possibly the ER where further monosaccharides are added, or endocytosed and trafficked through the endosomal compartments to the lysosome for degradation to more simple metabolites.

Degradation is essentially the reverse of biosynthesis, taking place in the lysosome, discovered by de Duve and colleagues in 1955 (De Duve et al., 1955). GSLs are degraded by water-soluble exoglycosidases, present in the lumen of the lysosome, that sequentially cleave the oligosaccharide residues to produce ceramide, which then undergoes deacylation to produce sphingosine. This lipid can then leave the lysosome and re-enter the biosynthetic pathway or be degraded further. Besides exohydrolases, sphingolipid activator proteins (SAPs) are required for the degradation of sphingolipids with short hydrophilic head groups. SAPs perturb the membrane and bind the lipid in order to present it to the soluble enzyme for digestion (Kolter and Sandhoff, 2005).

If there is an inherited defect in SAP or a lysosomal glycosidase, GSL accumulates within the lysosome upstream of the defective reactions, resulting in progressive disease pathology.

To date, most of the diseases resulting from defective GSL metabolism have been found to affect GSL degradation, and only GM3 synthase deficiency has been described thus far that affects GSL biosynthesis.

## 17.3 Defects in GSL Metabolism

### *17.3.1 Defects in GSL Biosynthesis: GM3 Synthase Deficiency*

Only one disease caused by a biosynthetic defect has been reported so far, GM3 synthase deficiency. An infantile-onset form of epilepsy, developmental stagnation, tonic-clonic seizures and blindness characterize this inherited disorder, and the underlying cause is a nonsense 649C→T substitution mutation in exon 8 of the SIAT9 gene on chromosome 2. This gene encodes GM3 synthase and the mutation results in premature termination of the production of this enzyme and thus a non-functional protein product (Proia, 2004; Simpson et al., 2004). GM3 synthase synthesizes ganglioside GM3 from LacCer, the first step in the synthesis of complex a- and b-series ganglioside species. It remains unknown whether this disease is caused by decreased levels of GM3 or other downstream gangliosides, or by accumulation of LacCer caused by lack of flux through the a- and b-series ganglioside pathways and/or by accumulation of the globoside or isogloboside 0-series via β1,3-*N*-acetylgalactosaminyltransferase (Fig. 17.2).

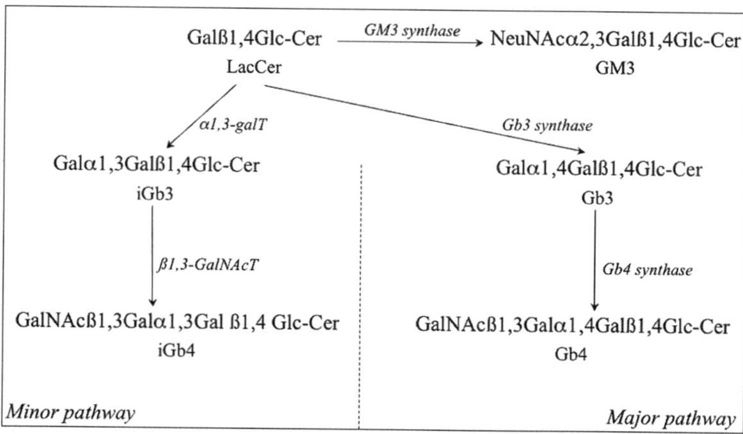


**Fig. 17.2** Overview of the biosynthetic pathway of globo-series neutral GSLs and gangliosides. α1,3-galT: α1,3-galactosyltransferase; β1,3-GalNAcT: β1,3-*N*-acetylgalactosaminyltransferase

GM3 synthase is a ubiquitously expressed protein present at particularly high levels in the central nervous system (CNS). The generation of a genetically engineered GM3 synthase knock-out mouse model led to the further understanding of the function of its product, GM3 ganglioside. It was found that these mice exhibited increased insulin sensitivity, suggesting GM3 as a negative regulator of insulin signalling and thus a potential therapeutic target for treating type II diabetes (Aerts et al., 2007).

Intriguingly however, this mouse model was phenotypically normal in contrast to the severe epileptic and developmental phenotype in humans; the reason as to how the pathological mechanisms differ amongst species is at present unclear (Yamashita et al., 2003).

Furthermore, in metastatic GM3 synthase deficient R3230AC cells the isogloboside (iGb4Cer) level is dramatically increased, suggesting that the presence of GM3 synthase prevents the formation of this metastasis-associated glycolipid in these cells. Therefore, GM3 synthase may play a crucial role in the progression of malignant cancer (Dumonceaux and Carlsen, 2001).

### *17.3.2 Defects in GSL Degradation*

In contrast to defects in glycolipid synthesis, diseases associated with defective degradation have been studied extensively and more than 40 lysosomal storage diseases are known of which at least ten are due to defective sphingolipid degradation (Table 17.1) (Futerman and van Meer, 2004). The frequency of individual diseases is not high but collectively they are a significant and severe group of disorders with a frequency of 1:7,700 live births in Australia and the most common cause of paediatric neurodegenerative diseases (Meikle et al., 1999). Some LSDs are more prevalent among specific ethnic groups, such as the high incidence of Tay-Sachs disease (1/3,900 compared to 1/200,000 in the general population) (Petersen et al., 1983) and Gaucher disease (1/855 compared to 1/100,000) among Ashkenazi Jews (Zimran, 1997).

The majority of LSDs are autosomal recessive inherited, with the exception of X-linked Fabry disease.

LSDs are normally classified according to the type of substrate that accumulates. Interestingly however, GSLs, such as GM2 and GM3 ganglioside, may also accumulate secondarily to accumulation of primary storage materials, as seen in Niemann-Pick disease. This additional accumulation, predominantly observed in axons of neurons and ectopic dendrites is essentially identical to that observed in the gangliosidoses (Walkley, 2004). Despite the underlying severe lesion in NPC does not necessarily involve defects in ganglioside metabolism, neuronal function in this LSD is impaired due to secondary storage of GSLs. The mechanism behind the alteration of these gangliosides in many types of lysosomal storage remains elusive.

**Table 17.1** Deficiencies and storage products in LSDs

| LSD | Enzyme/protein deficiency | Primary storage | Secondary storage |
|---|---|---|---|
| Farber* | Ceramidase | Ceramide | |
| Fucosidosis* | α-Fucosidase | Pentahexosylfuco-glycolipid | |
| Sialidosis* (mucolipidosis I) | Sialidase | GM3, sialyl-OS, sialoglycoproteins | GD3, GM4, LM1 |
| Metachromatic leukodystrophy* | Arylsulfatase A or saposin B | Cerebroside sulfate, 3-0-sulfogalactosyl-containing glycolipids | |
| Galactosialidosis* | Protective protein/ cathepsin A | Sialyl-OS | GM2, GM3, GM1, GD1a |
| Niemann-Pick | | | |
| A*, B | Sphingomyelinase | Sphingomyelin | GM2, GM3 |
| C* | Mutation in NPC1 or NPC2 gene | Cholesterol, bismono-acylglycerol phosphate | GM2, GM3, (GM1) |
| Krabbe* | Galactosylceramidase | Galactosylceramide | |
| Gaucher I, II* and III* | ß-glucocerebrosidase | GlcCer | GM2, GM3, GM1, GD3 |
| Fabry | β-galactosidase A | Gb3, Digalacto-sylceramide | |
| GM1 gangliosidosis* | ß-galactosidase | GM1, GA1, Gal-OS | GM2, GM3 |
| GM2 gangliosidosis*: | | | |
| Tay-Sachs | ß-hex A | GM2, GlcNAc-OS | Phospholipids, cholesterol |
| Sandhoff | ß-hex A, B | GM2, GA2, GlcNAc-OS | |
| GM2 activator | GM2 activator protein | GM2 | |

* Neurological involvement
Lysosomal storage diseases, their deficiency, primary storage, and secondary storage products are shown. OS: oligosaccharide.

Despite the different types of primary substrate that accumulate in different LSDs, all LSDs to date are characterized by macrophage activation following storage of GSLs by phagocytosis of senescent and apoptotic cells, and disruption of endocytic pathways targeted to the Golgi. However, the disease pathology, onset and severity are dependent on the enzyme deficiency and residual enzyme activity. Thus, every LSD exhibits pathological features characteristic for that particular disorder.

In this chapter the focus will be on the most common LSDs with neurological involvement, Niemann-Pick type C, types II and III Gaucher disease, GM1 and GM2 gangliosidosis.

#### 17.3.2.1 GM2 Gangliosidoses

GM2 gangliosidosis is a family of autosomal recessive disorders characterized by accumulation of GM2 ganglioside and its related glycolipids in the neuronal lysosome. It comprises GM2 activator protein deficiency, Tay-Sachs and Sandhoff disease, the latter of which are caused by a deficiency in the α-subunit or ß-subunit of ß-hexosaminidase, respectively (Gravel et al., 1995).

Sandhoff disease is a LSD that arises from a variety of point/substitution mutations in the HexB gene on chromosome 5 which encodes the ß-subunit present in ß-hexosaminidase A (a heterodimer consisting of an α- and ß-subunit) and B (ßß). ß-Hexosaminidase is a lysosomal enzyme required for the hydrolysis of the ß-glycosidic links of *N*-acetylgalactosamine (GalNAc) and *N*-acetylglucosamine (GlcNAc) terminating glycoconjugates.

Neutral substrates such as glycolipid GA2 can be catabolised by ß-hexosaminidase B as well as A, whereas only the α-subunit is able to hydrolyse negatively charged substrates, e.g. GM2 ganglioside, in the presence of its substrate-specific cofactor, GM2 activator protein (GM2AP) (Hou et al., 1996). Normally, membrane-bound lysosomal GSLs with short oligosaccharide chains, such as GM2, require a sphingolipid activator protein (SAP), such as GM2AP, to lift it out of the membrane and present it to lumenal ß-hexosaminidase A for degradation (Werth et al., 2001).

The third Hex isozyme, HexS, consisting of two α-subunits, does not occur in high quantities and was not thought to have a high catalytic activity. However, it has been shown that double knock-out mice that are totally deficient in Hex activity have a more severe phenotype than mice expressing only HexS as they display mucopolysaccharidosis (Sango et al., 1996) besides gangliosidosis and accumulate more anionic oligosaccharides. This demonstrates that HexS does contribute to the activity of ß-hexosaminidase, and has the potential to catabolise sulphated glycosaminoglycans, GSLs and water-soluble and amphiphilic glycoconjugates (Hepbildikler et al., 2002).

The deficiency in hexosaminidase A and B occurring in Sandhoff disease results in accumulation of GM2 and GA2 primarily in the lysosomes of neuronal cells (Fig. 17.3), leading to the formation of membranous cytoplasmic bodies (MCBs) (Tutor, 2004).

Furthermore, the levels of free oligosaccharides containing ß-linked *N*-acetylglucosamine at their non-reducing terminus, of which the formation is due to incomplete degradation of glycoproteins, are elevated (Fig. 17.4) (Warner et al., 1985; Winchester, 2005).

The main clinical features of this group of neurodegenerative diseases include startle reaction, hypotonia, psychomotor retardation, blindness and a greatly reduced life span of which the severity and disease onset are dependent on residual enzyme activity (Table 17.2) (Jeyakumar et al., 2002). All forms of acute, infantile GM2 gangliosidosis are characterized by a total absence of ß-hexosaminidase and thus no or hardly any residual enzyme activity, whereas the chronic, juvenile/adult forms have approximately 5%, and asymptomatic

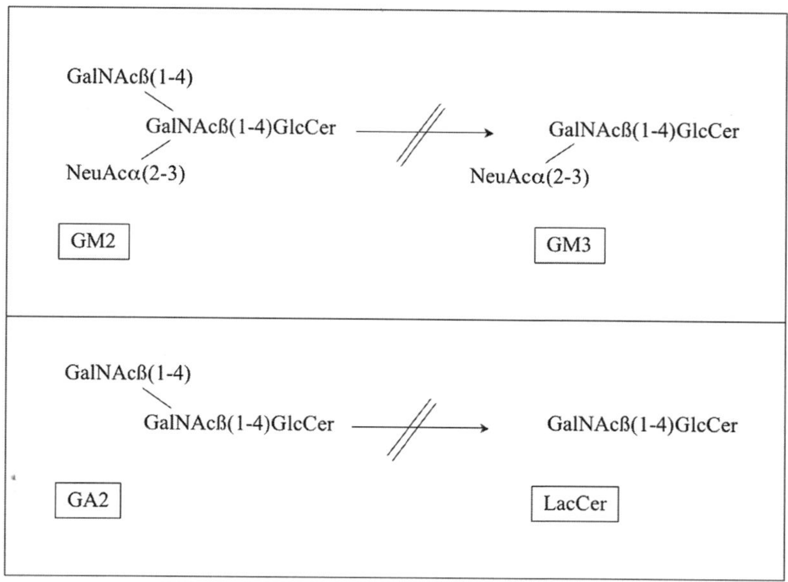


**Fig. 17.3** Structures and degradation of GM2 and GA2 glycolipid. In normal cells, *N*-acetylgalactosamine terminating GM2 ganglioside and its asialo-derivative GA2 are hydrolysed to GM3 and lactosylceramide respectively, by hexosaminidase. Deficiencies in hexosaminidase activity results in the accumulation of GM2 and GA2
GalNAc: *N*-acetylgalactosamine; Gal: galactose; Cer: ceramide; NeuAc: sialic acid

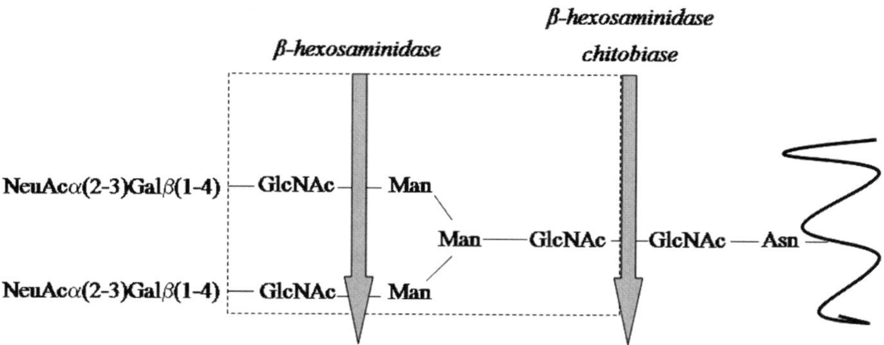


**Fig. 17.4** Simplified structure of a complex N-linked glycan and the sites of action of ß-hexosaminidase. Glycoproteins are degraded by specific enzymes from both directions. Lysosomal catabolism is terminated when an *N*-acetylglucosamine residue is encountered by the catalytically impaired enzyme ß-hexosaminidase, and thus the glycoprotein-derived oligosaccharide in the box accumulates in Sandhoff patients.
(NeuAc: sialic acid, Gal: galactose, Man: mannose, Asn: asparagine)

**Table 17.2** Major features of GM2 gangliosidoses (adapted from Jeyakumar et al., 2002)

| Clinical sign or symptom | Infantile | Juvenile | **Adult** |
|---|---|---|---|
| Motor dysfunction | + + | + | + |
| Hypotonia | + + | + | + |
| Blindness | + + | + | − |
| Seizures | + + | + | − |
| Macrocephaly | + | − | − |
| Cerebellar ataxia | − | + | + |
| Impaired cognitive function | + | + | + (−) |
| Psychosis | − | − | + (−) |
| Oligosaccharidurla (Sandhoff) | + | + | ? |
| Organomegaly (Sandhoff) | + | − | − |

heterozygotes have been found with just 10% of normal enzyme levels (Tropak et al., 2004).

While the GM2 gangliosidoses have become well understood at the biochemical level, the cause of neurodegeneration remains enigmatic. Initial studies suggested that GM2 ganglioside and/or its derivative may be potential inducers of apoptosis and thus progressive neurodegeneration (Huang et al., 1997). This finding prompted investigators to study the mechanism further, revealing that microglia activation, resulting in the production of inflammatory markers, precedes neuronal cell death (Fig. 17.5).

Bone marrow transplantation of Sandhoff disease mice does not reduce GM2 storage levels in the brain but does lead to the prevention of microglial activation and thus neuronal death (Norflus et al., 1998). Therefore, the inflammatory process may play a key role in the pathogenesis of GM2 gangliosidoses.

Several neurotoxic mediators have been found to contribute to the inflammatory response; for instance, increased tumour necrosis factor-α (TNFα) mRNA levels have been reported in Sandhoff mouse spinal cord, indicating transcriptionally-controlled up-regulation (Wada et al., 2000). Other

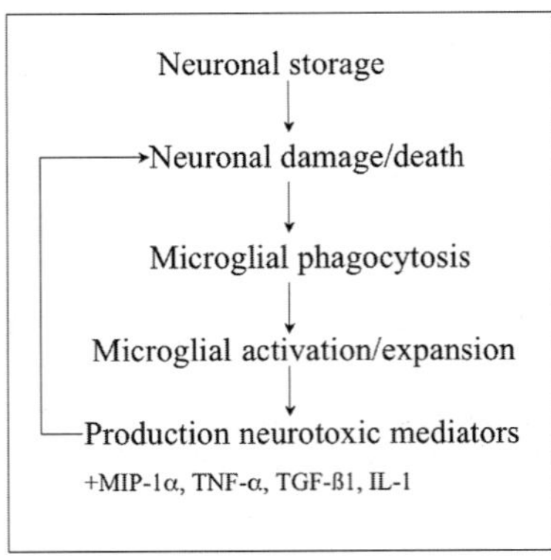


**Fig. 17.5** Schematic overview of cellular mechanism in which storage leads to inflammatory response

inflammation markers that have been found to be expressed were major histocompatability complex class II (MHC class II), nitric oxide, interleukin-1ß (IL-1ß) and transforming growth factor-ß1 (TGF-ß1), and their levels rise as the disease progresses. This suggests these markers are produced in response to, or contribute to, the local CNS immune activation (Jeyakumar et al., 2003).

Furthermore, macrophage inflammatory protein-1α (MIP-1α) induction has been suggested to correlate with the accumulation of *N*-acetylhexosaminyl glycoconjugates and not that of GM2 in specific brain regions (Tsuji et al., 2005). This study suggests that accumulation of glycoprotein-derived oligosaccharides as well as that of GM2, and potentially, GA2, may trigger an inflammatory response via different cascades dependent on the storage product. Deletion of this leukocyte chemokine in Sandhoff disease model mice decreases neuronal apoptosis and results in an improved and longer lifespan, further supporting the crucial role of inflammation in neurodegeneration and potentiating the use of anti-inflammatory drugs in the treatment of Sandhoff disease and possibly other LSDs (Wu and Proia, 2004).

### 17.3.2.2 GM1 Gangliosidosis

GM1 gangliosidosis is an inherited disorder caused by a deficiency in ß-galactosidase, resulting in lysosomal storage of GM1-ganglioside and its asialo-derivative GA1, primarily in the lysosome. Accumulation of these GSLs is widespread, with almost all neurons affected to a certain extent, and gradual deterioration of motor functions. A direct consequence of GM1 accumulation and depletion of ER calcium stores results in the activation of an unfolded protein response, triggering neuronal cell death in the ß-gal$^{-/-}$ mouse model of GM1-gangliosidosis (Tessitore et al., 2004).

A further pathological feature, so far thought to be exclusive to GM1 gangliosidosis, is vacuolisation of visceral cells and in particular liver parenchymal cells, spleen and the glomerular and renal tubular epithelial cells. These vacuoles have been found to contain high levels of highly water-soluble carbohydrate polymers, demonstrating a crucial role of ß-galactosidase in the degradation of these substrates(Wolfe et al., 1974).

Symptom onset and disease severity are dependent on residual enzyme activity, ranging from little or no activity in the infantile and juvenile forms, to measurable activity in adult-onset forms of GM1-gangiosidosis. Infantile-onset patients usually succumb to the disease in the first few months of life.

The potentially contributory role of the accumulating storage products to the pathology of GM1 gangliosidosis has been studied, and the inflammatory process, common to the symptomatic mouse model of GM1 storage appears to pre-date symptom onset, suggesting a potentially contributory role of GM1 in disease progression (Jeyakumar et al., 2003). Compared to wild-type mice, the expression of all apoptosis-mediators studied, MHC class II, Fas and tumour necrosis factor receptor 1 (TNFR1), are significantly elevated in storage regions of the brain, possibly as a result of progressive accumulation of GM1 and its related glycolipids.

The extent of inflammation correlates with disease progression with an age-dependent increase in microglial activation being observed which, interestingly, pre-dates the symptom onset and becomes more extensive as the disease progresses.

Furthermore, local neuroinflammation triggers activation of chemokines, such as stromal-cell-derived factor 1 (SDF-1), macrophage inflammatory protein 1-α (MIP-1α) and MIP-1β. Interestingly, administered β-galactosidase-active bone marrow cells migrate to the central nervous system (CNS), correcting the enzyme deficiency and restoring cytokine and GSL storage levels in GM1 gangliosidosis (Sano et al., 2005a).

#### 17.3.2.3 Gaucher Disease

Gaucher disease is an autosomal recessive progressive disorder and one of the most frequently observed sphingolipidoses. Pancytopenia, hepatosplenomegaly and skeletal complications are hallmarks of Gaucher disease (Zimran, 1997).

The enzyme responsible for degrading glucosylceramide (GlcCer), β-glucocerebrosidase, is deficient which causes progressive accumulation of substrate GlcCer, predominantly in large macrophage-derived cells with a characteristic morphology. These Gaucher cells have been thought to originate from the turn-over of cell membranes, for instance phagocytosed red and white blood cells, and secrete various factors involved in local tissue damage.

In normal situations, hydrolases destined for the lysosome are transported to this acidic compartment via the mannose 6-phosphate receptor. However, recent findings have suggested that β-glucocerebrosidase is trafficked to the lysosome through binding to a transmembrane lysosomal residents protein, LIMP-2, possibly in the endoplasmic reticulum (Reczek et al., 2007). This suggests a novel lysosomal trafficking pathway, independent of the mannose 6-phosphate receptor, and potentially a novel drug target for treating Gaucher disease.

Like many other LSDs, Gaucher disease is remarkably heterogeneous in terms of its clinical expression. Based on neurological involvement, disease onset and thus residual enzyme activity, three subsets can be distinguished, namely non-neuronopathic type I, and acute and sub-acute neuronopathic types II and III, respectively.

Specific mutations in the gene encoding ß-glucocerebrosidase are associated with specific clinical manifestations. The N370S mutation confers type I disease and the L444P mutation often correlates with neurological involvement in type III disease. The neurological manifestations of Gaucher disease have been suggested to be due to the loss of calcium homeostasis related to GlcCer accumulation within neurons (Futerman, 2007).

Recently, it has been shown that Gaucher type I disease can also be caused by a deficiency in Saposin-C, a lysosomal protein required for the degradation of glucosylceramide. Besides GlcCer storage, dramatically increased levels of chitotriosidase, a quantitative marker for the degree of macrophage activation, and chemokine CCL18, highly specific for Gaucher disease, are found in the plasma of Gaucher patients, the levels of which correlate to disease severity. In contrast, β-glucocerebrosidase activity is normal (Tylki-Szymanska et al., 2007). Accumulation of substrate within Gaucher cells appears to lead to elevation of pro-inflammatory cytokines IL-1β, IL-6, TNF-α, IL-10 and M-CSF, demonstrating direct correlation between GlcCer storage and disease progression (Tylki-Szymanska et al., 2007). Furthermore, the expression of macrophage inflammatory proteins MIP-1α and MIP-1β, both implicated in skeletal complications in multiple myeloma, are significantly elevated (9 and 12 fold respectively) in plasma of Gaucher type I patients. Effective treatment via enzyme replacement (ERT) results in restoration in the levels of these inflammatory markers (van Breemen et al., 2007).

Recent data analysing secondary metabolites also reveal the complexity of this monogenetic disorder that could influence pathogenesis. Macrophage models of Gaucher disease and patient plasma show elevations in ceramide, lactosylceramide, Gb3 and GM3 (Ghauharali-van der Vlugt et al., 2007; Hein et al., 2007). The increase in GM3 in particular may correlate with increased insulin resistance in Gaucher patients (Langeveld et al., 2007).

#### 17.3.2.4 Niemann-Pick Disease

Niemann-Pick disease comprises three subsets, A, B and C, of which the former two are caused by a genetic defect in lysosomal sphingomyelinase activity. Niemann-Pick type C (NPC) however is biochemically distinct from types A and B, due to a mutation in either the NPC1 gene (in more than 90% of the cases) or the NPC2 gene (Sun et al., 2001). The precise functions of NPC1 and NPC2 proteins remain at present unclear; it has been speculated that NPC1 might be implicated in the transport of low density lipoprotein (LDL) derived cholesterol out of the endosomal pathway. NPC2, present in the luminal lysosome, may bind cholesterol and cooperate with NPC1 in mediating the egress of cholesterol from the late endosome to the lysosomal compartment (Zhang et al., 2003).

The clinical manifestations, including vertical gaze palsy, ataxia, dystonia and seizures, usually become evident in early childhood and death typically occurs in adolescence. Although NPC is relatively rare (1 in 150,000 live births), the incidence in Yarmouth County, Nova Scotia, is 1% with an estimated carrier frequency of 10–25% (Winsor and Welch, 1978).

The most characteristic features of the brain tissue of NPC patients are neuronal cell loss, particularly Purkinje cells, swollen neurites and extensive demyelination (Higashi et al., 1993).

The underlying causes of this pathological phenotype have been found to be massive lysosomal accumulation of cholesterol and other lipids including bis-monoacylglyerol phosphate as well as gangliosides GM2 and GM3. The brain is the organ most enriched in cholesterol and as there is no access through the blood-brain barrier, its synthesis takes place in the CNS (Turley et al., 1998). Normally, the majority of the cholesterol in the brain resides in myelin; however in NPC brain tissue cholesterol is absent which is thought to lead to extensive demyelination (Takikita et al., 2004).

Mechanisms leading to pathology in NPC are only beginning to become understood; it has been speculated that the impaired LDL-derived cholesterol from the lysosome in NPC patients results in lysosomal cholesterol sequestration and thus an increase in hydrolases such as cathepsin D. High levels of cathepsin D are found to be toxic and lead to microglial activation (German et al., 2002) and up-regulation of interleukin-1ß confined to degenerating brain regions as a consequence of astroglial activation (Baudry et al., 2003), as well as autophagy (Liao et al., 2007). Autophagy has been found to be involved in neurodegeneration in the to NPC related Alzheimer disease (Nixon et al., 2005). These findings suggest that lysosomal dysfunction indirectly contributes to NPC pathogenesis and its progressive neurodegenerative feature.

## 17.4 Models for the Glycosphingolipidoses

Relatively recently, various models for sphingolipidoses have been developed in order to elucidate cellular functions of a glycolipid and to study the pathogenesis and potential approaches towards therapy of LSDs. Yeast models, Drosophila, cells obtained from patients, genetically engineered cells and animal models have all provided valuable insight.

An *in vitro* cellular model for Gaucher was developed by treating murine J774 macrophages with irreversible inhibitor conduritol B-epoxide (CBE) and feeding these cells red blood cell ghosts obtained from a Gaucher disease patient, in order to more closely mimic the disease state. CBE inhibits glucocerebrosidase, which is deficient in Gaucher disease. Interestingly, the activity of glucocerebrosidase could be reduced to 11–15% of the normal control level before storage of its substrate glucosylceramide occurred, thus demonstrating that reduction in enzyme activity does not automatically lead to storage and disease pathology (Schueler et al., 2004). This demonstrates that a critical threshold of residual enzyme activity must be reached to cause lysosomal storage (Conzelmann and Sandhoff, 1983).

Furthermore, the use of a similar chemically-derived *in vitro* cellular model led to the therapeutic application of a novel drug, *N*B-DNJ (miglustat, Zavesca®), a potent inhibitor of ceramide glucosyltransferase at a relatively low concentration (5–50 μM) in murine macrophages (Platt et al., 1994) (see Section 17.5.2).

Patient-derived cell lines or those obtained from mouse LSD models have also been used extensively to gain further insights in the disease pathology and potential therapeutics. For example, GM1 gangliosidosis has been studied in patient-derived fibroblasts and those obtained from ß-galactosidase deficient mice. The GM1 synthase activity in human fibroblasts was significantly reduced, whereas in murine fibroblasts the levels were even slightly increased. This suggests the occurrence of alternative metabolic pathways in mouse and man.

Moreover, the use of this model demonstrates that the massive intracellular accumulation of GM1 due to the loss of ß-galactosidase activity affected the biosynthetic pathway of this ganglioside, suggesting the occurrence of regulatory mechanisms that balance ganglioside biosynthesis and maintain cellular homeostasis (Sano et al., 2005b). Thus, it can be speculated that the turn-over of glycolipids is dependent on residual enzyme activity and therefore GSL metabolism in juvenile/adult patients occurs more rapidly than in infantile patients with no or hardly any activity (Leinekugel et al., 1992).

Genetically engineered mouse models for disease have also been used to develop authentic *in vitro* models. Aortic endothelial cells isolated from Fabry mice, which lack α-galactose A, retain the elevated levels of globotriaosylceramide (Gb3) observed in culture (Shu et al., 2005). This demonstrates that when LSD cells are removed from their native environment, the disease phenotype in terms of storage is still present. Thus, *in vitro* cell models for gangliosidoses are suitable for studying these storage diseases, as the reproducibility of cultured cells compared to multi-cellular organisms is higher due to controlled conditions, and the effects of higher concentrations of potential therapeutic drugs can be tested, as well as the performance of long-term experiments.

Most studies on GM2 gangliosidoses have been performed on mouse knockout models. However, while only the human phenotype slightly differs from that observed in these mice, the severity and disease course differs significantly, as the result of a sialidase, so far thought to be exclusively present in mice (Kolter and Sandhoff, 1998) (Fig. 17.6).

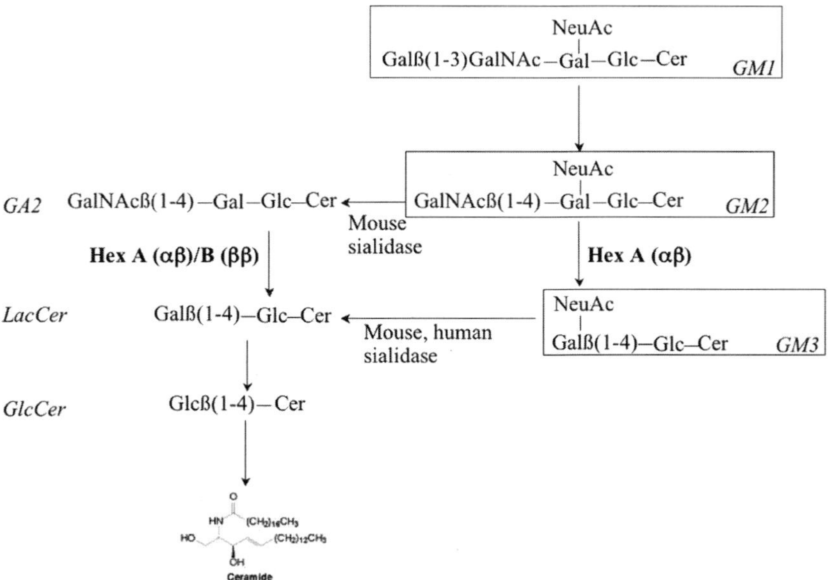


**Fig. 17.6** Partial GSL degradation pathway. GM2 is degraded by hexosaminidase A (Hex A) in cooperation with GM2 activator protein to form GM3. The presence of a specific sialidase in mice prevents accumulation of this glycolipid in GM2 gangliosidosis and degrades its substrate to GA2, which in turn can be degraded by Hex A or B. GM3 is acted on by sialidase, present in mouse and human, and lactosylceramide (LacCer) is formed which is then further degraded by specific hydrolases

For example, by contrast to human patients, Tay-Sachs knock-out mice (Hexa$^{-/-}$) show no neurological abnormalities despite the progressive accumulation of GM2 ganglioside in the central nervous system, whereas the Sandhoff knock-outs (Hexb$^{-/-}$) are severely affected (Sango et al., 1995). It has been proposed that this difference of the Tay-Sachs disease severity between mice and humans is due to the possibility that Hexa$^{-/-}$ mice, not humans, escape disease pathology through partial catabolism of GM2 via GA2 which can then be catabolised by active ß-hexosaminidase B (Phaneuf et al., 1996).

In our laboratory an authentic *in vitro* cellular model for Sandhoff disease has been generated upon treatment of RAW264.7 murine macrophages with an inhibitor of β-hexosaminidase (Boomkamp and Butters, unpublished data). High performance liquid chromatography (HPLC) analyses of extracted GSL oligosaccharides show that GA2 predominantly accumulates in inhibitor-treated RAW cells as opposed to a minor but significant increase in GM2 levels (Fig. 17.7). This difference in elevation is possibly due to the

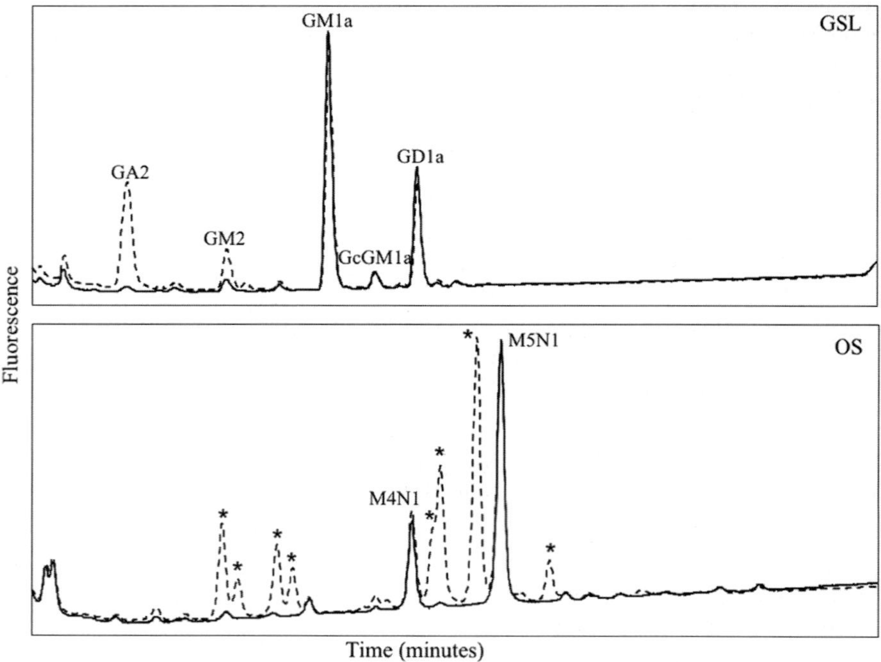


**Fig. 17.7** HPLC chromatograms of GSLs and OS in RAW264.7 cells prior and following treatment with ß-hexosaminidase inhibitor.
*Top panel*: Untreated RAW cells contain mainly GM1a, GcGM1a and GD1a (*solid line*) whereas upon treatment GA2 and GM2 glycolipid are additionally present (*dotted line*).
*Bottom panel*: Untreated RAW cells contain mainly polymannose M4N1 and M5N1 (*solid line*), whereas upon treatment the levels of GlcNAc-terminating structures (marked with an *asterisk*) are significantly increased (*dotted line*)

presence of a mouse-specific sialidase that provides a bypass mechanism by hydrolysing the sialic acid residue from GM2 to form GA2, as discussed previously.

Furthermore, glycoprotein-derived oligosaccharide (OS) purifications show that inhibitor treated cells exhibit a wide variety of GlcNAc-terminating structures, in addition to polymannose-type structures $Man_4GlcNAc_1$ (M4N1) and $Man_4GlcNAc_1$ (M5N1), also present in untreated cells which are products of ER associated degradation (Alonzi et al., 2008) (Fig. 17.7).

In urine and tissues of Sandhoff patients, OS are observed which serve as biological markers for disease. The main species, accounting for 70% of the oligosaccharides stored in the brain, has been shown to be a biantennary, bisected heptasaccharide, $Man_3GlcNAc_4$, (Warner et al., 1985) consistent with that seen in the Sandhoff mouse model (Lowe et al., 2005). This structure is also a major species in inhibitor-treated RAW cells. All other stored OS were also found to be GlcNAc-terminating, illustrating the deficiency/inhibition of ß-hexosaminidase and termination of further degradation of the partially hydrolysed glycoprotein.

Overall, the developed *in vitro* RAW model is an authentic, cellular representative of Sandhoff disease and may provide further insights into the disease pathology and for evaluating potential therapies for this disorder.

## 17.5 Therapeutic Options for the Neuronopathic Glycosphingolipidoses

Until recently few treatment options were available for LSDs with neurological involvement and the progressive nature of disease pathology in the most severe phenotypes allowed little time for clinical intervention. For those patients with juvenile and late-onset disease the potential for slowing the course of disease should have a dramatic outcome on quality of life measurements but the prospects for generating therapeutics for orphan diseases, where the population of affected individuals are extremely low, appeared to be poor. However, the success of enzyme supplementation for type 1 Gaucher disease (Brady, 2006) has provided a platform for pre-clinical studies for treating similar glycosphingolipid storage disorders.

The approaches to disease intervention can be separated into an enzyme augmentation strategy that includes direct enzyme infusion, pharmacological chaperoning, bone marrow replacement and gene transfer, and inhibition of substrate synthesis (Butters, 2007b). The first approach is dedicated to the precise enzyme deficiency in each disease whereas substrate inhibition is a generic therapy that can be applied to all glycosphingolipidoses that have a common biosynthetic pathway.

### 17.5.1 *Enzyme Augmentation*

#### 17.5.1.1 Enzyme Replacement Therapy (ERT)

Improving the catalytic competence in the lysosome can be achieved by direct infusion of an enzyme to provide enzyme augmentation or enzyme replacement therapy. For those diseases where neurological storage of the products of incomplete metabolism is the major biochemical phenotype, such as the gangliosidoses and types II and III Gaucher disease (Table 17.1), the blood-brain barrier restricts the access of intravenously delivered protein and results in a less efficient outcome for reducing the storage burden. Some success has been achieved by direct intracerebroventricular injection or high dose infusion (Vogler et al., 2005) but the clinical response may be insufficient to provide benefit.

#### 17.5.1.2 Molecular Chaperons

In many of the glycosphingolipidoses, amino acid changes as a result of gene mutations cause the protein to misfold during translation in the ER. Some of this protein is removed from the ER by a process of ER-associated degradation (ERAD) and other misfolded variants may be inefficiently trafficked to the lysosome (Ron and Horowitz, 2005). The net result is a reduction in lysosomal catalytic activity. The use of small molecules as chaperons to protect the protein from gross misfolding and ERAD, enhances the catalytic activity of enzyme in the lysosome. This novel approach, chaperon-mediated therapy (CMT), is in the early stages of clinical evaluation for the peripherally associated storage disorders, such as Gaucher and Fabry disease, following key pre-clinical studies (Butters, 2007b; Butters et al., 2005; Fan, 2003; Pastores and Sathe, 2006).

Many of the chaperons used have been imino sugars because of their tight-binding properties to the active site and at sub-inhibitory concentrations produce an enhancement of enzyme that may be sufficient to degrade the lysosomally stored material. For the neuronopathic disorders few candidates have been identified but a recent chemical library screen for potential hexosaminidase chaperons has revealed additional molecular frameworks that could be used that are outside the imino sugar structural motif (Tropak et al., 2007).

In the glycosphingolipidoses, the severity of disease is determined by the site(s) of mutation that critically affect protein folding. Whilst the mild to moderate disorders of protein folding, such as Gaucher and Fabry disease may be amenable to CMT, the more severe phenotypes may pose difficulty. In Gaucher disease, a common mutation predicting a neuronopathic course of disease is L444P, and the lack of ability of imino sugar chaperons to rescue enzyme activity and provide therapeutic potential may be limited (Butters, 2007a). Despite this, the identification of small molecules such as the imino sugars used for substrate reduction therapy (SRT) have promise since the blood-brain barrier is crossed, although rather inefficiently, and many of the safety and toxicity concerns for using these as pharmacological agents in man have been allayed. At the experimental level, partial correction of the cellular activity of the misfolded enzyme has been achieved for Gaucher ß-glucocerebrosidase (Compain et al., 2006; Sawkar et al., 2005; Steet et al., 2006; Yu et al., 2007; Zhu et al., 2005), Fabry α-galactosidase (Asano et al., 2000; Fan, 2003; Yam et al., 2005), GM1 gangliosidosis (Matsuda et al., 2003; Suzuki, 2006; Tominaga et al., 2001) and Tay-Sachs/Sandhoff ß-hexosaminidase (Tropak et al., 2004).

#### 17.5.1.3 Bone Marrow Transplantation and Gene Therapy

In mouse models for the gangliosidoses, bone marrow transplantation (BMT) extended the life span, slowed the progressive neurological deficit and reduced the peripheral cellular storage of glycoconjugates (Norflus et al., 1998). The extent to which donor-derived macrophages are able to repopulate the CNS and cross-correct the enzyme deficiency may be a limitation for this approach for severe disorders. A combination of approaches may be more successful and a synergistic improvement in the survival rate was demonstrated in the Sandhoff mouse when BMT was used in conjunction with SRT (Jeyakumar et al., 2001). In man, transplantation carries a number of risks associated with an invasive technique but offers a longer-term solution to enzyme infusion. The relatively few clinical reports suggest that this may be an option when other therapeutic approaches are unavailable.

Gene therapy is still at the pre-clinical stage and cellular and *in vivo* studies (Guidotti et al., 1998; Kyrkanides et al., 2005) support the utility of this approach for cross-correction of cells in privileged sites such as the brain. In the Sandhoff mouse model, enhanced survival following delivery to the brain via stereotaxic injection of adeno-associated viral vectors (Cachon-Gonzalez et al., 2006) was observed to be greater than a combination of BMT and SRT (Jeyakumar et al., 2001).

### *17.5.2 Substrate Reduction Therapy (SRT)*

Substrate reduction therapy is a strategy that aims to partially inhibit the biosynthetic cycle to reduce glycosphingolipid substrate influx into the catabolically compromised lysosome. A number of small molecules have been synthesised to inhibit ceramide-specific glucosyltransferase, the first enzyme in the biosynthetic pathway that glycosylates ceramide lipid, to balance synthesis with catabolism and restore homeostasis. *N*B-DNJ was shown to be an effective inhibitor of glycolipid biosynthesis and could reduce lysosomal storage in a chemically-derived tissue culture cell model of Gaucher disease (Platt et al., 1994). In mouse, models for disease were generated by deleting the gene encoding for lysosomal hexosaminidase and SRT using *N*B-DNJ was demonstrated to have biochemical efficacy in reducing neuronal cell ganglioside in the Tay-Sachs mouse (Platt et al., 1997) and improved the life span by 40% in the Sandhoff mouse (Jeyakumar et al., 1999). These data allowed a proof of principle study of *N*B-DNJ (miglustat) in man, and a 12-month assessment for efficacy in type I Gaucher disease showed improvements of organ volumes and haematological parameters (Cox et al., 2000). Miglustat gained approval in Europe, USA and Israel for use for type I patients who were unable or unwilling to take ERT. Clinical benefit was clearly demonstrated in this multi-centre trial and further trials to evaluate low dose administration (Heitner et al., 2002) and a 3-year continuation study (Elstein et al., 2004) have revealed significant improvements in all the major clinical endpoints. These data are consistent with the mechanism of action of miglustat as a drug dependent modulator of GSL biosynthesis since treatment was increasingly effective with time, as confirmed by further studies over 24 months (Pastores et al., 2005) and in combination with ERT (Elstein et al., 2007). Additional studies have demonstrated that when miglustat was administered for 12 months to a Spanish cohort of 25 patients, an equivalent improvement in hematologic parameters to ERT was observed (Giraldo et al., 2006). The access of small molecules such as miglustat to tissues that would be intractable to enzyme therapy, such as bone, results in an improvement in bone density and a reduction in pain (Pastores et al., 2007). A significant body of data now supports the use of this drug for SRT in Gaucher disease and other glycosphingolipidoses (Butters, 2007b).

In Niemann-Pick type C disease, where glycosphingolipid storage is a consequence of abnormal lipid trafficking and may contribute to the pathology observed, SRT has been shown to reduce biochemical markers of the lysosomal burden (Lachmann et al., 2004). In two recent studies, miglustat has been shown to stabilize the systemic disease in children with NPC (Chien et al., 2007; Patterson et al., 2007), implying that a reduction in CNS storage of glycosphingolipid can be of therapeutic value in this and similar neurodegenerative disorders.

## 17.6 Problems and Perspectives

Poor access via the blood-brain barrier is the major obstacle that prevents ERT from being applicable to the neuronopathic disorders. The inflammatory cascade that is mediated by CNS storage seen in the gangliosidoses for example, may respond to steroid treatment but unless the neuronal cell lysosomal burden can be reduced, this peripheral and neurological component of the phenotype could dominate the clinical outcome of disease.

Small molecule therapeutics are generally more able to access the neural tissue and the efficacy of miglustat to reduce neuronal cell ganglioside accumulation in murine models of Tay-Sachs and Sandhoff disease, indicates that a proportion of the plasma dose reaches the CNS. Direct measurement of miglustat in the cerebrospinal fluid (CSF) obtained from patients with either Niemann-Pick type C disease or Tay-Sachs disease following oral treatment have been obtained. These data reveal that 16–27% of the plasma concentration can be measured in the CSF at an oral dose of 100 mg/day (Bembi et al., 2006; Lachmann et al., 2004). Although the clinical response to treatment is variable and difficult to assess, as with many other progressive neurodegenerative disorders, the expected reduction in lipid burden offers potential for treating those disorders such as Gaucher type II/III, NP-C and the gangliosidoses, where no current successful therapy is available. Although the mechanism of action of miglustat is consistent with pre-clinical data showing a reduction of substrate, data in support of clinical efficacy for treatment of severe neuronopathic disorders is required. Recently, a two year study of a combination of ERT and miglustat administered to 20 type III Gaucher disease patients, aged between two and 20 years, did not demonstrate any improvement of the primary end-points (Butters, 2007a). The inability to reverse much of the existing pathology may be an overriding factor for treating many of the neuronopathic conditions, particularly for paediatric patients who may require high doses of a pharmacological agent to provide a more rapid response. The development of drugs for SRT or CMT with an improved efficacy and safety profile to allow either mono-therapy or in combination at doses required to deplete substrate and/or enhance enzyme activity significantly, may be a partial solution to these problems.

The dissection of the biochemical pathways that generate lysosomal storage of glycoconjugates has been a major achievement but our further understanding to allow the development of novel routes for intervention remains a challenge.

## References

Aerts, J. M., Ottenhoff, R., Powlson, A. S., Grefhorst, A., van Eijk, M., Dubbelhuis, P. F., Aten, J., Kuipers, F., Serlie, M. J., Wennekes, T., Sethi, J. K., O'Rahilly, S., and Overkleeft, H. S., Pharmacological inhibition of glucosylceramide synthase enhances insulin sensitivity, Diabetes 56 (2007) 1341–1349.

Alonzi, D. S., Neville, D. C., Lachmann, R. H., Dwek, R. A., and Butters, T. D., Glucosylated free oligosaccharides are biomarkers of endoplasmic- reticulum alpha-glucosidase inhibition, Biochem J 409 (2008) 571–580.

Asano, N., Ishii, S., Kizu, H., Ikeda, K., Yasuda, K., Kato, A., Martin, O. R., and Fan, J. Q., In vitro inhibition and intracellular enhancement of lysosomal alpha-galactosidase A activity in Fabry lymphoblasts by 1-deoxygalactonojirimycin and its derivatives, Eur J Biochem 267 (2000) 4179–4186.

Baudry, M., Yao, Y., Simmons, D., Liu, J., and Bi, X., Postnatal development of inflammation in a murine model of Niemann-Pick type C disease: immunohistochemical observations of microglia and astroglia, Exp Neurol 184 (2003) 887–903.

Bembi, B., Marchetti, F., Guerci, V. I., Ciana, G., Addobbati, R., Grasso, D., Barone, R., Cariati, R., Fernandez-Guillen, L., Butters, T., and Pittis, M. G., Substrate reduction therapy in the infantile form of Tay-Sachs disease, Neurology 66 (2006) 278–280.

Brady, R. O., Enzyme replacement for lysosomal diseases, Annu Rev Med 57 (2006) 283–296.

Butters, T. D., Gaucher disease, Curr Opin Chem Biol 11 (2007a) 412–418.

Butters, T. D., Pharmacotherapeutic strategies using small molecules for the treatment of glycolipid lysosomal storage disorders, Expert Opin Pharmacother 8 (2007b) 427–435.

Butters, T. D., Dwek, R. A., and Platt, F. M., Imino sugar inhibitors for treating the lysosomal glycosphingolipidoses, Glycobiology 15 (2005) 43R–52R.

Cachon-Gonzalez, M. B., Wang, S. Z., Lynch, A., Ziegler, R., Cheng, S. H., and Cox, T. M., Effective gene therapy in an authentic model of Tay-Sachs-related diseases, Proc Natl Acad Sci USA 103 (2006) 10373–10378.

Chien, Y. H., Lee, N. C., Tsai, L. K., Huang, A. C., Peng, S. F., Chen, S. J., and Hwu, W. L., Treatment of Niemann-Pick disease type C in two children with miglustat: initial responses and maintenance of effects over 1 year, J Inherit Metab Dis 30 (2007) 826.

Compain, P., Martin, O. R., Boucheron, C., Godin, G., Yu, L., Ikeda, K., and Asano, N., Design and synthesis of highly potent and selective pharmacological chaperones for the treatment of Gaucher's disease, Chembiochem 7 (2006) 1356–1359.

Conzelmann, E., and Sandhoff, K., Partial enzyme deficiencies: residual activities and the development of neurological disorders, Dev Neurosci 6 (1983) 58–71.

Cox, T., Lachmann, R., Hollak, C., Aerts, J., van Weely, S., Hrebicek, M., Platt, F., Butters, T., Dwek, R., Moyses, C., Gow, I., Elstein, D., and Zimran, A., Novel oral treatment of Gaucher's Disease with N-butyldeoxynojirimycin (OGT 918) to decrease substrate biosynthesis, Lancet 355 (2000) 1481–1485.

De Duve, C., Pressman, B. C., Gianetto, R., Wattiaux, R., and Appelmans, F., Tissue fractionation studies. 6. Intracellular distribution patterns of enzymes in rat-liver tissue, Biochem J 60 (1955) 604–617.

Dumonceaux, T., and Carlsen, S. A., Isogloboside biosynthesis in metastatic R3230AC cells results from a decreased GM3 synthase activity, Arch Biochem Biophys 389 (2001) 187–194.

Elstein, D., Dweck, A., Attias, D., Hadas-Halpern, I., Zevin, S., Altarescu, G., Aerts, J. F., van Weely, S., and Zimran, A., Oral maintenance clinical trial with miglustat for type I Gaucher disease: switch from or combination with intravenous enzyme replacement, Blood 110 (2007) 2296–2301.

Elstein, D., Hollak, C., Aerts, J. M., van Weely, S., Maas, M., Cox, T. M., Lachmann, R. H., Hrebicek, M., Platt, F. M., Butters, T. D., Dwek, R. A., and Zimran, A., Sustained therapeutic effects of oral miglustat (Zavesca, N -butyldeoxynojirimycin, OGT 918) in type I Gaucher disease, J Inherit Metab Dis 27 (2004) 757–766.

Fan, J. Q., A contradictory treatment for lysosomal storage disorders: inhibitors enhance mutant enzyme activity, Trends Pharmacol Sci 24 (2003) 355–360.

Futerman, A. H., Cellular pathology in Gaucher disease. In Gaucher disease (A. H. Futerman, and A. Zimran, Eds.), Taylor and Francis Group, Boca Raton, 2007, pp. 97–108.

Futerman, A. H., and van Meer, G., The cell biology of lysosomal storage disorders, Nat Rev Mol Cell Biol 5 (2004) 554–565.

German, D. C., Liang, C. L., Song, T., Yazdani, U., Xie, C., and Dietschy, J. M., Neurodegeneration in the Niemann-Pick C mouse: glial involvement, Neuroscience 109 (2002) 437–450.

Ghauharali-van der Vlugt, K., Langeveld, M., Poppema, A., Kuiper, S., Hollak, C. E., Aerts, J. M., and Groener, J. E., Prominent increase in plasma ganglioside GM3 is associated with clinical manifestations of type I Gaucher disease, Clin Chim Acta (2007) in press

Gillard, B. K., Clement, R. G., and Marcus, D. M., Variations among cell lines in the synthesis of sphingolipids in de novo and recycling pathways, Glycobiology 8 (1998) 885–890.

Giraldo, P., Latre, P., Alfonso, P., Acedo, A., Alonso, D., Barez, A., Corrales, A., Franco, R., Roldan, V., Serrano, S., and Pocovi, M., Short-term effect of miglustat in every day clinical use in treatment-naive or previously treated patients with type 1 Gaucher's disease, Haematologica 91 (2006) 703–706.

Gravel, R. A., Clarke, J. T. R., Kaback, M. M., Mahuran, D., Sandhoff, K., and Suzuki, K., The GM2 Gangliosidoses, 7 ed. In The Metabolic Bases of Inherited Disease (C. R. Scriver, A. L. Beaudet, W. S. Sly, and D. Valle, Eds.), Vol. II, McGraw-Hill, New York, 1995, pp. 2839–2879.

Guidotti, J., Akli, S., Castelnau-Ptakhine, L., Kahn, A., and Poenaru, L., Retrovirus-mediated enzymatic correction of Tay-Sachs defect in transduced and non-transduced cells, Hum Mol Genet 7 (1998) 831–838.

Hein, L. K., Meikle, P. J., Hopwood, J. J., and Fuller, M., Secondary sphingolipid accumulation in a macrophage model of Gaucher disease, Mol Genet Metab 92 (2007) 336–345.

Heitner, R., Elstein, D., Aerts, J., van Weely, S., and Zimran, A., Low-dose N-butyldeoxynojirimycin (OGT918) for type I Gaucher disease, Blood Cells Mol Dis 28 (2002) 127–133.

Hepbildikler, S. T., Sandhoff, R., Kolzer, M., Proia, R. L., and Sandhoff, K., Physiological substrates for human lysosomal beta -hexosaminidase S, J Biol Chem 277 (2002) 2562–2572.

Higashi, Y., Murayama, S., Pentchev, P. G., and Suzuki, K., Cerebellar degeneration in the Niemann-Pick type C mouse, Acta Neuropathol 85 (1993) 175–184.

Hou, Y., Tse, R., and Mahuran, D. J., Direct determination of the substrate specificity of the alpha-active site in heterodimeric beta-hexosaminidase A, Biochemistry 35 (1996) 3963–3969.

Huang, J. Q., Trasler, J. M., Igdoura, S., Michaud, J., Hanal, N., and Gravel, R. A., Apoptotic cell death in mouse models of GM2 gangliosidosis and observations on human Tay-Sachs and Sandhoff diseases, Hum Mol Genet 6 (1997) 1879–1885.

Jeyakumar, M., Butters, T. D., Dwek, R. A., and Platt, F. M., Glycosphingolipid lysosomal storage diseases: therapy and pathogenesis, Neuropathol Appl Neurobiol 28 (2002) 343–357.

Jeyakumar, M., Butters, T. D., CortinaBorja, M., Hunnam, V., Proia, R. L., Perry, V. H., Dwek, R. A., and Platt, F. M., Delayed symptom onset and increased life expectancy in Sandhoff disease mice treated with N-butyldeoxynojirimycin, Proc Nat Acad Sci USA 96 (1999) 6388–6393.

Jeyakumar, M., Norflus, F., Tifft, C. J., CortinaBorja, M., Butters, T. D., Proia, R. L., Perry, V. H., Dwek, R. A., and Platt, F. M., Enhanced survival in Sandhoff disease mice receiving a combination of substrate deprivation therapy and bone marrow transplantation, Blood 97 (2001) 327–329.

Jeyakumar, M., Thomas, R., Elliot-Smith, E., Smith, D. A., van der Spoel, A. C., d'Azzo, A., Perry, V. H., Butters, T. D., Dwek, R. A., and Platt, F. M., Central nervous system inflammation is a hallmark of pathogenesis in mouse models of GM1 and GM2 gangliosidosis, Brain 126 (2003) 974–987.

Kolter, T., and Sandhoff, K., Glycosphingolipid degradation and animal models of GM2-gangliosidoses, J Inherit Metab Dis 21 (1998) 548–563.

Kolter, T., and Sandhoff, K., Principles of lysosomal membrane digestion: stimulation of sphingolipid degradation by sphingolipid activator proteins and anionic lysosomal lipids, Annu Rev Cell Dev Biol 21 (2005) 81–103.

Kyrkanides, S., Miller, J. H., Brouxhon, S. M., Olschowka, J. A., and Federoff, H. J., beta-hexosaminidase lentiviral vectors: transfer into the CNS via systemic administration, Brain Res Mol Brain Res 133 (2005) 286–298.

Lachmann, R. H., Te Vruchte, D., Lloyd-Evans, E., Reinkensmeier, G., Sillence, D. J., Fernandez-Guillen, L., Dwek, R. A., Butters, T. D., Cox, T. M., and Platt, F. M., Treatment with miglustat reverses the lipid-trafficking defect in Niemann-Pick disease type C, Neurobiol Dis 16 (2004) 654–658.

Langeveld, M., Ghauharali, K. J., Sauerwein, H. P., Ackermans, M. T., Groener, J. E., Hollak, C. E., Aerts, H. J., and Serlie, M. J., Type I Gaucher disease, a glycosphingolipid storage disorder, is associated with insulin resistance, J Clin Endocrinol Metab (2007) in press.

Leinekugel, P., Michel, S., Conzelmann, E., and Sandhoff, K., Quantitative correlation between the residual activity of beta-hexosaminidase A and arylsulfatase A and the severity of the resulting lysosomal storage disease, Hum Genet 88 (1992) 513–523.

Liao, G., Yao, Y., Liu, J., Yu, Z., Cheung, S., Xie, A., Liang, X., and Bi, X., Cholesterol accumulation is associated with lysosomal dysfunction and autophagic stress in Npc1 –/– mouse brain, Am J Pathol 171 (2007) 962–975.

Lowe, J. P., Stuckey, D. J., Awan, F. R., Jeyakumar, M., Neville, D. C., Platt, F. M., Griffin, J. L., Styles, P., Blamire, A. M., and Sibson, N. R., MRS reveals additional hexose N-acetyl resonances in the brain of a mouse model for Sandhoff disease, NMR Biomed 18 (2005) 517–526.

Matsuda, J., Suzuki, O., Oshima, A., Yamamoto, Y., Noguchi, A., Takimoto, K., Itoh, M., Matsuzaki, Y., Yasuda, Y., Ogawa, S., Sakata, Y., Nanba, E., Higaki, K., Ogawa, Y., Tominaga, L., Ohno, K., Iwasaki, H., Watanabe, H., Brady, R. O., and Suzuki, Y., Chemical chaperone therapy for brain pathology in G(M1)-gangliosidosis, Proc Natl Acad Sci U S A 100 (2003) 15912–15917.

Meikle, P. J., Hopwood, J. J., Clague, A. E., and Carey, W. F., Prevalence of lysosomal storage disorders, J Am Med Assoc 281 (1999) 249–254.

Nixon, R. A., Wegiel, J., Kumar, A., Yu, W. H., Peterhoff, C., Cataldo, A., and Cuervo, A. M., Extensive involvement of autophagy in Alzheimer disease: an immuno-electron microscopy study, J Neuropathol Exp Neurol 64 (2005) 113–122.

Norflus, F., Tifft, C. J., McDonald, M. P., Goldstein, G., Crawley, J. N., Hoffmann, A., Sandhoff, K., Suzuki, K., and Proia, R. L., Bone marrow transplantation prolongs life span and ameliorates neurologic manifestations in Sandhoff disease mice, J Clin Invest 101 (1998) 1881–1888.

Pastores, G. M., and Sathe, S., A chaperone-mediated approach to enzyme enhancement as a therapeutic option for the lysosomal storage disorders, Drugs R D 7 (2006) 339–348.

Pastores, G. M., Barnett, N. L., and Kolodny, E. H., An open-label, noncomparative study of miglustat in type I Gaucher disease: efficacy and tolerability over 24 months of treatment, Clin Ther 27 (2005) 1215–1227.

Pastores, G. M., Elstein, D., Hrebicek, M., and Zimran, A., Effect of miglustat on bone disease in adults with type 1 Gaucher disease: a pooled analysis of three multinational, open-label studies, Clin Ther 29 (2007) 1645–1654.

Patterson, M. C., Vecchio, D., Prady, H., Abel, L., and Wraith, J. E., Miglustat for treatment of Niemann-Pick C disease: a randomised controlled study, Lancet Neurol 6 (2007) 765–772.

Petersen, G. M., Rotter, J. I., Cantor, R. M., Field, L. L., Greenwald, S., Lim, J. S., Roy, C., Schoenfeld, V., Lowden, J. A., and Kaback, M. M., The Tay-Sachs disease gene in North American Jewish populations: geographic variations and origin, Am J Hum Genet 35 (1983) 1258–1269.

Phaneuf, D., Wakamatsu, N., Huang, J. Q., Borowski, A., Peterson, A. C., Fortunato, S. R., Ritter, G., Igdoura, S. A., Morales, C. R., Benoit, G., Akerman, B. R., Leclerc, D., Hanai, N., Marth, J. D., Trasler, J. M., and Gravel, R. A., Dramatically different phenotypes in mouse models of human Tay-Sachs and Sandhoff diseases, Hum Mol Genet 5 (1996) 1–14.

Platt, F. M., Neises, G. R., Dwek, R. A., and Butters, T. D., N-Butyldeoxynojirimycin Is a Novel Inhibitor of Glycolipid Biosynthesis, J Biol Chem 269 (1994) 8362–8365.

Platt, F. M., Neises, G. R., Reinkensmeier, G., Townsend, M. J., Perry, V. H., Proia, R. L., Winchester, B., Dwek, R. A., and Butters, T. D., Prevention of lysosomal storage in Tay-Sachs mice treated with N-butyldeoxynojirimycin, Science 276 (1997) 428–431.

Proia, R. L., Glycosphingolipid functions: insights from engineered mouse models, Phil Trans R Soc Lond B 358 (2003) 879–883.

Proia, R. L., Gangliosides help stabilize the brain, Nat Genet 36 (2004) 1147–1148.

Reczek, D., Schwake, M., Schroder, J., Hughes, H., Blanz, J., Jin, X., Brondyk, W., Van Patten, S., Edmunds, T., and Saftig, P., LIMP-2 is a receptor for lysosomal mannose-6-phosphate-independent targeting of beta-glucocerebrosidase, Cell 131 (2007) 770–783.

Ron, I., and Horowitz, M., ER retention and degradation as the molecular basis underlying Gaucher disease heterogeneity, Hum Mol Genet 14 (2005) 2387–2398.

Sandhoff, R., Geyer, R., Jennemann, R., Paret, C., Kiss, E., Yamashita, T., Gorgas, K., Sijmonsma, T. P., Iwamori, M., Finaz, C., Proia, R. L., Wiegandt, H., and Grone, H. J., Novel class of glycosphingolipids involved in male fertility, J Biol Chem 280 (2005) 27310–27318.

Sango, K., Yamanaka, S., Hoffmann, A., Okuda, Y., Grinberg, A., Westphal, H., McDonald, M. P., Crawley, J. N., Sandhoff, K., Suzuki, K., and Proia, R. L., Mouse models of Tay-Sachs and Sandhoff diseases differ in neurologic phenotype and ganglioside metabolism, Nat Genet 11 (1995) 170–176.

Sango, K., McDonald, M. P., Crawley, J. N., Mack, M. L., Tifft, C. J., Skop, E., Starr, C. M., Hoffmann, A., Sandhoff, K., Suzuki, K., and Proia, R. L., Mice lacking both subunits of lysosomal beta-hexosaminidase display gangliosidosis and mucopolysaccharidosis, Nat Genet 14 (1996) 348–352.

Sano, R., Tessitore, A., Ingrassia, A., and d'Azzo, A., Chemokine-induced recruitment of genetically modified bone marrow cells into the CNS of GM1-gangliosidosis mice corrects neuronal pathology, Blood 106 (2005a) 2259–2268.

Sano, R., Trindade, V. M., Tessitore, A., d'Azzo, A., Vieira, M. B., Giugliani, R., and Coelho, J. C., G(M1)-ganglioside degradation and biosynthesis in human and murine G(M1)-gangliosidosis, Clin Chim Acta 354 (2005b) 131–139.

Sawkar, A. R., Adamski-Werner, S. L., Cheng, W. C., Wong, C. H., Beutler, E., Zimmer, K. P., and Kelly, J. W., Gaucher disease-associated glucocerebrosidases show mutation-dependent chemical chaperoning profiles, Chem Biol 12 (2005) 1235–1244.

Schueler, U. H., Kolter, T., Kaneski, C. R., Zirzow, G. C., Sandhoff, K., and Brady, R. O., Correlation between enzyme activity and substrate storage in a cell culture model system for Gaucher disease, J Inherit Metab Dis 27 (2004) 649–658.

Shu, L., Murphy, H. S., Cooling, L., and Shayman, J. A., An in vitro model of Fabry disease, J Am Soc Nephrol 16 (2005) 2636–2645.

Simpson, M. A., Cross, H., Proukakis, C., Priestman, D. A., Neville, D. C., Reinkensmeier, G., Wang, H., Wiznitzer, M., Gurtz, K., Verganelaki, A., Pryde, A., Patton, M. A., Dwek, R. A., Butters, T. D., Platt, F. M., and Crosby, A. H., Infantile-onset symptomatic epilepsy syndrome caused by a homozygous loss-of-function mutation of GM3 synthase, Nat Genet 36 (2004) 1225–1229.

Steet, R. A., Chung, S., Wustman, B., Powe, A., Do, H., and Kornfeld, S. A., The iminosugar isofagomine increases the activity of N370S mutant acid beta-glucosidase in Gaucher fibroblasts by several mechanisms, Proc Natl Acad Sci U S A 103 (2006) 13813–13818.

Sun, X., Marks, D. L., Park, W. D., Wheatley, C. L., Puri, V., O'Brien, J. F., Kraft, D. L., Lundquist, P. A., Patterson, M. C., Pagano, R. E., and Snow, K., Niemann-Pick C variant detection by altered sphingolipid trafficking and correlation with mutations within a specific domain of NPC1, Am J Hum Genet 68 (2001) 1361–1372.

Suzuki, Y., beta-Galactosidase deficiency: An approach to chaperone therapy, J Inherit Metab Dis 29 (2006) 471–476.

Takikita, S., Fukuda, T., Mohri, I., Yagi, T., and Suzuki, K., Perturbed myelination process of premyelinating oligodendrocyte in Niemann-Pick type C mouse, J Neuropathol Exp Neurol 63 (2004) 660–673.

Tessitore, A., del, P. M. M., Sano, R., Ma, Y., Mann, L., Ingrassia, A., Laywell, E. D., Steindler, D. A., Hendershot, L. M., and d'Azzo, A., GM1-ganglioside-mediated activation of the unfolded protein response causes neuronal death in a neurodegenerative gangliosidosis, Mol Cell 15 (2004) 753–766.

Tominaga, L., Ogawa, Y., Taniguchi, M., Ohno, K., Matsuda, J., Oshima, A., Suzuki, Y., and Nanba, E., Galactonojirimycin derivatives restore mutant human beta-galactosidase activities expressed in fibroblasts from enzyme-deficient knockout mouse, Brain Develop 23 (2001) 284–287.

Tropak, M. B., Reid, S. P., Guiral, M., Withers, S. G., and Mahuran, D., Pharmacological enhancement of beta-hexosaminidase activity in fibroblasts from adult Tay-Sachs and Sandhoff Patients, J Biol Chem 279 (2004) 13478–13487.

Tropak, M. B., Blanchard, J. E., Withers, S. G., Brown, E. D., and Mahuran, D., High-throughput screening for human lysosomal beta-N-Acetyl hexosaminidase inhibitors acting as pharmacological chaperones, Chem Biol 14 (2007) 153–164.

Tsuji, D., Kuroki, A., Ishibashi, Y., Itakura, T., Kuwahara, J., Yamanaka, S., and Itoh, K., Specific induction of macrophage inflammatory protein 1-alpha in glial cells of Sandhoff disease model mice associated with accumulation of N-acetylhexosaminyl glycoconjugates, J Neurochem 92 (2005) 1497–1507.

Turley, S. D., Burns, D. K., and Dietschy, J. M., Preferential utilization of newly synthesized cholesterol for brain growth in neonatal lambs, Am J Physiol 274 (1998) E1099–1105.

Tutor, J. C., Biochemical characterization of the GM2 gangliosidosis B1 variant, Braz J Med Biol Res 37 (2004) 777–783.

Tylki-Szymańska, A., Czartoryska, B., Vanier, M. T., Poorthuis, B. J., Groener, J. A., Lugowska, A., Millat, G., Vaccaro, A. M., and Jurkiewicz, E., Non-neuronopathic Gaucher disease due to saposin C deficiency, Clin Genet 72 (2007) 538–542.

van Breemen, M. J., de Fost, M., Voerman, J. S., Laman, J. D., Boot, R. G., Maas, M., Hollak, C. E., Aerts, J. M., and Rezaee, F., Increased plasma macrophage inflammatory protein (MIP)-1alpha and MIP-1beta levels in type 1 Gaucher disease, Biochim Biophys Acta 1772 (2007) 788–796.

van Meer, G., and Lisman, Q., Sphingolipid transport: rafts and translocators, J Biol Chem 277 (2002) 25855–25858.

Vogler, C., Levy, B., Grubb, J. H., Galvin, N., Tan, Y., Kakkis, E., Pavloff, N., and Sly, W. S., Overcoming the blood-brain barrier with high-dose enzyme replacement therapy in murine mucopolysaccharidosis VII, Proc Natl Acad Sci U S A 102 (2005) 14777–14782.

Wada, R., Tifft, C. J., and Proia, R. L., Microglial activation precedes acute neurodegeneration in Sandhoff disease and is suppressed by bone marrow transplantation, Proc Natl Acad Sci U S A 97 (2000) 10954–10959.

Walkley, S. U., Secondary accumulation of gangliosides in lysosomal storage disorders, Semin Cell Dev Biol 15 (2004) 433–444.

Warner, T. G., deKremer, R. D., Sjoberg, E. R., and Mock, A. K., Characterization and analysis of branched-chain N-acetylglucosaminyl oligosaccharides accumulating in Sandhoff disease tissue. Evidence that biantennary bisected oligosaccharide side chains of glycoproteins are abundant substrates for lysosomes, J Biol Chem 260 (1985) 6194–6199.

Werth, N., Schuette, C. G., Wilkening, G., Lemm, T., and Sandhoff, K., Degradation of membrane-bound ganglioside GM2 by beta -hexosaminidase A. Stimulation by GM2 activator protein and lysosomal lipids, J Biol Chem 276 (2001) 12685–12690.

Winchester, B., Lysosomal metabolism of glycoproteins, Glycobiology 15 (2005) 1R–15R.

Winsor, E. J., and Welch, J. P., Genetic and demographic aspects of Nova Scotia Niemann-Pick disease (type D), Am J Hum Genet 30 (1978) 530–538.

Wolfe, L. S., Senior, R. G., and Ng-Ying-Kin, N. M., The structures of oligosaccharides accumulating in the liver of G-M1-gangliosidosis, type I, J Biol Chem 249 (1974) 1828–1838.

Wu, Y. P., and Proia, R. L., Deletion of macrophage-inflammatory protein 1 alpha retards neurodegeneration in Sandhoff disease mice, Proc Natl Acad Sci USA 101 (2004) 8425–8430.

Yam, G. H., Zuber, C., and Roth, J., A synthetic chaperone corrects the trafficking defect and disease phenotype in a protein misfolding disorder, FASEB J 19 (2005) 12–18.

Yamashita, T., Wada, R., Sasaki, T., Deng, C. X., Bierfreund, U., Sandhoff, K., and Proia, R. L., A vital role for glycosphingolipid synthesis during development and differentiation, Proc Natl Acad Sci USA 96 (1999) 9142–9147.

Yamashita, T., Wu, Y. P., Sandhoff, R., Werth, N., Mizukami, H., Ellis, J. M., Dupree, J. L., Geyer, R., Sandhoff, K., and Proia, R. L., Interruption of ganglioside synthesis produces central nervous system degeneration and altered axon-glial interactions, Proc Natl Acad Sci USA 102 (2005) 2725–2730.

Yamashita, T., Hashiramoto, A., Haluzik, M., Mizukami, H., Beck, S., Norton, A., Kono, M., Tsuji, S., Daniotti, J. L., Werth, N., Sandhoff, R., Sandhoff, K., and Proia, R. L., Enhanced insulin sensitivity in mice lacking ganglioside GM3, Proc Natl Acad Sci USA 100 (2003) 3445–3449.

Yu, Z., Sawkar, A. R., Whalen, L. J., Wong, C. H., and Kelly, J. W., Isofagomine- and 2,5-anhydro-2,5-imino-D-glucitol-based glucocerebrosidase pharmacological chaperones for Gaucher disease intervention, J Med Chem 50 (2007) 94–100.

Zeller, C. B., and Marchase, R. B., Gangliosides as modulators of cell function, Am J Physiol 262 (1992) C1341–1355.

Zhang, M., Sun, M., Dwyer, N. K., Comly, M. E., Patel, S. C., Sundaram, R., Hanover, J. A., and Blanchette-Mackie, E. J., Differential trafficking of the Niemann-Pick C1 and 2 proteins highlights distinct roles in late endocytic lipid trafficking, Acta Paediatr Suppl 92 (2003) 63–73.

Zhu, X., Sheth, K. A., Li, S., Chang, H. H., and Fan, J. Q., Rational design and synthesis of highly potent beta-glucocerebrosidase inhibitors, Angew Chem Int Ed Engl 44 (2005) 7450–7453.

Zimran, A., Gaucher's Disease,Bailliere's Clinical Haematology. International Practice and Research, 10, Bailliere Tindall, London, 1997.

# Chapter 18
# Role of Neutral Sphingomyelinases in Aging and Inflammation

**Mariana Nikolova-Karakashian, Alexander Karakashian and Kristina Rutkute**

**Abstract** Aging is characterized by changes in the organism's immune functions and stress response, which in the elderly leads to increased incidence of complications and mortality following inflammatory stress. Alterations in the neuro-endocrine axes and overall decline in the immune system play an essential role in this process. Overwhelming evidence however suggests that many cellular cytokine signaling pathways are also affected, thus underscoring the idea that both, "cellular" and "systemic" changes contribute to aging. IL-1β for example, induces more potent cellular responses in hepatocytes isolated from aged animals then in hepatocytes from young rats. This phenomenon is referred to as IL-1β hyperresponsiveness and is linked to abnormal regulation of various acute phase proteins during aging.

Evidence has consistently indicated that activation of neutral sphingomyelinase and the resulting accumulation of ceramide mediate cellular responses to LPS, IL-1β, and TNFα in young animals. More recent studies identified the cytokine-inducible neutral sphingomyelinase with nSMase2 (smpd3) that is localized in the plasma membrane and mediates cellular responses to IL-1β and TNFα. Intriguingly, constitutive up-regulation of nSMase2 occurs in aging and it underlies the hepatic IL-1β hyperresponsiveness. The increased activity of nSMases2 in aging is caused by a substantial decline in hepatic GSH content linking thereby oxidative stress to the onset of pro-inflammatory state in liver. nSMase2 apparently follows a pattern of regulation consisting with "developmental-aging" continuum, since in animal models of delayed aging, like calorie-restricted animals, the aging-associated changes in NSMase activity and function are reversed.

**Keywords** IL-1β · oxidative stress · GSH · ceramide · nSMase2

---

M. Nikolova-Karakashian
University of Kentucky College of Medicine, Department of Physiology, Lexington, KY, USA

P.J. Quinn, X. Wang (eds.), *Lipids in Health and Disease*, 

## 18.1 The "Inflamm-Aging" Theory of Aging

The contemporary understanding of aging is based on a wealth of theories involving evolutionary, genetic, metabolic, and environmental principles. The "inflamm-aging" hypothesis (Chung et al., 2001), in particular, arises from some phenomenological clinical observations that identify inflammation not only as an underlying companion of aging-related diseases, but also as a factor that in itself often defines the systemic aging process (Han et al., 1995; Sly et al., 2001; Spaulding et al., 1997; Tang et al., 2000; Wu et al., 2003; Yamamoto et al., 2002). In youth, inflammation is a complex host defense mechanism against environmental stresses, like chemicals, drugs, oxidants, and microbial organisms. During aging however, a sustained elevation in a variety of inflammatory markers occurs even in the absence of clinically relevant stimulants of host defense, adversely affecting basal homeostasis of different organs and the ability to cope with environmental challenges. Such changes are well documented for (among others) the cardiovascular system (Yamamoto et al., 2002), macrophage functions (Claycombe et al., 2002; Tang et al., 2000; Wu et al., 2003), liver (Hsieh et al., 1998, 2003; Rabek et al., 1998; Suh, 2001) and brain (Kalehua et al., 2000; Sly et al., 2001; Suh, 2001).

The mechanisms underlying the onset of pro-inflammatory state in aged organisms are complex. Changes in the neuro-endocrine axis, overall decline in the adaptive component of the immune system, and upregulation of the innate immune response are all well established phenomena associated with the aging process that certainly are involved in onset of inflammatory state. The widely accepted "free radical theory of aging" postulates that aging-associated decay in the mitochondrial functions of various cells gradually leads to increased production of reactive oxygen species and to the activation of redox-sensitive pro-inflammatory molecules, mainly, Nuclear Factor κB (NFκB) and c-Jun N-Terminal Kinase (JNK) (Chung et al., 2001; Franceschi et al., 2000; Franceschi et al., 2007). The constitutive activation of these molecules causes sustained increase in the basal levels of pro-inflammatory cytokines, such as TNFα, IL-1β and IL-6, which consequently stimulate inflammatory reactions in various organs. However, for many organisms, including humans, aging-associated increases in cytokines never reach the levels necessary to evoke cellular responses in young organisms.

Recent studies have given indications to the fact that in many, but not all cell types, various signaling pathways are also affected by aging (Hsieh et al., 1998; Hsieh et al., 2003; Rutkute et al., 2007b; Tang et al., 2000; Wu et al., 2003). For example, when treated with bacterial endotoxin, lipopolysaccharide (LPS), tumor necrosis factor α (TNFα) or pharmacological inducers of oxidative stress, peritoneal macrophages (Tang et al., 2000; Wu et al., 2003), hepatocytes (Hsieh et al., 1998), and glial cells isolated from aged animals exhibit more severe and prolonged responses as compared to cells isolated from young animals. This is evidenced by substantial differences in the magnitude and temporal pattern of activation of cyclooxygenase-2 (Cox-2) (Wu et al., 2003), JNK (Hsieh et al., 2003; Rutkute et al., 2007b), NF-κB (Wu et al., 2003), CCAAT enhancer-binding protein (C/EBP) (Hsieh et al., 1998). Therefore, onset of the pro-inflammatory state in aging is apparently not only due to systemic factors, but also to changes in cellular responsiveness caused by, among others, a decreased capacity of anti-oxidant defense, up- or down-regulation of proteins functioning as rate-limiting factors in signaling cascades, and disruption of the balance between kinase and phosphatase activities.

## 18.2 Hepatic IL-1 β Signaling Pathway During Inflammation and Aging

IL-1β is a prototypic inflammatory cytokine that mediates the host response to infection, and its basal systemic levels are relatively constant with age (Di Iorio et al., 2003). Various IL-1β-related functions however, like the regulation of acute phase protein (APP) expression in liver appear to be age-dependent. The levels of acute phase reactants increase with age in a variety of species, from *D. melanogaster* (Zerofsky et al., 2005) to mammals (Berk et al., 1990; Rosenthal et al., 1975). Most importantly, these increases have been linked to aging-related disorders, like Alzheimer's disease (Abraham et al., 1988), rheumatoid arthritis (Rosenthal et al., 1975), cardiovascular diseases (Berk et al., 1990), impaired tissue renewal, frailty and reduced glucose tolerance (Ceda et al., 2005; Walston and Fried, 1999).

Cellular responses to IL-1β are mediated through the interleukin-1 receptor type I (IL-1RI). Together with TLR-4, the LPS receptor, IL-1RI is a prototypic member of the Toll-Like Receptor (TLR) family that shares a conserved signaling pathway (Fig. 18.1). Ligand binding to TLR results in the recruitment of several adaptor proteins including MyD88 (Wesche et al., 1997), followed by IL-1R-associated kinase-1 (IRAK-1) and IRAK-4 binding to the receptor complex. The subsequent IRAK-1 phosphorylation facilitates binding of tumor necrosis factor-associated factor-6 (TRAF-6) and the IRAK-1/IRAK-4/TRAF-6 complex then separates from the receptor and interacts with the transforming growth factor-β-activated kinase-1 (TAK-1) (Cao et al., 1996). Activation of TAK-1 apparently initiates the mitogen activated protein kinase kinase cascade, leading to the activation of transcription factors like activator protein-1 (AP-1) and NF-κB through JNK and inhibitory κB kinases respectively, resulting in the induction of APP mRNA transcription (Ninomiya-Tsuji et al., 1999).

IRAK-1 plays a central role in the TLR signaling cascade and it has recently been suggested that the rate of IRAK-1 degradation determines the magnitude of the response. Initially, phosphorylation of IRAK-1 is essential for its release from the receptor and the activation of downstream signaling molecules, but IRAK-1 phosphorylation also leads to its ubiquitination and proteasome-mediated

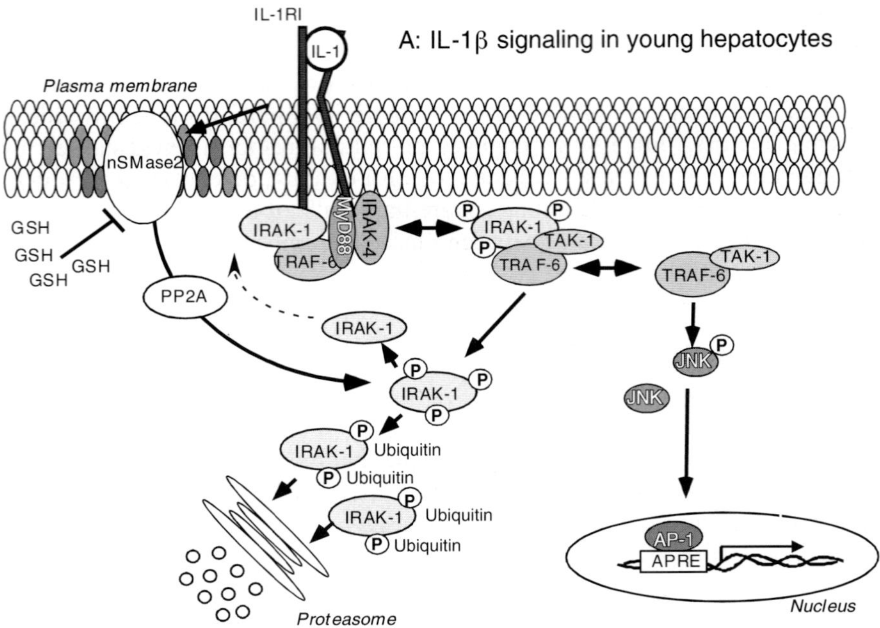


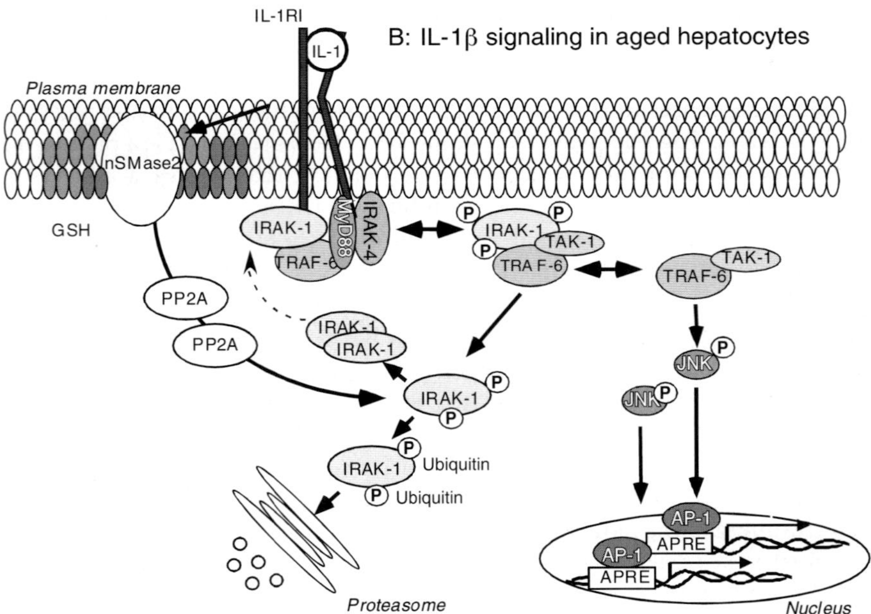


**Fig. 18.1** Mechanism for the age-related hyperresponsiveness to IL-1β. Binding of IL-1β to the IL-1RI induces the formation of a signaling complex containing IRAK-1, IRAK-4, MyD88 and other adapter molecules. IRAK-4 and other, unidentified kinases phosphorylate

degradation, resulting in the termination of the signaling cascade (Yamin and Miller, 1997). Suppression of IRAK-1 degradation in macrophages through the use of ubiquitin ligase inhibitors has been shown to potentiate the inflammatory response, while decreased stability of IRAK-1 is the hallmark of endotoxin tolerance exemplified by the muted response to secondary LPS administration (Cuschieri et al., 2004; Li et al., 2000). Interferon $\gamma$ and granulocyte-macrophage colony-stimulating factor can prevent this LPS tolerance by inhibiting IRAK-1 degradation (Adib-Conquy and Cavaillon, 2002).

Numerous studies have now confirmed that the IL-1β signaling cascade in liver is strongly affected by aging (Rutkute et al., 2007a,b; Rutkute and Nikolova-Karakashian, 2007c). The basal expression levels of IL1RI, IRAK-1, TAK-1 and JNK remain unchanged; yet stimulation with IL-1β evokes a more potent JNK phoshorylation in primary hepatocytes isolated from aged rats than in hepatocytes isolated from young ones. Importantly, concentrations of IL-1β that are too low to appreciably affect JNK activation in young animals are capable of inducing significant JNK phosphorylation in aged ones (Rutkute et al., 2007b). The increased phosphorylation of JNK leads to increased phosphorylation of c-jun, the major JNK substrate, and to more potent stimulation of expression of Insulin-Like Growth Factor Binding Protein 1 (IGFBP1) (Rutkute et al., 2007c). The latter is a hepatic acute phase protein and its production is potently upregulated in response to sepsis (Lang et al., 1996), endotoxin injection (Fan et al., 1994; Lang et al., 1997), and other inflammatory conditions. In young organisms increased circulating levels of IGFBP-1 during inflammation maintain the inflammatory catabolic state by diminishing the levels of bioavailable Insulin-Like Growth Factor-I (IGF-I) and thereby counteracting its anabolic effects (Lee et al., 1993). Notably, IGFBP1 serum concentrations are significantly elevated in aged organisms and are linked to declining levels of bioactive IGF-I (Yang et al., 2005), which in turn is

**Fig. 18.1** (continued) IRAK-1 at multiple residues, thus activating its own kinase activity leading to further autophosphorylation. The hyper-phosphorylation of IRAK-1 leads to disassociation of the complex from the IL-1β receptor and facilitates its interaction with TAK-1, which is upstream of JNK. Phosphorylation of JNK ultimately leads to activation of AP-1 transcription factor that regulates APP transcription through the acute phase response element (APRE) in the promoter(s). Phosphorylation of IRAK-1 also serves as a signal for its ubiquitination and rapid degradation in proteasomes, which effectively terminates the signaling cascade. IL-1β binding to its receptor however, also transiently activates the plasma membrane-localized nSMase2, which regulates the rate of IRAK degradation in a PP2A-dependent manner. Most likely, PP2A de-phosphorylates IRAK-1, thus preventing its ubiquitination and degradation and allowing further receptor binding of non-phosphorylated IRAK-1. In young hepatocytes, the high concentrations of GSH, which is a reversible inhibitor of nSMase2, limit the activity of nSMase2. During aging, depletion of cellular GSH content leads to constitutive activation of nSMase2, increased PP2A activity, and higher abundance of IRAK-1 available for receptor activation. These changes lead to higher abundance of phosphorylated JNK molecules and respectively more pronounced stimulation of gene expression associated with slow cell growth, impaired tissue renewal, frailty and reduced glucose tolerance (Ceda et al., 2005; Walston and Fried et al., 1999). IL-1β-induced hyper-production of IGFBP1can be substantially diminished *in vitro* by inhibiting JNK activity (Rutkute et al., 2007c). This indicates that differences in the magnitude of JNK phosphorylation in aged organisms could have important consequences for the hepatic physiological responses.

Recent studies into the mechanisms underlying aging-associated hyperresponsiveness to IL-1β in liver have revealed that the lipid-metabolizing enzyme, neutral sphingomyelinase (NSMase), and its product, ceramide, are implicated (Fig. 18.1). In young animals, NSMase is known to mediate the cellular response to IL-1β and other inflammatory cytokines; in aging, however, its activity is constitutively elevated in a redox-sensitive manner. The higher basal NSMase activity (and higher ceramide content at the plasma membrane) in hepatocytes from aged animals is responsible for slower IRAK-1 degradation, which keeps IRAK-1 available for recruitment to various signaling complexes, in turn causing more abundant activation of downstream signaling molecules, such as JNK. The experimental evidence for the role of NSMase, and more specifically the role of nSMase2 in the onset of hepatic IL-1β hyperresponsiveness during aging and the regulation of nSMase2 by oxidative stress and GSH are discussed below.

## 18.3 Functions of Neutral Sphingomyelinase-2 (NSMASE2) as Mediator of Inflammatory Responses in Young Organisms

The sphingomyelinase (nSMase2) family is a group of biochemically and genetically different enzymes all of which hydrolize Sphingomyelin (SM) to ceramide. SMase activities with neutral and acidic pH optima are found in most mammalian cells, and an enzyme active in alkaline pH is localized in the intestinal wall. Overwhelming evidence shows that pro-inflammatory cytokines like IL-1β and TNFα, as well as other inducers of host immune response, like LPS, activate either the neutral or acidic SMase activity causing transient elevation in the concentration of cellular ceramide. Ceramide has been shown to play an important role as a mediator of cellular responses to stress and its downstream targets are known to include kinases, phosphatases and transcription factors. Taken together, these observations have resulted in postulation of the "SM signaling pathway", which is essential part of the mechanism by which cells of young organisms cope with stress (Hannun and Obeid, 2002).

The cytokine-induced neutral SMase (NSMase) activity is associated with the plasma membrane and is Mg2+-dependent. However, a more detailed understanding of its role in signaling at the molecular level was hampered until recently, due to the lack of knowledge about the encoding gene(s). Studies from various groups have now identified the smpd3 gene as the one responsible for this NSMase activity. It encodes a 71 kDa protein termed nSMase2 with two putative transmembrane domains in the N-terminus and a catalytic domain within the C-terminal region (Hofmann et al., 2000). Extensive biochemical studies (reviewed recently by Clarke et al (Clarke et al., 2006)) have shown that nSMase2 is a *bona fide* NSMase that is Mg dependent and exclusively utilizes SM as a substrate (Luberto et al., 2002; Sawai et al., 1999). Moreover, it shows substrate specificity for very long chain SM, indicating that only a certain subset of cellular SM may be accessible as a substrate (Marchesini et al., 2004). Of particular interest is the observation that nSMase2 is activated by phosphatidylserine and other anionic phospholipids, a property that could be of importance in the regulation of NSMase activity taking into account the intrinsic asymmetry of the plasma membrane where phosphatidylserine is enriched in the inner leaflet. It is noteworthy that the plasma membrane asymmetry is disrupted during apoptosis and other pathophysiological conditions, as well as during aging.

Stoffel and co-workers reported that endogenous and overexpressed nSMase2 localized at the Golgi apparatus in several cell lines (Stoffel et al., 2005). However, studies with highly differentiated primary hepatocytes cultured in three-dimentional matrix of extracellular proteins, Matrigel™, in the absence of growth factors except insulin, have shown that the overexpressed nSMase2 is localized primary at the plasma membrane (Karakashian et al., 2004). Other studies have shown that nSMase2 probably translocates to the plasma membrane when cell confluence is reached (Marchesini et al., 2004) and that nSMase2 is required for the confluence-induced cell cycle arrest to ensue (Hayashi et al., 1997). Lending further support as to the role of nSMase2 in signaling has been the observation made in oligodendroma-derived cells of regulated translocation of nSMase2 to the caveolae, which are the signaling domains of the plasma membrane (Goswami et al., 2005).

Published data from different labs have unequivocally shown that nSMase2 is regulated by cytokines like IL-1β and TNFα, but also that it mediates some of the cytokine effects in young organisms (De Palma et al., 2006; Karakashian et al., 2004; Marchesini et al., 2004). Treatment of hepatocytes with IL-1β leads to increased cellular NSMase activity and accumulation of ceramide (Chen et al., 1995). Increases in the ceramide level on the other hand have been shown to stimulate expression of the hepatic APP, like α-1-acid glycoprotein and C-reactive protein (Chen et al., 1995; Lozanski et al., 1997). Ceramide accumulation is also linked to the activation of TAK-1, JNK and NF-κB, all of which have important roles in the IL-1β cascade (Shirakabe et al., 1997; Verheij et al., 1996; Westwick et al., 1995).

In hepatocytes, nSMase2 is one of the two so far described neutral, $Mg^{2+}$-dependent sphingomyelinases, the other being nSMase1. Specific silencing of nSMase2 with siRNA has only a minimal effect on the basal cellular NSMase activity, however it results in a complete inhibition of the cytokine-stimulated NSMase activity (Rutkute et al., 2007b). Therefore, nSMase2 is probably an inducible enzyme that contributes little to the basal turnover of SM, but at the same time it is also the only neutral SMase activated by IL-1β. The IL-1β-induced activation of NSMase2 is required for stimulation of JNK phosphorylation (Rutkute et al., 2007b); however, overexpression of NSMase2 is not sufficient to mimic IL-1β-induced JNK phosphorylation (Karakashian et al., 2004). This indicates that the role of nSMase2 in the IL-1β signaling cascade is rather complex. Indeed, it has been show that activation of nSMase2 regulates the extent of IRAK-1 phosphorylation and degradation following receptor activation. Increased NSMase-2 activity attenuates IL-1β-induced IRAK-1 phosphorylation, and its subsequent ubiquitination and degradation (Karakashian et al., 2004). In contrast, silencing of nSMase2 using siRNA has the opposite consequences and leads to more rapid degradation of IRAK-1. Also, these effects strictly correlate with higher, or respectively, lower JNK phosphorylation (Rutkute et al., 2007b). Apparently, the IL-β-induced stimulation of NSMase, which peculiarly happens after the initial burst in MAP kinase activation, serves to slow down IRAK-1 degradation, keeping IRAK-1 available for recruitment to various signaling complexes, leading to potentiation of JNK phosphorylation. Such a scenario is further supported by reports of increased JNK activation in the presence of inhibitors of proteasome functions which serves to slow down IRAK degradation (Cuschieri et al., 2004). nSMase2 induced stabilization of IRAK-1 most likely involves a phosphatase, like the ceramide-activated protein phosphatase 2A (Chalfant et al., 1999). Indications to that effect come from earlier studies where increased IRAK-1 stability in nSMase2 overexpressing cells was protein phosphatase 2A-dependent (Karakashian et al., 2004).

TNFα has also been shown to activate nSMase2 (Clarke et al., 2007; Marchesini et al., 2004; Tellier et al., 2007). TNFα-induced nSMase2 activation is a prerequisite for endothelial nitric oxide synthase activation in HUVEC cells (De Palma et al., 2006), as well as for the up-regulation in A549 lung epithelial cells of vascular cell adhesion molecule (VCAM) and intracellular adhesion molecule 1 (ICAM) (Clarke et al., 2007) all of which have prominent roles in vascular inflammatory responses. Taken together, the studies described above firmly establish nSMase2 and its product, ceramide, as important factors in the cytokine signaling networks of young organisms.

## 18.4 Hepatic Neutral Sphingomyelinase: The Link Between Oxidative Stress and IL-1β Hyperresponsiveness in Aging

The functions of ceramide as mediator of host response to infection and stress are well preserved among species, from drosophila and yeast to mice, and are important for organism survival; however, mounting evidence suggests that in the course of aging, these functions change. Studies in the late 80s and early 90s found increased basal levels of ceramide during aging in a variety of cells, including hepatocytes (Cutler and Mattson, 2001; Lightle et al., 2000; Petkova et al., 1988). Studies in cellular models of senescence using human diploid fibroblast related these increases to retardation in the rate of cell growth via inhibition of rB hyperphosphorylation and the appearance of markers of senescence, such as β-galactosidase (Mouton and Venable, 2000; Venable et al., 1995), suggesting a fundamental role for ceramide in the mechanisms leading to cellular aging. In rodents, ceramide concentrations also change following a "development-aging" continuum: normal aging is associated with accumulation of ceramide, while caloric restriction (CR) that extends the rat life span has been found to decrease the levels of ceramide (Algeri et al., 1991).

Mechanistic understanding of the role of ceramide in aging comes from studies in cultured hepatocytes isolated from young and old rats. In rodents, age-associated changes in ceramide are paralleled by a gradual increase in hepatic NSMase activity (Lightle et al., 2000; Petkova et al., 1988). These changes seem to be limited to the plasma membrane, implying that nSMase2 activity might be involved (Lightle et al., 2000). Indeed, suppression of nSMase2 in hepatocytes from aged animals, either by siRNA silencing or by pharmacological inhibitors, reduces NSMase activity to levels similar to those found in young animals (Rutkute et al., 2007b). Recent evidence has emerged suggesting that these age-associated changes in NSMase-2 activity are due to a decline in hepatic GSH content, providing a novel link between oxidative stress and inflammation during aging.

Oxidative stress is known to be a major factor leading to inflammation during aging. The onset of oxidative stress in aging animals is accompanied by a net decrease in the concentrations of GSH, the major scavenger of free radicals, in different organs, but most notably, in liver. In isolated primary hepatocytes, a dramatic age-dependent decline (40 to 70%) in GSH levels has been reported (Hagen et al., 2000; Vericel et al., 1994). This is noteworthy, because GSH has been found to be a reversible inhibitor of cellular NSMase activity (Liu and Hannun, 1997, 1998). The modulation of NSMase activity by GSH was first established in the context of regulation of TNFα signaling and apoptosis (Liu and Hannun, 1997, 1998). Later, the ability of GSH to affect the sensitivity of T47D/H3 breast cancer cells to doxorubicin was attributed to the inhibitory effect GSH has on NSMase activity (Gouaze et al., 2001). A correlation between oxidative stress and NSMase activity was also found in long-lived rats on vitamin Q10 enriched diet (Bello et al., 2005), and in astrocytes treated with vitamin E (Ayasolla et al., 2004). Finally, recent research has shown that specific downregulation of nSMase2 with siRNA blocks $H_2O_2$-induced apoptosis of human aortic endothelial cells, identifying nSMase2 as a redox-sensitive protein (Castillo et al., 2007).

Apparently, in the liver the activation of nSMase2 in the process of aging is linked to the substantial age-related decline in the hepatic GSH concentrations. Indeed, supplementation of aged hepatocytes with N-Acetyl cysteine (NAC), a precursor of GSH synthesis, significantly decreases the endogenous NSMase activity to levels typically found in young hepatocytes, while the addition of L-buthionine-S,R-sulfoximine (BSO), an inhibitor of GSH synthesis, to young hepatocytes lowers GSH levels causing sharp increases in NSMase activity (Rutkute et al., 2007a). Notably, GSH depletion exerts its effect on NSMase activity in a biphasic manner: experiments in hepatocytes from young animals, in which gradual depletion of GSH was achieved by varying the time and dose of BSO treatment, reveal that only when hepatic GSH concentration drops below 30% of its basal level (which coincides remarkably well with the decline typically found in aged animals), is a sharp, dose-dependent activation of NSMase observed. This is consistent with the existence of either a threshold for NSMase activation or of a specific GSH pool, possibly within the immediate surroundings of the enzyme, that influences NSMase activity. Since the direct, *in vitro,* effect of GSH on the overexpressed nSMase2 is also biphasic, the former possibility appears more likely. Finally, evidence supporting the role of GSH depletion in NSMase activation during aging comes also from experiments where the sensitivity of NSMase to inhibition by exogenously added GSH was tested *in vitro*. These studies found that NSMase activity in aged hepatocytes, as compared to young ones, is much more sensitive to direct inhibition by GSH, indicating that in hepatocytes from young rats, the activity is already inhibited by the endogenous GSH (Rutkute et al., 2007a). Interestingly, nSMase1, a NSMase that is genetically distinct from nSMase2, is also sensitive to changes in the GSH/GSSG ratio, but it is not affected by direct *in vitro* treatment with GSH (Martin et al., 2007). This suggests that redox sensitivity might be a common property of neutral sphingomyelinases. It should be noted that, while the increase in NSMase activity during aging is apparently caused by GSH depletion, the IL-1β-induced NSMase activation is GSH-independent (Rutkute et al., 2007a). This is consistent with the observations that IL-1β (unlike TNF-α) induces only modestly ROS production in the liver (Lang et al., 1999), without substantially altering the liver GSH content. Based on all these findings it could be hypothesized that in the process of aging, nSMase2 activity plays the role of a converging point that integrates the effects of oxidative stress into the IL-1β signaling cascade, and by doing so essentially modulates cellular responsiveness to IL-1β.

It has been reported that basal JNK phosphorylation in whole livers increases with aging due to increased oxidative stress (Bose et al., 2005). However, this observation does not hold true in primary hepatocyte cultures, where basal JNK phosphorylation is undetectable in cells from both young and aged animals, despite a significant difference in their GSH content. These observations suggest that at least in hepatocytes, oxidative stress alone is not sufficient to induce JNK activation. However, like the increase in nSMase2 activity, the drop in GSH content in aging is sufficient to stabilize IRAK-1 and potentiate JNK phosphorylation in the presence of external stimulus such as IL-1β, leading to IL-1β hyperresponsiveness (Rutkute et al., 2007a). The addition of BSO to young hepatocytes lowers their GSH levels causing sharp increase in NSMase activity and, notably, IL-1b hyperresponsiveness. Vice versa, restoration of normal GSH concentration and NSMase activity in aged hepatocytes by NAC supplementation also restores normal levels of IRAK-1 degradation and JNK phosphorylation. These results suggest that GSH depletion in aged hepatocytes might induce IL-1β hyperresponsiveness by activating nSMase2.

Accordingly, overexpression of nSMase2 in hepatocytes from young rats induces hyperresponsiveness to IL-1β stimulation; but more importantly, inhibition of nSMase2 activity in hepatocytes from aged animals, either by siRNA silencing or by pharmacological inhibitors, is sufficient to restore normal IL-1β response (Rutkute et al., 2007b). Finally, in NAC-treated aged hepatocytes that show "youthful" GSH levels, NSMase activity, and IL-1β responsiveness, the overexpression of NSMase-2 rescues the aging phenotype (Rutkute et al., 2007a). Thus, in the pathway responsible for hepatic IL-1β hyperresponsiveness during aging, nSMase2 is probably the only factor acting downstream of GSH depletion.

Important data supporting the hypothesis that NSMase is the link between oxidative stress/GSH depletion and IL-1β hyperresponsiveness in the liver comes from studies using calorie restricted aged rats (Rutkute et al., 2007a). Calorie restriction extends the lifespan of many species, from yeast to mammals. Animals, subjected to calorie-restricted diet at early maturity, exhibit delayed aging in terms of their physiological functions, and a marked decrease in the incidence of aging-related diseases. Notably, hepatocytes from aged, calorie restricted rats as compared to those from age-matched *ad libitum* fed rats, have attenuated IL-1β responsiveness resulting from higher GSH content and lower NSMase activity. These observations indicate that the IL-1β response and NSMase activity are modulated in age-specific manner in correlation with the lifespan of the organism.

## 18.5 Importance of Other Ceramide-Metabolizing Enzymes for Aging

### *18.5.1 Acid Sphingomyelinase and Brain Aging*

Acid sphingomyelinase (ASMase) is a lysosomal enzyme with some functional resemblance to nSMase2 in regards to its ability to generate ceramide and be activated during various stress conditions. Importantly, ASMase is one out of 15 genes that are up-regulated during senescence as determined by subtraction library studies of cultured young and senescent fibroblasts and of fibroblasts from normal and Werner syndrome subjects (Lecka-Czernik et al., 1996). The functional significance of these changes in the ASMase activity of brain is underscored by recent studies describing ASMase knockout mice as being protected against hypoxia-induced neuronal damage (Yu et al., 2000). This could indicate that ASMase is important in the onset of neuronal loss. Indeed, elevation in cellular ceramide has been implicated in the onset of pathophysiological amyloidosis associated with Alzheimer Disease (Puglielli et al., 2003) . Furthermore, increases in ASMase activity were found in the brain of a senescence–accelerated mouse, the enzyme activity correlating with the onset of premature aging (Kim et al., 1997).

### *18.5.2 De Novo Synthesis of Ceramide and Yeast Longevity*

The most direct evidence substantiating the importance of ceramide in the process of aging comes from studies identifying the yeast Longevity Assurance Gene-1 (LAG-1) and its homologue, LAC-1 as regulators of ceramide biosynthesis (Riebeling et al., 2003; Venkataraman and Futerman, 2002a). De novo synthesis is the major pathway for ceramide generation in yeast. The LAG-1 and LAC-1 proteins together with a third protein, Lip1, form a complex that acylates sphinganine (dihydrosphingosine) to form dihydroceramide and also acylates sphingosine to form ceramide (Pewzner-Jung et al., 2006). LAG-1 has previously been identified as a regulator of yeast life span and its deletion has been shown to prolong the life of S. Cerevisiae by 50%. The role of ceramide synthesis in aging is further underscored by the observation that the mammalian homology of LAG-1, UOG 1 (currently re-named to LASS1) which is known to regulate ceramide synthesis in mammalian cells can normalize the life span of Lag-1ΔLac-1Δ strain 85% as effective as LAG-1 itself (Venkataraman et al., 2002b). The mechanism by which LAG-1 affects longevity is unclear. However, *de novo* ceramide synthesis in yeast is a major signaling mechanism involved in regulating the yeast response to heat or osmotic chock (Jenkins and Hannun, 2001).

### *18.5.3 Ceramidase and Drosophila Retinal Degeneration*

A recent study in drosophila showed that modulation of cellular ceramide content might play an important role in the onset of aging associated retinal degeneration (Acharya et al., 2003). Reduction in the levels of ceramide through the targeted expression of the Drosophila Neutral Ceramidase (an enzyme that catalyzes the degradation of ceramide to sphingosine and fatty acid) led to the rescue of retinal degeneration, underlying a possible functional significance for the aging-associated accumulation of ceramide.

## 18.6 Concluding Remarks

NSMase2 has now been identified as the cytokine-regulated SMase that is required for hepatic acute phase response during infection. NSMase2 is not only a *bona fide* neutral SMase, but in the liver, it exhibits characteristics typical of a signaling enzyme: its very low basal activity is substantially increased by cytokine treatment. The role that nSMase2 activation plays in the IL-1β signaling cascades is rather complex: nSMase2 activity is required, but not sufficient for IL-1β induced activation of JNK. This indicates that nSMase2 activity and its product, ceramide, mediate a regulatory step in the IL-1β signaling pathway. Thus, it is plausible that nSMase2 is a converging point in the cell signaling cascades that integrates the effects of various stress stimuli, thereby modifying the magnitude of the response. This hypothesis is further supported by the reported role of nSMase2 during aging in the onset of the hepatic hyperresponsiveness to IL-1β, where the sensitivity of nSMase2 towards GSH provides a novel mechanism by which oxidative stress modifies the normal hepatic response to IL-1β.

Apparently, the age-related increases in NSMase activity observed in liver play a crucial role in the onset of the pro-inflammatory state that epitomizes the process of aging. The dual role of nSMase2: as an essential component of the host defense against infection in young organisms; and at the same time as facilitator of the onset of pro-inflammatory state in the elderly, fits with the "antagonistic pleiothropy" view of the aging process, according to which evolution only favors traits that are beneficial during the reproductive age, even though the same traits may be deleterious later in life (Bonsall, 2006; Kirkwood and Austad, 2000).

It is also quite possible that nSMase2 has much broader function in the process of aging. New evidence to that effect comes in the face of nSMase2 knockout mice (Stoffel et al., 2005) that develop a novel form of dwarfism, exhibit delayed puberty, and have reduced levels of IGF1.

## References


Abraham, C. R., Selkoe, D. J., and Potter, H. Immunochemical identification of the serine protease inhibitor alpha 1-antichymotrypsin in the brain amyloid deposits of Alzheimer's disease. Cell 52 (1988) 487–501.

Acharya, U., Patel, S., Koundakjian, E., Nagashima, K., Han, X., and Acharya, J. K. Modulating sphingolipid biosynthetic pathway rescues photoreceptor degeneration. Science 299 (2003) 1740–1743.

Adib-Conquy, M., and Cavaillon, J. M. Gamma interferon and granulocyte/monocyte colony-stimulating factor prevent endotoxin tolerance in human monocytes by promoting interleukin-1 receptor-associated kinase expression and its association to MyD88 and not by modulating TLR4 expression. J Biol Chem 277 (2002) 27927–27934.

Algeri, S., Biagini, L., Bianchi, S., Garofalo, P., Marconi, M., Pitsikas, N., Raimondi, L., Sacchetti, G., Tacconi, M. T., Bertani, T., and et al. The effect of caloric restriction on a rat model of aging: biological, pathological, biochemical and behavioral characterization. Aging (Milano) 3 (1991) 388–390.

Ayasolla, K., Khan, M., Singh, A. K., and Singh, I. Inflammatory mediator and beta-amyloid (25–35)-induced ceramide generation and iNOS expression are inhibited by vitamin E. Free Radic Biol Med 37 (2004) 325–338.

Bello, R. I., Gomez-Diaz, C., Buron, M. I., Alcain, F. J., Navas, P., and Villalba, J. M. Enhanced anti-oxidant protection of liver membranes in long-lived rats fed on a coenzyme Q10-supplemented diet. Exp Gerontol 40 (2005) 694–706.

Berk, B. C., Weintraub, W. S., and Alexander, R. W. Elevation of C-reactive protein in "active" coronary artery disease. Am J Cardiol 65 (1990) 168–172.

Bonsall, M. B. Longevity and ageing: appraising the evolutionary consequences of growing old. Philos Trans R Soc Lond B Biol Sci 361 (2006) 119–135.

Bose, C., Bhuvaneswaran, C., and Udupa, K. B. Age-related alteration in hepatic acyl-CoA: cholesterol acyltransferase and its relation to LDL receptor and MAPK. Mech Ageing Dev 126 (2005) 740–751.

Cao, Z., Xiong, J., Takeuchi, M., Kurama, T., and Goeddel, D. V. TRAF6 is a signal transducer for interleukin-1. Nature 383 (1996) 443–446.

Castillo, S. S., Levy, M., Thaikoottathil, J. V., and Goldkorn, T. Reactive nitrogen and oxygen species activate different sphingomyelinases to induce apoptosis in airway epithelial cells. Exp Cell Res 313 (2007) 2680–2686.

Ceda, G. P., Dall'Aglio, E., Maggio, M., Lauretani, F., Bandinelli, S., Falzoi, C., Grimaldi, W., Ceresini, G., Corradi, F., Ferrucci, L., Valenti, G., and Hoffman, A. R. Clinical implications of the reduced activity of the GH-IGF-I axis in older men. J Endocrinol Invest 28 (2005) 96–100.

Chalfant, C. E., Kishikawa, K., Mumby, M. C., Kamibayashi, C., Bielawska, A., and Hannun, Y. A. Long chain ceramides activate protein phosphatase-1 and protein phosphatase-2A. Activation is stereospecific and regulated by phosphatidic acid. J Biol Chem 274 (1999) 20313–20317.

Chen, J., Nikolova-Karakashian, M., Merrill, A. H., Jr., and Morgan, E. T. Regulation of cytochrome P450 2C11 (CYP2C11) gene expression by interleukin-1, sphingomyelin hydrolysis, and ceramides in rat hepatocytes. J Biol Chem 270 (1995) 25233–25238.

Chung, H. Y., Kim, H. J., Kim, J. W., and Yu, B. P. The inflammation hypothesis of aging: molecular modulation by calorie restriction. Ann N Y Acad Sci 928 (2001) 327–335.

Clarke, C. J., Snook, C. F., Tani, M., Matmati, N., Marchesini, N., and Hannun, Y. A. The Extended Family of Neutral Sphingomyelinases. Biochemistry 45 (2006) 11247–11256.

Clarke, C. J., Truong, T. G., and Hannun, Y. A. Role for neutral sphingomyelinase-2 in tumor necrosis factor alpha-stimulated expression of vascular cell adhesion molecule-1 (VCAM) and intercellular adhesion molecule-1 (ICAM) in lung epithelial cells: p38 MAPK is an upstream regulator of nSMase2. J Biol Chem 282 (2007) 1384–1396.

Claycombe, K. J., Wu, D., Nikolova-Karakashian, M., Palmer, H., Beharka, A., Paulson, K. E., and Meydani, S. N. Ceramide mediates age-associated increase in macrophage cyclooxygenase-2 expression. J Biol Chem 277 (2002) 30784–30791.

Cuschieri, J., Gourlay, D., Garcia, I., Jelacic, S., and Maier, R. V. Implications of proteasome inhibition: an enhanced macrophage phenotype. Cell Immunol 227 (2004) 140–147.

Cutler, R. G., and Mattson, M. P. Sphingomyelin and ceramide as regulators of development and lifespan. Mech Ageing Dev 122 (2001) 895–908.

De Palma, C., Meacci, E., Perrotta, C., Bruni, P., and Clementi, E. Endothelial Nitric Oxide Synthase Activation by Tumor Necrosis Factor {alpha} Through Neutral Sphingomyelinase 2, Sphingosine Kinase 1, and Sphingosine 1 Phosphate Receptors: A Novel Pathway Relevant to the Pathophysiology of Endothelium. Arterioscler Thromb Vasc Biol 26 (2006) 99–105.

Di Iorio, A., Ferrucci, L., Sparvieri, E., Cherubini, A., Volpato, S., Corsi, A., Bonafe, M., Franceschi, C., Abate, G., and Paganelli, R. Serum IL-1beta levels in health and disease: a population-based study. 'The InCHIANTI study'. Cytokine 22 (2003) 198–205.

Fan, J., Molina, P. E., Gelato, M. C., and Lang, C. H. Differential tissue regulation of insulin-like growth factor-I content and binding proteins after endotoxin. Endocrinology 134 (1994) 1685–1692.

Franceschi, C., Bonafe, M., Valensin, S., Olivieri, F., De Luca, M., Ottaviani, E., and De Benedictis, G. Inflamm-aging. An evolutionary perspective on immunosenescence. Ann N Y Acad Sci 908 (2000) 244–254.

Franceschi, C., Capri, M., Monti, D., Giunta, S., Olivieri, F., Sevini, F., Panourgia, M. P., Invidia, L., Celani, L., Scurti, M., Cevenini, E., Castellani, G. C., and Salvioli, S. Inflammaging and anti-inflammaging: a systemic perspective on aging and longevity emerged from studies in humans. Mech Ageing Dev 128 (2007) 92–105.

Goswami, R., Ahmed, M., Kilkus, J., Han, T., Dawson, S. A., and Dawson, G. Differential regulation of ceramide in lipid-rich microdomains (rafts): Antagonistic role of palmitoyl:-protein thioesterase and neutral sphingomyelinase 2. J Neurosci Res 81 (2005) 208–217.

Gouaze, V., Mirault, M. E., Carpentier, S., Salvayre, R., Levade, T., and Andrieu-Abadie, N. Glutathione peroxidase-1 overexpression prevents ceramide production and partially inhibits apoptosis in doxorubicin-treated human breast carcinoma cells. Mol Pharmacol 60 (2001) 488–496.

Hagen, T. M., Vinarsky, V., Wehr, C. M., and Ames, B. N. (R)-alpha-lipoic acid reverses the age-associated increase in susceptibility of hepatocytes to tert-butylhydroperoxide both in vitro and in vivo. Antioxid Redox Signal 2 (2000) 473–483.

Han, D., Hosokawa, T., Aoike, A., and Kawai, K. Age-related enhancement of tumor necrosis factor (TNF) production in mice. Mech Ageing Dev 84 (1995) 39–54.

Hannun, Y. A., and Obeid, L. M. The Ceramide-centric universe of lipid-mediated cell regulation: stress encounters of the lipid kind. J Biol Chem 277 (2002) 25847–25850.

Hayashi, Y., Kiyono, T., Fujita, M., and Ishibashi, M. cca1 Is Required for Formation of Growth-arrested Confluent Monolayer of Rat 3Y1 Cells. J Biol Chem 272 (1997) 18082–18086.

Hofmann, K., Tomiuk, S., Wolff, G., and Stoffel, W. Cloning and characterization of the mammalian brain-specific, Mg2+-dependent neutral sphingomyelinase. Proc Natl Acad Sci USA 97 (2000) 5895–5900.

Hsieh, C. C., Xiong, W., Xie, Q., Rabek, J. P., Scott, S. G., An, M. R., Reisner, P. D., Kuninger, D. T., and Papaconstantinou, J. Effects of age on the posttranscriptional regulation of CCAAT/enhancer binding protein alpha and CCAAT/enhancer binding protein beta isoform synthesis in control and LPS-treated livers. Mol Biol Cell 9 (1998) 1479–1494.

Hsieh, C. C., Rosenblatt, J. I., and Papaconstantinou, J. Age-associated changes in SAPK/JNK and p38 MAPK signaling in response to the generation of ROS by 3-nitropropionic acid. Mech Ageing Dev 124 (2003) 733–746.

Jenkins, G. M., and Hannun, Y. A. Role for de novo sphingoid base biosynthesis in the heat-induced transient cell cycle arrest of Saccharomyces cerevisiae. J Biol Chem 276 (2001) 8574–8581.

Kalehua, A. N., Taub, D. D., Baskar, P. V., Hengemihle, J., Munoz, J., Trambadia, M., Speer, D. L., De Simoni, M. G., and Ingram, D. K. Aged mice exhibit greater mortality concomitant to increased brain and plasma TNF-alpha levels following intracerebroventricular injection of lipopolysaccharide. Gerontology 46 (2000) 115–128.

Karakashian, A. A., Giltiay, N. V., Smith, G. M., and Nikolova-Karakashian, M. N. Expression of neutral sphingomyelinase-2 (NSMase-2) in primary rat hepatocytes modulates IL-beta-induced JNK activation. Faseb J 18 (2004) 968–970.

Kim, S., Kang, M., Choi, Y., Suh, Y., and Kim, D. Sphingomyelinase activity is enhanced in cerebral cortex of senescence-accelerated mouse-P/10 with advancing age. Biophys Biophim Res Comm 237 (1997) 583–587.

Kirkwood, T. B., and Austad, S. N. Why do we age? Nature 408 (2000) 233–238.

Lang, C. H., Fan, J., Cooney, R., and Vary, T. C. IL-1 receptor antagonist attenuates sepsis-induced alterations in the IGF system and protein synthesis. Am J Physiol 270 (1996) E430–437.

Lang, C. H., Pollard, V., Fan, J., Traber, L. D., Traber, D. L., Frost, R. A., Gelato, M. C., and Prough, D. S. Acute alterations in growth hormone-insulin-like growth factor axis in humans injected with endotoxin. Am J Physiol 273 (1997) R371–378.

Lang, C. H., Nystrom, G. J., and Frost, R. A. Regulation of IGF binding protein-1 in hep G2 cells by cytokines and reactive oxygen species. Am J Physiol 276 (1999) G719–727.

Lecka-Czernik, B., Moerman, E., Jones, R., and Goldstein, S. Identification of gene sequences overexpressed in senescent and Werner syndrome human fibroblasts. Exp. Gerontology. 31 (1996) 159–174.

Lee, A., Whyte, M. K., and Haslett, C. Inhibition of apoptosis and prolongation of neutrophil functional longevity by inflammatory mediators. J Leukoc Biol 54 (1993) 283–288.

Li, L., Cousart, S., Hu, J., and McCall, C. E. Characterization of Interleukin-1 Receptor-associated Kinase in Normal and Endotoxin-tolerant Cells. J Biol Chem 275 (2000) 23340–23345.

Lightle, S. A., Oakley, J. I., and Nikolova-Karakashian, M. N. Activation of sphingolipid turnover and chronic generation of ceramide and sphingosine in liver during aging. Mech Ageing Dev 120 (2000) 111–125.

Liu, B., and Hannun, Y. A. Inhibition of the neutral magnesium-dependent sphingomyelinase by glutathione. J Biol Chem 272 (1997) 16281–16287.

Liu, B., Andrieu-Abadie, N., Levade, T., Zhang, P., Obeid, L. M., and Hannun, Y. A. Glutathione regulation of neutral sphingomyelinase in tumor necrosis factor-alpha-induced cell death. J Biol Chem 273 (1998) 11313–11320.

Lozanski, G., Berthier, F., and Kushner, I. The sphingomyelin-ceramide pathway participates in cytokine regulation of C-reactive protein and serum amyloid A, but not alpha-fibrinogen. Biochem J 328 (Pt 1) (1997) 271–275.

Luberto, C., Hassler, D. F., Signorelli, P., Okamoto, Y., Sawai, H., Boros, E., Hazen-Martin, D. J., Obeid, L. M., Hannun, Y. A., and Smith, G. K. Inhibition of Tumor Necrosis Factor-induced Cell Death in MCF7 by a Novel Inhibitor of Neutral Sphingomyelinase. J Biol Chem 277 (2002) 41128–41139.

Marchesini, N., Osta, W., Bielawski, J., Luberto, C., Obeid, L. M., and Hannun, Y. A. Role for Mammalian Neutral Sphingomyelinase 2 in Confluence-induced Growth Arrest of MCF7 Cells. J Biol Chem 279 (2004) 25101–25111.

Martin, S. F., Sawai, H., Villalba, J. M., and Hannun, Y. A. Redox regulation of neutral sphingomyelinase-1 activity in HEK293 cells through a GSH-dependent mechanism. Arch Biochem Biophys 459 (2007) 295–300.

Mouton, R. E., and Venable, M. E. Ceramide induces expression of the senescence histochemical marker, beta-galactosidase, in human fibroblasts. Mech Ageing Dev 113 (2000) 169–181.

Ninomiya-Tsuji, J., Kishimoto, K., Hiyama, A., Inoue, J., Cao, Z., and Matsumoto, K. The kinase TAK1 can activate the NIK-I kappaB as well as the MAP kinase cascade in the IL-1 signalling pathway. Nature 398 (1999) 252–256.

Petkova, D. H., Momchilova-Pankova, A. B., Markovska, T. T., and Koumanov, K. S. Age-related changes in rat liver plasma membrane sphingomyelinase activity. Exp Gerontol 23 (1988) 19–24.

Pewzner-Jung, Y., Ben-Dor, S., and Futerman, A. H. When do Lasses (longevity assurance genes) become CerS (ceramide synthases)?: Insights into the regulation of ceramide synthesis. J Biol Chem 281 (2006) 25001–25005.

Puglielli, L., Ellis, B. C., Saunders, A. J., and Kovacs, D. M. Ceramide stabilizes beta-site amyloid precursor protein-cleaving enzyme 1 and promotes amyloid beta-peptide biogenesis. J Biol Chem 278 (2003) 19777–19783.

Rabek, J. P., Scott, S., Hsieh, C. C., Reisner, P. D., and Papaconstantinou, J. Regulation of LPS-mediated induction of C/EBP delta gene expression in livers of young and aged mice. Biochim Biophys Acta 1398 (1998) 137–147.

Riebeling, C., Allegood, J. C., Wang, E., Merrill, A. H., Jr., and Futerman, A. H. Two mammalian longevity assurance gene (LAG1) family members, trh1 and trh4, regulate dihydroceramide synthesis using different fatty acyl-CoA donors. J Biol Chem 278 (2003) 33735–33738.

Rosenthal, C. J., and Franklin, E. C. Variation with age and disease of an amyloid A protein-related serum component. J Clin Invest 55 (1975) 746–753.

Rutkute, K., Asmis, R. H., and Nikolova-Karakashian, M. N. Regulation of neutral sphingomyelinase-2 by GSH: a new insight to the role of oxidative stress in aging-associated inflammation. J Lipid Res 48 (2007a) 2443–2452.

Rutkute, K., Karakashian, A. A., Giltiay, N. V., Dobierzewska, A., and Nikolova-Karakashian, M. N. Aging in rat causes hepatic hyperresponsiveness to interleukin 1 b which is mediated by neutral sphingomyelinase-2. Hepatology 46 (2007b) 1166–1176.

Rutkute, K., and Nikolova-Karakashian, M. N. Regulation of insulin-like growth factor binding protein-1 expression during aging. Biochem Biophys Res Commun 361 (2007c) 263–269.

Sawai, H., Domae, N., Nagan, N., and Hannun, Y. A. Function of the Cloned Putative Neutral Sphingomyelinase as Lyso-platelet Activating Factor-Phospholipase C. J Biol Chem 274 (1999) 38131–38139.

Shirakabe, K., Yamaguchi, K., Shibuya, H., Irie, K., Matsuda, S., Moriguchi, T., Gotoh, Y., Matsumoto, K., and Nishida, E. TAK1 mediates the ceramide signaling to stress-activated protein kinase/c-Jun N-terminal kinase. J Biol Chem 272 (1997) 8141–8144.

Sly, L. M., Krzesicki, R. F., Brashler, J. R., Buhl, A. E., McKinley, D. D., Carter, D. B., and Chin, J. E. Endogenous brain cytokine mRNA and inflammatory responses to lipopolysaccharide are elevated in the Tg2576 transgenic mouse model of Alzheimer's disease. Brain Res Bull 56 (2001) 581–588.

Spaulding, C. C., Walford, R. L., and Effros, R. B. Calorie restriction inhibits the age-related dysregulation of the cytokines TNF-alpha and IL-6 in C3B10RF1 mice. Mech Ageing Dev 93 (1997) 87–94.

Stoffel, W., Jenke, B., Block, B., Zumbansen, M., and Koebke, J. Neutral sphingomyelinase 2 (smpd3) in the control of postnatal growth and development. Proc Natl Acad Sci USA 102 (2005) 4554–4559.

Suh, Y. Age-specific changes in expression, activity, and activation of the c-Jun NH(2)-terminal kinase and p38 mitogen-activated protein kinases by methyl methanesulfonate in rats. Mech Ageing Dev 122 (2001) 1797–1811.

Tang, Y., Di Pietro, L., Feng, Y., and Wang, X. Increased TNF-alpha and PGI(2), but not NO release from macrophages in 18-month-old rats. Mech Ageing Dev 114 (2000) 79–88.

Tellier, E., Negre-Salvayre, A., Bocquet, B., Itohara, S., Hannun, Y. A., Salvayre, R., and Auge, N. Role for furin in tumor necrosis factor alpha-induced activation of the matrix metalloproteinase/sphingolipid mitogenic pathway. Mol Cell Biol 27 (2007) 2997–3007.

Venable, M. E., Lee, J. Y., Smyth, M. J., Bielawska, A., and Obeid, L. M. Role of ceramide in cellular senescence. J Biol Chem 270 (1995) 30701–30708.

Venkataraman, K., and Futerman, A. H. Do longevity assurance genes containing Hox domains regulate cell development via ceramide synthesis? FEBS Lett 528 (2002a) 3–4.

Venkataraman, K., Riebeling, C., Bodennec, J., Riezman, H., Allegood, J. C., Sullards, M. C., Merrill, A. H., Jr., and Futerman, A. H. Upstream of growth and differentiation factor 1 (uog1), a mammalian homolog of the yeast longevity assurance gene 1 (LAG1), regulates N-stearoyl-sphinganine (C18-(dihydro)ceramide) synthesis in a fumonisin B1-independent manner in mammalian cells. J Biol Chem 277 (2002b) 35642–35649.

Verheij, M., Bose, R., Lin, X. H., Yao, B., Jarvis, W. D., Grant, S., Birrer, M. J., Szabo, E., Zon, L. I., Kyriakis, J. M., Haimovitz-Friedman, A., Fuks, Z., and Kolesnick, R. N. Requirement for ceramide-initiated SAPK/JNK signalling in stress-induced apoptosis. Nature 380 (1996) 75–79.

Vericel, E., Narce, M., Ulmann, L., Poisson, J. P., and Lagarde, M. Age-related changes in antioxidant defence mechanisms and peroxidation in isolated hepatocytes from spontaneously hypertensive and normotensive rats. Mol Cell Biochem 132 (1994) 25–29.

Walston, J., and Fried, L. P. Frailty and the older man. Med Clin North Am 83 (1999) 1173–1194.

Wesche, H., Henzel, W. J., Shillinglaw, W., Li, S., and Cao, Z. MyD88: an adapter that recruits IRAK to the IL-1 receptor complex. Immunity 7 (1997) 837–847.

Westwick, J. K., Bielawska, A. E., Dbaibo, G., Hannun, Y. A., and Brenner, D. A. Ceramide activates the stress-activated protein kinases. J Biol Chem 270 (1995) 22689–22692.

Wu, D., Marko, M., Claycombe, K., Paulson, K. E., and Meydani, S. N. Ceramide-induced and age-associated increase in macrophage COX-2 expression is mediated through up-regulation of NF-kappa B activity. J Biol Chem 278 (2003) 10983–10992.

Yamamoto, K., Shimokawa, T., Yi, H., Isobe, K., Kojima, T., Loskutoff, D. J., and Saito, H. Aging accelerates endotoxin-induced thrombosis : increased responses of plasminogen activator inhibitor-1 and lipopolysaccharide signaling with aging. Am J Pathol 161 (2002) 1805–1814.

Yamin, T. T., and Miller, D. K. The interleukin-1 receptor-associated kinase is degraded by proteasomes following its phosphorylation. J Biol Chem 272 (1997) 21540–21547.

Yang, J., Anzo, M., and Cohen, P. Control of aging and longevity by IGF-I signaling. Exp Gerontol 40 (2005) 867–872.

Yu, Z. F., Nikolova-Karakashian, M., Zhou, D., Cheng, G., Schuchman, E. H., and Mattson, M. P. Pivotal role for acidic sphingomyelinase in cerebral ischemia-induced ceramide and cytokine production, and neuronal apoptosis. J Mol Neurosci 15 (2000) 85–97.

Zerofsky, M., Harel, E., Silverman, N., and Tatar, M. Aging of the innate immune response in Drosophila melanogaster. Aging Cell 4 (2005) 103–108.

# Chapter 19
# Sphingolipid Metabolizing Enzymes as Novel Therapeutic Targets

**Andreas Billich and Thomas Baumruker**

**Abstract** Pharmacological interference with sphingolipid metabolizing enzymes promises to provide novel ways to modulate cellular pathways relevant in multiple diseases. In this review, we focus on two sphingolipid signaling molecules, sphingosine-1-phosphate (S1P) and ceramide, as they are involved in cell fate decisions (survival vs. apoptosis) and in a wide range of pathophysiological processes. For S1P, we will discuss sphingosine kinases and S1P lyase as the enzymes which are crucial for its production and degradation, respectively, emphasizing the potential therapeutic usefulness of inhibitors of these enzymes. For ceramide, we will concentrate on acid sphingomyelinase, and critically review the substantial literature which implicates this enzyme as a worthwhile target for pharmacological inhibitors. It will become clear that the task to validate these enzymes as drug targets is not finished and many questions regarding the therapeutic usefulness of their inhibitors remain unanswered. Still this approach holds promise for a number of totally new therapies, and, on the way, detailed insight into sphingolipid signaling pathways can be gained.

**Keywords** Acid sphingomyelinase · ceramide · sphingosine kinase · sphingosine-1-phosphate lyase · sphingosine-1-phosphate

**Abbreviations** ASMase: acid sphingomyelinase; bFGF: basic fibroblast growth factor; BMMCs: bone marrow derived mast cells; DC: dendritic cell; DHS: D,L-*threo*-dihydrosphingosine; DMS: *N,N*-dimethylsphingosine; NPD: Niemann-Pick disease; PMA: phorbol-12 myristate-13 acetate; S1P: sphingosine-1-phosphate; $S1P_1$: sphingosine-1-phosphate receptor type 1; siRNA: small interfering RNA; SPHK: sphingosine kinase; SPL: sphingosine-1-phosphate lyase; SPP1: S1P phosphatase type 1; VEGF: vascular endothelial growth factor

---

A. Billich
Novartis Institutes for BioMedical Research, Brunnerstrasse 59, A-1235 Vienna, Austria

P.J. Quinn, X. Wang (eds.), *Lipids in Health and Disease*,

## 19.1 Introduction

Sphingolipids are a family of lipids that play essential roles both as structural components of mammalian cells and in cell signaling. The signaling function, which has physiological relevance in many organ systems, comprises interaction of sphingolipids with cell surface receptors, modulation of lipid rafts in the plasma membrane, and intracellular actions. Pharmacological intervention with sphingolipid signaling has been achieved in two different ways, namely by interference with G-protein-coupled receptors and by modulating enzymes of sphingolipid metabolism thereby altering intracellular concentrations of sphingolipids. With respect to sphingolipid receptor modulation, FTY720, whose corresponding phosphate binds to sphingosine-1-phosphate (S1P) receptors, has recently shown clinical efficacy in a phase II trial in patients with multiple sclerosis (Baumruker et al., 2007) which greatly stimulated the interest in S1P receptor pharmacology. In contrast, inhibitors of sphingolipid metabolizing enzymes comprise various experimental compounds at preclinical stages; thus, the usefulness of this approach to interfere with sphingolipid signaling in order to treat diseases remains to be demonstrated.

Recent reviews have broadly covered the multitude of sphingolipid enzyme inhibitors (Nussbaumer et al. 2008; Delgado et al., 2006; Morales and Fernandez-Checa, 2007). Here, we will rather focus on two sphingolipid signaling molecules, S1P and ceramide, as they are involved in cell fate decisions and in a wide range of pathophysiological processes. For S1P, we will discuss sphingosine kinases and S1P lyase as the enzymes which are crucial for its production and degradation, respectively, emphasizing the potential therapeutic usefulness of inhibitors of these enzymes. For ceramide, we will concentrate on acid sphingomyelinase, and critically review the substantial literature which implicates this enzyme as a worthwhile target for pharmacological inhibitors.

## 19.2 Sphingosine Kinases as Novel Drug Targets

S1P exhibits a plethora of actions in normal physiology and in pathology (Hla, 2004), leading to an obvious interest in the enzymes which produce this important lipid mediator. The sphingosine kinases (SPHKs) type 1 and 2 are the only known enzymes relevant for the production of S1P in mammalian cells. Other potential isoforms delineated from biochemical studies, in particular in platelets (Banno et al., 1998), have not been characterized so far, and the production of S1P from sphingosylphosphorylcholine by autotaxin (Clair et al., 2003) appears to be not of physiological relevance in view of the high $K_m$ of that substrate. In fact, R. Proia and co-workers showed absence of detectable S1P levels in mouse embryos deficient in both SPHK1 and 2 (Mizugishi et al., 2005).

While SPHK1 and 2 both catalyze the formation of S1P from sphingosine and ATP with comparable $K_m$ and $V_{max}$ values (Billich et al., 2003), their relative contribution to S1P levels in blood and tissues and their differential role in various signaling pathways is a point of continued interest. Both SPHK1 and SPHK2 knock-out mice are viable and do not feature obvious abnormalities (Allende et al., 2004; Kharel et al. 2005; Mizugichi et al., 2005; Zemann et al., 2006); S1P levels in blood and tissues of SPHK1 knock-out mice are reduced by about 50%, while they are unchanged or even increased in different SPHK2 knock-out strains (Kharel et al. 2005; Zemann et al., 2006) (apparently due to a counterregulation by SPHK1), indicating an overlapping role of the two enzymes in contributing to S1P levels. In contrast, the above mentioned S1P-deficient SPHK1/2 double-KO mice (Mizugishi et al., 2005) are not viable, due to severely disturbed angiogenesis and neurogenesis, including neural tube closure, followed by death *in utero*. Based on the apparently crucial role of SPHK1/2 and S1P in embryonal development, one would suggest that blocking both SPHK1 and 2 simultaneously with a pharmacological inhibitor might give rise to severe toxicity. However, recent data (Pappu et al., 2007) on conditional SPHK1/2 double knock-out mice induced 3–5 days after birth have changed the picture: Those mice have systemic S1P levels reduced by $\sim$100-fold but seem to be perfectly healthy. Thus, a substantial reduction of systemic S1P levels by a dual SPHK1/2 inhibitor can be expected to be tolerated in an adult individual. As will be explained below, a dual-action SPHK inhibitor might be desirable for treatment of allergy and asthma where extracellular S1P levels appear to be involved. However, in many other instances it appears that SPHK1 is a specific and highly regulated element in cellular signaling cascades, relevant in multiple pathologies; therefore, inhibitors that are specific for that isoform should be aimed at.

### 19.2.1 SPHK and Cancer

The balance between S1P and ceramide is hypothesized to determine cell fate (survival *vs.* apoptosis), as originally proposed by S.Spiegel and co-workers (Maceyka et al., 2002). Therefore, measures to increase cellular levels of ceramide or to decrease S1P in tumor cells should be able to prevent tumor growth. An important element in understanding the S1P/ceramide rheostat comes from the observation that specific down-regulation of SPHK2 reduces conversion of sphingosine to ceramide in the recycling pathway of sphingolipid synthesis, while down-regulation of SPHK1 increases levels of pro-apoptotic ceramide via enhanced *de novo* synthesis (Maceyka et al., 2005; Le Stunff et al., 2007). These results demonstrate that SPHK1 and SPHK2 have opposing roles in the regulation of ceramide biosynthesis and (given the different subcellular localization of the two enzymes) suggest that the location of S1P production dictates its functions. Indeed, numerous studies have characterized SPHK1 as an anti-apoptotic protein: (i) SPHK1 siRNA is anti-proliferative/pro-apoptotic in several cancer cell lines (Akao et al., 2006; Billich et al., 2005; Döll et al., 2005, 2007; Murakami et al., 2007; Pchejetski et al., 2005; Sarkar et al., 2005; Taha et al., 2006) and in primary chronic myeloid leukemia cells (Li et al., 2007); (ii) SPHK1 overexpression stimulates cell growth in 3T3 fibroblasts (Olivera et al., 1999), colony formation of breast cancer cells in soft agar (Sukocheva et al., 2003), and promotes tumor growth in nude mice (Xia et al., 2000). By contrast, SPHK2 has been characterized as a pro-apoptotic protein (Liu et al., 2003; Sankala et al. 2007).

Our group found that SPHK1 mRNA is over expressed in renal tumors, while concomitantly SPHK2 mRNA levels were dramatically reduced (D.Mechtcheriakova, unpublished data). In two other studies, elevation of SPHK1 mRNA and protein in several solid tumors has also been shown (French et al., 2003; Johnson et al., 2005). Furthermore, high expression of SPHK1 was found to correlate with tumor aggressiveness and poor prognosis in glioblastoma patients (Van Brocklyn et al., 2005), and also with poor prognosis in breast cancer patients, where SPHK1 was most highly expressed in estrogen receptor negative tumors (Ruckhäberle et al., 2007). SPHK1 expression was also found to correlate with cancer in mouse models: it was found to be correlated with progression of murine erythroleukemia as a model for leukemic disorders in man (Le Scolan et al., 2005); furthermore, SPHK1 was upregulated in colon carcinogenesis in mice (Kawamori et al., 2006), and genetic deletion of SPHK1 in a mouse model of intestinal adenomas resulted in suppression in adenoma size and epithelial cell proliferation (Kohno et al., 2006). Importantly, SPHK inhibitors were shown to reduce tumor progression in a murine mammary adenocarcinoma model (French et al., 2006) and to inhibit growth of prostatic adenocarcinoma cells (Leroux et al., 2007).

To conclude, there is substantial evidence for an important role of SPHK1 in tumor cell survival, suggesting this enzyme as a drug target in anti-cancer therapy (see also Cuvillier, 2007, for a detailed review). To support this further, there is also literature demonstrating that SPHK1 inhibition can sensitize cells (i) towards the action of $\gamma$-irradiation (in the case of the prostate cancer cell line LNCaP (Nava et al., 2000)); (ii) to ceramide (melanoma cells (Bektas et al., 2005)); and (iii) to well-known chemotherapeutic drugs, such as doxorubicin and etoposide (in leukemia and breast cancer cells (Bonhoure et al., 2006; Olivera et al., 1999)), imatinib (leukemia cells (Baran et al., 2007)), and cisplatin (Min et al., 2005; Min et al. 2007). Finally, as COX-2 and VEGF play important roles in some types of cancers, the dependence of COX-2 expression and of VEGF-induced angiogenesis on SPHK1 (see below) is of significant interest.

## 19.2.2 SPHK and Angiogenesis

S1P is essential for vascular development, as demonstrated by the SPHK1/2 double-knockout mice which are devoid of S1P and feature severely disturbed angiogenesis (Mizugichi et al., 2005). In addition to this involvement in embryonal development of the vasculture, S1P is also an important factor inducing neo-angiogenesis in an adult organism which is of obvious pharmacological interest. Both the extracellular action of S1P via its receptors and the role of S1P/SPHK1 within cells have to be considered here. Extracellular S1P is a strong angiogenic factor and clearly mediates its effects via S1P receptors, primarily $S1P_1$. While S1P is constantly present in the blood stream, its local concentration at the endothelium can be increased by secretion, e.g., from platelets and mast cells; furthermore, the SPHK1a isoform is constitutively secreted by vascular endothelial cells and thus may contribute to extracellular S1P levels at the endothelium (Ancellin et al., 2002; Venkataraman et al., 2006).

For the intracellular role of S1P/SPHK in angiogenesis, a clear link of SPHK1 activation in the vascular endothelial growth factor (VEGF) signaling cascade has been deduced. VEGF, one of the major drivers in the process of angiogenesis, was shown to induce SPHK activity and raise S1P levels in the T24 bladder tumor cell line (Shu et al., 2002; Wu et al., 2003). SiRNA to SPHK1, but not SPHK2, and a catalytically inactive dominant-negative mutant of SPHK1 blocked the accumulation of VEGF induced Ras-GTP and phospho-ERK (Shu et al., 2002). Interestingly no effect of these agents was seen on an EGF induction using phospho-ERK1/2 as a readout, indicating a clear specificity.

This specificity, meaning that not all angiogenic factors are dependent on SPHK in their signaling, might be one of the reasons why the SPHK1 knockout mice have normal vasculature. However, that the involvement of SPHK1 is not restricted to VEGF was recently shown (Granata et al., 2007): IGFBP-3 induced network formation by human endothelial cells in a Matrigel assay and up-regulated proangiogenic genes (VEGF, MMP-2 and -9); decreasing SPHK1 expression by siRNA blocked the IGFBP-3-induced tube formation showing that IGFBP-3 also regulates angiogenesis through SPHK1/S1P activation.

Besides a number of growth factors, also phorbol-12 myristate-13 acetate (PMA) is known to promote blood vessel formation; the involved pathways are, however, ill defined. It was found (Taylor et al., 2006) that PMA-induced angiogenesis depends on protein kinase C and its downstream target SPHK1. The evidence for this notion is based on the use of the SPHK1-selective inhibitor SKI-II (see below).

Supporting evidence for a role of SPHK in angiogenesis was recently provided. (Maines et al., 2006) using prototype SPHK inhibitors SKI-II and ABC294640. SPHK activity was present in human and bovine retinal endothelial cells, and stimulated by VEGF. VEGF-induced formation of microvessels by these cells was inhibited by SKI-II *in vitro* due to an inhibition of their migration. In addition, the compound also blocked TNF-$\alpha$ induced expression of adhesion molecules. Dosage of the orally bioavailability inhibitor ABC294640 substantially reduced the levels of FITC-BSA leakage (as an indirect measure of neo-angiogenesis) in already strongly affected hyperglycemic rats, dramatically inhibiting the progression to retinopathy seen in the untreated controls.

Recent data from our group show that dermal microvascular endothelial cells isolated from SPHK1-deficient mice are defective in angiogenesis induced by VEGF and S1P (but not by FGF and TNF-α) *in vitro*, and that SPHK1 knock-out mice show defective neoangiogenesis *in vivo* (S.Niwa, manuscript in preparation).

Finally, SPHK has been shown to be a target of alphastatin, a 24–amino acid fragment of human fibrinogen acting as a potent inhibitor of activated endothelial cells *in vitro* and *in vivo* (Staton et al., 2004); it was found that alphastatin in a yet unexplored manner downregulates SPHK activity in endothelial cells (Chen et al., 2006).

Taken together, there is substantial evidence in published literature that the S1P/SPHK1 pathway plays a role in angiogenesis and vascular integrity; also, since neoangiogenesis plays an essential role in solid tumor progression and in several chronic inflammatory diseases (rheumatoid arthritis, multiple sclerosis, psoriasis) as well as in diabetic retinopathy, SPHK inhibitors appear as promising novel therapeutic option. It should be noted, however, that most likely there is a fine balance between inhibiting angiogenesis and reducing endothelial barrier function. S1P usually tightens the endothelial barrier and a permanent reduction is believed to be the first step in the leaking of plasma components such as LDL, an initiator in atherogenesis.

As angiogenesis is an important factor in establishment of solid tumors and of metastases, the essential role of SPHK1 in neo-angiogenesis provides a second mode-of-action for its inhibitors which otherwise have been proposed as potential anti-cancer drugs based on their pro-apoptotic effects (see above). Indirect support for this concept comes from studies on efficacy of an anti-S1P antibody in various tumor models (Visentin et al., 2006). This antibody has an affinity for S1P of ~100 pM which is about 100-fold higher than the affinity of S1P receptors for S1P and was active in three orthotopic models of breast carcinoma, ovarian cancer, and a subcutaneous xenograft models of lung adenocarcinoma. Tumor volumes and growth were inhibited between 40% and 62% in all models over the treatment period with comparable efficacy to paclitaxel. Mechanistically, the effect was assigned to an inhibition of the release and function of pro-angiogenic factors (such as VEGF, bFGF) and a direct effect on endothelial cells in neovascularization.

### *19.2.3 SPHK and Atherosclerosis*

In patients at risk for obstructive coronary artery disease, levels of S1P in the plasma were found to increase with severity and degree of stenosis from 634 (in control subjects from a different study) to 894 (in patients with mild symptoms) to 1035 pmol/ml (in severe stenosis) (n = 308) (Deutschman et al., 2003). S1P was found to be more predictive for stenosis than traditional risk factors, the S1P marker being the single most important independent predictor of stenosis. The question whether increased S1P level are a consequence of the disease or if they are causally involved in the pathogenesis (and therefore a target to be attacked with SPHK inhibitors), remains to be explored.

However, it appears that SPHK1 is an important element in endothelial cell activation, giving rise to expression of pro-atherogenic factors: (i) SPHK1 mediates TNF-α-induced MCP-1 gene expression through a p38 MAPK-dependent pathway and may participate in oscillatory flow-mediated proinflammatory signaling pathways in the vasculature (Chen et al., 2004); (ii) SPHK1 mediates expression of adhesion molecules like VCAM-1, ICAM-1 and E-selectin by TNF-α and by globular adiponectin (Xia et al., 1998; Kase et al., 2007); TNF-induced expression of such factors can be blocked by SPHK inhibitors (Xia et al., 1998; Kim et al., 2001); interestingly, it was shown that HDL profoundly inhibits TNF-stimulated SPHK activity in endothelial cells resulting in a decrease in S1P production and adhesion protein expression, suggesting a mechanism of protection against atherosclerosis (Xia et al., 1999). Clearly, more work is needed to further establish the role of SPHK1 in the pathogenesis of atherosclerosis.

### 19.2.4 Involvement of SPHK1 in COX-2 Activation

Y. Hannun and co-workers (Pettus et al., 2003) found that S1P induces both COX-2 and $PGE_2$ in A549 lung carcinoma cells and in murine fibroblasts; TNF induced a rapid and transient increase of S1P levels. SiRNAs directed against SPHK1 inhibited TNF-α-induced SPHK activity and almost completely abolished the ability of TNF to induce COX-2 or generate $PGE_2$. We extended these findings (Billich et al., 2005) by showing that indeed the transient activation of SPHK1 is required for cytokine-induced COX-2 transcription and $PGE_2$ production, since not only specific siRNA (abolishing both basal and induced SPHK1 enzyme activity), but also a dominant-negative SPHK1 mutant (suppressing induced SPHK1 activity only) reduced COX-2 and $PGE_2$. Furthermore, TNF-α- or IL-1β-induced transcription of selected cytokines, chemokines, and adhesion molecules (IL-6, RANTES, MCP-1, and VCAM-1) was found to require SPHK1 activation. Suppression of SPHK1 activation led to reduction of cytokine-induced IκBα phosphorylation and consequently diminished NFκB activity due to reduced nuclear translocation of RelA (p65), explaining the dependence of inflammatory mediator production on SPHK1 activation. These data are complemented by a recent report on SPHK inhibitors as suppressors of NFκB and $PGE_2$ (Maines et al., 2007).

We then proceeded to renal fibroblasts isolated from SPHK knock-out and wild-type mice: SPHK1 deficient cells did not upregulate COX-2 nor produced $PGE_2$ in response to TNF-α and IL-1β, in line with the studies summarized above (A.Billich, unpublished data). However, the hope that this might translate into a protection of SPHK1 knock-out mice in COX-2 dependent models of disease was not fulfilled: the SPHK1-deficient animals reacted similar to wild-type controls in models of arthritis induced by antigen or serum transfer (J.Dawson, R.Feifel; unpublished data); also, Hla and co-workers (Michaud et al., 2006) reported lack of protection of SPHK1 knock-out animals in collagen-induced arthritis. On the other hand, we showed that the SPHK1 inhibitor SKI-II blocks $PGE_2$ production in human synoviocytes derived from patients with rheumatoid arthritis (A.Billich, unpublished data). Taken together, we assume that compensatory mechanisms in the SPHK1-deficient animals might lead to a circumvention of the need for SPHK1 in COX-2 activation observed in cultured cells. Therefore, experiments with pharmacological inhibitors of SPHK1 in models of arthritis are mandatory to define the role of this enzyme as potential target in this disease. Interestingly, orally bioavailable SPHK inhibitors were recently shown to be active in a mouse model of colitis, another inflammatory disease in which COX-2 plays a pathogenetic role (Maines et al., 2008).

### *19.2.5 SPHK1 and Neutrophil Functions*

Transient increase of intracellular SPHK activity and S1P levels implied a role of S1P within neutrophils (Alemany et al., 1999; Ibrahim et al., 2004; MacKinnon et al., 2002). A central role of SPHK in neutrophil activation was suggested based on experiments with *N,N*-dimethylsphingosine (DMS), which blocks both SPHK1 and 2, and in addition may have other targets (see 19.2.9). More recently, Melendez and co-workers demonstrated that neutrophil activation by anaphylatoxin C5a is dependent on SPHK1, based on experiments with specific anti-sense oligonucleotides (Ibrahim et al., 2004). Furthermore, they showed that pretreatment of mice with DMS prevented C5a-induced peritonitis (Vlasenko and Melendez, 2005). Finally, inhibition of SPHK1 by SKI-II was shown to prevent neutrophil activation in a model of hemorrhagic shock in rats (Lee et al., 2004). In sharp contrast, Hla and co-workers (Michaud et al., 2006) reported a normal response of the SPHK1-deficient mice in a model of thioglycollate-induced peritonitis, and normal superoxide generation by neutrophils in response to fMLP. To get more insight into the role of the type 1 and 2 isoforms of SPHK in neutrophil functions, we isolated neutrophils from the bone marrow of SPHK1 and 2 knock-out mice (Zemann et al., 2007); the cells showed normal increase of intracellular $Ca^{2+}$ when stimulated *in vitro* by fMLP, platelet-activating factor, C5a, or ATP, and normal migration towards fMLP and C5a. Also, recruitment of neutrophils into the peritoneum towards the chemokines KC and MIP-2 or to LPS, and into the peripheral blood after fMLP injection was similar in SPHK knockout strains and wild-type animals. An in vivo model of bacterial lung infection revealed an accelerated progression of disease in SPHK2 (but not SPHK1) knockout mice as compared to wild-type controls. However, effector functions of SPHK-deficient neutrophils, such as superoxide production, β-glucuronidase release and their capacity to kill bacteria were unchanged as compared to wild-type cells.

To conclude, the data derived from SPHK knockout mice do not support the hypothesis that any of the two lipid kinases plays a crucial role in signaling downstream of various neutrophil stimuli; SPHKs appear not to be essential for neutrophil recruitment and effector functions.

## 19.2.6 SPHK1 in Macrophages and Dendritic Cells

Involvement of SPHK1 in macrophage activation and functions was shown by Melendez and Ibrahim, 2004, using antisense oligonucleotides to SPHK1 in human monocyte-derived macrophages and C5a as a stimulus. Intracellular $Ca^{2+}$ mobilization was inhibited by 90% followed by partial inhibition of degranulation, chemotaxis, and inhibition of the cytokines TNF-α, IL-6, and IL-8; using human monocytes the same group showed that SPHK1 also plays an essential role in the TNF-α triggered intracellular $Ca^{2+}$ signals, degranulation, cytokine production, and activation of NFκB in this cell type (Zhi et al., 2006). In contrast, unpublished data from our group show that responses of peritoneal macrophages from SPHK1 knock-out mice to C5a and a number of other stimuli, including LPS, are indistinguishable from responses of wild-type cells; furthermore, the inhibitor SKI-II did not attenuate the macrophage responses in wild-type cells. Moreover, TNF-α injection into SPHK1 knock-out mice gave rise to a normal elevation of IL-6 in the blood. Thus, monocyte/macrophages are, in our view, not an established target for SPHK1 inhibition with the aim to inhibit cytokine secretion by these cells. Wu et al. (2004) reported that SPHK1 protects macrophages from LPS-induced apoptosis, in line with the general pro-survival functions of the enzyme, and suggested that it may be a target to prevent hyperimmune responses induced by gram-negative bacteria.

Interesting data on a role of SPHK1 in dendritic cells (DCs) come from the group of Y.M. Park (Jung et al., 2007b): SKI-II inhibits their migration towards CCL19 through the down-regulation of CCR7, possibly mediated through inhibition of p38 mitogen-activated protein kinase (MAPK). Furthermore, the SPHK1 inhibitor significantly downregulated co-stimulatory molecules in DCs, suppressed their IL-12 production and concomittantly IFN-γ production by T cells (Jung et al., 2007a). In addition, the compound blocked LPS-induced translocation of NFκB in dendritic cells, whereas it did not affect the degradation of IL-1 receptor-associated kinase-1 by LPS (Jung et al., 2007b). In parallel studies, Eigler and co-workers (Eigenbrod et al., 2006) showed that the SPHK1 inhibitor dihydrosphingosine (DHS) reduced migration of immature but not of mature DCs, indicating that the SPHK1/S1P may play a role in accumulation of peripheral immature DC at the site of inflammation and subsequent antigen-uptake. Since there are open questions regarding the specificity of both SKI-II and DHS (see below), studies using specific siRNA or SPHK1 knock-out mice will be important to further validate the role of the enzyme in dendritic cells.

## *19.2.7 SPHK in Mast Cells and in Airway Inflammation*

In 1996, Choi et al. (1996) showed that DHS inhibits $Ca^{2+}$ rise and S1P generation after FcϵRI stimulation in RBL-2H3 cells and hypothesized that S1P acts as an alternative second messenger to inositol-1,4,5-trisphosphate. Melendez and co-workers (Melendez and Khaw, 2002) applied antisense technology to human bone marrow derived mast cells (BMMCs) and identified a SPHK1-mediated fast and transient $Ca^{2+}$ release from intracellular stores that precedes a second, slower wave of $Ca^{2+}$ release from intracellular stores by phospholipase Cγ1; the first wave was shown to regulate degranulation in mast cells. Using exogenously supplied sphingosine and S1P, we could show that in mouse mast cell lines and mouse BMMCs a rheostat of these two sphingolipids determines the excitability: sphingosine is inhibiting and S1P is supporting an activation via the FcϵRI, with SPHK serving as a permissive switch for activation (Prieschl et al., 1999). We speculate that in the course of mast cell activation SPHK deactivates the inhibitory potential of sphingosine on external $Ca^{2+}$ influx and simultaneously generates S1P as a new second messenger to open internal $Ca^{2+}$ stores.

Importantly, Rivera and coworkers (Olivera et al., 2007) demonstrated in a model of passive systemic anaphylaxis that SPHK1 knock-out mice show about 35% reduced histamine release; SPHK2 deficient animals react normally. Anaphylaxis-reduction in SPHK1 knock-out mice correlated with reduced concentration of circulating S1P relative to the wild-type counterparts, while degranulation of mast cells derived from these mice was completely normal when tested *in vitro*. This suggested an extrinsic role of S1P, generated through SPHK1 in a cell type other than mast cells, in regulating allergic responsiveness *in vivo*. The normal reaction in $SPHK2^{-/-}$ mice seemed to be the sum of two opposite effects in these animals: (i) their mast cells produced little S1P and featured decreased calcium influx, activation of protein kinase C, cytokine production, and degranulation; but (ii) they featured high concentrations of circulating S1P (150% of wild-type). $SPHK1^{+/-}$ $SPHK2^{-/-}$ mice, unlike their $SPHK1^{+/+}$ $SPHK2^{-/-}$ counterparts, had normal amounts of circulating S1P; however, these mice were now resistant to anaphylaxis. This characterized a contribution of SPHK2 to allergic reactivity intrinsic to the mast cells. Our own unpublished data (using SPHK knock-out mice on the Balb/c background as opposed to the C57Bl/6 strain used by Rivera and co-workers) confirmed the partial suppression (35–50%) of response in the PCA model, both in SPHK1 and SPHK2 deficient animals, the only difference to the data by Rivera being that SPHK2 deficient animals themselves already showed the defective response. Taken together these data imply that a dual SPHK1/SPHK2 inhibitor may be suited to treat type I allergic diseases, as inhibition of one or the other isoform may not be sufficient to have a therapeutic effect.

However, a number of confounding findings remain: (i) The report of S. Spiegel and co-workers (Jolly et al., 2005) that over-expression of SPHK1 in RBL-2H3 mast cells (with consequently higher S1P levels) impairs degranulation. (ii) The existence of extracellular components in S1P-triggering of mast cell activation via the $S1P_1$ and $S1P_2$ receptors (Jolly et al., 2004). (iii) The controversial data from Melendez and co-workers (Melendez and Khaw, 2002) in human mast cells (high-lighting intrinsic SPHK1 to be important in mast cell activation) which stand in contrast to the data from Rivera and from our group in the mouse. More work in this area to consolidate the picture is certainly required.

Finally, in this context it is worth mentioning that S1P levels in bronchoalveolar lavage fluid collected from asthmatic subjects challenged with allergen were significantly increased, while S1P levels in control subjects were unaffected (Ammit et al., 2001). The increased levels in the asthmatic group correlated with the degree of inflammation reflected in eosinophil number and protein influx. Thus, also based on these correlative data SPHK inhibition might be viewed as a therapeutic option.

### *19.2.8 SPHK and T and B Cells*

In contrast to the vast amount of information on S1P receptor modulators in the field of B- and T-cells, only few studies on these cell types relate to SPHKs. It was found (Yang et al., 2005) that inhibition of SPHK1 in naïve human $CD4^+CD45RA^+/RO^-$ cells by siRNA increased IL-2, INF-$\gamma$ and TNF-$\alpha$ by 30–50% while overexpression of SPHK1 in mouse DO11.10 Th0/Th1 cells resulted in their downregulation. This characterizes SPHK1 as a negative regulator of $CD4^+$ Th1 cells; thus SPHK1 inhibitors could be utilized to attenuate Th2-type reactions, as they occur in allergy and asthma.

Lymphoblastoid cells lines (LCLs) generated from patients with rheumatoid arthritis display a high constitutive SPHK activity, increased SPHK1 mRNA expression levels and increased levels of S1P; also SPHK1 mRNA levels in synovial tissue from arthritis patients are increased (Pi et al., 2006; Tan et al., 2007). The enhanced S1P levels are hypothesized to be the cause of the impairment of Fas-mediated apoptosis with consequent resistance to cell death of disease-specific B-cells observed in rheumatoid arthritis.

### *19.2.9 Inhibitors of SPHKs*

A number of experimental SPHK inhibitors have been used in the literature (see Fig. 19.1 for selected structures), but the specificity of these compounds has not

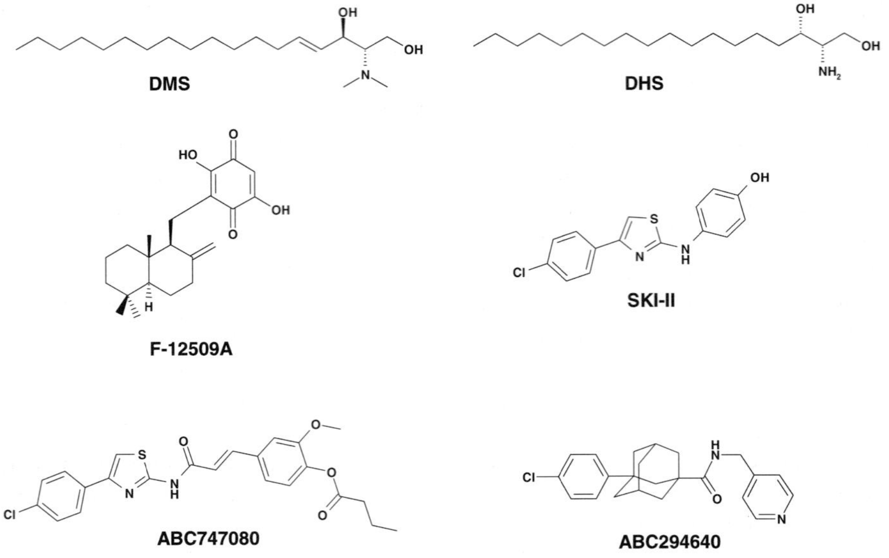

**Fig. 19.1** Chemical structures of selected SPHK1 inhibitors

been thoroughly explored in most cases. DMS, the most widely used compound, acts as competitive inhibitor for SPHK1 and as uncompetitive inhibitor for SPHK2, with $K_i$ values in the low micromolar range for both enzymes (Liu et al., 2000). Curiously, a recent report showed inhibition of SPHK1 from rat heart by DMS, while SPHK2 from the same source was activated (Vessey et al., 2007). DMS was originally described as inhibitor of protein kinase C (Igarashi et al., 1989), but the relevance of this finding has been debated (Edsall et al., 1998). Importantly, protein kinases that are activated by sphingosine and DMS have been described (Megidish et al., 1998). Therefore, at least some of the activities ascribed to SPHK inhibition by DMS in cellular assays may be "off-target" effects. DHS (safingol) (Buehrer and Bell, 1992) is also of limited use, as it blocks SPHK1 at low micromolar levels but is a substrate for SPHK2, thus being converted to DHS-1-phosphate which has biological activity on its own. Some authors used the sesquiterpene F-12509A, a competitive inhibitor of both SPHK1 ($K_i = 4\ \mu M$ ) and SPHK2 ($K_i = 5.5\ \mu M$) (Kono et al., 2002), with unexplored specificity beyond these lipid kinases.

In recent literature, compound SKI-II (Fig. 19.1), identified by high-throughput screening (French et al., 2003) as SPHK inhibitor ($IC_{50} = 0.5\ \mu M$) has been frequently used. In our hands, SKI-II is a selective SPHK1 inhibitor, with no measurable inhibition of SPHK2 and a series of 19 protein kinases (Jung et al., 2007a). The compound acts as non-competitive inhibitor both with respect to ATP (French et al., 2003) and to sphingosine (A.Billich, unpublished data). SKI-II has poor oral bioavailability and short plasma half-life in rodents (our unpublished data), yet it showed anti-tumor efficacy in mice

(Maines et al., 2006). Furthermore, compounds ABC294640 and ABC747080 (Maines et al. 2006; Maines et al. 2007) are reported to be orally active in blocking SPHK in rodents; $IC_{50}$ values towards SPHKs have not been published so far, but blockade of S1P levels in cells needs concentrations in the range of 1–100 μM (Maines et al., 2007). In general, it appears that the pharmacological effects of SKI-II, ABC294640, and ABC747080 such as anti-angiogenic activity (Maines et al., 2006), inhibition of COX-2 and NFκB (Maines et al., 2007), and inhibition of tumor growth (French et al., 2006) are congruent with those generated by siRNA or genetic knock-down of SPHK1. However, in view of the relatively modest potency of the compounds towards SPHK activity, in particular in cellular assays, one may still suspect that these compounds could have off-target effects which bring about the activity; therefore, much closer scrutiny of their specificity is needed before drawing any conclusions.

Besides blocking the enzyme activity of SPHK1 with compounds that bind directly to the protein, there might be other possibilities to interfere with SPHK1 signaling. SPHK1 activation was shown to be a consequence of extracellular signal-regulated kinase 1/2-mediated phosphorylation at Ser225 which increases catalytic activity and is responsible for the agonist-induced translocation of SPHK1 to the plasma membrane. Importantly, a phosphorylation-deficient SPHK1 mutant, while still enzymatically actives, looses its oncogenic potential (Pitson et al., 2005). Interestingly, two anti-angiogenic compounds, phenoxidiol and glabridin (Gamble et al., 2006; Kang, et al., 2006) have been reported to block the activation of SPHK1 after TNF-α stimulation of endothelial cells.

Another currently hypothetical way to interfere with SPHK signaling is to block its interaction with proteins to which the enzyme binds within cells; numerous binding partners have been identified, but in most cases the functional relevance of these interactions is not defined. As an exception, the interaction of SPHK1 with TRAF2 could indeed be of functional relevance in the well-established role of the enzyme down-stream of the TNF-receptor (Xia et al., 2002); thus, interference with this protein-protein interaction by low-molecular weight compounds may be a way to abrogate this signalling pathway.

### *19.2.10 Perspective*

Today, there is substantial evidence that SPHK1 qualifies as a pharmacological target, in particular in cancer and in angiogenesis. In other areas, such as inflammation, conflicting data make it too early to finally assess if SPHK-directed treatment would be a viable option. Involvement of SPHK1 in additional signaling pathways continues to be identified, recent examples being the involvement of SPHK1 activation by TGF-β1 in the lung fibrogenic process (Kono et al., 2007), by TNF and IL-1β in the pathological response of pancreatic islet cells to cytokines (Mastrandrea et al., 2005), and by extracellular nucleotides in renal disease (Klawitter et al., 2007). On the other hand, involvement of SPHKs in pathways that one does not want to inhibit also becomes known, such as the role of SPHK1 as a survival factor in cardiomyocytes (Kacimi et al., 2007) and in insulin signaling (Ma et al., 2007), and the function of SPHK2 as a suppressor of IL-2 expression (Samy et al., 2007). As far as we see, SPHK1 inhibitors which qualify as drug candidates have not yet been identified; however, the picture on the validity of SPHK1 as a pharmacological target can be expected to become clearer once nanomolar inhibitors with defined specificity become available.

In contrast to SPHK1 inhibitors, a utility of selective SPHK2 inhibitors has not been defined. As SPHK2 acts in concert with S1P phosphatase type 1 (SPP1) to regulate recycling of sphingosine into ceramide (Le Stunff et al., 2007), we hypothesize that inhibitors of SPHK2 (and of SPP1) could be used in diseases featuring excessive apoptosis, such as in liver diseases and in neurodegeneration (see also section on acid sphingomyelinase). Finally, as outlined above, dual SPHK1/2 inhibitors may be of use in treating allergy and asthma.

## 19.3 Sphingosine-1-Phosphate Lyase as Novel Target in Immunmodulation

Sphingosine-1-phosphate lyase (SPL) is the enzyme responsible for the irreversible cleavage of S1P in a retroaldol reaction to yield hexadecenal and phosphoethanolamine as degradation products. The enzyme, therefore, is in a critical position to regulate levels of S1P (Bandhuvula and Saba, 2007). Here, we will focus on recent evidence that SPL may serve as a target for modulation of immune responses in transplantation settings and in the treatment of autoimmune disease.

The egress of activated lymphocytes from secondary lymphoid organs is dependent on expression of S1P receptor type 1 ($S1P_1$) on the cell surface. Downregulation of this receptor by FTY720 leads to retention of the lymphocytes which thus cannot reach sites of inflammation, an effect that is desired in the treatment of autoimmune diseases. Long before FTY720 was discovered, two structurally distinct compounds – the vitamin B6 antagonist 4′-deoxypyridoxine (DOP) and the food colorant 2-acetyl-4-tetrahydroxybutylimidazole (THI) (see Fig. 19.2) – were known to cause a reduction of lymphocyte numbers in the circulation when given to rodents. In 2005, the group of J. Cyster found that DOP and THI inhibit SPL thereby causing a steep increase of S1P concentrations in tissues, including the lymphoid organs (Schwab et al., 2005). This leads to downregulation of $S1P_1$ on the lymphocytes in the secondary lymphoid organs which thus do not migrate out into the circulation. It appears that SPL establishes an S1P gradient between tissues (low S1P concentrations) and circulation (high S1P, probably due to absence of SPL from erythrocytes (Ito et al., 2007)); this gradient is disrupted by inhibition of SPL.

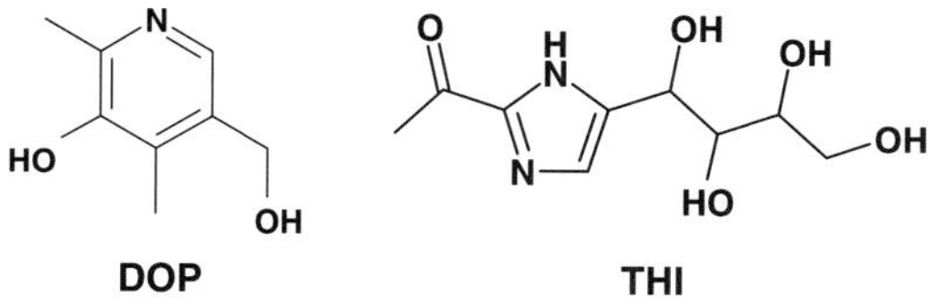

DOP THI

**Fig. 19.2** Chemical structure of the SPL inhibitors 4′-deoxypyridoxine (DOP) and 2-acetyl-4-tetrahydroxybutylimidazole (THI)

Based on these observations, one may speculate that SPL inhibitors could be useful in inflammatory or autoimmune diseases by preventing lymphocyte recruitment to diseased tissues. However, the known inhibitors, DOP and THI, do not qualify as lead compounds for drug design: DOP is acting by replacing the co-factor pyridoxal phosphate on many enzymes, not only SPL, which leads to long-term toxicity; THI does not block SPL activity in cellular assays, but only exhibits its effect *in vivo*, pointing to a molecular target indirectly affecting SPL (A.Billich, unpublished observations). Therefore, active-site directed inhibitors of SPL need to be identified to explore their utility as potential drugs.

The group of J.Saba reported that SPL is inhibited by FTY720 and speculated that at least part of the action of the drug might be due to inhibition of that enzyme (Bandhuvula et al., 2005, 2007). However, FTY720 inhibits SPL only partially at high concentrations (20–70% at 30 μM (Bandhuvula et al., 2007)) which are $10^4$-fold above the $EC_{50}$ for FTY720 phosphate binding to $S1P_1$. In fact, in tissues (including brain) of normal rats treated with doses of FTY720 that are protective in experimental autoimmune encephalomyelitis, we observed no reduction of the baseline SPL activity; also S1P concentrations in the tissues were unchanged (A.Billich, unpublished observations). Therefore, the action of FTY720 as an enzyme inhibitor of SPL does not appear to be of pharmacological significance.

Since S1P as the substrate of SPL plays an important role in cell differentiation and survival, it is not surprising that silencing of SPL has a variety of consequences (see Bandhuvula and Saba, 2007, for review). These include reduced apoptotic responses to serum deprivation and chemotherapeutic drugs (Bandhuvula and Saba, 2007; Oskouian et al., 2006) and increased responsiveness of COX-2 transcription to TNF-α stimulation (Pettus et al., 2003). SPL expression was seen to be downregulated in colon carcinoma (Oskouian et al., 2006), but upregulated in ovarian cancer (Hibbs et al., 2004) and atopic dermatitis (Seo et al., 2006). Most importantly, SPL knockout mice feature postnatal mortality due to multiple congenital abnormalities, in particular of the vasculature (Schmahl et al., 2007). Whether these findings would limit the usefulness of S1P lyase inhibitors as immunomodulatory drugs remains to be explored.

## 19.4 Acid Sphingomyelinase as Novel Drug Target

Acid sphingomyelinase (ASMase) is one of the enzymes producing ceramide from sphingomyelin. The enzyme plays a fundamental "house-keeping" function in the turnover of sphingomyelin: in patients with genetic deficiency of active ASMase, sphingomyelin accumulates in the lysosomes giving rise to Niemann-Pick disease (NPD) types A and B (Schuchman, 2007). Interestingly, however, there are at least four lines of evidence suggesting that ASMase could be regarded as a therapeutic target, namely in prevention of apoptosis and of certain infections, in endotoxemia, and in atherosclerosis. These topics will be discussed in the subsequent paragraphs, followed by an assessment of currently available inhibitors.

### *19.4.1 Role of ASMase in Apoptosis*

Today it is widely accepted that ceramide is an almost universal mediator of programmed cell death (see Thevissen et al., 2006, for a recent review). Its formation is triggered by a multitude of pro-apoptotic stimuli, and it activates both the extrinsic (receptor-mediated) and the intrinsic (mitochondrial) pathways of apoptosis via interaction with multiple downstream targets. Proapoptotic ceramide is formed in stimulated cells either by *de novo* synthesis (either from palmitoyl-CoA and serine via dihydrosphingosine, or via the sphingolipid salvage pathway from sphingosine), or by the action of neutral sphingomyelinase(s) (NSMase) or of ASMase on preexisting pools of sphingomyelin. The relative contribution of the different modes of ceramide formation to apoptosis appears to vary between stimuli and cell types; cases have been described where solely ASMase-derived ceramide (Kashkar et al., 2005; Kolesnick and Fuks, 2003) or *de novo* ceramide synthesis (Seumois et al., 2007) play a role, where ASMase plus the *de novo* pathway are important (Petrache et al., 2005), or where ASMase and NSMase are both stimulated (Malaplate-Armand et al., 2006)

Some years ago, the involvement of ASMase in apoptosis mediated through the Fas/CD95-receptor had been discussed controversially in the literature (Bezombes et al., 2001; Cock et al., 1998; De Maria et al., 1998; Kirschnek et al., 2000). However, the work of R.Kolesnick and co-workers (Lin et al., 2000) clarified by *in vivo* experiments on ASMase-deficient mice that ASMase-derived ceramide mediates Fas-mediated apoptosis in some cell types (in particular hepatocytes), but not in others (e.g., thymocytes, T-lymphocytes). This explains why ASMase deficient mice are less susceptible to lethality and liver apoptosis induced by the anti-Fas antibody Jo-2, but do not develop lymphoproliferative disease like lpr mice that lack functional CD95. We note, however, that the reason for the observed cell type specificity is not known, and that

CD95-induced apoptosis in the human T-cell line Jurkat can be readily inhibited by blocking aSMase (A.Billich, unpublished data).

The functional link between Fas-induced apoptosis, ceramide and ASMase was proven by showing that supplementation with natural ceramide can overcome the resistance of ASMase-deficient hepatocytes to anti-Fas (Paris et al., 2001). Subsequent studies by the groups of E. Gulbins and R. Kolesnick showed that Fas-mediated apoptosis requires ceramide-mediated receptor clustering that depends on ASMase and caspase-8 (Cremesti et al., 2001; Grassmé et al., 2001, 2003b; Rotolo et al., 2005); very recently, Häussinger and coworkers showed that in rat hepatocytes CD95L induced endosomal acidification, ceramide-formation and downstream events, such as p47$^{phox}$-phosphorylation, ROS-formation, CD95-activation and apoptosis all of these responses were abolished after knock-down of ASMase (Reinehr et al., 2007). While earlier studied had relied on the use of the anti-Fas antibody Jo-2, only, more recent work showed that the Fas-mediated hepatocyte apoptosis induced by either bile acids (Gupta et al., 2004; Becker et al., 2007) or ethanol (Deaciuc et al., 2000) also requires activation of ASMase.

The role of ASMase in TNF-α-induced hepatocyte apoptosis has been studied in some detail by J. Fernández-Checa and coworkers. They showed that administration of TNF-α or LPS to galactosamine-pretreated ASMase$^{-/-}$ mice led to minimal hepatocellular injury, while severe damage was induced in wild-type animals (Garcia-Ruiz et al., 2003; Mari et al., 2004). In a murine model of warm hepatic ischemia/reperfusion injury administration of the ASMase inhibitor, imipramine, or ASMase knockdown by siRNA decreased ceramide generation, and attenuated serum levels of alanine aminotransferase, hepatocellular necrosis, cytochrome c release, and caspase-3 activation; survival of animals was enhanced when dosed with the ASMase inhibitor (Llacuna et al., 2006). As to the mechanism of ASMase involvement in TNF-induced hepatocyte apopotosis, the authors propose two routes: ASMase-generated ceramide is coverted into gangliosides (in particular GD3) which targets the mitochondria to induce apoptosis; in parallel, ceramide-induced downregulation of the liver specific methionine adenosyl transferase 1A expression leads to glutathione depletion in mitochondria, thus inducing apoptosis (Mari et al., 2004). Notably, in a study on mouse hepatocytes *in vitro*, D.A. Brenner and Y.Hannun confirmed that activation of ASMase and generation of ceramide (specifically of chain length C16) contributes to TNF-α-induced hepatocyte apoptosis (Osawa et al., 2005), but they did not see an involvement of downstream GD3 formation in their model. Another observation to be mentioned in this context is that TNF-α-induced apoptosis in the human myelogenous leukemia cells was blocked by the ASMase inhibitor SR33557 (Higuchi et al., 1996).

In addition to FasL and TNF-α, another TNF-family member, namely TRAIL (= TNF-related apoptosis-inducing ligand) has recently been shown to also depend on ASMase activity in its capability to induce apoptosis. E.Gulbins and co-workers showed that TRAIL activates ASMase in splenocytes via a redox mechanism resulting in release of ceramide and formation of ceramide-enriched membrane platform (Dumitru and Gulbins, 2006). Our group recently showed that TRAIL-induced apoptosis in hepatocytes also depends on ASMase (A.Billich, unpublished data).

Recently, a functional role of ASMase and ceramide in copper-induced apoptosis of hepatocytes and erythrocytes was demonstrated by F. Lang, E. Gulbins, and co-workers (Lang et al., 2007). This is of functional significance since in patients with Wilson's disease, featuring a defect in a $Cu^{2+}$ secretase (ATP7B) and therefore have elevated serum and tissue levels of that cation, apoptosis of these cell types is of importance in the pathology. The authors show that genetic deficiency or pharmacological inhibition of ASMase prevented $Cu^{2+}$-induced hepatocyte apoptosis and protected rats, genetically prone to develop Wilson disease, from acute liver failure and death. While the authors discuss a possible direct activation of ASMase by $Cu^{2+}$- induced reactive oxygen species (ROS), others had previously suggested dependence of $Cu^{2+}$-induced hepatocyte apoptosis on CD95 (Strand et al., 1998). As CD95-induced apoptosis depends on ASMase (see above), it appears that ASMase is involved in any case, regardless of the precise mechanism.

To summarize, based on the available data, pharmacological inhibition of ASMase might be an opportunity to prevent apoptosis, in particular in the liver, without influencing T-cell apoptosis. In fact, hepatocyte apoptosis is an important element in a number of liver diseases, ranging from cholestatic and alcoholic liver disease to autoimmune and viral hepatitis (Eichhorst, 2005). Furthermore, a role of ASMase-produced ceramide in neuronal and myocardial apoptosis has been implicated from studies in models of rat cerebral and rabbit heart ischemia, respectively (Yu et al., 2000; Argaud et al., 2004); the criticism to the latter two studies is, however, that their conclusion rely on the use of the xanthogenate D609, which inhibits ASMase in an indirect way but certainly is not specific for that enzyme.

Two studies have implicated a role of ASMase in pathological conditions that involve lung cell apoptosis: (i) S. Uhlig and collaborators (Goggel et al., 2004) showed that lung emphysema induced by platelet-activating factor, as clinically relevant in severe lung injury, depends on ceramide and ASMase activation; these authors used both ASMase-deficient cells and D609 as pharmacological tool to arrive at their conclusion. (ii) I. Petrache and co-workers (Petrache et al., 2005) demonstrated that alveolar cell apoptosis is involved in the pathogenesis of emphysema (e.g., in chronic obstructive pulmonary disease induced by cigarette smoking) and that *de novo* ceramide synthesis as well as a feed-forward mechanism mediated by activation of ASMase is crucial in induction of apoptosis. Finally, apoptosis in airway epithelial cells *in vitro* was shown to depend on sphingomyelinase activity: using specific siRNAs, T.Goldkorn and coworkers (Castillo et al., 2007) showed that silencing neutral sphingomyelinase type 2 prevented $H_2O_2$-induced apoptosis, while silencing of ASMase impaired peroxynitrite-induced apoptosis. Interestingly, also treatment of retinal photoreceptor cells with a nitric oxide donor activated ASMase, produced ceramide and induced apoptosis, leading the authors of the study to propose ASMase as a potential therapeutic target for the treatment of retinal pathologies (Sanvicens and Cotter, 2006).

### *19.4.2 ASMase in Endotoxemia and Inflammation*

ASMase-deficient mice are protected in models of lethal endotoxic shock induced by LPS alone (Haimovitz-Friedman et al., 1997) or LPS after galactosamine pre-treatment (Garcia-Ruiz et al., 2003). Also, treatment of mice with the ASMase inhibitor NB6 enhances survival in endotoxemia (Claus et al., 2005). These findings may be readily explained by the role of ASMase in apoptotic cell death, in particular of the endothelium and of hepatocytes.

Injection of LPS or of the cytokines TNF-α or IL-1β into mice increases the activity of $Zn^{2+}$-dependent ASMase in the serum (Wong et al., 2000). Interestingly, serum ASMase activity was seen to be elevated in patients under septic shock, and was found to be negatively correlated with survival (Claus et al., 2005). It has to be noted that serum ASMase, which differs from the lysosomal form in its posttranslational processing (Ferlinz et al., 1997; Hurwitz et al., 1994), shows only marginal catalytic activity in its natural environment (i) the near neutral pH of the serum is suboptimal for catalytic activity; (ii) serum ASMase is $Zn^{2+}$-dependent, its activity being revealed only upon supplementation with that cation. In contrast, the lysosomal form – being exposed to saturating concentrations of $Zn^{2+}$ in the lysosomes – is constitutively active and operates at optimal pH (Schissel et al., 1998a). Therefore, the relevance of the increased secreted ASMase in endotoxemia is unclear; it might be merely a consequence of apoptotic cell death in the liver, which also gives rises to increased serum levels of other enzymes, such as the transaminases. Interestingly, secreted ASMase was reported to be also elevated in chronic heart failure (Doehner et al., 2007); if this is a trivial consequence of apoptosis in myocardial cells or if any causal relation exists, remains to be dissected.

While ASMase-deficient mice survive lethal doses of LPS (see above), the serum levels of TNF-α-induced by sublethal doses of LPS are similar to those in WT mice (Haimovitz-Friedman et al., 1997); isolated ASMase-deficient macrophages respond to LPS with a normal pattern of cytokine secretion (Manthey and Schuchman, 1998). Thus, there is no evidence that ASMase is important for LPS-induced inflammatory responses. A recent report showed that imipramine, an inhibitor of cellular ASMase activity, blocked LPS-induced TNF-α secretion from differentiated human monocyte-like THP-1 cells (Cuschieri et al., 2007). However, in our hands imipramine has no effect on cytokine secretion from these cells or from monocytes isolated from human blood (A.Billich, unpublished observations).

Recently, it was published (Sakata et al., 2007a,b) that a sphingomyelinase inhibitor has a beneficial effect in a murine model of DSS-induced colitis, an experimental disease where TNF-α plays a major role. However, the compound used appears to be a weak and unspecific inhibitor of both ASMase and NSMase – any conclusion about a role for ASMase in colitis based on these data is not possible in our opinion.

Taken together, ASMase appears to be important in endotoxemia based on its involvement in ceramide-induced apoptosis; in contrast, any solid evidence for a role of the enzyme in inflammatory responses is not available.

### *19.4.3 ASMase as a Target in Atherosclerosis*

I. Tabas and co-workers have proposed a role of secretory ASMase in atherosclerosis (Tabas, 1999; Tabas et al., 2007) They reported that the enzyme can be secreted from cultured vascular endothelial cells in a form that is only partially dependent on $Zn^{2+}$ (Marathe et al., 1998) and can cleave sphingomyelin on the surface of atherogenic lipoproteins even at pH 7.5, leading to fusion and aggregation of the lipoprotein particles (Schissel et al., 1998b). It is not known how this form of the enzyme may differ on the molecular level from the strictly Zn-dependent ASMase circulating in human blood mentioned above.

Aggregation and subsequent fusion of lipoproteins after they enter the subendothelium can increase the size of the particles to the point where exit from the arterial wall is prohibited; actually, sphingomyelinase-to-LDL molar ratio determines LDL aggregation size (Oorni et al., 2000), and the hydrolysis of sphingomyelin associated with low-density lipoproteins (LDL) increases LDL affinity for arterial wall proteoglycans (Guarino et al., 2006). Aggregated forms of LDL, including those induced by secretory ASMase, are taken up by macrophages and are potent inducers of macrophage foam cell formation (Marathe et al., 1999, 2000). Secretory ASMase is found in atheromata and lipoproteins extracted from animal and human atherogenic lesions have increased ceramide (Jiang et al., 2000). Importantly, in a recent review article (Tabas et al., 2007) mentioned unpublished data showing that $ApoE^{-/-}$ mice lacking the secreted form of ASMase have decreased development of early atherosclerotic lesions and decreased retention of atherogenic lipoproteins compared with $ApoE^{-/-}$ mice matched for similar plasma lipoprotein levels. Finally, studies have shown an association between high sphingomyelin content in circulating lipoproteins and an increased risk for aortic atherosclerosis in mice and coronary artery disease in humans (Jiang et al., 2000); also, an association of ceramides in human plasma with risk factors of atherosclerosis was reported (Ichi et al., 2006).

Importantly, Marcil and co-workers (Lee et al., 2006) recently reported that increased sphingomyelin content impairs high-density lipoprotein (HDL) biogenesis and maturation in human Niemann-Pick disease type B patients which feature highly reduced ASMase activity. Since HDL is atheroprotective, it remains to be seen whether pharmacological intervention with sphingomyelin hydrolysis by secreted ASMase in order to prevent LDL aggregation would have a beneficial effect in prevention or treatment of atherosclerosis.

### *19.4.4 ASMase and Infection*

Ceramide plays an important role in the infection of mammalian cells with at least some pathogens. It was demonstrated by E. Gulbins and co-workers that internalization of *N. gonorrhoeae*, *P. aeruginosa*, and rhinoviruses is mediated by ASMase (Grassmé et al., 1997, 2003a, 2005; Hauck et al., 2000) other researchers showed that cell entry by Sindbis virus (Jan et al., 2000) and cellular invasion by the parasite *Cryptosporidium parvum* depends on ASMase (Nelson et al., 2006). The pathogens rapidly activate the enzyme, induce a rapid surface translocation of the acid sphingomyelinase onto the cell surface and trigger the release of ceramide and, thus, the formation of ceramide-enriched membrane platforms, which seem to be central for the uptake of the pathogens.

In the case of rhinovirus, it was convincingly demonstrated that genetic deficiency or pharmacological inhibition of the acid sphingomyelinase prevented infection of human epithelial cells (Grassmé et al., 2005); if this *in vitro* finding can be translated into a useful effect in human rhinovirus infection remains to be explored. Surprisingly, in the case of Sindbis virus, an increase, rather than the expected decrease, in susceptibility of ASMase deficient mice to infection was seen as the result of more-rapid replication and spread in the nervous system and increased neuronal death (Ng and Griffin, 2006). Nevertheless, further exploration of the role of ASMase in viral infection would be of interest.

### *19.4.5 Inhibitors of ASMase*

Several physiological inhibitors of ASMase have been described. Kölzer et al. reported L-α-phosphatidyl-D-myoinositol-3,5-bisphosphate as a specific ASMase inhibitor ($K_i = 0.53\ \mu M.$) (Kölzer et al., 2003). Testai et al. showed that the corresponding 3,4,5-triphosphate is a non-competitive inhibitor of ASMase from human oligodendroglioma cells (Testai et al., 2004), and speculate that this inhibition is a way to regulate ASMase activity within rafts involved in receptor clustering and capping. Furthermore, the sphingolipid phosphates S1P and C1P both inhibit ASMase (Gómez-Muñoz et al., 2003; 2004); for C1P inhibition of ASMase activity in cell lysates was shown, while S1P inhibited the enzyme only in intact cells, suggesting an indirect mode of inhibition. As hypothesized by Gómez-Muñoz and co-workers, inhibition of ASMase by the lipid phosphates may have a regulatory function in macrophage apoptosis (Gómez-Muñoz et al., 2003, 2004).

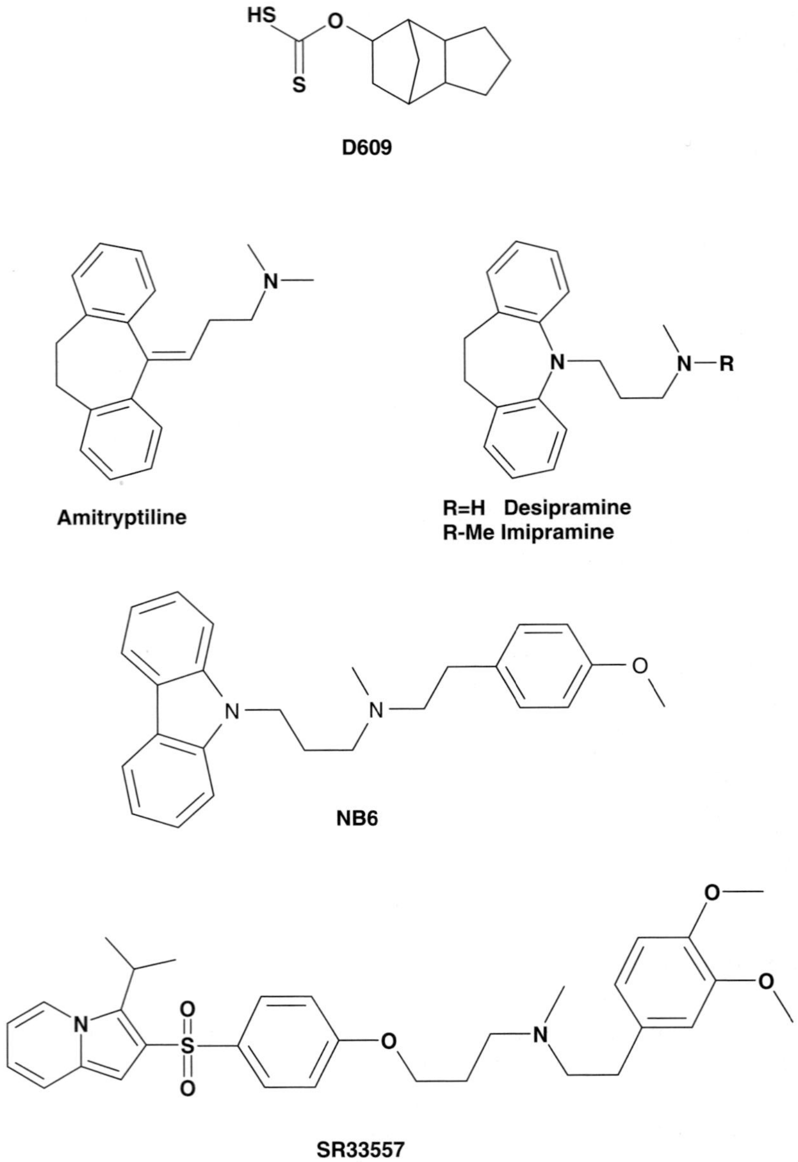

D609

Amitryptiline

R=H Desipramine
R-Me Imipramine

NB6

SR33557

**Fig. 19.3** Chemical structure of selected ASMase inhibitors

Currently known non-natural ASMase inhibitors (see Fig. 19.3 for selected structures) may be classified as follows: (i) compounds that block AMSase activity by direct interaction with the enzyme; (ii) compounds that block ASMase activity in cells in an indirect way. The first class is the most interesting from the pharmaceutical point of view, but only few examples of such ASMase inhibitors can be found in the literature. Compounds that block ASMase as apparent substrate analogs include compound SMA-7, the weak and unspecific inhibitor sphingomyelinase inhibitor mentioned above [45,46]; a thiourea derivative of sphingomyelin (AD2765) that also blocks sphingomyelin synthesis (Darroch et al., 2005) Finally, carnitine was described as non-competitive inhibitor of ASMase (Andrieu-Abadie et al., 1999), and α-mangostin and derivatives as sphingomyelinase inhibitors with a preference for ASMase (Okudaira et al., 2000; Hamada et al., 2003).

The second class of compounds, inhibitors of cellular ASMase activity, contains compounds such as curcumin (Cheng et al., 2007) and D609 which block the enzyme in an indirect way. The xanthogenate D609 has been used by some authors as "acid sphingomyelinase inhibitor" but blocks also phosphatidylcholine-specific phospholipase C/sphingomyelin synthase (Luberto and Hannun, 1998). More importantly, a group of cationic amphiphilic drugs has been shown to efficiently inhibit ASMase activity in cells and in tissues *in vivo*. The compounds include the tricyclic antidepressants imipramine, desipramine, and amitryptiline; NB6, a structurally related compound (Claus et al., 2005); and the calcium channel blocker SR33557. More recently, it was shown that a wide variety of drugs classifed as weak organic bases (e.g., doxepine, fluoxetine, maprotiline, paroxetine, sertraline, suloctidil and terfenadine) act as functional ASMase inhibitors (Kornhuber et al., 2007). The mode of action of such lysosomotrophic compounds has been elucidated by K. Sandhoff and co-workers for the case of desipramine (Kölzer et al., 2004): the compound, by virtue of its cationic moiety, displaces ASMase from the lysosomal membrane and thus induces proteolytic digestion of ASMase in the lumen of the lysosomes. Given the mode of action, the question about specificity of the cationic amphiphilic ASMase inhibitors arises. For SR33557 it was shown that ASMase is selectively inhibited (β-hexosaminidase, α- and β-galactosidases, β-glucocerebrosidase, and arylsulfatase A were not blocked) (Jaffrézou et al., 1991). However, J.Norris and collaborators showed that also acid ceramidase activity in cells vanishes following treatment with desipramine (Elojeimy et al., 2006); thus, the compound depletes lysosomes of both a ceramide generating and consuming enzyme (and few authors have studies the net effect on cellular ceramide levels when using such inhibitors). Therefore, the many studies in the literature using this class of ASMase inhibitors have to be taken with caution; firm conclusions can only be reached when effects seen with these compounds are backed up by experiments with genetically ASMase deficient cells or cells depleted for ASMase by using siRNA.

To conclude, ASMase appears as an interesting potential drug target, in particular in prevention of apoptosis. It can be expected that more selective and more potent inhibitors with appropriate pharmacokinetic properties would be excellent tools to explore the scope of pharmacological effects caused by ASMase inhibition. At least one pharmaceutical company, Berlex, performed high-throughput screening for ASMase inhibitors – but was unsuccessful in finding tractable lead structures (Mintzer et al., 2005). Rational design of inhibitors is difficult today, since the crystal structure of the enzyme is not available, but a similarity model of the protein based on the structure of a distantly related phosphatase may offer a starting point (Seto et al., 2004).

## 19.5 Perspective

Sphingolipid metabolizing enzymes have only recently been proposed as novel drug targets; however, as this review focussed on three members of this class of enzymes shows, they hold considerable promise to be useful in the treatment of a variety of diseases. Other enzymes of pharmacological interest not covered here include the neutral and alkaline sphingomyelinases, acid ceramidase, and glucosyl ceramide synthase (Nussbaumer, 2007; Delgado et al., 2006; Morales and Fernandez-Checa, 2007); new additions to this list are still being made, such as the S1P phosphatase type 2 which is upregulated in psoriatic lesions and is essential for IL-1β production in endothelial cells (Mechtcheriakova et al., 2007). Validation of all of these enzymes as drug targets is certainly incomplete – genuine validation is only reached when an inhibitors shows efficacy in patients. One example where, according to our data, even preclinical validation failed is ceramide kinase, as mice deficient in this enzyme do not feature the expected phenotype in immunological disorders apart from an impaired response to bacterial infection (Graf et al., 2008). However, it is important to recognize that study of sphingolipid metabolism continues to reveal novel signaling pathways, which – beyond the focus on individual enzymes – may yield possibilities for pharmacological interference.

## References

Akao, Y., Banno, Y., Nakagawa, Y., Hasegawa, N., Kim, T.J., Murate, T., Igarashi, Y., and Nozawa, Y., High expression of sphingosine kinase 1 and S1P receptors in chemotherapy-resistant prostate cancer PC3 cells and their camptothecin-induced up-regulation, Biochem Biophys Res Commun, 342 (2006) 1284–1290.

Alemany, R., Meyer zu Heringdorf, D., van Koppen, C.J., and Jakobs K.H., Formyl peptide receptor signaling in HL-60 cells through sphingosine kinase, J Biol Chem, 274 (1999) 3994–3999.

Allende, M.L., Sasaki, T., Kawai, H., Olivera, A., Mi, Y., van Echten-Deckert, G., Hajdu, R., Rosenbach, M., Keohane, C.A., Mandala, S., Spiegel, S., and Proia, R.L., Mice deficient in sphingosine kinase 1 are rendered lymphopenic by FTY720, J Biol Chem, 279 (2004) 52487–52492.

Ammit, A.J., Hastie, A.T., Edsall, L.C., Hoffman, R.K., Amrani, Y., Krymskaya, V.P., Kane, S.A., Peters, S.P., Penn, R.B., Spiegel, S., and Panettieri, RA Jr., Sphingosine 1-phosphate modulates human airway smooth muscle cell functions that promote inflammation and airway remodeling in asthma, FASEB J, 15 (2001) 1212–1214.

Ancellin, N., Colmont, C., Su, J., Li, Q., Mittereder, N., Chae, S.S., Stefansson, S., Liau, G., and Hla, T., Extracellular export of sphingosine kinase-1 enzyme. Sphingosine 1-phosphate generation and the induction of angiogenic vascular maturation, J Biol Chem, 277 (2002) 6667–6675.

Andrieu-Abadie, N., Jaffrezou, J.P., Hatem, S., Laurent, G., Levade, and T., Mercadier, J.J., L-carnitine prevents doxorubicin-induced apoptosis of cardiac myocytes: role of inhibition of ceramide generation, FASEB J, 13 (1999) 1501–1510.

Argaud, L., Prigent, A.F., Chalabreysse, L., Loufouat, J., Lagarde, M., and Ovize, M., Ceramide in the antiapoptotic effect of ischemic preconditioning, Am J Physiol Heart Circ Physiol, 286 (2004) H246–251.

Bandhuvula, P., Tam, Y.Y., Oskouian, B., and Saba, J.D., The immune modulator FTY720 inhibits sphingosine-1-phosphate lyase activity, J Biol Chem, 280 (2005) 33697–36700.

Bandhuvula, P., Fyrst, H., and Saba, J.D., A rapid fluorescent assay for sphingosine-1-phosphate lyase enzyme activity, J Lipid Res, 48 (2007) 2769–2778.

Bandhuvula, P., and Saba, J.D., Sphingosine-1-phosphate lyase in immunity and cancer: silencing the siren, Trends Mol Med, 13 (2007) 210–217.

Banno, Y., Kato, M., Hara, A., and Nozawa, Y., Evidence for the presence of multiple forms of Sph kinase in human platelets, Biochem J, 335 (1998) 301–304.

Baran, Y., Salas, A., Senkal, CE., Gunduz, U., Bielawski, J., Obeid, L.M., and Ogretmen, B. Alterations of ceramide/sphingosine 1-phosphate rheostat involved in the regulation of resistance to imatinib-induced apoptosis in K562 human chronic myeloid leukemia cells, J Biol Chem, 282 (2007) 10922–10934.

Baumruker, T., Billich, A., and Brinkmann, V., FTY720, an immunomodulatory sphingolipid mimetic: translation of a novel mechanism into clinical benefit in multiple sclerosis, Expert Opin Investig Drugs, 16 (2007) 283–289.

Becker, S., Reinehr, R., Grether-Beck, S., Eberle, A., and Haussinger, D., Hydrophobic bile salts trigger ceramide formation through endosomal acidification, Biol Chem, 388 (2007) 185–196.

Bektas, M., Jolly, P.S., Müller, C., Eberle, J., Spiegel, and S., Geilen, C.C., Sphingosine kinase activity counteracts ceramide-mediated cell death in human melanoma cells: role of Bcl-2 expression, Oncogene, 24 (2005) 178–187.

Bezombes, C., Segui, B., Cuvillier, O., Bruno, AP., Uro-Coste, E., Gouaze, V., Andrieu-Abadie, N., Carpentier, S., Laurent, G., Salvayre, R., Jaffrezou, J.P., and Levade, T., Lysosomal sphingomyelinase is not solicited for apoptosis signaling, FASEB J, 15 (2001) 297–299.

Billich, A., Bornancin, F., Dévay, P., Mechtcheriakova, D., Urtz, N., and Baumruker, T., Phosphorylation of the immunomodulatory drug FTY720 by sphingosine kinases, J Biol Chem, 278 (2003) 47408–47415.

Billich, A., Bornancin, F., Mechtcheriakova, D., Natt, F., Huesken, D., and Baumruker, T., IL-1β and TNF-α induced production of inflammatory mediators in A549 epithelial lung carcinoma cells is dependent on sphingosine kinase 1 activity, Cell Signal, 17 (2005) 1203–1217.

Bonhoure, E., Pchejetski, D., Aouali, N., Morjani, H., Levade, T., Kohama, T., and Cuvillier, O., Overcoming MDR-associated chemoresistance in HL-60 acute myeloid leukemia cells by targeting sphingosine kinase-1, Leukemia, 20 (2006) 95–102.

Buehrer, B.M., and Bell, R.M., Inhibition of sphingosine kinase in vitro and in platelets, Implications for signal transduction pathways, J Biol Chem, 267 (1992) 3154–3159.

Castillo, S.S., Levy, M., Thaikoottathil, J.V., and Goldkorn, T., Reactive nitrogen and oxygen species activate different sphingomyelinases to induce apoptosis in airway epithelial cells, Exp Cell Res, 313 (2007) 2680–2866.

Chen, L., Li, T., Li, R., Wei, and B., Peng, Z., Alphastatin downregulates vascular endothelial cells sphingosine kinase activity and suppresses tumor growth in nude mice bearing human gastric cancer xenografts, World J Gastroenterol, 12 (2006) 4130–4136.

Chen, X.L., Grey, J.Y., Thomas, S., Qiu, F.H., Medford, R.M., Wasserman, M.A., and Kunsch, C., Sphingosine kinase-1 mediates TNF-alpha-induced MCP-1 gene expression in endothelial cells: upregulation by oscillatory flow, Am J Physiol Heart Circ Physiol, 287 (2004) H1452–1458.

Cheng, Y., Kozubek, A., Ohlsson, L., Sternby, B., and Duan, R.D., Curcumin decreases acid sphingomyelinase activity in colon cancer Caco-2 cells, Planta Med, 73 (2007) 725–730.

Choi, O.H., Kim, J.H., and Kinet, J.P., Calcium mobilization via sphingosine kinase in signalling by the Fc epsilon RI antigen receptor, Nature, 380 (1996) 634–636.

Clair, T., Aoki, J., Koh, E., Bandle, R.W., Nam, SW., Ptaszynska, M.M., Mills, G.B., Schiffmann, E., Liotta, L.A., and Stracke, M.L., Autotaxin hydrolyzes sphingosylphosphorylcholine to produce the regulator of migration, sphingosine-1-phosphate cancer Res, 63 (2003) 5446–5453.

Claus, R.A., Bunck, A.C., Bockmeyer, C.L., Brunkhorst, F.M., Losche, W., Kinscherf, R., and Deigner, H.P., Role of increased sphingomyelinase activity in apoptosis and organ failure of patients with severe sepsis, FASEB J, 19 (2005) 1719–1721.

Cock, J.G., Tepper, A.D., de Vries, E., van Blitterswijk, WJ., and Borst, J., CD95 (Fas/APO-1) induces ceramide formation and apoptosis in the absence of a functional acid sphingomyelinase, J Biol Chem, 273 (1998) 7560–7565.

Cremesti, A., Paris, F., Grassme, H., Holler, N., Tschopp, J., Fuks, Z., Gulbins, E., and Kolesnick, R., Ceramide enables fas to cap and kill, J Biol Chem, 276 (2001) 23954–23961.

Cuschieri, J., Bulger, E., Billgrin, J., Garcia, I., and Maier, R.V., Acid sphingomyelinase is required for lipid Raft TLR4 complex formation, Surg Infect, 8 (2007) 91–106.

Cuvillier, O., Sphingosine kinase-1 – a potential therapeutic target in cancer, Anticancer Drugs, 18 (2007) 105–110.

Darroch, P.I., Dagan, A., Granot, T., He, X., Gatt, S., and Schuchman, E.H., A lipid analogue that inhibits sphingomyelin hydrolysis and synthesis, increases ceramide and leads to cell death, J Lipid Res, 46 (2005) 2315–2324.

Deaciuc, I.V., Nikolova-Karakashian, M., Fortunato, F., Lee, E.Y., Hill, D.B., and McClain, C. J., Apoptosis and dysregulated ceramide metabolism in a murine model of alcohol-enhanced lipopolysaccharide hepatotoxicity, Alcohol Clin Exp Res, 24 (2000) 1557–1565.

Delgado, A., Casas, J., Llebaria, A., Abad, J.L., and Fabrias, G., Inhibitors of sphingolipid metabolism enzymes, Biochim Biophys Acta, 1758 (2006) 1957–1977.

De Maria, R., Rippo, M.R., Schuchman, E.H., and Testi, R., Acidic sphingomyelinase (ASM) is necessary for fas-induced GD3 ganglioside accumulation and efficient apoptosis of lymphoid cells, J Exp Med, 187 (1998) 897–902.

Deutschman, D.H., Carstens, J.S., Klepper, R.L., Smith, W.S., Page, M.T., Young, T.R., Gleason, L.A., Nakajima, N., and Sabbadini, R.A., Predicting obstructive coronary artery disease with serum sphingosine-1-phosphate, Am Heart J, 146 (2003) 62–68.

Doehner, W., Bunck, A.C., Rauchhaus, M., von Haehling, S., Brunkhorst, F.M., Cicoira, M., Tschope, C., Ponikowski, P., Claus, R.A., and Anker, S.D., Secretory sphingomyelinase is upregulated in chronic heart failure: a second messenger system of immune activation relates to body composition, muscular functional capacity and peripheral blood flow, Eur Heart J, 28 (2007) 821–828.

Döll, F., Pfeilschifter, J., Huwiler, and A., Prolactin upregulates sphingosine kinase-1 expression and activity in the human breast cancer cell line MCF7 and triggers enhanced proliferation and migration., Endocr Relat Cancer, 14 (2007) 325–335.

Döll, F., Pfeilschifter, J., and Huwiler, A., The epidermal growth factor stimulates sphingosine kinase-1 expression and activity in the human mammary carcinoma cell line MCF7, Biochim Biophys Acta, 1738 (2005) 72–81.

Dumitru, C.A., and Gulbins, E., TRAIL activates acid sphingomyelinase via a redox mechanism and releases ceramide to trigger apoptosis, Oncogene, 25 (2006) 5612–5625.

Edsall, L.C., Van Brocklyn, J.R., Cuvillier, O., Kleuser, B., and Spiegel, S., N,N-Dimethylsphingosine is a potent competitive inhibitor of sphingosine kinase but not of protein kinase C: modulation of cellular levels of sphingosine 1-phosphate and ceramide, Biochemistry, 37 (1998) 12892–12898.

Eichhorst, S.T., Modulation of apoptosis as a target for liver disease, Expert Opin Ther Targets, 9 (2005) 83–99.

Eigenbrod, S., Derwand, R., Jakl, V., Endres, and S., Eigler, A., Sphingosine kinase and sphingosine-1-phosphate regulate migration, endocytosis and apoptosis of dendritic cells, Immunol Invest, 35 (2006) 149–165.

Elojeimy, S., Holman, D.H., Liu, X., El-Zawahry, A., Villani, M., Cheng, .J.C., Mahdy, A., Zeidan, Y., Bielwaska, A., Hannun, Y.A., and Norris, J.S., New insights on the use of desipramine as an inhibitor for acid ceramidase, FEBS Lett, 580 (2006) 4751–4756.

Ferlinz, K., Hurwitz, R., Moczall, H., Lansmann, S., Schuchman, E.H., and Sandhoff, K., Functional characterization of the N-glycosylation sites of human acid sphingomyelinase by site-directed mutagenesis, Eur J Biochem, 243 (1997) 511–517.

French, K.J., Schrecengost, R.S., Lee, B.D., Zhuang, Y., Smith, S.N., Eberly, J.L., Yun, J.K., and Smith, C.D., Discovery and evaluation of inhibitors of human sphingosine kinase, Cancer Res, 63 (2003) 5962–5969.

French, K.J., Upson, J.J., Keller, S.N., Zhuang, Y., Yun, J.K., and Smith, C.D., Antitumor activity of sphingosine kinase inhibitors, J Pharmacol Exp Ther, 318 (2006) 596–603.

Gamble, J.R., Xia, P., Hahn, C.N., Drew, J.J., Drogemuller, C.J., Brown, D., and Vadas, M.A., Phenoxodiol, an experimental anticancer drug, shows potent antiangiogenic properties in addition to its antitumour effects, Int J Cancer, 118 (2006) 2412–2420.

Garcia-Ruiz, C., Colell, A., Marí, M., Morales A., Calvo M., Enrich C., and Fernández-Checa, J.C., Defective TNF-alpha-mediated hepatocellular apoptosis and liver damage in acidic sphingomyelinase knockout mice, J Clin Invest, 111 (2003) 197–208.

Goggel, R., Winoto-Morbach, S., Vielhaber, G., Imai, Y., Lindner, K., Brade, L., Brade, H., Ehlers, S., Slutsky, A.S., Schutze, S., Gulbins, and E., Uhlig, S., PAF-mediated pulmonary edema: a new role for acid sphingomyelinase and ceramide, Nat Med, 10 (2004) 155–160.

Gómez-Muñoz, A., Kong, JY., Salh, B., and Steinbrecher, U.P., Ceramide-1-phosphate blocks apoptosis through inhibition of acid sphingomyelinase in macrophages., J Lipid Res, 45 (2004) 99–105

Gómez-Muñoz, A., Kong, J., Salh, B., and Steinbrecher U.P., Sphingosine-1-phosphate inhibits acid sphingomyelinase and blocks apoptosis in macrophages, FEBS Lett, 539 (2003) 56–60.

Graf, C., Zemann, B., Rovina, P., Urtz, N., Schanzer, A., Mechtcheriakova, D., Müller, M., Reuschel, R., Fischer, E.,Van Overloop, H., Reichel, C., Huber, S., Dawson, J., Meingassner, J., Billich, A., Niwa, S., Badegruber, R., Van Veldhoven, P.P., Kinzel B., Baumruker, T., and Bornancin, F., Neutropenia with impaired immune response to Streptococcus pneumoniae in ceramide-kinase-deficient mice, J Immunol, 180 (2008) 3457–3466.

Granata, R., Trovato, L., Lupia, E., Sala, G., Settanni, F., Camussi, G., and Ghidoni, R., Ghigo E.Insulin-like growth factor binding protein-3 induces angiogenesis through IGF-I- and SphK1-dependent mechanisms, J Thromb Haemost, 5 (2007) 835–845.

Grassmé, H., Gulbins, E., Brenner, B., Ferlinz, K., Sandhoff, K., Harzer, K., Lang, F., and Meyer, T.F., Acidic sphingomyelinase mediates entry of N. gonorrhoeae into nonphagocytic cells, Cell, 91 (1997) 605–615.

Grassmé, H., Jekle, A., Riehle, A., Schwarz, H., Berger, J., Sandhoff, K., Kolesnick, R., and Gulbins, E., CD95 signaling via ceramide-rich membrane rafts, J Biol Chem, 276 (2001) 20589–20596.

Grassmé, H., Jendrossek, V., Riehle, A., von Kürthy, G., Berger, J., Schwarz, H., Weller, M., Kolesnick, R., and Gulbins, E., Host defense against Pseudomonas aeruginosa requires ceramide-rich membrane rafts, Nat Med, 9 (2003a) 322–330.

Grassmé, H., Cremesti, A., Kolesnick, R., and Gulbins, E., Ceramide-mediated clustering is required for CD95-DISC formation, Oncogene, 22 (2003b) 5457–5470.

Grassmé, H., Riehle, A., Wilker, B., and Gulbins, E. Rhinoviruses infect human epithelial cells via ceramide-enriched membrane platforms, J Biol Chem, 280 (2005) 26256–26262.

Guarino, A.J., Tulenko, T.N., and Wrenn, S.P., Sphingomyelinase-to-LDL molar ratio determines low density lipoprotein aggregation size: biological significance, Chem Phys Lipids, 142 (2006) 33–42.

Gupta, S., Natarajan, R., Payne, S.G., Studer, E.J., Spiegel, S., Dent, P., and Hylemon, P.B., Deoxycholic acid activates the c-Jun N-terminal kinase pathway via FAS receptor activation in primary hepatocytes, Role of acidic sphingomyelinase-mediated ceramide generation in FAS receptor activation, J Biol Chem, 279 (2004) 5821–5828.

Haimovitz-Friedman, A., Cordon-Cardo, C., Bayoumy, S., Garzotto, M., McLoughlin, M., Gallily, R., Edwards, C.K, 3rd., Schuchman, E.H., Fuks, Z., and Kolesnick, R., Lipopolysaccharide induces disseminated endothelial apoptosis requiring ceramide generation, J Exp Med, 186 (1997) 1831–1841.

Hamada, M., Iikubo, K., Ishikawa, Y., Ikeda, A., Umezawa, K., and Nishiyama, S., Biological activities of alpha-mangostin derivatives against acidic sphingomyelinase, Bioorg Med Chem Lett, 13 (2003) 3151–3153.

Hauck, CR., Grassmé, H., Bock, J., Jendrossek, V., Ferlinz, K., Meyer, T.F., and Gulbins, E., Acid sphingomyelinase is involved in CEACAM receptor-mediated phagocytosis of Neisseria gonorrhoeae, FEBS Lett, 478 (2000) 260–266.

Hibbs, K., Skubitz, K.M., Pambuccian, S.E., Casey, R.C., Burleson, K.M., Oegema, T.T.jr., Thiele, J.J., Grindle, S.M., Bliss, R.L., and Skubitz, A.P., Differential gene expression in ovarian carcinoma: identification of potential biomarkers, Am J Pathol, 165 (2004) 397–414.

Higuchi, M., Singh, S., Jaffrezou, J.P., and Aggarwal, B.B., Acidic sphingomyelinase-generated ceramide is needed but not sufficient for TNF-induced apoptosis and nuclear factor-kappa B activation, J Immunol, 157 (1996) 297–304.

Hla, T., Physiological and pathological actions of sphingosine 1-phosphate, Semin Cell Dev Biol, 15 (2004) 513–520.

Hurwitz, R., Ferlinz, K., Vielhaber, G., Moczall, H., and Sandhoff, K., Processing of human acid sphingomyelinase in normal and I-cell fibroblasts, J Biol Chem, 269 (1994) 5440–5445.

Ibrahim, F.B., Pang, S.J., and Melendez, A.J., Anaphylatoxin signaling in human neutrophils, A key role for sphingosine kinase, J Biol Chem, 279 (2004) 44802–44811.

Ichi, I., Nakahara, K., Miyashita, Y., Hidaka, A., Kutsukake, S., Inoue, K., Maruyama, T., Miwa, Y., Harada-Shiba, M., Tsushima, and M., Kojo, S., Kisei Cohort Study Group, Association of ceramides in human plasma with risk factors of atherosclerosis, Lipids, 41 (2006) 859–863.

Igarashi, Y., Hakomori, S., Toyokuni, T., Dean, B., Fujita, S., Sugimoto, M., Ogawa, T., el-Ghendy, K., and Racker, E., Effect of chemically well-defined sphingosine and its N-methyl derivatives on protein kinase C and src kinase activities, Biochemistry, 28 (1989) 6796–6800.

Ito, K., Anada, Y., Tani, M., Ikeda, M., Sano, T., Kihara, A., and Igarashi, Y., Lack of sphingosine 1-phosphate-degrading enzymes in erythrocytes, Biochem Biophys Res Commun, 357 (2007) 212–217.

Jaffrézou, J.P., Herbert, JM., Levade, T., Gau, M.N., Chatelain, P., and Laurent, G., Reversal of multidrug resistance by calcium channel blocker SR33557 without photoaffinity labeling of P-glycoprotein, Biol Chem, 266 (1991) 19858–19864.

Jan, J.T., Chatterjee, S., and Griffin, D.E., Sindbis virus entry into cells triggers apoptosis by activating sphingomyelinasd, leading to the release of ceramide, J Virol, 74 (2000) 6425–6432.

Jiang, X.C., Paultre, F., Pearson, T.A., Reed, R.G., Francis, C.K., Lin, M., Berglund, L., and Tall, A.R., Plasma sphingomyelin level as a risk factor for coronary artery disease, Arterioscler Thromb Vasc Biol, 20 (2000) 2614–2618.

Johnson, K.R., Johnson, K.Y., Crellin, H.G., Ogretmen, B., Boylan, A.M., Harley, R.A., and Obeid, L.M., Immunohistochemical distribution of sphingosine kinase 1 in normal and tumor lung tissue, J Histochem Cytochem.,53 (2005) 1159–1166.

Jolly, P.S., Bektas, M., Olivera, A., Gonzalez-Espinosa, C., Proia, R.L., Rivera, J., Milstien, S., and Spiegel, S., Transactivation of sphingosine-1-phosphate receptors by FcepsilonRI triggering is required for normal mast cell degranulation and chemotaxis, J Exp Med, 199 (2004) 959–970.

Jolly, P.S., Bektas, M., Watterson, K.R., Sankala, H., Payne, S.G., Milstien, S., Spiegel S., Expression of SphK1 impairs degranulation and motility of RBL-2H3 mast cells by desensitizing S1P receptors, Blood, 105 (2005) 4736–4742.

Jung, I.D., Lee, J.S., Kim, Y.J., Jeong, Y.I., Lee, C.M., Baumruker, T., Billich, A., Banno, Y., Lee, M.G., Ahn, S.C., Park, W.S., Han, J., and Park, Y.M., Sphingosine kinase inhibitor suppresses a Th1 polarization via the inhibition of immunostimulatory activity in murine bone marrow-derived dendritic cells, Int Immunol, 19 (2007a) 411–426.

Jung, I.D., Lee, J.S., Kim, Y.J., Jeong, Y.I., Lee, C.M., Lee, M.G., Ahn, S.C., and Park, Y.M., Sphingosine kinase inhibitor suppresses dendritic cell migration by regulating chemokine receptor expression and impairing p38 mitogen-activated protein kinase. Immunology, 121 (2007b) 533–544.

Kacimi, R., Vessey, DA., Honbo, N., and Karliner, JS., Adult cardiac fibroblasts null for sphingosine kinase-1 exhibit growth dysregulation and an enhanced proinflammatory response, J Mol Cell Cardiol, 43 (2007) 85–91.

Kang, JS., Yoon, Y.D., Han, M.H., Han, S.B., Lee, K., Lee, K.H., Park, S.K., and Kim, H.M., Glabridin suppresses intercellular adhesion molecule-1 expression in tumor necrosis factor-alpha-stimulated human umbilical vein endothelial cells by blocking sphingosine kinase pathway: implications of AkT., extracellular signal-regulated kinasE., and nuclear factor-kappaB/Rel signaling pathways, Mol Pharmacol, 69 (2006) 941–949.

Kase, H., Hattori, Y., Jojima, T., Okayasu, T., Tomizawa, A., Suzuki, K., Banba, N., Monden, T., Satoh, H., Akimoto, K., and Kasai, K., Globular adiponectin induces adhesion molecule expression through the sphingosine kinase pathway in vascular endothelial cells, Life Sci, 81 (2007) 939–943.

Kashkar, H., Wiegmann, K., Yazdanpanah, B., Haubert, and D., Kronke, M., Acid sphingomyelinase is indispensable for UV light-induced Bax conformational change at the mitochondrial membrane, J Biol Chem, 280 (2005) 20804–20813.

Kawamori, T., Osta, W., Johnson, K.R., Pettus, B.J., Bielawski, J., Tanaka, T., Wargovich, M.J., Reddy, B.S., Hannun, Y.A., Obeid, L.M., and Zhou, D., Sphingosine kinase 1 is up-regulated in colon carcinogenesis, FASEB J, 20 (2006) 386–388.

Kharel, Y., Lee, S., Snyder, A.H., Sheasley-O'Neill, S.L., Morris, M.A., Setiady, Y., Zhu, R., Zigler, M.A., Burcin, T.L., Ley, K., Tung, K.S., Engelhard, V.H., Macdonald, T.L., Pearson-White, S., and Lynch, K.R., Sphingosine kinase 2 is required for modulation of lymphocyte traffic by FTY720, J Biol Chem, 280 (2005) 36865–36872.

Kim, I., Moon, S.O., Kim, S.H., Kim, H.J., Koh, Y.S., and Koh, G.Y., Vascular endothelial growth factor expression of intercellular adhesion molecule 1 (ICAM-1), vascular cell adhesion molecule 1 (VCAM-1), and E-selectin through nuclear factor-kappa B activation in endothelial cellS, J Biol Chem, 276 (2001) 7614–7620.

Kirschnek, S., Paris, F., Weller, M., Grassme, H., Ferlinz, K., Riehle, A., Fuks, Z., Kolesnick, R., and Gulbins, E., CD95-mediated apoptosis in vivo involves acid sphingomyelinase, J Biol Chem, 275 (35) (2000) 27316–27323.

Klawitter, S., Hofmann, L.P., Pfeilschifter, J., and Huwiler, A., Extracellular nucleotides induce migration of renal mesangial cells by upregulating sphingosine kinase-1 expression and activity, Br J Pharmacol, 150 (2007) 271–280.

Kölzer, M., Arenz, C., Ferlinz, K., Werth, N., Schulze, H., Klingenstein, R., and Sandhoff, K., Phosphatidylinositol-3,5-Bisphosphate is a potent and selective inhibitor of acid sphingomyelinase, Biol Chem, 384 (2003) 1293–1298.

Kölzer, M., Werth, N., and Sandhoff, K., Interactions of acid sphingomyelinase and lipid bilayers in the presence of the tricyclic antidepressant desipramine, FEBS Lett, 559 (2004) 96–98.

Kohno, M., Momoi, M., Oo, M.L., Paik, J.H., Lee, Y.M., Venkataraman, K., Ai, Y., Ristimaki, A.P., Fyrst, H., Sano, H., Rosenberg, D., Saba, J.D., Proia, .R.L., and Hla, T., Intracellular role for sphingosine kinase 1 in intestinal adenoma cell proliferation, Mol Cell Biol, 26 (2006) 7211–7223.

Kolesnick, R., and Fuks, Z., Radiation and ceramide-induced apoptosis, Oncogene, 22 (2003) 5897–5906.

Kono, K., Sugiura, M., and Kohama, T., Inhibition of recombinant sphingosine kinases by novel inhibitors of microbial origin, F-12509A and B-5354c, J Antibiot, 55 (2002) 99–103.

Kono, Y., Nishiuma, T., Nishimura, Y., Kotani, Y., Okada, T., Nakamura, S., and Yokoyama, M., Sphingosine kinase 1 regulates differentiation of human and mouse lung fibroblasts mediated by TGF-beta1, Am J Respir Cell Mol Biol, 37 (2007) 395–404.

Kornhuber, J., Tripal, P., Reichel, M., Terfloth, L., Bleich, S., Wiltfang, J., and Gulbins, E., Identification of new functional inhibitors of acid sphingomyelinase using a structure-property-activity relation model, J Med Chem, 51 (2007) 219–237.

Lang, P.A., Schenck, M., Nicolay, J.P., Becker, J.U., Kempe, D.S., Lupescu, A., Koka, S., Eisele, K., Karl, B.A., Rübben, H., Schmid, K.W., Mann, K, Hildenbrand, S., Hefter, H., Huber, S.M., Wieder, T., Erhardt, A., Häussinger, D., Gulbins, E., and Lang, F., Liver cell death and anemia in Wilson disease involve acid sphingomyelinase and ceramide, Nat Med, 13 (2007) 164–170.

Lee, C., Xu, D.Z., Feketeova, E., Kannan, K.B., Yun, J.K., Deitch, E.A. Fekete, Z., Livingston, D.H, and Hauser, C,J. Attenuation of shock-induced acute lung injury by sphingosine kinase inhibition, J Trauma, 57 (2004) 955–960.

Lee, C.Y., Lesimple, A., Denis, M., Vincent, J., Larsen, A., Mamer, O., Krimbou, L., Genest, J., and Marcil, M., Increased sphingomyelin content impairs HDL biogenesis and maturation in human Niemann-Pick disease type B, J Lipid Res, 47 (2006) 622–632.

Le Scolan, E., Pchejetski, D., Banno, Y., Denis, N., Mayeux, P., Vainchenker, W., Levade, and T., Moreau-Gachelin, F. Overexpression of sphingosine kinase 1 is an oncogenic event in erythroleukemic progression., Blood, 106 (2005) 1808–1816.

Le Stunff, H., Giussani, P., Maceyka, M., Lépine, S., Milstien, S., and Spiegel, S., Recycling of sphingosine is regulated by the concerted actions of sphingosine-1-phosphate phosphohydrolase 1 and sphingosine kinase 2, J Biol Chem, 282 (2007) 34372–34380.

Leroux, M.E., Auzenne, E., Evans, R., Hail, N. Jr., Spohn, W., Ghosh, S.C., Farquhar, D., McDonnell, T., and Klostergaard, J. Sphingolipids and the sphingosine kinase inhibitor, SKI II, induce BCL-2-independent apoptosis in human prostatic adenocarcinoma cells, Prostate, 67 (2007) 1699–1717

Li, Q.F., Huang, WR., Duan, H.F., Wang, H., Wu, C.T., and Wang, L.S., Sphingosine kinase-1 mediates BCR/ABL-induced upregulation of Mcl-1 in chronic myeloid leukemia cells, Oncogene, 26 (2007) 7904–7908.

Lin, T., Genestier, L., Pinkoski, MJ., Castro, A., Nicholas, S., Mogil, R., Paris, F., Fuks, Z., Schuchman, E.H., Kolesnick, R.N., and Green, D.R., Role of acidic sphingomyelinase in Fas/CD95-mediated cell death, J Biol Chem, 275 (2000) 8657–8663.

Llacuna, L., Mari, M., Garcia-Ruiz, C., and Fernandez-Checa, J.C., Morales A. Critical role of acidic sphingomyelinase in murine hepatic ischemia-reperfusion injury, Hepatology, 44 (2006) 561–572.

Liu, H., Sugiura, M, Nava, V.E., Edsall, L.C., Kono, K., Poulton, S., Milstien, S., Kohama, T., and Spiegel, S., Molecular cloning and functional characterization of a novel mammalian sphingosine kinase type 2 isoform, J Biol Chem, 275 (2000) 19513–19520.

Liu, H., Toman, R.E., Goparaju, S.K., Maceyka, M., Nava, VE., Sankala, H., Payne, S.G., Bektas, M., Ishii, I., Chun, J., Milstien, S., and Spiegel, S., Sphingosine kinase type 2 is a putative BH3-only protein that induces apoptosis, J Biol Chem, 278 (2003) 40330–40336.

Luberto, C., and Hannun, Y.A., Sphingomyelin synthase, a potential regulator of intracellular levels of ceramide and diacylglycerol during SV40 transformation, Does sphingomyelin synthase account for the putative phosphatidylcholine-specific phospholipase C, J Biol Chem, 273 (1998) 14550–14559.

Ma, M.M., Chen, J.L., Wang, G.G., Wang, H., Lu, Y., Li, J.F., Yi, J., Yuan, Y.J., Zhang, Q.W., Mi J., Wang, L.S.H., Duan, H.F., and Wu, C.T., Sphingosine kinase 1 participates in insulin signalling and regulates glucose metabolism and homeostasis in KK/Ay diabetic mice, Diabetologia, 50 (2007) 891–900.

Maceyka, M., Payne S.G., Milstien, S., and Spiegel, S., Sphingosine kinase, sphingosine-1-phosphate, and apoptosis, Biochim Biophys Acta, 1585 (2002) 193–201.

Maceyka, M., Sankala, H., Hait, N.C., Le Stunff, H., Liu, H., Toman, R., Collier, C., Zhang, M., Satin, LS., Merrill, AH JR., Milstien, S., and Spiegel, S., SphK1 and SphK2, sphingosine kinase isoenzymes with opposing functions in sphingolipid metabolism, J Biol Chem, 280 (2005) 37118–37129.

MacKinnon, A.C., Buckley, A., Chilvers, E.R., Rossi, A.G., Haslett, C., and Sethi, T., Sphingosine kinase: a point of convergence in the action of diverse neutrophil priming agents, J Immunol, 169 (2002) 6394–6400.

Maines, L.W., French, K.J., Wolpert, E.B., Antonetti, D.A., and Smith, C.D., Pharmacologic manipulation of sphingosine kinase in retinal endothelial cells: implications for angiogenic ocular diseases, Invest Ophthalmol Vis ScI, 47 (2006) 5022–5031.

Maines, L.W., Fitzpatrick, L.R., French, K.J., Zhuang, Y., Xia, Z., Keller, S.N., Upson, J.J., and Smith, C.D., Suppression of ulcerative colitis in mice by orally available inhibitors of sphingosine kinase, Dig Dis Sci, 2007 Dec 4; [Epub ahead of print]

Malaplate-Armand, C., Florent-Bechard, S., Youssef, I., Koziel, V., Sponne, I., Kriem, B., Leininger-Muller, B., Olivier, J.L., Oster, T., and Pillot, T., Soluble oligomers of amyloid-beta peptide induce neuronal apoptosis by activating a $cPLA_2$-dependent sphingomyelinase-ceramide pathway, Neurobiol Dis, 23 (2006) 178–189.

Manthey, C.L., and Schuchman, E.H., Acid sphingomyelinase-derived ceramide is not required for inflammatory cytokine signalling in murine macrophages, Cytokine, 10 (1998) 654–661.

Marathe, S., Schissel, S.L., Yellin, MJ., Beatini, N., Mintzer, R., Williams, K.J., Tabas, I., Human vascular endothelial cells are a rich and regulatable source of secretory sphingomyelinase, Implications for early atherogenesis and ceramide-mediated cell signaling, J Biol Chem, 273 (1998) 4081–4088.

Marathe, S., Kuriakose, G., Williams, and K.J., Tabas, I., Sphingomyelinase, an enzyme implicated in atherogenesis, is present in atherosclerotic lesions and binds to specific components of the subendothelial extracellular matrix, Arterioscler Thromb Vasc Biol, 19 (1999) 2648–2658.

Marathe, S., Choi, Y., Leventhal, A.R., and Tabas, I., Sphingomyelinase converts lipoproteins from apolipoprotein E knockout mice into potent inducers of macrophage foam cell formation, Arterioscler Thromb Vasc Biol, 20 (2000) 2607–2613.

Mari, M., Colell, A., Morales, A., Pañeda, C., Varela-Nieto, I., García-Ruiz, C., Fernández-Checa JC, Acidic sphingomyelinase downregulates the liver-specific methionine adenosyltransferase 1A, contributing to tumor necrosis factor-induced lethal hepatitis, J Clin Invest, 113 (2004) 895–904.

Mastrandrea, L.D., Sessanna, S.M., Laychock, S.G., Sphingosine kinase activity and sphingosine-1-phosphate production in rat pancreatic islets and INS-1 cells: response to cytokines, Diabetes, 54 (2005) 1429–1436.

Mechtcheriakova, D., Wlachos, A., Sobanov, J., Kopp, T., Reuschel, R., Bornancin, F., Cai, R., Zemann, B., Urtz,. N., Stingl, G., Zlabinger, G., Woisetschläger, M., Baumruker, T., and Billich, A. Sphingosine 1-phosphate phosphatase 2 is induced during inflammatory responses, Cell Signal, 19 (2007) 748–760.

Megidish, T., Cooper, J., Zhang, L., Fu, H., and Hakomori, S., A novel sphingosine-dependent protein kinase (SDK1) specifically phosphorylates certain isoforms of 14-3-3 protein, J Biol Chem, 273 (1998) 21834–21845.

Melendez, A.J., and Khaw, A.K., Dichotomy of Ca2+ signals triggered by different phospholipid pathways in antigen stimulation of human mast cells, J Biol Chem, 277 (2002) 17255–17262.

Melendez, A.J., and Ibrahim, F.B., Antisense knockdown of sphingosine kinase 1 in human macrophages inhibits C5a receptor-dependent signal transduction, Ca2+ signals, enzyme release, cytokine production, and chemotaxis, J Immunol, 173 (2004) 1596–1603.

Michaud, J., Kohno, M., Proia, R.L., and Hla, T., Normal acute and chronic inflammatory responses in sphingosine kinase 1 knockout mice, FEBS Lett, 580 (2006) 4607–4612.

Min, J., Traynor, D., Stegner, A.L., Zhang, L., Hanigan, M.H., Alexander, H., and Alexander, S., Sphingosine kinase regulates the sensitivity of Dictyostelium discoideum cells to the anticancer drug cisplatin, Eukaryot Cell, 4 (2005) 178–819.

Min, J., Mesika, A., Sivaguru, M., Van Veldhoven, P.P., Alexander, H., Futerman, A.H., andAlexander, S., (Dihydro) ceramide synthase 1 regulated sensitivity to cisplatin is associated with the activation of p38 mitogen-activated protein kinase and is abrogated by sphingosine kinase 1. Mol Cancer Res, 5 (2007) 801–812.

Mintzer, R.J., Appell, K.C., Cole, A., Johns, A., Pagila, R., Polokoff, M.A., Tabas, I., Snider, R.M., and Meurer-Ogden, J.A., A novel high-throughput screening format to identify inhibitors of secreted acid sphingomyelinase, J Biomol Screen, 10 (2005) 225–234.

Mizugishi, K., Yamashita, T., Olivera, A., Miller, G.F., Spiegel, S., and Proia, RL., Essential role for sphingosine kinases in neural and vascular development, Mol Cell Biol, 25 (2005) 11113–11121.

Morales, A., and Fernandez-Checa, J.C., Pharmacological modulation of sphingolipids and role in disease and cancer cell biology, Mini Rev Med Chem, 7 (2007) 371–382.

Murakami, M., Ichihara, M., Sobue, S., Kikuchi, R., Ito, H., Kimura, A., Iwasaki, T., Takagi A., Kojima, T., Takahashi, M., Suzuki, M., Banno, Y., Nozawa, and Y., Murate, T., Reand T signaling-induced SPHK1 gene expression plays a role in both GDNF-induced differentiation and MEN2-type oncogenesis, J Neurochem, 102 (2007) 1585–1594.

Nava, V.E., Cuvillier, O., Edsall L.C., Kimura, K., Milstien, S., Gelmann, E.P., and Spiegel, S., Sphingosine enhances apoptosis of radiation-resistant prostate cancer cellS., Cancer Res, 60 (2000) 4468–4474.

Nelson, J.B., O'Hara, S.P., Small, A.J., Tietz, P.S., Choudhury, AK., Pagano, R.E., Chen, X.M., and LaRusso, N.F., Cryptosporidium parvum infects human cholangiocytes via sphingolipid-enriched membrane microdomains, Cell Microbiol, 8 (2006) 1932–1945.

Ng, C.G., and Griffin, D.E., Acid sphingomyelinase deficiency increases susceptibility to fatal alphavirus encephalomyelitis, J Virol, 80 (2006) 10989–10999.

Nussbaumer, P., Medicinal chemistry aspects of drug targets in sphingolipid metabolism, Chem Med Chem, 3 (2008) 543–551.

Okudaira, C., Ikeda, Y., Kondo, S., Furuya, S., Hirabayashi, Y., Koyano, T., Saito, Y., and Umezawa, K., Inhibition of acidic sphingomyelinase by xanthone compounds isolated from Garcinia speciosa, J Enzyme Inhib, 15 (2000) 129–138.

Olivera, A., Kohama, T., Edsall, L., Nava, V., Cuvillier, O., Poulton, S., and Spiegel, S., Sphingosine kinase expression increases intracellular sphingosine-1-phosphate and promotes cell growth and survival, J Cell Biol, 147 (1999) 545–558.

Olivera, A., Mizugishi, K., Tikhonova, A., Ciaccia, L., Odom, S., Proia, R.L., and Rivera, J., The sphingosine kinase-sphingosine-1-phosphate axis is a determinant of mast cell function and anaphylaxis, Immunity, 26 (2007) 287–297.

Oorni, K., Pentikainen, M.O., Ala-Korpela, M., Kovanen, P.T., Aggregation, fusion, and vesicle formation of modified low density lipoprotein particles: molecular mechanisms and effects on matrix interactions, J LipidRes, 41 (2000) 1703–1714

Osawa, Y., Uchinami, H., Bielawski, J., Schwabe, RF., Hannun, YA., Brenner, DA., Roles for C16-ceramide and sphingosine 1-phosphate in regulating hepatocyte apoptosis in response to tumor necrosis factor-alpha, J Biol Chem, 280 (2005) 27879–27887.

Oskouian, B., Sooriyakumaran, P., Borowsky, AD., Crans, A., Dillard-Telm, L., Tam, Y.Y., Bandhuvula, P., and Saba, JD., Sphingosine-1-phosphate lyase potentiates apoptosis via p53- and p38-dependent pathways and is down-regulated in colon cancer, Proc Natl Acad Sci USA, 103 (2006) 17384–17389.

Pappu, R., Schwab, S.R., Cornelissen, I., Pereira, J.P., Regard, J.B., Xu, Y., Camerer, E., Zheng, Y.W., Huang, Y., Cyster, J.G., and Coughlin, S.R., Promotion of lymphocyte egress into blood and lymph by distinct sources of sphingosine-1-phosphate, Science, 316 (2007) 295–298.

Paris, F., Grassme, H., Cremesti, A., Zager, J., Fong, Y., Haimovitz-Friedman, A., Fuks, Z., Gulbins, E., and Kolesnick, R., Natural ceramide reverses Fas resistance of acid sphingomyelinase(–/–) hepatocytes, J Biol Chem, 276 (2001) 8297–8305.

Pchejetski, D., Golzio, M., Bonhoure, E., Calvet, C., Doumerc, N., Garcia, V., Mazerolles, C. , Rischmann, P., Teissié, J., Malavaud, B., and Cuvillier, O., Sphingosine kinase-1 as a chemotherapy sensor in prostate adenocarcinoma cell and mouse models, Cancer Res, 65 (2005) 11667–11675.

Petrache, I., Natarajan, V., Zhen, L., Medler, T.R., Richter, A.T., Cho, C., Hubbard, W.C., Berdyshev, E.V., and Tuder, R.M., Ceramide upregulation causes pulmonary cell apoptosis and emphysema-like disease in mice, Nat Med, 11 (2005) 491–498.

Pettus, B.J., Bielawski, J., Porcelli, A.M., Reames, D.L., Johnson, K.R., Morrow, J., Chalfant, C.E., Obeid, L.M., and Hannun, Y.A., The sphingosine kinase 1/sphingosine-1-phosphate pathway mediates COX-2 induction and $PGE_2$ production in response to TNF-alpha, FASEB J, 17 (2003) 1411–1421.

Pi, X., Tan, S.Y., Hayes, M, Xiao, L, Shayman, J.A., Ling, S., and Holoshitz, J. Sphingosine kinase 1-mediated inhibition of Fas death signaling in rheumatoid arthritis B lymphoblastoid cellS., Arthritis Rheum, 54 (2006) 754–764.

Pitson, S.M, Xia, P., Leclercq, T.M, Moretti, P.A., Zebol, J.R., Lynn, H.E., Wattenberg, B. W., and Vadas, M.A. Phosphorylation-dependent translocation of sphingosine kinase to the plasma membrane drives its oncogenic signalling. J Exp Med, 201 (2005) 49–54.

Prieschl, E.E., Csonga, R.,, Novotny, V., Kikuchi, G.E., and Baumruker, T., The balance between sphingosine and sphingosine-1-phosphate is decisive for mast cell activation after Fc epsilon receptor I triggering. J Exp Med, 190 (1999) 1–8.

Reinehr, R., Sommerfeld, A., Keitel, V., Grether-Beck, S., and Haussinger, D. Amplification of CD95 activation by caspase 8-induced endosomal acidification in rat hepatocytes, J Biol Chem, 283 (2007) 2211–2222.

Rotolo, J.A., Zhang, J., Donepudi, M, Lee, H., Fuks, Z., and Kolesnick, R., Caspase-dependent and -independent activation of acid sphingomyelinase signaling. J Biol Chem, 280 (2005) 26425–26434.

Ruckhäberle, E., Rody, A., Engels, K, Gaetje, R., von Minckwitz, G., Schiffmann, S., Grösch, S., Geisslinger, G., Holtrich, U., Karn, T., and Kaufmann, M., Microarray analysis ofaltered sphingolipid metabolism reveals prognostic significance of sphingosine kinase 1 in breast canceR., Breast Cancer Res Treat, 2007 Dec 4. [Epub ahead of print]

Sakata A., Yasuda K, Ochiai T., Shimeno H., Hikishima S., Yokomatsu T., Shibuya S., Soeda S., Inhibition of lipopolysaccharide-induced release of interleukin-8 from intestinal epithelial cells by SMA., a novel inhibitor of sphingomyelinase and its therapeutic effect on dextran sulphate sodium-induced colitis in mice. Cell Immunol, 245 (2007a) 24–31.

Sakata, A., Ochiai, T., Shimeno, H., Hikishima, S., Yokomatsu, T., Shibuya, S., Toda, A., Eyanagi, R., Soeda, S., Acid sphingomyelinase inhibition suppresses lipopolysaccharide-mediated release of inflammatory cytokines from macrophages and protects against disease pathology in dextran sulphate sodium-induced colitis in mice. Immunology, 122 (2007b) 54–64.

Samy, E.T., Meyer, C.A., Caplazi, P., Langrish, C.L, Lora J.M, Bluethmann, H., and Peng S.L., Modulation of intestinal autoimmunity and IL-2 signaling by sphingosine kinase 2 independent of sphingosine 1-phosphate. J Immunol, 179 (2007) 5644–5648.

Sankala, H.M, Hait N.C., Paugh, S.W., Shida, D., Lépine, S., Elmore, L.W., Dent, P., Milstien, S., and Spiegel, S., Involvement of sphingosine kinase 2 in p53-independent induction of p21 by the chemotherapeutic drug doxorubicin., Cancer Res, 67 (2007) 10466–10474.

Sanvicens, N., and Cotter, T.G. Ceramide is the key mediator of oxidative stress-induced apoptosis in retinal photoreceptor cells., J Neurochem, 98 (2006) 1432–1444.

Sarkar, S., Maceyka, M, Hait, N.C., Paugh, S.W., Sankala, H., Milstien, S., and Spiegel, S., Sphingosine kinase 1 is required for migration, proliferation and survival of MCF-7 human breast cancer cells., FEBS Lett, 579 (2005) 5313–5317.

Schmahl, J., Raymond, C.S., and Soriano, P., PDGF signaling specificity is mediated through multiple immediate early genes, Nat Genet, 39 (2007) 52–60.

Schissel, S.L, Keesler, G.A., Schuchman, E.H., Williams, K.J., and Tabas, I. The cellular trafficking and zinc dependence of secretory and lysosomal sphingomyelinase, two products of the acid sphingomyelinase gene, J Biol Chem, 273 (1998a) 18250–18259.

Schissel, S.L, Jiang, X,C., Tweedie-Hardman, J., Jeong, T.S., Camejo, E.H., Najib, J., Rapp J.H., Williams, K.J., and Tabas, I. Secretory sphingomyelinase, a product of the acid sphingomyelinase gene, can hydrolyze atherogenic lipoproteins at neutral pH: implications for atherosclerotic lesion development, J Biol Chem, 273 (1998b) 2738–2746.

Schuchman, E.H. The pathogenesis and treatment of acid sphingomyelinase-deficient Niemann-Pick disease. J Inherit Metab Dis, 30 (2007) 654–663.

Schwab, S.R., Pereira, J.P., Matloubian, M., Xu, Y., Huang, Y., and Cyster, J.G. Lymphocyte sequestration through S1P lyase inhibition and disruption of S1P gradient, Science, 309 (2005) 1735–1739.

Seo, E.Y., Park, G.T., Lee, K.M, Kim, J.A., Lee, J.H., and Yang, J.M., Identification of the target genes of atopic dermatitis by real-time PCR., J Invest Dermatol, 126 (2006) 1187–1189.

Seto, M., Whitlow, M., McCarrick, M.A., Srinivasan, S., Zhu, Y., Pagila, R., Mintzer, R., Light, D., Johns, A., and Meurer-Ogden, J.A., A model of the acid sphingomyelinase phosphoesterase domain based on its remote structural homolog purple acid phosphatase, Protein Sci, 13 (2004) 3172–3186.

Seumois, G., Fillet, M., Gillet, L., Faccinetto, C., Desmet, C., Francois, C., Dewals, B., Oury, C., Vanderplasschen, A., Lekeux, P., and Bureau, F., De novo C16- and C24-ceramide generation contributes to spontaneous neutrophil apoptosis, J Leukoc Biol, 81 (2007) 1477–1486.

Shu, X., Wu, W., Mosteller, R.D., and Broek D. Sphingosine kinase mediates vascular endothelial growth factor-induced activation of ras and mitogen-activated protein kinases, Mol Cell Biol, 22 (2002) 7758–7768.

Staton, C.A., Brown, N.J., Rodgers, G.R., Corke, K.P., Tazzyman, S., Underwood, J.C., and Lewis, C.E., Alphastatin, a 24-amino acid fragment of human fibrinogen, is a potent new inhibitor of activated endothelial cells in vitro and in vivo, Blood, 103 (2004) 601–606.

Strand, S., Hofmann, W.J., Grambihler, A., Hug, H., Volkmann, M., Otto, G., Wesch, H., Mariani, S.M., Hack, V., Stremmel, W., Krammer, P.H., and Galle, P.R., Hepatic failure and liver cell damage in acute Wilson's disease involve CD95 (APO-1/Fas) mediated apoptosis, Nat Med, 4 (1998) 588–593.

Sukocheva, O.A., Wang, L., Albanese, N., Pitson, S.M., Vadas, M.A., and Xia, P., Sphingosine kinase transmits estrogen signaling in human breast cancer cells, Mol Endocrinol, 17 (2003) 2002–2012.

Tabas, I., Williams, K.J., and Boren, J., Subendothelial lipoprotein retention as the initiating process in atherosclerosis: update and therapeutic implications, Circulation, 116 (2007) 1832–1844.

Tabas, I., Secretory sphingomyelinase, Chem Phys Lipids, 102 (1999) 123–130.

Taha, T.A., Kitatani, K., El-Alwani, M., Bielawski, J., Hannun, Y.A., Obeid, L.M., Loss of sphingosine kinase-1 activates the intrinsic pathway of programmed cell death: modulation of sphingolipid levels and the induction of apoptosis, FASEB J, 20 (2006) 482–484.

Tan, S.Y., Xiao, L., Pi, X., and Holoshitz, J., Aberrant Gi protein coupled receptor-mediated cell survival signaling in rheumatoid arthritis B cell lines, Front Biosci, 12 (2007) 1651–1660.

Taylor, C.J., Motamed, K, and Lilly, B., Protein kinase C and downstream signaling pathways in a three-dimensional model of phorbol ester-induced angiogenesis, Angiogenesis, 9 (2006) 39–51.

Testai, F.D., Landek, M.A., Goswami, R., Ahmed, M., and Dawson, G., Acid sphingomyelinase and inhibition by phosphate ion: role of inhibition by phosphatidyl-myo-inositol 3,4,5-triphosphate in oligodendrocyte cell signaling, J Neurochem, 89 (2004) 636–644.

Thevissen, K., Francois, I.E., Winderickx, J., Pannecouque, C., and Cammue, B.P., Ceramide involvement in apoptosis and apoptotic diseases., Mini Rev Med Chem, 6 (2006) 699–709.

Van Brocklyn, J.R., Jackson, C.A., Pearl, D.K., Kotur, M.S., Snyder, P.J., and Prior, T.W., Sphingosine kinase-1 expression correlates with poor survival of patients with glioblastoma multiforme: roles of sphingosine kinase isoforms in growth of glioblastoma cell lines, J Neuropathol Exp Neurol, 64 (2005) 695–705.

Venkataraman, K., Thangada, S., Michaud, J., Oo, M.L., Ai, Y., Lee, Y.M., Wu, M., Parikh, N.S., Khan, F., Proia, R.L., and Hla, T. Extracellular export of sphingosine kinase-1a contributes to the vascular S1P gradient, Biochem J, 397 (2006) 461–471.

Vessey, D.A., Kelley, M., Zhang, J., Li, L., Tao, R., and Karliner, J.S., Dimethylsphingosine and FTY720 inhibit the SK1 form but activate the SK2 form of sphingosine kinase from rat hearT., J Biochem Mol Toxicol, 21 (2007) 273–279.

Visentin, B., Vekich, J.A., Sibbald, B.J., Cavalli, A.L, Moreno, K.M., Matteo, R.G., Garland, W.A., Lu, Y., Yu, S., Hall, H.S., Kundra, V., Mills, G.B., and Sabbadini, R.A., Validation of an anti-sphingosine-1-phosphate antibody as a potential therapeutic in reducing growth, invasion, and angiogenesis in multiple tumor lineages, Cancer Cell, 9 (2006) 225–238.

Vlasenko, L.P., and Melendez, A.J., A critical role for sphingosine kinase in anaphylatoxin-induced neutropenis, peritonitis, and cytokine production in vivo, J Immunol, 174 (2005) 6456–6461.

Wong, M.L., Xies B., Beatinis N., Phu, P., Marathe, S., Johns, A., Gold, P.W., Hirsch, E., Williams, K.J., Licinio, J., and Tabas, I., Acute systemic inflammation up-regulates secretory sphingomyelinase in vivo: a possible link between inflammatory cytokines and atherogenesis, Proc Natl Acad Sci USA., 97 (2000) 8681–8686.

Wu, W., Shu, X., Hovsepyan, H., Mosteller, R.D., and Broek, D., VEGF receptor expression and signaling in human bladder tumor, Oncogene, 22 (2003) 3361–3370.

Wu, W., Mosteller, R.D., and Broek, D., Sphingosine kinase protects lipopolysaccharide-activated macrophages from apoptosis, Mol Cell Biol, 24 (2004) 7359–7369.

Xia, P., Gamble, J.R., Rye, K.A., Wang, L., Hii, C.S., Cockerill, P., Khew-Goodall, Y., Bert, A.G., Barter, P.J., and Vadas, M.A. Tumor necrosis factor-alpha induces adhesion molecule expression through the sphingosine kinase pathway, Proc Natl Acad Sci USA, 95 (1998) 14196–14201.

Xia, P., Vadas, M.A., Rye, K.A., Barter, P.J., and Gamble, J.R., High density lipoproteins (HDL) interrupt the sphingosine kinase signaling pathway, A possible mechanism for protection against atherosclerosis by HDL, Biol Chem, 274 (1999) 33143–33147.

Xia, P., Gamble, J.R., Wang, L., Pitson, S.M., Moretti, P.A., Wattenberg, B.W., D'Andrea, R.J., and Vadas, M.A. An oncogenic role of sphingosine kinase, Curr Biol, 10 (2000) 1527–1530.

Xia, P., Wang, L., Moretti, P.A., Albanese, N., Chai, F., Pitson, S.M., D'Andrea, R.J., Gamble, J.R., and Vadas, M.A., Sphingosine kinase interacts with TRAF2 and dissects tumor necrosis factor-alpha signaling, J Biol Chem, 277 (2002) 7996–8003.

Yang, J., Castle, B.E., Hanidu, A., Stevens, L., Yu, Y., Li, X., Stearns, C., Papov, V., Rajotte D., Li, J., Sphingosine kinase 1 is a negative regulator of CD4+ Th1 cell, J Immunol, 175 (2005) 6580–6588.

Yu, Z.F., Nikolova-Karakashian, M., Zhou, D., Cheng, G., Schuchman E.H., and Mattson, M.P., Pivotal role for acidic sphingomyelinase in cerebral ischemia-induced ceramide and cytokine production, and neuronal apoptosis, J Mol Neurosci, 15 (2000) 85–97.

Zemann, B., Kinzel, B., Müller, M., Reuschel, R., Mechtcheriakova, D., Urtz, N., Bornancin, F., Baumruker, T., and Billich A., Sphingosine kinase type 2 is essential for lymphopenia induced by the immunomodulatory drug FTY720.Blood, 107 (2006) 1454–1458.

Zemann, B., Urtz, N., Reuschel, R., Mechtcheriakova, D., Bornancin, F., Badegruber, R., Baumruker, T., and Billich, A., Normal neutrophil functions in sphingosine kinase type 1 and 2 knockout mice, Immunol Lett, 109 (2007) 56–63.

Zhi, L., Leung, B.P., and Melendez, A.J., Sphingosine kinase 1 regulates pro-inflammatory responses triggered by TNFalpha in primary human monocytes, J Cell Physiol, 208 (2006) 109–115.

# Chapter 20
# Ceramide-Enriched Membrane Domains in Infectious Biology and Development

**Katrin Anne Becker, Alexandra Gellhaus, Elke Winterhager and Erich Gulbins**

**Abstract** Ceramide has been shown to be critically involved in multiple biological processes, for instance induction of apoptosis after ligation of death receptors or application of gamma-irradiation or UV-A light, respectively, regulation of cell differentiation, control of tumor cell growth, infection of mammalian cells with pathogenic bacteria and viruses or the control of embryo and organ development to name a few examples. Ceramide molecules form distinct large domains in the cell membrane, which may serve to re-organize cellular receptors and signalling molecules. Thus, in many conditions, ceramide may be involved in the spatial and temporal organisation of specific signalling pathways explaining the pleiotrophic effects of this lipid. Here, we focus on the role of ceramide and ceramide-enriched membrane domains, respectively, in bacterial infections, in particular of the lung, and sepsis. We describe the role of ceramide for infections with *Neisseriae gonorhoeae*, *Staphylococcus aureus* and *Pseudomonas aeruginosa*. Finally, we discuss newly emerging aspects of the cellular function of ceramide, i.e. its role in germ line and embryo development.

**Keywords** Ceramide · platforms · development · cystic fibrosis · infection

## 20.1 Biophysical Principles

The fluid mosaic model of the cell membrane proposed by Singer and Nicolson in 1972 suggesting a random distribution of lipids and proteins and, thus, a liquid disordered phase (Singer and Nicolson, 1972) of the cell membrane was significantly revised in the last 10 years. Biophysical studies indicated that in particular sphingolipids and cholesterol form distinct membrane domains that may exist in a liquid-ordered phase (Simons and Ikonen, 1997; Brown and London, 1998; Kolesnick et al., 2000). Hydrophilic interactions between the headgroups of (glyco)sphingolipids and the hydroxy-group of cholesterol and

K.A. Becker
Department of Molecular Biology, University of Duisburg-Essen, Hufelandstrasse 55, 45122 Essen, Germany

P.J. Quinn, X. Wang (eds.), *Lipids in Health and Disease*, 

hydrophobic van der Waal interactions between the acyl-chains of sphingolipids and the sterol ring system result in a tight interaction of these lipids and spontaneous formation of gel-like domains with a much higher melting temperature than other phospholipids in the cell membrane (Simons and Ikonen, 1997; Brown and London, 1998; Kolesnick et al., 2000). Cholesterol seems to fill the void spaces between bulky sphingolipids and stabilize these domains (Xu et al., 2001; London et al., 2006). These interactions may result in the lateral organization of the cell membrane and the formation of distinct sphingolipid- and cholesterol-enriched membrane domains, named rafts (Simons and Ikonen, 1997), since they float in the ocean of other phospholipids in the cell membrane. The existence of rafts in the cell membrane at physiological temperatures still requires direct proof, while biophysical studies in model membranes proved the existence of distinct membranes composed of sphingolipids and cholesterol (Brown and London, 1998; Xu et al., 2001; London et al., 2006). Many studies employed non-ionic detergents to extract lipid domains and the proteins therein from cells, since it is proposed that the strong interactions between lipids may result in a highly ordered spatial structure and, thus, detergent resistance of these domains at low temperatures.

Sphingolipids are asymmetrically distributed in the cell membrane and are predominantly present in the outer leaflet of the cell membrane. Thus, sphingolipid- and cholesterol-enriched membrane domains exist in the outer leaflet, while at present it can be only speculated whether similar structures are also present in the inner leaflet of the cell membrane. In particular, long acyl-chains of sphingolipids are certainly able to interact with farnesylated, geranylated or palmitoylated proteins integrated in the inner leaflet of the cell membrane, and, furthermore, to influence the structure and lateral distribution of the inner membrane. However, the mechanisms that mediate an organization of the inner membrane leaflet similar to the structure of the outer membrane leaflet still need to be defined.

## 20.2 Ceramide-Enriched Membrane Domains/Platforms

The most prevalent membrane sphingolipid present in the outer leaflet of the cell membrane is sphingomyelin. Sphingomyelin consists of a D-erythro-sphingosine connected via an amide ester bond with a fatty acid containing 2–28 carbon atoms in the acyl chain that together form the hydrophobic ceramide moiety and a hydrophilic phosphorylcholine headgroup (Hakomori, 1983). Physiological ceramides display $C_{16}$- through $C_{32}$-chains, although short chain ceramides such as $C_2$-ceramide seem to be also present in cells. Ceramide is released in cellular membranes by the activity of sphingomyelinases that are characterized by their pH optimum (Quintern et al., 1989; Kitatani et al., 2007; Gulbins and Kolesnick, 2003). Thus, acid sphingomyelinases hydrolyze sphingomyelin to ceramide at an acidic pH, although other lipids critically influence the $K_m$ of the enzyme to its substrate (Schissel et al., 1996, 1998).

The acid sphingomyelinase has been shown to reside within vesicles, mostly lysosomes and secretory lysosomes, in mammalian cells (Grassme et al., 2001a). Depending on the glycosylation pattern cells also express a secretory acid sphingomyelinase that is secreted upon stimulation for instance via Interleukin 1 receptors (Schissel et al., 1996, 1998). The mobilization of intracellular vesicles results in exposure of the acid sphingomyelinase on the outer leaflet of the cell membrane, a process that has been shown to occur for instance after CD95, DR5 or CD40-receptor stimulation, but also after some bacterial and viral infections or stress stimuli (Grassme et al., 2001b, 2002, 2003a,b; Cremesti et al., 2001; Dumitru and Gulbins, 2006; Lang et al., 2007). The translocation of the acid sphingomyelinase onto the extracellular leaflet brings the enzyme into close contact with its substrate sphingomyelin resulting in the formation of ceramide within the extracellular leaflet of the cell membrane. In addition to translocation many stress stimuli also trigger an activation of the enzyme indicated by an increase of $V_{max}$ (Gulbins and Kolesnick, 2000). Ceramide molecules generated by the acid sphingomyelinase in the outer leaflet of the cell membrane alter the biophysical properties of the membrane, finally resulting in a marked lateral organization within the membrane. Ceramide molecules associate with each other resulting in the formation of small ordered membrane domains (Kolesnick et al., 2000). These small domains have the spontaneous tendency to fuse and to form larger ceramide-enriched membrane domains, also named platforms that can reach a diameter of up to 5 micrometer. In addition, ceramide molecules are very hydrophobic and, thus, the accumulation of ceramide in distinct domains results in the formation of highly hydrophobic platforms (Kolesnick et al., 2000; London et al., 2006). Ceramide molecules are tightly packed and, thus, ceramide-enriched membrane domains are in a gel-like, highly ordered status compared to other domains of the cell membrane (Kolesnick et al., 2000; London et al., 2006). Finally, ceramide molecules may exclude cholesterol from rafts, at least in model membranes, and this might further alter the composition of pre-formed rafts (London and London, 2004). In conclusion, the generation of ceramide in the outer (or any anti-cytoplasmatic) leaflet of cell membranes results in a membrane platform that is suitable to reorganize receptor and intracellular signaling molecules and, thus, to facilitate signal transduction. Consistent with this notion it was shown that receptors such as CD95 (Grassme et al., 2001b; Cremesti et al., 2001), CD40 (Grassme et al., 2002), DR5 (Dumitru and Gulbins, 2006) or CFTR (Pier et al., 1996) or signaling molecules such as NADPH-oxidase (Zhang et al., 2001, 2003), caspase 8 (Eramo et al., 2004) or Kv1.3 (Bock et al., 2003) to name a few, are concentrated and clustered within ceramide-enriched membrane domains, a process that is critically required for signal transduction via these molecules.

Several methods were applied to show the formation of ceramide-enriched membrane platforms: For instance, unilamellar vesicles composed of phosphatidylcholine/sphingomyelin were treated with immobilized sphingomyelinase, which resulted in rapid formation of large ceramide-enriched membrane platforms visualized by fluorescence microscopy of giant vesicles (Holopainen et al., 1998; Nurminen et al., 2002). Further studies on model membranes indicate that incorporation of even low amounts of ceramide results in transition of fluid phospholipid layers into a gel-like phase (Veiga et al., 1999) suggesting that ceramide promotes the formation of stable domains in bilayers. Phase separation was also confirmed by atomic force microscopy experiments on artificial $C_{16}$-ceramide-enriched, glycerol-phospholipid/cholesterol membranes (ten Grotenhuis et al., 1996). Our own studies employed fluorescent-labelled anti-ceramide-antibodies to demonstrate the formation of ceramide-enriched membrane domains after CD95 and DR5-stimulation or infection with some pathogenic bacteria (Grassme et al., 2001b, 2003b; Dumitru and Gulbins, 2006).

A recent study demonstrated that ceramide generated by the neutral sphingomyelinase also plays a critical role in membrane fusion (Trajkovic et al., 2008). These studies show that neutral sphingomyelinases generate ceramide in exosomes that mediates the fusion of these exosomes with the endosomes to form multivesicular endosomes and, thus, the secretion of the exosome cargo into the endosome (Trajkovic et al., 2008).

## 20.3 Ceramide in Bacterial Infections

Studies on *Neisseriae gonorrhoeae* (*N. gonorrhoeae*) established an activation of the acid sphingomyelinase upon bacterial infection (Grassme et al., 1997; Hauck et al., 2000). Ceramide is required for internalisation of *N. gonorrhoeae* as shown by studies on acid sphingomyelinase-deficient cells or after treatment of cells with an inhibitor of the acid sphingomyelinase (Grassme et al., 1997; Hauck et al., 2000). Further studies extended the concept that sphingolipids are involved in bacterial infections to *Staphylococcus aureus* (*S. aureus*) and demonstrated an activation of the acid sphingomyelinase with a concomitant release of ceramide upon infection of endothelial cells with *S. aureus* (Esen et al., 2001). In addition to mediating internalisation of pathogens, ceramide is also critical to trigger death of *S. aureus*-infected endothelial cells. However, the most detailed studies on the role of sphingolipids in infectious biology are published for *Pseudomonas aeruginosa* (*P. aeruginosa*) (Grassme et al., 2003b). Infection of murine or human respiratory epithelial cells with *P. aeruginosa* results in rapid activation of the acid sphingomyelinase, release of ceramide and the formation of large ceramide-enriched membrane platforms (Grassme et al., 2003b) (Fig. 20.1). *P. aeruginosa* localizes to these domains and infects the cells via ceramide enriched-membrane platforms (Fig. 20.2). These platforms are also positive for cholera toxin suggesting that they were formed from small rafts (Fig. 20.1). Genetic deficiency of the acid sphingomyelinase abrogates the generation of ceramide-enriched membrane platforms and prevents hallmarks of the infection, i.e. internalisation of the bacteria, induction of death in infected cells and a controlled release of cytokines in the infected lung. These data indicate that ceramide-enriched membrane platforms are critically involved in

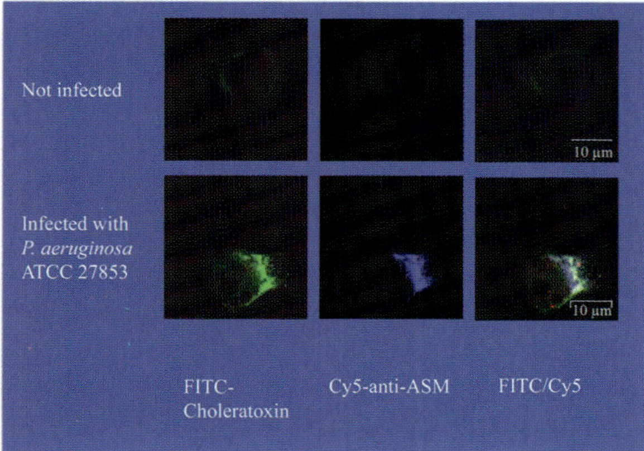


**Fig. 20.1** Infection with *P. aeruginosa* results in surface exposure of the acid sphingomyelinase and the formation of distinct membrane domains. Epithelial cells were infected with *P. aeruginosa*, the cells were fixed and stained with Cy5-coupled anti-acid sphingomyelinase antibodies and FITC-choleratoxin to visualize membrane domains. The results reveal the formation of a large membrane platform that co-localizes with surface acid sphingomyelinase

the infection of mammalian epithelial cells with *P. aeruginosa*. Studies by Pier et al. demonstrated that *P. aeruginosa* binds to the cystic fibrosis transmembrane conductance regulator (CFTR), which results in internalization of the bacteria (Pier et al., 1996). Further, cellular infection with *P. aeruginosa* was also shown to result in an activation of the CD95/CD95 ligand system and induction of cell death (Grassme et al., 2000). Both, CD95 and CFTR cluster in

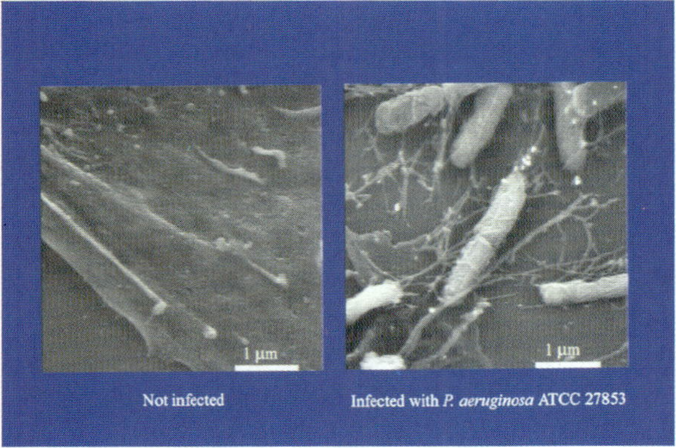


**Fig. 20.2** Surface acid sphingomyelinase co-localizes with *P. aeruginosa* on the infection site. Epithelial cells were infected with *P. aeruginosa*, the cells were fixed and stained with nano-Gold-coupled anti-acid sphingomyelinase antibodies to determine the exact localization of the acid sphingomyelinase on the cell surface. The studies demonstrate a localization of the acid sphingomyelinase at the site of the bacterial infection on the cell surface

ceramide-enriched membrane domains upon infection suggesting that these domains might promote the re-organisation of cellular receptors and signalling molecules that stimulate the infected cell and, thus, are required for an adequate response of the infected host cell to the pathogen.

## 20.4 Ceramide in Sepsis

Initial studies by Haimovitz-Friedman (Haimovitz-Friedman et al., 1997) demonstrated that injection of LPS into wildtype mice results in death of endothelial cells, the development of a sepsis and finally death of the mice. In contrast, acid sphingomyelinase-deficient mice are resistant to the infection, endothelial cells are protected from cell death and the mice are able to survive a systemic injection of LPS. These data clearly indicate that ceramide generated via the acid sphingomyelinase by bacterial LPS exposure triggers a dramatic systemic inflammation, a deterioration of the microcirculation by induction of endothelial cell death and finally death of the animal. These studies also suggest that an inhibition of the acid sphingomyelinase might be a useful concept to prevent or ameliorate septic symptoms.

The notion that the acid sphingomyelinase plays an important role in sepsis was further supported by elegant studies by Claus et al. (2005), who described an increase of the acid sphingomyelinase-activity in the blood of patients with sepsis correlating with poor clinical outcome of these patients. The acid sphingomyelinase and ceramide might trigger deterioration of organ function by mediating a massive release of pro-inflammatory mediators such as Interleukin 1 and TNF-alpha. In animal experiments application of a low molecular weight inhibitor of the acid sphingomyelinase protected the animals from severe sepsis further supporting the concept that the acid sphingomyelinase may play a critical role in the development of sepsis and organ failure.

Further studies by Göggel et al. demonstrate an important role of the acid sphingomyelinase in organ failure during sepsis (Göggel et al., 2004). These authors demonstrated that platelet activating factor (PAF) induces lung edema and injury via activation of an acid sphingomyelinase- and cyclooxygenase-dependent pathway presumably in endothelial cells. Ceramide elevates vascular permeability, which is antagonized by sphingosine-1-phosphate (S1P) (Göggel et al., 2004; Lindner et al., 2005). Inhibition of either the acid sphingomyelinase using D609 or cycloxygenase or genetic deficiency of the acid sphingomyelinase partially protect the animals from PAF-induced edema, while a complete abrogation of PAF-effects is obtained by the simultaneous inhibition of both pathways. Recently published data extended the role of the acid sphingomyelinase in acute lung dysfunction and demonstrate that a washout of surfactant from the lungs of newborn piglets or inhalation of LPS into the lung of mice activate the acid sphingomyelinase, negatively affect lung function and induce pulmonary edema (von Bismarck et al., 2008). These adverse events on lung function are prevented by inhibition of the acid sphingomyelinase with imipramine or by genetic deficiency of the acid sphingomyelinase.

## 20.5 Ceramide in Cystic Fibrosis

Pulmonary infections with *P. aeruginosa* are most important in children with cystic fibrosis, very often result in a destruction of the lung and are critical for the life quality and expectance of these patients. Although it is long known that cystic fibrosis is caused by a mutation of the Cystic Fibrosis Transmembrane Conductance Regulator (CFTR) (Riordan et al., 1989), the molecular mechanisms leading to the chronic infection with *P. aeruginosa* are still unknown. At present, one of the most commonly discussed concepts suggests a chronic inflammation in the lung of cystic fibrosis patients finally resulting in high susceptibility to *P. aeruginosa* infection (Zahm et al., 1997; Joseph et al., 2005). However, several studies demonstrated that the pH in lysosomes and other acidic compartments of cells deficient for functional Cftr is alkalanized (Barasch et al., 1991; Di et al., 2006). Cftr is a chloride channel and deficiency of the protein prevents sufficient chloride-ion influx into lysosomes, which are required as counter ions for $H^+$ to acidify these compartments. In a recent study we demonstrated that the increase of the pH in normally acidic compartments in *Cftr*-deficient cells results in an imbalance of the activities of the acid sphingomyelinase and the acid ceramidase (Teichgräber et al., 2008). While an increase of the pH to a value of approximately 6 results in almost complete inhibition of the acid ceramidase activity, the activity of the acid sphingomyelinase is only reduced by 30–40%. This imbalance resulted in accumulation of ceramide in bronchial epithelial cells in the lung of *Cftr*-deficient mice, which triggers an inadequate death of epithelial cells, the release of DNA into the bronchial lumen, the release of pro-inflammatory cytokines and an accumulation of macrophages and to a lesser degree also neutrophils. These changes are age-dependent and increase with age of the mice. Most important pharmacological inhibition of the acid sphingomyelinase or genetic heterozygosity of the enzyme in *Cftr*-deficient mice prevented the accumulation of ceramide in lungs of *Cftr*-deficient mice and, the hypersusceptibility of these mice to develop *P. aeruginosa* infections. Since a similar accumulation of ceramide was observed in human nasal epithelial cells and lung specimen, it might be possible to develop a novel treatment strategy for cystic fibrosis based on inhibition of the acid sphingomyelinase, although this treatment must be carefully adapted to prevent a complete inhibition of the enzyme and, thus, the failure to adequately respond to *P. aeruginosa* infections. Furthermore, these data indicate that a constitutive increase of organ concentrations of ceramide might be involved in the development of other chronic disorders and/or a high susceptibility to infections.

## 20.6 Ceramide in Viral and Parasitic Infections

The exact function of ceramide in viral and parasitic infections is much less defined than for bacterial infections. However, studies on HIV demonstrated that an increase of ceramide, for instance triggered by treatment of cells with sphingomyelinase, reduced infection with HIV (Finnegan et al., 2004; Finnegan and Blumenthal, 2006). Ceramide did not alter the binding of the virus or distribution of viral receptors, but reduced infectivity of the virus by redirecting the virus into the endocytic fraction, at least in vitro (Finnegan et al., 2004; Finnegan and Blumenthal, 2006). Studies on rhinoviruses demonstrated an activation of the acid sphingomyelinase and a release of ceramide upon infection with the viruses (Grassme et al., 2005). Rhinoviruses associate with ceramide-enriched membrane domains and genetic deficiency or pharmacological inhibition of the acid sphingomyelinase prevented uptake of at least some rhinoviral strains. Sindbis viruses extend the list of viruses that employ the acid sphingomyelinase for infection of mammalian cells. Infection with Sindbis virus triggers cell death during viral uptake, a process that requires activation of the acid sphingomyelinase and is prevented by acid ceramidase-mediated degradation of ceramide (Jan et al., 2000). Interestingly, even some parasites, for instance *Plasmodium falciparum (P. falciparum)*, seem to employ sphingomyelinases and ceramide for host cell invasion. *P. falciparum* express a sphingomyelinase that was shown to be critical for the uptake of the parasite by erythrocytes (Hanada et al., 2002), although it is unknown whether this also involves ceramide-enriched membrane domains.

## 20.7 Sphingolipids in Germ Cell Development

### 20.7.1 Sphingolipids in Female Germ Line Development

In the female germ cell line the oocyte development starts from proliferating primordial cells, which migrate into the gonads and differentiate into approximately 7 million oogonia prenatally. Around birth most of these oocytes are depleted by apoptosis and they cease proliferation with entering into the first meiotic division. Oocytes remain arrested in the prophase of division 1 until puberty. In humans, meiosis is completed in a cohort of mature oocytes in two steps first after ovulation and second – if fertilization occurs – meiosis 2 is terminated upon sperm entrance. Interestingly from the remaining 1–2 million primary oocytes at birth only few hundred thousands survive until puberty. During the reproductive phase only 400–500 follicles develop into mature oocytes and ovulate. All others undergo controlled atresia in the different steps of follicular developmental stages. Taken together, there is a continuous depletion of proliferating fetal oogonia, primary oocytes and after puberty of all follicular stages by apoptosis. While age-related alterations at the level of the hypothalamus-pituitary appear to determine the timing of menarche, the number of follicles remaining in the mature ovary is the major determinant of the timing of both the peri-menopause and the menopause. In addition programmed cell death seems to be increased in immature oocytes of elderly women compared to young individuals concomitant with decreasing oocyte maturation (Wu et al., 2000), a mechanism which enhances the depletion process before menopause.

Here, we will focus on the role of the acid sphingomyelinase in the control of the balance of apoptosis in germ line development. The acid sphingomyelinase has been shown to be critically involved in the regulation of primordial oocyte cell death (Morita et al., 2000). Deficiency of the acid sphingomyelinase and the concomitant reduction of the conversion of spingolipids into ceramide reduce fetal germ cell apoptosis leading to ovarian hyperplasia at birth (Morita et al., 2000). Furthermore, acid sphingomyelinase-deficiency or supplementation with S1P, an inhibitor of ceramide-promoted cell death, protected oocytes from radiation and chemotherapy (doxorubicin)-mediated apoptosis indicating that ceramide released by the acid sphingomyelinase is key for the induction of cell death in oocytes by these forms of stress (Morita et al., 2000). Thus, acid sphingomyelinase-deficient mice give strong evidence that ceramide promotes survival of oocytes and protects oocytes from radiation- and chemotherapy-induced apoptosis (Morita et al., 2000).

These effects of ceramide are not only related to cell death in the fetal germ cells and primary oocytes, but also to the age-related increase in oocyte apoptosis. Recently, Perez et al. (2005) described that ceramide is predominantly found in cumulus cells of aged mice and only weakly in those of young animals. Ceramide trafficking from the cumulus cells into the oocyte is dependent on the intact cumulus-oocyte complex with a physiologically intact gap junctional communication. This trafficking is accompanied by induction of oocyte apoptosis. It has to be considered, however, that the water filled pore of connexin channels consisting of connexin37 or connexin43 (Kidder and Mhawi, 2002) are not able to exchange hydrophobic long chain ($C_{16}$ or longer acyl chains) ceramides, a phenomenon that requires further clarification. In any case, a manipulation of ceramide within the system of oocytes and cumulus cells may be a promising new therapeutic approach to prevent premature oocyte loss or to extend ovarian function.

### *20.7.2 Sphingolipids in Male Germ Line Development*

Major differences exist between oogenesis and spermatogenesis: While oogonia proliferate only in the fetus, which results in a limited stock, in males sperms proliferate continuously from puberty onwards. As a consequence apoptosis occurs in the testis as an important physiological mechanism to limit the number of germ cells in the seminiferous epithelium and thus is central for normal gametogenesis in testis (Lee et al., 1997; Dunkel et al., 1997; Print and Loveland, 2000). It has been long known that neutral and acid sphingomyelinase activities are high in the testis (Spence et al., 1979; Hinkovska et al., 1987), but only recently more information has been published about a direct role of ceramide or sphingosine derivatives for sperm differentiation and apoptosis in testis.

Acid sphingomyelinase-deficient males exhibit reduced fertility. Investigation of testis and sperms reveals accumulation of sphingomyelin and cholesterol in spermatozoa of these mice, which results in severe disruptions of the sperm cell membrane as well as the acrosome membranes concomitantly with an impaired capacitation. The mutant sperm regain normal morphology by adding detergents reducing the overhead of lipids (Butler et al., 2002). Mice heterozygous for acid sphingomyelinase reveal two distinct populations of sperms, which represent the morphology of mutant and wild type animals. Fertilization with the healthy ones after sperm sorting results in a higher proportion of wild type pups. Heterogeneity in the morphological shape might thus be a possibility for couples heterozygous for Niemann-Pick disease to perform assisted reproduction to avoid a transfer of this genetic disease (Butler et al., 2007).

Similar to the findings on ovaries, Otala et al. (2005) demonstrated that irradiation induced apoptosis in male germ cells is mediated by the acid sphingomyelinase and prevented by deficiency of the enzyme or treatment with S1P in vivo (Otala et al., 2005).

## 20.8 Sphingolipids in Implantation

S1P does not only prevent oocyte apoptosis, but is also required for successful implantation. S1P is synthezised from sphingosine by sphingosine-kinase SPHK1 and SPHK2. Though deletion of *Sphk1* or *Sphk2* does not influence fertility (Kono et al., 2004; Mizugishi et al., 2005) the double mutants missing completely S1P were embryonic lethal due to severe malformations in neural and vascular development (Mizugishi et al., 2005). Interestingly double mutants with heterozygosity of *sphk2* (*sphk1* −/− *sphk2* +/−) exhibit significantly reduced levels of S1P, but are phenotypically normal and viable. However, female mice are infertile due to early pregnancy loss (Mizugishi et al, 2007). Embryos do implant, but get absorbed between days 7.5 and 9.5 post conceptionem. The reason for these resorptions is the impaired decidualization process of the stromal compartment surrounding the implanting blastocyst. Though the decidualization process is induced by the implanting blastocyst, the balance between proliferation and apoptosis of decidual cells in normal implantation as well as in artificial decidualization is disturbed in favour to cell death in these mice. Moreover, decidual blood vessels loose stability leading to hemorrhages causing malnutrition and death of the embryo *in utero* in these mice.

## 20.9 Sphingolipids in Early Embryonic Development

### *20.9.1 Pre-implantation Embryos*

During pre-implantation mammalian embryos undergo a series of divisions to generate a blastocyst competent for implantation. Within this context maternal factors regulate the plan of divisions by unevenly distribution of cytoplasm. In addition with several key transcriptional factors the different cell lineages, inner cell mass, the trophoblast and the primitive endoderm are probably generated by this mechanism (for review see Rossant 2004).

The cleavage steps require in addition to all molecules involved in cytokinesis insertion of additional plasma membrane material to provide sufficient membranes for both daughter cells. It has been shown that accumulation of phosphatidylethanolamine in the cleavage furrow membrane is crucial for division (Emoto and Umeda, 2000; Emoto et al., 2005). Furthermore, Comiskey and Warner (2007) demonstrate that membrane cholesterol and spingolipid-enriched rafts identified by labelled choleratoxin ß subunits are highly present in mouse oocyte and zygote membranes and associate with the membrane furrows during pre-implantation. Cholesterol depletion by methyl-ß-cyclodextrin impairs development of pre-implantation embryos. Very recent studies provide direct evidence for a critical role of ceramide in embryonic development prior to implantation. These studies reveal that deficiency of the acid ceramidase (*Asah1*$^{-/-}$ mice) (Li et al., 2002; Eliyahu et al., 2007) results in embryonic lethality, because the pre-implantation embryos do not develop beyond the two-cell stage. Expression studies indicate that the acid ceramidase is already expressed in the two-cell stage. Supplementation of the two-cell stage with S1P leads to a further development into the 4–8 cell stage, but not into blastocysts. The molecular details of ceramide-induced death of the two-cell stage embryos remain to be clarified. However, these studies clearly indicate that ceramide levels must be tightly controlled to permit embryonic development.

### *20.9.2 Sphingolipids in Embryogenesis*

Ceramide including glycosphingolipids are also central for embryogenesis as shown by gene targeting of mice. Thus, mice deficient in the ceramide synthase pathway die before E15. Deleting the major synthesis pathway for the glycosphingolipids, i.e. the glucosylceramide synthetase, results in even earlier death directly after gastrulation around E7.5 upon massive apoptosis in all embryonic layers (Yamashita et al., 1999). Glucosylceramide synthetase-deficient ES cells developed into endodermal, mesodermal and ectodermal progenitors, but did not differentiate further into tissues.

Surprisingly, ceramide does not only exhibit pro-apoptotic functions in development, but also promotes morphogenesis in the primitive ectoderm

(Krishnamurthy et al., 2007). This study indicates that ceramide and ceramide-enriched membrane domains are critical for the development of the primitive ectoderm into all embryonic tissues. Primitive ectoderm cells displayed increased levels of ceramide that showed a polar, apicolateral distribution. The GTPase Cdc42, the protein kinase C-zeta/lambda and F-actin co-localized with these ceramide-enriched domains. Inhibition of ceramide synthesis using myriocin or RNA-interference impaired primitive ectoderm morphogenesis proving the significance of ceramide for primitive ectoderm development. Morphogenesis was re-established when embryoid bodies were supplemented with ceramide. Thus, ceramide seems to bind to PKC-zeta/lambda at the cell membrane and may form a complex with polarity proteins, i.e. Cdc42, followed by activation of PKCzeta/lambda, which in turn deactivates GSK-3beta. Loss of this polarized stabilisation may result in a loss of ectodermal cell polarity and finally cellular apoptosis.

Taken together, in embryo development ceramide seems to be a morphogenetic lipid by organisation of defined cell membrane domains, which in turn leads to a polar organisation of the embryo required for normal development.

**Acknowledgments** Parts of the work described in the present review was supported by German Research Foundation grants DFG 774/21-2 and DFG 774/22-1 to E.W. and DFG 335/16-1 to E.G.

## References

Barasch, J., Kiss, B., Prince, A., Saiman, L., Gruenert, D. and al-Awqati, Q. Defective acidification of intracellular organelles in cystic fibrosis. Nature 352 (1991) 70–3.

Bock, J., Szabo, I., Gamper, N., Adams, C. Gulbins, E. Ceramide inhibits the potassium channel Kv1.3 by the formation of membrane platforms. Biochem Biophys Res Commun 305 (2003) 890–7.

Brown, D. A. London, E. Functions of lipid rafts in biological membranes. Annu Rev Cell Dev Biol 14 (1998) 111–36.

Butler, A., He, X., Gordon, R. E., Wu, H. S., Gatt, S. Schuchman, E. H. Reproductive pathology and sperm physiology in acid sphingomyelinase-deficient mice. Am J Pathol 161 (2002) 1061–75.

Butler, A., Gordon, R. E., Gatt, S. Schuchman, E. H. Sperm abnormalities in heterozygous acid sphingomyelinase knockout mice reveal a novel approach for the prevention of genetic diseases. Am J Pathol 170 (2007) 2077–88.

Claus, R. A., Bunck, A. C., Bockmeyer, C. L., Brunkhorst, F. M., Losche, W., Kinscherf, R. Deigner, H. P. Role of increased sphingomyelinase activity in apoptosis and organ failure of patients with severe sepsis. Faseb J 19 (2005) 1719–21.

Comiskey, M. Warner, C. M. Spatio-temporal localization of membrane lipid rafts in mouse oocytes and cleaving preimplantation embryos. Dev Biol 303 (2007) 727–39.

Cremesti, A., Paris, F., Grassme, H., Holler, N., Tschopp, J., Fuks, Z., Gulbins, E. Kolesnick, R. Ceramide enables fas to cap and kill. J Biol Chem 276 (2001) 23954–61.

Di, A., Brown, M. E., Deriy, L. V., Li, C., Szeto, F. L., Chen, Y., Huang, P., Tong, J., Naren, A. P., Bindokas, V., Palfrey, H. C. Nelson, D. J. CFTR regulates phagosome acidification in macrophages and alters bactericidal activity. Nat Cell Biol 8 (2006) 933–44.

Dumitru, C. A. Gulbins, E. TRAIL activates acid sphingomyelinase via a redox mechanism and releases ceramide to trigger apoptosis. Oncogene 25 (2006) 5612–25.
Dunkel, L., Hirvonen, V. Erkkila, K. Clinical aspects of male germ cell apoptosis during testis development and spermatogenesis. Cell Death Differ 4 (1997) 171–9.
Eliyahu, E., Park, J. H., Shtraizent, N., He, X. Schuchman, E. H. Acid ceramidase is a novel factor required for early embryo survival. Faseb J 21 (2007) 1403–9.
Emoto, K. Umeda, M. An essential role for a membrane lipid in cytokinesis. Regulation of contractile ring disassembly by redistribution of phosphatidylethanolamine. J Cell Biol 149 (2000) 1215–24.
Emoto, K., Inadome, H., Kanaho, Y., Narumiya, S. Umeda, M. Local change in phospholipid composition at the cleavage furrow is essential for completion of cytokinesis. J Biol Chem 280 (2005) 37901–7.
Eramo, A., Sargiacomo, M., Ricci-Vitiani, L., Todaro, M., Stassi, G., Messina, C. G., Parolini, I., Lotti, F., Sette, G., Peschle, C. De Maria, R. CD95 death-inducing signaling complex formation and internalization occur in lipid rafts of type I and type II cells. Eur J Immunol 34 (2004) 1930–40.
Esen, M., Schreiner, B., Jendrossek, V., Lang, F., Fassbender, K., Grassme, H. Gulbins, E. Mechanisms of Staphylococcus aureus induced apoptosis of human endothelial cells. Apoptosis 6 (2001) 431–9.
Finnegan, C. M., Rawat, S. S., Puri, A., Wang, J. M., Ruscetti, F. W. Blumenthal, R. Ceramide, a target for antiretroviral therapy. Proc Natl Acad Sci USA 101 (2004) 15452–7.
Finnegan, C. M. Blumenthal, R. Fenretinide inhibits HIV infection by promoting viral endocytosis. Antiviral Res 69 (2006) 116–23.
Göggel, R., Winoto-Morbach, S., Vielhaber, G., Imai, Y., Lindner, K., Brade, L., Brade, H., Ehlers, S., Slutsky, A. S., Schutze, S., Gulbins, E. Uhlig, S. PAF-mediated pulmonary edema: a new role for acid sphingomyelinase and ceramide. Nat Med 10 (2004) 155–60.
Grassme, H., Gulbins, E., Brenner, B., Ferlinz, K., Sandhoff, K., Harzer, K., Lang, F. Meyer, T. F. Acidic sphingomyelinase mediates entry of *N. gonorrhoeae* into nonphagocytic cells. Cell 91 (1997) 605–15.
Grassme, H., Kirschnek, S., Riethmueller, J., Riehle, A., von Kürthy, G., Lang, F., Weller, M. Gulbins, E. Host defense to *Pseudomonas aeruginosa* requires CD95/CD95 ligand interaction on epithelial cells. Science 290 (2000) 527–30.
Grassme, H., Schwarz, H. Gulbins, E. Molecular mechanisms of ceramide-mediated CD95 clustering. Biochem Biophys Res Commun 284 (2001a) 1016–30.
Grassme, H., Jekle, A., Riehle, A., Schwarz, H., Berger, J., Sandhoff, K., Kolesnick, R. Gulbins, E. CD95 signaling via ceramide-rich membrane rafts. J Biol Chem 276 (2001b) 20589–96.
Grassme, H., Jendrossek, V., Bock, J., Riehle, A. Gulbins, E. Ceramide-rich membrane rafts mediate CD40 clustering. J Immunol 168 (2002) 298–307.
Grassme, H., Cremesti, A., Kolesnick, R. Gulbins, E. Ceramide-mediated clustering is required for CD95-DISC formation. Oncogene 22 (2003a) 5457–70.
Grassme, H., Jendrossek, V., Riehle, A., von Kurthy, G., Berger, J., Schwarz, H., Weller, M., Kolesnick, R. Gulbins, E. Host defense against *Pseudomonas aeruginosa* requires ceramide-rich membrane rafts. Nat Med 9 (2003b) 322–30.
Grassme, H., Riehle, A., Wilker, B. Gulbins, E. Rhinoviruses infect human epithelial cells via ceramide-enriched membrane platforms. J Biol Chem 280 (2005) 26256–62.
Gulbins, E. Kolesnick, R. Measurement of sphingomyelinase activity. Methods Enzymol 322 (2000) 382–8.
Gulbins, E. Kolesnick, R. Raft ceramide in molecular medicine. Oncogene 22 (2003) 7070–7.
Haimovitz-Friedman, A., Cordon-Cardo, C., Bayoumy, S., Garzotto, M., McLoughlin, M., Gallily, R., Edwards, C. K., 3rd, Schuchman, E. H., Fuks, Z. Kolesnick, R. Lipopolysaccharide induces disseminated endothelial apoptosis requiring ceramide generation. J Exp Med 186 (1997) 1831–41.

Hakomori, S. Chemistry of Glycosphingolipids. In: Kanfer JN, Hakomori S, editors. Sphingolipid Biochemistry. New York and London: Plenum Press (1983) 1–165.

Hanada, K., Palacpac, N. M., Magistrado, P. A., Kurokawa, K., Rai, G., Sakata, D., Hara, T., Horii, T., Nishijima, M. Mitamura, T. Plasmodium falciparum phospholipase C hydrolyzing sphingomyelin and lysocholinephospholipids is a possible target for malaria chemotherapy. J Exp Med 195 (2002) 23–34.

Hauck, C. R., Grassme, H., Bock, J., Jendrossek, V., Ferlinz, K., Meyer, T. F. Gulbins, E. Acid sphingomyelinase is involved in CEACAM receptor-mediated phagocytosis of Neisseria gonorrhoeae. FEBS Lett 478 (2000) 260–6.

Hinkovska, V. T., Petkova, D. H. Koumanov, K. S. A neutral sphingomyelinase in spermatozoal plasma membranes. Biochem Cell Biol 65 (1987) 525–8.

Holopainen, J. M., Subramanian, M. Kinnunen, P. K. Sphingomyelinase induces lipid microdomain formation in a fluid phosphatidylcholine/sphingomyelin membrane. Biochemistry 37 (1998) 17562–70.

Jan, J. T., Chatterjee, S. Griffin, D. E. Sindbis virus entry into cells triggers apoptosis by activating sphingomyelinase, leading to the release of ceramide. J Virol 74 (2000) 6425–32.

Joseph, T., Look, D. Ferkol, T. NF-kappaB activation and sustained IL-8 gene expression in primary cultures of cystic fibrosis airway epithelial cells stimulated with Pseudomonas aeruginosa. Am J Physiol Lung Cell Mol Physiol 288 (2005) L471–9.

Kidder, G. M. Mhawi, A. A. Gap junctions and ovarian folliculogenesis. Reproduction 123 (2002) 613–20.

Kitatani, K., Idkowiak-Baldys, J. Hannun, Y. A. The sphingolipid salvage pathway in ceramide metabolism and signaling. Cell Signal (2007) PMID 18191382

Kolesnick, R. N., Goni, F. M. Alonso, A. Compartmentalization of ceramide signaling: physical foundations and biological effects. J Cell Physiol 184 (2000) 285–300.

Kono, M., Mi, Y., Liu, Y., Sasaki, T., Allende, M. L., Wu, Y. P., Yamashita, T. Proia, R. L. The sphingosine-1-phosphate receptors S1P1, S1P2, and S1P3 function coordinately during embryonic angiogenesis. J Biol Chem 279 (2004) 29367–73.

Krishnamurthy, K., Wang, G., Silva, J., Condie, B. G. Bieberich, E. Ceramide regulates atypical PKCzeta/lambda-mediated cell polarity in primitive ectoderm cells. A novel function of sphingolipids in morphogenesis. J Biol Chem 282 (2007) 3379–90.

Lang, P. A., Schenck, M., Nicolay, J. P., Becker, J. U., Kempe, D. S., Lupescu, A., Koka, S., Eisele, K., Klarl, B. A., Rubben, H., Schmid, K. W., Mann, K., Hildenbrand, S., Hefter, H., Huber, S. M., Wieder, T., Erhardt, A., Häussinger, D., Gulbins, E. Lang, F. Liver cell death and anemia in Wilson disease involve acid sphingomyelinase and ceramide. Nat Med 13 (2007) 164–70.

Lee, J., Richburg, J. H., Younkin, S. C. Boekelheide, K. The Fas system is a key regulator of germ cell apoptosis in the testis. Endocrinology 138 (1997) 2081–8.

Li, C. M., Park, J. H., Simonaro, C. M., He, X., Gordon, R. E., Friedman, A. H., Ehleiter, D., Paris, F., Manova, K., Hepbildikler, S., Fuks, Z., Sandhoff, K., Kolesnick, R. Schuchman, E. H. Insertional mutagenesis of the mouse acid ceramidase gene leads to early embryonic lethality in homozygotes and progressive lipid storage disease in heterozygotes. Genomics 79 (2002) 218–24.

Lindner, K., Uhlig, U. Uhlig, S. Ceramide alters endothelial cell permeability by a nonapoptotic mechanism. Br J Pharmacol 145 (2005) 132–40.

London, M. London, E. Ceramide selectively displaces cholesterol from ordered lipid domains (rafts): implications for lipid raft structure and function. J Biol Chem 279 (2004) 9997–10004.

London, M., Bakht, O. London, E. Cholesterol precursors stabilize ordinary and ceramide-rich ordered lipid domains (lipid rafts) to different degrees. Implications for the Bloch hypothesis and sterol biosynthesis disorders. J Biol Chem 281 (2006) 21903–13.

Mizugishi, K., Yamashita, T., Olivera, A., Miller, G. F., Spiegel, S. Proia, R. L. Essential role for sphingosine kinases in neural and vascular development. Mol Cell Biol 25 (2005) 11113–21.

Mizugishi, K., Li, C., Olivera, A., Bielawski, J., Bielawska, A., Deng, C. X. Proia, R. L. Maternal disturbance in activated sphingolipid metabolism causes pregnancy loss in mice. J Clin Invest 117 (2007) 2993–3006.

Morita, Y., Perez, G. I., Paris, F., Miranda, S. R., Ehleiter, D., Haimovitz-Friedman, A., Fuks, Z., Xie, Z., Reed, J. C., Schuchman, E. H., Kolesnick, R. N. Tilly, J. L. Oocyte apoptosis is suppressed by disruption of the acid sphingomyelinase gene or by sphingosine-1-phosphate therapy. Nat Med 6 (2000) 1109–14.

Nurminen, T. A., Holopainen, J. M., Zhao, H. Kinnunen, P. K. Observation of topical catalysis by sphingomyelinase coupled to microspheres. J Am Chem Soc 124 (2002) 12129–34.

Otala, M., Pentikainen, M. O., Matikainen, T., Suomalainen, L., Hakala, J. K., Perez, G. I., Tenhunen, M., Erkkila, K., Kovanen, P., Parvinen, M. Dunkel, L. Effects of acid sphingomyelinase deficiency on male germ cell development and programmed cell death. Biol Reprod 72 (2005) 86–96.

Perez, G. I., Jurisicova, A., Matikainen, T., Moriyama, T., Kim, M. R., Takai, Y., Pru, J. K., Kolesnick, R. N. Tilly, J. L. A central role for ceramide in the age-related acceleration of apoptosis in the female germline. Faseb J 19 (2005) 860–2.

Pier, G. B., Grout, M., Zaidi, T. S., Olsen, J. C., Johnson, L. G., Yankaskas, J. R. Goldberg, J. B. Role of mutant CFTR in hypersusceptibility of cystic fibrosis patients to lung infections. Science 271 (1996) 64–7.

Print, C. G. Loveland, K. L. Germ cell suicide: new insights into apoptosis during spermatogenesis. Bioessays 22 (2000) 423–30.

Quintern, L. E., Schuchman, E. H., Levran, O., Suchi, M., Ferlinz, K., Reinke, H., Sandhoff, K. Desnick, R. J. Isolation of cDNA clones encoding human acid sphingomyelinase: occurrence of alternatively processed transcripts. Embo J 8 (1989) 2469–73.

Riordan, J. R., Rommens, J. M., Kerem, B., Alon, N., Rozmahel, R., Grzelczak, Z., Zielenski, J., Lok, S., Plavsic, N., Chou, J. L. and et al. Identification of the cystic fibrosis gene: cloning and characterization of complementary DNA. Science 245 (1989) 1066–73.

Rossant, J. Lineage development and polar asymmetries in the peri-implantation mouse blastocyst. Semin Cell Dev Biol 15 (2004) 573–81.

Schissel, S. L., Schuchman, E. H., Williams, K. J. Tabas, I. $Zn^{2+}$-stimulated sphingomyelinase is secreted by many cell types and is a product of the acid sphingomyelinase gene. J Biol Chem 271 (1996) 18431–6.

Schissel, S. L., Keesler, G. A., Schuchman, E. H., Williams, K. J. Tabas, I. The cellular trafficking and zinc dependence of secretory and lysosomal sphingomyelinase, two products of the acid sphingomyelinase gene. J Biol Chem 273 (1998) 18250–9.

Simons, K. Ikonen, E. Functional rafts in cell membranes. Nature 387 (1997) 569–72.

Singer, S. J. Nicolson, G. L. The fluid mosaic model of the structure of cell membranes. Science 175 (1972) 720–31.

Spence, M. W., Burgess, J. K. Sperker, E. R. Neutral and acid sphingomyelinases: somatotopographical distribution in human brain and distribution in rat organs. A possible relationship with the dopamine system. Brain Res 168 (1979) 543–51.

Teichgräber, V., Ulrich, M., Endlich, N., Riethmüller, J., Wilker, B., De Oliveira-Munding, C., van Heeckeren, A. M., Barr, M. L., von Kurthy, G., Schmid, K. W., Weller, M., Tümmler, B., Lang, F., Grassme, H., Döring, G. Gulbins, E. Ceramide accumulation mediates inflammation, cell death and infection susceptibility in cystic fibrosis. Nat Med 14 (2008) 382–91.

ten Grotenhuis, E., Demel, R. A., Ponec, M., Boer, D. R., van Miltenburg, J. C. Bouwstra, J. A. Phase behavior of stratum corneum lipids in mixed Langmuir-Blodgett monolayers. Biophys J 71 (1996) 1389–99.

Trajkovic, K., Hsu, C., Chiantia, S., Rajendran, L., Wenzel, D., Wieland, F., Schwille, P., Brugger, B. Simons, M. Ceramide triggers budding of exosome vesicles into multivesicular endosomes. Science 319 (2008) 1244–7.

Veiga, M. P., Arrondo, J. L., Goni, F. M. Alonso, A. Ceramides in phospholipid membranes: effects on bilayer stability and transition to nonlamellar phases. Biophys J 76 (1999) 342–50.

von Bismarck, P., Garcia Wistadt, C. F., Klemm, K., Winoto-Morbach, S., Uhlig, U., Schutze, S., Adam, D., Lachmann, B., Uhlig, S. Krause, M. F. Improved pulmonary function by acid sphingomyelinase inhibition in a newborn piglet lavage model. Am J Respir Crit Care Med (2008) PMID 18310483.

Wu, J., Zhang, L. Wang, X. Maturation and apoptosis of human oocytes in vitro are age-related. Fertil Steril 74 (2000) 1137–41.

Xu, X., Bittman, R., Duportail, G., Heissler, D., Vilcheze, C. London, E. Effect of the structure of natural sterols and sphingolipids on the formation of ordered sphingolipid/sterol domains (rafts). Comparison of cholesterol to plant, fungal, and disease-associated sterols and comparison of sphingomyelin, cerebrosides, and ceramide. J Biol Chem 276 (2001) 33540–6.

Yamashita, T., Wada, R., Sasaki, T., Deng, C., Bierfreund, U., Sandhoff, K. Proia, R. L. A vital role for glycosphingolipid synthesis during development and differentiation. Proc Natl Acad Sci USA 96 (1999) 9142–7.

Zahm, J. M., Gaillard, D., Dupuit, F., Hinnrasky, J., Porteous, D., Dorin, J. R. Puchelle, E. Early alterations in airway mucociliary clearance and inflammation of the lamina propria in CF mice. Am J Physiol 272 (1997) C853–9.

Zhang, D. X., Zou, A. P. Li, P. L. Ceramide reduces endothelium-dependent vasodilation by increasing superoxide production in small bovine coronary arteries. Circ Res 88 (2001) 824–31.

Zhang, D. X., Zou, A. P. Li, P. L. Ceramide-induced activation of NADPH oxidase and endothelial dysfunction in small coronary arteries. Am J Physiol Heart Circ Physiol 284 (2003) H605–12.

# Part IV
# Lipidomics

# Chapter 21
# MALDI-TOF MS Analysis of Lipids from Cells, Tissues and Body Fluids

**Beate Fuchs and Jürgen Schiller**

**Abstract** Many diseases as atherosclerosis and metabolic dysfunctions are known to correlate with changes of the lipid profile of tissues and body fluids. Therefore, the importance of reliable methods of lipid analysis is obvious. Although matrix-assisted laser desorption and ionization time-of-flight mass spectrometry (MALDI-TOF MS) was so far primarily used for protein analysis, this method has itself proven to be very useful in lipid analysis, too. This review provides an overview of applications of MALDI-TOF MS in lipid analysis and summarizes the specific advantages and drawbacks of this modern soft-ionization method. The focus will be on the analysis of body fluids and cells as well as the diagnostic potential of the method in the lipid field. It will be shown that MALDI-TOF mass spectra can be recorded in a very short time and provide important information on the lipid as well as the fatty acyl composition of the lipids of an unknown sample. However, it will also be shown that only selected lipid classes (in particular those with quaternary ammonia groups as phosphatidylcholine) are detected if crude mixtures are analyzed as they are more sensitively detectable than other ones. This review ends with a short outlook emphasizing current methodological developments.

**Keywords** Lipids · phospholipids · lipid analysis · MALDI-TOF MS · lipid extracts · body fluids · cells · tissues

**Abbreviations** APCI: Atmospheric Pressure Chemical Ionization; DAG: Diacylglycerols; DE: Delayed Extraction; DHB: 2,5-Dihydroxybenzoic Acid; EI: Electron Impact; ESI: Electrospray Ionisation; FACS: Fluorescence-activated cell sorting; GC: Gas Chromatography; HDL: High Density Lipoprotein; HPLC: High-Performance Liquid Chromatography; IR: Infrared; LDL: Low Density Lipoprotein; LOD: Level of Detection; LOQ: Level of Quantification; LPC: Lyso-Phosphatidylcholine; LPL: Lyso-Phospholipid; MALDI: Matrix-

B. Fuchs
University of Leipzig, Medical Department, Institute of Medical Physics and Biophysics, Härtelstr. 16/18, D-04107 Leipzig, Germany

P.J. Quinn, X. Wang (eds.), *Lipids in Health and Disease*,

Assisted Laser Desorption and Ionization; MS: Mass Spectrometry; m/z: mass over charge; NMR: Nuclear Magnetic Resonance; PA: Phosphatidic Acid; PC: Phosphatidylcholine; PE: Phosphatidylethanolamine; PG: Phosphatidylglycerol; PI: Phosphatidylinositol; PIP: Phosphatidylinositol-Phosphate; PL: Phospholipid; $PLA_2$: Phospholipase $A_2$; PNA: Para-Nitroaniline; ppm: Parts per Million; PPI: (Poly-)Phosphoinositides; PS: Phosphatidylserine; PSD: Post Source Decay; ROS: Reactive Oxygen Species; SM: Sphingomyelin; S/N: Signal to Noise; sn: Stereospecific Numbering; TAG: Triacylglycerol; TFA: Trifluoroacetic Acid; TLC: Thin-Layer Chromatography; TOF: Time-of-Flight; UV: Ultraviolet.

## 21.1 Survey of Methods Used for the Analysis of Lipids and Phospholipids

Due to the (recently) recognized importance of lipids, many different analytical methods to assess the composition of an unknown sample (e.g. an organic extract of a body fluid) are available. Based on the available information, all these techniques can be sorted into methods providing primarily (a) chemical or (b) physical information (Christie 2003). Chemical (compositional) information is primarily provided by methods based on chromatography and both, liquid chromatography (in particular high-performance liquid chromatography (HPLC)) as well as thin-layer chromatography (TLC), are widely used for lipid analysis. This important field was recently comprehensively reviewed (Peterson and Cummings 2006). Although HPLC is a powerful tool for lipid analysis, TLC is also widely used because it provides many advantages: For instance, TLC is convenient, can be performed very rapidly and, in particular, does not provide any "memory" effects as a completely new stationary is used in all cases. Surveys dealing with lipid applications of TLC are provided in (Touchstone 1995) and of HPLC in Clejan (1998). Although a more comprehensive discussion is clearly outside the scope of this review, it should be mentioned that normal phase chromatography is regularly used for the separation of individual lipid classes, whereas reversed phase chromatography is the method of choice to monitor differences in fatty acyl compositions. Besides separation, the visualization of individual PL classes is also an important matter. Although some nondestructive dyes as primuline (White et al. 1998) were recently propagated to enable the quantitative determination of absolute PL concentrations, these dye-based methods are not yet widely accepted and the determination of the PL concentrations by the phosphate content according to Barlett is still a very common assay (Bartlett 1959). Spectroscopic methods are often used to determine the chemical composition of PL mixtures as well as the physical (membrane) structures of PL. Due to the limited space of this review only nuclear magnetic resonance (NMR) will be shortly discussed here as it provides both, compositional and structural, information. For instance, $^{31}P$ NMR lineshape (from solid-state NMR spectra) analysis is very useful to characterize PL phases in an aqueous environment and crystalline, lamellar, hexagonal, cubic and micellar phases of PL can be easily differentiated (Huster 2005). Additionally it is also possible to study the interaction between proteins and lipids and this is very important for the structural characterization of membrane proteins (Huster 2005). Last but not least, high resolution $^{31}P$ NMR is an established method for the determination of the PL composition of complex mixtures (Schiller and Arnold 2002) that can be very easily quantified by using the integral intensities of the individual resonances. Both aspects of $^{31}P$ NMR were recently reviewed (Schiller et al. 2007a).

### *21.1.1 Methods of Lipid Analysis Based on Mass Spectrometry (ESI, APCI, MALDI)*

Nearly all known ionization methods of mass spectrometry (including electron impact, laser desorption and fast atom bombardment) were already successfully applied to lipids. However, many ionization techniques are not very suitable for the analysis of complex PL mixtures as they provide considerable amounts of fragment ions. Therefore, only three "soft-ionization" methods play nowadays a major role in lipid analysis. Beside atmospheric pressure chemical ionization (APCI) (Byrdwell 2001), electrospray ionization (ESI) (Pulfer and Murphy 2003) is nowadays considered to be the method of choice for lipid analysis. Both methods are widely used in "lipidomic" studies (Wenk 2005) as they allow the detection of the intact lipid molecules and give only a small extent of fragmentation. However, it must be explicitly stated that conventional EI mass spectra provide more reliable quantitative information than soft ionization techniques as the ion yield of EI depends primarily on the ionization potential of the functional groups of a molecule. Using soft-ionization MS, however, quasimolecular ions are generated, the yield of which depends significantly on the acidic / basic properties of the analyte. Before describing how MALDI MS works, it should still be mentioned that the inventors of MALDI and ESI were awarded the Nobel Prize for Chemistry in 2002 (Cho and Normile 2002).

### *21.1.2 Matrix-Assisted Laser Desorption and Ionization Time-of-Flight (MALDI-TOF MS)*

Matrix-assisted laser desorption and ionization mass spectrometry was developed at the end of the 80s of the last century independent by Japanese and German scientists and is nowadays primarily used in protein and peptide research, but also for the analysis of carbohydrates and DNA. A detailed survey of the different applications of MALDI-TOF MS is provided in the excellent book by Hillenkamp and Peter-Katalinić (2007). Lipids and phospholipids were analyzed by MALDI-TOF MS so far only to a minor extent. However, the interest in lipid analysis by MALDI-TOF MS is nowadays continuously increasing (Schiller et al. 2007b). One potential reason for that growing interest is coming from MALDI-TOF imaging that is currently experiencing great interest (McDonnell and Heeren 2007). As all tissues contain cells and cells possess a membrane composed of significant quantities of PLs, lipids are easily detected in all MALDI imaging experiments of tissues (Schiller et al. 2004).

#### 21.1.2.1 How Does MALDI-TOF MS Work?

MALDI-TOF MS is based on the utilization of an (in most cases) ultraviolet-absorbing matrix. Although there are nowadays also IR lasers in use that require completely different matrix compounds (e.g. glycerol), only UV matrices will be discussed here because nearly all commercially available MALDI devices are equipped with UV lasers. The matrix fulfils two important tasks: (a) to absorb the energy emitted by the laser and (b) to prevent aggregation of the analyte molecules, that would result in cluster ion formation (Schiller et al. 2007b). When the laser beam hits the sample (co-crystals from the matrix and the lipid), its energy is primarily absorbed by the matrix that is present in a vast excess over the lipid (a 100 100.000 fold excess of matrix is typically used). Consequently, the matrix is vaporized, carrying intact analyte molecules into the vapor phase. During the expanding process of this gas cloud, ions (e.g. $H^+$ and $Na^+$) are exchanged between the matrix and the lipid, leading to the formation of charged analyte molecules. These ions are called “adducts” or “quasimolecular ions”. Besides cation generation, anions can also be generated by abstracting $H^+$ or $Na^+$ from the analyte. The ratio between cations and anions is determined by the (gas phase) acidities of the analyte and the matrix (Schiller et al. 2007c). The most important difference between “molecular (radical) ions” (generated in conventional EI spectra) and “quasimolecular ions” is the mass. As molecular ions are generated by the abstraction of one electron from the analyte, the mass of the generated ion fits perfectly to the mass of the analyte. In contrast, quasimolecular ions are generated by the addition of a cation to the analyte. Therefore, the mass of the quasimolecular ions is higher in comparison to the analyte molecule. Unfortunately, despite of its profound importance, the process of ion generation in MALDI-TOF MS is only poorly understood (Knochenmuss and Zehobi 2003) and many papers are currently dealing with this important topic. For instance, one recent issue of European Journal of Mass Spectrometry was exclusively dedicated to “Mechanisms of MALDI”. Despite this obvious lack of knowledge, it is sure that singly-charged ions are primarily generated (Karas et al. 2000) and, therefore, the actual measured quantity, the mass-to-charge ratio (m/z) may be replaced directly by the monoisotopic mass of the analyte molecule–plus or minus the mass of the ion required to generate a charge. After being formed, ions are accelerated in a strong electric field (typically of the order of 20.000 V). After passing a charged grid, the ions are drifting freely over a field-free space where mass separation is achieved: Low mass ions arrive at the detector in a shorter time than high mass ions (Schiller et al. 2004). It should be noted that there is no absolute need to combine a MALDI ion source with a "TOF" detector. However, as MALDI is often used for the detection of large molecules, the TOF detector is very popular because it has a nearly unlimited mass range. An additional reason is the pulsed ion generation of MALDI that is most suitable for the TOF detector (Hillenkamp and Peter-Katalinic 2007). Although the detection of positive ions is much more common, negative ions may also be easily detected: Positive and negative ions can be simply differentiated by inverting the direction of the applied electric field. Due to the limited available space only a rough description of the ion generating process can be provided. However, further details are available in the excellent book by Hillenkamp and Karas (2007).

#### 21.1.2.2 Advantages and Drawbacks of Lipid Analysis by MALDI-TOF MS

As initially stated, MALDI is just one selected MS method. Therefore, the question arises why lipids should be analyzed by MALDI-TOF MS. There are at least two important advantages of MALDI: (a) MALDI analysis is very fast and simple and one sample can be investigated in less than one minute and (b) MALDI tolerates higher amounts of impurities than other MS methods, minimizing the need of sample purification (Schiller and Arnold 2000, Fukuzawa et al. 2005). Interfering polar compounds are often already sufficiently removed by the lipid extraction process.

Therefore, spectra of lipids are characterized by a high reproducibility and can be directly used for quantitative analysis (Cohen and Gusev 2002). Additionally, lipids possess a molecular weight (the majority of lipids have masses between about 500 and 1500 Da) that is ideal for MALDI-TOF MS analysis. On the one hand, it is small enough that high ion yields are obtained (the tendency of molecules to get into the gas phase correlates reciprocally with their molecular weights). On the other hand, the molecular weight of lipids is sufficiently high that interferences with background signals from the matrix do regularly not play a major role (Schiller et al. 2004). The most important advantage of lipids is the even co-crystallization with the matrix: Since the lipid and the matrix represent relatively apolar molecules, both are highly soluble in organic solvents and there is (in contrast to water-soluble analytes) no need of solvent mixtures. Nevertheless, the sample preparation also confers disadvantages: In MALDI-TOF MS, a solid but not a liquid sample (as in all other MS methods) is used. The co-crystals between matrix and analyte are never completely homogeneous but possess irregularities. Therefore, the achievable intensities of the signals depend on the position where the laser hits the sample/matrix crystal and fluctuations from shot to shot occur. Although this problem may be minimized to some extent by averaging a larger number of laser shots on different positions, quantitative evaluation of MALDI spectra is still rather difficult.

### Improving Reproducibility of MALDI Mass Spectra

Different methods to improve the reproducibility of MALDI spectra were suggested and the majority of them are based on improving the homogeneity of the matrix analyte cocrystals by optimizing the sample preparation method. Using electrospray sample deposition seems extraordinarily helpful in this respect (Wetzel et al. 2004).

### Quantitative Aspects

As the majority of analytical methods, MALDI-TOF MS is also characterized by two different limits, namely (a) the level of detection (LOD) and (b) the level of quantification (LOQ). Obviously the LOD requires lower amounts of lipids than the LOQ. It is rather difficult to provide absolute data as the detection limits differ from lipid to lipid class but the LOD is of the order of 20 picograms (Gellermann et al. 2006) corresponding to about 25 femtomole if 800 g/mole is assumed as the typical molecular weight of a lipid. These data hold for 2,5 dihydroxybenzoic acid (DHB) as matrix and an isolated lipid sample in pure (nearly salt free and in the absence of impurities) organic solvents. The presence of salts, detergents, other lipids, etc. may significantly increase the detection levels (Zschörnig et al. 2006). Basically there are three different approaches how absolute quantitative information may be obtained from a mass spectrum:

1. Internal standards: The comparison of the intensity of an unknown compound with the intensity of a known compound is likely the most suitable approach and was already applied to the analysis of diacylglycerols (Benard et al. 1999) and phosphatidylcholines (Zschörnig et al. 2006): By relating the intensity of the analyte signal to a known reference peak, the impact of further parameters (differences in the applied laser strength, presence of impurities, etc.) may be reduced. On the other hand, the addition of internal standards requires some prior knowledge about the sample composition as the standard must be provided in a suitable concentration, i.e. the peak of the standard should be clearly detectable, but must not dominate the mass spectrum because this might result in the suppression of the peaks of interest.
2. Comparison to a defined matrix peak: Although common matrices have normally much lower masses than lipids, nearly all matrices tend to undergo photochemical reactions upon laser irradiation and under gas phase conditions leading to peaks at higher m/z ratios (Schiller et al. 2007c). Although some applications of this method to apolar lipids as triacylglycerols (Schiller et al. 2004) were described, the applicability of this method seems rather limited.
3. Using the signal-to-noise (S/N) ratio: The achievable S/N ratio increases with lipid concentration. Although this method seems to be applicable for all substances, it was so far only used for rather polar species, in particular lysophospholipids (Petković et al. 2001) and phosphoinositides (Müller et al. 2001).

The absolute quantification of PLs in complex mixtures is generally difficult and previous separation is normally required (Schiller et al. 2007b).

## 21.2 Characteristics of MALDI-TOF Mass Spectra of Lipids

Although the focus of this report is the analysis of physiologically relevant lipid mixtures, some comments on typical patterns of MALDI-TOF mass spectra of lipids will be given on the hand of defined lipids.

### *21.2.1 "Non-Phospholipids"*

Beside the physiologically most relevant PLs, there is also a large variety of non-phospholipids. In biological samples there are primarily free fatty acids, di-, and triacylglycerols as well as cholesterol and cholesteryl esters, the properties of which will be initially shortly discussed.

#### 21.2.1.1 Cholesterol and cholesteryl esters

Cholesterol is an important constituent of biological membranes, whereas cholesteryl esters occur in vast amounts in the lipoproteins of blood (Zschörnig et al. 2006). In comparison to cholesterol that possesses amphiphilic properties due to its hydroxyl group, cholesteryl esters are completely apolar molecules. As typical of apolar molecules, cholesteryl esters (as well as di- and triacylglycerols) are exclusively detected as $Na^+$ but never as $H^+$ adducts, whereas cholesterol–as a primary alcohol–is exclusively detected as protonated molecule subsequent to water elimination, i.e. cholesterol is detected at $m/z = 369.3$, although its monoisotopic mass is 386.3 (Schiller et al. 2000). Some additional important aspects will be discussed below in the context of blood.

#### 21.2.1.2 Free Fatty Acids, Diacyl- and Triacylglycerols

Although fatty acyl residues are present in all lipids, the analysis of free fatty acids by MALDI-TOF MS is rather difficult as they show–due to their relatively low molecular weights–a significant overlap with the matrix. This particularly holds if fatty acids at low concentrations have to be analyzed. One possibility to overcome this problem is the use of mesotetrakis(pentafluorophenyl)porphyrin (Ayorinde et al. 1999) as matrix: It could be shown that the analysis of free fatty acids obtained by alkaline saponification of different plant oils is possible by this matrix (Ayorinde et al. 2000). As an excess of sodium acetate was added, exclusively the $Na^+$ adducts of the sodium salts of the fatty acids were detected by positive ion MALDI-TOF MS. Therefore, problems with peak assignments did not occur. Unfortunately, this approach works only with completely saturated fatty acids: In the presence of unsaturated fatty acids, all peaks - even that of saturated fatty acids–are shifted for 14 Da to higher m/z values in comparison to the expected molecular weights.

Also this is highly speculative, an oxidation of a methylene (14) to a carbonyl group (28) might have occurred. Therefore, the established GC/MS approach is still the method of choice for the analysis of free fatty acids (Marriot et al. 2002). Triacylglycerols are important for the storage of energy (fat tissue), while diacylglycerols are highly relevant second messengers (Wakelam et al. 1998). Both, DAG (Benard et al. 1999) as well as TAG (Reid Asbury et al. 1999) appear exclusively as $Na^+$ adducts and are characterized by significant yields of fragment ions according to the loss of one sodiated acyl residue. As a typical example the positive ion spectrum of triolein (TAG 3×18:1) is shown in Fig. 21.1a: The peak at m/z = 907.7 corresponds to the $Na^+$ adduct of triolein, whereas the peak at m/z = 603.5 corresponds to sodium oleate (304) elimination and the very small peak at m/z = 797.6 to a cleavage of one double bond under

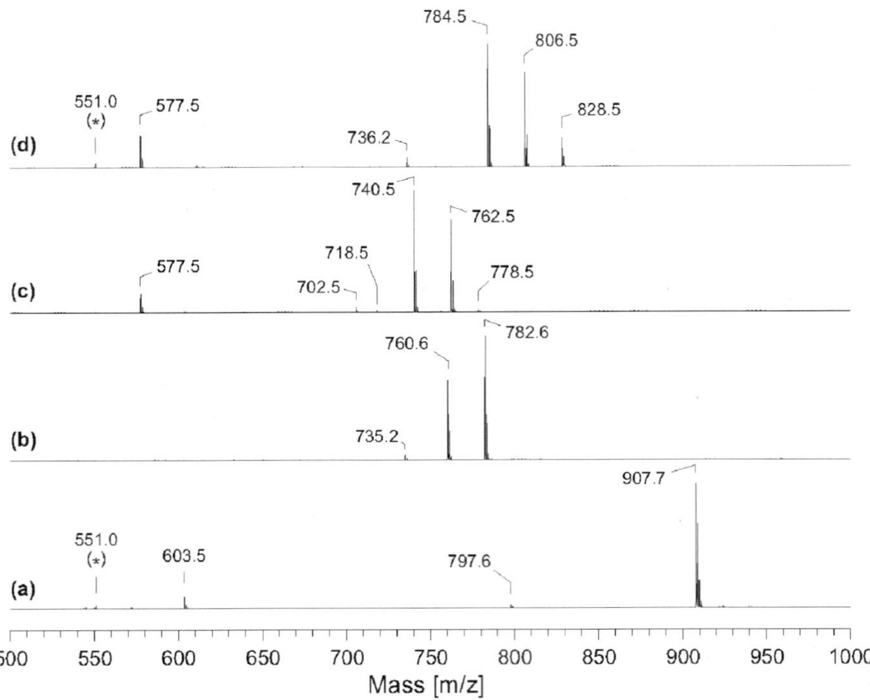


**Fig. 21.1** Positive ion MALDI-TOF mass spectra of triolein (**a**), 1-palmitoyl-2-oleoyl-sn-phosphatidylcholine (**b**), 1-palmitoyl-2-oleoyl-sn-phosphatidylethanolamine (**c**) and 1-palmitoyl-2-oleoyl-sn-phosphatidylserine (**d**). In all cases a 0.5 M solution of DHB in methanol was used as matrix and the sample solutions (0.2 mg/ml) diluted 1:1 (v/v) with the matrix. All peaks are marked according to their m/z ratio and the DHB matrix peaks are marked with asterisks

formation of an aldehyde (most probably via the decay of a peroxide) (Schiller et al. 2004). The DHB matrix itself gives the peak at m/z = 551.0 (Schiller et al. 2007c). In a brand-new work (Gidden et al. 2007) it has been shown that the extent of fragment ion generation can be significantly reduced under alkaline conditions. It was suggested that the observed fragments actually arise from unseen protonated TAGs as their fragmentation occurs so rapidly and completely that protonated TAGs are not normally observed. If the pH is increased the $H^+$ concentration is simultaneously decreased leading to reduced $H^+$ adduct generation and, accordingly, to a lower yield of fragmentation products (Gidden et al. 2007).

## *21.2.2 Phospholipids*

Phospholipids are of paramount interest as they make out a significant moiety of the lipids of the cellular membrane and are of high relevance as second messenger molecules, for instance, phosphoinositides, lysophospholipids and phosphatidic acids (Exton et al. 1994). Therefore, these molecules are in the focus of current research (Schiller et al. 2007b) as well as "Lipidomic" studies (Wenk 2005).

### 21.2.2.1 Zwitterionic Phospholipids

In organic extracts of tissues and body fluids significant amounts of phosphatidylcholine (PC), phosphatidylethanolamine (PE) and phosphatidylserine (PS) occur that all represent zwitterionic PLs. Due to their significance they were among the first PLs investigated by MALDI-TOF MS. In order to illustrate the typical patterns of MALDI-TOF mass spectra, the positive ion spectra of PC 16:0/18:1 (b) PE 16:0/18:1 (c) and PS 16:0/18:1 (d) are shown in Fig. 21.1. These PLs were chosen because they occur in high quantities in cells and are useful to demonstrate the influence of different head structures. All spectra were recorded with 2,5-dihydroxybenzoic acid (DHB) as matrix. Although many other matrix compounds were suggested in the past, DHB is still the most versatile matrix for lipid analysis (Schiller et al. 2007c). PC 16:0/18:1 (1b) provides two peaks at m/z = 760.6 and 782.6 corresponding to the $H^+$ and $Na^+$ adduct of the neutral PC molecule. Additionally, there is one further, but much smaller peak at m/z = 735.2 that indicates the cleavage of the quaternary ammonia headgroup (Al-Saad et al. 2003a). This is a common fragmentation mechanism of all PLs containing quaternary ammonia groups, e.g. PC, LPC and SM. Although not yet completely clarified, the extent of fragmentation may be influenced by the applied laser intensity and the content of impurities, in particularly the salt content of the sample (Zschörnig et al. 2006). In contrast, the spectrum of PE 16:0/18:1 (1c) is more difficult to interpret. This is coming from the $-NH_3^+$ group of PE with its exchangeable protons: The peaks at m/z = 718.5 and 740.5 in (1c) correspond to the $H^+$ and $Na^+$ adducts of the PE as in the case of PC. In contrast, the peak at m/z = 762.5 corresponds to the $Na^+$ adduct subsequent to an exchange of one $H^+$ by $Na^+$. One should note that upon moderate laser irradiation, PC does not give intense fragment ions (1b), whereas the PE has a much higher tendency to loose its head group leading to the peak at m/z = 577.5 (Al-Saad et al. 2003b). All lipids discussed so far give much stronger positive than negative ion signals. The tendency to give negative ions increases in the order triolein < PC < PE. Triolein is not detectable at all as negative ion, whereas PC is not detected as $PC-H^+$, too, but may be detected as negatively charged cluster ion with the matrix (Schiller et al. 2002). Finally PE is less sensitively detected as negative ion ($PE-H^+$, see below) and PS is most sensitively detected as negative ion. So far, DHB was primarily used for lipid analysis by MALDI-TOF MS since this compound provides the best results and gives only weak signals of its own (cf. the weak peak at m/z = 551) (Schiller et al. 2007c). Nevertheless, common MALDI matrices as α-cyano-hydroxycinnamic acid or sinapinic acid (typically used for protein analysis) can also be used (Harvey 1995). In order to further improve the homogeneity between the matrix and the PL, the use of liquid crystalline matrices was also recently suggested (Li et al. 2005).

#### 21.2.2.2 Acidic Phospholipids

Basically the same MS patterns as discussed above may be expected for acidic PLs. For instance, the positive ion spectrum of PS 16:0/18:1 is shown in (1d) and can be explained in the same way as the PE (see above). It is also obvious that the same fragment (m/z = 577.5) corresponding to the cleavage of the headgroup is observed in both cases. However, the most interesting aspect of acidic lipids is their tendency to give intense negative ion signals. In Fig. 21.2 the negative ion spectra of PE 16:0/18:1 (a), PS 16:0/18:1 (b) and dipalmitoyl-phosphatidylinositol-3-phosphate (c) are shown. It is obvious that the DHB matrix (marked with asterisks) provides much more intense signals in the negative than in the positive ion spectra. It is also evident that the signal of the PE subsequent to $H^+$ abstraction (m/z = 716.5) is very small, whereas a very intense signal corresponding to a matrix cluster is also detected (m/z = 892.5). Although PE may be detected as negative ion if DHB is used as matrix, it was recently shown that the application of p-nitroaniline (PNA) gives much better results because PNA possesses an enhanced basicity (Estrada and Yappert 2004). Using the so-called "post-source decay" technique (an MS/MS method typical of TOF devices) (Hillenkamp and Peter-Katalinic 2007), it turned out that the negative ion fragmentation pattern is extraordinarily helpful to assign the fatty acyl positions within a given PE (Fuchs et al. 2007a). Basically the same peak pattern is detected in the case of the PS (2b). One should note that the typical mass differences in comparison to the positive ion spectrum (1d) of 24 ($H^+/Na^+$) and 46 (2 $Na^+$) are often useful in order to confirm peak assignments. Finally, it is also evident from the spectrum of the PIP (2c) that under negative ion conditions, no cleavage of the headgroup, but of the individual fatty acyl residues

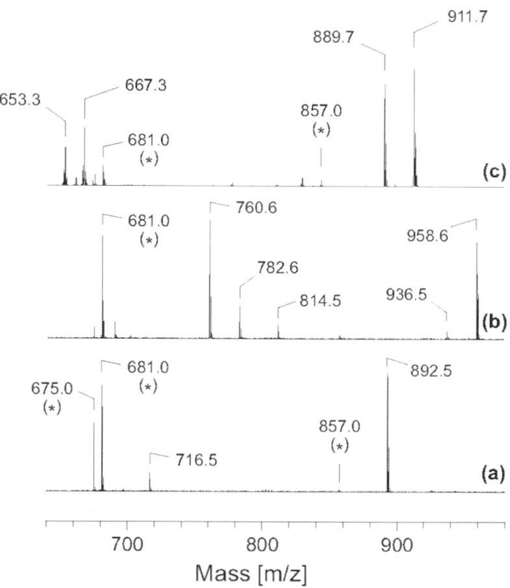


**Fig. 21.2** Negative ion MALDI-TOF mass spectra of 1-palmitoyl-2-oleoyl-sn-phosphatidylethanolamine (**a**), 1-palmitoyl-2-oleoyl-sn-phosphatidylserine (**b**) and 1,2-dipalmitoyl-sn-phosphatidylinositol-3-phosphate (**c**). In all cases a 0.5 M solution of DHB in methanol was used as matrix and the sample solutions (0.5 mg/ml) diluted 1:1 (v/v) with the matrix. All peaks are marked according to their m/z ratio and DHB matrix peaks are marked with asterisks. Please note the considerable intensity of the cluster ions between DHB and the different analytes. Please also note that the contribution of matrix peaks is much higher than in the positive ion spectra

predominates (m/z = 653.4 and 667.4). The MALDI-TOF MS analysis of acidic PLs is, however, aggravated by three facts: (a) they normally occur only in rather low amounts in cells and tissues, (b) with increasing charge density their detectability is continuously reduced (Müller et al. 2001) and (c) other more abundant membrane lipids, in particular PC, may suppress their detection (Schiller et al. 2007b). However, a rather simple approach based on simply filtering the lipid sample through a column was recently introduced to remove interfering PC components (Johanson et al. 2007).

## 21.3 Selected Applications of MALDI-TOF MS

Although the majority of applications of MALDI-TOF MS are still dedicated to the analysis of polar biopolymers, there is an increasing number of reports dealing with lipid analysis. Due to the limited space, however, only a few selected applications are discussed.

### 21.3.1 Determination of Enzyme Activities

One of the most interesting enzymes in the context of PLs is for sure phospholipase $A_2$ ($PLA_2$). This enzyme cleaves the fatty acyl residue in sn-2 position of PLs and many different types of $PLA_2$ with different PL specificities are nowadays known. A significant contribution of $PLA_2$ to many pathologies seems also obvious and, therefore, there is increasing diagnostic interest in this enzyme (Heller et al. 1998). The determination of $PLA_2$ activity is regularly performed by spectrophotometric assays based on the release of a fatty acid with high UV absorption, particularly arachidonic acid from the PL of interest. It is obvious that this assay does not work if saturated lipids are to be analyzed. A generally applicable assay based on MALDI-TOF MS was recently introduced (Petković et al. 2002). The special advantage of this approach is that both, the substrate (e.g. PC) as well as the product (e.g. LPC) can be simultaneously monitored by a single measurement and that there is no need of using an artificially labeled lipid. Therefore, virtually all lipids can be used as substrates. Of course, $PLA_2$ is not the only enzyme, the activity of which can be determined by MALDI-TOF MS and related investigations were also performed on the hand of cholesterol esterase (Zschörnig et al. 2005). A more general paper on the determination of enzyme activities by MALDI-TOF MS was recently published (Liesener and Karst 2005).

### 21.3.2 Analysis of Body Fluids from Patients and Healthy Controls

Due to its high sensitivity and the fast performance, MALDI-TOF MS attracts also significant interest in clinical chemistry for disease monitoring. Many different body fluids are of diagnostic interest and only a few of them will be discussed here. Blood or blood plasma is easily available and it is nowadays accepted that the analysis of the lipoproteins of blood is of great interest regarding the diagnosis of atherosclerosis (Zschöring et al. 2006). Lipids of the individual lipoprotein fractions of human blood were already investigated by MALDI-TOF MS and it could be clearly shown that they can be easily differentiated by their LPC and SM contents (Schiller et al. 2001a). Typical positive ion MALDI-TOF mass spectra of a chloroform-methanol extract of LDL (a) and HDL (b) are shown in Fig. 21.3. It is obvious that the achievable resolution - even on routine devices - is absolutely sufficient (cf. the insert in trace 3b) for the discrimination of differently saturated lipids (one double bond corresponds to a mass shift of 2 Da) and that both lipoprotein fractions differ significantly in their SM ($m/z = 703.6$ and 725.6) and LPC ($m/z = 496.3$ and 524.3) contents (Schiller et al. 2001a). LPC is generated under inflammatory conditions from PC, and, therefore, an important marker of oxidative stress (Arnhold et al. 2002). A very recent paper confirmed and extended these data:

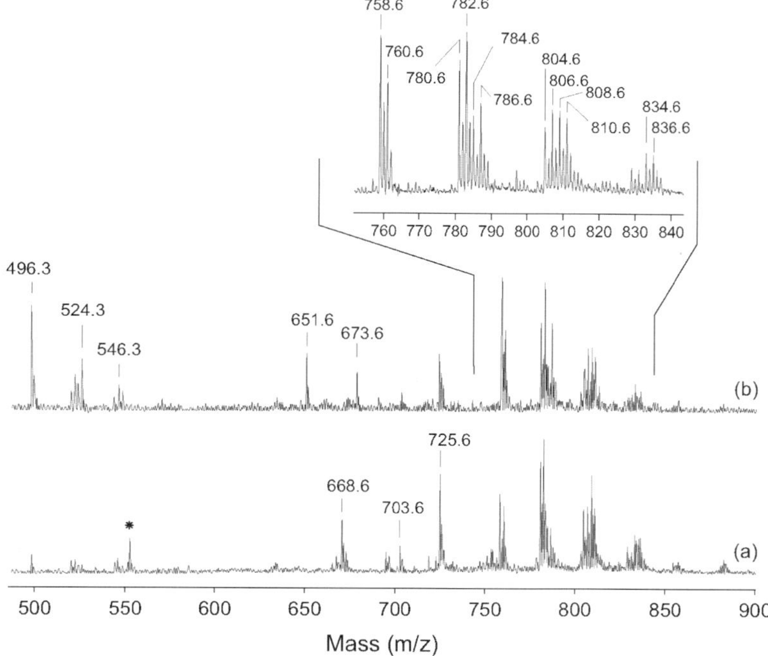


**Fig. 21.3** Typical positive ion MALDI-TOF mass spectra of a chloroform-methanol extract of LDL (**a**) and HDL (**b**). The insert emphasizes the most relevant mass range of the HDL sample. The peak labeled with an asterisk is caused by the matrix. Reprinted with modification from Journal of Lipid Research 42 (2001) 1501–1508

Even absolute quantitative data can be derived from the simple mass spectra of lipoproteins if a suitable internal standard is used (Hidaka et al. 2007). Nevertheless, great care is needed if MALDI-TOF mass spectra are to be quantitatively analyzed and this warning is illustrated in Fig. 21.4. In this figure the organic extracts from human whole blood plasma were characterized by positive ion MALDI-TOF MS. Trace 4a represents the sample that was used without further dilution. It is evident that under these conditions PC species are exclusively detected–primarily as $H^+$ adducts (the peak at m/z = 760.6 corresponds to PC 16:0/18:1). There are not even minor peaks of triacylglycerols (TAG). However, TAG signals become clearly detectable at higher dilutions accompanied by increased intensities of $Na^+$ adducts of PC (m/z = 782.6). This indicates that dilution of the sample is advisable if the sample of interest contains significant amounts of PC in order to check if there are further lipid species that are suppressed by the PC at higher concentrations. A similar problem regarding the detection of lysolipid species was already reported (Petković et al. 2001). Rheumatic diseases are extremely widespread in the industrialized countries and of significant socio-economic significance (Schiller et al. 2006). Much diagnostic information is available from the analysis of the

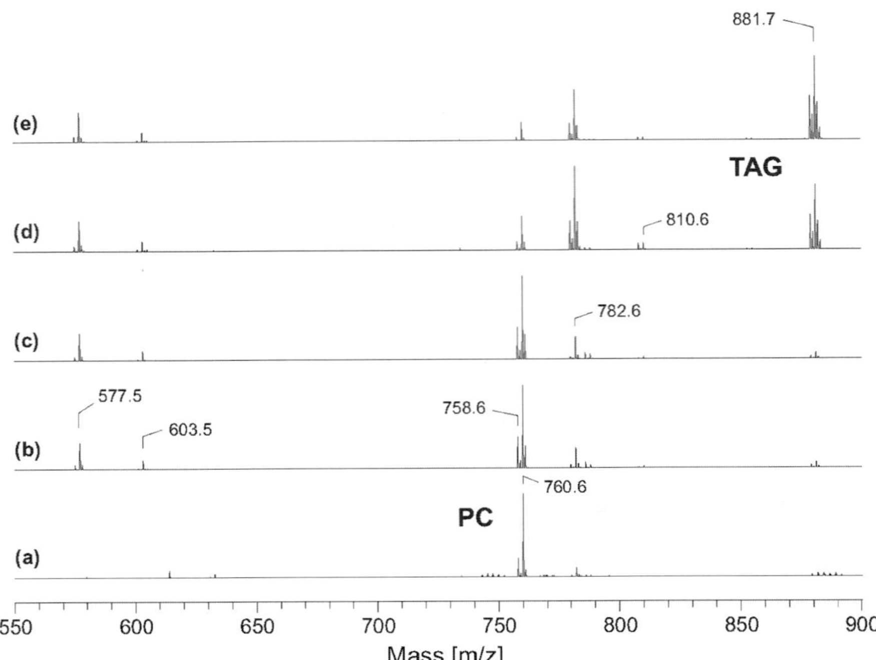


**Fig. 21.4** Positive ion MALDI-TOF mass spectra of an organic extract of human blood plasma recorded with DHB as matrix. Trace (**a**) represents the original extract, whereas all other samples were diluted with $CHCl_3$ prior to mixing with matrix. The following dilutions were used: 1:5 (**b**), 1:10 (**c**), 1:25 (**d**) and 1:50 (**e**). The peaks at m/z = 577.5 and 603.5 are caused by the loss of one acyl residue from the TAG. Please note the arising of triacylglycerol peaks and the reduction of PC at higher dilutions

joint (synovial) fluids from patients suffering from rheumatic diseases. It could be shown that the content of lysophosphatidylcholine (LPC) correlates with the severity of rheumatoid arthritis and that successful drug treatment results in an increased PC/LPC ratio (Fuchs et al. 2005) either by reduced $PLA_2$ activity or by increased re-acylation of LPC to PC. As demonstrated in Fig. 21.5, it could also be shown that nearly the same results are obtained equally if the synovial fluid or the blood from the same patient is analyzed. This is a surprising, but important result as blood can be more easily obtained than synovial fluids (Fuchs et al. 2005). MALDI-TOF MS was also successfully applied to many further body fluids, including the bronchoalveolar lavage fluid (the "surfactant" of lungs) (Schiller et al. 2001b). The vast majority of the PLs of this body fluid is represented by PC and PG, which can both be easily identified from the positive and the negative ion spectra of the total extracts, respectively. However, it also turned out that some minor species as PE or PI could only be monitored subsequent to previous separation into the individual lipid classes (Sommerer et al. 2004).

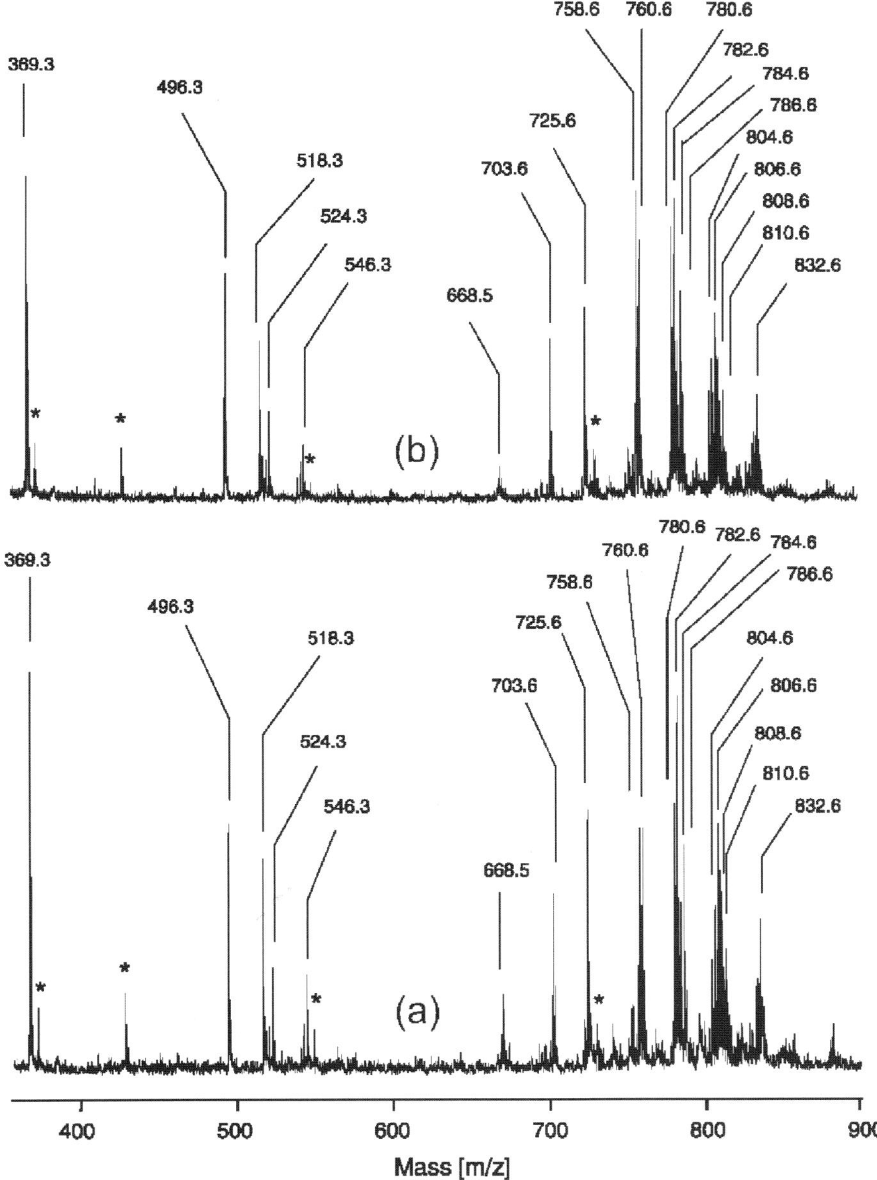


**Fig. 21.5** Representative MALDI-TOF MS spectra (positive ion mode) of an organic extract of synovial fluid (**a**) and plasma (**b**) from a selected patient suffering from rheumatoid arthritis. Peaks are labeled according to their m/z ratio and characteristic DHB matrix peaks are labeled with asterisks. Reprinted with modifications from Clinical Biochemistry, 38, Fuchs et al., The phosphatidylcholine/lysophosphatidylcholine ratio in human plasma is an indicator of the severity of rheumatoid arthritis: investigations by $^{31}P$ NMR and MALDI-TOF MS, 925-933, Copyright (2005) with permission from Elsevier

### *21.3.3 Analysis of Cells and Cellular Metabolism*

The lipid composition of a large variety of cells–obtained from biological fluids as well as from cell culture–was already successfully investigated by MALDI-TOF MS. The special advantage of MALDI-TOF MS is that no extensive sample work-up is required as some important information can be already obtained from the analysis of the total cell extracts although not all lipid classes are detectable under these conditions. At the first glance, this seems as a disadvantage. However, this fact simplifies data interpretation significantly as only a limited number of lipid classes has to be considered at all. For instance, it could be shown by this approach that the lipid composition of human neutrophilic granulocytes changes significantly upon stimulation (Schiller et al. 1999). Human and animal spermatozoa were also intensively investigated by MALDI-TOF MS. Using human spermatozoa it could be shown that changes upon cryopreservation can be easily determined (Schiller et al. 2000) because upon the freezing process the spermatozoa membrane is at least partially destroyed and phospholipases as well as sphingomyelinase are released. Under the influence of these enzymes ceramide (from SM) as well as LPC (from PC) are generated that are both useful markers of sperm quality. It was also shown that the LPC content is massively enhanced in apoptotic spermatozoa (Glander et al. 2002) that can be separated from intact spermatozoa by annexin-V binding. Annexin-V binds to phosphatidylserine that is normally exclusively located in the inner leaflet of the membrane but gets detectable in the outer leaflet of apoptotic cells. A comparison of the positive ion MALDI-TOF mass spectra of annexin-V negative (a) and positive (b) human spermatozoa is shown in Fig. 21.6 and the tremendous increase of LPC 16:0 ($m/z = 496.3$ and 518.3) and LPC 22:6 ($m/z = 568.4$ and 590.4) under simultaneous reduction of PC 16:0/22:6 is obvious (Glander et al. 2002). It must be emphasized that LPC may be generated under the influence of $PLA_2$ as well as reactive oxygen species (ROS) and that a free radical theory of sterility was already introduced (Aitken 1994): As spermatozoa contain significant amounts of highly unsaturated fatty acyl residues, in particular docosahexaenoyl (22:6), the ROS-induced LPC generation is likely to play a major role in these cells. It was shown by treating isolated PCs with a certain number of double bonds with HOCl that the LPC yield is the larger the higher the degree of unsaturation of the lipid (Arnhold et al. 2002). In comparison to man, spermatozoa from breeding animals (e.g. bull) contain a high portion of plasmalogen PLs that are characterized by the presence of one alkenyl ether residue in sn-1 position instead of the acyl linkage in common lipids. This alkenyl ether is extremely sensitive towards even traces of acids, and, therefore, the matrix used for MALDI-TOF MS must not contain trifluoroacetic acid (a common additive in order to enhance the yield of protonated species) the presence of which would lead to the hydrolysis of plasmalogens under generation of LPC (Schiller et al. 2003). Therefore, if a sample contains unexpected

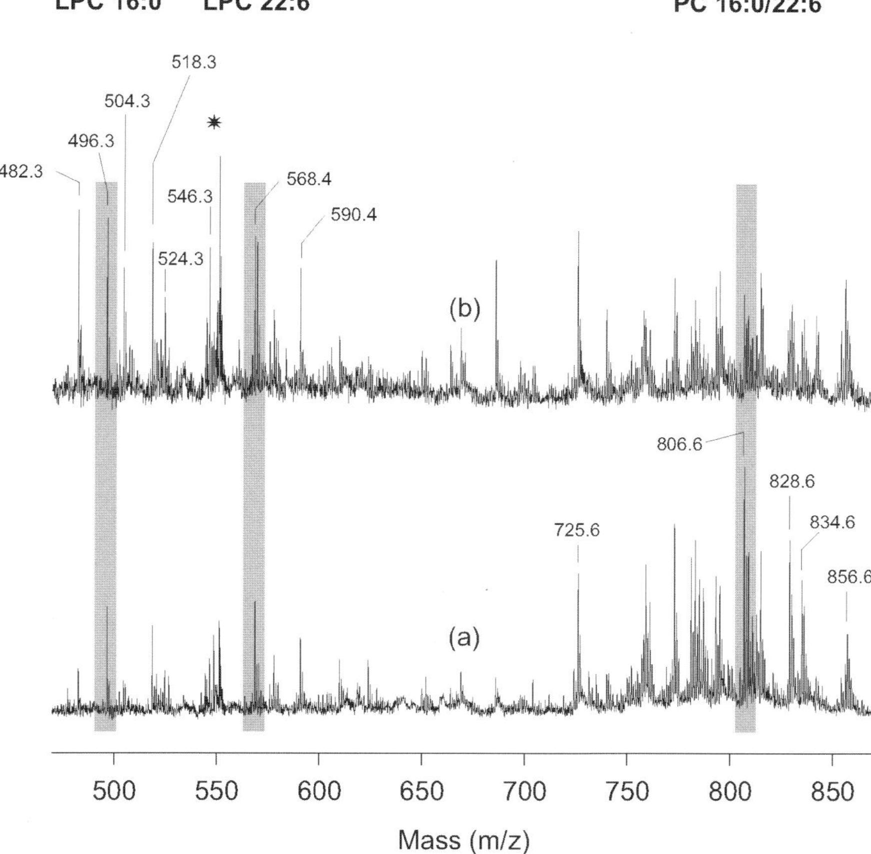


**Fig. 21.6** Typical positive ion MALDI-TOF mass spectra of organic extracts of spermatozoa from a healthy volunteer. Before MS analysis, spermatozoa were separated into annexin V-negative (**a**) and annexin V-positive (**b**) spermatozoa. Peaks are labeled according to the corresponding m/z ratios. The most indicative peak groups are marked with grey bars. The asterisk indicates a typical DHB matrix peak (m/z = 551). Abbreviations: LPC, lyso-phosphatidylcholine; PC, phosphatidylcholine. Reprinted from Glander et al., Deterioration of spermatozoal plasma membrane is associated with an increase of sperm lyso-phosphatidylcholines, Andrologia, 2002 with permission from Blackwell

high amounts of lysolipids, the presence of plasmalogens may be assumed (Murphy 2002). Finally, it could also be shown that formyl-LPC is a suitable marker of animal spermatozoa quality that allows the assessment of ROS-induced damages (Fuchs et al. 2007b). It was also shown very recently in a combined MS and $^{31}$P NMR study that MALDI-TOF MS is a useful method for the determination of apoptotic stem cells (Fuchs et al. 2007c). As illustrated in Fig. 21.7, the apoptotic cells (b) are characterized by a lipid and fatty acyl composition that differs significantly from the intact cells (a). Changes from

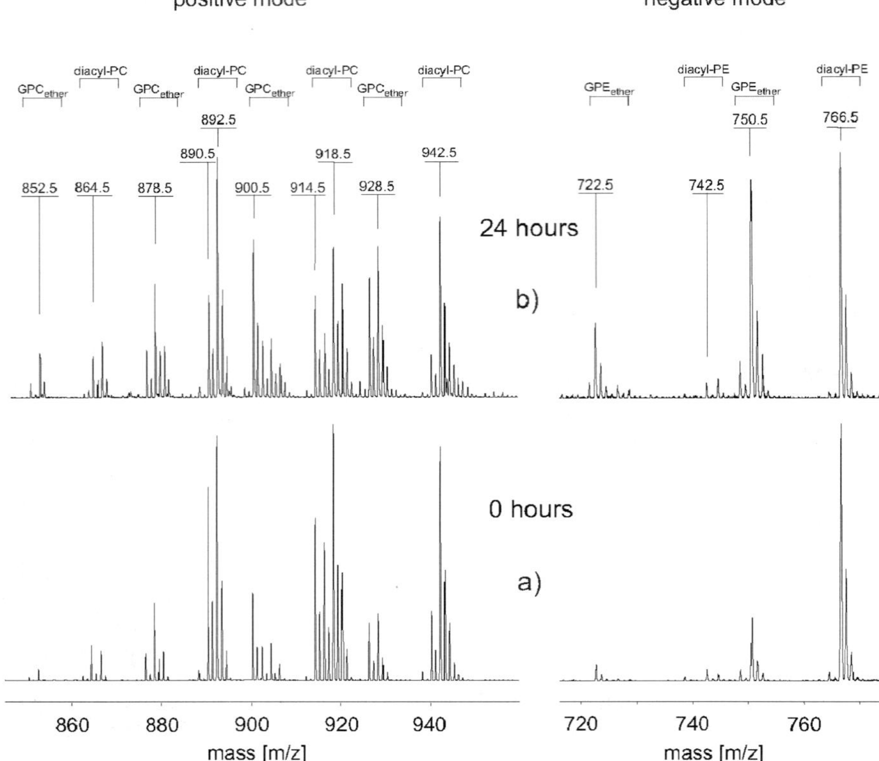


**Fig. 21.7** Positive (left) and negative (right) ion MALDI-TOF mass spectra of the organic phase of FDCPmix stem cells after proliferation in medium containing the growth factor interleukin 3 (IL-3) (**a**) and subsequent to the withdrawal of IL-3 after 24 hours (**b**). By growth factor withdrawal apoptosis was induced. Selected peaks are labeled according to the corresponding m/z ratios. Please note that the $Cs^+$ adducts were used exclusively for the analysis of the positive ion spectra. Abbreviations: GPCether, ether-linked glycerophosphocholine; diacyl-PC, diacyl-phosphatidylcholine; GPEether, ether-linked glycerophosphoethanolamine; diacyl-PE, diacyl-phosphatidylethanolamine. Reprinted from Chemistry and Physics of Lipids, 150, Fuchs et al., Apoptosis-associated changes in the glycerophospholipid composition of hematopoietic progenitor cells monitored by $^{31}$P NMR spectroscopy and MALDI-TOF mass spectrometry, 229-238, Copyright (2007) with permission from Elsevier

normal to apoptotic cells primarily concern the content of ether-linked PLs that are present in much smaller amounts in living cells. On the left hand, the positive and on the right hand, the negative ion spectra are shown. The positive ion spectra are dominated by PC, whereas the negative ion spectra exhibit primarily PE signals. This clear differentiation was achieved by using PNA that is a less acidic matrix than commonly used DHB (Estrada and Yappert 2004). Please also note that the positive ion spectra were recorded in the presence of an excess of $Cs^+$ in order to avoid overlap of $H^+/Na^+$ adducts on the one hand and differences in acyl compositions on the other hand. The data obtained by MS approach correlate closely with the relative contents of apoptotic cells determined by FACS analysis (Fuchs et al. 2007c). However, FACS analysis is much more expensive due to the need of fluorescently-labeled lipids.

### *21.3.4 Whole Tissue Analysis and MALDI Imaging*

Although common (commercially available) nitrogen lasers penetrate only a few micrometers into a solid sample (Hillenkamp and Peter-Katalinić 2007), thin tissue (for instance histological) slices can be directly analyzed by MALDI-TOF MS. The simplest approach is to fix the tissue slice onto the MALDI target and to cover it with matrix. The homogeneous spotting of the matrix is, however, the most critical point and many efforts for its improvements are nowadays undertaken (Hankin et al. 2007). The PL compositions of many different tissues were already investigated by MALDI-TOF MS (Jones et al. 2006) and the obtained data compared with the results from established biochemical methods. A surprisingly good agreement between both methods could be found. Therefore, crude tissue samples, for instance brain, may be analyzed directly by MALDI-TOF MS without sample workup (Jackson et al. 2005). This is a very important prerequisite for using MALDI-TOF MS as an imaging technique (Rujoi et al. 2004), i.e. to obtain spatially-resolved information about the PL distribution within a given sample, although the achievable resolution is still only of the order of 60 μm (Hillenkamp and Peter-Katalinic 2007). Although the lipid compositions of many different tissues (e.g. brain (Jackson et al. 2005), eye lenses (Rujoi et al. 2004), dystrophic muscle (Touboul et al. 2004), etc.) were already examined by MALDI-TOF MS, the most intriguing success of MALDI imaging was to prove that it is possible to differentiate tumour tissue from the ischemic and necrotic areas of the lesion (Stoeckli et al. 2001). This selected example already clearly indicates that MALDI imaging will have great future development because it might be useful for clinical diagnostics. A comprehensive review dealing with MS imaging is available in (McDonnell and Heeren 2007).

## 21.4 Current Developments

Hopefully, we were able to show on the hand of the few selected examples discussed above that MALDI-TOF MS represents a powerful research method. Even if the detailed mechanisms of ion generation are still unknown (Knochenmuss and Zenobi 2003). Although a lot of methodological work is currently performed, only two selected aspects will discussed here due to their significant practical interest.

### *21.4.1 Coupling MALDI-TOF MS with Chromatographic Methods*

One serious disadvantage of MALDI-TOF MS is coming from the strongly varying detectabilities of the individual PL classes: In the presence of compounds with quaternary ammonia groups (e.g. PC, SM, LPC), further less sensitively detectable lipids may be suppressed (Petković et al. 2001). Therefore, chromatographic separation of total lipid extracts is normally required if a detailed analysis has to be performed. The coupling between LC and MALDI-TOF MS is already well established although this coupling is primarily used in proteomic studies. As TLC is a common method of lipid analysis (Peterson and Cummings 2006), the direct analysis of lipids separated on a TLC plate by MALDI-TOF MS was recently attempted by two different methods. One approach is based on the use of an infrared laser and glycerin as matrix (Rohlfing et al. 2007). This approach has the significant advantage that IR radiation penetrates deeper into the sample plate, but also confers the disadvantage that abundant glycerine adducts are detected. Another problem is that IR lasers are barely available so far. Therefore, another approach used a readily available $N_2$ laser and standard DHB as matrix (Fuchs et al. 2007d). Regarding detection limits, both approaches provided comparable results (about 400 pmol) and, therefore, both approaches might be useful for routine lipid analysis. This in particular holds as surprisingly good mass accuracies (about 100 ppm) and mass resolutions (about 3000) could be obtained.

### *21.4.2 Automated Spectra Acquisition*

Automation is one important prerequisite for high-throughput methods. Automated routines are nowadays available from nearly all companies. This equipment also comprises automatic pipeting as well as automatic sample changers. As this is particularly helpful for proteomics studies where–due to the digestion with different proteases–a large number of samples is obtained, the clear focus is so far on proteins and no commercial "Lipidomics" MALDI-based platform is available so far (Wenk 2005).

## 21.5 Summary

There is considerable interest in lipid analysis and it is expected that this interest will increase in the future since an increasing number of diseases is recognized to be accompanied by alterations of lipid compositions. Although not yet commonly accepted, MALDI-TOF MS represents a reliable method of lipid analysis. In our opinion, MALDI-TOF MS is a very suitable method because measurements can be performed in a relatively short time and a very convenient way. Only minor time-consuming sample purification is required since considerable amounts of impurities are tolerated. Therefore, MALDI-TOF MS is useful for the routine analysis of a large number of samples. MALDI analyses are also more inexpensive than established biochemical assays. Quantitative analysis of MALDI-TOF mass spectra is, however, still the critical point since individual PL classes lead to different peak intensities: The concentration of compounds with quaternary ammonia groups like PC or SM is overestimated, whereas the PE concentration is underestimated. This problem is overcome by previous separation of the sample into the individual lipid classes. Nevertheless, in many cases it is sufficient to be able to quantify some selected, well-known metabolites, e.g. one PL and one certain LPL if one is interested, for instance, in the determination of enzyme activities.

**Acknowledgments** The authors wish to thank all colleagues and friends who helped them in writing this review. Especially we wish to thank Dr. Holger Spalteholz, Dr. Jacqueline Leßig, Dr. Matthias Müller and Ms Rosmarie Süß. The kind and helpful advice of Dr. Suckau and Dr. Schürenberg (Bruker Daltonics, Bremen) is also gratefully acknowledged. This work was supported by the German Research Council (DFG Schi 476/5-1 and Schi 476/7-1 as well as the former HBFG program enabling the purchase of a Bruker "Autoflex" device) and the Federal Ministry of Education and Research (Grant BMBF 0313836).

## References

Aitken, R.J., A free radical theory of male infertility, Reprod Fertil Dev, 6 (1994) 19–23.

Al-Saad, K.A., Siems, W.F., Hill, H.H., Zabrouskov, V. and Knowles, N.R., Structural analysis of phosphatidylcholines by post-source decay matrix-assisted laser desorption/ ionization time-of-flight mass spectrometry, J Am Soc Mass Spectrom, 14 (2003a) 373–382.

Al-Saad, K.A., Zabrouskov, V., Siems, W.F., Knowles, N.R., Hannan, R.M. and Hill, H.H. Jr., Matrix-assisted laser desorption/ionization time-of-flight mass spectrometry of lipids: ionization and prompt fragmentation patterns, Rapid Commun Mass Spectrom, 17 (2003b) 87–96.

Arnhold, J., Osipov, A.N., Spalteholz, H., Panasenko, O.M. and Schiller, J., Formation of lysophospholipids from unsaturated phosphatidylcholines under the influence of hypochlorous acid, Biochim Biophys Acta, 1572 (2002) 91–100.

Ayorinde, F.O., Garvin, K. and Saeed, K., Determination of the fatty acid composition of saponified vegetable oils using matrix-assisted laser desorption/ionization time-of-flight mass spectrometry, Rapid Commun Mass Spectrom, 14 (2000) 608–615.

Ayorinde, F.O., Hambright, P., Porter, T.N. and Keith, Q.L. Jr., Use of mesotetrakis(pentafluorophenyl)porphyrin as a matrix for low molecular weight alkylphenol ethoxylates in laser desorption/ ionization time-of-flight mass spectrometry, Rapid Commun Mass Spectrom, 13 (1999) 2474–2479.

Bartlett, G.R., Phosphorus assay in column chromatography, J Biol Chem, 234 (1959) 466–468.

Benard, S., Arnhold, J., Lehnert, M., Schiller, J. and Arnold, K., Experiments towards quantification of saturated and polyunsaturated diacylglycerols by matrix-assisted laser desorption and ionization time-of-flight mass spectrometry (MALDI-TOF MS), Chem Phys Lipids, 100 (1999) 115–125.

Byrdwell, W.C., Atmospheric pressure chemical ionization mass spectrometry for analysis of lipids, Lipids 36 (2001) 327–346.

Cho, A. and Normile, D., Nobel Prize in chemistry. Mastering macromolecules, Science, 298 (2002) 527–528.

Christie W. W., Lipid Analysis, Oily Press, Bridgwater, 2003.

Clejan, S., HPLC analytical methods for the separation of molecular species of fatty acids in diacylglycerol and cellular phospholipids, Methods Mol Biol, 105 (1998) 255–274.

Cohen, L.H. and Gusev, A.I., Small molecule analysis by MALDI mass spectrometry, Anal Bioanal Chem, 373 (2002) 571–586.

Estrada, R. and Yappert, M.C., Alternative approaches for the detection of various phospholipid classes by matrix-assisted laser desorption/ionization time-of-flight mass spectrometry, J Mass Spectrom, 39 (2004) 412–422.

Exton, J.H., Messenger molecules derived from membrane lipids, Curr Opin Cell Biol, 6 (1994) 226–229.

Fuchs, B., Müller, K., Göritz, F., Blottner, S. and Schiller, J., Characteristic oxidation products of choline plasmalogens are detectable in cattle and roe deer spermatozoa by MALDI-TOF mass spectrometry, Lipids, 42 (2007b) 991–998.

Fuchs, B., Schiller, J. and Cross, M.A., Apoptosis-associated changes in the glycerophospholipid composition of hematopoietic progenitor cells monitored by $^{31}$P NMR spectroscopy and MALDI-TOF mass spectrometry, Chem Phys Lipids, 150 (2007c) 229–238.

Fuchs, B., Schiller, J., Süß, R., Schürenberg, M. and Suckau, D., A direct and simple method of coupling matrix-assisted laser desorption and ionization time-of-flight mass spectrometry (MALDI-TOF MS) to thin-layer chromatography (TLC) for the analysis of phospholipids from egg yolk, Anal Bioanal Chem, 389 (2007d) 827–834.

Fuchs, B., Schiller, J., Wagner, U., Häntzschel, H. and Arnold, K., The phosphatidylcholine/lysophosphatidylcholine ratio in human plasma is an indicator of the severity of rheumatoid arthritis: investigations by $^{31}$P NMR and MALDI-TOF MS, Clin Biochem, 38 (2005) 925–933.

Fuchs, B., Schober, C., Richter, G., Süß, R. and Schiller, J., MALDI-TOF MS of phosphatidylethanolamines: different adducts cause different post source decay (PSD) fragment ion spectra, J Biochem Biophys Methods, 70 (2007a) 689–692.

Fukuzawa, S., Asanuma, M., Tachibana, K. and Hirota, H., On-probe sample preparation without washes for matrix-assisted laser desorption/ionization mass spectrometry using an anion exchange medium, Anal Chem, 77 (2005) 5750–5754.

Gellermann, G.P., Appel, T.R., Davies, P. and Diekmann, S., Paired helical filaments contain small amounts of cholesterol, phosphatidylcholine and sphingolipids. Biol Chem, 387 (2006) 1267–1274.

Gidden, J., Liyanage, R., Durham, B. and Lay, J.O. Jr., Reducing fragmentation observed in the matrix-assisted laser desorption/ionization time-of-flight mass spectrometric analysis of triacylglycerols in vegetable oils, Rapid Commun Mass Spectrom, 21 (2007) 1951–1957.

Glander, H.J., Schiller, J., Süß, R., Paasch, U., Grunewald, S. and Arnhold, J., Deterioration of spermatozoal plasma membrane is associated with an increase of sperm lyso-phosphatidylcholines, Andrologia, 34 (2002) 360–366.

Hankin, J.A., Barkley, R.M. and Murphy, R.C., Sublimation as a method of matrix application for mass spectrometric imaging, J Am Soc Mass Spectrom, 18 (2007) 1646–1652.

Harvey, D.J., Matrix-assisted laser desorption/ionization mass spectrometry of phospholipids, J Mass Spectrom, 30 (1995) 1333–1346.

Heller, A., Koch, T., Schmeck, J. and van Ackern, K., Lipid mediators in inflammatory disorders, Drugs 55 (1998) 487–496.

Hidaka, H., Hanyu, N., Sugano, M., Kawasaki, K., Yamauchi, K. and Katsuyama, T., Analysis of human serum lipoprotein lipid composition using MALDI-TOF mass spectrometry, Ann Clin Lab Sci, 37 (2007) 213–221.

Hillenkamp, F. and Karas, M., The MALDI process and method, in: Hillenkamp, F. and Peter-Katalinić, J. (Eds), MALDI MS, Wiley-VCH, Weinheim, 2007, pp. 1–28.

Hillenkamp, F. and Peter-Katalinić, J., MALDI MS -A Practical Guide to Instrumentation, Methods and Application, Wiley-VCH, Weinheim, 2007.

Huster, D., Investigations of the structure and dynamics of membrane-associated peptides by magic angle spinning NMR, Progr Nucl Magn Reson Spectrosc, 46 (2005) 79–107.

Jackson, S.N., Wang, H.Y. and Woods, A.S., Direct profiling of lipid distribution in brain tissue using MALDI-TOF MS, Anal Chem, 77 (2005) 4523–4527.

Johanson, R.A., Buccafusca, R., Quong, J.N., Shaw, M.A. and Berry, G.T., Phosphatidylcholine removal from brain lipid extracts expands lipid detection and enhances phosphoinositide quantification by matrix-assisted laser desorption/ionization time-of-flight (MALDI-TOF) mass spectrometry, Anal Biochem, 362 (2007) 155–167.

Jones, J.J., Borgmann, S., Wilkins, C.L. and O'Brien, R.M., Characterizing the phospholipid profiles in mammalian tissues by MALDI FTMS, Anal Chem, 78 (2006) 3062–3071.

Karas, M., Glückmann, M. and Schäfer, J., Ionization in matrix-assisted laser desorption/ionization: singly charged molecular ions are the lucky survivors. J Mass Spectrom, 35 (2000) 1–12.

Knochenmuss, R. and Zenobi, R., MALDI ionization: the role of in-plume processes, Chem Rev, 103 (2003) 441–452.

Li, Y.L., Gross, M.L. and Hsu, F.F., Ionic-liquid matrices for improved analysis of phospholipids by MALDI-TOF mass spectrometry, J Am Soc Mass Spectrom, 16 (2005) 679–682.

Liesener, A. and Karst, U., Monitoring enzymatic conversions by mass spectrometry: a critical review, Anal Bioanal Chem, 382 (2005) 1451–1464.

Marriot, P.J., Shellie, R. and Cornwell, C., Gas chromatographic technologies for the analysis of essential oils, J Chromatogr A, 936 (2002) 1–22.

McDonnell, L.A. and Heeren, R.M., Imaging mass spectrometry, Mass Spectrom Rev, 26 (2007) 606–643.

Müller, M., Schiller, J., Petković, M., Oehrl, W., Heinze, R., Wetzker, R., Arnold, K. and Arnhold, J., Limits for the detection of (poly-)phosphoinositides by matrix-assisted laser desorption and ionization time-of-flight mass spectrometry (MALDITOF MS), Chem Phys Lipids, 110 (2001) 151–164.

Murphy, R.C., Mass spectrometry of phospholipids: Tables of molecular and productions, Illuminati Press, Denver, 2002.

Peterson, B.L. and Cummings, B.S., A review of chromatographic methods for the assessment of phospholipids in biological samples, Biomed Chromatogr, 20 (2006) 227–243.

Petković, M., Müller, J., Müller, M., Schiller, J., Arnold, K. and Arnhold, J., Application of matrix-assisted laser desorption/ionization time-of-flight mass spectrometry for monitoring the digestion of phosphatidylcholine by pancreatic phospholipase $A_2$, Anal Biochem, 308 (2002) 61–70.

Petković, M., Schiller, J., Müller, J., Müller, M., Arnold, K. and Arnhold, J., The signal-to-noise ratio as the measure for the quantification of lysophospholipids by matrix-assisted laser desorption/ionization time-of-flight mass spectrometry, Analyst, 126 (2001) 1042–1050.

Pulfer, M. and Murphy, R.C., Electrospray mass spectrometry of phospholipids. Mass Spectrom Rev, 22 (2003) 332–364.

Reid Asbury, G., Al-Saad, K., Siems, W.F., Hannan, R.M. and Hill, H.H. Jr., Analysis of triacylglycerols and whole oils by matrix-assisted laser desorption/ionization time of flight mass spectrometry, J Am Soc Mass Spectrom, 10 (199) 983–991.

Rohlfing, A., Müthing, J., Pohlentz, G., Distler, U., Peter-Katalinić, J., Berkenkamp, S. and Dreisewerd, K., IR-MALDI-MS analysis of HPTLC-separated phospholipid mixtures directly from the TLC plate, Anal Chem, 79 (2007) 5793–5808.

Rujoi, M., Estrada, R. and Yappert, M.C., In situ MALDI-TOF MS regional analysis of neutral phospholipids in lens tissue, Anal Chem, 76 (2004) 1657–1663.

Schiller, J. and Arnold, K., Application of high resolution $^{31}$P NMR spectroscopy to the characterization of the phospholipid composition of tissues and body fluids -a methodological review, Med Sci Monit, 8 (2002) MT205–222.

Schiller, J. and Arnold, K., Mass spectrometry in structural biology, in: Meyers, R.A. (Ed), Encyclopedia of Analytical Chemistry, John Wiley & Sons Ltd., Chichester, 2000, pp. 559–585.

Schiller, J., Arnhold, J., Benard, S., Müller, M., Reichl, S. and Arnold, K., Lipid analysis by matrix-assisted laser desorption and ionization mass spectrometry: A methodological approach, Anal Biochem, 267 (1999) 46–56.

Schiller, J., Arnhold, J., Glander, H.-J. and Arnold, K., Lipid analysis of human spermatozoa and seminal plasma by MALDI-TOF mass spectrometry and NMR spectroscopy–effects of freezing and thawing, Chem Phys Lipids, 106 (2000) 145–156.

Schiller, J., Fuchs, B. and Arnold, K., The molecular organization of polymers of artilage: An overview of health and disease, Curr Org Chem, 10 (2006) 1771–1789.

Schiller, J., Hammerschmidt, S., Wirtz, H., Arnhold, J. and Arnold, K., Lipid analysis of bronchoalveolar lavage fluid (BAL) by MALDI-TOF mass spectrometry and $^{31}$P NMR spectroscopy, Chem Phys Lipids, 112 (2001b) 67–79.

Schiller, J., Müller, K., Süβ, R., Arnhold, J., Gey, C., Herrmann, A., Leβig, J., Arnold. K. and Müller, P., Analysis of the lipid composition of bull spermatozoa by MALDITOF mass spectrometry–a cautionary note, Chem Phys Lipids, 126 (2003) 85–94.

Schiller, J., Müller, M., Fuchs, B., Arnold, K. and Huster, D., $^{31}$P NMR spectroscopy of phospholipids: From micelles to membranes. Curr Anal Chem, 3 (2007a) 283–301.

Schiller, J., Süβ, R., Arnhold, J., Fuchs, B., Leβig, J., Müller, M., Petković, M., Spalteholz, H., Zschörnig, O. and Arnold, K., Matrix-assisted laser desorption and ionization time-of-flight (MALDI-TOF) mass spectrometry in lipid and phospholipid research, Prog Lipid Res, 43 (2004) 449–488.

Schiller, J., Süβ, R., Fuchs, B., Müller, M., Petković, M., Zschörnig, O. and Waschipky, H., The suitability of different DHB isomers as matrices for the MALDITOF MS analysis of phospholipids: which isomer for what purpose? Eur Biophys J, 36 (2007c) 517–527.

Schiller, J., Süβ, R., Fuchs, B., Müller, M., Zschörnig, O. and Arnold, K., MALDITOF MS in lipidomics. Front Biosci, 12 (2007b) 2568–2579.

Schiller, J., Süss, R., Petković, M., Zschörnig, O. and Arnold, K., Negative-ion matrix-assisted laser desorption and ionization time-of-flight mass spectra of complex phospholipids mixtures in the presence of phosphatidylcholine: a cautionary note on peak assignment, Anal Biochem, 309 (2002) 311–314.

Schiller, J., Zschörnig, O., Petković, M., Müller, M., Arnhold, J. and Arnold, K., Lipid analysis of human HDL and LDL by MALDI-TOF mass spectrometry and $^{31}$P NMR, J Lipid Res, 42 (2001a) 1501–1508.

Sommerer, D., Süβ, R., Hammerschmidt, S., Wirtz, H., Arnold, K. and Schiller, J., Analysis of the phospholipids composition of bronchoalveolar lavage (BAL) fluid from man and minipig by MALDI-TOF mass spectrometry in combination with TLC, J Pharm Biomed Anal, 35 (2004) 199–206.

Stoeckli, M., Chaurand, P., Hallahan, D.E. and Caprioli, R.M., Imaging mass spectrometry: a new technology for the analysis of protein expression in mammalian tissues, Nat Med, 7 (2001) 493–496.

Touboul, D., Piednoël, H., Voisin, V., De La Porte, S., Brunelle, A., Halgand, F. and Laprévote, O., Changes of phospholipid composition within the dystrophic muscle by matrix-assisted laser desorption/ionization mass spectrometry and mass spectrometry imaging. Eur J Mass Spectrom, 10 (2004) 657–664.

Touchstone, J.C., Thin-layer chromatographic procedures for lipid separation, J Chromatogr B, 671 (1995) 169–195.

Wakelam, M.J., Diacylglycerol-when is it an intracellular messenger? Biochim Biophys Acta, 1436 (1998) 1117–1126.

Wenk, M.R., The emerging field of lipidomics, Nat Rev Drug Discov, 4 (2005) 594–610.
Wetzel, S.J., Guttman, C.M. and Flynn, K.M., The influence of electrospray deposition in matrix-assisted laser desorption/ionization mass spectrometry sample preparation for synthetic polymers, Rapid Commun Mass Spectrom, 18 (2004) 1139–1146.
White, T., Bursten, S., Frederighi, D., Lewis, R.A. and Nudelman, E., High-resolution separation and quantification of neutral lipid and phospholipids species in mammalian cells and sera by multi-one-dimensional thin-layer chromatography, Anal Biochem, 10 (1998) 109–117.
Wolf, C. and Quinn, P.J., Lipidomics: Practical aspects and applications, Progr Lipid Res, 2007, in press.
Zschörnig, O., Pietsch, M., Süβ. R., Schiller, J. and Gütschow, M., Cholesterol esterase action on human high density lipoproteins and inhibition studies: detection by MALDI-TOF MS, J Lipid Res, 46 (2005) 803–811.
Zschörnig, O., Richter, V., Rassoul, F., Süβ, R., Arnold, K. and Schiller, J., Analysis of human blood plasma by MALDI-TOF MS-Evaluation of critical parameters, Anal Lett, 39 (2006) 1101–1113.

# Chapter 22
# Lipidomics in Diagnosis of Lipidoses

**C. Wolf and P.J. Quinn**

**Abstract** A review is presented of the major clinical features of a number of glycolipidoses including Fabry, Gaucher, Tay-Sachs, metachromatic leukodystrophy as well as CeroidLipofucinosis and Sjogren-Larsson syndrome. The possibilities offered by lipidomics for diagnosis and follow-up after enzyme replacement therapy are presented from a practical perspective. The contribution of HPLC coupled with tandem mass spectrometry has considerably simplified the detection and assay of abnormal metabolites. Corresponding internal standards consisting of weighed mixtures of the stable-isotope labeled metabolites required to calibrate and quantitate lipid components of these orphan diseases standards have yet to become commercially available. A lipidomics approach has been found to compare favorably with DNA-sequence analysis for the rapid diagnosis of pre-birth syndromes resulting from these multiple gene defects. The method also seems to be suitable for screening applications in terms of a high throughput combined with a low rate of false diagnoses based on the wide differences in metabolite concentrations found in affected patients as compared with normal subjects. The practical advantages of handling samples for lipidomic diagnoses as compared to enzyme assay are presented for application to diagnosis during pregnancy.

**Keywords** Lipidomics · fabry · gaucher · Tay-Sachs · lipofucinosis · Sjogren-Larsson syndrome

## 22.1 Introduction

An interrogation of the Pubmed database for 2006–7 returns nearly 1000 articles and 132 reviews for "lipidoses". Of these the predominant subject is the glycolipidoses excluding sterol disorders (e.g., Niemann-Pick type C and

C. Wolf
Mass Spectrometry Unit, INSERM U538, Faculte de Medecine P. et M. Curie, University Paris-6, 27 rue de Chaligny, Paris 75012, France

P.J. Quinn, X. Wang (eds.), *Lipids in Health and Disease*,

deficits of cholesterol biosynthesis) and perturbations of energy-generating neutral glyceride metabolism which are not in the scope of this volume.

In the recent literature, Fabry disease, a deficiency of the lysosomal α-galactosidase A (EC 3.2.1.22) and subsequent accumulation in various tissues of globotriosylceramide (Gb3, Galα-(1→4)Galβ-(1→4)Glc-ceramide) and Gaucher disease, a deficit of glucocerebrosidase (GBA) are represented by 248 and 345 references, respectively. Other glycolipidosis frequently documented are Tay-Sachs disease, a deficiency of Hexosaminidase A causing the accumulation of gangliosides GM2 and metachromatic leukodystrophy, another lysosomal storage disorder caused by the deficiency in the sulfolipid degrading enzyme arylsulfatase A (ASA).

Of the deficits unrelated to the glycolipids which show an abundant "clinical" literature but less interest manifested by scientists in the field of lipidomics are ceroid lipofuscinosis (the deficiency in palmitoyl-protein thioesterase (PPT)) and the Sjogren-Larsson syndrome, a defect of ALDH3A2 gene which encodes fatty aldehyde dehydrogenase (FALDH). It is pertinent to question the reason for such differences between clinical and basic scientific interests reflected in literature citations.

The shift in the interest of many clinicians to the subject can easily be explained by the recent development of enzyme replacement therapy for some of these diseases. During the last decade this therapeutic strategy and the possibility of its use in prenatal diagnosis have excited interest in syndromes described more than 130 years ago. A current advance in therapy is the sophisticated and successful gene therapy developed recently for a frequent peroxisomal deficit of lipid metabolism, X linked adrenoleucodystrophy (Sevin et al., 2007). This method exploits an opportunity of gene transfection mediated by lentivirus. This is one of many treatments which have been critically reviewed by Beutler (Beutler, 2006) on the basis of their cost ($100–200 k/year) and efficacy. Other treatments at a lower cost, such as substrate depletion and chaperone therapy, are also examined in the review. As a result of these developments it is expected that basic studies focusing on deficits of lipid degradation and toxicity of accumulated by-products will be accelerated to reconcile the diverging views of pediatrians with health insurance.

We present in this Chapter an overview on these orphan diseases from the perspective of opportunities offered by developments in lipidomics. The severity of these conditions, which are usually progressively debilating, become manageable by therapies that considerably delay the appearance of the consequent irreversible damage.

The wide differences in the mutation rate inherent in these disorders govern the efforts of national associations of patients affected by lipidoses and that of enzyme replacement therapeutic industries. In turn, the clinical "everyday" practice triggers campaigns for the detection by genotyping combined with biochemical lipidomic methods which establish the screening efficiency in neonates. The history of cholesterol deficits such as the Smith-Lemli-Opitz syndrome (SLOS) serves to illustrate how long it may take before the clinical and scientific aspects of disorders to be married. The clinical description of SLOS was first reported in 1964 (Smith et al., 1964). In the same year experimental teratologists established the activity of inhibitors of cholesterol synthesis for the mid-brain development and masculinization in rat embryos (Roux, 1964). Nevertheless, it has taken 30 years to recognize that blockade of cholesterol synthesis was the cause of SLOS (Tint, 1993). For this particular deficit a wide geographical distribution without any remarkably high incidence in small groups has most likely been responsible for a lack of an understanding of the underlying biochemical causes of the malformations. However most lipidoses presented in Table 22.1 have now received a clear biochemical explanation even if the detailed mechanisms of cell toxicity remain in many cases partially obscure.

**Table 22.1** Summary of genetic defects, incidences and therapies of lipidoses

| Syndrome | Enzyme deficit | Abnormal metabolite | Defective gene | Estimated incidence | Recent change in the status |
|---|---|---|---|---|---|
| Fabry | α-galactosidase A | Globotrihexosyl-ceramide (Gb3) | GLA (X-linked) | 1:40000-1:55000 | ERT |
| Gaucher | acid β-glucocerebrosidase | Glucoceramide | GBA or PSAP (saposin C, activor) | carrier up to ~ 4% (Ashkenazi Jews) | ERT |
| Tay-Sachs | hexosaminidase A (α subunit) | Ganglioside GM2 | HEXA | carrier ~ 4% (in Ashkenazi Jews, French Canadians) to 0.4% | ERT |
| Metachromatic Leucodystrophy | arylsulfatase A | Cerebroside sulfate | ARSA | 1:40000 | LCMS2 screening of urinary sulfatides |
| Ceroid Lipofucinosis | palmitoyl-protein thioesterase (PPT 1) | granular osmiophilic deposits of acylated proteins | CNL1 (for PPT1) CNLx>10 | 1:12500 | fluorimetric enzyme assay of PPT |
| Sjogren-Larson | fatty aldehyde dehydrogenase | Fatty alcohol | FALDH3 | 1:12000 north-east Sweden | GC assay of plasma aliphatic alcool |

## 22.2 Fabry's Disease

The disorder results in an accumulation of globotriosylceramide (Gb3) in tissues due to deficiency of α-galactosidase activity.

### *22.2.1 Biochemical Features of Fabry's Disease*

Gb3 which accumulates in cells of patients affected with Fabry's disease fulfils a number of regulatory processes. It has been known for many years that the glycosphingolipids act as a blood-group antigen, designated pk antigen (Naiki and Marcus, 1974). More recently,Gb3 has been shown to have a role as a receptor for shiga toxins (Lingwood, 1996), as a marker for the germinal centre stage of B-cell development (CD77) (Mangeney et al., 1993) and as an antigen associated with Burkett's lymphoma where it may be required for antigen presentation in malignant B-cells (George et al., 2001). The involvement of Gb3 in signaling *via* interferon-alpha pathways (Khine and Lingwood, 2000), CD19-mediated cell adhesion (Maloney and Lingwood, 1994) and apoptosis (Taga et al., 1997) has also been reported.

The role of Gb3 in its action as a plasma membrane receptor for internalization of Shiga-like toxins has been investigated in some detail. Binding of the toxin results in endocytosis and translocation via the endosomal system and *trans*-Golgi network in such a manner as to avoid the late endosomal compartment. Transport through the Golgi proceeds independently of coatormer protein-1 to the endoplasmic reticulum. There is evidence that Gb3 receptor functions are performed by the translocation of the glycosphingolipids into membrane microdomains on the cell surface. Thus treatment of cells in tissue culture with N-butyldeoxygalactonojirimycin, a specific inhibitor of ceramide glucosyl transferase, completely prevents uptake of toxin and induction of intracellular toxicity by denuding detergent-insoluble membrane fractions of Gb3 (Smith et al., 2006). These studies indicate that Gb3 function in cells is mediated by its sequestration in membrane rafts.

One of the attendant complications of Gb3 accumulation is the propensity towards vascular thrombosis (Utsumi et al., 1997). This is manifest in animal models such as the *Gla*-deficient mouse (Eitzman et al., 2003) but the connection is as yet unclear. This question has been examined by an analysis of tissue fibrin deposition in mice and thrombosis (Shen et al., 2006). The model confirmed a synergistic interaction between Gb3 and clotting factor V in tissue deposition suggesting an underlying explanation for susceptibility of stroke in Fabry's patients. The prothrombic state is associated with increased expression of the integrin CD11b on monocytes indicating leucocyte and endothelial activation (DeGraba et al., 2000). An elevation of blood levels of myeloperoxidase in Fabry's disease patients also contributes to risk factors in formation of atherosclerotic plaques generated, in part, by the production of reactive oxygen species (Kaneski et al., 2006). Cerebral involvement in Fabry's disease is also believed to be due to vascular pathologies especially in young male patients (Moore et al., 2001, Moore et al., 2001). This is manifest as white matter lesions identified in cranial scans of both male and female patients (Fellgiebel et al., 2005)

### 22.2.2 Treatment and Monitoring

A concise presentation of this disorder has been given by European Fabry Outcome Survey group (Cybulla and Neumann, 2007): "Fabry's disease is a rare, X-chromosome linked recessive lysosomal storage disorder. In its course multiple organ damage occurs, e.g. in skin, nerves, kidneys and heart. If untreated the disease not only markedly impairs the quality of life but also shortens life expectancy. As it is a rare and not widely known disease with considerable variability of its symptoms it is often not or only belatedly diagnosed. Since 2001, enzyme replacement has become available as an option in the causal treatment." The opening lines clearly emphasize the critical role for clinicians to recognizing the condition as early as possible so as to trigger an appropriate diagnostic test. It also stresses that starting enzyme replacement therapy at an early stage of onset is imperative. However, the Survey Group indicates that usually the diagnosis is only made in adults after a long and deleterious period of time (Cybulla and Neumann, 2007) "Fabry Outcome Survey data bank for the documentation of the disease's clinical course shows on 262 patients (130 males, 132 females) a mean age 37.5 and 34 years, respectively, on entry in the FOS Typical symptoms". This is in contrast with the usual course where – acroparesthesias, joint pain, hypohidrosis, fever and angiokeratoma – have their onset in childhood (mean age nine years). The time interval is about 15 years between onset of the first symptoms and establishment of the diagnosis and the severity of the clinical picture correlates significantly with age ($p = 0.0001$). "Main causes of morbidity and death in Fabry's disease is involvement of the kidneys or heart, the one or other occurring in 75% of patients." What was a late diagnosis for a non-treatable deficit should now be changed into an early and widely available biochemical test since ERT is available.

A recent clinical trial of ERT (Banikazemi et al., 2007) shows that "Agalsidase-beta therapy slowed progression to the composite clinical outcome of renal, cardiac, and cerebrovascular complications and death compared with placebo in patients with advanced Fabry disease. Therapeutic intervention before irreversible organ damage may provide greater clinical benefit." The results of ERT are even better in children and emphasizes the importance of early detection prescribed by the pediatrician (Ries et al., 2006) "The boys showed a significant reduction in plasma globotriaosylceramide on treatment. Three patients (out of 24) with anhidrosis, as determined by quantitative sudomotor axon reflex testing, developed sweating. Six of 11 patients could reduce or cease their use of antineuropathic analgesics." The results are appealing for a simple specific lipidomics method to detect an abnormal accumulation of globotriosylceramide.

In a pioneering study it was shown that infusion of α-galactosidase A reduced the tissue globotriaosylceramide storage in patients with Fabry disease (Schiffmann et al., 2000). Two methods were proposed in the trial with a potential diagnosis application: the assay of "total" Gb3 concentration by HPLC and a α -galactosidase A kinetics using the fluorogenic substrate 4-methylumbelliferyl- α -D-galactopyranoside. HPLC assay of perbenzoylated Gb3 (Ullman and McCluer, 1985; Ullman et al., 1985) can serve also to assay the distinct molecular species of glycolipids altogether. The procedure involves the conversion of gangliosides to their perbenzoyl derivatives, isolation of derivatives on a C18-reversed-phase cartridge, separation of the derivatives on a column (straight phase silica) maintained at an elevated temperature, and UV detection of the derivatives at 230 nm. The aspects of the procedure which contribute to its utility are a convenient isolation of derivatives and chromatographic conditions that provide the baseline resolution of derivatives. However the resolution of molecular species can also be obtained by LC coupled to tandem MS (Nelson et al., 2004; Roddy et al., 2005). "The accurate measurement of Gb3 in biological samples, i.e., plasma, is not trivial due to the inherent heterogeneity and amphiphilic nature of the Gb3 molecule. The structure of both the sphingoid base region (long-chain base), as well as the fatty-acyl chain region, can exhibit heterogeneity which increases the overall complexity of Gb3 measurements; the additional measurement complexity is attributed to the need to measure a large number of possible Gb3 isoforms". Indeed it has been suggested that monitoring of individual isoforms or specific isoform ratios in addition to the quantification of total Gb3 could yield an improved effectiveness of enzyme replacement therapy. A deuterated synthetic internal standard is now available for accurate quantitation of Gb3 with LC-MS (Mills et al., 2002). A non-invasive screening method for Fabry disease by measuring globotriaosylceramide in whole urine samples is now available using tandem MS. For instance the method allows a sensitive detection including cases in hemizygote girls (Kitagawa et al., 2005). Interference of the assay with the renal condition was noted in the reported results.

## 22.3 Gaucher's Disease

Gaucher's disease is an inborn error of sphingolipid metabolism caused by a deficiency of the lysosomal enzyme, acid β-glucosidase (EC 3.2.1.45) which is responsible for cleaving glucosylceramide (GlcCer) into glucose and ceramide, a critical terminal step in many glycolipids catabolism (Brady et al., 1965, Brady et al., 1965).

### 22.3.1 Subcellular Changes

The subcellular catabolism of GlcCer has been examined using fluorescent derivatives of lactosylceramide. It was reported that lactosylceramide was targeted to late endosomes and lysosomes in fibroblasts and macrophage models of GD rather than to the Golgi in normal cells (Sillence et al., 2002). This effect was observed at levels of accumulated GlcCer well below that required to produce pathological signs of GD. Moreover, there was a concomitant increase in cholesterol content of the cells in these models inferring that cholesterol may be involved in miss-targeting because depletion of intracellular cholesterol restored the normal subcellular trafficking of lactosylceramide to the Golgi (Puri et al., 1999).

Globotriaosyl ceramide is required not only for Shiga toxin binding to cells, but also for its intracellular trafficking. Shiga toxin induces globotriaosyl ceramide recruitment to detergent-resistant membranes (Smith et al., 2006), and subsequent internalization of the lipid (Falguieres et al., 2006). The globotriaosyl ceramide pool at the plasma membrane is then replenished from internal stores. Whereas endocytosis is not affected in the recovery condition, retrograde transport of Shiga toxin to the Golgi apparatus (Falguieres et al., 2001) and the endoplasmic reticulum is strongly inhibited. This effect is specific, as cholera toxin trafficking on GM(1) and protein biosynthesis are not impaired. The differential behavior of both toxins is also paralleled by the selective loss of Shiga toxin association with detergent-resistant membranes in the recovery condition, and comparison of the molecular species composition of plasma membrane globotriaosyl ceramide indicates subtle changes in favor of unsaturated fatty acids.

More recently the effect of GlcCer accumulation on intracellular trafficking of other sphingolipids and phospholipids in the macrophage model of Gaucher's disease has been investigated (Hein et al., 2007). A 12-fold excess of GlcCer was induced in macrophages cultured in the presence of conduricol B-epoxide and the distribution of GlcCer in subcellular membranes was determined. It was found that initially GlcCer accumulates in lysosomes but as the content of the glycosphingolipids increases it becomes distributed relatively evenly throughout all the subcellular fractions. A similar pattern of accumulation of ceramide, di- and trihexosylceramides and phosphatidylglycerol was also observed in these cells suggesting that as the lysosomes become saturated with lipid the excess is shunted off to other subcellular compartments. The consequences of the excess glycosphingolipids and phosphatidylglycerol were said to interfere with biochemical pathways at these extralysosomal sites resulting in cell dysfunction and manifestation of the typical signs of Gaucher's disease.

### 22.3.2 Clinical Features

The deficiency of acid β-glucosidase manifests primarily in the macrophage. This is because macrophages acquire exogenously amounts of derived lipids from ingested senescent and apoptotic red and white blood cells. Therefore, the residual level of enzyme activity in the macrophage is insufficient to meet the needs of relatively large GlcCer turnover in these cells. Consequently GlcCer accumulates in the lysosomes of the mononuclear lineage throughout the reticulo-endothelial system including the liver, bone marrow, spleen and lung. Hematological involvement (thrombocytopenia and anemia), bone lesions and neurological impairment occur early in patients with certain severe genotypes (Koprivica et al., 2000). The prediction of disease progression in patients with Gaucher disease is difficult. The clinical manifestations of Gaucher disease are highly variable and, although certain genotypes are often associated with mild symptoms (N370S substitution) or severe (L444P), a simple correlation between genotype and phenotype does not exist because modifier genes play a modulatory role. However a relationship was found to exist among the 16:0-glucosylceramide/16:0-lactosylceramide ratio (glucosylceramide level is increased but lactosylceramide is decreased in plasma of more severely affected patients), LAMP-1 (lysosomal-associated membrane protein-1) and saposin C levels (non specific markers for lysosomal storage disorders) and the patient phenotype. The refinement of the genotype-phenotype correlation finding has major implications for the diagnosis, prediction of disease severity and monitoring of therapy in patients with Gaucher disease (Whitfield et al., 2002).

The size of the spleen has been shown to increase 25-fold in patients with Gaucher's disease. However it has been recently shown that a secondary overall sphingolipid accumulation in a macrophage model explains such an increase, GlcCer representing only 2% of the lipid mass. There is clear evidence from studies of the mouse model of Gaucher's disease, in which GlcCer accumulation is induced by inhibition of GlcCerase by conduritol B-epoxide, of perturbations in metabolism of phospholipids (Bodennec et al., 2002). The enlargement of cells with fibrillar and vacuolated cytoplasm, irregular nuclei and atypical subcellular membrane structures is believed to be associated with increased GlcCer content but it may not be entirely responsible for the observed phenotype. More recent studies of human macrophages treated with GlcCerase inhibitor showed additionally that the rate of synthesis of phosphatidylcholine was increased in these cells (Trajkovic-Bodennec et al., 2004). The accelerated rate of phosphatidylcholine synthesis is apparently due to an increase in activity of the key enzyme controlling the rate of synthesis, CTP:phosphocholine cytidylyltransferase (Kacher et al., 2007). The molecular mechanisms responsible for increased activity of CTP:phosphocholine transferase is presently unknown. Neither is it clear whether the increased synthesis of phosphatidylcholine resulting from GcCer accumulation contributes directly to the pathology of the disease.

### 22.3.3 Diagnosis and Treatment Monitoring

Biochemical identification and quantification of glucosyl- and lactosyl-ceramide has widely benefited of LC tandem MS method. A general method using 1-phenyl-3-methyl-5-pyrazolone derivatives (Ramsay et al., 2005) of urinary oligosaccharides prior to analysis by electrospray ionization-tandem MS has been adapted to enable assay of large number of samples. PMP derivatization method was initially described by Honda (Honda et al., 1989). It can readily be extended to other oligosaccharidurias as a general method to monitor the levels of different oligosaccharide in patients receiving ERT. It has also potential for incorporation into a newborn screening program. The method semiquantifies urinary oligosaccharides from patients suffering from a variety of oligosaccharidurias including the Gaucher disease and GM1-GM2 gangliosidosis. The oligosaccharides are referenced against a single internal standard, methyl lactose, to produce ratios for a semiquantitative comparison with control samples. Elevations in specific urinary oligosaccharides were significantly indicative of lysosomal disease. The defective catabolic enzyme needs for specific assay a deuterated internal standard of the glycolipid to be incorporated in the sample before extraction.

## 22.4 Tay-Sachs Disease

Tay-Sachs disease is caused by the mutation of the alpha subunit of hexosaminidase A gene (HEXA). Deficitated hexosaminidases A and B produce 3 distinct clinical forms of ganglioside GM2 storage disease-Tay-Sachs disease, Sandhoff disease, and juvenile GM2-gangliosidosis. Hexosaminidase-A has a structure comprised of alpha-beta subunits and Tay-Sachs disease is the alpha-minus mutation, whereas Sandhoff disease is a beta-minus mutation (Beutler and Kuhl, 1975; Beutler et al., 1975). Subunit alpha is mapped to chromosome 15 (and beta to chromosome 5). Different levels of residual activities are correlated with the age of clinical onset: Tay-Sachs disease, 0.1% of normal hexosaminidase; late infantile, 0.5%; adult GM2-gangliosidosis, 2–4%; healthy persons with 'low hexosaminidase,' 11% and 20% (Conzelmann et al., 1983).

### 22.4.1 Genetic Distinction

Accumulation of GM2 is understood as a neurodegeneration that results in an excessive inflammatory reaction. Serial analysis of gene expression (SAGE) determined gene expression profiles in cerebral cortex from a Tay-Sachs patient, a Sandhoff disease patient and a pediatric control. Examination of genes that showed altered expression in both patients revealed molecular details of the pathophysiology of the disorders relating to neuronal dysfunction and loss. A large fraction of the elevated genes in the patients could be attributed to activated macrophages/microglia and astrocytes, and included class II histocompatability antigens, the pro-inflammatory cytokine osteopontin, complement components, proteinases and inhibitors, galectins, osteonectin/SPARC, and prostaglandin D2 synthase (Myerowitz et al., 2002).

Enzyme screening of the serum remains an essential component of carrier screening in non-Jewish carriers because of uncommon mutations. DNA screening can be best used as an adjunct to enzyme testing to exclude known HEXA pseudo-deficiency alleles (Akerman et al., 1997). Tay-Sachs disease is approximately 100 times more common in infants of Ashkenazi Jewish ancestry (central-eastern Europe) than in non-Jewish infants (Rimoin et al., 1977). The most frequent (>80%) DNA lesion in Tay-Sachs disease in Ashkenazi Jews is a 4-bp insertion in exon 11 of the HEXA gene (Myerowitz and Costigan, 1988). These data strongly support the use of DNA testing alone as the most cost-effective and efficient approach to carrier screening for Tay-Sachs disease in individuals of confirmed Ashkenazi Jewish ancestry (Bach, et al., 2001).

### *22.4.2 Enzyme Screening and Lipidomics*

Methods for the assay of hexosaminidase A and total hexosaminidase activities in dried blood spots on filter paper offer considerable advantages for screening (Chamoles et al., 2002). The deficient activity of the lysosomal enzymes hexosaminidase A and total hexosaminidase (hexosaminidase A plus B) are usually measured in plasma or extracts of leukocytes. To tubes containing a 3-mm-diameter blood spot, elution liquid and substrate solution were added. After incubation at 37°C, the amount of hydrolyzed product was compared with a calibrator to allow the quantification of enzyme activity. The method was proven reliable even after storage for up to 38 months at room temperature. For total hexosaminidase, the substrate was 4-methyl-umbelliferyl-2-acetamido-2-deoxy-β-D-glucopyranoside (neutral derivative) and for hexosaminidase A, 4-methyl-umbelliferyl-β-D-*N*-acetyl-glucopyranoside-6-sulfate. For ERT therapeutic evaluation of GM2 gangliosidoses ELISA using anti-GM2 ganglioside antibodies are now used for intracellular quantification of GM2 (Tsuji et al., 2007). Another method for profiling gangliosides extracted from animal tissues is using ESI-MS/MS with high through-put potential (Tsui et al., 2005). This system utilizes specific detection of a precursor ion (m/z 290), a derivative of *N*-acetylneuraminic acid comprised in gangliosides. The method includes the enrichment of gangliosides in the aqueous phase from total cellular lipid extracts which eliminates the damping effect of phospholipids and permits direct precursor scan.

## 22.5 Metachromatic Leucodystrophy

Metachromatic leukodystrophy is another lysosomal storage disorder caused by a deficiency of arylsulfatase A which leads to the accumulation of 3-O-sulfogalactosylceramide. The defect results in severe demyelination. The disease often takes a presentation of an under- and mis-diagnosed psychiatric affection long before neurological symptoms appear and MRI displays the anatomical lesions. MRI reveals a diffuse demyelination, bilateral and often symmetrical, initially limited to the periventricular areas.

### *22.5.1 Genetic Prevelance*

In a recent multicenter study two mutations were found the most prevalent amongst the European population: "c.459 + 1G>A and p.P426L, in 384 unrelated European patients presenting with different types of metachromatic leukodystrophy were found. In total, c.459 + 1G>A was found 194 times among the 768 investigated ARSA alleles (25%), whereas P426L was identified 143 times (18.6%). Thus, these two mutations accounted for 43.8% of investigated MLD alleles." As a function of age and geographical distribution the 2 mutations were found very different (Lugowska et al., 2005) "Mutation c.459 + 1G>A was most frequent in late-infantile MLD patients (40%), while P426L was most frequent in adults (42.5%). Mutation c.459 + 1G>A is more frequent in countries situated at the western edges of Europe, i.e., in Great Britain and Portugal, and also in Belgium, Switzerland, and Italy, which is visible as a strand ranging from North to South, and additionally in Czech and Slovak Republics. Mutation P426L is most prevalent in countries assembled in a cluster containing the Netherlands, Germany, and Austria."

### *22.5.2 Biochemical Characteristics*

Studies of the animal model have established a link between biosynthesis of sulfogalactosylceramides, accumulation and demyelination. The sulfatide storage pattern in ASA-deficient [ASA(−/−)] mice is comparable to humans, but the mice do not mimic the myelin pathology. It was assumed that increasing sulfatide storage in this animal model might provoke demyelination. Transgenic ASA(−/−) mice overexpressing the sulfatide-synthesizing enzyme galactose-3-O-sulfotransferase-1 [tg/ASA(−/−)] were prepared. Indeed, these tg/ASA(−/−) mice displayed a significant increase in sulfatide storage in brain and peripheral nerves and older than 1 year mice developed severe neurological symptoms related to demyelination (Eckhardt et al., 2007; Ramakrishnan et al., 2007). The accretion of sulfatides is directly correlated with the neurological phenotype. In generated transgenic ASA-deficient [ASA(−/−)] mice overexpressing the sulfatide synthesizing enzymes UDP-galactose:ceramide galactosyltransferase

(CGT) and cerebroside sulfotransferase (CST) neuronal lipid storage was provoked. CGT-transgenic ASA(−/−) [CGT/ASA(−/−)] mice showed an accumulation of C18:0 fatty acid-containing sulfogalactosylceramide in the brain. Histochemicaly, an increase in sulfolipid storage could be detected in central and peripheral neurons of both CGT/ASA(−/−) and CST/ASA(−/−) mice compared with ASA(−/−) mice. CGT/ASA(−/−) mice developed severe neuromotor coordination deficits and weakness of hindlimbs and forelimbs.

### 22.5.3 Clinical Signs

The biochemical abnormalities as a function of the presentation have been recently scrutinized by Bauman's group (Colsch et al., 2007) "During adolescence and/or adulthood, there are 2 clinical presentations. It may be that of a degenerative disease of the central nervous system with mainly spastic manifestations or a spino-cerebellar ataxia, or that of a psychosis. As several lines of evidence indicate that the psychotic form of MLD could be a model of psychosis, we decided to do a pluridisciplinary study on 11 psycho-cognitive cases involving mental and psychiatric testing, in comparison with 5 adult motor cases. However a biochemical study with enzyme assays and quantitative mass spectrometry of urinary sulfatides, so as to determine whether there were biochemical particularities related to the psychotic forms does not show any difference." These biochemical data have yet to be reconciled with previous observations where a phenotype-genotype correlation was described (Rauschka et al., 2006) "P426L homozygotes principally presented with progressive gait disturbance caused by spastic paraparesis or cerebellar ataxia; mental disturbance was absent or insignificant at the onset of disease but became more apparent as the disease evolved. In contrast, compound heterozygotes for I179S presented with schizophrenia-like behavioral abnormalities, social dysfunction, and mental decline, but motor deficits were scarce. Reduced peripheral nerve conduction velocities and less residual arylsulfatase A activity were present in P426L homozygotes *vs* I179S".

### 22.5.4 Lipidomics

Sulfatides (3-sulfogalactosylceramides) were initially detected by mass spectrometry using FAB ionization but spectrum showed a number of fragmentations which complicates considerably the method (Ohashi and Nagai, 1991). More recently structural characterization of sulfatides by collisional-activated dissociation (CAD) was described in quadrupole ion-trap tandem mass spectrometric methods with electrospray ionization. With the method [M – H]- ions of sulfatides yield abundant structurally informative ions that permit unequivocal assignments of the long-chain base and fatty acid constituent including the location of double bond (Hsu and Turk, 2004). The major sulfatide molecular species are quantified similarly in the 2 clinical forms (motor and psycho-cognitive adult forms) with the following fatty acids and sphingoid bases: C22:1/d18:1 and /or C22:0/d18:2 (m/z 862.5), C22:0 (OH)/d18:1 (m/z 878.5), C24:0/d18:1 and / or C24:0/C23:1(OH)/d18:2 (m/z 890.3), C24:0 (OH)/d18:1(m/z 906.5) (Colsch et al., 2007). Because the diagnosis may be complicated in cases of arylsulfatase A pseudodeficiency and sphingolipid activator protein deficiency, this measurement of sulfatide in the urinary sediment of affected individuals by a rapid, sensitive, and specific mass spectrometric method has been long wanted (Whitfield et al., 2001). Urinary sulfatides are now commonly detected using electrospray ionization-tandem mass spectrometry by means of the precursor ion scan 97. Levels are considerably increased to X20–30 folds as compared to controls which allows the rapid screening of a large number of samples.

## 22.6 Ceroid Lipofuscinoses

The condition is caused by deficiencies of palmitoyl protein thioesterase 1 (PPT1) (or tripeptidyl peptidase 1 (TPP1) and possibly other enzymes resulting in the same clinical presentation of NCL (for *N*eural *C*eroid *L*ipofuscinose)). PPT1 cleaves long-chain fatty acids from S-acylated proteins within the lysosome (Lu et al., 1996). How the loss of this activity causes the death of central nervous system neurons is not known.

### *22.6.1 Clinical Aspects*

A strong interest for this group of disorders was recently raised after a clinical trial (Steiner et al., 2007) has been conducted to evaluate the safety and preliminary efficacy (phase 1) of human central nervous stem cells (HuCNS-SC) implanted into the cortex and lateral ventricles of patients with advanced neuronal ceroid lipofuscinoses. It is known that the accumulation of undigested substrates leads to the formation of neuronal storage bodies that are associated with the clinical symptoms. With the same clinical presentation, TPP1 gene mutation is related to the deficiency of tripeptidyl peptidase 1. The corresponding CLN2 gene product is synthesized as an inactive proenzyme that is autocatalytically converted to an active serine protease (Lin et al., 2001). The deficit of this lysosomal protease causes neuronal ceroid lipofuscinoses, a group of inherited, neurodegenerative, lysosomal-storage disorders characterized by intracellular accumulation of autofluorescent ceroid lipofuscin storage material in neurons and other cells. The patients experience progressive cognitive and motor deterioration, blindness, seizures (as early as 2–3 yrs) and early death.

The two subtypes of NCLs are due to deficiencies either in the palmitoyl protein thioesterase 1 (PPT1) protein or in the tripeptidyl peptidase 1 (TPP1) protein. The trial aforementioned by a cell therapy was comprised of neural stem/progenitor cells which constitutively synthesize and secrete both the PPT1 and TPP1 enzymes. In culture, these secreted enzymes are internalized by fibroblasts from patients with the PPT1 or TTP1 deficiency. Transplantation of HuCNS-SCs into PPT1 knockout immunodeficient mice leads to global engraftment, provides PPT1 enzyme, reduces storage material, neuroprotects host neurons, and extends survival of host mice.

### 22.6.2 *Enzymology*

The recent development of simple, fluorogenic enzyme assay using 4-methylumbelliferyl-6-thiopalmitoyl-ß-glucoside (MUTG) for infantile and late infantile neuronal ceroid lipofuscinosis has greatly facilitated the diagnostic process for these diseases (Young et al., 2001). In leucocytes and fibroblasts from infantile patients profound deficiencies of palmitoyl-protein thioesterase 1 (PPT1 ) are found, the residual activity being $< 5\%$ of mean control. The feasibility of a reliable prenatal enzyme analysis was tested successfully using the fluorogenic substrate. In fibroblasts from late infantile lipofuscinoses patients a similar syndrome, the deficiency of tripeptidyl-peptidase I activity (TPP-I) was frequently found. More than 30 mutations have been reported altogether in PPT1 and TPP1 genes, rendering the molecular genetic analysis impractical as a primary means of diagnosis. Electron microscopy of characteristic cellular inclusions remains an important diagnostic method, but it is also tedious and not readily available. A simple assay for the determination of tripeptidyl peptidase and palmitoyl protein thioesterase activities in dried blood spots is now available (Lukacs, et al., 2003). The clinical presentation of ceroid lipofuscinoses is especially difficult to correlate with a single biochemical mechanism. Indeed after reviewing 319 patients with NCL, the authors (Wisniewski et al., 2001) found that 64 (20%) did not fit into gene defects for *CLN1* and *CLN2* which encode lysosomal palmitoyl protein thioesterase and tripeptidyl peptidase 1, respectively. Eight NCL forms are now considered which result from 100 different mutations.

## 22.7 Sjogren-Larsson Syndrome

The syndrome is caused by a particular mutation in the gene encoding fatty aldehyde dehydrogenase (gene FALDH3A2). About 1.3% of the population of northern Sweden is heterozygous for the defective gene.

### 22.7.1 *Clinical Aspects*

Sjogren and Larsson described in 1957 patients with spastic diplegia or tetraplegia, low grade oligophrenia, and ictyosiform erythrodermia developing during infancy. Many patients have also characteristically retinal glistering spots. The ophtalmological abnormalities are often severe in the syndrome. Detailed information is given on this complication in (Aslam and Sheth, 2007; Romanes, 1968). About half the affected children have an early pigmentary degeneration of the retina (juvenile macular dystrophy) (Willemsen et al., 2000). Most of the patients never walk and about half the patients have seizures.

## *22.7.2 Genetic and Enzymic Characteristics*

The evidence that Sjögren-Larsson syndrome is genetically homogeneous was critically analyzed by two references (Pigg et al., 1999; Rogers et al., 1995). It is assumed that Swedish soldiers bivouacking in Germany during the 30-year war in the 17th century could have introduced the Sjogren-Larsson gene into the German population. A missense mutation in the FALDH gene was maped on chromosome 17 in Sjögren-Larsson syndrome patients originating from the northern part of Sweden. Fatty alcohol:$NAD^+$ oxidoreductase, the enzyme catalyzing the oxidation of hexadecanol or octadecanol to the corresponding fatty acid is deficient in fibroblasts (mean activity at 13% of that in normal fibroblasts). Fibroblasts from heterozygotes show intermediate levels of activity (Rizzo et al., 1987; Rizzo et al., 1989).

Fatty alcohol:$NAD^+$ oxidoreductase is a complex enzyme which consists of two separate proteins that sequentially catalyze the oxidation of fatty alcohol to fatty aldehyde and to fatty acid that is to say a fatty alcohol dehydrogenase and a fatty aldehyde dehydrogenase (FALDH) activity. Sjögren-Larsson cells were selectively deficient in the FALDH component and had normal activity of fatty alcohol dehydrogenase. Intact fibroblast oxidized octadecanol to fatty acid at $<10\%$ of the normal rate but oxidized octadecanal normally confirming that FALDH is specifically affected (Rizzo and Craft, 1991).

Sjögren-Larsson can be diagnosed prenatally using enzymatic methods (Rizzo et al., 1994). Fatty alcohols in cultured cells and plasma were analyzed as acetate derivatives using capillary column gas chromatography. By this method, cultured skin fibroblasts from Sjögren-Larsson patients were found to have 7- and 8-fold elevations in the mean content of hexadecanol (16:0-OH) and octadecanol (18:0-OH), respectively. Most importantly, the mean plasma 16:0-OH and 18:0-OH concentrations in Sjögren-Larsson patients ($n = 11$) were 9- and 22-fold higher than in normal controls (about 10 ng/ml), respectively. In fibroblasts, most of the fatty alcohol (59%) that accumulated was free rather than esterified alcohol, whereas free alcohol accounted for 23% of the total alcohol in normal cells indicating that elevations in free fatty alcohols provide a sensitive marker for Sjögren-Larsson syndrome. Furthermore, dietary or cell media sources may contribute to the tissue or cell content of specific fatty alcohols.

## 22.8 Lipidomics Applied to Diagnosis

The characteristical lipid metabolites (glycolipids or fatty alcohols) which accumulate in lipidoses are reviewed above. None represents a particular difficulty to authentify or assay quantitatively as regard to the multiple lipids which have already been successfully studied by the lipidomics. The method for usual lipids is commonly understood as a process comprising extraction, solvent partition and a critical HPLC-tandem MS step to ascertain the metabolite structure and allow a (semi)-quantitative assay. For glycolipids and for highly polar gangliosides (TSD) or cerebroside sulfate (MLD), specifically, the partition between alcoholic water layer and chloroformic lower layer (Bligh and Dyer, 1959; Folch, et al., 1957) cannot be applied as it is for most of the lipids with a moderate or no hydrophilicity at neutral or acidic pH. With the perspective of clinical application a variety of methods were challenged for gangliosides from red blood cells (Wang and Gustafson, 1995). Results show that ganglioside extraction is unfavourably affected by the addition of the solvents as a mixture and by the use of less polar solvents and by a lower total solvent-to-sample ratio. The distribution of gangliosides could be uneven in an apparently monophasic extraction solvent mixture. The uneven distribution occurred during and also after the extraction (in filtration and centrifugation). In the recommended method using 19 volumes of methanol/chloroform (2:1) solvent in a one-step extraction, the above disadvantages in ganglioside extraction and quantification are kept under control. This method appears simple and it gives a high recovery of gangliosides. Pre-analytical steps in the perspective of HPLC-tandem MS serves to separate the metabolite of interest from compounds which may suppress ion formation in the ESI source. Usually suppression is caused by abundant ions with a charge similar with the metabolite.

In the case of acidic glycolipids the relative proton affinity of chemicals can shift the balance for negative ionisation in favor of co-eluted compounds. Pre-analytical separation under acidic conditions serves also to reduce as much as possible the "dispersion" in the MS spectrum of the metabolite into multiple m/z representing the various adducts of counterions $Na^+$, $K^+$, $NH_4{}^+$, organic amines$^+$,... which improves sensitivity of the test. Sulfatides are lost during the partition between the hexane and the methanol/water phase. The analysis of sulfatides involves the isolation of the glycosphingolipid fraction and the subsequent separation of sulfatides from neutral lipids by chromatography on DEAE-sephadex or DEAE-cellulose column (the variety of methods are referenced in the website CyberLipid (http://www.cyberlipid.org/).

The contribution of lipidomics for the diagnosis of any severe defect related to a metabolic inherited abnormality should be judged taking into account 2 criteria; the 1srt is the delay required for the laboratory response given to the obstetrician after sampling (amniotic fluid or chorionic villosity) and the 2nd criterion is the difficulty for an alternative molecular DNA diagnostic when no specific but multiple mutations have to be searched for along a multi-kb gene.

Based on everyday practice lipidomics methods are well able to provide the ascertained diagnostic without false positive or negative on very long periods of time (>15 yrs). Answers are usually given within few days and delay can be furthermore reduced after the sample transfer from the diagnostic center is optimized and the electronically certified results are directly mailed to the clinician in charge with the patient. Lipids metabolites as compared to most other diagnostics requiring enzyme assay do not require any particular precautions in terms of sample transfer. Postal mail with no cooling or freezing of the biological sample can be achieved with the advantages of a modest cost for transportation and the critical possibility to collect samples for a regional or national lipidomics unit appropriately trained. The reduction of delay for a positive ascertained diagnostic leaves a possibility for a medical abortion if required. If the severe defect is evoked by sound-examination at the term of 10–12 weeks of gestation mifepristone abortifacient may be applied instead of the surgical uterine revision. By difference DNA analysis of multi-kb genes (such as CNL or TSD in non-Ahskenazi population) and multiple polymorphisms may imply time-consuming PCR amplification and tedious analysis of the sequence.

**Acknowledgments** The review was supported by a grant from the Human Frontier Science Programme (RGP16/2005).

## References

Akerman B.R., Natowicz M.R., Kaback M.M., Loyer M., Campeau E., Gravel R.A., Novel mutations and DNA-based screening in non-Jewish carriers of Tay-Sachs disease, American journal of human genetics 60 (1997) 1099–1106.

Aslam S.A., Sheth H.G., Ocular features of Sjogren-Larsson syndrome, Clinical & experimental ophthalmology 35 (2007) 98–99.

Bach G., Tomczak J., Risch N., Ekstein J., Tay-Sachs screening in the Jewish Ashkenazi population: DNA testing is the preferred procedure, American journal of medical genetics 99 (2001) 70–75.

Banikazemi M., Bultas J., Waldek S., Wilcox W.R., Whitley C.B., McDonald M., Finkel R., Packman S., Bichet D.G., Warnock D.G., Desnick R.J., Agalsidase-beta therapy for advanced Fabry disease: a randomized trial, Annals of internal medicine 146 (2007) 77–86.

Beutler E., Lysosomal storage diseases: natural history and ethical and economic aspects, Molecular genetics and metabolism 88 (2006) 208–215.

Beutler E., Kuhl W., Subunit structure of human hexosaminidase verified: interconvertibility of hexosaminidase isozymes, Nature 258 (1975) 262–264.

Beutler E., Kuhl W., Comings D., Hexosaminidase isozyme in type O Gm2 gangliosidosis (Sandhoff-Jatzkewitz disease), American journal of human genetics 27 (1975) 628–638.

Bligh E.G., Dyer W.J., A Rapid Method of Total Lipid Extraction and Purification, Canadian Journal of Biochemistry and Physiology 37 (1959) 911–917.

Bodennec J., Pelled D., Riebeling C., Trajkovic S., Futerman A.H., Phosphatidylcholine synthesis is elevated in neuronal models of Gaucher disease due to direct activation of CTP:phosphocholine cytidylyltransferase by glucosylceramide, Faseb J 16 (2002) 1814–1816.

Brady R.O., Kanfer J., Shapiro D., The Metabolism of Glucocerebrosides. I. Purification and Properties of a Glucocerebroside-Cleaving Enzyme from Spleen Tissue, The Journal of biological chemistry 240 (1965) 39–43.

Brady R.O., Kanfer J.N., Shapiro D., Metabolism of Glucocerebrosides. Ii. Evidence of an Enzymatic Deficiency in Gaucher's Disease, Biochem Biophys Res Commun 18 (1965) 221–225.

Chamoles N.A., Blanco M., Gaggioli D., Casentini C., Tay-Sachs and Sandhoff diseases: enzymatic diagnosis in dried blood spots on filter paper: retrospective diagnoses in newborn-screening cards, Clinica chimica acta; international journal of clinical chemistry 318 (2002) 133–137.

Colsch B., Afonso C., Turpin J.C., Portoukalian J., Tabet J.C., Baumann N., Sulfogalactosylceramides in motor and psycho-cognitive adult metachromatic leukodystrophy: relations between clinical, biochemical analysis and molecular aspects, Biochim Biophys Acta BBA 1780 (2008) 434–440.

Conzelmann E., Kytzia H.J., Navon R., Sandhoff K., Ganglioside GM2 N-acetyl-beta-D-galactosaminidase activity in cultured fibroblasts of late-infantile and adult GM2 gangliosidosis patients and of healthy probands with low hexosaminidase level, American journal of human genetics 35 (1983) 900–913.

Cybulla M., Neumann H.P., [Fabry disease. An interdisciplinary challenge], Deutsche medizinische Wochenschrift (1946) 132 (2007) 2271–2277.

DeGraba T., Azhar S., Dignat-George F., Brown E., Boutiere B., Altarescu G., McCarron R., Schiffmann R., Profile of endothelial and leukocyte activation in Fabry patients, Annals of neurology 47 (2000) 229–233.

Eckhardt M., Hedayati K.K., Pitsch J., Lullmann-Rauch R., Beck H., Fewou S.N., Gieselmann V., Sulfatide storage in neurons causes hyperexcitability and axonal degeneration in a mouse model of metachromatic leukodystrophy, Journal of Neuroscience 27 (2007) 9009–9021.

Eitzman D.T., Bodary P.F., Shen Y., Khairallah C.G., Wild S.R., Abe A., Shaffer-Hartman J., Shayman J.A., Fabry disease in mice is associated with age-dependent susceptibility to vascular thrombosis, Journal of the American Society of Nephrology 14 (2003) 298–302.

Falguieres T., Mallard F., Baron C., Hanau D., Lingwood C., Goud B., Salamero J., Johannes L., Targeting of Shiga toxin B-subunit to retrograde transport route in association with detergent-resistant membranes, Molecular Biology of the Cell 12 (2001) 2453–2468.

Falguieres T., Romer W., Amessou M., Afonso C., Wolf C., Tabet J.C., Lamaze C., Johannes L., Functionally different pools of Shiga toxin receptor, globotriaosyl ceramide, in HeLa cells, FEBS Journal 273 (2006) 5205–5218.

Fellgiebel A., Muller M.J., Mazanek M., Baron K., Beck M., Stoeter P., White matter lesion severity in male and female patients with Fabry disease, Neurology 65 (2005) 600–602.

Folch J., Lees M., Stanley G.H.S., A Simple Method for the Isolation and Purification of Total Lipides from Animal Tissues, Journal of Biological Chemistry 226 (1957) 497–509.

George T., Boyd B., Price M., Lingwood C., Maloney M., MHC class II proteins contain a potential binding site for the verotoxin receptor glycolipid CD77, Cell Mol Biol (Noisy-le-grand) 47 (2001) 1179–1185.

Hein L.K., Meikle P.J., Hopwood J.J., Fuller M., Secondary sphingolipid accumulation in a macrophage model of Gaucher disease, Molecular genetics and metabolism 92 (2007) 336–345.

Honda S., Akao E., Suzuki S., Okuda M., Kakehi K., Nakamura J., High-performance liquid chromatography of reducing carbohydrates as strongly ultraviolet-absorbing and electrochemically sensitive 1-phenyl-3-methyl-5-pyrazolone derivatives, Analytical biochemistry 180 (1989) 351–357.

Hsu F.F., Turk J., Studies on sulfatides by quadrupole ion-trap mass spectrometry with electrospray ionization: structural characterization and the fragmentation processes that include an unusual internal galactose residue loss and the classical charge-remote fragmentation, Journal of the American Society for Mass Spectrometry 15 (2004) 536–546.

Kacher Y., Golan A., Pewzner-Jung Y., Futerman A.H., Changes in macrophage morphology in a Gaucher disease model are dependent on CTP:phosphocholine cytidylyltransferase alpha, Blood cells, molecules & diseases 39 (2007) 124–129.

Kaneski C.R., Moore D.F., Ries M., Zirzow G.C., Schiffmann R., Myeloperoxidase predicts risk of vasculopathic events in hemizgygous males with Fabry disease, Neurology 67 (2006) 2045–2047.

Khine A.A., Lingwood C.A., Functional significance of globotriaosyl ceramide in interferon-alpha(2)/type 1 interferon receptor-mediated antiviral activity, Journal of Cellular Physiology 182 (2000) 97–108.

Kitagawa T., Ishige N., Suzuki K., Owada M., Ohashi T., Kobayashi M., Eto Y., Tanaka A., Mills K., Winchester B., Keutzer J., Non-invasive screening method for Fabry disease by measuring globotriaosylceramide in whole urine samples using tandem mass spectrometry, Molecular genetics and metabolism 85 (2005) 196–202.

Koprivica V., Stone D.L., Park J.K., Callahan M., Frisch A., Cohen I.J., Tayebi N., Sidransky E., Analysis and classification of 304 mutant alleles in patients with type 1 and type 3 Gaucher disease, American journal of human genetics 66 (2000) 1777–1786.

Lin L., Sohar I., Lackland H., Lobel P., The human CLN2 protein/tripeptidyl-peptidase I is a serine protease that autoactivates at acidic pH, The Journal of biological chemistry 276 (2001) 2249–2255.

Lingwood C.A., Role of verotoxin receptors in pathogenesis, Trends Microbiol 4 (1996) 147–153.

Lu J.Y., Verkruyse L.A., Hofmann S.L., Lipid thioesters derived from acylated proteins accumulate in infantile neuronal ceroid lipofuscinosis: correction of the defect in lymphoblasts by recombinant palmitoyl-protein thioesterase, Proceedings of the National Academy of Sciences of the United States of America 93 (1996) 10046–10050.

Lugowska A., Amaral O., Berger J., Berna L., Bosshard N.U., Chabas A., Fensom A., Gieselmann V., Gorovenko N.G., Lissens W., Mansson J.E., Marcao A., Michelakakis H., Bernheimer H., Ol'khovych N.V., Regis S., Sinke R., Tylki-Szymanska A., Czartoryska B., Mutations c.459+1G>A and p.P426L in the ARSA gene: prevalence in metachromatic leukodystrophy patients from European countries, Molecular genetics and metabolism 86 (2005) 353–359.

Lukacs Z., Santavuori P., Keil A., Steinfeld R., Kohlschutter A., Rapid and simple assay for the determination of tripeptidyl peptidase and palmitoyl protein thioesterase activities in dried blood spots, Clinical chemistry 49 (2003) 509–511.

Maloney M.D., Lingwood C.A., CD19 has a potential CD77 (globotriaosyl ceramide)-binding site with sequence similarity to verotoxin B-subunits: implications of molecular mimicry for B cell adhesion and enterohemorrhagic Escherichia coli pathogenesis, Journal of Experimental Medicine 180 (1994) 191–201.

Mangeney M., Lingwood C.A., Taga S., Caillou B., Tursz T., Wiels J., Apoptosis induced in Burkitt's lymphoma cells via Gb3/CD77, a glycolipid antigen, Cancer Research 53 (1993) 5314–5319.

Mills K., Johnson A., Winchester B., Synthesis of novel internal standards for the quantitative determination of plasma ceramide trihexoside in Fabry disease by tandem mass spectrometry, FEBS letters 515 (2002) 171–176.

Moore D.F., Herscovitch P., Schiffmann R., Selective arterial distribution of cerebral hyperperfusion in Fabry disease, Journal of Neuroimaging 11 (2001) 303–307.

Moore D.F., Scott L.T., Gladwin M.T., Altarescu G., Kaneski C., Suzuki K., Pease-Fye M., Ferri R., Brady R.O., Herscovitch P., Schiffmann R., Regional cerebral hyperperfusion and nitric oxide pathway dysregulation in Fabry disease: reversal by enzyme replacement therapy, Circulation 104 (2001) 1506–1512.

Myerowitz R., Costigan F.C., The major defect in Ashkenazi Jews with Tay-Sachs disease is an insertion in the gene for the alpha-chain of beta-hexosaminidase, The Journal of biological chemistry 263 (1988) 18587–18589.

Myerowitz R., Lawson D., Mizukami H., Mi Y., Tifft C.J., Proia R.L., Molecular pathophysiology in Tay-Sachs and Sandhoff diseases as revealed by gene expression profiling, Human molecular genetics 11 (2002) 1343–1350.

Naiki M., Marcus D.M., Human erythrocyte P and Pk blood group antigens: identification as glycosphingolipids, Biochemical and Biophysical Research Communications 60 (1974) 1105–1111.

Nelson B.C., Roddy T., Araghi S., Wilkens D., Thomas J.J., Zhang K., Sung C.C., Richards S.M., Globotriaosylceramide isoform profiles in human plasma by liquid chromatography-tandem mass spectrometry, Journal of chromatography 805 (2004) 127–134.

Ohashi Y., Nagai Y., Fast-atom-bombardment chemistry of sulfatide (3-sulfogalactosylceramide), Carbohydrate research 221 (1991) 235–243.

Pigg M., Annton-Lamprecht I., Braun-Quentin C., Gustavson K.H., Wadelius C., Further evidence of genetic homogeneity in Sjogren-Larsson syndrome, Acta dermato-venereologica 79 (1999) 41–43.

Puri A., Hug P., Jernigan K., Rose P., Blumenthal R., Role of glycosphingolipids in HIV-1 entry: requirement of globotriosylceramide (Gb3) in CD4/CXCR4-dependent fusion, Bioscience reports 19 (1999) 317–325.

Ramakrishnan H., Hedayati K.K., Lullmann-Rauch R., Wessig C., Fewou S.N., Maier H., Goebel H.H., Gieselmann V., Eckhardt M., Increasing sulfatide synthesis in myelin-forming cells of arylsulfatase A-deficient mice causes demyelination and neurological symptoms reminiscent of human metachromatic leukodystrophy, Journal of Neuroscience 27 (2007) 9482–9490.

Ramsay S.L., Meikle P.J., Hopwood J.J., Clements P.R., Profiling oligosaccharidurias by electrospray tandem mass spectrometry: quantifying reducing oligosaccharides, Analytical biochemistry 345 (2005) 30–46.

Rauschka H., Colsch B., Baumann N., Wevers R., Schmidbauer M., Krammer M., Turpin J.C., Lefevre M., Olivier C., Tardieu S., Krivit W., Moser H., Moser A., Gieselmann V., Zalc B., Cox T., Reuner U., Tylki-Szymanska A., Aboul-Enein F., LeGuern E., Bernheimer H., Berger J., Late-onset metachromatic leukodystrophy: genotype strongly influences phenotype, Neurology 67 (2006) 859–863.

Ries M., Clarke J.T., Whybra C., Timmons M., Robinson C., Schlaggar B.L., Pastores G., Lien Y.H., Kampmann C., Brady R.O., Beck M., Schiffmann R., Enzyme-replacement therapy with agalsidase alfa in children with Fabry disease, Pediatrics 118 (2006) 924–932.

Rimoin D.L., Greenwald S., Nathan T.J., Kaback M.M., Unique considerations for genetic counseling in community-based carrier screening programs, Progress in clinical and biological research 18 (1977) 297–304.

Rizzo W.B., Craft D.A., Sjogren-Larsson syndrome. Deficient activity of the fatty aldehyde dehydrogenase component of fatty alcohol:NAD+ oxidoreductase in cultured fibroblasts, The Journal of clinical investigation 88 (1991) 1643–1648.

Rizzo W.B., Craft D.A., Dammann A.L., Phillips M.W., Fatty alcohol metabolism in cultured human fibroblasts. Evidence for a fatty alcohol cycle, The Journal of biological chemistry 262 (1987) 17412–17419.

Rizzo W.B., Craft D.A., Kelson T.L., Bonnefont J.P., Saudubray J.M., Schulman J.D., Black S.H., Tabsh K., Dirocco M., Gardner R.J., Prenatal diagnosis of Sjogren-Larsson syndrome using enzymatic methods, Prenatal diagnosis 14 (1994) 577–581.

Rizzo W.B., Dammann A.L., Craft D.A., Black S.H., Tilton A.H., Africk D., Chaves-Carballo E., Holmgren G., Jagell S., Sjogren-Larsson syndrome: inherited defect in the fatty alcohol cycle, The Journal of pediatrics 115 (1989) 228–234.

Roddy T.P., Nelson B.C., Sung C.C., Araghi S., Wilkens D., Zhang X.K., Thomas J.J., Richards S.M., Liquid chromatography-tandem mass spectrometry quantification of globotriaosylceramide in plasma for long-term monitoring of Fabry patients treated with enzyme replacement therapy, Clinical chemistry 51 (2005) 237–240.

Rogers G.R., Rizzo W.B., Zlotogorski A., Hashem N., Lee M., Compton J.G., Bale S.J., Genetic homogeneity in Sjogren-Larsson syndrome: linkage to chromosome 17p in families of different non-Swedish ethnic origins, American journal of human genetics 57 (1995) 1123–1129.

Romanes G.J., Sjogren-Larson syndrome, The British journal of ophthalmology 52 (1968) 174–177.

Roux C., Action teratogène du triparanol chez l'animal, Archives françaises de pédiatrie 21 (1964) 451–464.

Schiffmann R., Murray G.J., Treco D., Daniel P., Sellos-Moura M., Myers M., Quirk J.M., Zirzow G.C., Borowski M., Loveday K., Anderson T., Gillespie F., Oliver K.L., Jeffries N.O., Doo E., Liang T.J., Kreps C., Gunter K., Frei K., Crutchfield K., Selden R.F., Brady R.O., Infusion of alpha-galactosidase A reduces tissue globotriaosylceramide storage in patients with Fabry disease, Proceedings of the National Academy of Sciences of the United States of America 97 (2000) 365–370.

Sevin C., Verot L., Benraiss A., Van Dam D., Bonnin D., Nagels G., Fouquet F., Gieselmann V., Vanier M.T., De Deyn P.P., Aubourg P., Cartier N., Partial cure of established disease in an animal model of metachromatic leukodystrophy after intracerebral adeno-associated virus-mediated gene transfer, Gene therapy 14 (2007) 405–414.

Shen Y., Bodary P.F., Vargas F.B., Homeister J.W., Gordon D., Ostenso K.A., Shayman J.A., Eitzman D.T., Alpha-galactosidase A deficiency leads to increased tissue fibrin deposition and thrombosis in mice homozygous for the factor V Leiden mutation, Stroke 37 (2006) 1106–1108.

Sillence D.J., Puri V., Marks D.L., Butters T.D., Dwek R.A., Pagano R.E., Platt F.M., Glucosylceramide modulates membrane traffic along the endocytic pathway, Journal of lipid research 43 (2002) 1837–1845.

Smith D.C., Sillence D.J., Falguieres T., Jarvis R.M., Johannes L., Lord J.M., Platt F.M., Roberts L.M., The association of Shiga-like toxin with detergent-resistant membranes is modulated by glucosylceramide and is an essential requirement in the endoplasmic reticulum for a cytotoxic effect, Molecular Biology of the Cell 17 (2006) 1375–1387.

Smith D.W., Lemli L., Opitz J.M., A Newly Recognized Syndrome of Multiple Congenital Anomalies, The Journal of pediatrics 64 (1964) 210–217.

Steiner R., Koch T., Al-Uzri A., Uchida N., Tamaki S., Tsukamoto A., Guillaume D., Selden N., Molecular Genetics and Metabolism A phase I clinical study of human CNS stem cells (HUCNS-SC) in patients with neuronal ceroid lipofuscinosis, in: Elsevier (Ed.), Molecular genetics and metabolism, Elsevier, 2007, p. 17.

Taga S., Carlier K., Mishal Z., Capoulade C., Mangeney M., Lecluse Y., Coulaud D., Tetaud C., Pritchard L.L., Tursz T., Wiels J., Intracellular signaling events in CD77-mediated apoptosis of Burkitt's lymphoma cells, Blood 90 (1997) 2757–2767.

Tint G.S., Cholesterol defect in Smith-Lemli-Opitz syndrome, American Journal of Medical Genetics 47 (1993) 573–574.

Trajkovic-Bodennec S., Bodennec J., Futerman A.H., Phosphatidylcholine metabolism is altered in a monocyte-derived macrophage model of Gaucher disease but not in lymphocytes, Blood cells, molecules & diseases 33 (2004) 77–82.

Tsui Z.C., Chen Q.R., Thomas M.J., Samuel M., Cui Z., A method for profiling gangliosides in animal tissues using electrospray ionization-tandem mass spectrometry, Analytical biochemistry 341 (2005) 251–258.

Tsuji D., Higashine Y., Matsuoka K., Sakuraba H., Itoh K., Therapeutic evaluation of GM2 gangliosidoses by ELISA using anti-GM2 ganglioside antibodies, Clinica chimica acta; international journal of clinical chemistry 378 (2007) 38–41.

Ullman M.D., McCluer R.H., Quantitative analysis of brain gangliosides by high performance liquid chromatography of their perbenzoyl derivatives, Journal of lipid research 26 (1985) 501–506.

Ullman S., Nelson L.B., Jackson L.G., Prenatal diagnostic techniques. Chorionic villus sampling, Survey of ophthalmology 30 (1985) 33–40.

Utsumi K., Yamamoto N., Kase R., Takata T., Okumiya T., Saito H., Suzuki T., Uyama E., Sakuraba H., High incidence of thrombosis in Fabry's disease, Internal Medicine 36 (1997) 327–329.

Wang W.Q., Gustafson A., Ganglioside extraction from erythrocytes: a comparison study, Acta Chemica Scandinavica 49 (1995) 929–936.

Whitfield P.D., Nelson P., Sharp P.C., Bindloss C.A., Dean C., Ravenscroft E.M., Fong B.A., Fietz M.J., Hopwood J.J., Meikle P.J., Correlation among genotype, phenotype, and biochemical markers in Gaucher disease: implications for the prediction of disease severity, Molecular genetics and metabolism 75 (2002) 46–55.

Whitfield P.D., Sharp P.C., Johnson D.W., Nelson P., Meikle P.J., Characterization of urinary sulfatides in metachromatic leukodystrophy using electrospray ionization-tandem mass spectrometry, Molecular genetics and metabolism 73 (2001) 30–37.

Willemsen M.A., Cruysberg J.R., Rotteveel J.J., Aandekerk A.L., Van Domburg P.H., Deutman A.F., Juvenile macular dystrophy associated with deficient activity of fatty aldehyde dehydrogenase in Sjogren-Larsson syndrome, American journal of ophthalmology 130 (2000) 782–789.

Wisniewski K.E., Zhong N., Philippart M., Pheno/genotypic correlations of neuronal ceroid lipofuscinoses, Neurology 57 (2001) 576–581.

Young E.P., Worthington V.C., Jackson M., Winchester B.G., Pre- and postnatal diagnosis of patients with CLN1 and CLN2 by assay of palmitoyl-protein thioesterase and tripeptidyl-peptidase I activities, European Journal of Paediatric Neurology 5 Suppl A (2001) 193–196.

# Index